Extreme-Temperature and Harsh-Environment Electronics (Second Edition)

Physics, technology and applications

Online at: https://doi.org/10.1088/978-0-7503-5072-3

Extreme-Temperature and Harsh-Environment Electronics (Second Edition)

Physics, technology and applications

Vinod Kumar Khanna
Independent Researcher, Chandigarh, India
Retired Chief Scientist, CSIR-Central Electronics Engineering Research Institute, Pilani, India

IOP Publishing, Bristol, UK

ISBN 978-0-7503-5072-3 (ebook)
ISBN 978-0-7503-5070-9 (print)
ISBN 978-0-7503-5073-0 (myPrint)
ISBN 978-0-7503-5071-6 (mobi)

DOI 10.1088/978-0-7503-5072-3

Version: 20230701

IOP ebooks

British Library Cataloguing-in-Publication Data: A catalogue record for this book is available from the British Library.

Published by IOP Publishing, wholly owned by The Institute of Physics, London

IOP Publishing, No.2 The Distillery, Glassfields, Avon Street, Bristol, BS2 0GR, UK

US Office: IOP Publishing, Inc., 190 North Independence Mall West, Suite 601, Philadelphia, PA 19106, USA

To my late father Shri Amarnath Khanna

For his earnest endeavors to shape my educational career.

To my late mother Smt. Pushpa Khanna

For her love and blessings to guide me on the path of life.

To my grandson Hansh and daughter Aloka

For bringing joy and happiness in the family.

To my wife Amita

For her unflinching and unfailing support.

Contents

Part II Harsh-environment electronics
Sub-part IIA General considerations

Part II Harsh-environment electronics
Sub-part IIB Application-specific robust electronics techniques

Preface to the revised edition

The first edition of this popular reference book presented a new perspective on electronic applications catering to hostile environments, and filled a long-felt need for an advanced reference book on the subject. This second edition provides fully updated content, including new references and developments during the years since the first edition was published. New material in this edition includes overall updating of chapter contents, an extensively upgraded bibliography, addition of sections on GaAs, SiC, GaN and diamond sensors (sections 7.15, 8.12, 9.10, and 10.11) together with incorporation of five new chapters added as sub-part IIB providing expanded discussion of electromagnetic interference and compatibility issues anticipated with the colossal spread of communication links and power infrastructure, sensors for hostile environments, medical electronic devices implanted in the human body for corrective therapy, electronic equipment operation in the vacuum and radiation environments experienced during space odysseys, and last but not the least, protection from dangers faced from malicious jamming and hacking attacks. The second edition of the book is organized in two parts I (chapters 1–14) and II (chapters 15–25), with parts I and II further subdivided into sections A and B. Their content coverage is concisely spelt out as follows:

Part I: Environmental hazards and extreme-temperature electronics (chapters 1–14)

'Sub-part IA: Environmental hazards (chapters 1 and 2)' gives an overview of the hazardous environments of operation of electronics.

'Sub-part IB: Extreme-temperature electronics (chapters 3–14)' covers the effects of temperature on semiconductors, silicon bipolar devices and circuits, silicon MOS devices and circuits, and SiGe heterojunction bipolar transistors. This part presents gallium arsenide electronics, silicon carbide electronics, gallium nitride electronics, diamond electronics, passive components, interconnections and packaging, superconductive electronics, superconductor-based microwave circuits, and high temperature superconductor-based delivery of power.

Part II: Harsh-environment electronics (chapters 15–25)

'Sub-part IIA: Harsh-environment electronics (chapters 15–20): General considerations' describes humidity and contamination effects on electronics, moisture and waterproof electronics, chemical corrosion and radiation effects on electronics, radiation-hardened electronics, and vibration-tolerant electronics.

'Sub-part IIB: Harsh-environment electronics: Application-specific robust electronics techniques (chapters 21–25)' explains electromagnetic interference and methods to achieve electromagnetic compatibility, sensors for aggressive environments, implantable medical electronics, space electronics, electronic jamming mitigation and assurance of cyber security.

The electronics engineer faces numerous challenges to develop electronic components and devices that can operate in difficult environmental conditions or situations where long-term reliability is critical and where mission failure will lead to human safety risks besides incurring substantial financial losses. The precautionary measures begin right at the outset during the conception phase of an electronic circuit in reference to conditions under which it is planned to function. They must be addressed at all stages commencing from device or circuit design to its fabrication and packaging, including proper choice of constructional materials.

It is hoped that the revised edition encapsulating latest information on the subject will serve as a treasure-trove of knowledge immensely useful to researchers, postgraduate students, practising engineers and other concerned stakeholders working with electronic devices and circuits under extreme temperatures and harsh environments, including the automotive, avionics, oil and nuclear power industries. Like its predecessor, the new edition will motivate and inspire all readers tackling environmental perils confronting electronics to strive with greater zeal and enthusiasm towards the goal of rugged, durable and reliable electronics.

Vinod Kumar Khanna
Chandigarh, India

Preface to first edition

Customarily, electronic devices and circuits are required to operate at room temperature. This is more a matter of convenience and convention than optimization. On lowering the temperature, the performance of electronic devices is improved two-fold or by several orders of magnitude. This upgradation is observed in various forms, e.g. increased speed of digital systems, a better signal-to-noise ratio and greater bandwidth for analog systems, improved sensitivity for sensors, greater precision and range for measuring instruments, and overall deceleration of the ageing process of materials. However, low temperature is not always beneficial, e.g. the current gain and breakdown voltage of bipolar transistors are degraded with decreasing temperature. Broadly speaking, low-temperature electronics (LTE) has two offshoots: semiconductor-based electronics and superconductor-based electronics. Semiconductor-based electronics pertains to electronics that can be made to operate at any temperature from room temperature, or even much higher, down to the lowest cryogenic temperatures, 1 K or below. On the other hand, present-day systems based on superconductivity are confined to operation at low cryogenic temperatures, below about 10 K, which has been a serious impediment to their widespread use. The advent of high-temperature superconductors and associated systems appears to hold some promise.

The other side of the story is high-temperature electronics (HTE). 'High temperature' is any temperature above 125 °C. This cut-off point is frequently specified as the upper limit at which standard commercial silicon devices are supposed to function properly. However, tests on standard commercial components indicate that even temperatures up to 150 °C may be applied to selected silicon components. Certain niche applications have to operate beyond the melting point of many materials used in present-day industrial electronics. Examples are monitors and down-hole well-drilling tools for energy exploration. Aircraft and turbine engine controls also have to withstand high temperatures.

The scope of extreme-temperature electronics (ETE) encompasses operation in temperature ranges outside the traditional commercial, industrial or military ranges, i.e. −55 °C/−65 °C to +125 °C. There are three categories of ETE, known as HTE referring to temperatures over +125 °C, LTE for temperatures below −55 °C/−65 °C, and cryogenic temperature electronics, below −150 °C.

Apart from the extreme-temperature applications discussed above, mention may be made of chemically corrosive conditions. Excessively high humidity causes corrosion in electronic devices. Low humidity favors the building up of static electricity. Atmospheric corrosion, an electrochemical process that occurs on metals covered by a thin film of water and ions, is often responsible for damage to electrical and electronic components leading to premature failures even in indoor ambient conditions.

Ionizing radiation consisting of electromagnetic waves such as x-rays and γ-rays, and particulate radiation, i.e. protons, electrons, neutrons, etc, causes malfunctions and failures in electronic components and circuits. The extent of damage depends on the type of radiation, its intensity or energy, the time of exposure and hence the dose, in addition to the distance between the radiation source and the electronic equipment.

Unconventional electronics is subdivided into different areas, e.g. electronics capable of operating at high temperatures for deep well and geothermal logging, lightweight ground and air vehicles, and space exploration; electronics benefitting from low-temperature operation as well as electronics able to withstand low temperatures for infrared systems, satellite communications and medical equipment, and newer opportunities in wireless and mobile communications, computers, and measurement and scientific equipment; electronics capable of countering the detrimental effects of humid and chemically corrosive environments for use in tropical climates and industries such as pulp and paper processing, oil and petroleum refining, mining, foundry, chemicals, etc; and radiation environment electronics for the space, medical and nuclear power industries.

This book has three objectives:

(i) to explore the beneficial/harmful impacts of extreme temperatures on electronic devices and circuits, as well as to enquire into the complexities introduced by harsh conditions such as damp, dirty and radiation-filled environments;
(ii) to describe the techniques adopted to utilize the advantages of these unconventional situations; and
(iii) to present the remedial measures taken to counteract and deal with these unfavorable circumstances.

This book is written for graduate and research students in electrical and electronic engineering. It will serve as a useful supplement to microelectronics course material, treating this specialized discipline with breadth and depth. The book answers several questions that come to mind when one starts thinking and imagining beyond a normal electronics course. The book is also written to fulfill the needs of electronic device and process design engineers, as well as circuit and system developers engaged in this fast-moving field, covering both fundamentals and applications. Scientists and professors engaged in this field will also find it useful as a comprehensive guide to the state-of-the-art electronic technologies for hostile environments.

Vinod Kumar Khanna
CSIR-CEERI, Pilani, India

Acknowledgements

Above all, I must thank Almighty God for giving me the vigor and wisdom to complete this work. I am grateful to all my colleagues and Director, CSIR-CEERI, Pilani for constant encouragement in my efforts. I wish to thank my editor and editorial assistant for rejuvenating my interest and boosting my confidence from time to time. Their kind co-operation and support led to timely completion of the work. I owe a profound debt of gratitude to all authors/editors of research papers, magazine articles and web pages, on whose work this book is based. Most of these excellent works are cited in the bibliographies at the end of each chapter. However if anyone has inadvertently escaped mention, I hope it may be forgiven. Last but not least, I am grateful to my family for providing the serene atmosphere and relieving me of many domestic responsibilities so that I could devote more time to book writing. To all the above, I express my sincere thanks.

Vinod Kumar Khanna
CSIR-CEERI, Pilani, India

About this book

Extreme-Temperature and Harsh-Environment Electronics (Second Edition): Physics, technology and applications presents a unified perspective combining the impact of extremely high and exceedingly low temperatures on the operation of semiconductor electronics, in addition to the influence of hostile environments such as high-humidity conditions, as well as surroundings contaminated by chemical vapors, nuclear radiation or those disturbed by mechanical shocks and vibrations. Incorporating the preliminary background material and thus laying down foundations for easy understanding, the progress in mainstream silicon, silicon-on-insulator and gallium arsenide electronics is sketched. Contemporary wide bandgap semiconductor technologies such as silicon carbide, gallium nitride and diamond electronics are explored. After a brief treatment of superconductivity, concepts of superconductive electronics are introduced. Progress in Josephson junctions, SQUIDs and RSFQ logic circuits is highlighted. The state of the art in high-temperature superconductor-based power delivery is surveyed. Succeeding chapters look at various protection schemes that have been devised to shield electronic circuits and equipment from adverse ambient conditions. These conditions range from the presence of high moisture concentrations in the atmosphere, to showers of high-energy particles such as the alpha particles, protons and nuclei of heavy elements that flood the atmosphere from outer space. A particularly attractive area is the dampening of vibration effects to protect electronics from the quivering disturbances or jerks that are always present near large machines or during accidental falls of electronic equipment. Specialized topics explored include electromagnetic interference and compatibility, use of sensors in aggressive environments, implantable medical electronics, space electronics, and protection of electronics from jamming and hacking.

In this book, a lucid description of this vast panorama of topics is reinforced by an elegant mathematical treatment. Broad in scope, this comprehensive treatise provides a coherent, well-organized and amply illustrated exposition of the subject which will be immensely useful for graduate and research students, professional engineers and researchers engaged in this frontier of technology. Its three-pronged approach encompassing physical aspects, technological breakthroughs and application examples will make reading interesting and enjoyable.

Author biography

Vinod Kumar Khanna

Introduction

Vinod Kumar Khanna is an independent researcher at Chandigarh, India. He is a retired Chief Scientist from Council of Scientific and Industrial Research (CSIR)-Central Electronics Engineering Research Institute (CEERI), Pilani, India, and a retired Professor from Academy of Scientific and Innovative Research (AcSIR), Ghaziabad, India. He is a former Emeritus Scientist, CSIR and Professor Emeritus, AcSIR, India. His broad areas of research were the design, fabrication and characterization of power semiconductor devices, micro- and nanosensors.

Academic qualifications

He received the MSc Degree in Physics with specialization in Electronics from University of Lucknow in 1975 and PhD degree in Physics from Kurukshetra University in 1988 for the thesis entitled, 'Development, characterization and modeling of the porous alumina humidity sensor'.

Work experience and accomplishments

His research experience spans over a period of 40 years from 1977 to 2017. He started his career as a Research Assistant in the Department of Physics, University of Lucknow from 1977 to 1980. He joined CSIR-Central Electronics Engineering Research Institute, Pilani (Rajasthan) in April 1980. At CSIR-CEERI he worked on several CSIR-funded as well as sponsored research and development projects. His major fields of research included power semiconductor devices and microelectronics/MEMS and nanotechnology-based sensors and dosimeters.

In power semiconductor devices area, he worked on the high-voltage and high-current rectifier (600 A, 4300 V) for railway traction, high voltage TV deflection transistor (5 A, 1600 V), power Darlington transistor for AC motor drives (5 A, 1600 V), fast-switching thyristor (1300 A, 1700 V), power DMOSFET and IGBT. He contributed towards the development of sealed tube Ga/Al diffusion for deep junctions, surface electric field control techniques using edge beveling and contouring of large-area devices, and floating field limiting ring design; and characterization of minority-carrier lifetime as a function of process steps. He also contributed towards the development of P–I–N diode neutron dosimeter and PMOSFET-based gamma ray dosimeter.

In the area of sensor technology, he worked on the nanoporous aluminum oxide humidity sensor, ion-sensitive field-effect transistor-based microsensors for biomedical, food and environmental applications; microheater embedded gas sensor for automotive electronics, MEMS acoustic sensor for launch vehicles and capacitive MEMS ultrasonic transducer for medical applications.

Semiconductor facility creation and maintenance

He was responsible for setting up and looking after diffusion/oxidation facilities, edge beveling and contouring, reactive sputtering and carrier lifetime measurement facilities. As the Head of MEMS and microsensors group, he looked after the maintenance of six-inch MEMS fabrication facility for R&D projects as well as augmentation of processing equipment under this facility at CSIR-CEERI.

Scientific positions held

During his tenure of service at CSIR-CEERI from April 1980 till superannuation in November 2014, he was promoted to various positions including one merit promotion. He retired as Chief Scientist and Professor (AcSIR: Academy of Scientific and Innovative Research), and Head of MEMS and Microsensors Group. Subsequently, he worked for three years as Emeritus Scientist, CSIR and Emeritus Professor, AcSIR from November 2014 to November 2017. After completion of the emeritus scientist scheme, he now lives at Chandigarh. He is a passionate author, and enjoys reading and writing.

Membership of professional societies

He is a Fellow and life member of the Institution of Electronics and Telecommunication Engineers (IETE), India. He is a life member of Indian Physics Association (IPA), Semiconductor Society, India (SSI) and Indo-French Technical Association (IFTA).

Foreign travel

He is widely travelled. He participated in and presented research papers at IEEE Industry Application Society (IEEE-IAS) Annual Meeting at Denver Colorado, USA, in September–October, 1986. His short-term research assignments include deputations to Technische Universität Darmstadt, Germany in 1999, at Kurt-Schwabe-Institut fur Mess- und Sensortechnik e.V., Meinsberg, Germany in 2008 and at Fondazione Bruno Kessler, Trento, Italy in 2011 under collaborative program. He was a member of the Indian Delegation to Institute of Chemical Physics, Novosibirsk, Russia in 2009.

Scholarships and awards

He was awarded National Scholarship by the Ministry of Education and Social Welfare, Government of India, on the basis of Higher Secondary result, 1970; CEERI Foundation Day Merit Team Award for projects on fast-switching thyristor (1986); power Darlington transistor for transportation (1988), P–I–N diode neutron dosimeter (1992); and high-voltage TV deflection transistor (1994); Dr N G Patel Prize for best poster presentation in 12th National Seminar on Physics and Technology of Sensors 2007, BARC, Mumbai; CSIR-DAAD Fellowship in 2008 under Indo-German Bilateral Exchange Programme of Senior scientists, 2008. He is featured in the Stanford–Elsevier prestigious list of world top 2% scientists (2022, Elsevier Data Repository, V4, doi: 10.17632/btchxktzyw.4).

Research publications and books

He has published 194 research papers in leading peer-reviewed national/international journals and conference proceedings. He has authored 19 books, and has also contributed six chapters in edited books. He has five patents to his credit, including two US patents.

Abbreviations, acronyms, chemical symbols and mathematical notation

A	ampere
AC	alternating current
ACE	arbitrary code execution
ACK	acknowledge
Ag	silver (argentum)
AgCl	silver chloride
Ag_2O	silver oxide
Ag_2S	silver sulfide
AH	absolute humidity
AI	artificial intelligence
Al	aluminum
ALD	atomic layer deposition
AlGaAs	aluminum gallium arsenide
AlGaN	aluminum gallium nitride
AlN	aluminum nitride
Al_2O_3	aluminum oxide, alumina
AO	atomic oxygen
APT	advanced persistent threat
Ar	argon
As_2O_3	arsenic trioxide
As_2O_5	arsenic pentoxide
ASTM	American society for testing and materials
Au	gold (aurum)
$BaCO_3$	barium carbonate
BCS	Bardeen–Cooper–Schrieffer (theory of superconductivity)
Be	beryllium
BFSK	binary frequency-shift keying
BGJFET	buried-grid JFET
B_2H_6	diborane
Bi_2Te_3	bismuth telluride
BiCMOS	bipolar CMOS
BJT	bipolar junction transistor
BLS	board level shielding
BN	boron nitride
B_2O_3	boron trioxide
BOX	buried oxide
BPF	bandpass filter
BPSK	binary phase-shift keying
BSCCO	bismuth–strontium–calcium–copper oxide
BSG	borosilicate glass
C	carbon

°C	degrees centigrade
Ca^{2+}	calcium cation
cal g^{-1}	calorie $gram^{-1}$
CaO	calcium oxide
$Ca_3(PO_4)_2$	calcium phosphate
$Ca_{10}(PO_4)_6(OH)_2$	calcium hydroxyapatite
CBMA	carboxybetaine methacrylate
CCD	charge-coupled device
CEN	Comité Européen de Normalisation
CeO_2	cerium (IV) oxide
CF-PEEK	carbon fiber reinforced PEEK (polyetheretherketone)
CH_4	methane
$(CH_3)_3B$	trimethylborane
$C_{16}H_{14}Cl_2$	parylene C
$-(C_2H_2F_2)_n-$	PVDF
CHFET	complementary heterojunction FET
$C_7H_7N_3$	tolytriazole
$C_{17}H_{16}N_2O_4$	polyurethane
$C_{12}H_{24}N_2O_2$	dicyclohexylammonium nitrite
$C_{35}H_{28}N_2O_7$	polyimide resin
$C_6H_8O_7$	citric acid
$C_6H_{10}O_4$	adipic acid
$C_6H_{10}O_6$	glucono-delta-lactone or gluconolactone
$C_6H_{12}O_6$	glucose
$(C_2H_6OSi)_n$	polydimethylsiloxane (PDMS)
CISPR	Comité International Spécial des Perturbations Radioélectriques
Cl^-	chloride anion
CMOS	complementary metal–oxide–semiconductor (FET)
CO	carbon monoxide
Co	cobalt
CO_2	carbon dioxide
COB	chip-on-board (assembly)
CO-OP	controlled over pressure (process)
$CoSi_2$	cobalt silicide
CP-Ti	commercially pure titanium
CPU	central processing unit
CQT	cascaded quadruplet trisection coupling structure (filter geometry)
Cr	chromium
Cr_2O_3	chromium (III) oxide
CRT	cathode ray tube
CSFs	colony-stimulating factors
3C-SiC	cubic silicon carbide
CSMA	carrier sense multiple access (protocol)
CT	computed tomography
CTE	coefficient of thermal expansion
Cu	copper
CuO	copper oxide
Cu_2S	copper sulfide
CVD	chemical vapor deposition
CZ	Czochralski (single-crystal silicon)

DAC	discretionary access control
dB	decibel
DBE	droplet backside exposure
DC	direct current
DCF	Distributed coordination function
DD	displacement damage
DDoS	distributed denial-of-service (attack)
2DEG	two-dimensional electron gas
DGVTJFET	dual-gate vertical channel trench JFET
2DHG	two-dimensional hole gas
DI-BSCCO	dynamically innovative-BSCCO
DICE	dual interlocked storage cell
DICHAN	dicyclohexylammonium nitrite
DIFS	DCF interframe space interval
DMOSFET	double-diffused MOSFET
DMVTJFET	depletion-mode VTJFET
DNS	domain name system
DOM	document object model
DoS	denial-of-service
DRAM	dynamic random-access memory
DSSS	direct sequence spread spectrum
E	east
e^-	electron
e-beam	electron beam
ECF	extracellular fluid compartment
ECOG	electrocorticography
EDR	endpoint detection and response (tool)
EEG	electroencephalogram
EEPROM	electrically-erasable programmable read-only memory
EG	electronic grade (polysilicon)
EHP	electron hole pair
eHTP	extremely high temperature package
EIA	Electronics Industries Association
EL Ni	electroless nickel
EMC	electromagnetic compatibility
EMF	electromotive force
EMI	electromagnetic interference
EMVTJFET	enhancement-mode VTJFET
ESD	electrostatic discharge
ETO	ethylene oxide
eV	electronvolt
2FA	two-factor authentication
FBGCs	foreign body giant cells
FBR	foreign body reaction
FC/APC	fiber connector with angled physical contact
FCC	Federal Communications Commission
fcc	face-centered cubic
FCL	fault current limiter
FET	field-effect transistor
fF Pa^{-1}	femtofard $pascal^{-1}$
FHSS	frequency hopping spread system

FIRST	Far Infra-Red Space Telescope
FR-4	flame retardant-4, a grade designation by National Electrical Manufacturers Association for glass epoxy laminate
FSK	frequency-shift keying
Ga	gallium
GaN	gallium nitride
Ga_2O_3	gallium trioxide
$GaPO_4$	gallium phosphate or gallium orthophosphate
Ge	germanium
GEO	geosynchronous equatorial orbit
GHz	gigahertz
GM	Gifford–McMahon (cryocoolers)
Gox	glucose oxidase
GPS	global positioning system
GSFC	Goddard Space Flight Center
GTO	gate turn-off (thyristor)
Gy	gray
h	hour
H^+	hydrogen cation
HA	hydroxyapatite
HAF	high-attenuation fiber
HASL	hot air solder leveling
HA/Ti-6Al-4V	hydroxyapatite reinforced Ti-6Al-4V
HBT	heterojunction bipolar transistor
HCO_3^-	bicarbonate anion
HCs	hydrocarbons
HEMT	high electron mobility transistor
HFET	heterojunction FET
HfO_2	hafnium oxide
Hg	mercury (hydrargyrum)
H_2O	water, dihydrogen monoxide
H_2O_2	hydrogen peroxide
HPHT	high-pressure and high-temperature
6H-SiC	hexagonal silicon carbide
HT	high-temperature (sensors)
HTCC	high temperature co-fired ceramic
HTE	high-temperature electronics
HTML	hypertext markup language
HTS	high-temperature superconductor
HTTP	hypertext transfer protocol
HVAC	high-voltage alternating current
Hz	hertz
IBAD	ion beam-assisted deposition
IC	integrated circuit
I^2C	inter-integrated circuit
ICF	intracellular fluid (compartment)
ICMP	internet control message protocol
ICQ	short form of the phrase, 'I seek you', it is an internet instant cross-platform messenger app supporting audio/video chats, text messaging to cellular network phones, emails and file transfers

iCVD	initiated chemical vapor deposition
IDS/IPS	intrusion detection system/intrusion prevention system
IDT	interdigitated (electrodes)
IEEE	Institute of Electrical and Electronic Engineers
IF	intermediate frequency
i-GaN	intrinsic gallium nitride
IGBT	insulated gate bipolar transistor
IL-3, IL-4, IL-10, etc	interleukin-3, interleukin-4, interleukin-10, etc
IM Au	immersion gold
InAlN	indium aluminum nitride
InAs	indium arsenide
InGaAs	indium gallium arsenide
InP	indium phosphide
InSb	indium antimonide
Intelsat	International Telecommunications Satellite Organization
IP	internet protocol
IPS	intrusion prevention system
Ir	iridium
ISOPHOT	Infrared Space Observatory Photometer
ITO	indium tin oxide
JFET	junction FET
JJ	Josephson junction
JTE	junction termination extension
K	kelvin (scale of temperature), key
K^+	potassium cation
KCl	potassium chloride
keV	kilo-electronvolt
kg	kilogram
kHz	kilohertz
km	kilometer
kPa	kilopascal
K_{Public}	public key
K_{Secret}	secret key
kV	kilovolt
kW	kilowatt
L	liter
$LaAlO_3$	lanthanum aluminate
LaB_6	lanthanum hexaboride
Langasite	lanthanum gallium silicate, $La_3Ga_5SiO_{14}$
LAO	lanthanum aluminate
LC	Inductance–capacitance (circuit)
LCD	liquid crystal display
LCJFET	lateral channel JFET
LCPs	liquid crystal polymers
LDPC	low density parity coding
LEC	liquid encapsulated Czochralski
LED	light-emitting diode
LEO	low earth orbit
LFI	local file inclusion

LGS	lanthanum gallium silicate, $La_3Ga_5SiO_{14}$
Li	lithium
LO	local oscillator
LPCVD	low-pressure chemical vapor deposition
LPF	low-pass filter
LSN	low-stress silicon nitride
LTCC	low temperature co-fired ceramic
LTE	low-temperature electronics
LTO	low-temperature oxide
M1	classically-activated macrophages
M2	wound-healing macrophages
m	meter
mA	milli-ampere
MAC	molecular absorber coating, managed/mandatory access control
MACOR	trademark for a machinable glass ceramic: fluorophlogopite mica (55%) + borosilicate glass (45%)
MAG	maximum available gain
$M_xAl_xSi_{1-x}O_2 \cdot yH_2O$	zeolite, M is a metal or hydrogen ion, $0 < x < 1$ and y is the number of water molecules
MDR	managed detection and response
MEA	more electric aircraft
MEMS	microelectromechanical systems
MESFET	metal–semiconductor FET
meV	milli-electronvolt
Mg	magnesium
Mg^{2+}	magnesium cation
MG	metallurgical grade (polysilicon)
$MgCr_2O_4$–TiO_2	magnesium chromite–titanium dioxide
MgO	magnesium oxide
MHz	megahertz
MISFET	metal–insulator–semiconductor FET
mK	millikelvin
MLI	multilayer insulation
mm	millimeter
MMIC	monolithic microwave IC
Mn	manganese
$Mn(NO_3)_2$	manganese nitrate
MnO_2	manganese dioxide
mΩ	milli-ohm
MΩ-cm	megaohm-centimeter
Mo	molybdenum
MOCVD	metal–organic CVD
MOSFET	metal–oxide–semiconductor FET
$MoSi_2$	molybdenum disilicide
MPa	megapascal
MPC	2-methacryloyloxyethyl phosphorylcholine
MPCVD	microwave plasma-enhanced CVD
MPS	merged PiN/Schottky (diode)
Mregs	regulatory macrophages
MRI	magnetic resonance imaging

mS	millisiemen
MSOC	modern security operations center
MTTF	mean-time-to-failure
MV	megavolt
mV	millivolt
N	nitrogen, north
n	number, neutron
Na^+	sodium cation
Na_2O	sodium oxide
NaOH	sodium hydroxide
NASA	National Aeronautics and Space Administration
Nb	niobium
nC	nanocoulomb
NH_3	ammonia
Ni	nickel
NiCr	nickel–chromium, nichrome
NiO	nickel (II) oxide
nm	nanometer
NMOS	n-channel MOSFET
NO	nitric oxide
NO_2	nitrogen dioxide gas
NO_x	oxides of nitrogen
NP0	negative positive zero
ns	nanosecond
O_2	oxygen
OCVD	open circuit voltage decay
OH^-	hydroxide anion
ONO	silicon oxide–silicon nitride–silicon oxide
OP-AMP	operational amplifier
OSI	open system interconnection (model)
OSP	organic solderability preservative
pA	picoampere
Parylene C	derivative of parylene in which one hydrogen atom in the aryl ring is replaced with chlorine
Pb	lead (plumbum)
pBN	pyrolytic boron nitride
pC	pico Coulomb
PCB	printed circuit board
PCBA	printed circuit board assembly
PCM	phase-change material
Pd	palladium
PDMS	polydimethylsiloxane
PDR	packet delivery ratio
PECVD	plasma-enhanced CVD
PEEK	polyetheretherketone
PEG	polyethylene glycol
pH	potential hydrogen
PH_3	phosphine
pHEMA	poly (2-hydroxyethyl methacrylate)
PKI	public key infrastructure

PLD	pulsed laser deposition
PLGA	poly(lactic-co-glycolic) acid
PMOS	p-channel MOSFET
PMU	power management unit
PO_4^{3-}	phosphate anion
P_2O_5	phosphorous pentoxide
$POCl_3$	phosphorous oxychloride
Poly-SiC	polycrystalline silicon carbide
ppm K^{-1}	parts per million $Kelvin^{-1}$
ps	picosecond
PSG	phosphosilicate glass
Pt	platinum
PTFE	polytetrafluoroethylene
PTH	plated-through-hole
PVA	poly(vinyl alcohol)
PVDF	polyvinylidene fluoride
QUAD	quadruple
rad	radiation absorbed dose
RADAR	radio detection and ranging
RAM	random-access memory
RBAC	role-based access control
RCE	remote code execution
ReBCO	rare-earth barium copper oxide
RF	radio frequency
RFBG	regenerated fiber Bragg grating
RFI	remote file inclusion
RFID	radio-frequency identification
RH	relative humidity
RHBD	radiation hardening by design
RHBP	radiation hardening by process
RoHS	restriction of hazardous substances
RPCVD	reduced pressure CVD
R–S	reset–set (flip-flop)
RSFQ	rapid single flux quantum
R_2SiO	structural unit of silicone where R is an organic group
RTA	rapid thermal annealing
RTP	reversible thermal panel
RTV	room temperature vulcanization
Ru	ruthenium
s	second
S	sulfur, siemen
SAE International	Society of Automotive Engineers International
Sapphire	Al_2O_3
SAW	surface acoustic wave
SBD	Schottky barrier diode
SBMA	sulfobetaine methacrylate
SCPs	satellite control processors
SCS	switched capacitor system
Se	selenium

SEB	single event burnout
SEE	single event effect
SEFI	single event functional interrupt
SEGR	single event gate rupture
SEI	Sumitomo Electric Industries Ltd
SEJFET	static expansion channel JFET
SEL	single event latchup
SES	single event snapback
SET	single event transient
SEU	single event upset
SFBG	sapphire fiber Bragg grating
SFCL	superconducting FCL
Si	silicon
SIAFET	static induction-injected accumulated FET
SiC	silicon carbide
SIEM	security information and event management
SiGe	silicon–germanium (alloy)
$Si_{1-x}Ge_x$	silicon–germanium (alloy) where x is the mole fraction of germanium in the alloy with a value from 0 to 1
SiH_4	silane
$SiHCl_3$	trichlorosilane
SiH_2Cl_2	dichlorosilane
SiO_2	silicon dioxide
SMD	surface mount device
Si_3N_4	silicon nitride
$Si(OC_2H_5)_4$	tetraethylorthosilicate (TEOS)
$Si_xO_yN_z$	silicon oxynitride
SISO	spiral in/spiral out (resonator)
SLD	superluminescent diode
Sm	samarium
$s\ m^{-1}$	second meter^{-1}
SMF-28	single mode optical fiber-28, a standard telecom fiber
SMLI	stateful multilayer inspection (firewall)
Sn	tin (stannum)
$SO_4{}^{2-}$	sulphate anion
SOI	silicon-on-insulator
SPF	sender policy framework
SQUID	superconducting quantum interference device
SRH	Shockley–Read–Hall (recombination)
SS	subthreshold swing
STI	shallow trench isolation
STP	shielded twisted pair (cable)
SYN	synchronize
SYN-ACK	synchronize-acknowledge
T	tesla
Ta	tantalum
TaN	tantalum nitride
Ta_2O_5	tantalum pentoxide
$TaSi_2$	tantalum disilicide
TBCCO	thallium–barium–calcium–copper oxide

TC	temperature coefficient
TCP SYN	transmission control protocol synchronize
TCR	TC of resistance
TCS	trichlorosilane
Te	tellurium
TEOS	tetraethylorthosilicate
Ti	titanium
Ti-6Al-4V	90% Ti, 6% aluminum and 4% vanadium alloy
TID	total ionizing dose
TiN	titanium nitride
TiO_2	titanium dioxide
TiO_2–ZrO_2	titanium dioxide–zirconium dioxide
$TiSi_2$	titanium disilicide
TMB	trimethylboron
TMR	triple modular redundancy
TP	twisted pair (cable)
TPTS	two-phase thermal switch
TPU	thermoplastic polyurethane
TTL	transistor–transistor logic
TV	television
UHVCVD	ultra-high vacuum CVD
UMOSFET	U-shaped MOSFET
URL	uniform resource locator
UTP	unshielded twisted pair (cable)
UV	ultraviolet
V	volt
VAPT	vulnerability assessment and penetration testing (tool)
VCI	volatile corrosion inhibitor (coating)
V_2O_5	vanadium pentoxide
VTJFET	vertical trench JFET
W	tungsten, watt
Wb	weber, the SI unit of magnetic flux. Weber per second = Volt
WI-FI	not the short form of 'wireless fidelity'; a registered trademark of WI-FI alliance; a family of wireless protocols based on IEEE 802.11 standard; a local area networking technology for high-speed data exchange between digital devices over short distances through radio communication by internet
WN_x	tungsten nitride
WSi_2	tungsten disilicide
w/v	weight by volume concentration
w/w	weight by weight concentration
XML	extensible markup language
XSS	cross-site scripting
$YBa_2Cu_3O_7$	YBCO
YBCO	yttrium barium cuprate
$YCa_4O(BO_3)_3$	yttrium calcium oxoborate
YCOB	$YCa_4O(BO_3)_3$
YIG	yttrium iron garnet
Y_2O_3	yttrium oxide
YSZ	yttria stabilized zirconia

Z	atomic number
Zn	zinc
Zr	zirconium
ZrO_2	zirconium oxide, zirconia
ZTC	zero temperature coefficient biasing point of a MOSFET

Roman alphabet symbols

a	depth of the active region in MESFET
A	parameter in the model of Quay *et al*, area, amplitude of signal
$\mathbf{A}$	magnetic vector potential
A^*	effective Richardson constant
A, B, C	parameters of the Bludau *et al* model
A_V	voltage gain
b	parameter in Arora–Hauser–Roulston equation, parameter in Chynoweth equations
B	bandwidth of the original signal in FHSS
$B(0)$	magnetic induction at the surface
B_{FHSS}	bandwidth determined by the spacing between the M carrier frequencies of FHSS signal
b_i	the ith data bit
BV_{CBO}	collector–base breakdown voltage of a bipolar transistor with emitter open
BV_{CEO}	collector–emitter breakdown voltage of a bipolar transistor with base open
BV_{DSS}	drain–source breakdown voltage of a MOSFET with gate shorted to source
$B(x)$	magnetic induction along the x-direction
C_1, C_2, C_3	parameters in the model of threshold ionization energy
C	capacitance
c	velocity of light, damping coefficient (vibration theory)
C_{ds}, C_{DS}	drain–source capacitance
C_{gd}	gate–drain capacitance
C_{iss}	intrinsic capacitance
C–L	capacitance–inductance
C–L–C	capacitance–inductance–capacitance
C_{rss}	reverse transfer capacitance
C_{ox}	oxide capacitance per unit area
$c(t)$	chipping signal
D	diffusion coefficient of carrier, diffusion coefficient of dopant
d	diameter, thickness, length, deformation
$d\mathbf{l}$	linear element
D_{nB}	diffusion constant of electrons in the p-base
$\widetilde{\boldsymbol{D}}_{nB}$	position-averaged diffusion coefficient across the base profile
$D_{nB}(x)$	position-dependent diffusion coefficient of electrons in the base
D_{pB}	diffusion constant of holes in the base
D_{pE}	diffusion constant of holes in the n^+-emitter
$d\mathbf{s}$	areal element
$d(t)$	discrete function
E	electric field
e	electronic charge
E_C	energy of conduction band edge
E_g, E_G	energy bandgap
E_{gB0}	silicon bandgap at zero doping of the base layer (in an HBT)

$\Delta E_{gB,A}$	bandgap narrowing of base layer due to acceptor impurity doping effect (in an HBT)
$\Delta E_{gBGe}(x = 0)$	bandgap offset of base layer at $x = 0$ (in an HBT)
$\Delta E_{gB,Ge}(x = W_B)$	bandgap offset of base layer at $x = W_B$ (in an HBT)
$E_{gB}(x)$	position-dependent energy bandgap of SiGe base layer (in an HBT)
E_F	Fermi energy level
E_{fn}	quasi-Fermi level of electrons
E_{fp}	quasi-Fermi level of holes
E_g	energy bandgap
$E_g(0)$	energy bandgap at 0 K
$\langle E_p \rangle$	average energy loss due to phonon scattering
E_V	energy of valence band edge
F	force, free energy
f	frequency of the signal
F_0	peak value of the force waveform
f_0	frequency of the carrier signal in FHSS
F1, F2, F33, F4, F5, F6, F7, F8	frequencies corresponding to the 8 k-bit patterns: 000, 001, 010, 011, 100, 101, 110, 111
$f_1 = f_0$	low frequency for binary 0 in FHSS
$f_2 = f_0 + \Delta f$	high frequency for binary 1 in FHSS
f_c	carrier frequency in MFSK-FHSS, frequency of the carrier wave in BPSK
f_d	difference frequency in MFSK-FHSS
f_i	frequency of the spreading signal during ith hopping period in FHSS
f_{max}	maximum frequency of oscillation
f_n	natural frequency
f_T	transition frequency (unity-gain frequency)
G	processing gain
g	acceleration due to gravity
g_d	output conductance
g_m	transconductance of MOSFET
g_{m0}	transconductance value (maximum) at $V_{GS} = 0$
g_{mb}	body-effect conductance
g_{ms}	transconductance of MOSFET in saturated condition
h	Planck's constant
$\hbar$	reduced Planck's constant = $h/2\pi$
H_C	critical magnetic field of a superconductor
h_{FE}	current gain of a bipolar transistor in a common-emitter connection
I, i	current
I_A	current in circuit A
I_B	base current of a bipolar transistor, current flowing through load resistance of circuit B
I_b	biasing current (superconductor, Josephson junction, SQUID)
I_C	collector current of a bipolar transistor, critical current (superconductor)
I_{CBO}	collector–base reverse current of a bipolar transistor with emitter open

I_{CCH}	current drawn from the supply during logic high output state
I_{CEO}	collector–emitter reverse current of a bipolar transistor with base open
I_d	current in SBD
I_{D0}	reverse saturation current (leakage current) of SBD
I_{DS}	drain–source current of a MOSFET
I_{DSS}	drain–source leakage current of a MOSFET with the gate shorted to the source, the saturated drain current of a MESFET at $V_{GS} = 0$
I_E	emitter current of a bipolar transistor
I_F	forward current
I_{fc}	full saturation current
I_{IL}	input low current
I_{nB}	electron current from the n^+-emitter to the p-base
I_{OFF}	off-state current
I_{ON}	on-state current
I_p	persistent current (superconductor)
I_{pE}	hole current from the p-base to the n^+-emitter of a bipolar transistor
I_R	reverse current
I_S	saturation current, screening current (superconductor)
J, j	current density
J_B	base current density
J_C, j_C	collector current density of a bipolar transistor, critical current density of a superconductor
J_n	Bessel functions of the first kind
K	stiffness, spring constant
k	number of bits in a binary word
K, K_1, K_2	constants
k_B	Boltzmann constant
K_{FE}	damage coefficient related to current gain
K_τ	damage coefficient associated with carrier lifetime
L	length, diffusion length of carrier, channel length of an FET, inductance, number of bits per signal element (in FHSS)
L_{nB}	diffusion length of electrons in the base
L_{pE}	diffusion length of holes in the emitter
M	number of equivalent valleys in the conduction band of SiC, collector multiplication factor, a digit for the number of levels or groupings likely for an assigned number of binary variables, number of signal elements in MFSK-FHSS, index in MOSFET drain current equation in Shoucair's analysis
m_0	rest mass of electron = 9.11×10^{-31} kg
m_n^*, m_p^*	effective masses of electrons and holes for density-of-state calculations
N	dopant concentration
n	electron concentration, ideality factor of SBD
$\widetilde{N}$	position-averaged ratio of effective densities of states in SiGe and Si, across the base profile
N_A	total acceptor impurity concentration
N_A^-	ionized acceptor impurity concentration
N_{AB}	acceptor concentration in the p-base layer

N_C, N_V	effective densities of states in the conduction and valence bands of a semiconductor
N_{crit}	critical impurity density
$N_{C,SiGe}(x)$	position-dependent effective density of states in the conduction band of SiGe
N_D	total donor impurity concentration
N_D^+	ionized donor impurity concentration
N_{DE}	doping concentration in the n-emitter layer
$N_{D(g)}$	doping concentration of polysilicon gate
N_I	density of charged impurities, trap density
n_i	intrinsic carrier concentration (of a semiconductor)
$n_{iB}(x)$	position-dependent intrinsic carrier concentration in the base
n_{pB}	number of electrons in the p-base
$N(t)$, $n(t)$	noise as a function of time
$N_{V,SiGe}(x)$	position-dependent effective density of states in the valence band of SiGe
P	pressure
$\mathbf{p}$	canonical momentum of a classical particle
p	hole concentration
P_0	amplitude of the transmitted force
p_0	TC of threshold voltage
$p_B(x)$	position-dependent hole concentration in the base varying with position x
p_{H_2O}	partial pressure of water vapor present in air
$p^*_{H_2O}$	equilibrium vapor pressure of water vapor
P_I	incident power
P_J	jammer power
p_{nE}	number of holes in the n^+-emitter
P_T	transmitted power
$p(t)$	product or spread signal in FHSS
Q	total electronic charge
q	electronic charge
Q_0	heat of compression
q_0	parameter in Shoucair's threshold voltage equation (the value of V_{Th} at 0 K, as found by extrapolation)
$Q_a(T)$	conducting channel charge
Q_C	heat absorbed from the cold environment at temperature T_C
Q_f	fixed charge in the silicon dioxide
R	resistance, the bit rate in the input information signal in DSSS
R_1, R_2, R_3	three reflected beams
R_1, R_3	resistors formed at the junctions between the beams and the support frame
R_2, R_4	resistors made at the junctions between the beams and the proof mass
R_A	resistance connected to voltage source V_A
R–C	resistance–capacitance
R_c	bit rate in the spread signal in DSSS
$R_{CHANNEL}$	resistance of channel region of a MOSFET
$r_{Corrected}(t)$	received corrected signal
R_{DRIFT}	resistance of drift region of a MOSFET

R_D^{Si}	on-resistance of Si diode
R_D^{SiC}	on-resistance of SiC diode
$R_{DS(ON)}$	drain–source on-resistance of a MOSFET
R_G	resistance of the current path to ground
R_{LA}	resistance of load resistance of circuit A
R_{LB}	resistance of load resistance of circuit B
R_n	external resistor shunting a JJ
R_{Return}	resistance of return path of current
R_{SA}	resistance of voltage source of circuit A
R_{SB}	resistance of voltage source of circuit B
R_s, R_S	series resistance of the source
$r(t)$	received signal
S, s	cross-section, area
$s_{Corrected}(t)$	corrected signal $s(t)$ in DSSS (by including noise)
$s_{Corrected\text{-}Jamming}(t)$	corrected signal $s(t)$ in DSSS (by including noise as well as jamming)
$s_d(t)$	transmitted digital signal with respect to time, BPSK signal
$s_i(t)$	transmitted digital signal for one signal element with respect to time (in MFSK-FHSS)
S_{Jammer}	power of jammer
$S_{Jamming(Filter)}$	power of the jamming signal entering the filter
$s_{Jamming}(t)$	jamming signal as a function of time
$s(t)$	spread signal in FHSS
T	temperature in the Kelvin scale, periodic time of clock signal, transmissibility, bit period in MFSK-FHSS, bit width in DSSS
t	time
T_0	ambient temperature
T_C	temperature of the cold environment
T_c	time taken for translation of the MFSK signal to a new frequency in MFSK-FHSS, bit width of the pseudorandom noise signal in DSSS
t_H	hop period
T_L	lattice temperature
t_{ox}	oxide thickness
T_s	time occupied by each signal element in MFSK-FHSS
U_1, U_2	states of lowest energy on the opposite sides of the tunnel barrier
V, v	voltage
v	velocity of electron
V_0	amplitude of sinusoidal voltage
V_+, V_-	two signals equal in magnitude but opposite in phase (ϕ_+, ϕ_-)
V_1, V_2	voltage magnitudes
V_a	early voltage
V_A, V_B	voltage sources of circuits A and B
V_a, V_b, V_c	potentials at the ground points a, b, c of circuits A, B, C
V_{AC}	AC voltage
v_{BC}	base–collector voltage
v_{BE}	base–emitter voltage
V_{bi}	built-in potential
V_{CE}	collector–emitter voltage, voltage conversion efficiency
v_{CES}	collector–emitter voltage in saturation mode

V_{CM}	common-mode voltage
V_d	voltage across SBD
V_{DC}	DC voltage
V_{DD}	drain supply
V_{DM}	differential-mode voltage
V_{DS}	drain–source voltage
V_D^{Si}	forward voltage of Si diode
V_D^{SiC}	forward voltage of SiC diode
V_{FB}	flatband voltage
V_G	voltage drop produced across resistor R_G
V_{GS}	gate–source voltage
V_{GS} (ZTC)	V_{GS} at ZTC point
V_{IN+}, V_{IN-}	input voltage terminals of the Wheatstone bridge
v_{nB}	velocity of electrons at the emitter end of the base
V_{NMH}	high-level noise margin
V_{NML}	low-level noise margin
V_{OH}	output high voltage
V_{OL}	output low voltage
V_{OUT}	output voltage of the Wheatstone bridge
V_{OUT+}, V_{OUT-}	output voltage terminals
V_{Output}	voltage of the output signal
V_{peak}	peak voltage
V_{po}	pinch-off voltage of MESFET
v_{pE}	velocity of holes at the base end of the emitter
V_R	reverse bias
V_{RLA}	voltage across the load resistance R_{LA} of the source circuit A
V_{RLB}	voltage across the load resistance R_{LB} of the receptor circuit B
V_{SA}	voltage source of circuit A
V_{SB}	substrate bias, voltage source of circuit B
V_{Signal}	voltage of the signal
V_{SS}	source supply
$V(t)$	time-dependent voltage
V_{sub}, $V_{substrate}$	substrate voltage
V_{Th}	threshold voltage of a MOSFET
$V_{Thermal}$	thermal voltage
v_{sat}	saturation velocity of carrier
v_n^{sat}	saturation velocity of electron
v_p^{sat}	saturation velocity of hole
W	bandwidth of the input information signal in DSSS
w	depletion layer width
W, W_B	base width of a bipolar transistor, channel width of an FET
W_0	work done at room temperature
W_C	work done by the gas during expansion in the engine at temperature T_C
W_D	depletion region thickness at the drain
W_d	bandwidth of the input data signal in MFSH-FHSS
W_I	threshold ionization energy
W_S	depletion region thickness at the source
W_s	bandwidth of the spread signal in MFSH-FHSS, the bandwidth of the spread signal in DSSS

X_C	capacitive reactance
X_L	inductive reactance
$Xtalk_{BA}$	crosstalk parameter between receptor circuit B and source circuit A
$Z_A, Z_B, Z_C, \ldots, Z_N$	impedances of wires W_A, W_B, W_C, …, W_N

Greek/other symbols

α	fitting parameter in the Varshini equation, index of temperature for mobility variation, ionization coefficient, current gain of a bipolar transistor in the common-base configuration, an empirical constant determining the saturation voltage of the drain current of a MESFET
α_{it}	coefficient for interface states
α_n	ionization coefficient of an electron
α_{ot}	coefficient for oxide-trapped charges
α_p	ionization coefficient of a hole
α_T	base transport factor
β	fitting parameter in the Varshini equation, current gain of a bipolar transistor in the common-emitter configuration, transconductance parameter of a MESFET containing the electron mobility μ_n
β_{max}	maximum current gain
β_{NPN}	current gain of n–p–n transistor
β_{PNP}	current gain of p–n–p transistor
β_{Si}	current gain of Si BJT
$\beta_{Si/SiGe}(T)$	current gain of Si/SiGe HBT as a function of temperature T
γ	emitter injection efficiency of a bipolar transistor, the effective threshold voltage displacement with V_{DS} for a MESFET
δ	phase difference
ΔC	change in capacitance
ΔE_A	activation energy of acceptor impurity
ΔE_D	activation energy of donor impurity
ΔE_g	energy bandgap difference between the emitter and base semiconductor materials
ΔE_{gB}	bandgap narrowing of the base
ΔE_{gE}	bandgap narrowing of the emitter
$(\Delta E_g)_{Si/SiGe}$	difference between energy bandgaps of the Si emitter and Si_xGe_{1-x} base
ΔE_V	valence band offset
Δf	difference between frequencies f_1 and f_2
ΔN_{it}	interface trap density
ΔN_{ot}	oxide-trapped charge density
ΔR	change in resistance
Δt	time interval
$\Delta\xi_g$, $\Delta\xi_{gBE}$	decrease in bandgap of the emitter relative to base
$\Delta\Phi$	phase difference between junctions A, B in a SQUID
ε	dielectric constant
ε_0	permittivity of free space
ε_{ox}	relative permittivity of silicon dioxide
ε_s	relative permittivity of silicon
η	emission coefficient or ideality factor in diode equation, shielding efficiency
η_A	efficiency of absorption
η_a, η_b	empirical constants
η_{MR}	efficiency of multiple reflections
η_R	efficiency of reflection
θ	gauge-invariant phase difference, contact angle at solid–liquid interface
$\theta(\mathbf{r})$	phase of the wave function

κ	TC of threshold voltage
λ	mean free path of carrier, channel length modulation parameter of MESFET, wavelength of the wave, air–fuel ratio in the combustion chamber of automobile
λ or λ_L	London penetration length/depth
λ_0	high-energy low-temperature asymptotic phonon mean free path
μ	mobility of carrier, permeability of the material
μ_0	vacuum permeability
μA	microampere
μ_I	ionized impurity scattering-limited mobility
μ_i	initial mobility
μ_f	final mobility
μJ	microJoule
μ_L	lattice scattering-limited mobility
μm	micron (micrometer)
μ_n	mobility of an electron
μ_{nB}	mobility of electrons in the base
μ_p	mobility of a hole
μ_{pE}	mobility of holes in the emitter
μV	microvolt
μW	microwatt
ξ	coherence length (for Cooper pairs)
ξ_g	bandgap in the emitter
Ξ	an integer
ρ	resistivity of a substance
ρ_1, ρ_2	densities of the Cooper pairs
ρ_{ox}	charge density in the oxide
$\rho(\mathbf{r})$	Cooper pair concentration in the superconductor where **r** is the position vector
σ	conductivity
σ_v	an empirical constant
τ	carrier lifetime, time constant
τ_B	base transit time
τ_E	emitter transit time
τ_F	forward transit time, final lifetime
τ_{HL}	high-level lifetime
τ_I	initial lifetime
τ_{LL}	low-level lifetime
τ_p	lifetime of holes
τ_{pE}	lifetime of holes in the emitter
Φ	magnetic flux threading a loop
Φ_0	magnetic flux quantum (fluxon)
ϕ	fluence, phase difference between the wavefunctions ψ_1 and ψ_2; $\phi = \phi_2 - \phi_1$
ϕ_0	phase difference when $V = 0$
ϕ_1, ϕ_2	phases of the wavefunctions of Cooper pairs, phases of two signals V_1, V_2
Φ_{A1}, Φ_{A2}	phases adjacent to the junction A in a SQUID
Φ_{B1}, Φ_{B2}	phases adjacent to the junction B in a SQUID
Φ_A, Φ_B	phase differences across the JJs A and B in a SQUID
ϕ_B	Schottky barrier height
Φ_{bn}	Schottky barrier height of metal–semiconductor interface
$\Phi_{External}$	external magnetic flux

ϕ_F	bulk potential
ϕ_{ms}	metal–semiconductor work function
ψ	wavefunction
ψ_1 and ψ_2	wavefunctions of Cooper pairs
$\psi(\mathbf{r})$	wavefunction of Cooper pair electron where **r** is the position vector
ω	angular frequency
Ω	ohm
Ω-cm	ohm-centimeter
ω_n	natural angular frequency
Ω_s	excitation frequency
∇	nabla (differential operator)

Part I

Environmental hazards and extreme-temperature electronics

Sub-part IA

Environmental hazards

Chapter 1

Introduction and overview

Electronic equipment is often required to work in environments which are very difficult compared to their normal working conditions. These environments involve high temperatures and harsh conditions. Effects of temperature on silicon- and wide bandgap semiconductor-based electronics as well as on passive components are discussed. Superconductive electronics offers several opportunities. Among harsh environment effects, humidity, chemical corrosion, radiation, vibrations and shocks are prominent. Electromagnetic interference calls for building systems compliant with electromagnetic compatibility. Medical electronic devices implanted in the human body must be benevolent to the body, and therefore need special attention to packaging. Normal sensors cannot work in aggressive conditions. Electronics built for space missions makes various precautionary measures essential for reliable long-lasting operations. No less important is the protection of electronics from mischievous activities of jammers and hackers. All these topics to be described in the book are touched upon, and an organizational plan of the book is sketched.

1.1 Reasons for moving away from normal practices in electronics

Extreme-temperature and harsh-environment electronics begins where routine conventional electronics designed for operation in a room temperature friendly environments ends. It breaks away from the traditional treatment of electronics to cover aspects which may, at times, appear less friendly and more antagonistic to electronic circuit operation, although there are exceptions to this rule, as we shall see below. Necessity is the mother of invention. The rationale for taking excursions from routine electronics is that in many applications it becomes imperative to build electronic systems that have

doi:10.1088/978-0-7503-5072-3ch1

to perform satisfactorily and reliably for long durations in unfavorable circumstances (Werner and Fahmer 2001, Alhendi *et al* 2022). Such conditions may prevail when we dig deep into the Earth, when we move out into space, when electronic equipment is placed near nuclear reactors and particle accelerators, when the operation of heavy machinery creates vibrations in buildings, when equipment has to withstand high humidity, rainy and stormy weather, to name just a few of the aggressive situations one can contemplate. So, the need for deviation from conventional electronics is primarily driven by the increasing demands from users and customers who work in such hostile conditions (Johnson *et al* 2004).

But there is a secondary reason as well. This reason originates from the realization that many physical phenomena, such as superconductivity, take place only at temperatures which are far below room temperature. To apply such phenomena for human use, the temperature has to be deliberately decreased close to absolute zero, or at least to the vaporization temperature of liquid nitrogen. Here, the basic operational principle of electronic devices and circuits imposes the requirement of breaking the norms of working at normal temperatures. This applies not only from the viewpoint of superconductivity phenomena; many electrical parameters of semiconductor devices show improvement as the temperature is decreased. Thus one should not consider that deviations from the norm will always lead to an aggressive and incompatible situation. It may be a fortunate situation as well.

In fact, the properties of semiconductors change over a very wide range as one increases the temperature from −273 K to 1000 K. This variation of properties of semiconductors is visible in the form of changes in the electrical behavior of the electronic devices fabricated from them (Stoica *et al* 2016). Some electrical parameters tend to improve at low or high temperatures, while others show deterioration. The comprehensive study of these behavioral trends helps scientists take advantage of the changes that are beneficial in the utilization of electronic circuits when we are operating outside the recommended range of temperatures.

In response to both types of situations portrayed above, one stemming from application-specific requirements and the other from phenomenological needs, one has to move away from regular practice and deal with challenging situations to meet one's aims.

1.2 Organization of the book

This book is subdivided into 25 chapters. The book (chapters 1–25) is grouped into two parts I and II (see figure 1.1). Part I, consisting of chapters 1–14, is divided into two sub-parts IA and IB.

Sub-part IA (present chapter 1 and chapter 2) deals with environmental hazards, namely, very high and very low temperatures, humidity, chemical corrosion, radiation, vibration, etc. It elaborates on the motivations for departure from treading the established path of electronics. The reader will come across many situations and applications that show the deficiencies of conventional electronics in solving problems. These examples also illustrate the need to shift away from the normal course to benefit from the advantages of utilizing phenomena that take place only under special conditions.

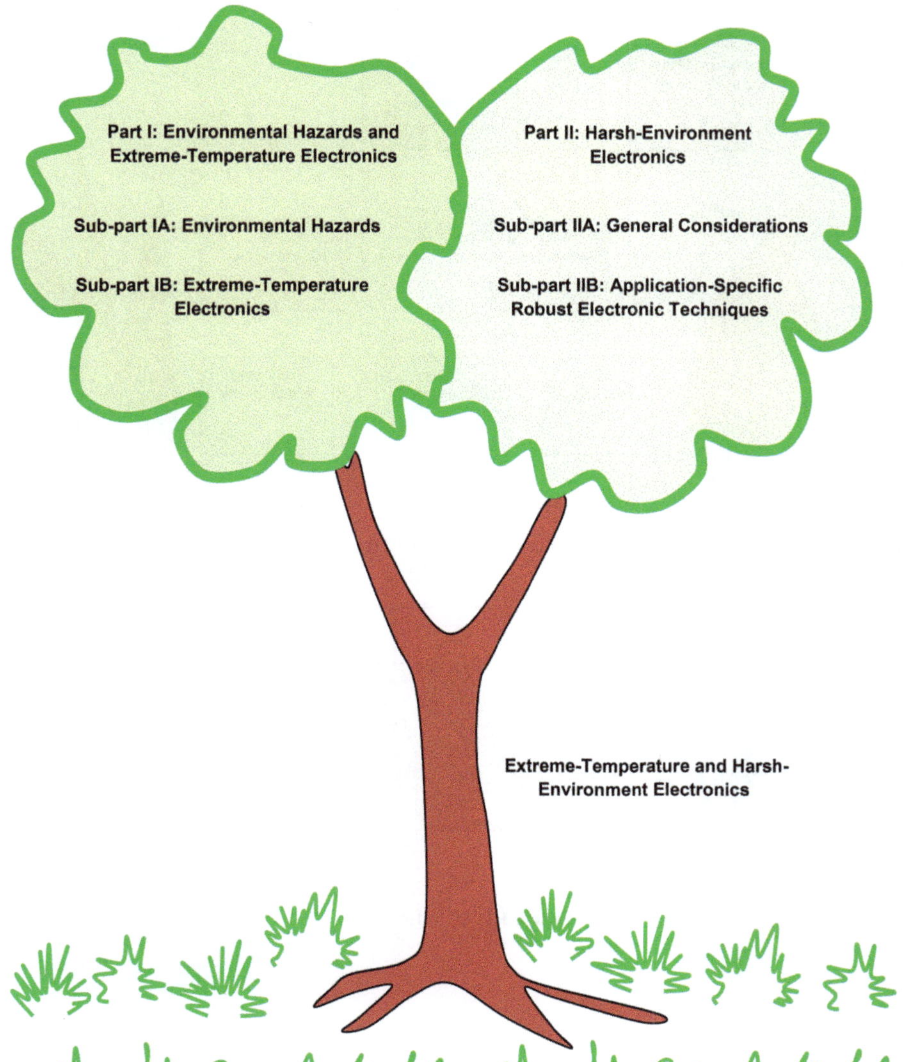

(a)

Figure 1.1. Organization of the book: (a) subdivision of 'extreme-temperature and harsh-environment electronics' tree into parts I and II perched on the branches of the tree, (b) explanation of part I, (c) details of sub-part IIA, and (d) exposition of sub-part IIB.

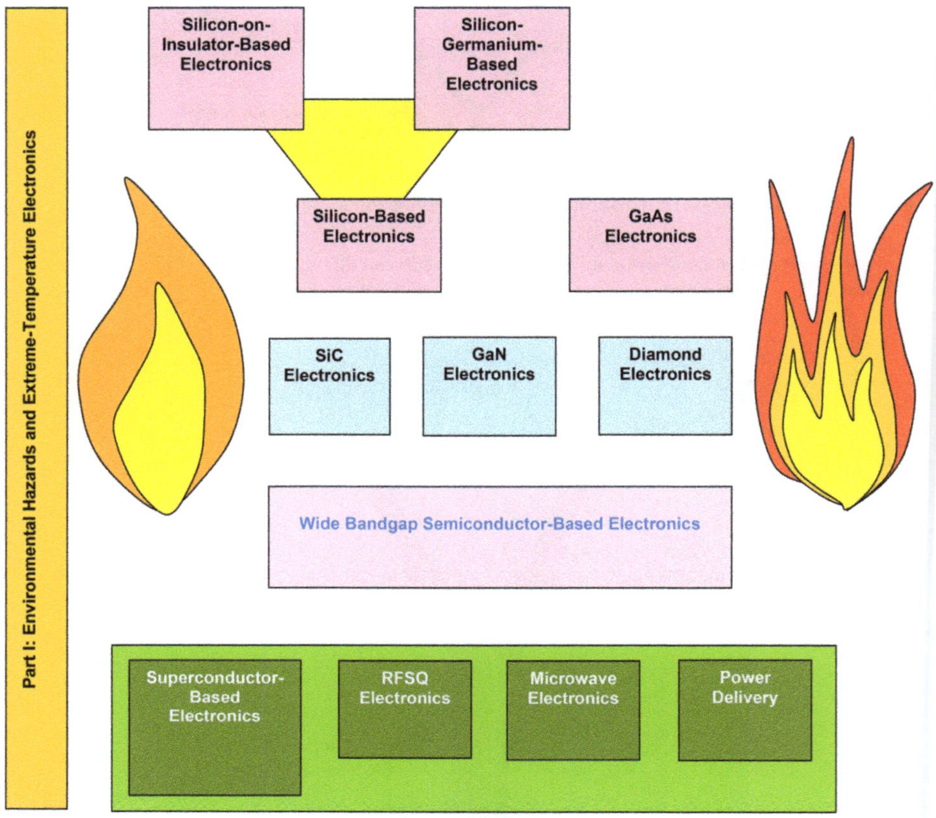

(b)

Figure 1.1. (Continued.)

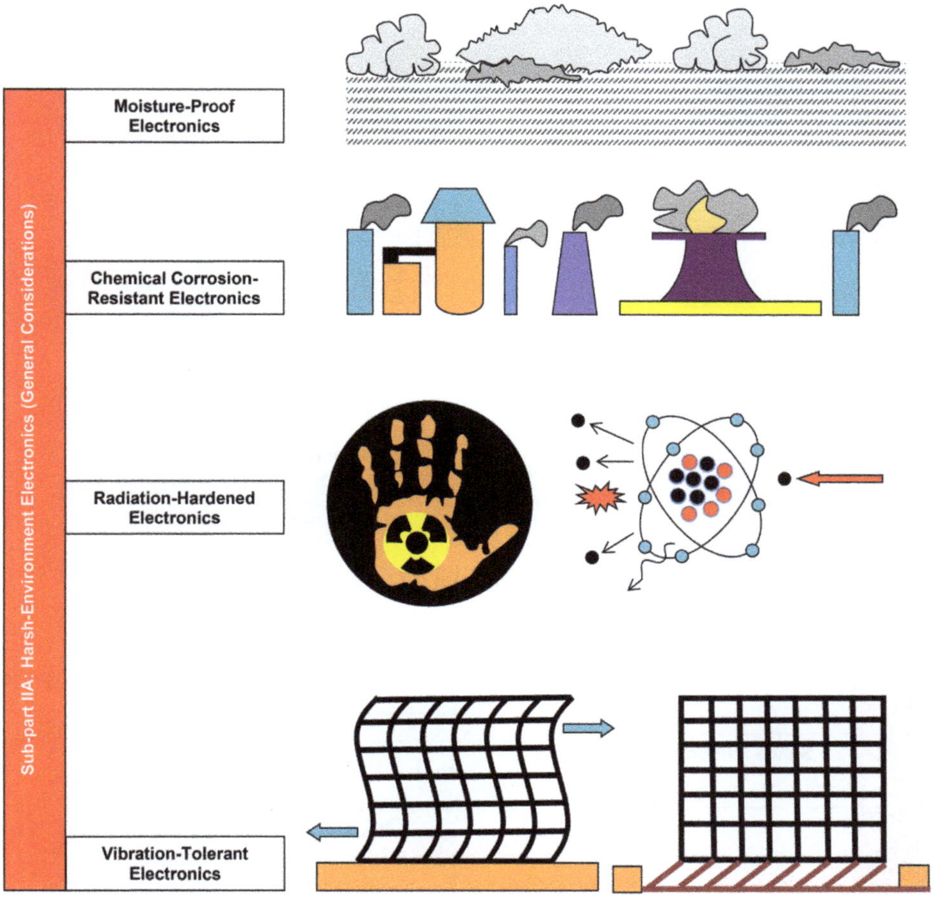

(c)

Figure 1.1. (Continued.)

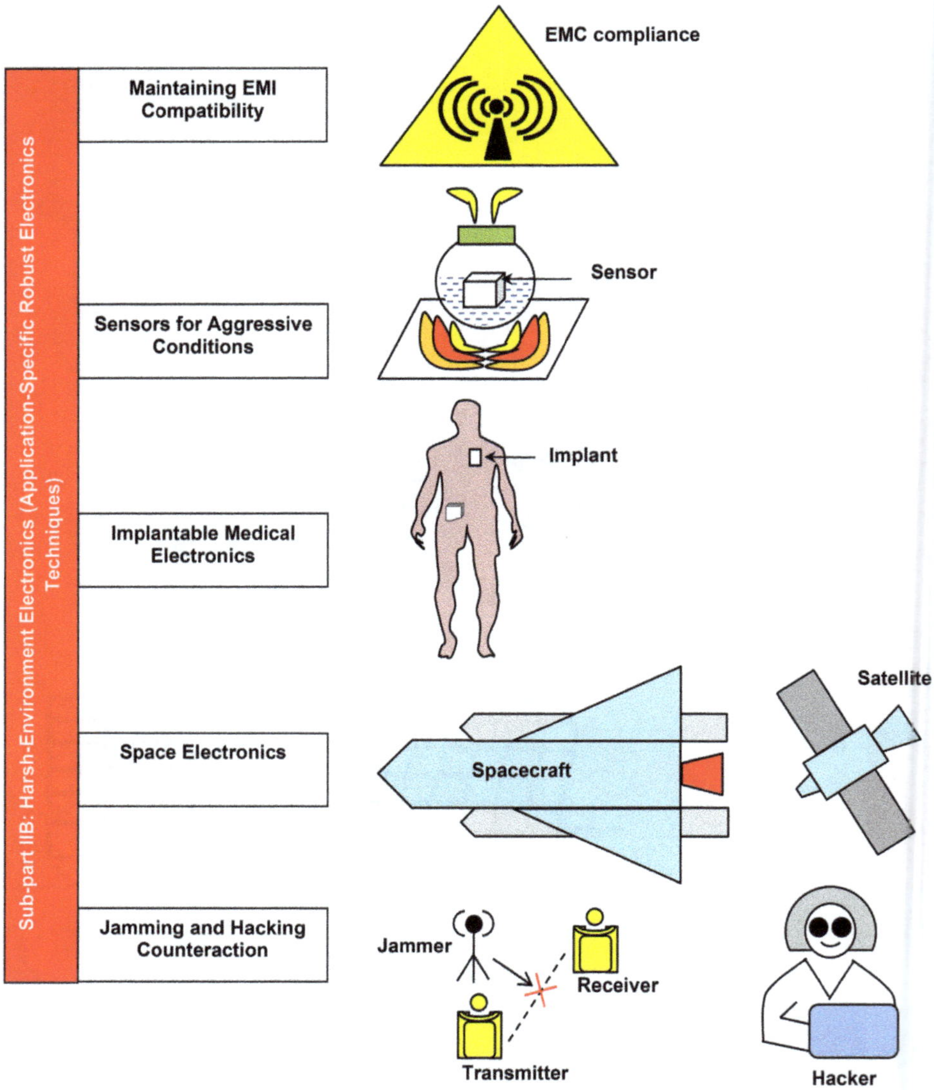

(d)

Figure 1.1. (Continued.)

Sub-part IB (chapters 3–14) treats extreme-temperature electronics (ETE). It examines the measures adopted in countering the harmful effects of very high temperatures as well as the processes developed for taking advantage of the beneficial effects of very low temperatures, as in superconductive electronics. It also includes utilization of the characteristics of semiconductor devices which improve with a fall in temperature, e.g., the leakage currents. Thus, temperature effects manifest themselves as both a curse and a boon for electronics exhibiting hostile as well as friendly behavior.

Part II comprising chapters 15–25 treats harsh environment electronics (HEE). Part II consists two sub-parts, IIA and IIB. Chapters 15–20 fall under sub-part IIA while chapters 21–25 come under sub-part IIB.

Sub-part IIA covers the general aspects of phenomena which are detrimental to electronic circuit operation, such as: highly humid climatic conditions; corrosive environments inside or in the neighborhood of chemical factories; in radiation-contaminated areas, such as near nuclear power stations, x-ray or gamma-ray equipment in hospitals; and in vibrating buildings amidst the busy and heavy traffic of metropolitan cities or where heavy machinery is being used, making the surroundings noisy and shaky.

Sub-part IIB concentrates on building rugged electronics for specific applications. Note that part I and sub-part IIA are focused on conventional classification of electronics into thermal and harsh segments. But there may be cases where there may be mixtures of variables, e.g. both thermal and harsh such as in the case of sensors, which may be required to operate both in hot and corroding ambience. In space flights, several difficult situations like vibrations, high/low temperatures, outgassing, etc are simultaneously faced. There are also distinctive circumstances in which the environment does not appear harsh in the classical sense, e.g., one may require satisfactory functioning of electronic equipment in places afflicted with electromagnetic noise pollution or vacuum conditions in outer space which do not seem so harsh at the first instance but are unsuitable for electronics. The biological environment of the human body looks so benign but may reject an implantable electronic device as a foreign entity. Last but not the least, it is possible to block the operation of electronic instrumentation by eavesdropping, jamming or similar inimical activities. This is an example of a hostile environment. All such special environments, which have not been covered in part I and sub-part IIA, constitute the contents of the sub-part IIB (chapters 21–25).

1.3 Temperature effects

1.3.1 Silicon-based electronics

Chapter 3 explains how the properties of semiconductors change as a function of temperature. Since variation of these properties is reflected in the thermal behavior of devices, a thorough understanding of this chapter lays a firm foundation for understanding the contents of ensuing chapters. The upper temperature limit of a semiconductor material is fixed by its bandgap energy. Device operation is governed by the concentration of free carriers in a pure semiconductor, known as the intrinsic carrier concentration. The intrinsic carrier concentration is an exponential function of temperature. An increase in temperature augments the energy of electrons in the valence band of a semiconductor material. At a particular temperature called the intrinsic temperature, the thermal energy of electrons exceeds the bandgap energy of the semiconductor. Then the electrons are promoted from the valence band to the conduction band. The number of thermally generated carriers becomes equal to the number of free carriers due to impurity doping, either n- or p-type. Then there are no longer any distinguishable n- or p-type regions. The p–n junction reduces to a resistor and its function is paralyzed.

On the opposite side, as the temperature of a semiconductor decreases towards absolute zero, the ionization of impurity atoms to release free carriers ceases. Then there are no, or very few, carriers available for conduction. In this carrier freeze-out regime as well, the normal operation of a semiconductor device stops. In fact, a semiconductor is an insulator at 0 K.

Apart from the liberation of electrons from their bonds rendering possible their availability for conduction of electrical current, other noteworthy phenomena include the scattering of carriers by lattice atoms and impurity ions, which affect the ease of movement of carriers through the crystal lattice, i.e. the carrier mobility. Mobility is a strong function of temperature. As lattice atoms vibrate with greater amplitudes with rising temperature, the electrons undergo more collisions on their paths and mobility decreases. Impurity scattering limited mobility varies in the opposite way, because the increased vibration of impurity ions makes them less effective in influencing electron motion. The temperature dependence of carrier mobilities is also affected by whether a semiconductor is non-degenerately or degenerately doped.

From chapter 4 onwards up to chapter 10 we move towards discussion, assessment of the capabilities and appraisal of the critical issues concerning electronic devices and circuits made from semiconductor materials of progressively increasing energy bandgap, and hence intrinsic temperature. In this sequence, the starting material of interest is silicon, which has been the favorite of electronic engineers for a very long time and has reigned as the king of electronic materials. Silicon electronics forms the contents of chapters 4 and 5. In chapter 4, bipolar silicon devices are addressed. Chapter 5 focuses on MOS silicon devices. Silicon electronics has two forms: bulk silicon and silicon-on-insulator (SOI) technologies. The objectives of these chapters are to present simple analytical formulae for the temperature coefficients (TCs) of silicon bipolar and MOS devices. These derivations help in appreciating the degradation or upgradation in electrical characteristics of bipolar/MOS silicon devices and circuits subjected to constantly rising temperatures.

The forward voltage drop across a p–n diode or Schottky diode decreases with temperature, the current gain of a bipolar transistor increases with temperature and the breakdown voltage of a diode increases with temperature. In almost all circuit applications, the leakage currents of p–n junctions should be kept infinitesimally small with respect to the signal currents. These leakage currents increase exponentially with temperature. In complementary metal–oxide semiconductor (CMOS) structures, junction leakage occurs at the junctions between source/substrate and drain/substrate due to the minority carrier diffusion current near the depletion region, together with electron–hole pair (EHP) generation inside the depletion region. Gate leakage increases with thinning of the gate oxide as MOSFETs are scaled towards smaller dimensions. The threshold voltage decreases linearly with rise in temperature and hot carrier effects become less pronounced with increasing temperature. High-cost SOI technology considerably helps in obviating the leakage current issues in bulk silicon devices at high temperatures.

Silicon and germanium have been two key materials forming the backbone of electronics. Silicon–germanium, an alloy of silicon with germanium, combines the best properties of both materials. Silicon/silicon–germanium heterojunction bipolar transistors (HBTs) overcome the problem of a rapid fall in the current gain and switching frequency of silicon bipolar temperatures at cryogenic temperatures. Chapter 6 treats the mathematical theory of the HBT and shows how this device serves as a replacement for bipolar transistors, showing much better performance at these temperatures.

Silicon was the first-generation material of the twentieth century. It played a leading role in ushering in the microelectronics revolution. Time and again, it has appeared that silicon electronics has hit a wall, prompting the search for alternative materials. Silicon was followed by gallium arsenide in the second generation of semiconductors. Together with silicon, gallium arsenide set off the information technology and wireless revolution around the turn of the twenty-first century. Chapter 7 deals with gallium arsenide, which is superior to silicon in the fabrication of ultra-fast radio-frequency (RF) devices, and is also suitable for making opto-electronic devices such as LEDs and laser diodes. In contrast to silicon technology, where the primary devices are bipolar transistors and MOSFETs, GaAs relies on MESFETs and HBTs.

1.3.2 Wide bandgap semiconductors

The wide bandgap semiconductors, silicon carbide and gallium nitride, belong to the third-generation semiconductors. They heralded the optoelectronics and high-temperature electronics (HTE) era at the beginning of the twenty-first century. The capability of silicon carbide and gallium nitride chips to operate at higher temperatures, voltages and frequencies promoted the research interest in these materials. Approximately three times the energy used in the case of silicon is needed to transport an electron from the conduction to valence band in SiC and GaN, which makes both these materials ideal candidates for realizing high-temperature devices with high breakdown strength. Naturally, the power electronic modules built from these materials are significantly more energy efficient.

The manufacturers of silicon carbide wafers have been able to minimize the defect density considerably as well as increase the size of the wafers. Technological breakthroughs in SiC materials and process techniques have led to the realization of several devices such as p–n diodes, Schottky barrier diodes (SBDs), JFETs, bipolar transistors, etc SiC JFETs are very attractive for HTE, but SiC MOSFETs still need a lot of improvement. In SiC, the interface state density is high and the carrier mobility is very low. Scrupulous efforts are being made in the development of SiC thyristors, IGBTs and gate turn-off thyristors (GTOs). Chapter 8 provides glimpses into achievements in silicon carbide electronics.

Gallium nitride is an excellent material for the fabrication of light-emitting diodes and power transistors. Advancing from gallium nitride wafers on sapphire substrates to free-standing gallium nitride wafers with low dislocation density is likely to boost confidence in GaN technology. GaN MESFETs and HEMTs have been developed and tested at high temperatures. Chapter 9 takes a look at some of the GaN devices.

After silicon carbide and gallium nitride, the superior properties of diamond make it suitable to catapult another major upheaval in microelectronics. Diamond has been crowned as the ultimate semiconductor material due to its spectacular combination of properties, including high thermal conductivity and radiation hardness. Diamond electronics is still immature because the full capabilities of diamond have so far evaded utilization. The ability to synthesize diamond from the vapor phase to produce large-area films has generated a lot of interest in diamond-based devices, and the scenario seems to be gradually changing as single-crystal electronic grade (EG) diamond is becoming commercially available. Chapter 10 deals with the synthesis, processing and characteristics of diamond films and related devices.

1.3.3 Passive components and packaging

The aforementioned few chapters concentrate on wide-bandgap semiconductors and the active devices fabricated from them, the aim being to fabricate semiconductor devices that are able to withstand successively higher temperatures. Apart from semiconductors, the other important materials used in electronic devices are the metals used for contact electrodes (Sharif 2019). No electronic circuit can be fabricated without passive components. Furthermore, the devices have to be packaged within safe enclosures to protect them from mechanical damage, as well as atmospheric and weather effects. The electronic circuit will fail at elevated temperatures if the active or passive components, metallization or packaging are not up to the mark. From these considerations, chapter 11 diverts attention from semiconductors towards resistors, capacitors, metal interconnections and packaging. Carbon resistors show good thermal stability. Diamond resistors are fabricated from CVD diamond on an aluminum nitride substrate. Teflon capacitors can be used up to 200 °C and mica capacitors up to 260 °C. Diamond capacitors are based on a dielectric film of diamond with Au contacts. Low dielectric loss and constant capacitance are observed up to 450 °C.

In addition to good adhesion with the underlying silicon, the metallization must be able to withstand thermal cycling and must not decompose or undergo chemical reactions at high temperatures. Films of refractory metals and refractory metal silicides formed by chemical vapor deposition (CVD) serve as a useful metallization for high temperatures. This method provides selective deposition over silicon regions, leaving aside oxides and insulating areas. Electrolytic Cr–Ni–Au metallization is a robust scheme for wire bonding applications exposed to high temperatures (250 °C).

During die attachment, care must be taken to match the coefficients of thermal expansion of the die, die-attach and substrate. This avoids any mechanical stressing or fracturing of the die during thermal cycling. Die-attach materials proven for room-temperature operation cannot be used owing to their low glass transition temperature.

For the reliability of wire bonds at high temperatures, the metals used for the wire and the metallization of bonding pad must be mutually compatible. Poor compatibility of metals involved in wire bonding leads to two types of problem. Either an intermetallic compound is formed at the interface between the metals causing brittleness of the bond and breakability, or voids are formed at the interface by diffusion, an effect called the Kirkendall effect, thereby weakening the bond. Inopportunely, the common Au–Al combination between the Au wire and Al metallization pad is prey to such phenomena. These arguments entice us to use the same metal for the wire bond and the bond pad.

For high-temperature operation, hermetic ceramic packages are far better than plastic packages. They also serve as moisture and contamination impenetrable barriers. Limiting the ingress of moisture and dirt prevents corrosion. Regrettably ceramic packages are larger, heavier and costlier than plastic packages; the latter, however, cannot be used past 150 °C. High melting point solders with a melting point >250 °C must be used with ceramic packaging.

1.3.4 Superconductivity

The following three chapters (chapters 12–14) are concerned with superconductivity, both low-temperature and high-temperature. Superconducting films exhibit low resistance even at frequencies of approximately a few hundred GHz, paving the way for their utilization in magnetometry using superconducting quantum interference devices (SQUIDs), microwave filters, transmission lines, etc. Rapid single-flux-quantum (RSFQ) logic electronics uses Josephson junctions (JJs) to perform logic operations based on the quantization of magnetic flux. HTS-based power transmission uses cables comprising hundreds of strands of HTS wire with a cryogenic cooling system to maintain the required low temperature. In dense urban localities, power substations often reach capacity limits. HTS systems bind these substations together circumventing expensive transformer upgradation. HTS power delivery is used to load pockets in high-demand metropolitan areas with a saturated grid.

1.4 Harsh environment effects

1.4.1 Humidity and corrosion effects

Temperature is an important parameter, and has been the focus of the attention of electronic engineers because by increasing/decreasing the temperature, both advantageous and disadvantageous results can be produced on electronic equipment. In addition to temperature, humidity is another vital parameter. Humidity-related failures and remedial schemes are discussed in chapters 15 and 16. Through moisture condensation, water droplets accumulate on the surfaces of semiconductor devices, producing ionic current flow. The effects of humidity depend on the materials used, the dimensions of the components and their layout. Humidity accelerates the corrosion rate. Corrosion is the deterioration of the materials constituting the electronic devices and circuits under the influence of reactive gases in the environment. The principal reason for the vulnerability of electronic products to corrosion is the large variety of metals and alloys used in the electronics industry. A few of these are aluminum, copper and its alloys, silver and its alloys, tin and its alloys, titanium, chromium, nickel, gold, platinum, palladium, tungsten, etc. Apart from humidity, corrosion depends on inorganic and organic contaminants, atmospheric pollutants, salt spray, noxious gases, and residues from soldering and other assembly processes. Corrosion increases the contact resistances between joints and leakage currents between wires and decay products. Corrosion by water, dust and gases causes short circuits and produces ugly surfaces. It can initiate cross-talks. As a consequence of the reduced spacing between components in the wake of miniaturization of circuits, a material loss of a few pictograms due to corrosion is adequate to spawn a fault. Therefore, the impact of corrosion becomes more magnified in damaging electronic circuits. Chapter 17 suggests ways to mitigate corrosion problems.

1.4.2 Radiation effects

Next to temperature and humidity comes the radiation from space as well as terrestrial sources. These types of radiation have a negative influence on electronic circuits located on Earth, those placed in orbiting satellites or used in long-distance space flights. These defects range from performance degradation to complete loss of functionality. Due to these radiation-induced failures, satellite lifetimes are shortened and space missions are disrupted. Radiation effects can be one or more of the following types: single-event effects (SEEs), displacement damage (DD) and total ionizing dose (TID) in the form of a cumulative effect over a long time span. Radiation-hardened circuits are modified versions of non-hardened equivalents, incorporating revised designs for fault-tolerance and software approaches to deal with disturbances, along with suitable manufacturing process amendments. Bipolar ICs are more radiation-hard than CMOS circuits. Radiation-hard circuits are fabricated in SOI or silicon-on-sapphire substrates, instead of the common bulk silicon wafers, because of the higher leakage currents produced by radiation in junction isolated devices made on bulk Si. Chapter 18 describes the detrimental effects of radiation exposure on electronic circuits. Feasible ways to thwart the disturbances and damage produced by radiation in electronic circuits are suggested in chapter 19.

1.4.3 Vibration and mechanical shock effects

Last but not the least, the damaging effects of vibrations, impacts, kicks, drops and shocks, as caused by acceleration/deceleration and impulsive forces on electronic circuitry cannot be ignored. Chapter 20 describes the common techniques that must be adopted for protection of electronics from vibrations.

1.4.4 Electronics in electromagnetic interference environments

With the expansion of cyber physical systems, electrical transportation and smart grids, the shrinking distances between power conditioning and logic circuits lead to electromagnetic interference concerns. Measures to deal with these situations are discussed in chapter 21.

1.4.5 Sensors in aggressive environments

While standard electronic devices and circuits are carefully packaged to protect them from environmental degradation, sensors are an exclusive class of devices, which must closely interact with the environment and often cannot be protected. Technological innovations have paved the way towards development of novel sensors that can be used in these conditions. These devices are treated in chapter 22.

1.4.6 Medical implant electronics

Frequently, medical electronics devices are implanted in the human body for correction of erroneous functions or supplementing the biological activities. The body has a tendency to reject foreign objects. Foreign body response is a primary issue in such cases. But the situation is not so simple. Neither the toxic materials in the electronics device should seep into the body to harm the patient nor should the body fluids leak into the package to damage the electronics. Chapter 23 looks into devices built for these applications.

1.4.7 Space environment electronics

Manned missions in outer space bring problems, which are altogether different from those in terrestrial settings. Multifarious problems confronting electronics in space flights form the content of chapter 24. These include mechanical shocks and vibrations, wide temperature excursions, radiation bombardment, outgassing, tin whisker growth and a plethora of unforeseen difficulties.

1.4.8 Jamming attacks prevention, and cyber security

Electronic signals are prone to attacks by jammers. Electronic communication can be disrupted by jamming causing great inconvenience to users. Computers are vulnerable to hacking threats from people with mala fide intentions. Therefore, electronic telecommunication and computing facilities must be built to withstand such threats. Chapter 25 presents methods to defend against them to be able to work in a hostile environment.

1.5 Discussion and conclusions

The temperature effects on electronic circuits must be carefully understood, taking due consideration of their pros and cons. Pernicious effects must be suppressed. Beneficial effects must be gainfully exploited to execute utilitarian functions, as an enhancement to system functionality. The effects of humidity, corrosion, radiation and vibration need to be addressed according to the specific application. They have to be dealt with on a case-by-case basis. Some ideas of this chapter are succinctly described in the following poem:

Safeguarding electronics

Extreme temperatures are sometimes hostile, sometimes cozy;
Sometimes thorny, sometimes rosy;
Away from room temperature, some device parameters downgrade,
Other parameters upgrade.
But high humidities and aggressive chemicals
Are always detrimental.
Vibrations too are harmful.
So, please be careful
To make electronic design fault-tolerant,
And package environment-resistant.
Wide bandgap semiconductors are promising,
Superconductive electronics is amazing.
To ensure success of mission,
Choose electronic design, materials and fabrication
According to application;
Provide proper surface passivation
To protect from moisture invasion.
Prevent chemical corrosion.
Take precautions for radiation
And cushion electronic product against vibration.
Keep in touch with new developments and innovations.

Review exercises

1.1. Contemplate two situations, one negative or unfavorable situation and one positive or favorable situation, which urge electronic engineers to develop circuits that can work in non-conventional conditions.
1.2. What property of a semiconductor material fixes the upper permissible limit of temperature up to which an electronic device made from it may be operated?
1.3. Why is operation of a semiconductor device not possible at a temperature exceeding its intrinsic temperature?
1.4. Give some examples illustrating the beneficial effects of low temperatures for electronic device operation.
1.5. Explicate, with reference to carrier mobility, the outcomes of scattering of electrons by lattice atoms and impurity ions. Elucidate the influence of temperature on mobility variations caused by the two types of scattering.
1.6. Cite three examples showing the variation of the most important electrical parameters of semiconductor devices with temperature.
1.7. Does temperature affect the leakage current of a p–n junction? If so, why does it happen? Describe the various types of leakage currents that a CMOS structure is susceptible to.
1.8. In what manner is a silicon–germanium heterojunction transistor superior to a silicon bipolar transistor for operation at cryogenic temperatures?

1.9. Mention two application areas where gallium arsenide devices find wide usage.
1.10. In what ways can silicon carbide and gallium nitride outperform silicon and gallium arsenide devices?
1.11. Is it possible to synthesize diamond artificially?
1.12. Give examples of resistors and capacitors developed for high-temperature applications.
1.13. What kind of contact metallization can be used in HTE?
1.14. What precautions are necessary for selecting appropriate die-attach materials to be used at high temperatures?
1.15. What problems occur due to poor compatibility of the wire metal and metallization pad?
1.16. What are the advantages and disadvantages of hermetic ceramic packages?
1.17. Give three applications of superconductivity.
1.18. What is the physical basis of RSFQ logic?
1.19. Why is HTS-based power delivery beneficial in upgrading dense, congested urban power network?
1.20. Why are electronic products prone to corrosion effects? How do humid environments aggravate corrosion effects? How does corrosion impair the performance of electronic devices?
1.21. What are the three types of radiation effects on electronic circuits? What are the approaches followed for countering these effects?
1.22. Which type of ICs are more radiation-hard: bipolar or CMOS?
1.23. Which type of silicon wafers are used for the fabrication of radiation-hard circuits: bulk silicon wafers or SOI wafers?
1.24. Mention one adversary of electronic circuits, apart from temperature, humidity, corrosion and radiation.

References

Alhendi M, Alshatnawi F, Abbara E M, Sivasubramony R, Khinda G, Umar A I, Borgesen P, Poliks M D and Shaddock D *et al* 2022 Printed electronics for extreme high temperature environments *Addit. Manuf.* **54** 102709

Johnson R W, Evans J L, Jacobsen P, Thompson J R and Christopher M 2004 The changing automotive environment: high-temperature electronics *IEEE Trans. Electron. Packag. Manuf.* **27** 164–76

Sharif A (ed) 2019 *Harsh Environment Electronics: Interconnect Materials and Performance Assessment* (Weinheim: Wiley-VCH)

Stoica L, Riches S and Johnston C 2016 *High Temperature Electronics Design for Aero Engine Controls and Health Monitoring* (Gistrup: River Publishers)

Werner M R and Fahmer W R 2001 Review on materials, microsensors, systems and devices for high-temperature and harsh-environment applications *IEEE Trans. Ind. Electron.* **48** 249–57

Chapter 2

Operating electronics beyond conventional limits

Temperatures vary widely from place to place on the Earth. Electronics working satisfactorily at a given place may fail when taken to another place. Moreover, electronic functionality must be ensured, either by chip design/fabrication or packaging in chemically corrosive or humid environments or in laboratories where radiation exposure is expected. Electronics for space instrumentation or for journeys to distant planets demands extra care not only because of temperature effects, but also the requirements to withstand radiation hazards and several damaging phenomena. In view of these difficulties, special technologies dealing with electronic operations at regions exceeding conventional limits have been developed. This chapter outlines various situations where extreme-temperature and harsh-environment electronics is the answer to counter failure possibilities, and trepidations.

Degradation effects on the performance characteristics of semiconductor devices at elevated temperatures are described. The reason why the thermal management approach is insufficient to deal with the menace of high temperatures is explained, as are the beneficial and disadvantageous effects of low temperatures on device operation. Many interesting areas for the application of high-temperature electronics (HTE) are highlighted, notably in the automobile, aerospace and well-logging industries. The effects of humidity, radiation and vibration on device behavior are described. The terminology of low-temperature, high-temperature, extreme-temperature and harsh-environment electronics is elaborated. A useful strategy to overcome the problems faced consists in addressing all of the issues, starting from the chip design through process steps up to the packaging stage.

doi:10.1088/978-0-7503-5072-3ch2

2.1 Life-threatening temperature imbalances on Earth and other planets

Very hot temperatures are uncomfortable for human beings. Very cold temperatures are no less excruciating. At times, such temperatures can prove perilous to human health. For those working in a prickly hot environment, the most serious concern is posed by heat stroke, causing loss of consciousness or fainting (syncope). During heat stroke, the temperature-regulating mechanism of the body malfunctions. Relatively less serious is heat exhaustion due to overheating. This leads to heavy sweating accompanied by a fast pulse rate.

In very cold temperatures, the most serious danger is posed by hypothermia. It is a precarious overcooling of the body. In this condition, the body loses heat more speedily than it produces heat. The body temperature may fall below 35 °C. Another serious effect of exposure to severe cold is frostbite (Gupta *et al* 2021). This results in injury to the skin and underlying tissues of the extremities of the body, such as the fingers, toes, nose and ear lobes, at temperatures near or below the freezing point of water.

It is interesting to know the temperatures of some of the hottest and coldest places on Earth. Table 2.1 gives an idea of some such places, both hot and cold (Tavanaei 2013). One can easily imagine how demanding it is to live and survive in these places. The very thought of these temperatures makes us tremble and chill with fear. But temperatures on other planets of the Solar System are even more extreme (Williams 2014, Howell 2014). Depending on their respective distances from the Sun, the surface temperatures of the planets vary from >400 °C on Mercury and Venus to < −200 °C on the distant planets (figure 2.1). One can imagine the conditions on these planets from table 2.2. These temperatures may be considered as one of the reasons for the non-existence of living organisms on these planets.

Table 2.1. The six hottest and six coldest places on Earth.

Sl. No.	Hottest places		Coldest places	
	Place	Temperature (°C)	Place	Temperature (°C)
1.	Lut Desert in Iran	+70.7	Vostok Station, Antarctica	−89.2
2.	Death Valley, California, North America	+56.7	Oymyakon, Russia	−71.2
3.	Al'Aziziyah, Northwest Libya, Africa	+57.8	Verkhoyansk, Russia	−69.8
4.	Ghudamis, Libya, Africa	+55	North Ice, Greenland	−66
5.	Kebili, Tunisia, Africa	+55	Snag, Yukon, Canada	−63
6.	Timbuktu, Mali, West Africa	+54.5	Prospect Creek, Alaska, USA	−62

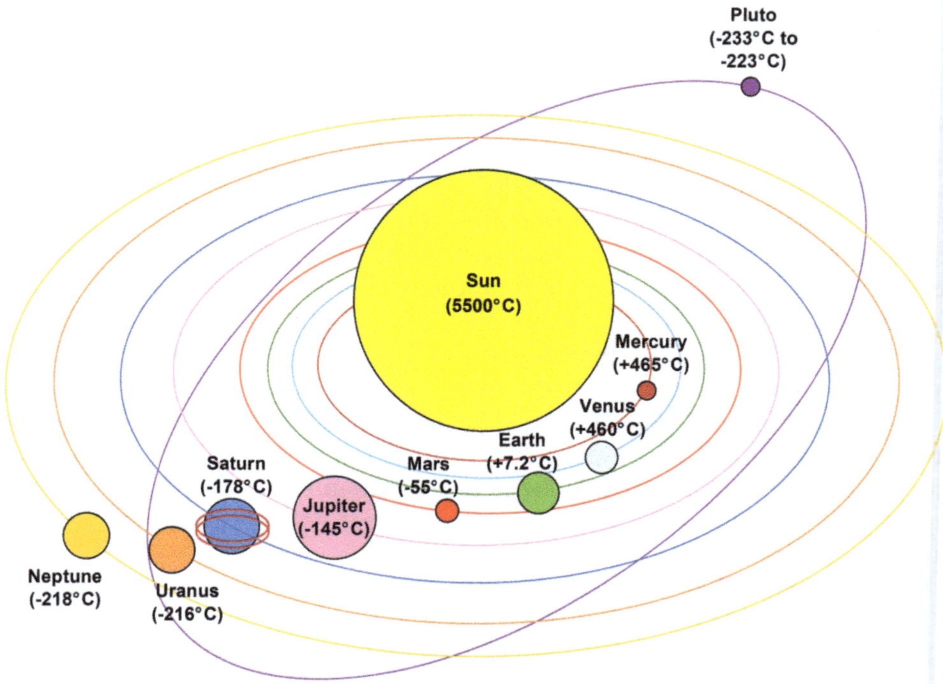

Figure 2.1. Illustrating the wide temperature variation in the Solar System.

Table 2.2. Average surface temperatures of planets.

Sl. No.	Planet	Average distance from Sun (km)	Temperature (°C)
1	Mercury	5.7×10^7	+465 (side exposed to Sun); −184 (dark side)
2	Venus	1.08×10^8	+460
3	Earth	1.50×10^8	+7.2 varying from +70.7 (the deserts of Iran) to −89.2 (Antarctica)
4	Mars	2.28×10^8	−55, ranging between as high as +20 at the equator during midday, to as low as −153 at the poles
5	Jupiter	7.79×10^8	−145
6	Saturn	1.43×10^9	−178
7	Uranus	2.88×10^9	−216
8	Neptune	4.50×10^9	−218
9	Pluto	5.91×10^9	−233 to −223

2.2 Temperature disproportions for electronics

What is true about humans holds also for electronics. But in electronics, the scales of intolerable temperatures are different. In electronics, the traditional range of operating temperatures lies between −55 °C/−65 °C at the lower extremity to +125 °C on the higher side. Temperatures falling outside these classical boundaries are known as extreme temperatures. ETE is a branch of electronics that deals with the properties and operation of electronic materials, devices, circuits and systems under the severe thermal conditions lying above or below the limits specified above (Kirschman 2012). It has two sub-branches:

(a) Electronics addressing operation above +125 °C to as high as electronics can be made to operate. This sub-branch is known as HTE.
(b) Electronics concerned with operation below −55 °C/−65 °C to −273 °C or 0 K. This sub-branch is called low-temperature electronics (LTE).

2.3 High-temperature electronics

The first query that can be raised is the following: why should one worry about abnormal situations, which may constitute a small segment of the electronics industry? This remark is valid because electronics is generally required to operate under conditions favorable to human life. The answer is that this segment, even though small, is vital strategically. It comprises several important sectors in which HTE has firmly strengthened its roots. One of these sectors is the automotive industry, i.e., motor vehicles. Another sector is the aerospace industry, dealing with travel in the Earth's atmosphere or in the region above it. The final sector is the well-logging industry, which records the properties of subsurface rocks and fluids by drilling a bore hole into the Earth's interior. In each of these sectors, specific problems are encountered. These issues cannot be solved without the help of HTE.

The second question that arises is: what will happen if traditional electronics is used in combination with active or passive cooling when designing electronics intended to function outside the normal temperature ranges? Yes, customarily engineers have followed this approach. Thermal management systems serve as useful alternatives to high-temperature devices. But these systems add undesired weight and volume to the total system. Additional overhead is introduced in the form of longer wires, extra connectors and/or cooling systems. As a result, the power-to-volume and power-to-weight ratios are reduced. These can negatively affect the advantages endowed by electronics with respect to the overall system operation. Thermal management systems also increase the potential for failure. By removing the heat sink and long interconnects, savings in the overall mass and volume of the power electronic modules can be achieved to the extent of one order of magnitude.

In certain circumstances, the use of cooling systems is impossible, principally when compactness is desired. In other applications, it is more appealing to operate electronics at high temperatures with the intent of increasing either the reliability of the system or to reducing the total expenditure. When objectives of this nature become primary, multiple challenges are faced in building the electronic systems: the design techniques are dramatically altered; the choice of semiconductor material is of paramount significance; the selection of other materials such as those used in metallization becomes a decisive determinant of performance; packaging materials and technologies differ from routine practice; and new qualification criteria and methodologies must be developed.

Needless to say, in the absence of HTE the cost of managing and monitoring hot environments becomes astronomically high. The direct costs escalate due to the increased system complexity and the specialized cabling requirements. Indirect costs also mount up through the increase in the weight of the electronic system due to cooling paraphernalia. The systems operated with these additional cooling jackets are also less reliable than the electronic systems which confront the intricacies of high temperatures *ab initio* and throughout the system (McCluskey *et al* 1997). Examples of applications demanding HTE are given in the following subsections (Delatte 2010). See also figure 2.2.

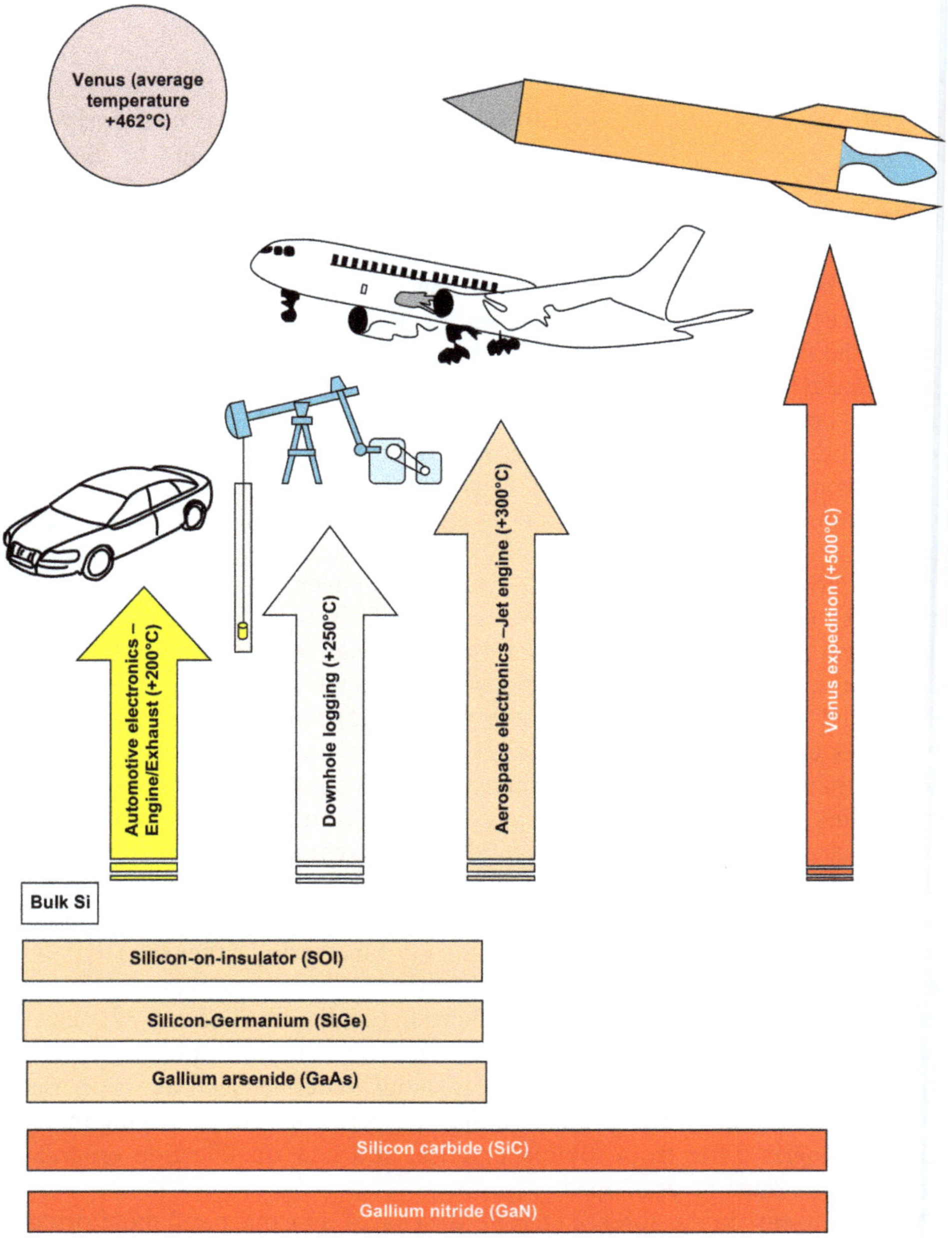

Figure 2.2. Applications of HTE and available materials/technologies.

2.3.1 The automotive industry

The ignition circuit used in a car is shown in figure 2.3. It consists of a battery, a switching transistor, a step-up transformer and a spark plug. A voltage ~20–50 kV is produced across the gap in the spark plug when the current flowing through the power transistor stops, such that the magnetic field around the primary coil collapses, inducing high voltage in the secondary winding.

Figure 2.4 shows how a car engine is electronically controlled using various types of sensors, namely the throttle position sensor, fuel pressure sensor, air flow sensor, knock sensor, temperature sensor and oxygen sensors (Bhatt *et al* 2019). The electronics provide an optimum timing cycle of ignition for all speeds and widely different loading conditions of the engine. Compared to a mechanical ignition circuit, electronic ignition provides superior starting and smooth running. It minimizes fuel consumption and thus causes reduced atmospheric pollution. In particular at very low and very high speeds, it provides a far-better performance. Servicing costs are also reduced.

In the automotive industry, engineers designing the body of a motor vehicle are continuously persuaded to augment the space inside the cabin, and thereby the capacity to carry more passengers and load. Cabin space can be increased by squeezing the engine and power module into smaller cubicles. This in turn necessitates the reduction of the size of electronic systems and wiring. Therefore,

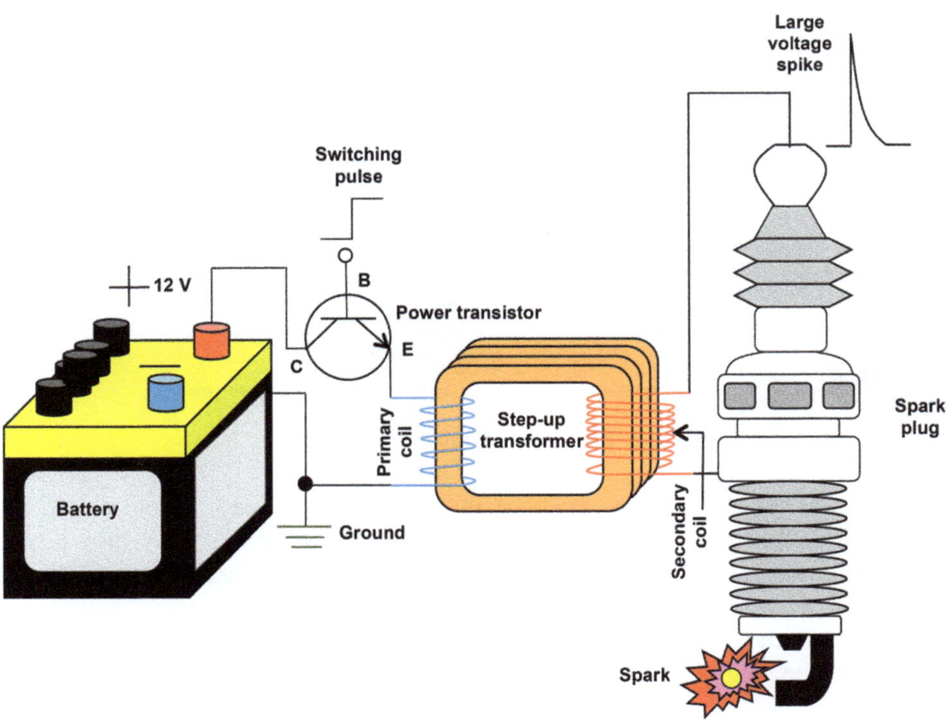

Figure 2.3. Car ignition circuit.

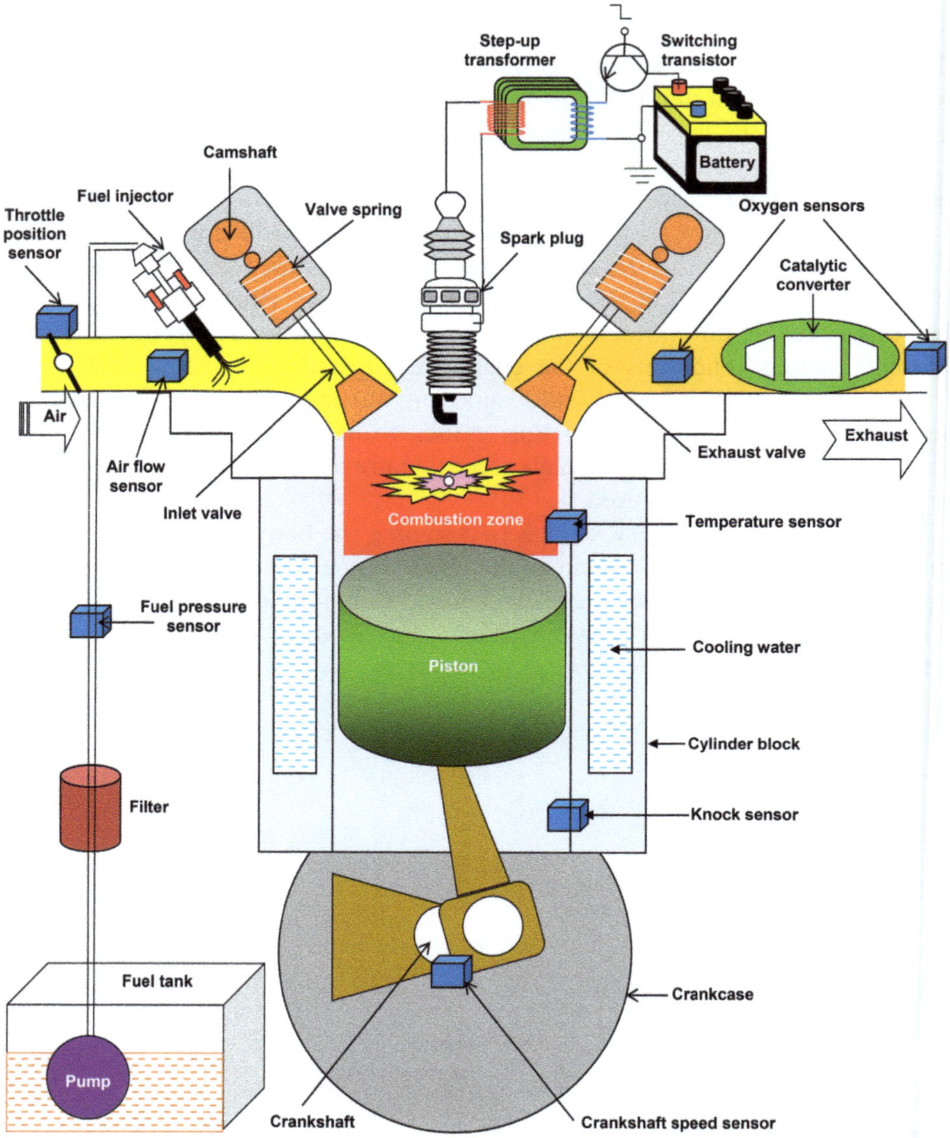

Figure 2.4. Electronic control of a car engine.

the electronic control systems must be mounted as proximate as possible to the engine. They may be placed in locations such as the gearbox. They may also be placed in transmission components. Consequently, a close-packed system is obtained. This kind of system can work satisfactorily if the electronic boards and components in the system can perform accurately at the approximate temperatures reached in the engine or gearbox. Additionally, smaller engine compartments impose stricter margins on heat sink sizes. Hence, vehicle aerodynamics is modified. This modified aerodynamics is able to provide less cooling air flow to the radiator.

An increase in temperature under the hood invariably follows. In all the above circumstances, the capabilities provided by HTE help to protect the electronic system from adverse thermal effects.

Another aspect of the advancement of electronics in the automotive industry relates to a recent paradigm shift. There is a strong desire to move away from purely mechanical and hydraulic systems to electromechanical or mechatronics schemes. This migration is essential to improve reliability. At the same time maintenance costs are reduced. But this requires the placement of sensors, signal conditioning and control electronics adjoining heat sources which produce a high-temperature environment. The maximum temperature and exposure time vary according to the type of vehicle and the location of the electronics on the vehicle. In some vehicles, a higher degree of integration of electrical and mechanical systems is achieved, such as collocation of the transmission and transmission controller. This leads to simplification of the manufacturing, testing and maintenance of automotive subsystems, but is associated with an increase in temperature. Here also, HTE is resorted to.

Remarkable developments have taken place in the hybrid automotive industry. These developments have generated an enormous need for different power electronic modules. The desired modules are direct current (DC)/DC converters and DC/alternating current (AC) inverters capable of working at elevated temperatures. On average, the junction temperatures for ICs are 10 °C–15 °C higher than the ambient temperature. For power devices, they are around 25 °C above the ambient temperature. Hence, the electronics used in the automotive industry, particularly those placed close to the engine, must be able to work at temperatures above 150 °C + 25 °C = 175 °C.

In electric cars and those based on a hybrid approach, power electronics is used for motor drives by integration of power converters and smart power devices into the drive train. This requirement increases the temperature requirements for automotive-qualified semiconductor devices from 150 °C to a 200 °C peak value (Huque *et al* 2008). This again is an application area of HTE.

With increase in the availability and reduction in price of HTE, its applications in automotive systems will continue to grow.

2.3.2 The aerospace industry

The aerospace industry entails civil and military aviation, and space flights and missions. Figures 2.5–2.7 illustrate different types of aircraft engines. Piston-engine powered aircraft obtain their thrust from a propeller driven by the engine. Turbopropeller aircraft obtain their thrust from a propeller driven by a gas turbine, while turbofan aircraft are similar but obtain their thrust from the internal fan driven by the turbine. In both turbopropeller and turbofan aircraft, a small amount of thrust is also obtained from the hot exhaust gases. The propeller/fan sucks in air, which is passed through a compressor and sent to a combustion chamber where it is ignited. While part of the air passes through the engine core where combustion occurs, the remaining air, known as the bypass air, moves around the core through a duct, producing additional thrust. The hot air moves the turbine blades which are

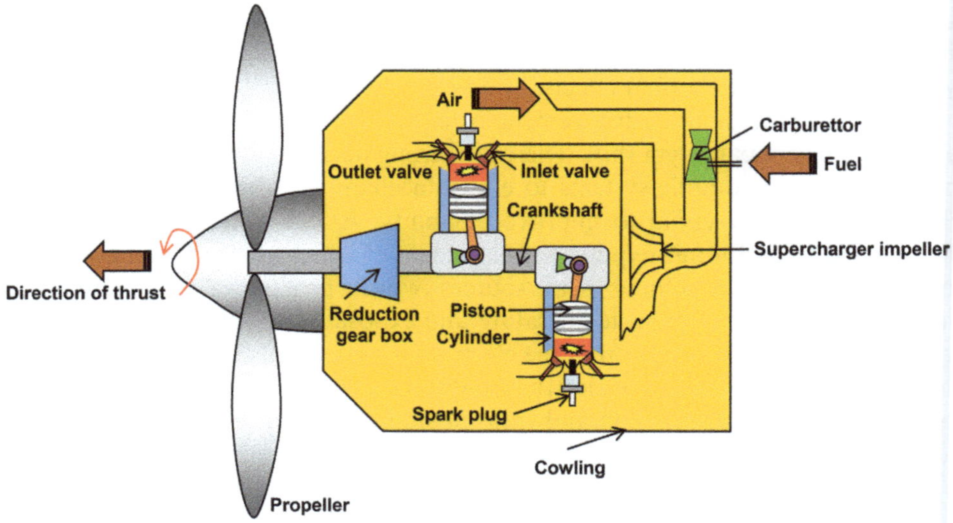

Figure 2.5. Aircraft propeller piston engine.

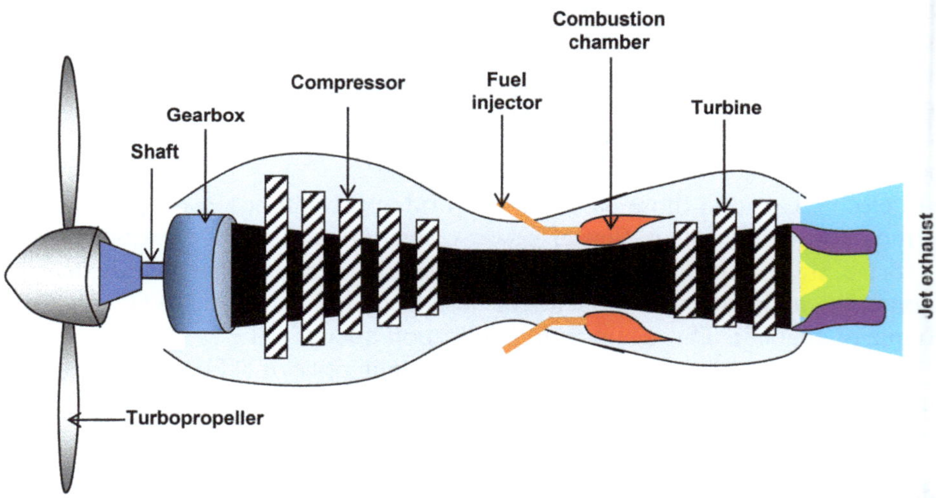

Figure 2.6. Aircraft turbopropeller jet engine.

coupled to the propeller/fan and help in its rotation, sucking more air. A rocket engine is shown in figure 2.8. It forms its exhaust jet using only the fuel stored inside. It also has an oxygen tank because it has to travel in space where an oxygen supply is not available. In all the above engines, whether aircraft or rockets, the fuel ignition is electronically monitored.

The ignition system of an aircraft is shown in figure 2.9. The dual magnetos used in an aircraft engine provide redundancy. In the case that one magneto fails, the other magneto serves as a backup. In addition, the ignition efficiency is improved by ensuring thorough combustion of the air–fuel mixture from both sides towards the center.

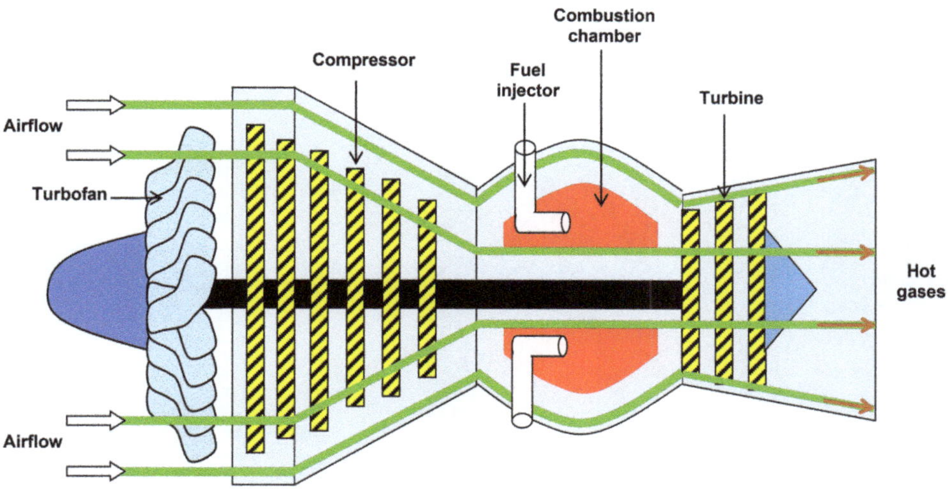

Figure 2.7. Aircraft turbofan jet engine.

Currently the aerospace industry is following in the footsteps of the automotive industry, in so far as changeover to newer electromechanical technologies is concerned. Like the automotive industry, there is a propensity for substitution of outdated hydraulic actuators with lighter and more cost-effective electrical equivalents. Engine components and braking systems are the focus of attention. In the lubrication subsystems of turbine engines, mechanical pumps are being replaced by electrical oil pumps. Consequently, motor-drive electronics is moving into close vicinity with lubricants. The lubricants operate at temperatures over 200 °C. Thus, in the aerospace industry, HTE assists in the deployment of electronic systems closer to heat sources, which are operating beyond the usual temperature boundaries laid down for electronic devices.

Furthermore, in the aerospace industry, a growing tendency is observed towards the inclusion of more electrical/electronic systems inside the aircraft. There is a clear-cut trend to make the aircraft more electric. This initiative of 'more electric aircraft' (MEA) partly seeks to eliminate traditional centralized engine controllers. They are replaced with distributed control systems. Centralized control requires large, heavy wire connections. These connections contain hundreds of conductors. Multiple connector interfaces are also used. Moving to a distributed control scheme brings the engine controls closer to the engine. The complexity of the interconnections is reduced by a factor of 10. Through these efforts, tons of aircraft weight is set aside. As a result, the reliability of the system is enhanced, but all these alterations are possible only if the electronic systems can bear up against the higher temperatures near the engine.

2.3.3 Space missions

In satellite applications (figure 2.10), the temperature of electronic assemblies may be controlled.

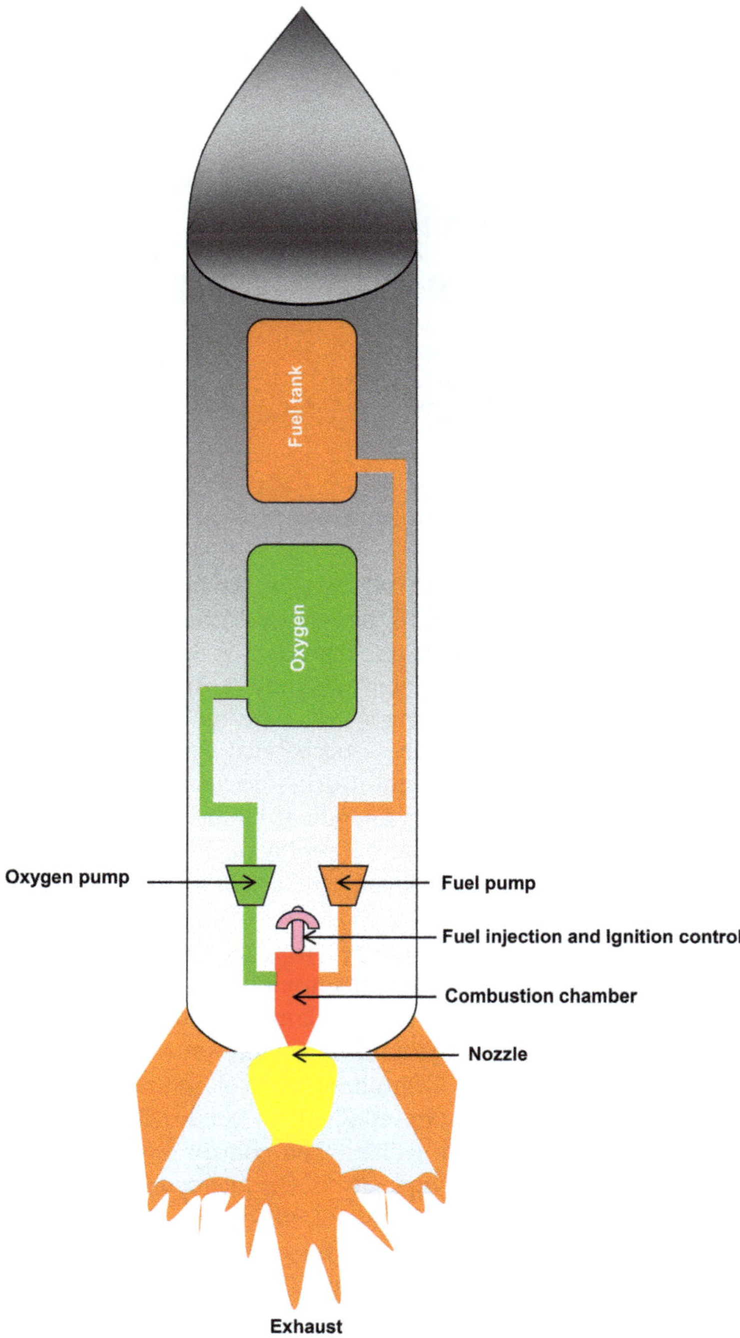

Figure 2.8. Liquid fuel rocket engine.

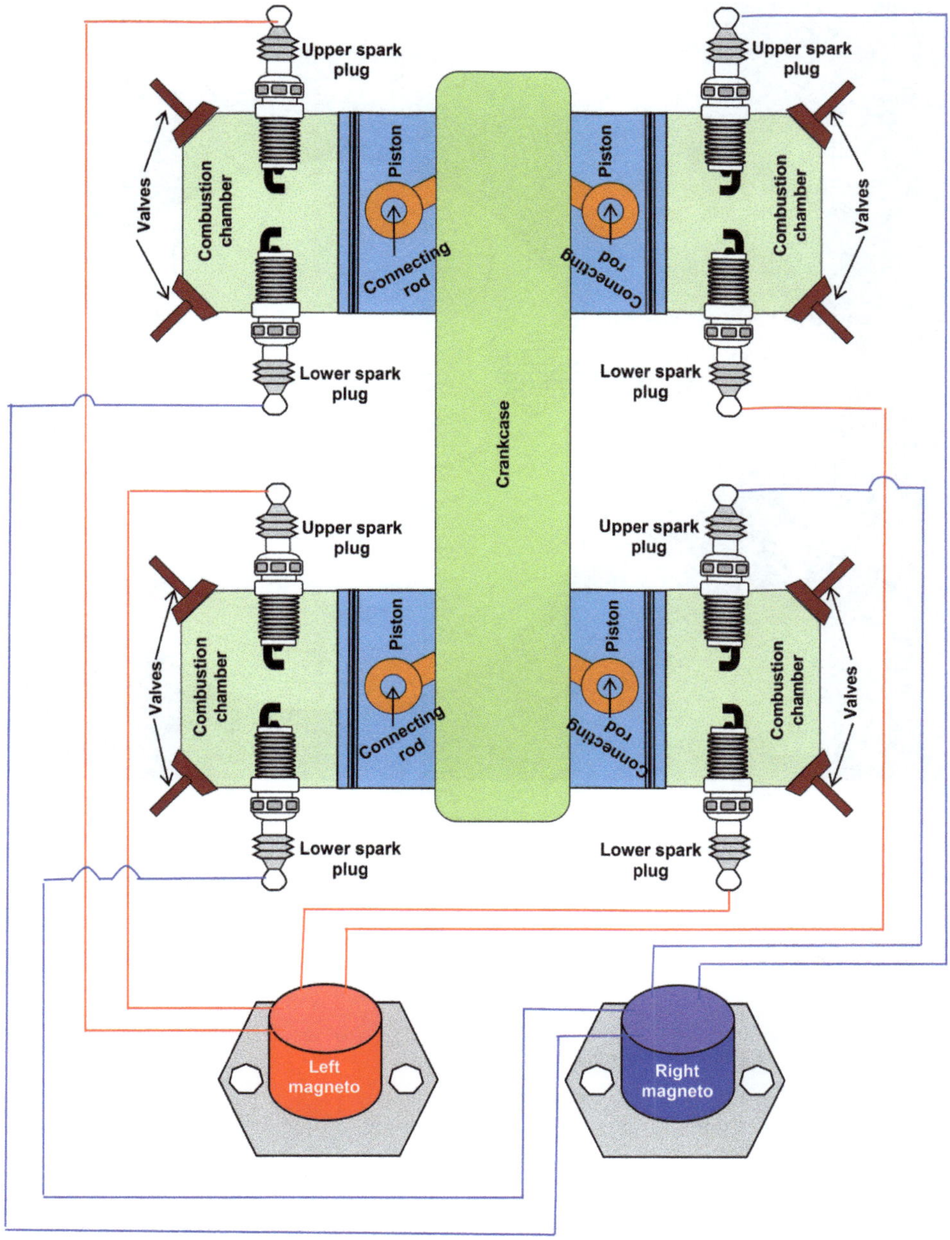

Figure 2.9. Wiring diagram of a dual magneto-based aircraft ignition system with two spark plugs in each cylinder for redundancy.

Space missions from the Earth to the Moon (figure 2.11) must take into consideration the wide temperature fluctuations that need to be tolerated on board the space flight and at the lunar surface. The temperature on the Moon plunges down to −181 °C during the lunar night and rises up to 101 °C during the lunar day.

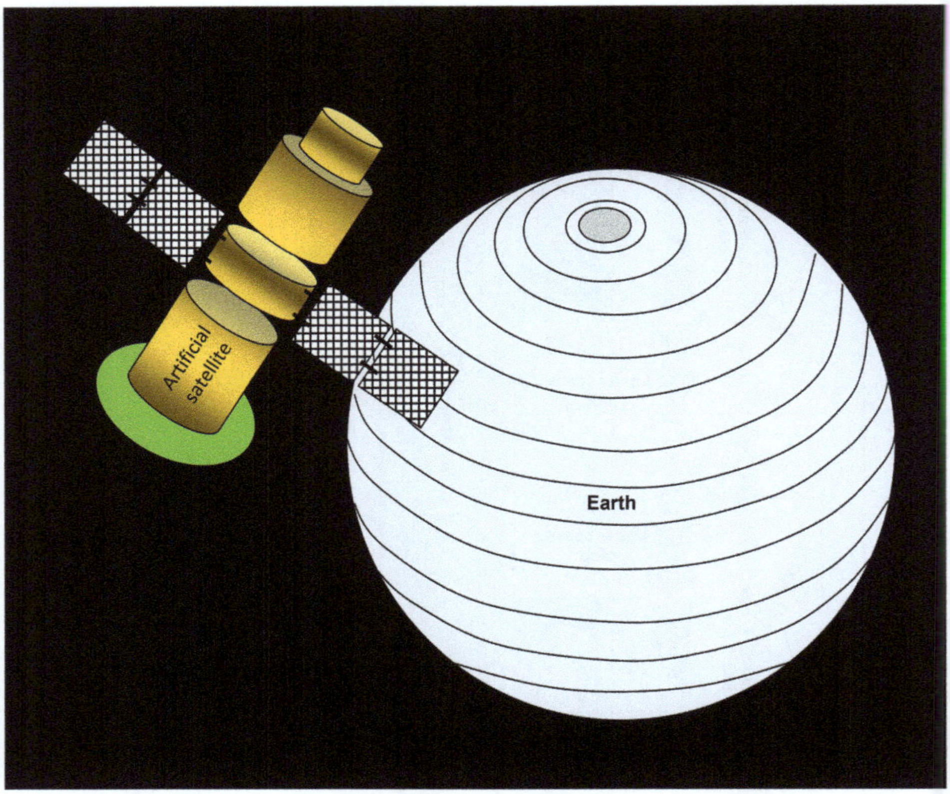

Figure 2.10. An artificial satellite orbiting the Earth.

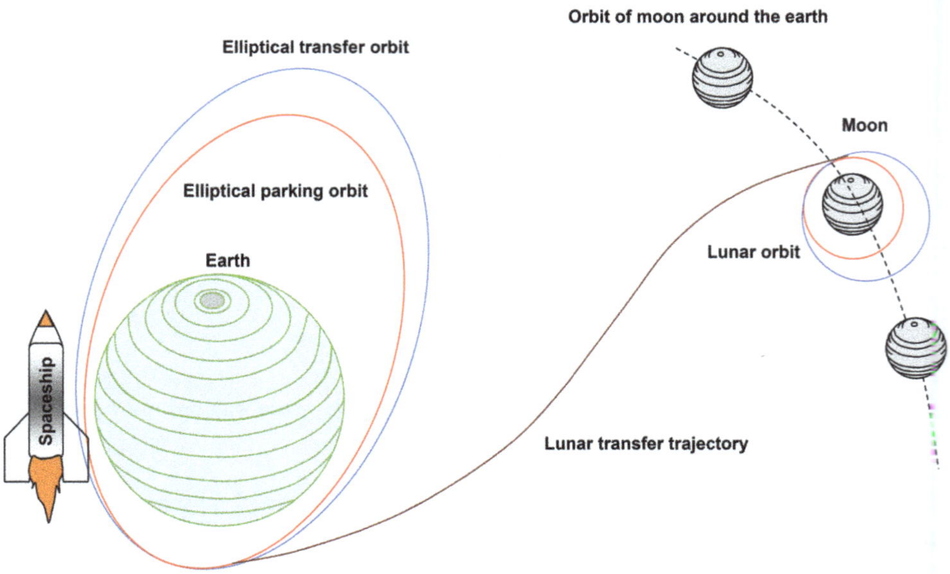

Figure 2.11. A space mission from Earth to the Moon.

This happens because the Moon has no atmosphere, unlike the Earth. On Earth, the atmosphere acts as a blanket, and traps the heat received from the Sun making its escape slower. During the day, sunlight passes through the atmosphere and warms up the soil. At night, the energy is emitted by the soil as infrared radiation, but it cannot escape through the atmosphere easily; hence the planet warms up. Nights are colder than days but unlike the Moon, the night temperature does not fall to a very low level. To deal with the extensive range of temperature variations on the Moon, spacesuits are heavily insulated with layers of fabric and are also covered with reflective outer layers. In addition, they have internal heaters and cooling systems. They use liquid heat exchange pumps to remove excess heat.

In deep-space missions such as planetary expeditions to Mercury and Venus, the ability to operate at very high temperatures up to +470 °C is a mandatory requirement.

Electronics can also be made to function by regulating the temperature within acceptable levels. However, in comparison to temperature regulation, the greater reliability and stable performance of high-temperature components provides more robust systems. It helps to remove the financial burden of temperature regulation and cooling. Thus, HTE becomes a necessity in space expeditions.

At the opposite extreme temperature condition, space systems are also exposed to extremely cold conditions, e.g., for Mars and Jupiter, which falls under the domain of LTE, figure 2.12. They must be able to work at −55 °C at Mars and −145 °C at Jupiter.

2.3.4 Oil well logging equipment

Petrochemical exploration companies are striving to exploit natural resources lying underground at depths of several kilometers beneath the surface of the Earth. Well

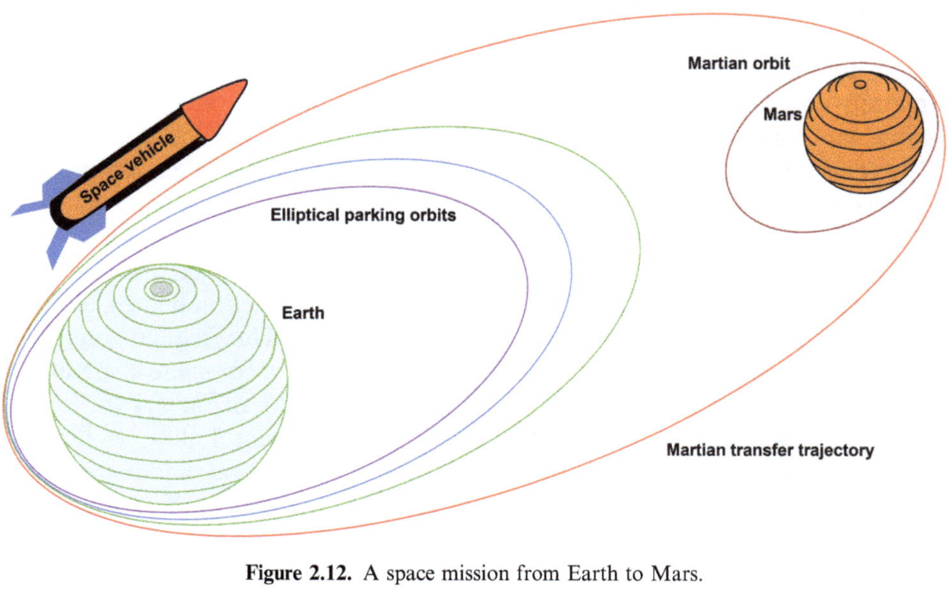

Figure 2.12. A space mission from Earth to Mars.

logging or bore hole logging is the practice of chronicling a detailed record of the geological rock strata, using comparable physical characteristics, when penetrated by a bore hole. In this application, the electronics are required to operate at a temperature which varies with the depth of the well below ground level. A crude estimate of this temperature is obtained from the geothermal gradient. This gradient is the increase in temperature per unit kilometer increase in depth into the Earth's interior. Near the surface and away from the tectonic plate boundaries, this gradient is 25 °C km^{-1}–30 °C km^{-1}. The tectonic plates are made up of the layers of the Earth's rocky crust, the uppermost envelope, together called the lithosphere.

In the past, explorations have been terminated at temperatures around 150 °C–175 °C. At present, such easily accessible natural resources have been exhausted. Moreover, technological advancements have encouraged deeper drilling operations. Terrestrial regions of higher thermal gradients must also be explored. In these regions, temperatures exceeding 200 °C at pressures >25 psi are encountered in the deepest wells, more than 5 km deep. In such hostile environments, cooling techniques are neither viable nor effective.

The use of HTE in the exploration sector is multifaceted (figure 2.13). The parameters measured vary widely. The resistivity of the rocks indicates their electrical properties. Radioactive decay describes the type of radioactivity in the material. Similarly, acoustic travel time, magnetic resonance and other properties are measured to ascertain the general characteristics of the geological formation. Apart from these, porosity, permeability and water/hydrocarbon saturation properties are determined. Gross physical characteristics such as the mineral composition, color, texture, grain size, etc, of a rock come under its lithology. The data thus gathered allow the geologist to determine the kinds of rock in the formation, the types of fluids present, their location, and the extractability of adequate amounts of hydrocarbons from fluid-bearing zones. Accuracy is essential for all measurement and data-acquisition systems, because data taken from deep bores play a decisive role in the analysis of formations leading to identification of deposits such as oil or gas. From the data accumulated, the most promising localities to commence extraction are agreed upon and extraction is performed only at these sites. Because large monetary investments are at stake and rely on the data collected, the systems must operate accurately at these exceptionally high temperatures, emphasizing the importance of HTE.

The role of HTE does not end with well exploration but extends much further. It is also required post-exploration, in the equipment used to complete the bore and in the extraction of resources. During the completion and extraction phases, electronic systems monitor the pressure, temperature and vibrations. They regulate multiphase flow by actively controlling valves. To cater to these needs, a complete chain of high-performance components is put into service. The reliability of the system is of the utmost priority. The cost of downtime due to equipment failure is high. An electronic assembly failure on a drill string operating several kilometers underground can take more than a day to restore, a Herculean task. If replacement is necessary, enormous expenditure may be incurred.

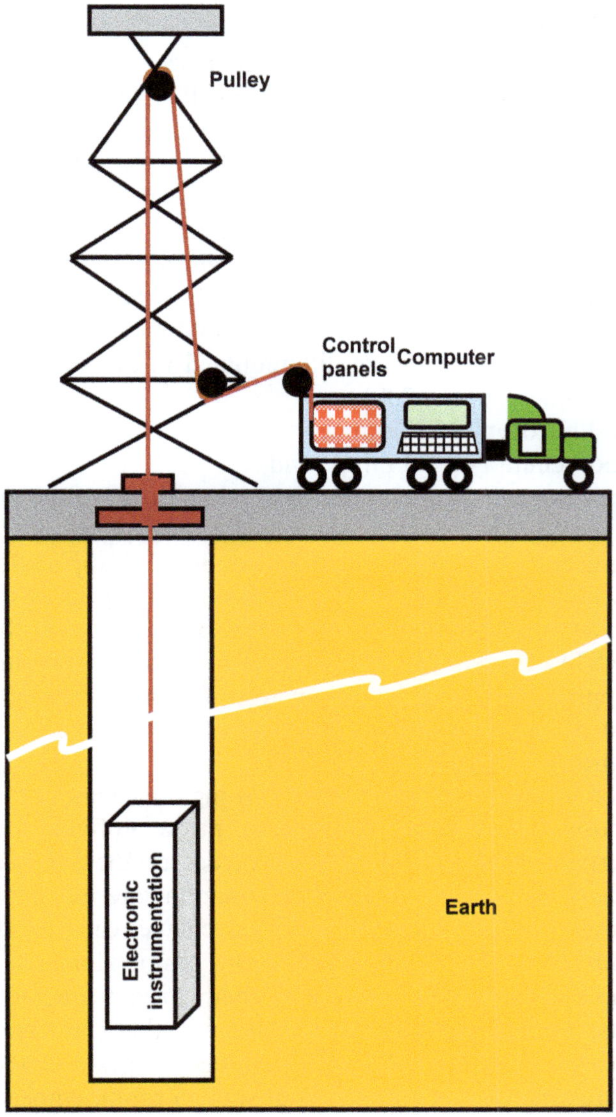

Figure 2.13. Data logging by suspending the electronic measuring unit inside the bore well and pulling it out.

2.3.5 Industrial and medical systems

These systems present special requirements for electronic modules. To fulfill industrial and medical needs, these systems must offer the specified exactness, constancy and permanence at high temperatures in the range of 175 °C. In industrial process control, electronic control and processing modules must be mounted close to sensors and actuators. Such mounting prevents noise from entering the system through long wiring. Medical electronic systems need HTE to, for example, take advantage of the benefits of high-temperature monitoring of sterilization systems.

2.4 Low-temperature electronics

LTE is referred to by different names such as cold electronics, cryogenic electronics or cryoelectronics (Gutiérrez *et al* 2001). It involves the operation of electronic materials, devices, circuits and systems at temperatures significantly below the conventional range, which conventionally ends at −55 °C (Kirschman 1990). Electronic devices and systems have been operated at temperatures down to within a degree of absolute zero (0 K = −273 °C). Because of the variety of materials, devices and effects, no single temperature can be given as the boundary between the conventional and the low-temperature ranges. However, the cryogenic range is considered to start from about 100 K (−173 °C) and move towards the colder side.

In general, three main regions are differentiated in discussions on low-temperature operation (Gutiérrez *et al* 2001):

(a) the liquid-nitrogen (77 K) range;
(b) the liquid-helium (4.2 K) range; and
(c) the deep cryogenic temperatures descending down to the millikelvin range.

The first region may lead to more or less commercial applications. The second region is mainly used for cold electronics associated with space missions such as the Infrared Space Observatory Photometer (ISOPHOT) and Far Infra-Red Space Telescope (FIRST). Extremely low temperatures are found in the world of astrophysical applications, such as bolometers. The lower the operating temperature, the greater is the transference of activities from potential industrial applications to research-oriented fields.

The motivation to study LTE lies in the following three reasons (Clark *et al* 1992). First and foremost, it is clear that temperature has a profound impact on several important properties of materials. Notable properties are the drift velocity of carriers and conductance of substances. In ICs, noise margins are influenced. Keeping the effect of temperature on such properties under surveillance can provide useful insights into the behavior of electronic circuits at room temperature and liquid-nitrogen temperature. This could potentially lead to the ability to counteract intrinsic problems, such as reliability, through methods that would have otherwise remained unknown.

The second reason why LTE must be investigated is to explore the areas in which a low-temperature ambience is essential and wherein room temperature electronics cannot penetrate. In this area falls electronics which cannot operate at room temperature, but become feasible only at low temperatures. This area is exemplified by two well-known phenomena, namely superconductance and the Josephson effect. They occur only at significantly low (77 K) temperatures. Furthermore, transistor characteristics, such as the threshold voltage of MOSFETs, do not scale properly with temperature. Thus, understanding the effects of low temperatures is upgraded from a simple option to a mandatory requirement.

Lastly, the performance of existing CMOS technology can be further improved if transformed into LTE, i.e. operated under low-temperature conditions as opposed to room temperature (Peeples *et al* 2000). Low-temperature conditions have also eliminated hazardous problems such as latchup. They have significantly benefitted

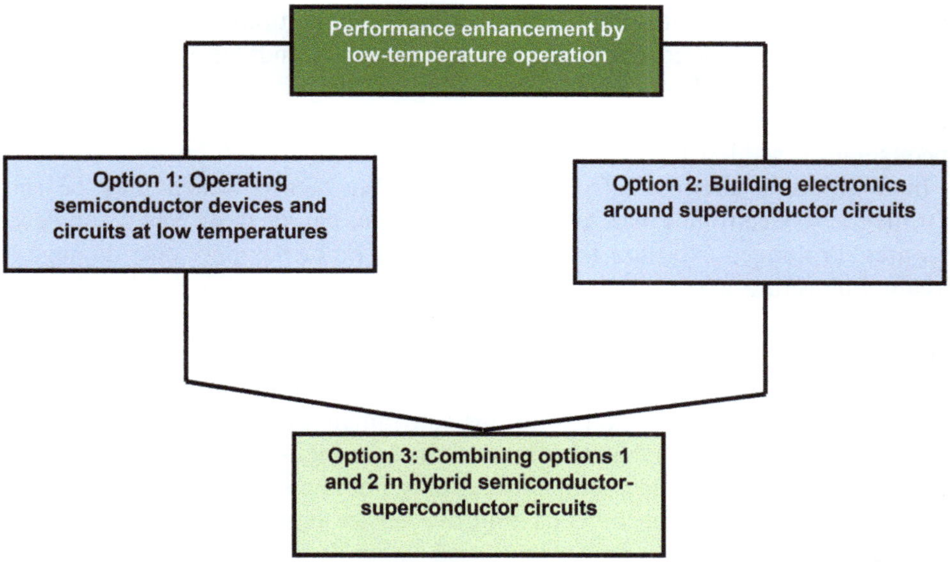

Figure 2.14. Cold electronics using semiconductors and superconductors.

applications such as dynamic random-access memory (DRAM) (Henkels *et al* 1989). The drawback, however, is the enormous cost necessary to maintain the temperature at a sufficiently low magnitude so that the desired phenomena can take place. Figure 2.14 illustrates how LTE can be utilized as a boon to electronics, helping in significant improvement in the performance of existing technologies and also affording the development of new technologies, notably by exploiting superconductivity.

For the operation of silicon devices there is a minimum temperature, called the freeze-out temperature, at which the thermal energy inside the semiconductor is too small and is inadequate to activate the donor or acceptor impurity atoms, depending on whether the material is n-type or p-type. Freeze-out is the condition when dopants are no longer ionized and the semiconductor behaves as an insulator. The freeze-out temperature depends on the semiconductor doping concentration. At an elevated level of doping, which is much higher than that used in standard silicon technology, one can operate devices at temperatures close to 0 K. On the other side, consumer electronic devices are clearly not aimed at operation below 233 K. However, exceptional, specific applications exist. The main snags with cryogenic design arise from the carrier freeze-out phenomena and changes in carrier mobilities that occur near absolute zero. These effects must be fully understood to correctly model devices. They must be incorporated accurately into electronic design software.

2.5 The scope of extreme-temperature and harsh-environment electronics

A harsh environment is described as any location or surroundings in which survival is arduous or outside the bounds of possibility. For a human being, a harsh or inhospitable environment is an unwelcoming set of conditions that can cause harm to the body over a period of time. The concept is extendable to electronics. An electrical circuit is easily spoiled and ceases to function if it is exposed to excessively high temperatures and fast temperature cycling. But temperature is not the only uncongenial variable which is to be blamed for failure.

In addition to thermal effects, there are other factors which may be more detrimental to electronics than temperature. Systems can be destroyed if submerged in water. Prolonged exposure to a high-humidity ambience may also damage the circuit. Subjecting a circuit to ingress of particulate matter is no less harmful. Electrostatic discharge (ESD) effects are known to cause dielectric breakdown of MOS devices if suitable precautions are not taken. Electromagnetic interference (EMI) upsets the normal functioning of circuits. Apart from these factors, vibrations, physical shocks and collisions can impair the functionality of devices and circuits. Therefore, harsh environments include high-temperature atmospheres as a sub-component. But low temperatures may not necessarily fall in this domain. A few low-temperature phenomena improve device characteristics, a few others degrade them. Nevertheless, they always need elaborate arrangements to provide the cooling which make the systems very bulky. Therefore, in this book temperature effects are treated separately from harsh-environment effects, such as those of humidity, radiation and shocks. Thermal effects can be both disadvantageous and advantageous.

2.5.1 High-temperature operation: a serious vulnerability

Compared to the low-temperature end, the high-temperature end causes more problems in electronics. Frequently, the high-temperature restriction is not determined by the inherent limitations of the semiconductor materials, but by the properties of the materials used for interconnection between devices and packaging of finished chips.

2.5.2 Upgradation/degradation of performance by cooling

Contrary to human beings, very low temperatures may not be always punitive to electronics. Instead, they may often lead to better performance of devices and circuits. In general, as temperature decreases, the electrical characteristics of a component can undergo a gradual or an abrupt change. It may even stop operating completely. At lower temperatures, field-effect transistors (FETs) exhibit increased gain and speed, and unwanted leakage current is decreased. A MOSFET or CMOS device can operate below the liquid-nitrogen temperature (−196 °C). Its performance is upgraded. On the other hand, a silicon bipolar transistor discontinues functioning at about −150 °C. Essentially, it offers such low gain that it becomes unworkable. Band-structure changes are responsible for these malfunctions, not the freeze-out phenomenon. Freeze-out does not occur in silicon until about −230 °C (Titus 2012).

With falling temperature, parasitic resistances and capacitances in interconnections decrease. Heat transfer improves and the noise level decreases. An important low-temperature application includes pacing up digital electronics and another application is noise reduction in microwave pre-amplification.

2.5.3 Corrosion: humidity and climatic effects

When the atmospheric relative humidity (RH) is very high, >80%, a lot of water vapor condenses on the metal surfaces of semiconductor devices. Optimal performance of computers is achieved by maintaining the humidity levels tightly between 45% and 55% RH limits. The reliability of computers is also assured at these humidities. The recommendation is for warning alerts to be triggered when humidity falls below 40% RH or rises above 60% RH. Critical alerts may be sounded at humidity levels less than 30% RH and above 70% RH (Grundy 2005).

Condensation of moisture molecules in the superincumbent air on metallic surfaces leads to the formation of a thin film of water on these surfaces. Application of an electric field to the metal results in the elution of metal ions, which is the process of stripping of ions as if by washing with a solvent. The neighboring conductor pulls the eluted metal ions. In this way, current can flow between adjoining conductors through the film formed by eluted metal ions, short-circuiting the device and thereby causing its failure. Often, the metal film resembles a dendrite of a human nerve cell with threadlike extensions. At the time of short circuit, it disappears because of its brittleness. Device failure induced by migration of metal ions is a cumbersome process, which is not reproducible after its occurrence (Apiste Corporation 2015).

Corrosion induced by climatic effects increases the contact resistances of joints. Leakage currents between wires are enhanced. It also causes materials to decay. The surfaces of electronic devices acquire an ugly appearance. Different types of operational faults are observed (Hienonen and Lahtinen 2000).

2.5.4 Deleterious effects of nuclear and electromagnetic radiations on electronic systems

We dwell in a world which is filled with radiations of different kinds. The radiation content in our environment varies dramatically from place to place. It is also variable with respect to time (Boscherini *et al* 2003). While civilians reside in a comparatively safe environment, military and space systems are confronted with consistent threats of assorted types of radiation of high intensities. Many times, military operations have to withstand man-made nuclear radiation. Catastrophic effects are produced in electronic systems exposed to such hazardous radiation-contaminated environments. The impact of this radiation on electronic systems may be apparent in various forms. It may lead to momentary cessation of services or to erasure of chip memory, or ultimately to death of the device by burning out. As radiation effects are primarily issues faced by military and space engineers, manufacturers of commercial consumer electronics goods hardly ever pay any attention to these problems. As a result, consumer electronic systems have no built-in features safeguarding them against the injurious effects of radiation. They often fail prematurely in such environments (Bagatin and Gerardin 2016).

Considering the types of radiation encountered by electronic systems, there are two broad categories, namely particulate type radiation and electromagnetic waves or photons. Particulate radiation consists of charged and neutral particles, such as electrons or beta particles, protons or hydrogen nuclei, alpha particles or helium nuclei, ions resulting from fission reactions called fission fragments, neutrons and heavy ions. Photonic radiation comprises gamma and x-rays. Radiation damage is the general term describing the detrimental consequences of radiations of all types.

Radiation can be distinguished into two other types, those containing charged particles (α-particles, β-rays, protons) and those of a neutral nature (neutrons and γ-rays). Fundamentally, exposure to the two types of radiation produces different kinds of effects on electronic devices and systems. These effects are subdivided into three broad categories:

(a) *Single event effects (SEEs).* Here passage of a single, highly ionizing particle alters the operational state of a semiconductor device in a single event, e.g. it may change the logic state of a transistor.

(b) *Total ionizing dose (TID) effects.* These are long-term effects which take place on prolonged exposure to radiation. They cause build-up of charge at the boundaries or interfacial regions between different layers in semiconductor devices.

(c) *Displacement damage (DD) effects.* Collisions take place between the irradiating particles and lattice atoms. These collisions displace the lattice atoms from their positions, resulting in the formation of Schottky or Frenkel defects in the crystal lattice. The crystal lattice is permanently damaged. Secondary particles liberated as a result of the collisions may cause further displacement of atoms, activating a cascading of collisions. Several defects in the form of vacancies and interstitials can be found along the tracks of the incoming particles. Such defects may occur in clusters at the ends of the tracks.

2.5.5 Vibration and shock effects

When first considering the harsh environmental conditions to be encountered by an electronic product, a design engineer is likely to overlook the effects of vibrations and shocks. However, vibration and shock effects represent a major cause of failure in many systems. During the lifecycle of a product, it may have to sustain such effects at many stages. Conceivably, it may experience shocks during shipping or transportation when the cargo is emplaned or deplaned, or carried in a vehicle for delivery to the customer. During everyday use, it may sometimes fall from its support. If it is too fragile, it will stop working. In many applications, ranging from automobile, train and aerospace systems, to oil drilling apparatus/hardware, powerhouses and power generating plants, the product has to face vibrational stresses of moderate or severe magnitudes (Askew 2015).

Protection from random vibrations is provided by mounting the electronic systems on resilient or elastic supports. The trade-off between the damping and stiffness of supporting platforms is based on the dynamic responses of the resulting electronic enclosures subject to the maximum rattle space provided in a given application. Thus, the design rule for resilient mounts is to optimize the isolation between vibration and the electronic system within the limits of the maximum deflections permitted. By pursuing this approach, reliable vibration isolators can be built in view of the intensity/amplitude of vibrations experienced (Veprik 2003).

2.5.6 Special environments

Besides the general scenario of harsh conditions sketched in the preceding sections, there are some specific circumstances in which electronics is required to perform its normal jobs. A notable situation is an environment in which electromagnetic interference from an electronic circuit itself or a nearby circuit does not allow its normal functioning. Corrective circuit design is essential. Another example is a sensor that has to directly interface with the environment which it is designed to sense, and this environment is damaging to the sensor. In such cases, regular sensors cannot be used and special sensors must be fabricated to withstand the operating environment. A series of electronic equipment has been developed for reliable operation inside the human body. Such equipment is carefully encapsulated for satisfactory long-term operation in the presence of biological fluids. Electronics aboard space flights has to work in a mixture of austere conditions which are not faced during terrestrial working. Last but not the least, electronics communication is open to jamming attacks and electronic computer engineers work under the constant apprehension of hackers. All these situations make electronics operation extremely difficult and critical.

2.6 Discussion and conclusions

It is evident that electronic systems designed to function under normal operational conditions cannot be expected to perform well under extreme-temperature and harsh environmental conditions, of which a few representative examples were presented in this chapter. The deleterious effects on electronic devices can occur due to the following reasons:

(a) operation at very high temperatures for which the device has not been fabricated;
(b) operation at very low temperatures for which the device was not planned;
(c) exposure to very wet conditions;
(d) exposure to radiations of different varieties;
(e) subjecting the device to vibrations and shocks;
(f) electromagnetic interference from a circuit's own functioning or from nearby circuits;
(g) damaging effects of sensed environment on a sensor device;
(h) effects of human body fluids on implanted electronics;
(i) hazards of vacuum environment in outer space during space missions; and
(j) electronic jamming and cybersecurity issues.

In this list, low temperatures are known to influence device operation in both directions, i.e., in upgradation as well as degradation. High temperatures too can be beneficial on some occasions. Therefore, temperature effects are considered as a separate subject which can be harmful or beneficial. The rest of the subject is treated under harsh-environment effects. The field of electronic devices and systems intended to operate in these abnormal circumstances is therefore a subject in itself, and this subject constitutes the focal theme of this book. Various interesting facets, issues and challenges faced by engineers and scientists in overcoming the problems will be highlighted and their solutions will be addressed as the reader progresses through the text. The utilization of the helpful effects of low-temperature operation will also be explained.

As the reader browses through chapters of this book, it will be evident that necessary precautionary measures need to be taken right at the outset, i.e., from product conceptualization for a particular application. The design of the product must take into consideration all adverse situations that the electronic device is likely to be confronted with during its lifetime. This practice should be adhered to during planning of its fabrication and packaging steps, right up to system assembly (Watson and Castro 2012).

Review exercises

2.1. Explain the following terms: (i) sun stroke, (ii) hypothermia and (iii) frostbite.

2.2. Which is the hottest place on the Earth's surface? What is its temperature?

2.3. Which is the coldest place on the Earth's surface? What is its temperature?

2.4. Name the hottest and coldest planets in the Solar System. What are their temperatures?

2.5. Explain the following terms: (i) HTE, (ii) LTE, (iii) ETE and (iv) harsh-environment electronics.

2.6. Why is using traditional electronics combined with active or passive cooling inadequate to meet the challenges posed by high temperatures?

2.7. Give three reasons explaining the need for HTE in the automotive industry.

2.8. Explain the significance of HTE in aerospace industries.

2.9. Why does the temperature on Earth not fall to very low magnitudes during the night? Explain the role of HTE in space missions.

2.10. What is the value of the thermal gradient per kilometer below the Earth's surface? Explain the need for developing high-temperature electronic devices for deep oil well exploration.

2.11. Give one example of the use of HTE in medicine.

2.12. Write three alternative names for LTE.

2.13. Mention the three major regions of low-temperature operation of semiconductor devices. Indicate the targeted applications of each region.

2.14. Name two properties of semiconductors which are benefitted by low-temperature operation.

2.15. Name two phenomena which take place only at low temperatures.

2.16. What is meant by carrier freeze-out? What are the minimum temperatures up to which traditional MOS and bipolar devices can operate? Does reduction of the gain of a bipolar transistor at low temperatures result from carrier freeze-out or some other phenomenon?

2.17. Between what limits of humidity is the performance of computers optimized? What happens when moisture condenses on the metal surfaces of semiconductor devices and a voltage is applied to them?

2.18. What are the two main classes of radiation which affect the performance of semiconductor devices?

2.19. Name and discuss the three types of effects that radiation produces in semiconductor devices.

2.20. How is a semiconductor device protected from the effects of vibrations and shocks?

References

Apiste Corporation 2015 Effect of humidity on electronic devices *Apiste Corporation* www.apiste-global.com/enc/technology_enc/detail/id=1263

Askew D 2015 Vibration protection of electronic components in harsh environments *Mouser Electronics Inc.* www.mouser.in/applications/harsh-environment-vibration-protection/

Bagatin M and Gerardin S (ed) 2016 *Ionizing Radiation Effects in Electronics: From Memories to Imagers* (Boca Raton, FL: CRC Press)

Bhatt G, Manoharan K, Chauhan P S and Bhattacharya S 2019 MEMS sensors for automotive applications: a review *Sensors for Automotive and Aerospace Applications. Energy, Environment, and Sustainability* ed S Bhattacharya, A Agarwal, O Prakash and S Singh (Singapore: Springer), ch 2 pp 223–39

Boscherini M, Adriani O, Bongi M, Bonechi L, Castellini G, D'Alessandro R and Gabbanini A *et al* 2003 Radiation damage of electronic components in space environment *Nucl. Instrum. Methods Phys. Res.* A **514** 112–6

Clark W F, El-Kareh B, Pires R G, Titcomb S L and Anderson R L 1992 Low temperature CMOS—a brief review *IEEE Trans. Compon. Hybrids Manuf. Technol.* **15** 397–403

Delatte P 2010 Designing high-temp electronics for auto and other apps *EE Times* www.eetimes.com/document.asp?doc_id=1272966&

Grundy R 2005 Recommended data centre temperature and humidity: preventing costly downtime caused by environment conditions *AVTECH News* www.avtech.com/About/Articles/AVT/NA/All/-/DD-NN-AN-TN/Recommended_Computer_Room_Temperature_Humidity.htm

Gupta A, Soni R and Ganguli M 2021 Frostbite—manifestation and mitigation *Burns Open* **5** 96–103

Gutiérrez D E A, Deen M J and Claeys C 2001 *Low Temperature Electronics: Physics, Devices, Circuits and Applications* (New York: Academic)

Henkels W H, Lu N C C, Hwang W, Rajeevakumar T V, Franch R L, Jenkins K A, Bucelot T J, Heidel D F and Immediato M J 1989 A 12-ns low-temperature DRAM *IEEE Trans. Electron Devices* **36** 1414–22

Hienonen R and Lahtinen R 2000 *Corrosion and Climatic Effects in Electronics* (Espoo: Technical Research Centre of Finland)

Howell E 2014 How far are the planets from the Sun? *Universe Today* www.universetoday.com/15462/how-far-are-the-planets-from-the-sun/

Huque M A, Islam S K, Blalock B J, Su C, Vijayaraghavan R and Tolbert L M 2008 Silicon-on-insulator based high-temperature electronics for automotive applications *IEEE Int. Symp. on Industrial Electronics (Cambridge, 30 June–2 July)* (Piscataway, NJ: IEEE), pp 2538–43

Kirschman R K 1990 Low-temperature electronics *IEEE Circuits Devices* **6** 12–24

Kirschman R K 2012 *Extreme-Temperature Electronics, Tutorials* www.extremetemperatureelectronics.com

McCluskey F P, Grzybowski R and Podlesak T 1997 *High Temperature Electronics* (New York: CRC Press)

Peeples J W, Little W, Schmidt R and Nisenoff M 2000 Low temperature electronics workshop *16th Annual Semiconductor Thermal Measurement and Management Symp. (San Jose, CA, 21–23 March)* (Piscataway, NJ: IEEE), pp 107–8

Tavanaei G 2013 The five coldest and hottest places on Earth *Epoch Times* www.theepochtimes.com/n3/101262-the-five-coldest-and-hottest-places-on-earth/

Titus J 2012 Design electronics for cold environments *ECN Mag.* www.ecnmag.com/articles/2012/12/design-electronics-cold-environments

Veprik A M 2003 Vibration protection of critical components of electronic equipment in harsh environmental conditions *J. Sound Vib.* **259** 161–75

Watson J and Castro G 2012 High-temperature electronics pose design and reliability challenges *Analog Dialogue* **46** 1–7

Williams M 2014 What is the average surface temperature of the planets in our solar system? *Universe Today* www.universetoday.com/35664/temperature-of-the-planets/

Part I

Environmental hazards and extreme-temperature electronics

Sub-part IB

Extreme-temperature electronics

Chapter 3

Temperature effects on semiconductors

The impact of temperature on the important properties of semiconducting materials used for electronic devices and circuit fabrication is examined, with a focus on silicon. The properties considered are the energy bandgap (the Varshini and Blaudau *et al* models), intrinsic carrier concentration and saturation velocity of carriers (the Quay model, and Ali-Omar and Reggiani model). Various mobility equations are discussed, e.g., the Arora–Hauser–Roulston equation, Klaassen equations and those in the MINIMOS model. The differences between uncompensated and compensated semiconductors regarding the temperature dependence of the mobility and carrier concentration are described. The ionization regimes of semiconductors are also described, namely the carrier freeze-out regime, extrinsic or saturation regime, and intrinsic regime. The conceptual development in this chapter paves the way for the temperature-related discussions in forthcoming chapters.

3.1 Introduction

A proper understanding of the operation of semiconductor devices at very low as well as at very high temperatures can only be obtained on the basis of a general comprehension of the physical properties of semiconductor materials, together with a correct perception of their variation under extreme thermal conditions. These properties often change drastically compared to the more familiar room temperature behavior. The present chapter investigates how the properties of a semiconductor change with temperature. Throughout, silicon is taken as the focal material, but the treatment of silicon helps us to interpret the behavior of other materials, taking into consideration the relevant differences between material properties.

doi:10.1088/978-0-7503-5072-3ch3

3.2 The energy bandgap

The energy bandgap is a fundamental property of a semiconductor which determines the electrical characteristics of the devices fabricated from it. For practical purposes, such as the simulation of semiconductor components, the variation of the bandgap value with the temperature and doping concentration, must be accurately known to the device designer (Stefanakis and Zekentes 2014).

The energy bandgap of semiconductors always diminishes when its temperature is increased (Van Zeghbroeck 2011). This behavior can be appreciated if one considers that the interatomic spacing becomes larger when the amplitude of the atomic vibrations increases. This increase in interatomic spacing is caused by the enhancement in thermal energy by intensification of thermal motion at higher temperatures. The effect is quantified by the linear expansion coefficient of a material. With an increase of interatomic spacing, the potential seen by the electrons in the material decreases. It is this decreased potential which is responsible for the reduction of the energy bandgap. A direct modulation of the interatomic distance also brings about bandgap changes. By applying high compressive stress the bandgap increases, whereas on applying a tensile stress it decreases.

A semi-empirical relationship for the variation of the energy gap (E_g) of semiconductors with temperature (T) was proposed by Varshini (1967):

$$E_g = E_g(0) - \frac{\alpha T^2}{T + \beta}, \tag{3.1}$$

where $E_g(0)$ (eV), α (eV K^{-1}) and β (K) are the fitting coefficients of the model. The symbol $E_g(0)$ represents the bandgap of the material at 0 K, and T is the temperature in the Kelvin scale. Values of the parameters α and β for Si, GaAs and 4H-SiC are compiled in table 3.1. The equation has been found to satisfactorily represent the experimental data for diamond, Ge, Si, 6H-SiC, GaAs, InP and InAs.

Applying the Varshini model, variation of the energy bandgaps of common semiconductors is presented in table 3.2.

For Si, a precise assessment of the bandgap energy E_g between 2 K and 300 K (Bludau *et al* 1974) led to the approximate formula as follows:

$$E_g(T) = A + BT + CT^2. \tag{3.2}$$

Table 3.1. Varshini fitting parameters.

Material	$E_g(0)$ (eV)	α (eV K^{-1})	β (K)	Reference
Germanium	0.7437	4.774×10^{-4}	235	Sze *et al* (2021)
Silicon	1.1695	4.73×10^{-4}	636	Singh (1993), Ioffe Institute (2015b)
Gallium arsenide	1.521	5.58×10^{-4}	220	Wilkinson and Adams (1993)
4H-SiC	3.285	3.3×10^{-4}	240	Stefanakis and Zekentes (2014)

Table 3.2. Bandgaps at different temperatures according to the Varshini model.

Material	4.2 K	77.2 K	300 K	600 K
Germanium	0.743 66	0.7346	0.663 39	0.537 87
Silicon	1.169 486	1.165 55	1.124 02	1.031 733
Gallium arsenide	1.520 956	1.509 81	1.424 42	1.276 024
4H-SiC	3.284 976	3.2788	3.23	3.143 571

Table 3.3. Bandgaps of silicon at different temperatures according to the Bludau *et al* model.

Temperature	4.2 K	77.2 K	100 K	200 K	300 K
Bandgap	1.170	1.167	1.165	1.148	1.124

The values of parameters A, B and C are provided for two temperature zones. In the temperature range $0 < T \leqslant 190$ K,

$$A = 1.170 \text{ eV}, B = 1.059 \times 10^{-5} \text{ eV K}^{-1} \text{ and } C = -6.05 \times 10^{-7} \text{ eV K}^{-2}, \quad (3.2a)$$

while for the temperature interval $150 \leqslant T \leqslant 300$ K,

$$\begin{array}{l} A = 1.1785 \text{ eV}, B = -9.025 \times 10^{-5} \text{ eV K}^{-1} \text{ and} \\ C = -3.05 \times 10^{-7} \text{ eV K}^{-2} \end{array}. \quad (3.2b)$$

Table 3.3 shows the changes in the Si bandgap with temperature on the basis of the Bludau *et al* model.

3.3 Intrinsic carrier concentration

The intrinsic carrier concentration of a semiconductor is a key material property which occurs every now and then in the formulae describing the operations of various devices (Couderc *et al* 2014). This basic parameter exhibits a strong dependence on temperature. Further, the nature of variation is vastly different for different semiconductors, as shown in figure 3.1.

Under thermal equilibrium conditions, the intrinsic carrier concentration in a semiconductor is given by

$$n_{\text{i}} = \sqrt{N_{\text{C}} N_{\text{V}}} \exp\left(-\frac{E_{\text{g}}}{2k_{\text{B}}T}\right), \quad (3.3)$$

where N_C, N_V are the effective densities of states in the conduction and valence bands, respectively. N_C, N_V are expressed as

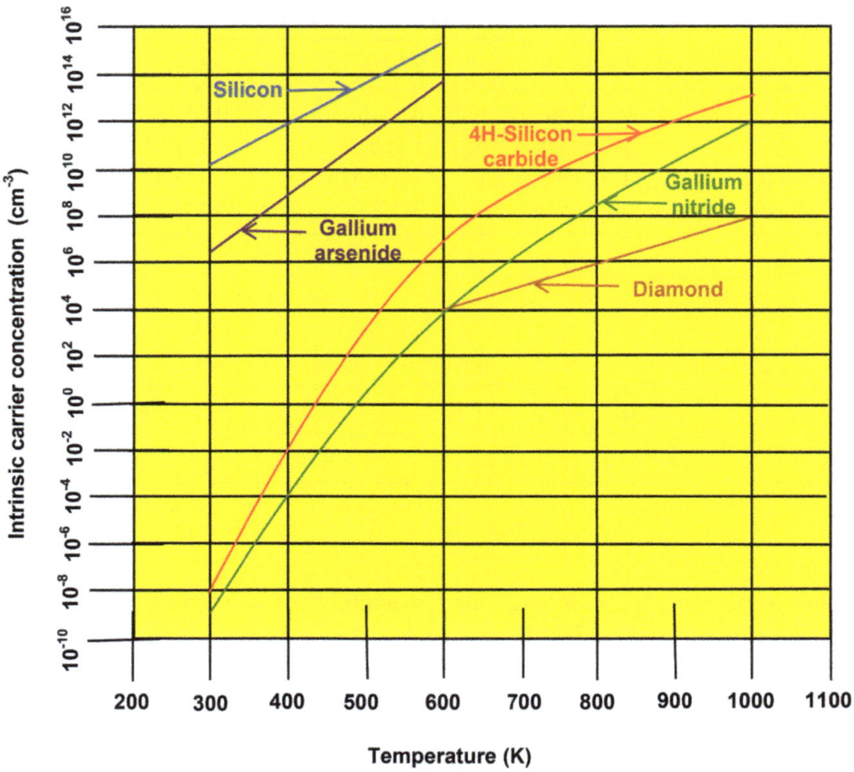

Figure 3.1. Change in the intrinsic carrier concentration of semiconductors on raising the temperature above room temperature (300 K).

Table 3.4. Effective densities of states (Van Zeghbroeck 1997).

Material	m_n^*/m_0	m_p^*/m_0	N_C (cm^{-3})	N_V (cm^{-3})
Germanium	0.56	0.29	$2.023 \times 10^{15}T^{1.5}$	$7.54 \times 10^{14}T^{1.5}$
Silicon	1.08	0.81	$5.42 \times 10^{15}T^{1.5}$	$3.52 \times 10^{15}T^{1.5}$
Gallium arsenide	0.067	0.47	$8.37 \times 10^{13}T^{1.5}$	$1.56 \times 10^{15}T^{1.5}$

m_0 = free electron rest mass = 9.11×10^{-31} kg.

$$N_C = 2\left(\frac{2\pi m_n^* k_B T}{h^2}\right)^{1.5} \tag{3.4}$$

$$N_V = 2\left(\frac{2\pi m_p^* k_B T}{h^2}\right)^{1.5}, \tag{3.5}$$

where m_n^*, m_p^* are the effective masses of electrons and holes for density of states calculations, k_B is the Boltzmann constant (= 1.381×10^{-23} m^2 kg s^{-2} K^{-1}), h is Planck's constant (= 6.626×10^{-34} m^2 kg s^{-1}). The calculated values of N_C, N_V for different semiconductors are tabulated in table 3.4.

In the above calculations, the masses m_n^*, m_p^* are presupposed to be constant with respect to temperature. But in reality, this assumption is not true. Careful review and correlation of experimental data on density of states effective masses m_n^*, m_p^*, and on the intrinsic concentration in silicon have revealed that the effective masses are, in fact, temperature- and energy-dependent (Barber 1967). Based on the measured temperature dependence of the energy gap, the obvious temperature variation of both the hole and electron effective masses was approximated in a first-order evaluation. Substitution of these temperature-dependent effective masses into the theoretical expression for intrinsic carrier concentration yielded a close agreement with reported measurements of n_i, confined within the bounds of experimental error. To perform a more rigorous calculation including the temperature-dependent effective masses of carriers as well as the energy bandgap of the semiconductor, we write

$$n_i = 2\left(\frac{2\pi k_B}{h^2}\right)^{1.5}\left(m_n^* m_p^*\right)^{0.75} T^{1.5} \exp\left(-\frac{E_g}{2k_B T}\right) \tag{3.6}$$

$$= 2\left\{\frac{2 \times 3.14 \times 1.381 \times 10^{-23}}{(6.626 \times 10^{-34})^2}\right\}^{1.5}\left(\frac{m_n^* m_p^*}{m_0^2} \times m_0^2\right)^{0.75}$$

$$T^{1.5} \exp\left(-\frac{E_g}{2k_B T}\right)$$

$$= 2\left(\frac{8.672\,68 \times 10^{-23}}{4.390\,3876 \times 10^{-67}}\right)^{1.5} \times (m_0^2)^{0.75} \times \left(\frac{m_n^* m_p^*}{m_0^2}\right)^{0.75}$$

$$T^{1.5} \exp\left(-\frac{E_g}{2k_B T}\right)$$

$$= 5.552\,72 \times 10^{66} \times m_0^{1.5} \times \left(\frac{m_n^* m_p^*}{m_0^2}\right)^{0.75} T^{1.5} \exp\left(-\frac{E_g}{2k_B T}\right) \tag{3.7}$$

$$= 5.552\,72 \times 10^{66} \times (9.11 \times 10^{-31})^{1.5} \times \left(\frac{m_n^* m_p^*}{m_0^2}\right)^{0.75}$$

$$T^{1.5} \exp\left(-\frac{E_g}{2k_B T}\right)$$

$$= 4.828\,1787 \times 10^{21} \times \left(\frac{m_n^* m_p^*}{m_0^2}\right)^{0.75} T^{1.5} \exp\left(-\frac{E_g}{2k_B T}\right) \text{ m}^{-3}$$

$$= 4.83 \times 10^{15} \times \left(\frac{m_n^* m_p^*}{m_0^2}\right)^{0.75} T^{1.5} \exp\left(-\frac{E_g}{2k_B T}\right) \text{ cm}^{-3}.$$

For silicon, the equations describing the temperature-dependent effective masses m_n* and m_p* in terms of free electron rest mass m_0 are (Caiafa *et al* 2003):

$$m_n^* = (-1.084 \times 10^{-9}T^3 + 7.580 \times 10^{-7}T^2 + 2.862 \times 10^{-4}T + 1.057)m_0 \quad (3.8)$$

$$m_p^* = \left(1.872 \times 10^{-11}T^4 - 1.969 \times 10^{-8}T^3 + 5.857 \times 10^{-6}T^2 + 2.712 \times 10^{-4}T + 0.584\right)m_0. \quad (3.9)$$

The temperature dependence of the bandgap energy is included through the equations of Bludau *et al* (1974) as follows:

$$E_g = 1.17 + 1.059 \times 10^{-5}T - 6.05 \times 10^{-7}T^2 \text{ for } T \leqslant 190 \text{ K} \quad (3.10)$$

$$E_g = 1.1785 - 9.025 \times 10^{-5}T - 3.05 \times 10^{-7}T^2 \text{ for } 300 \text{ K} \geqslant T \geqslant 190 \text{ K}. \quad (3.11)$$

At $T = 4.2$ K,

$$\begin{aligned} m_n^* &= \{-1.084 \times 10^{-9}(4.2)^3 + 7.580 \times 10^{-7}(4.2)^2 \\ &\quad + 2.862 \times 10^{-4}(4.2) + 1.057\}m_0 \\ &= \{-8.031 \times 10^{-8} + 1.337 \times 10^{-5} + 1.202 \times 10^{-3} + 1.057\}m_0 \\ &= 1.058m_0 \end{aligned} \quad (3.12)$$

$$\begin{aligned} m_p^* &= \{1.872 \times 10^{-11}(4.2)^4 - 1.969 \times 10^{-8}(4.2)^3 \\ &\quad + 5.857 \times 10^{-6}(4.2)^2 + 2.712 \times 10^{-4}(4.2) + 0.584\}m_0 \\ &= \{5.825 \times 10^{-9} - 1.459 \times 10^{-6} + 1.033 \times 10^{-4} \\ &\quad + 1.139 \times 10^{-3} + 0.584\}m_0 = 0.585m_0 \end{aligned} \quad (3.13)$$

$$\begin{aligned} E_g &= 1.17 + 1.059 \times 10^{-5}(4.2) - 6.05 \times 10^{-7}(4.2)^2 \\ &= 1.17 + 4.4478 \times 10^{-5} - 1.067 \times 10^{-5} \\ &= 1.17\text{eV} \end{aligned} \quad (3.14)$$

$$\begin{aligned} \therefore n_i &= 4.83 \times 10^{15} \times \left(\frac{1.058m_0 \times 0.585m_0}{m_0^2}\right)^{0.75} (4.2)^{1.5} \\ &\quad \exp\left(-\frac{1.17}{2 \times 8.617 \times 10^{-5} \times 4.2}\right) \text{cm}^{-3} \\ &= 4.83 \times 10^{15} \times 0.6978 \times 8.6074 \times 1.0087 \times 10^{-702} \text{ cm}^{-3} \\ &= 4.83 \times 10^{15} \times 6.0585 \times 10^{-702} \text{ cm}^{-3} = 2.926 \times 10^{-686} \text{ cm}^{-3}. \end{aligned} \quad (3.15)$$

Similar calculations are performed for $T = 77.2$ K and $T = 300$ K. Above 300 K, bandgaps are calculated by Varshini's formula, e.g. at $T = 600$ K, the computational procedure is given below:

$$
\begin{aligned}
m_n^* &= \{-1.084 \times 10^{-9}(600)^3 + 7.580 \times 10^{-7}(600)^2 + 2.862 \times 10^{-4}(600) \\
&\quad + 1.057\}m_0 \\
&= \{-0.234\,144 + 0.272\,88 + 0.171\,72 + 1.057\}m_0 = 1.267\,456m_0
\end{aligned} \tag{3.16}
$$

$$
\begin{aligned}
m_p^* = \{&1.872 \times 10^{-11}(600)^4 - 1.969 \times 10^{-8}(600)^3 + 5.857 \times 10^{-6}(600)^2 \\
&+ 2.712 \times 10^{-4}(600) + 0.584\}m_0
\end{aligned}
$$

$$
= (2.426\,112 - 4.253\,04 + 2.108\,52 + 0.162\,72 + 0.584)m_0 = 1.028\,312m_0 \tag{3.17}
$$

$$
\begin{aligned}
E_g = E_g(0) - \frac{\alpha T^2}{T + \beta} &= 1.1695 - \frac{4.73 \times 10^{-4}(600)^2}{600 + 636} \\
= 1.1695 - 0.1377\,6699 &= 1.031\,73\text{eV}
\end{aligned} \tag{3.18}
$$

$$
\begin{aligned}
\therefore n_i &= 4.828\,1787 \times 10^{15} \times \left(\frac{1.267\,456m_0 \times 1.0283\,12m_0}{m_0^2}\right)^{0.75} \\
&\quad (600)^{1.5} \exp\left(-\frac{1.031\,73}{2 \times 8.617 \times 10^{-5} \times 600}\right) \text{cm}^{-3} \\
&= 4.83 \times 10^{15} \times 1.2198 \times 14\,696.938\,46 \times 4.642\,56 \times 10^{-5} \text{ cm}^{-3} \\
&= 4.0199 \times 10^{15} \text{ cm}^{-3}.
\end{aligned} \tag{3.19}
$$

This calculation procedure is followed up to 1000 K. Table 3.5 lists the computed values of n_i.

Figure 3.2 presents the calculated change in the intrinsic carrier concentration of silicon with increase in temperature.

Table 3.5. Intrinsic carrier concentrations in silicon at different temperatures. m_0 = free electron rest mass = 9.11×10^{-31} kg.

Temperature (K)/ parameter	4.2	77.2	300	600	800 K	900 K	1000 K
m_n^*/m_0	1.058	1.0831	1.1818	1.2675	1.216	1.1383	1.017
m_p^*/m_0	0.585	0.6315	0.812 49	1.0283	2.1359	3.5	5.74
E_g (eV)	1.17	1.167	1.124	1.0317	0.9587	0.92	0.88
n_i (cm^{-3})	2.93×10^{-686}	1.954×10^{-20}	8.81×10^{9}	4.02×10^{15}	2.135×10^{17}	9.76×10^{17}	3.47×10^{18}

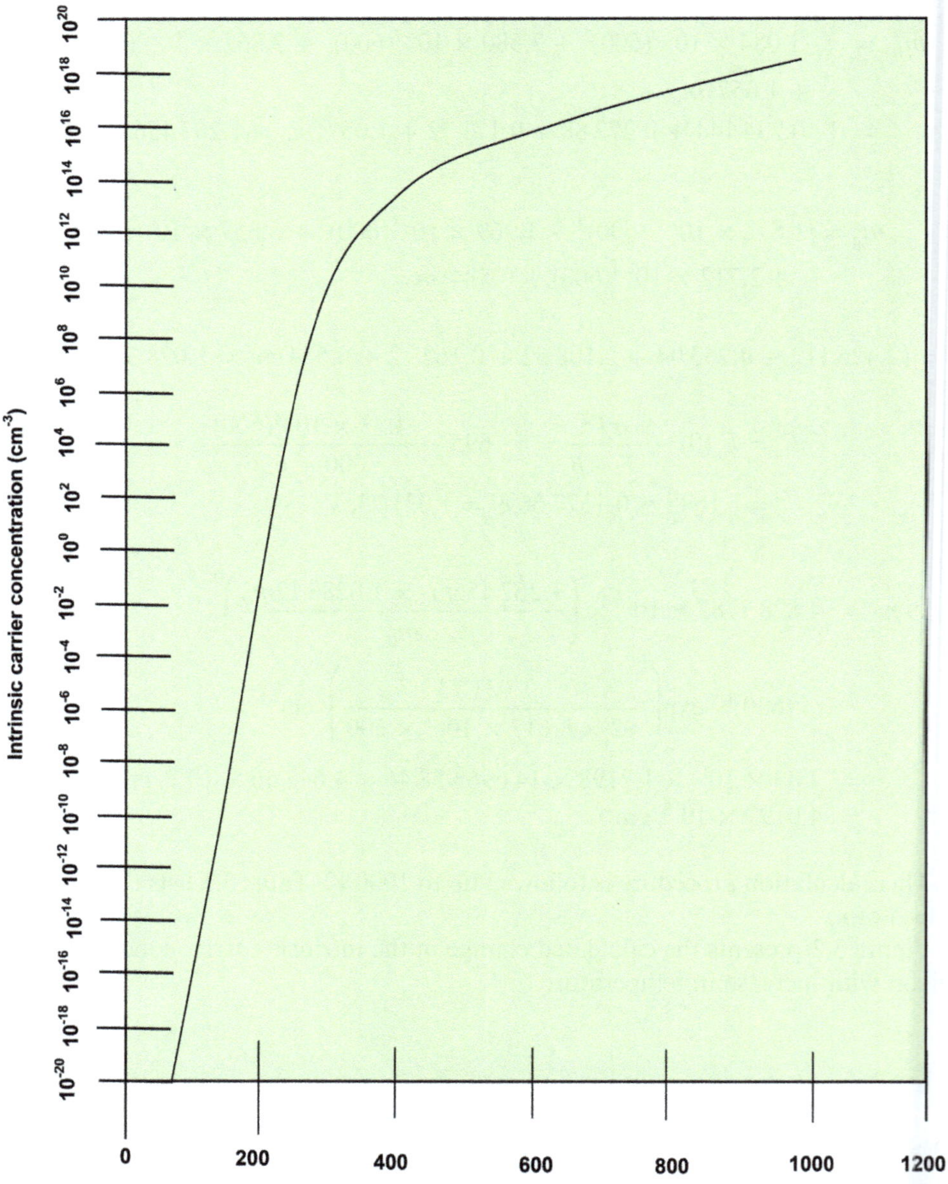

Figure 3.2. Increase in intrinsic carrier concentration of silicon with increase in temperature.

3.4 Carrier saturation velocity

The proportionality relationship between the average velocity of carriers and the applied electric field is violated at high electric fields. At these values of electric field, the carrier velocity attains a maximum value for both electrons and holes.

Quay *et al* put forward a simple and precise temperature-dependent model for the saturation velocity as a function of temperature in semiconductors (Quay *et al* 2000)

$$v_{sat}(T_L) = \frac{v_{sat}(300)}{(1 - A) + A(T_L/300)}. \tag{3.20}$$

This model is a two-parameter model. The first parameter v_{sat} (300) is the saturation velocity at the lattice temperature $T_L = 300$ K. The second parameter A is the TC portraying the temperature dependence of the different materials complying with the model. In the case of silicon, v_{sat} (300) = 1.02×10^7 cm s^{-1}, $A = 0.74$ for electrons; and v_{sat} (300) = 0.72×10^7 cm s^{-1}, $A = 0.37$ for holes. The model was demonstrated for the v_{sat} data of electrons in silicon in the temperature range 0 K–500 K by Jacoboni *et al* (1977). However, the model applies in general to a large number of technologically relevant semiconductor materials from 200 K to 500 K. Table 3.6 presents the calculated values from the model of Quay *et al* (2000) for silicon.

After Ali-Omar and Reggiani (1987),

$$v_n^{sat} = 1.45 \times 10^7 \sqrt{\tanh\left(\frac{155K}{T}\right)} \text{ cm s}^{-1} \tag{3.21}$$

$$v_p^{sat} = 9.05 \times 10^6 \sqrt{\tanh\left(\frac{312K}{T}\right)} \text{ cm s}^{-1}. \tag{3.22}$$

The values obtained from this model are listed in table 3.7.

Table 3.6. Electron and hole saturation velocities in silicon from the model of Quay *et al* (2000).

Temperature (K)	4.2	77.2	300	500
v_{sat} (T_L), cm s^{-1} for electrons	3.77×10^7	2.265×10^7	1.02×10^7	6.83×10^6
v_{sat} (T_L), cm s^{-1} for holes	1.13×10^7	9.928×10^6	7.2×10^6	5.775×10^6

Table 3.7. Electron and hole saturation velocities in silicon from the Ali-Omar and Reggiani model (Ali-Omar and Reggiani 1987).

Temperature (K)	4.2	77.2	300	500
v_{sat} (T_L), cm s^{-1} for electrons	1.45×10^7	1.424×10^7	9.995×10^6	7.948×10^6
v_{sat} (T_L), cm s^{-1} for holes	9.05×10^6	9.047×10^6	7.982×10^6	6.735×10^6

3.5 Electrical conductivity of semiconductors

The conductivity of a semiconductor is obtained by adding the contributions from the electron and hole populations as

$$\sigma = q\left(\mu_n n + \mu_p p\right), \tag{3.23}$$

where q is the charge of the electron, n and p stand for the densities of electrons and holes, and μ_n and μ_p refer to the mobilities of the electrons and holes, respectively. In a doped semiconductor under equilibrium conditions, the number of majority carriers greatly exceeds that of minority carriers. Then the above equation reduces to a single term involving the majority carriers.

The main point to be noted is that the conductivity of a semiconductor is determined by two factors, namely, the concentration of charge carriers moving freely and at disposal to transmit current, and the mobility or freedom of movement of these carriers (Sze *et al* 2021). Mobility determines the extent to which the free carriers are affected by electric fields. It is defined as the average drift velocity acquired by the carriers in unit electric field strength. In a semiconductor, both the carrier concentration and mobility are temperature-dependent. Thus, it is important to view the conductivity as a function of temperature. This assertion is expressed as

$$\sigma = q\left\{\mu_n(T)n(T) + \mu_p(T)p(T)\right\}. \tag{3.24}$$

To understand how the conductivity of a semiconductor changes with temperature and what its value will be at low temperatures relative to room temperature, it is necessary to know the influence of temperature on carrier concentrations and mobilities.

3.6 Free carrier concentration in semiconductors

Discrete power semiconductor devices as well as ICs are designed and fabricated to operate between designated temperature limits. These temperature limits are specified by manufacturers. The standard practice is that the device/circuit designer selects the doping level or levels. It is generally assumed that the dopants are approximately 100% ionized. It is taken for granted that the dopant atoms are completely exhausted by conversion into ions through release of free carriers from them. Also, it is postulated that the operating temperature is neither too high nor too low relative to room temperature. If this assumption regarding temperature ceases to hold, a profound impact on the values of a plethora of device parameters may be noticed. The depletion width or threshold voltage of an FET may drastically change.

Exponential temperature dependence dominates the temperature variation of intrinsic carrier concentration $n_i(T)$. In order to determine the total carrier concentration, space-charge neutrality must be taken into account. Hence,

$$n(T) = N_D^+(T) - N_A^-(T) + \frac{n_i^2(T)}{n(T)} \tag{3.25}$$

$$p(T) = N_A^-(T) - N_D^+(T) + \frac{n_i^2(T)}{p(T)}, \tag{3.26}$$

where $N_D^+(T)$ is the ionized donor concentration and $N_A^-(T)$ is the ionized acceptor concentration, which are clearly different from the number of neutral donor and acceptor atoms, N_D, N_A. Both $N_D^+(T)$ and $N_A^-(T)$ are temperature-dependent.

With temperature falling to very low values in the vicinity of 0 K (large $1/T$), n_i becomes infinitesimally small. Then, in an intrinsic material, the number of EHPs plummets to negligible proportions. The donor electrons are bound to the corresponding electron donating atoms. The holes are also fastened to the acceptor atoms concerned. When the temperature is too low, the percentage ionization of the dopant or dopants is appreciably less than 100%. This turnaround of the percentage of ionization of the dopants as opposed to their near 100% room temperature values is called freeze-out.

3.7 Incomplete ionization and carrier freeze-out

Impurity freeze-out is modeled using Fermi–Dirac statistics and the degeneracy factors associated with the conduction and valence energy bands. The ionized donor and acceptor concentrations $N_D{}^+$, $N_A{}^-$ are expressed in terms of the total donor and acceptor concentrations N_D, N_A as (Cole and Johnson 1989, Silvaco 2000)

$$\frac{N_D^+}{N_D} = \left[1 + g_{CD}\ \exp\left\{\left(\frac{\Delta E_D}{k_B T}\right) + \left(\frac{E_{fn} - E_C}{k_B T}\right)\right\}\right]^{-1} \tag{3.27}$$

$$\frac{N_A^-}{N_A} = \left[1 + g_{VD}\ \exp\left\{\left(\frac{\Delta E_A}{k_B T}\right) + \left(-\frac{E_{fp} - E_V}{k_B T}\right)\right\}\right]^{-1}, \tag{3.28}$$

where g_{CD}, g_{VD} are the degeneracy factors for conduction and valence bands, respectively, with the usually assumed values $g_{CD} = 2$ and $g_{VD} = 4$; E_{fn}, E_{fp} are the electron and hole quasi-Fermi energy levels; $(E_{fn} - E_C)$ is the position of the electron quasi-Fermi level relative to the conduction band edge for phosphorous-doped silicon; $(E_{fp} - E_V)$ is the position of the hole quasi-Fermi level relative to the valence band edge for boron-doped silicon; E_C, E_V are the energies of the conduction and valence band edges; $\Delta E_D = E_C - E_D$ is the activation energy for the phosphorous impurity in silicon (= 0.045 eV); and $\Delta E_A = E_A - E_V$ is the activation energy for the boron impurity in silicon (= 0.045 eV).

The temperature dependence of the activation energies of shallow donor and acceptor levels is not considered significant (Jonscher 1964).

At 4.2 K,

$$\frac{\Delta E_D}{k_B T} = \frac{0.045}{8.617 \times 10^{-5} \times 4.2} = 124.3389. \tag{3.29}$$

For $n = 1 \times 10^{15}$ cm^{-3}, the electron Fermi level E_{fn} is related to the electron concentration n as

$$\left(\frac{E_{\mathrm{fn}} - E_{\mathrm{C}}}{k_{\mathrm{B}}T}\right)_{4.2\ \mathrm{K}} = \left\{\ln\left(\frac{n}{N_{\mathrm{C}}}\right)\right\}_{4.2\ \mathrm{K}} = \ln\left\{\frac{1 \times 10^{15}}{N_{\mathrm{C}}}\right\}, \tag{3.30}$$

where N_{C} is the effective density of states in the conduction band at 4.2 K:

$$\begin{aligned}(N_{\mathrm{C}})_{4.2\ \mathrm{K}} &= 2\left(\frac{2\pi k_{\mathrm{B}}T}{h^2}\right)^{1.5}\left\{(m_{\mathrm{n}}^*)_{4.2\ \mathrm{K}}\right\}^{1.5} \\ &= 2\left\{\frac{2 \times 3.14 \times 1.381 \times 10^{-23} \times 4.2}{(6.626 \times 10^{-34})^2}\right\}^{1.5} \\ &\quad \times (1.058 \times 9.11 \times 10^{-31})^{1.5} \\ &= 2\left(\frac{2 \times 3.14 \times 1.381 \times 10^{-23} \times 4.2}{4.3904 \times 10^{-67}}\right)^{1.5} \\ &\quad \times (1.058 \times 9.11 \times 10^{-31})^{1.5} \\ &= 2\left(\frac{3.6425 \times 10^{-22}}{4.3904 \times 10^{-67}}\right)^{1.5} \times (1.058 \times 9.11 \times 10^{-31})^{1.5} \\ &= 2\{8.2965 \times 10^{44}\}^{1.5} \times (9.638 \times 10^{-31})^{1.5} \\ &= 2 \times 2.3897 \times 10^{67} \times 9.46 \times 10^{-46} \\ &= 4.521 \times 10^{22}\ \mathrm{m}^{-3} = 4.521 \times 10^{16}\ \mathrm{cm}^{-3}\end{aligned} \tag{3.31}$$

$$\therefore \left(\frac{E_{\mathrm{fn}} - E_{\mathrm{C}}}{k_{\mathrm{B}}T}\right)_{4.2\ \mathrm{K}} = \ln\left(\frac{1 \times 10^{15}}{4.521 \times 10^{16}}\right) = \ln\left(2.2119 \times 10^{-2}\right) = -3.8113. \tag{3.32}$$

Therefore,

$$\begin{aligned}\frac{N_{\mathrm{D}}^{+}}{N_{\mathrm{D}}} &= [1 + 2\ \exp\{(124.3389) + (-3.8113)\}]^{-1} = [1 + 2\ \exp(120.5276)]^{-1} \\ &= [1 + 2 \times 2.21 \times 10^{52}]^{-1} = (4.42 \times 10^{52})^{-1} = 2.262 \times 10^{-53}.\end{aligned} \tag{3.33}$$

At $T = 40$ K

$$\begin{aligned}m_{\mathrm{n}}^* &= \{-1.084 \times 10^{-9}(40)^3 + 7.580 \times 10^{-7}(40)^2 \\ &\quad + 2.862 \times 10^{-4}(40) + 1.057\}m_0 \\ &= \{-0.000\,069\,376 + 0.001\,2128 + 0.011\,448 + 1.057\}m_0 \\ &= 1.069\,59m_0\end{aligned} \tag{3.34}$$

$$\frac{\Delta E_{\mathrm{D}}}{k_{\mathrm{B}}T} = \frac{0.045}{8.617 \times 10^{-5} \times 40} = 13.0556. \tag{3.35}$$

For $n = 1 \times 10^{15}$ cm^{-3}, the quasi-Fermi level E_{fn} is related to the electron concentration n as

$$\left(\frac{E_{fn} - E_C}{k_B T}\right)_{40\ K} = \left\{\ln\left(\frac{n}{N_C}\right)\right\}_{40\ K} = \ln\left\{\frac{1 \times 10^{15}}{N_C}\right\}, \tag{3.36}$$

where N_C is the effective density of states in the conduction band at 40 K:

$$\begin{aligned}(N_C)_{40\ K} &= 2\left(\frac{2\pi k_B T}{h^2}\right)^{1.5}\left\{(m_n^*)_{40\ K}\right\}^{1.5} \\ &= 2\left\{\frac{2 \times 3.14 \times 1.381 \times 10^{-23} \times 40}{(6.626 \times 10^{-34})^2}\right\}^{1.5} \times (1.069\,59 \times 9.11 \times 10^{-31})^{1.5} \\ &= 2\left(\frac{2 \times 3.14 \times 1.381 \times 10^{-23} \times 40}{4.390\,388 \times 10^{-67}}\right)^{1.5} \times (1.069\,59 \times 9.11 \times 10^{-31})^{1.5} \\ &= 2\left(\frac{3.4691 \times 10^{-21}}{4.390\,388 \times 10^{-67}}\right)^{1.5} \times (1.069\,59 \times 9.11 \times 10^{-31})^{1.5} \\ &= 2\{7.9016 \times 10^{45}\}^{1.5} \times (9.743\,965 \times 10^{-31})^{1.5} \\ &= 2 \times 7.0238 \times 10^{68} \times 9.618\,42 \times 10^{-46} \\ &= 1.35 \times 10^{24}\ \text{m}^{-3} = 1.35 \times 10^{18}\ \text{cm}^{-3}\end{aligned} \tag{3.37}$$

$$\therefore \left(\frac{E_{fn} - E_C}{k_B T}\right)_{4.2\ K} = \ln\left(\frac{1 \times 10^{15}}{1.35 \times 10^{18}}\right) = \ln\left(7.4 \times 10^{-4}\right) = -7.208\,86. \tag{3.38}$$

Therefore,

$$\begin{aligned}\frac{N_D^+}{N_D} &= [1 + 2\ \exp\{(13.0556) + (-7.208\,86)\}]^{-1} = [1 + 2\ \exp(5.846\,74)]^{-1} \\ &= [1 + 2 \times 346.1]^{-1} = (693.2)^{-1} = 1.443 \times 10^{-3}.\end{aligned} \tag{3.39}$$

Similarly, N_D^+/N_D calculations are performed at 77.2 K, 300 K and 600 K. These values are given in table 3.8.

Table 3.8. Fractional ionization at different temperatures.

Temperature (K)	4.2	40	77.2	300	600
N_D^+/N_D	2.262×10^{-53}	1.443×10^{-3}	0.58	0.999 64	0.999 978
N_D^+ (cm^{-3})	2.262×10^{-38}	1.443×10^{-12}	5.8×10^{14}	9.9964×10^{14}	$9.999\,78 \times 10^{14}$

3.8 Different ionization regimes

For a semiconductor, three definite regimes of operation are observed in relation to temperature, as shown in figure 3.2. These regimes are distinctly discernible in figure 3.3. The following sections will bring out the salient features of these regimes.

3.8.1 At temperatures $T < 100$ K: carrier freeze-out or incomplete ionization regime

At temperatures which are not too low but still low enough, e.g. <100 K (=−173 °C), the thermal energy within a semiconductor is lacking in intensity to cause activation of the full donor and acceptor impurity content. As a result, the carrier concentration is not equal to the concentration of dopant atoms. This region of operation below 100 K, in which the thermal energy within the silicon is inadequate to fully ionize the impurity atoms, is known as the freeze-out regime. The statement about the availability of insufficient energy to ionize all impurity atoms is only partially true. In the quest for the truth, one must delve deeper into the actual situation. Two cases must be clearly distinguished.

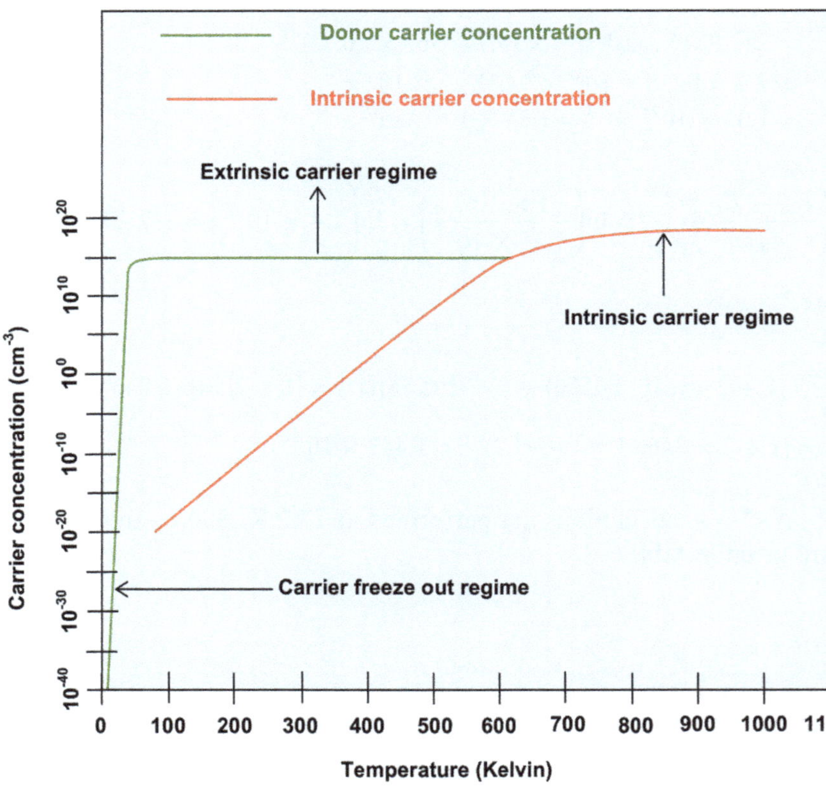

Figure 3.3. Diagram showing the appearance of the three operational regimes in an n-type semiconductor with $N_D = 1 \times 10^{15}$ cm^{-3} with increase of temperature: carrier freeze-out, extrinsic and intrinsic carrier concentration regimes.

Case I: intrinsic and low-doped or non-degenerate semiconductors. For a semiconductor in thermal equilibrium, the occupation of all electronic levels, both free and localized, is determined by the Fermi–Dirac distribution function $f(E)$ symbolizing the probability that an electronic state at energy E is occupied by an electron

$$f(E) = \left\{\exp\left(\frac{E - E_F}{k_B T}\right) + 1\right\}^{-1}, \tag{3.40}$$

where k_B is the Boltzmann constant ($=8.617 \times 10^{-5}$ eV K^{-1}) and E_F is a parameter known as the Fermi energy, which is the energy at which the probability of occupation by an electron is exactly one-half. The function $\{1 - f(E)\}$ is the probability that an electron is not found at energy E, i.e., the probability of finding a hole there.

For silicon, the range of doping considered here extends from 1.45×10^{10} cm^{-3} to 1×10^{18} cm^{-3}. At $T = 0$ K, when $E < E_F$, i.e. for energy levels lying below E_F, $(E - E_F)$ is a negative quantity and hence,

$$f(E) = \{\exp(-\infty) + 1\}^{-1} = \{0 + 1\}^{-1} = 1 \tag{3.41}$$

meaning that all the states below the Fermi level are filled; hence all the electrons are present in the valence band. In other words, there are no holes available for conduction.

At $T = 0$ K, when $E > E_F$, i.e. for energy levels lying above E_F, $(E - E_F)$ is a positive quantity, so that

$$f(E) = \{\exp(+\infty) + 1\}^{-1} = \{\infty + 1\}^{-1} = 0, \tag{3.42}$$

i.e. all the states above the Fermi level are empty so that there are no free electrons in the conduction band.

At $T = 0$ and at all temperatures, when $E = E_F$,

$$f(E) = \{\exp(0) + 1\}^{-1} = \{1 + 1\}^{-1} = 0.5. \tag{3.43}$$

Thus, the graph of $f(E)$ against $(E - E_F)$ is a step function in which, for $(E - E_F) < 0$, $f(E) = 1$, but when $(E - E_F) > 0$, $f(E) = 0$. The probability of finding an electron with energy equal to the Fermi energy in a semiconductor is ½ at any temperature including absolute zero. The graph of $f(E)$ versus $(E - E_F)$ is symmetrical around the Fermi level E_F. This graph is sketched in figure 3.4.

Let us now plot a graph between $f(E)$ and $(E - E_F)$ at the temperature $T = 10$ K (figure 3.5). We note that for energy $E - E_F = -3k_BT$, the exponential term in the equation for function $f(E)$ becomes <0.05. Hence the Fermi–Dirac distribution function increases to ~1. Also, for energy $E - E_F = 3k_BT$, the exponential term in the equation for function $f(E)$ becomes >20. Hence the Fermi–Dirac distribution function decays to ~0. Therefore, the $f(E)$ values will be determined for three negative values $E - E_F = -k_BT, -2k_BT, -3k_BT$ and three positive values of $E - E_F = k_BT, 2k_BT, 3k_BT$.

For negative values of $E - E_F$, it is found that when $E - E_F = -k_BT, -2k_BT, -3k_BT = -8.617 \times 10^{-4}, -1.723 \times 10^{-3}, -2.585 \times 10^{-3}$ eV, $f(E) = 0.731, 0.881,$

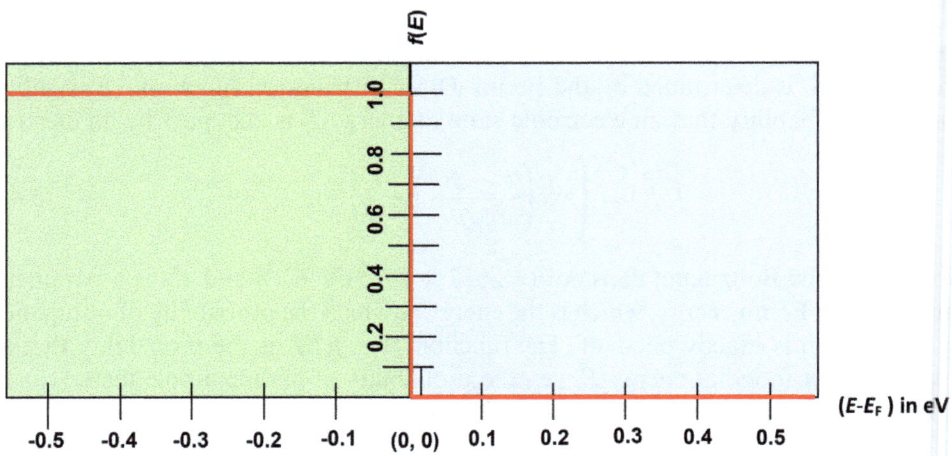

Figure 3.4. Plot of Fermi–Dirac distribution function $f(E)$ with respect to $(E - E_F)$ at $T = 0$ K.

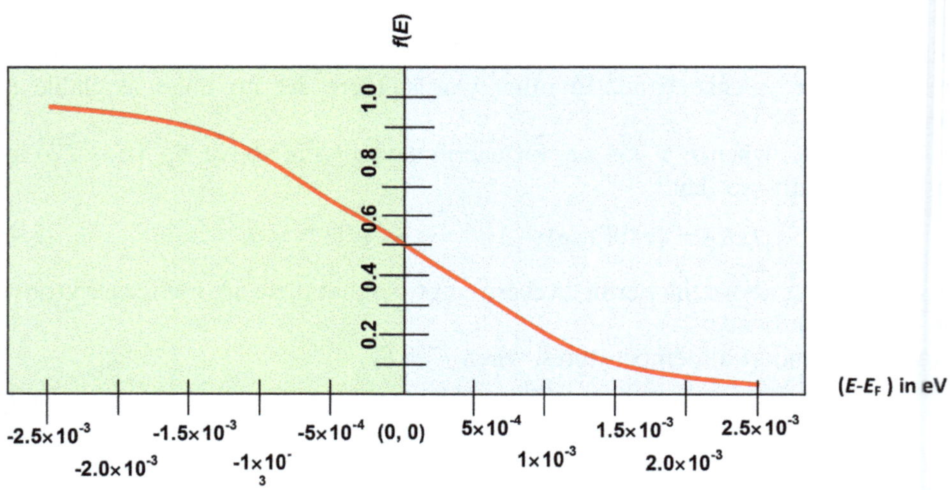

Figure 3.5. Plot of Fermi–Dirac distribution function $f(E)$ with respect to $(E - E_F)$ at $T = 10$ K.

0.9526. Thus, as $E - E_F$ decreases from $-k_BT$ to $-3k_BT$, $f(E)$ increases from 0.731 to 0.9526. Similarly for positive values of $E - E_F = k_BT$, $2k_BT$, $3k_BT = 8.617 \times 10^{-4}$, 1.723×10^{-3} eV, 2.585×10^{-3} eV, $f(E) = 0.269$, 0.119, 0.0474, respectively. Thus, as $E - E_F$ increases from k_BT to $3k_BT$, $f(E)$ decreases from 0.269 to 0.0474. Hence, the transition from $f(E) = 1$, corresponding to nearly full occupation of levels by electrons, to $f(E) = 0$, corresponding to almost empty levels, occurs in the range of energies $\pm 3k_BT = \pm 0.0026$ eV. For the very narrow range of energy levels enclosed within ± 0.0026 eV on either side of E_F, there is a very small, although finite, probability that some states are vacant on the valence band side and some states are filled on the conduction band side.

The gist of the discussion is that at temperatures near absolute zero, the transition from $f(E) = 1$ to $f(E) = 0$ being very rapid, the range of energies in which the probability of free carriers is non-zero is very small, so the number of free carriers in the semiconductor is insignificant. Inasmuch as the number of carriers is negligibly small due to the cooling effect, it appears as if the carriers have been completely frozen out. The freeze-out effect is a serious problem hampering the operation of semiconductor devices at cryogenic temperatures. Figure 3.6 showing the $f(E)$ versus $(E - E_F)$ plots in the temperature range 0 K–400 K corroborates the above assertion.

Case II: degenerate semiconductors. In a low-doped semiconductor, the Fermi level is inside the bandgap. For an n-type semiconductor, it is above the intrinsic Fermi level E_i and closer to the edge of the conduction band. For a p-type semiconductor it is below the intrinsic Fermi level E_i and closer to the valence band edge. Depending on whether the semiconductor is doped with donor or acceptor impurities, the higher the doping concentration is, the closer the Fermi level is to the edge of the conduction band or valence band. A non-degenerate semiconductor is defined as a semiconductor for which the Fermi energy is at least $3k_BT$ energy units away from either band edge.

As the doping level is continuously increased, at a particular stage, the Fermi level moves within an allowed band. It moves either inside the conduction or valence band. The semiconductor is then said to be degenerate. In this condition, the carrier densities no longer obey classical statistics. Thus a degenerate semiconductor is defined as a heavily doped semiconductor. In such a semiconductor, the Fermi level lies either in the conduction band or in the valence band, causing the material to behave as a metal. However, a degenerate semiconductor still has far fewer charge carriers than a true metal. So, its behavior is in many ways intermediary between a semiconductor and a metal.

At high impurity concentrations, the individual impurity atoms become such close neighbors that their doping levels merge into an impurity band. In other words, the onset of impurity interaction causes the broadening of the discrete energy levels of isolated impurities into an impurity band, with the resulting reduction of the effective activation energy below the value appropriate to high-purity material. This

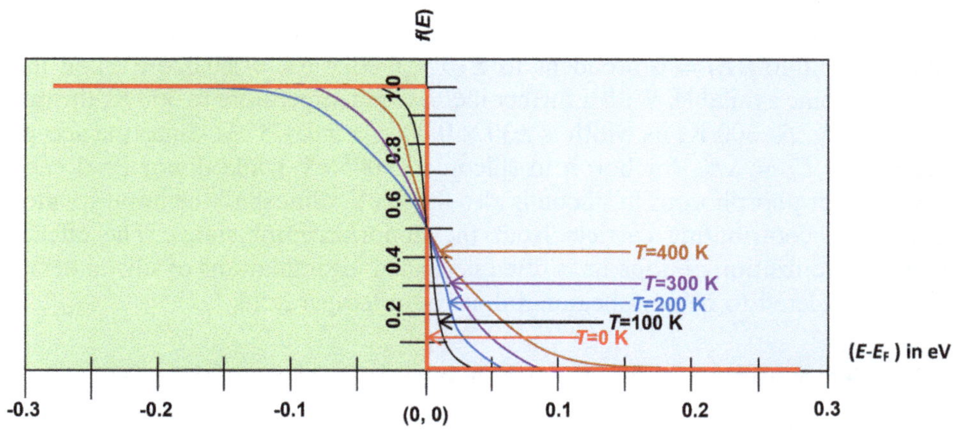

Figure 3.6. Graphs of the Fermi–Dirac distribution function with respect to energy difference $(E - E_F)$ from Fermi level E_F at different temperatures.

reduces somewhat the tendency to freeze-out carriers at any given temperature, and also produces some conduction in the broadened impurity band.

At a sufficiently high density of impurities, to the increasing broadening of the impurity band is added a shift downwards of the conduction band or upwards of the valence band, which is due to carrier–carrier interaction at high carrier densities. At this stage there is no longer any clear-cut distinction between the localized levels and the free bands; the activation energy becomes zero and the carrier density ceases to be a function of temperature. The important inference and verdict from the point of view of device operation is that there is no freeze-out of carriers even at the lowest temperatures; therefore the conductivity remains high. So, the statement about carrier freeze-out is true for a non-degenerate semiconductor. It is not true for a degenerate semiconductor.

3.8.2 At temperatures $T \sim 100$ K, and within $100 \text{ K} < T < 500$ K: extrinsic or saturation regime

A progressive increase in ionization takes place as the temperature is raised. At a temperature around 100 K, a large fraction of the donor/acceptor atoms has undergone ionization. At this point, the carrier concentration becomes a function of doping. It can be stated that at temperatures between 100 K and 500 K, i.e. from −173 °C to 227 °C, plentiful thermal energy resides within the silicon crystal. This thermal energy can ionize the impurity atoms. The region of operation where available dopants have been ionized and free carriers are liberated is known as the extrinsic regime or saturation region. It is the region where

$$N_D^+(T) = N_D \tag{3.44}$$

$$N_A^-(T) = N_A \tag{3.45}$$

$$n_i(T) \ll | N_D - N_A |. \tag{3.46}$$

Recalling the discussion regarding the $f(E)$ with $E - E_F$ plot at 0 K and 10 K, now increasing the temperature to 100 K, the range of energies $(E - E_F)$ for transition from $f(E) = 1$ to $f(E) = 0$ broadens to $\pm 10 \times 0.0026$ eV $= \pm 0.026$ eV and more carriers become available. With a further increase in temperature to 300 K, the range spreads more. At 300 K, its width is $\pm 30 \times 0.0026 = \pm 0.078$ eV. Since the acceptor level energy $E_a = \Delta E_A$ for boron in silicon is 0.045 eV while donor level energy $E_d = \Delta E_D$ for phosphorous in silicon is also 0.045 eV, almost all the impurity atoms are ionized, contributing carriers from the donor/acceptor states. The effect of incomplete ionization of dopants is often neglected in simulations of silicon devices, as it is considered to be non-meaningful at room temperature.

3.8.3 At temperatures $T > 500$ K: intrinsic regime

If the temperature is too high, the thermal generation effect causes the majority carrier concentration to become excessively higher than the dopant concentration in what is called the intrinsic temperature regime. As the temperature increases beyond

550 K (=277 °C), the intrinsic carrier concentration approaches and then exceeds the impurity concentration. At these high temperatures, the thermally generated intrinsic carriers outnumber the dopant-produced carriers, and the silicon returns to intrinsic-type behavior. In the intrinsic regime, the majority carrier concentration is nearly equal to the intrinsic concentration, n_i.

$$n_i(T) > | N_D - N_A |. \tag{3.47}$$

In this intrinsic region, the carrier concentration increases with temperature. The exhaustion regime lies between these two extremes, intrinsic and freeze-out (Pieper and Michael 2005).

At 600 K, the range of energies $E - E_F$ for $f(E) = 1$ to $f(E) = 0$ increases to $\pm 60 \times 0.0026 = \pm 0.156$ eV. As the temperature increases towards 600 K, electrons are dislodged from silicon atoms by thermal excitation, creating EHPs, and a large concentration of free carriers is built up, much greater than that due to impurity doping.

Figure 3.7 provides an explanation of the carrier concentration changes in a semiconductor from 0 K to 600 K in terms of the thermally induced EHP generation from the perspective of the energy band model.

3.8.4 Proportionality to bandgap at $T \geqslant 400$ K

One particularly interesting case occurs at high temperatures (above 400 K = 127 °C or higher) when mobility is dominated by lattice scattering ($\mu \propto T^{-3/2}$). In such cases, the conductivity can easily be shown to vary with temperature as

$$\sigma \propto \exp\left(-\frac{E_g}{2k_B T}\right). \tag{3.48}$$

In this case, conductivity depends only on the semiconductor bandgap and the temperature, as in an intrinsic semiconductor.

3.9 Mobilities of charge carriers in semiconductors

Carrier mobility is a phenomenological parameter whose value critically controls the performance characteristics of numerous semiconductor devices. As examples of these devices, mention may be made of diodes, and bipolar and FETs. Principally, the mobility of electrons and holes in a semiconductor is decided by two disparate types of scattering mechanisms. These mechanisms are: lattice scattering and impurity scattering (Van Zeghbroeck 2011).

3.9.1 Scattering by lattice waves

In scattering by lattice waves, the absorption or emission of either acoustical or optical phonons occurs. The density of phonons in a solid is known to increase with temperature. As a consequence, the scattering time due to this mechanism will decrease with temperature, as will the mobility. Theoretical calculations indicate that *acoustic* phonon interaction dominates the mobility in non-polar

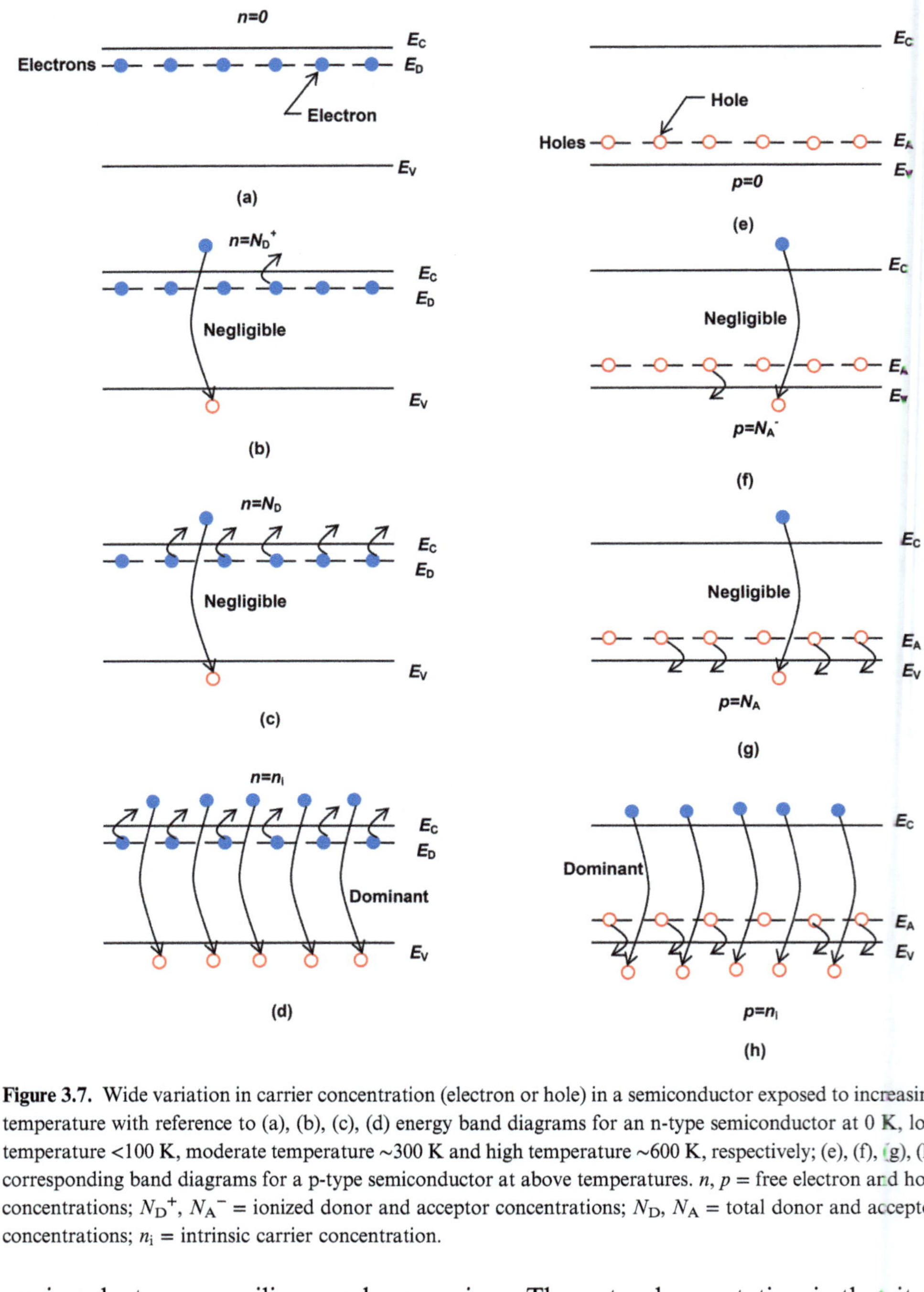

Figure 3.7. Wide variation in carrier concentration (electron or hole) in a semiconductor exposed to increasing temperature with reference to (a), (b), (c), (d) energy band diagrams for an n-type semiconductor at 0 K, low temperature <100 K, moderate temperature ~300 K and high temperature ~600 K, respectively; (e), (f), (g), (h) corresponding band diagrams for a p-type semiconductor at above temperatures. n, p = free electron and hole concentrations; N_D^+, N_A^- = ionized donor and acceptor concentrations; N_D, N_A = total donor and acceptor concentrations; n_i = intrinsic carrier concentration.

semiconductors, e.g. silicon and germanium. The natural expectation is that it is proportional to $T^{-3/2}$. On the other hand, the mobility due to *optical* phonon scattering only is expected to be proportional to T^{-2}. The effect of temperature on lattice mobility μ_L is conveyed by the equation

$$\mu_{\mathrm{L}} \propto T^{-\alpha}, \tag{3.49}$$

where $1.5 < \alpha < 2.5$. The lower limit takes account of acoustic phonon scattering, whereas the upper limit takes account of optical phonon scattering.

3.9.2 Scattering by ionized impurities

The extent of scattering due to electrostatic forces between the carrier and the ionized impurity depends on two factors:

(i) the number of impurity ions, and
(ii) the interaction time between the carriers and the impurity ions.

Larger impurity concentrations cause more scattering of carriers and thereby lead to a lower mobility. The interaction time is directly linked to the relative velocity of the carrier and the impurity. So, it is related to the thermal velocity of the carriers. As this thermal velocity increases with the ambient temperature, the interaction time decreases with an increase in temperature. Therefore, the amount of scattering decreases. The result is a mobility increase with temperature. To first order, the mobility due to impurity scattering varies directly as $T^{3/2}/N_{\mathrm{I}}$. The denominator N_{I} represents the density of charged impurities, regardless of sign. Hence, we can write

$$\mu_{\mathrm{I}} \propto T^{3/2}/N_{\mathrm{I}}. \tag{3.50}$$

The temperature dependences of mobility at low temperatures due to ionized impurity scattering and that at high temperatures caused by lattice scattering are shown in figure 3.8.

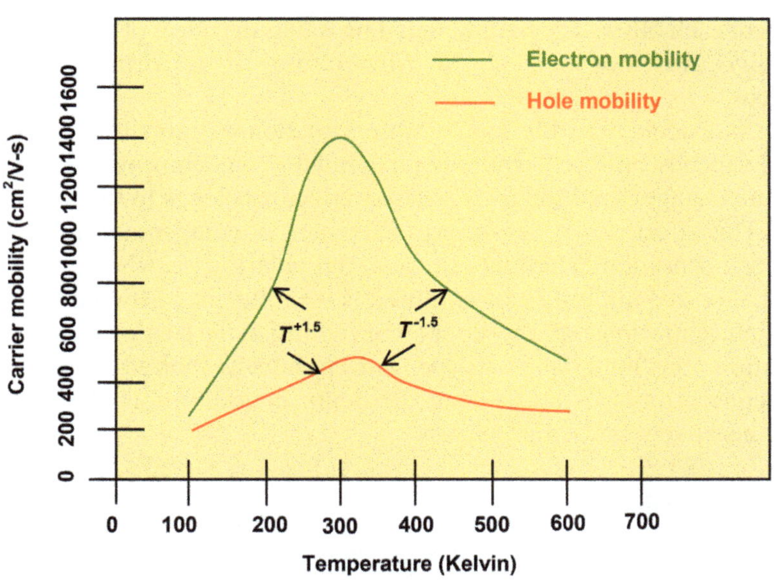

Figure 3.8. Variation of impurity scattering-controlled electron and hole mobilities in silicon at low temperatures and lattice scattering determined electron and hole mobilities at high temperatures in accordance with the $T^{1.5}$ and $T^{-1.5}$ laws, respectively.

3.9.3 Mobility in uncompensated and compensated semiconductors

There are two cases of interest: uncompensated and compensated semiconductors. In uncompensated material, the density of charged impurities is equal to the density of ionized shallow donors or acceptors, which decreases as the carrier density goes down with temperature. Shallow impurities are impurities which require little energy, typically around the thermal energy at room temperature or less, to ionize. Because of the decrease in density of charge centers with falling temperature, the effect on mobility due to impurity scattering becomes temperature-dependent in an uncompensated semiconductor.

Suppose a semiconductor contains both types of impurities: shallow donors and shallow acceptors. Also, let the concentrations of donors and acceptors be equal. Such a semiconductor is called a compensated semiconductor. It is so-named because in this semiconductor, equal amounts of donors and acceptors compensate each other, yielding no free carriers. The simultaneous presence of shallow donors and shallow acceptors in a semiconductor causes the electrons ejected from the donor to be captured by the acceptor state. This ionizes the acceptor without producing a free electron or hole. The resulting density of charged centers in compensated material, which contains equal quantities of both shallow donors and shallow acceptors, is equal to double the minority compensating density ($2N_A$ in an n-type material). Due to the large number of charged centers present, the scattering increases and the mobility in a compensated material is less than in an uncompensated material.

However, in this type of compensated material, the density of charged centers is virtually independent of temperature because approximately equal numbers of donor and acceptor atoms are oppositely ionized at a given temperature, cancelling the effects of each other. The natural outcome is that the effect of impurity scattering on mobility does not depend on temperature. It is virtually temperature-independent.

Thus it is evident that the temperature variation of mobility due to impurity scattering depends on whether it is compensated or uncompensated. The effect of temperature changes accordingly if a semiconductor belongs to the former or latter category. This effect is very sensitive to the extent of compensation. The degree or extent of compensation determines the material polarity, i.e. whether it is n-type or p-type. The ionized impurity concentration is the difference between the donor and acceptor concentrations. If the donor concentration is larger than the acceptor concentration, the material is n-type with positively ionized impurities. If the acceptor concentration is larger than the donor concentration, the material is p-type with negatively ionized impurities.

3.9.4 Resultant mobility

The observed carrier mobility in a semiconductor is the combined effect of the lattice-scattering limited mobility and the impurity-scattering limited mobility. The temperature dependence of mobility due to lattice scattering is approximately $T^{-\alpha}$. The temperature dependence of mobility due to impurity scattering is $T^{+3/2}$. Hence, carrier mobility is found to follow the trend of increasing with diminishing temperature if lattice scattering plays a dominant role, but decreasing with falling temperature if impurity scattering overrides lattice scattering. At lower temperatures, carrier movements become feebler and slower. Thus carriers have more time to interact with charged impurities. Consequently, as the temperature is lowered, impurity scattering increases. In consequence, the mobility decreases. This is just the reverse of the effect of lattice scattering.

As temperature goes down, the upward trend of mobility due to the ionized impurity scattering mechanism is counterbalanced by the downward trend of mobility due to lattice scattering. The resultant mobility μ is the harmonic mean of the components due to the two mechanisms μ_L, μ_I

$$\mu = \frac{\mu_L \mu_I}{\mu_L + \mu_I}, \tag{3.51}$$

i.e., it is effectively determined by the smaller of the two mobilities.

Case I: pure materials. At cryogenic temperatures in reasonably pure materials, the ionized impurity scattering is normally dominant, which is responsible for the decrease in mobility below 100 K.

Case II: degenerate semiconductors. The situation described above changes considerably in several respects in semiconductor materials containing sufficiently high densities of impurities to give rise to interaction between neighboring impurity centers. As a very rough estimate, the critical impurity density N_{crit} for shallow 'hydrogenic' centers corresponds to a mean inter-impurity distance equal to the radius of the Bohr orbit of the ground state a_1. Thus, $N_{crit} = a_1^{-3}$ depends on the effective mass and the dielectric constant K of the material or, what amounts to the same in this simple model, on the impurity activation energy. As to order of magnitude, the critical density is 10^{18} cm^{-3} in silicon, 10^{16} cm^{-3} in germanium and 10^{13} cm^{-3} in n-type InSb (Jonscher 1964).

The mobility is affected to a large extent by degeneracy. In the first place, the ionized impurity scattering depends on the effective energy of the carriers contributing to conduction and in degenerate systems this is the temperature-independent Fermi energy. Second, if the free carriers are present in reasonably high densities, their own screening action reduces the range of action of ionized centers. As a result, one can observe relatively high mobilities down to very low temperatures, even in highly impure materials.

3.10 Equations for mobility variation with temperature

3.10.1 Arora–Hauser–Roulston equation

Arora *et al* proposed the following empirical equations for electron/hole mobility in silicon up to a dopant concentration N of 10^{20} cm^{-3} for the temperature range $T = 200$–500 K (Arora *et al* 1982):

$$\mu_n(N, T) = 88\left(\frac{T}{300}\right)^{-0.57} + \frac{7.4 \times 10^8 T^{-2.33}}{1 + 0.88\left\{\frac{N}{1.26 \times 10^{17}(T/300)^{2.4}}\right\}(T/300)^{-0.146}} \tag{3.52}$$

$$\mu_p(N, T) = 54.3\left(\frac{T}{300}\right)^{-0.57} + \frac{1.36 \times 10^8 T^{-2.23}}{1 + 0.88\left\{\frac{N}{2.35 \times 10^{17}(T/300)^{2.4}}\right\}(T/300)^{-0.146}}. \tag{3.53}$$

The above equations were derived by them on the basis of experimental data and the amended Brooks–Herring theory of mobility. They found that the electron mobility data in the range 200 K–500 K fitted satisfactorily with the lattice scattering limited mobility μ_{Ln} formula

$$\mu_{Ln} = 8.56 \times 10^8 T^{-2.33} \tag{3.54}$$

and ionized impurity scattering μ_{In} formula

$$\mu_{In} = (7.3 \times 10^{17} T^{1.5})/[N_I\{\ln(b + 1) - b/(b + 1)\}], \tag{3.55}$$

where N_I is the number of ionized impurity atoms

$$b = \frac{1.52 \times 10^{15} T^2}{n\{2 - (n/N)\}} \tag{3.56}$$

n is the electron concentration per cm^3. The two mobilities μ_{Ln}, μ_{In} were combined together by the mixed scattering formula to arrive at equation (3.52).

Similarly, the hole mobility data for the 150 K–400 K range could be fitted to

$$\mu_{Lp} = 1.58 \times 10^8 T^{-2.23} \tag{3.57}$$

$$\mu_{Ip} = (5.6 \times 10^{17} T^{1.5})/[N_I\{\ln(b + 1) - b/(b + 1)\}], \tag{3.58}$$

$$b = \frac{2.5 \times 10^{15} T^2}{p\{2 - (p/N)\}} \tag{3.59}$$

where p is the hole concentration per cm^3. As before, the two mobilities μ_{Lp}, μ_{Ip} were combined together by the mixed scattering formula to obtain equation (3.53).

3.10.2 Klaassen equations

A physics-based analytical model was presented by Klaassen expressing the temperature-dependent mobility of charge carriers in terms of donor, acceptor, as well as electron and hole concentrations (Klaassen 1992a).

For lattice scattering,

$$\mu_{i,L} = \mu_{max}\left(\frac{300}{T}\right)^{\theta_i} \tag{3.60}$$

where $\theta_n = 2.285$ for electrons, $\theta_p = 2.247$ for holes and μ_{max} values are given with impurity scattering for different dopants.

For majority impurity scattering of electrons/holes,

$$\mu_{i,\,I}(N_I,\, c) = \mu_{i,\,N}\left(\frac{N_{ref,\,1}}{N_I}\right)^{\alpha_1} + \mu_{i,\,c}\left(\frac{c}{N_I}\right), \tag{3.61}$$

where (i, I) → (n, D) or (p, A) (n for electron, p for hole, D for donor and A for acceptor), c is the carrier concentration and

$$\mu_{i,\,N} = \frac{\mu_{max}^2}{\mu_{max} - \mu_{min}}\left(\frac{T}{300}\right)^{3\alpha_1 - 1.5} \tag{3.62}$$

$$\mu_{i,\,c} = \frac{\mu_{max}\,\mu_{min}}{\mu_{max} - \mu_{min}}\left(\frac{300}{T}\right)^{0.5}. \tag{3.63}$$

For phosphorous, $\mu_{max} = 1414.0$ cm^2 V^{-1} s^{-1}, $\mu_{min} = 68.5$ cm^2 V^{-1} s^{-1}, $N_{ref,1} = 9.2 \times 10^{16}$ cm^{-3}, $\alpha_1 = 0.711$. For boron, $\mu_{max} = 470.5$ cm^2 V^{-1} s^{-1}, $\mu_{min} = 44.9$ cm^2 V^{-1} s^{-1}, $N_{ref,1} = 2.23 \times 10^{17}$ cm^{-3}, $\alpha_1 = 0.719$.

3.10.3 MINIMOS mobility model

As in other models, the lattice mobility $\mu_{n,\,p}^{L}$ is modeled as a simple power law. In the MINIMOS 4 mobility model, the equations for mobilities of electrons and holes are (Selberherr *et al* 1990):

$$\mu_n^L = 1430(T/300\text{ K})^{-2}\text{ cm}^2/\text{V} - \text{s} \tag{3.64}$$

$$\mu_p^L = 460(T/300\text{ K})^{-2.18}\text{ cm}^2/\text{V} - \text{s}. \tag{3.65}$$

The ionized impurity mobility $\mu_{n,\,p}^{LI}$ is modeled by using the Caughey–Thomas equation

$$\mu_{n,\,p}^{LI} = \mu_{n,\,p}^{min} + \frac{\mu_{n,\,p}^{L} - \mu_{n,\,p}^{min}}{1 + \left(N_I/N_{n,\,p}^{ref}\right)^{\alpha_{n,\,p}}} \tag{3.66}$$

where

$$\mu_n^{min} = 80(T/300\text{ K})^{-0.45}\text{ cm}^2/\text{V} - \text{s for } T \geqslant 200\text{ K} \tag{3.67}$$

$$\mu_n^{min} = 80(200\text{ K}/300\text{ K})^{-0.45}(T/200\text{ K})^{-0.15}\text{ cm}^2/\text{V} - \text{s for } T < 200\text{ K} \tag{3.68}$$

$$\mu_p^{min} = 45(T/300\text{ K})^{-0.45}\text{ cm}^2/\text{V} - \text{s for } T \geqslant 200\text{ K} \tag{3.69}$$

$$\mu_p^{min} = 45(200\text{ K}/300\text{ K})^{-0.45}(T/200\text{ K})^{-0.15}\text{ cm}^2/\text{V} - \text{s for } T < 200\text{ K} \tag{3.70}$$

$$N_{\mathrm{n}}^{\mathrm{ref}} = 1.12 \times 10^{17}\left(\frac{T}{300}\right)^{3.2} \tag{3.71}$$

$$N_{\mathrm{p}}^{\mathrm{ref}} = 2.23 \times 10^{17}\left(\frac{T}{300}\right)^{3.2} \tag{3.72}$$

$$\alpha_{\mathrm{n,\,p}} = 0.72\left(\frac{T}{300}\right)^{0.065}. \tag{3.73}$$

3.11 Mobility in MOSFET inversion layers at low temperatures

The mobility of electrons in the conducting channel formed under the surface of the device monotonically increases as temperature falls to 5 K (Hairapetian *et al* 1989). Considering the electric field dependence of mobility, this increase of mobility takes place at all values of the electric field, as evidenced from measurements at all gate voltages. This increase in the inversion layer mobility of electrons is greatly beneficial to device operation at low temperatures. In NMOS devices, at an electric field of 3.5×10^5 V cm^{-1}, the experimentally measured electron mobility in the inversion layer was found to increase from approximately 450 cm^2 V^{-1} s^{-1} at 293 K to 2800 cm^2 V^{-1} s^{-1} at 77 K, 4700 cm^2 V^{-1} s^{-1} at 25 K and 5800 cm^2 V^{-1} s^{-1} at 5 K. In a PMOS device, at an electric field of 2.0×10^5 V cm^{-1}, the hole mobility was 160 cm^2 V^{-1} s^{-1} at 293 K, but increased to 400 cm^2 V^{-1} s^{-1} at 100 K, 620 cm^2 V^{-1} s^{-1} at 25 K and 680 cm^2 V^{-1} s^{-1} at 5 K.

Both electron and hole mobilities increased with decreasing temperature when the respective NMOS or PMOS devices were operating under strong inversion. But a difference was noticed between the behavior of electron and hole mobilities, when the device was operating near the threshold voltage instead of the strong inversion regime. For operation near the threshold voltage, the PMOS device exhibited a peak value of hole mobility around 50 K. The hole mobility started to decrease at temperatures <50 K. However, the NMOS device did not show such a peak. The mobility continued to increase up to 5 K.

3.12 Carrier lifetime

Minority-carrier lifetime or carrier lifetime is an important property of semiconductors that critically impacts the performance characteristics of electronic devices, e.g., solar cells and bipolar junction transistors (Murphy *et al* 2022). The carrier lifetime is defined as the average time during which an excess minority carrier, electron or hole, undergoes recombination (Park 2004). It is determined by three recombination mechanisms:

(i) band-to-band radiative recombination;
(ii) trap-assisted (or Shockley–Read–Hall (SRH)) recombination; and
(iii) Auger recombination.

Mechanisms (i) and (ii) involve only two carriers, but mechanism (iii) is a three-carrier process in which the energy liberated during electron–hole recombination is imparted to a third carrier, either an electron or a hole. It is not released in the form of heat or light. Amongst these three mechanisms, band-to-band recombination has a relatively inconsequential effect in silicon because the radiative lifetime in silicon is extremely large. The remaining two mechanisms (ii) and (iii) govern the electrical properties of silicon devices. SRH recombination is correlated with the concentration of deep energy-level metallic impurities contaminating the silicon wafer and also the lattice defects created in the silicon. Lifetime degradation occurs during crystal growth as well as in the device fabrication steps, mainly through impurities in processing chemicals, gases, quartzware and containers, and also through thermal shocks in high-temperature diffusion and oxidation furnaces. The lifetime killing causes the device characteristics to deteriorate. Although Auger lifetime is not related to impurity density, it varies inversely with the carrier concentration.

When a semiconductor device is in the operating condition, the minority carrier density in the active device layer is the decisive factor affecting the value of recombination lifetime in silicon. Two interesting situations shall be considered. At low minority carrier density, also called low-level injection, the non-equilibrium minority-carrier concentration is smaller than the majority-carrier concentration at thermal equilibrium. Then the SRH recombination mechanism is preponderant and so relevant for carrier lifetime in silicon. At high carrier density, also known as high-level injection, the non-equilibrium minority-carrier concentration is larger than the equilibrium majority carrier concentration. In this case, the Auger recombination mechanism predominates and hence influences the recombination lifetime.

It is found that the efficiency of recombination centers is high at lower temperatures (Hudgins *et al* 1994). Evidence for this statement comes from the decrease in the effective carrier lifetime at these temperatures. It is reported that for recombination centers that are poised at shallow locations in the forbidden energy gap, the capture cross-section increases very rapidly with falling temperature (Jonscher 1964). Interpretation for this steep rise in capture cross-section has been provided in terms of the popular 'giant trap' model. In this model, the carriers are not captured directly into the ground state of the capturing center. They are first trapped in one of the higher excited states. Afterwards, they cascade down to the ground state. The cascading resembles a waterfall or a series of waterfalls. The excited states provide a much larger capture cross-section than would be expected from the ground state. Extensive and colossal cross-sections in the range of 10^{-15} cm^2 to 10^{-13} cm^2 have been observed for a wide variety of traps in Si and Ge. Some of these traps involve binding energies several times larger than the Debye energy (Lax 1959).

Recombination is an inelastic process. In this process, a major chunk of the binding energy is carried away by acoustic phonons (Park *et al* 1988). The impurity ionization energy E_I is much greater than the energy of an acoustic phonon, i.e., $E_I \gg k_BT$. Naturally, direct recombination into the ground state involves several phonons. The multiphonon procedure has equivalently a very small cross-section. Therefore, the controlling activity is captured into highly excited states. The capturing is ensued by a cascade of single-phonon emissions and absorptions. These happen as the carrier slowly diffuses into the ground state.

The giant-trap model applies, however, only to the shallow 'hydrogenic' impurities. It is not clear how the theory can be extended to deeper levels. Regrettably, these are the levels which are normally associated with recombination.

Focusing attention on the temperature range between room temperature and about 90 K (Ichimura *et al* 1998), a very frail temperature dependence is observed in the carrier lifetime of as-polished wafers. In contrast, the carrier lifetime decreases sharply with decreasing temperature for oxidized wafers. In all specimens, an initial hasty decay occurs in the photoconductivity decay curves followed by a sluggish component in the photoconductivity decay curves at temperatures below 150 K. From numerical simulations, the inference was drawn that the decrease in the carrier lifetime with decreasing temperature was caused by recombination through shallow recombination centers. These centers had an energy level within 0.15 eV from the band edge. Moreover, the slow component was due to minority carrier traps with a small majority carrier capture cross-section.

The low-level lifetime is of no interest at low temperatures, since injected densities are invariably high in comparison with equilibrium densities (Jonscher 1964). The temperature dependence of high-level lifetime is given by a simple power law (Palmer *et al* 2003)

$$\tau_{\rm HL} = 5 \times 10^{-7}\left(\frac{T}{300}\right)^{1.5} \tag{3.74}$$

At $T = 4.2$ K,

$$\tau_{\rm HL} = 5 \times 10^{-7}\left(\frac{4.2}{300}\right)^{1.5} = 8.28 \times 10^{-10}{\rm s}. \tag{3.75}$$

Similarly, $\tau_{\rm HL}$ values are calculated at 77 K, 300 K and 600 K. These are tabulated in table 3.9.

A simple temperature exponent of 1.5 does not address the competing mechanisms of SRH recombination (carrier lifetime increases as temperature increases) and Auger recombination (carrier lifetime decreases as temperature increases) under the conditions of medium-to-large carrier concentrations. For rigorous modeling applications, the effects of intrinsic lifetime, SRH recombination, and Auger processes must be included, assuming high-level injection and quasi-neutrality (electron concentration $n \approx$ hole concentration p) (Klaassen 1992b).

Table 3.9. High-level carrier lifetimes at different temperatures.

Temperature (K)	4.2	77.2	300	600
Carrier lifetime(s)	8.2×10^{-10}	6.53×10^{-8}	5×10^{-7}	1.41×10^{-6}

3.13 Wider bandgap semiconductors than silicon

3.13.1 Gallium arsenide

The energy gap of GaAs obeys the formula (Ioffe Institute 2015a)

$$E_g(T) = 1.519 - 5.405 \times 10^{-4}T^2/(T + 204) \text{ eV for } 0 < T < 1000 \text{ K}. \quad (3.76)$$

Intrinsic carrier concentration in GaAs follows the relation (Madelung *et al* 2002)

$$n_i(T) = 1.05 \times 10^{16}T^{3/2} \exp\{-1.604/(2k_BT)\} \text{ for } 33 < T < 475 \text{ K}. \quad (3.77)$$

3.13.2 Silicon carbide

For 4H-SiC, the energy gap is (Ioffe Institute 2015d)

$$E_g(T) = E_g(0) - 6.5 \times 10^{-4}T^2/(T + 1300), \quad (3.78)$$

where $E_g(0) = 3.23$ eV is the bandgap at 300 K.

For 4H-SiC in the intrinsic carrier concentration formula, the effective density of states in the conduction band is

$$N_c(T) \approx 4.82 \times 10^{15}M(m_c/m_0)^{1.5}T^{1.5} \text{ (cm}^{-3}) \approx 4.82 \times 10^{15}(m_{cd}/m_0)^{1.5}T^{1.5} = 3.25 \times 10^{15}T^{1.5} \text{ (cm}^{-3}), \quad (3.79)$$

where $M = 3$ denotes the number of equivalent valleys in the conduction band. The symbol $m_c = 0.37m_0$ represents the effective mass of the density of states in one valley of the conduction band and $m_{cd} = 0.77\, m_0$ is the effective mass of the density of states.

The effective density of states in the valence band conforms to the equation

$$N_v(T) = 4.8 \times 10^{15}T^{1.5} \text{ (cm}^{-3}). \quad (3.80)$$

3.13.3 Gallium nitride

The energy gap–temperature relationship for gallium nitride is (Ioffe Institute 2015a)

$$E_g(T) = E_g(0) - 7.7 \times 10^{-4}T^2/(T + 600) \text{ eV}, \quad (3.81)$$

where

$$E_g(0) = 3.47 \text{ eV for wurtzite structure}, \quad (3.81a)$$

and

$$E_g(0) = 3.28 \text{ eV for zinc blende structure}. \quad (3.81b)$$

The Varshini equation for GaN is

$$E_g(T) = E_g(0) - 9.39 \times 10^{-4}T^2/(T + 772) \text{ eV where } E_g(0) = 3.427 \text{ eV}. \quad (3.82)$$

Given below are the equations for effective densities of states in conduction and valence bands for wurtzite and zinc blende GaN structures to be used for calculating the intrinsic carrier concentration

$$N_C \approx 4.3 \times 10^{14}T^{1.5} \text{ (cm}^{-3}\text{) for wurtzite,} \quad (3.83)$$

$$N_C \approx 2.3 \times 10^{14}T^{1.5} \text{ (cm}^{-3}\text{) for zinc blende,} \quad (3.84)$$

$$N_V \approx 8.9 \times 10^{15}T^{1.5} \text{ (cm}^{-3}\text{) for wurtzite,} \quad (3.85)$$

$$N_V \approx 8.0 \times 10^{15}T^{1.5} \text{ (cm}^{-3}\text{) for zinc blende.} \quad (3.86)$$

3.13.4 Diamond

At 300 K, the bandgap variation with temperature is given by (Ioffe Institute 2015c)

$$dE_g/dT = -(5.4 \pm 0.5) \times 10^{-5} \text{ eV K}^{-1}. \quad (3.87)$$

3.14 Discussion and conclusions

When considering operation of semiconductor devices at low temperatures, if we look at the temperature dependence of free carrier concentrations and mobilities, it is necessary to distinguish between two distinct temperatures. These are the liquid-nitrogen temperature (77 K) and liquid-helium temperature (4.2 K). Investigations have shown that carrier concentrations and mobilities do not experience any drastic changes up to the liquid-nitrogen temperature. But below this temperature, the scenario is dissimilar. As the temperature is decreased further down from 77 K, different low-temperature phenomena come into action, notably carrier freeze-out and impurity scattering. Again, when considering the effects of these phenomena on carrier concentrations and mobilities, one must take note of the purity of the semiconductor material, whether undoped or intrinsic, low-doped or heavily doped with impurities. Consideration must also be given to the extent of compensation of the semiconductor, i.e. whether it is totally uncompensated, partially compensated or fully compensated. Knowledge of these aspects is vital because in a pure material, carrier concentration and mobility decrease at low temperatures, the former due to carrier freeze-out and the latter by ionized impurity scattering, even though the amount of impurities present is low. In a heavily doped degenerate material, carrier concentration and mobility are temperature-independent. Uncompensated and compensated semiconductors differ regarding the effect of temperature on mobility. In an uncompensated semiconductor, mobility decreases as the temperature is decreased. But in a compensated semiconductor, there is scarcely any influence of temperature on mobility.

Review exercises

3.1. Why does the energy bandgap of a semiconductor material decrease when the temperature falls, and conversely? Write the equations of Varshini for variation of the bandgap with temperature. Explain the meanings of the symbols used in these equations. What are the values of the fitting coefficients α and β in the model for 4H-SiC?

3.2. Write down the equation for the intrinsic carrier concentration of a semiconductor and indicate the physical parameters in this equation which depend on temperature. What are the effective masses of electrons and holes in gallium arsenide to be used in effective densities of states in conduction and valence bands? Using these values, obtain the equation for temperature dependence of intrinsic carrier concentration in gallium arsenide.

3.3. Consider silicon. Calculate the intrinsic carrier concentration in it at 4.2 K and 600 K. Treat the effective masses of charge carriers and bandgap energy as temperature-dependent.

3.4. Write the equation given by Quay *et al* for the variation of saturation velocity of charge carriers with temperature and mention the values of the two parameters in the model for silicon.

3.5. Write the equation for conductivity of a semiconductor. What parameters in the equation for conductivity change with temperature?

3.6. Write the equations for temperature-dependent free electron concentration $n(T)$ and temperature-dependent free hole concentration $p(T)$ in terms of ionized donor concentration $N_D^+(T)$, ionized acceptor concentration $N_A^-(T)$ and intrinsic carrier concentration $n_i(T)$. Express the ionized donor and acceptor concentrations N_D^+, N_A^- in terms of the total donor and acceptor concentrations N_D, N_A.

3.7. Write the equation for the Fermi–Dirac distribution function $f(E)$. What does this function say regarding the occupancy of an electronic state at energy E by an electron at temperature T? Provide the interpretation for this function at $T = 0$ K for $E = E_F$, $E < E_F$ and $E > E_F$ where E_F is the Fermi energy.

3.8. Applying the Fermi–Dirac distribution function to a non-degenerate semiconductor, show that at a temperature close to absolute zero, there exists an insignificant density of free carriers. Hence, explain the occurrence of carrier freeze-out in a semiconductor at such a low temperature. Does the same phenomenon take place in a degenerate semiconductor? If not, what is the difference between a degenerate and a non-degenerate semiconductor with regard to carrier freeze-out? Give the explanation for the difference.

3.9. Why does carrier freeze-out take place in a non-degenerate semiconductor whereas in a degenerate semiconductor the free carrier density is not a function of temperature?

3.10. What is the temperature regime $100\text{ K} \leqslant T \leqslant 500\text{ K}$ called in the context of ionization of impurity atoms in a semiconductor? For this regime of operation, write the relation between the concentration of ionized donor impurity atoms and total concentration of donor impurity atoms.

3.11. What is meant by the intrinsic regime of ionization of impurity atoms in a semiconductor? How does it differ from the extrinsic regime?

3.12. How do lattice and impurity scattering phenomena affect carrier mobility in a semiconductor? How do the effects on mobility produced by these scattering mechanisms depend on temperature? How are the mobilities influenced by the two mechanisms combined together to obtain a resultant mobility of carriers in a semiconductor?

3.13. How does acoustic phonon scattering differ from optical phonon scattering in temperature dependence? Does impurity scattering vary with impurity concentration?

3.14. Why does the density of charged impurities in an uncompensated semiconductor decrease with temperature, whereas in a compensated semiconductor, the same is practically unaffected by temperature? What is the implication of this behavioral difference between uncompensated and compensated materials concerning the mobility variation with temperature?

3.15. How is ionized impurity scattering able to decrease the carrier mobility in a non-degenerate semiconductor at low temperatures? Why are high carrier mobilities observed in degenerate semiconductors even at very low temperatures?

3.16. Write down the Arora–Hauser–Roulston equations for electron/hole mobility in a semiconductor and explain the symbols used. What is the temperature range of their application?

3.17. How is lattice scattering mobility represented in: (a) Klaassen equations and (b) the MINIMOS 4 model?

3.18. How does carrier mobility in a MOSFET channel change as the temperature is lowered? Is this variation of mobility with temperature beneficial or disadvantageous for device operation?

3.19. Why do carrier concentration and mobility decease with temperature in a pure semiconductor, whereas in a degenerate semiconductor they are not affected by temperature?

3.20. Comment on how the variation of carrier concentration and mobility with temperature differ in the following temperature ranges: (a) room temperature to liquid-nitrogen temperature and (b) liquid-nitrogen temperature to liquid-helium temperature.

3.21. Define recombination lifetime. What are the three recombination mechanisms in semiconductors? Which of these mechanisms determines the carrier lifetime in silicon at low carrier concentrations? Which one plays a decisive role in controlling lifetime at high carrier densities? Which mechanism is inconsequential for silicon?

3.22. How is the decrease in the recombination lifetime at low temperatures explained in terms of the giant-trap model? What is the typical order of magnitude of the capture cross-sections in this model? To which type of impurity levels, shallow or deep, does this model apply?

3.23. Why is the low-level lifetime not of interest at low temperatures? Write the power law giving the temperature dependence of the high-level lifetime. What are the limitations of using a simple power exponent of 1.5 in the formula?

References

Ali-Omar M and Reggiani L 1987 Drift and diffusion of charge carriers in silicon and their empirical relation to the electric field *Solid-State Electron.* **30** 693–7

Arora N D, Hauser J R and Roulston D J 1982 Electron and hole mobilities in silicon as a function of concentration and temperature *IEEE Trans. Electron Devices* **29** 292–5

Barber H D 1967 Effective mass and intrinsic concentration in silicon *Solid-State Electron* **10** 1039–51

Bludau W, Onton A and Heinke W 1974 Temperature dependence of the band gap of silicon *J. Appl. Phys.* **45** 1846–8

Caiafa A, Wang X, Hudgins J L, Santi E and Palmer P R 2003 Cryogenic study and modeling of IGBTs *34th Annual IEEE Power Electronics Specialists Conf. (Acapulco, Mexico, June 2003)* vol 4 1897–903 Paper 59_02

Cole D C and Johnson J B 1989 Accounting for incomplete ionization in modeling silicon-based semiconductor devices *Proc. Workshop Low Temperature Semiconductor Electronics (7–8 August)* (Piscataway, NJ: IEEE) pp 73–7

Couderc R, Amara M and Lemiti M 2014 Reassessment of the intrinsic carrier density temperature dependence in crystalline silicon *J. Appl. Phys.* **115** 093705-1–5

Hairapetian A, Gitlin D and Viswanathan C R 1989 Low-temperature mobility measurements on CMOS devices *IEEE Trans. Electron Devices* **36** 1448–55

Hudgins J L, Godbold C V, Portnoy W M and Mueller O M 1994 Temperature effects on GTO characteristics *IEEE–IAS Annual Meeting Rec. (2–6 October, Denver, CO)* pp 1182–6

Ichimura M, Tajiri H, Ito T and Arai E 1998 Temperature dependence of carrier recombination lifetime in Si wafers *J. Electrochem. Soc.* **145** 3265–71

Ioffe Institute 2015a GaAs—gallium arsenide: band structure and carrier concentration *Ioffe Institute* www.ioffe.ru/SVA/NSM/Semicond/GaAs/bandstr.html#Temperature

Ioffe Institute 2015b Physical properties of silicon (Si) *Ioffe Institute* www.ioffe.rssi.ru/SVA/NSM/Semicond/Si/

Ioffe Institute 2015c C—diamond: band structure and carrier concentration *Ioffe Institute* www.ioffe.ru/SVA/NSM/Semicond/Diamond/bandstr.html#Temperature

Ioffe Institute 2015d SiC—silicon carbide: band structure and carrier concentration *Ioffe Institute* www.ioffe.ru/SVA/NSM/Semicond/SiC/bandstr.html#DependenceTemperature

Jacoboni C, Canali C, Ottaviani G and Alberigi Quaranta A 1977 A review of some charge transport properties of silicon *Solid-State Electron.* **20** 77–89

Jonscher A K 1964 Semiconductors at cryogenic temperatures *Proc. IEEE* **52** 1092–104

Klaassen D B M 1992a A unified mobility model for device simulation—I. Model equations and concentration dependence *Solid-State Electron.* **35** 953–9

Klaassen D B M 1992b A unified mobility model for device simulation—II. Temperature dependence of carrier mobility and lifetime *Solid-State Electron.* **35** 961–7

Lax M 1959 Giant traps *J. Phys. Chem. Solids* **8** 66–73

Madelung O, Rössler U and Schulz M 2002 Gallium arsenide (GaAs), intrinsic carrier concentration, electrical and thermal conductivity *Group IV Elements, IV–IV and III–V*

Compounds. Part B—Electronic, Transport, Optical and Other Properties (Landolt–Börnstein—Group III Condensed Matter vol 41A1b) (Berlin: Springer)

Murphy J D, Grant N E, Pain S L, Niewelt T, Wratten A, Khorani E and Markevich V P *et al* 2022 Carrier lifetimes in high-lifetime silicon wafers and solar cells measured by photoexcited muon spin spectroscopy *J. Appl. Phys.* **132** 065704-1–15

Palmer P R, Santi E, Hudgins J L, Kang X, Joyce J C and Eng P Y 2003 Circuit simulator models for the diode and IGBT with full temperature dependent features *IEEE Trans. Power Electron.* **18** 1220–9

Park I S, Haller E E, Grossman E N and Watson D M 1988 Germanium: gallium photoconductors for far infrared heterodyne detection *Appl. Opt.* **27** 4143–50

Park J-M 2004 Novel power devices for smart power applications *Dissertation* (Sydney: Macquarie University) Section 3.1.4

Pieper R J and Michael S 2005 An exact analysis for freeze-out and exhaustion in single impurity semiconductors *Proc. 2005 American Society of Engineering Education Annual Conf. and Exposition* (Washington, DC: American Society for Engineering Education)

Quay R, Moglestue C, Palankovski V and Selberherr S 2000 A temperature dependent model for the saturation velocity in semiconductor materials *Mater. Sci. Semicond. Process.* **3** 149–55

Selberherr S, Hänsch W, Seavey M and Slotboom J 1990 The evolution of the MINIMOS mobility model *Solid-State Electron.* **33** 1425–36

Silvaco 2000 Simulation standard: simulating impurity freeze-out during low temperature operation *Silvaco* www.silvaco.com/tech_lib_TCAD/simulationstandard/2000/nov/a1/a1.html

Singh J 1993 *Physics of Semiconductors and Their Heterostructures* (New York: McGraw-Hill)

Stefanakis D and Zekentes K 2014 TCAD models of the temperature and doping dependence of the bandgap and low field carrier mobility in 4H-SiC *Microelectron. Eng.* **116** 65–71

Sze S M, Li Y and Ng K K 2021 *Physics of Semiconductor Devices* 4th edn (New York: Wiley)

Van Zeghbroeck B J 1997 Effective mass in semiconductors http://ecee.colorado.edu/~bart/book/effmass.htm

Van Zeghbroeck B J 2011 *Principles of Semiconductor Devices* (Boulder, CO: University of Colorado) https://ecee.colorado.edu/~bart/book/

Varshini P 1967 Temperature dependence of the energy gap in semiconductors *Physica* **34** 149–54

Wilkinson V and Adams A 1993 The effect of temperature and pressure on InGaAs band structure *Properties of Lattice-Matched and Strained Indium Gallium Arsenide (EMIS Data Reviews Series vol 8)* ed P K Bhattacharya (Berlin: Wiley) pp 70–5

Chapter 4

Temperature dependence of the electrical characteristics of silicon bipolar devices and circuits

Silicon is the mainstay of discrete electronic devices and ICs, supported by gallium arsenide and wide bandgap materials such as silicon carbide, gallium nitride and others. This chapter recapitulates the essentials of silicon technology, and then makes an enquiry into the variation of vital electrical parameters of common devices of the bipolar family, such as p–n junctions, Schottky diodes and bipolar junction transistors (BJTs). Parameters of interest include the forward voltage drop, reverse breakdown voltage and leakage current of diodes and transistors. Special attention is paid to the current gain of a bipolar transistor. The switching parameters of devices are also delved into. Investigational studies carried out by researchers on bipolar analog and digital circuits are described.

4.1 Properties of silicon

Silicon is a mainstream electronics material (Zhang 2014, Hu 2021, Pandey *et al* 2022). In compliance with the requirements of many military systems, the uppermost borderline signifying thermal cut-off for routinely used silicon devices is 125 °C. However, this does not necessarily imply that silicon devices cease to function at this temperature. Instead, they become more susceptible to failure mechanisms when this temperature is reached. Failure of a device is a complex process involving edge terminations, biasing voltages and leakage current levels, in addition to temperature

doi:10.1088/978-0-7503-5072-3ch4

Table 4.1. Properties of silicon (Si).

Property	Value
Atomic number	14
Atomic mass	28
Classification	Metalloid
Crystal structure	Two interpenetrating face-centered cubic lattices
Color	Gray
Density at 300 K (g cm^{-3})	2.329
Number of atoms cm^{-3}	5×10^{22}
Lattice constant (Å)	5.43
Melting point (°C)	1410
Dielectric constant	11.7
Thermal conductivity (W cm K^{-1})	1.5
Energy bandgap E_g (eV) at 300 K	1.12
Electrical breakdown field (V cm^{-1})	3×10^5
Intrinsic carrier concentration (cm^{-3})	1×10^{10}
Intrinsic resistivity (Ω cm)	3.2×10^5
Electron mobility (cm^2 V^{-1} s^{-1})	1400
Hole mobility (cm^2 V^{-1} s^{-1})	450
Electron diffusion coefficient (cm^2 s^{-1})	36
Hole diffusion coefficient (cm^2 s^{-1})	12
Electron saturated velocity (cm s^{-1})	2.3×10^7
Hole saturated velocity (cm s^{-1})	1.65×10^7
Minority-carrier lifetime (s)	10^{-6}

effects. The thermal boundary of bipolar and CMOS circuits is easily extendable to 150 °C, and further up to 200 °C–250 °C by reducing supply voltages. The adoption of dielectric isolation (section 4.3.13) reduces leakage currents significantly, extending the operating range up to 300 °C. The properties of silicon relevant to semiconductor device operation are listed in table 4.1. The crystal structure of silicon is depicted in figure 4.1.

4.2 Intrinsic temperature of silicon

To calculate the intrinsic temperature of silicon, the approximate equation for intrinsic carrier concentration (3.7) is written down, assuming temperature-independent effective masses for the carriers and temperature-independent semiconductor bandgap:

$$n_i = 4.83 \times 10^{15} \times \left(\frac{m_n^* m_p^*}{m_0^2}\right)^{0.75} T^{1.5} \exp\left(-\frac{E_g}{2k_B T}\right) \text{cm}^{-3}$$
$$= K m^* T^{1.5} \exp\left(-\frac{E_g}{2k_B T}\right), \tag{4.1}$$

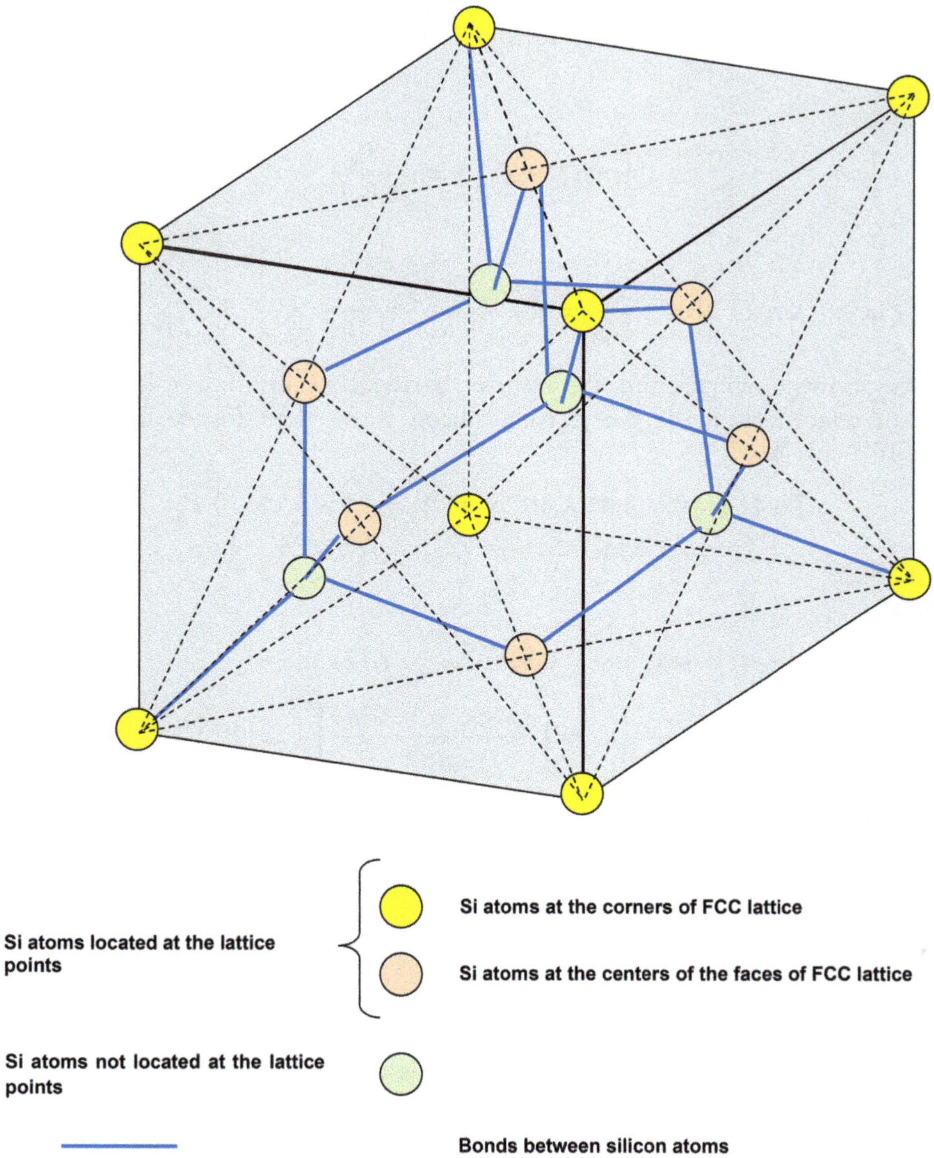

Figure 4.1. Structure of silicon crystal.

where K is a constant = 4.83×10^{15}; m^* is expressed in terms of electron effective mass m_n^*, hole effective mass m_p^* and electron rest mass m_0 as

$$m^* = \frac{m_n^* m_p^*}{m_0^2} \tag{4.2}$$

where T is temperature, E_g is the bandgap and k_B is the Boltzmann constant. Taking the natural logarithm of both sides of equation (4.1), we obtain

$$\ln(n_i) = \ln(Km^*T^{1.5}) - \frac{E_g}{2k_BT} \tag{4.3}$$

or

$$\ln\{n_i/(Km^*T^{1.5})\} = -\frac{E_g}{2k_BT} \tag{4.4}$$

or

$$T\ln\{n_i/(Km^*T^{1.5})\} = -\frac{E_g}{2k_B} = -\frac{E_g}{2 \times 8.617 \times 10^{-5}} = -5.8025 \times 10^3 E_g. \tag{4.5}$$

At the intrinsic temperature, the intrinsic carrier concentration n_i is taken as the dopant concentration of the starting silicon wafer used for device fabrication (1×10^{15} cm^{-3})

$$\begin{aligned} T\ln\{1 \times 10^{15}/(4.83 \times 10^{15} m^* T^{1.5})\} &= -5.8025 \times 10^3 E_g \\ \therefore T\ln\{0.207/(m^* T^{1.5})\} &= -5.8025 \times 10^3 E_g. \end{aligned} \tag{4.6}$$

For Si,

$$\begin{aligned} &m_n^* = 1.08 m_0,\ m_p^* = 0.81 m_0,\ E_g = 1.12\ \text{eV} \\ &\therefore m^* = \left(\frac{m_n^* m_p^*}{m_0^2}\right)^{0.75} = \left(\frac{1.08 m_0 \times 0.81 m_0}{m_0^2}\right)^{0.75} = 0.9046. \end{aligned} \tag{4.7}$$

Hence,

$$T\ln\{0.207/(0.9046 T^{1.5})\} = -5.8025 \times 10^3 \times 1.12 \tag{4.8}$$

or

$$T\ln(0.229/T^{1.5}) = -6498.8$$

or

$$\begin{aligned} T\ln 0.229 - T\ln T^{1.5} &= -6498.8 \\ \therefore -1.474T - T\ln T^{1.5} &= -6498.8. \end{aligned} \tag{4.9}$$

If we take,

$$\begin{aligned} T &= 588\ \text{K} \\ \text{left-hand side(lhs)} &= -1.474 \times 588 - 588\ln 588^{1.5} \\ &= -866.71 - 5624.27 = -6490.98. \end{aligned} \tag{4.10}$$

If

$$\begin{aligned} &T = 589\ \text{K} \\ &\text{lhs} = -1.474 \times 589 - 589\ln 589^{1.5} = -868.19 - 5635.34 \\ &\qquad = -6503.53. \end{aligned} \tag{4.11}$$

When

$$\begin{aligned} T &= 588.6\text{ K} \\ \text{lhs} &= -\ 1.474 \times 588.6 - 588.6 \ln 588.6^{1.5} \\ &= -\ 867.60 - 5630.91 = -6498.51. \end{aligned} \tag{4.12}$$

$$\begin{aligned} \therefore T \approx 588.63\text{ K for which lhs} &= -\ 1.474 \times 588.63 - 588.635 \\ &\quad \ln 588.63^{1.5} \\ &= -\ 867.64 - 5631.245 \\ &= -6498.885. \end{aligned} \tag{4.13}$$

This value represents an approximation to the intrinsic temperature because of the main assumptions underlying the calculations.

4.3 Recapitulating single-crystal silicon wafer technology

4.3.1 Electronic grade polysilicon production

Metallurgical grade (MG) polysilicon containing ~2% impurities is obtained by reducing silica. Silica reduction is performed by heating it with carbon in an electric arc furnace at 1900 °C (Fisher *et al* 2012). MG polysilicon is purified by conversion to trichlorosilane (TCS) with a boiling point 31.8 °C. Conversion to TCS is carried out by reaction with HCl in a fluidized bed reactor at 300 °C:

$$\text{Si (MG)} + 3\text{HCl} \rightarrow \text{SiHCl}_3\text{ (liquid)} + \text{H}_2. \tag{4.14}$$

The TCS is distilled yielding a highly pure liquid. Electronic grade (EG) polysilicon is produced by reaction between purified TCS and hydrogen in a chemical vapor depostion (CVD) reactor at 1000 °C–1200 °C:

$$\text{SiHCl}_3\text{ (liquid)} + \text{H}_2 \rightarrow \text{Si (EG)} + 3\text{HCl}. \tag{4.15}$$

An intermediate compound such as SiH_4 can also be decomposed to produce EG polysilicon:

$$\text{SiH}_4 + \text{H}_2 \rightarrow \text{Si (EG)} + 3\text{H}_2. \tag{4.16}$$

4.3.2 Single-crystal growth

4.3.2.1 The Czochralski method

In the Czochralski (CZ) method, polysilicon (EG) obtained as above is melted in a quartz crucible (figure 4.2). A seed crystal is immersed into molten polysilicon (EG). It is slowly pulled upwards. Simultaneously with the pulling, it is also rotated. Through crystal pulling/rotating from the melt, a cylindrical ingot of monocrystalline silicon is formed. Vital parameters that must be stringently monitored during this operation are: the temperature gradient, the pulling rate and speed of rotation of the crystal. The process is carried out in an argon atmosphere maintained inside a quartz chamber (Friedrich *et al* 2015).

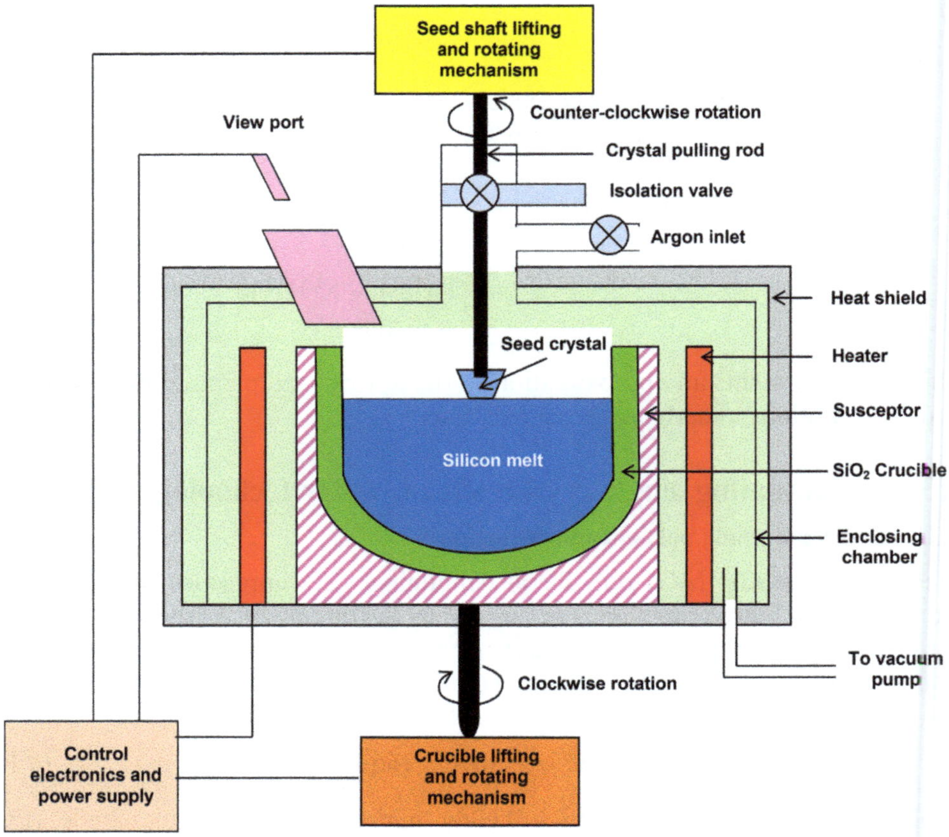

Figure 4.2. Silicon crystal growth by the CZ process.

4.3.2.2 Float-zone method

A molten region is slowly moved along a silicon bar (figure 4.3). Impurities segregate towards the liquefied side leaving the solidified portion in a purified state (Muiznieks *et al* 2015).

4.3.3 Photolithography

Several kinds of photolithographic exposure and mask aligning systems are available (Madou 2002, Zant 2014). Optical lithography has borne the major burden of silicon device and IC manufacturing. This technique is widely used for defining micron and sub-micron geometries up to the wavelength of light that imposes restrictions on minimum feature size. Two types of photosensitive materials called positive and negative photoresists are used. The positive photoresist becomes soft on exposure while the negative photoresist becomes hard in exposed regions. One type of optical mask aligner, namely a projection system, is shown in figure 4.4(a). In the nanometer range, electron-beam (e-beam) lithography (figure 4.4(b)) is employed.

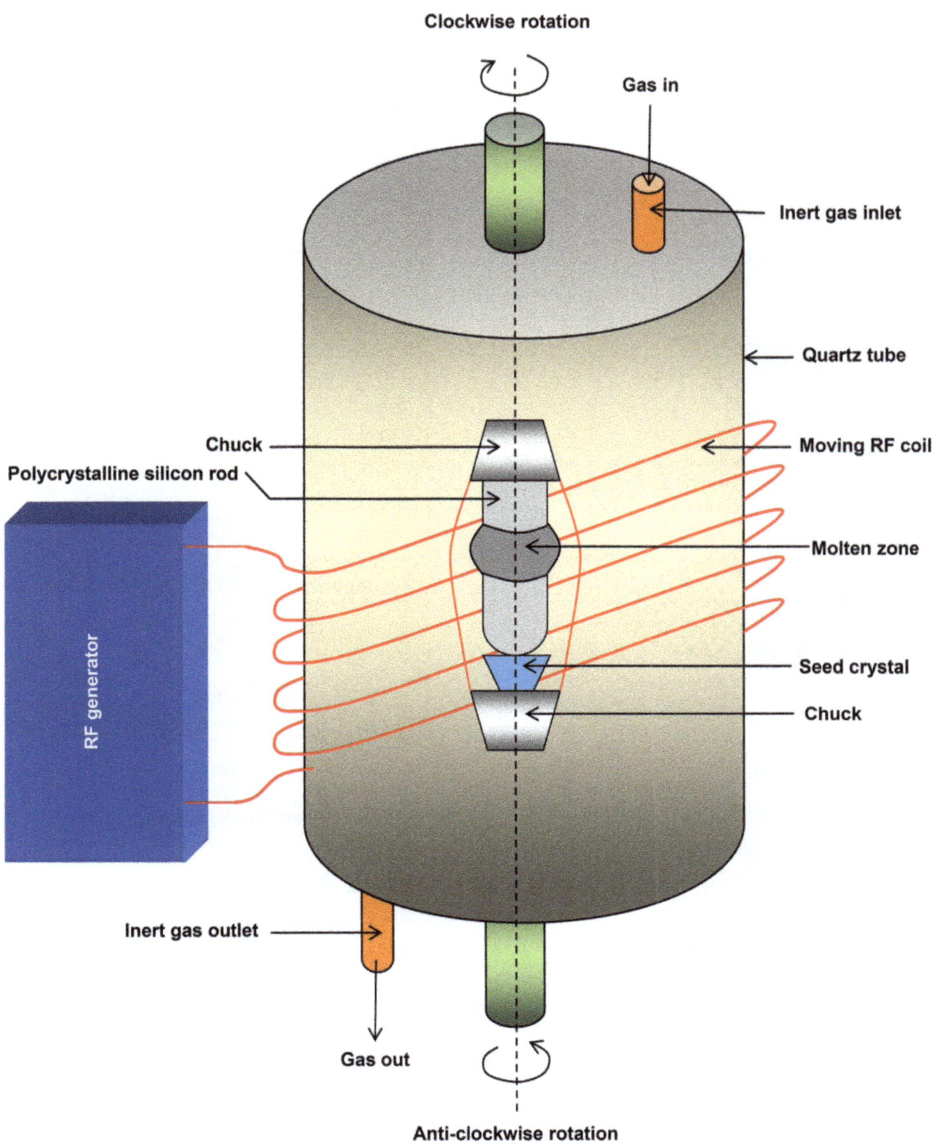

Figure 4.3. Silicon crystal growth by the float zone method.

4.3.4 Thermal oxidation of silicon

By exposing silicon wafers to oxygen, either pure oxygen gas or by bubbling oxygen through heated water in a flask and transporting the water vapors to a furnace at a high temperature of ~900 °C–1200 °C, good quality silicon dioxide films are formed on the wafer surface (Griffin *et al* 2014). When oxygen is used alone, the process is called dry oxidation. If oxygen is passed through hot water, the process is known as wet oxidation. These oxide films serve as insulators. They are used as diffusion

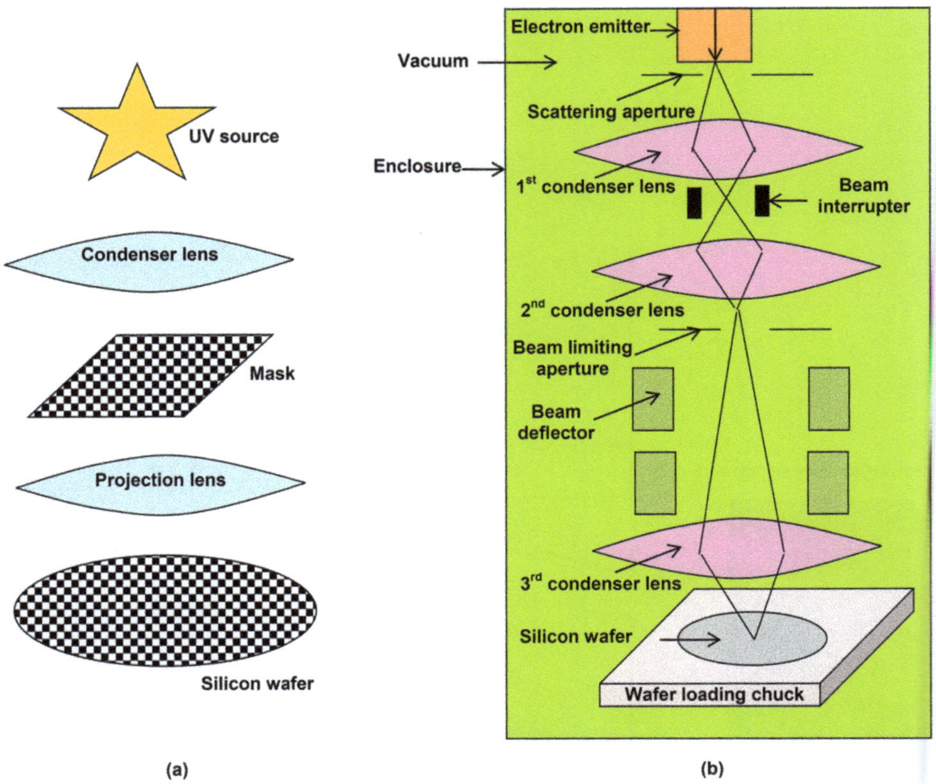

Figure 4.4. Photolithography: (a) optical projection lithography and (b) e-beam lithography.

masks to produce windows through which selective diffusion can be performed in defined areas:

$$Si + O_2 = SiO_2, \tag{4.17}$$

$$Si + 2H_2O = SiO_2 + 2H_2\ (g). \tag{4.18}$$

4.3.5 n-Type doping of silicon by thermal diffusion

In figure 4.5, the energy band diagram of silicon is drawn showing the energy levels of the impurities commonly used as dopants for producing n-type and p-type materials (Jones 2008).

Phosphorous oxychloride (figure 4.6) is a liquid dopant source used to form phosphosilicate glass (PSG) on the silicon wafer at 900 °C–1000 °C:

$$4POCl_3 + 3O_2 \rightarrow 2P_2O_5 + 6Cl_2; 2P_2O_5 + 5Si \rightarrow 4P + 5SiO_2. \tag{4.19}$$

A gaseous source (phosphine or PH_3) may also be used:

$$2PH_3 + 4O_2 \rightarrow P_2O_5 + 3H_2O. \tag{4.20}$$

The glass layer acts as a source of impurity for doping the semiconductor.

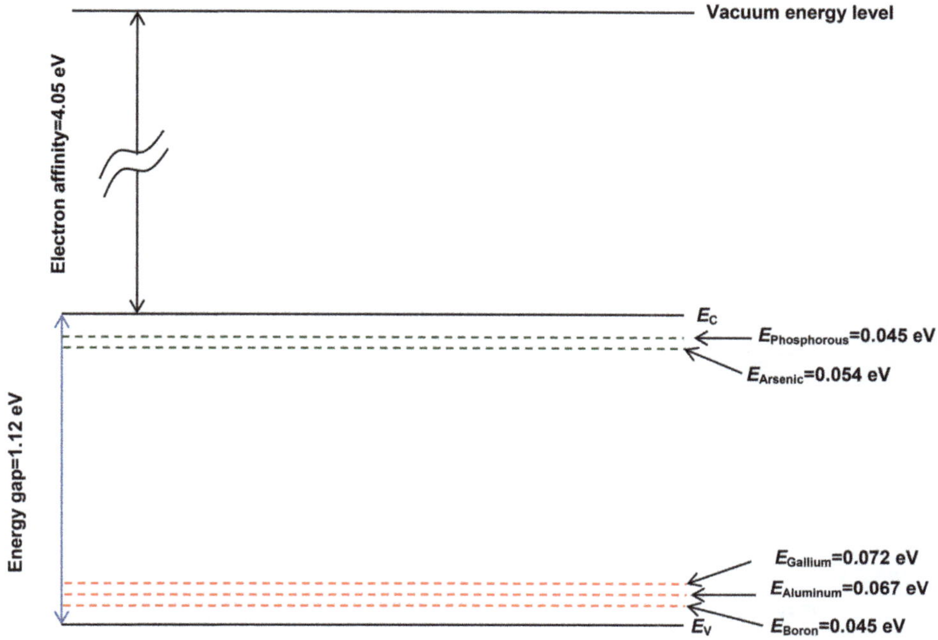

Figure 4.5. Energy band diagram of silicon showing the energy levels of commonly used impurities for n- (P and As) and p-doping (B, Al, Ga).

4.3.6 p-Type doping of silicon by thermal diffusion

The boron source is a ceramic wafer of boron nitride. This wafer is oxidized at 1000 °C in an oxygen atmosphere to form B_2O_3 glass. During boron deposition, B_2O_3 evaporates from the boron nitride wafer and condenses on the silicon wafer as a glassy layer (borosilicate glass (BSG)) serving as a source of elemental boron:

$$2B_2O_3 + 3Si \rightarrow 4B + 3SiO_2. \tag{4.21}$$

A gaseous source such as diborane (B_2H_6) is sometimes used to form B_2O_3 glass:

$$B_2H_6 + 3O_2 \rightarrow B_2O_3 + 3H_2O. \tag{4.22}$$

4.3.7 Impurity doping by ion implantation

As an alternative to diffusion, pre-deposition is performed by ion implantation (figure 4.7). It is a room temperature technique in which a beam of ions obtained from a gaseous source, e.g. boron trifluoride (BF_3) for boron, phosphine (PH_3) for phosphorous, etc, is separated by the charge-to-mass ratio of the ions, focused, accelerated by an electric field and scanned across the silicon wafer (Nastasi and Mayer 2006). The energy of the ion beam determines the depth of penetration of ions into the wafer, and the dose measured in ions cm^{-2} decides their concentration.

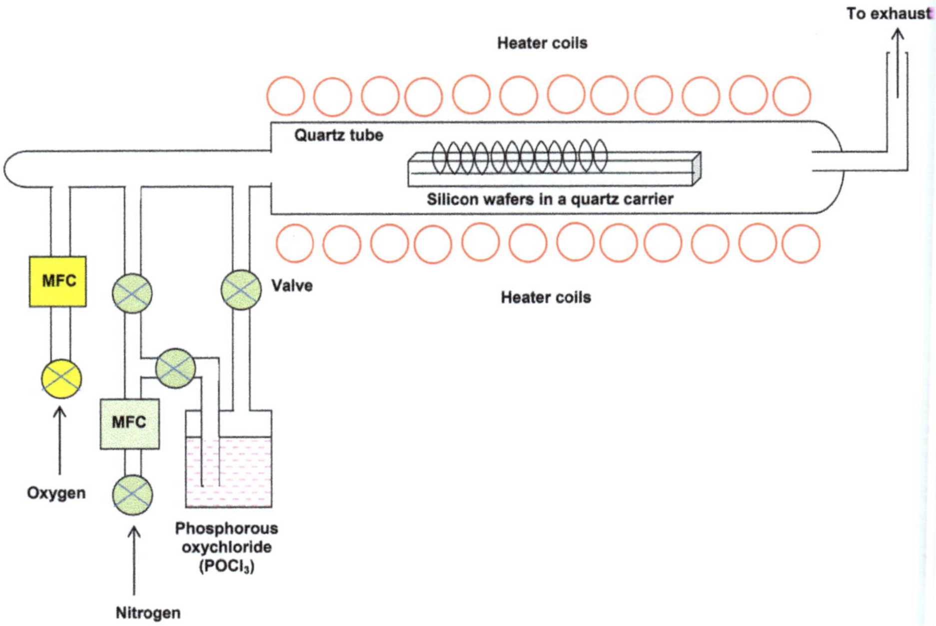

Figure 4.6. Phosphorous diffusion using a liquid $POCl_3$ source.

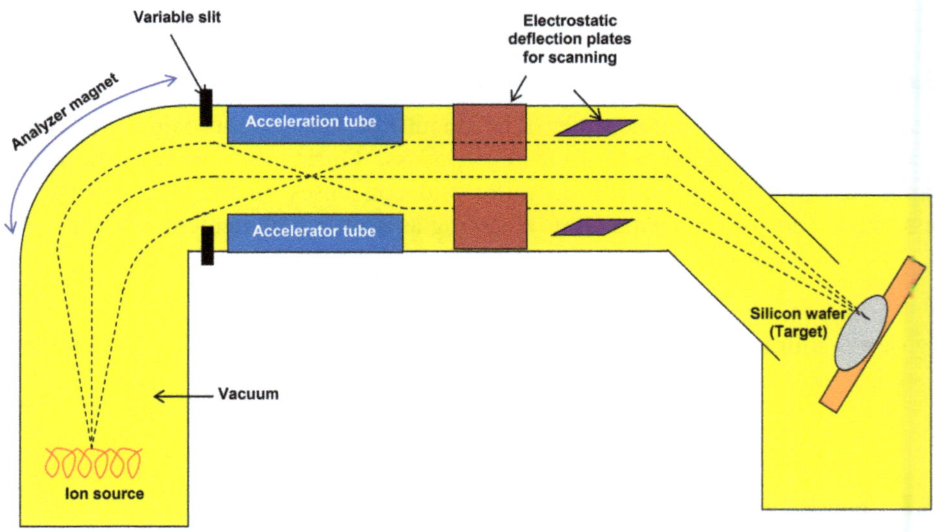

Figure 4.7. Ion implanter.

A thermal annealing step at 900 °C for 30 min almost always follows the ion implantation. This step is essential for two reasons. First, to anneal the damage caused by ion bombardment and second, to transfer the impurity ions from interstitial to substitutional sites. In order to diffuse the implanted ions deeper into the semiconductor crystal, a high-temperature drive-in step at 1100 °C–1200 °C may be required. Typical parameters for ion implantation are: accelerator energy = 1–500 keV, dose = 10^{11}–10^{16} cm^{-2}. Ion implanters are classified according to the magnitudes of beam currents used by them. Medium current implanters range in ion beam currents from ~10 μA to 2 mA; high current types can be up to 30 mA.

4.3.8 Low-pressure chemical vapor deposition

For low-pressure CVD (LPCVD) the starting pressure in the quartz tube (figure 4.8) is 0.1 Pa and the working gas (dilution gas + reactive species) is introduced at 10–1000 Pa (Vergara-Irigaray *et al* 2012). It is used for depositing polysilicon, silicon nitride and doped oxides such as BSG and PSG. Temperatures range from 600 °C to 660 °C for polysilicon to 850 °C–950 °C for BSG and 950 °C–1100 °C for PSG.

Polysilicon is obtained by pyrolysis of silane (SiH_4):

$$SiH_4 = Si + 2H_2, \tag{4.23}$$

or from trichlorosilane ($SiHCl_3$):

$$SiHCl_3 + H_2 = Si + 3HCl. \tag{4.24}$$

Silicon dioxide is deposited using silane (SiH_4), dichlorosilane (SiH_2Cl_2) or tetraethylorthosilicate, (TEOS, $Si(OC_2H_5)_4$):

$$SiH_4 + O_2 = SiO_2 + 2H_2 \tag{4.25}$$

$$SiH_2Cl_2 + 2N_2O = SiO_2 + 2N_2 + 2HCl \tag{4.26}$$

$$Si(OC_2H_5)_4 + 12O_2 = SiO_2 + 10H_2O + 8CO_2. \tag{4.27}$$

The reactions for silicon nitride (Si_3N_4) deposition are:

$$3SiH_4 + 4NH_3 = Si_3N_4 + 12H_2 \tag{4.28}$$

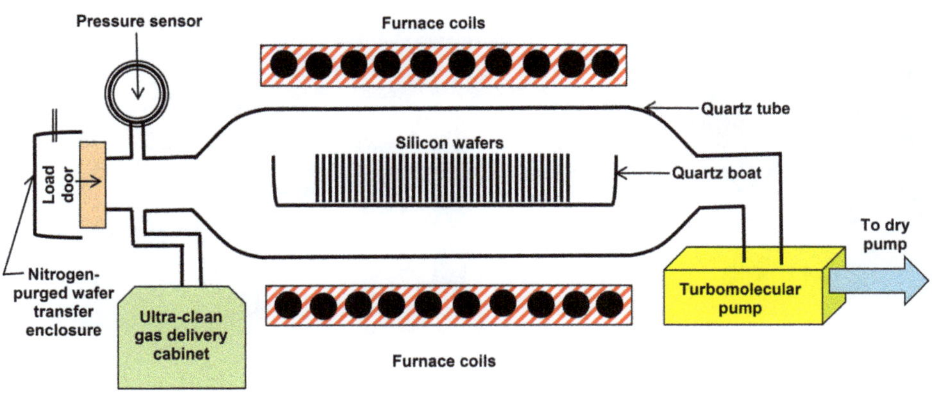

Figure 4.8. LPCVD system.

$$3SiH_2Cl_2 + 4NH_3 = Si_3N_4 + 6HCl + 6H_2. \quad (4.29)$$

Tungsten can be deposited using tungsten hexafluoride (WF_6) and silane:

$$WF_6 + 2SiH_4 = W + 2SiHF_3 + 3H_2. \quad (4.30)$$

4.3.9 Plasma-enhanced chemical vapor deposition

In plasma-enhanced CVD (PECVD), employing a parallel plate electrode configuration, plasma is excited by a RF signal (figure 4.9). The energy of electrons in cold thermal plasma helps PECVD to operate at relatively lower temperatures (100 °C–400 °C) than LPCVD. It is widely used for deposition of silicon dioxide (SiO_2), silicon nitride (Si_3N_4) and silicon oxynitride ($Si_xO_yN_z$) films (Tedrow and Reif 1994).

4.3.10 Atomic layer deposition

Atomic layer deposition (ALD) may be referred to as a subclass of CVD or an advanced thin-film coating technique. In this method, ultrathin, conformal, dense, defect-free films (mainly oxides such as Al_2O_3, HfO_2, Ta_2O_5, V_2O_5, ZrO_2, etc)

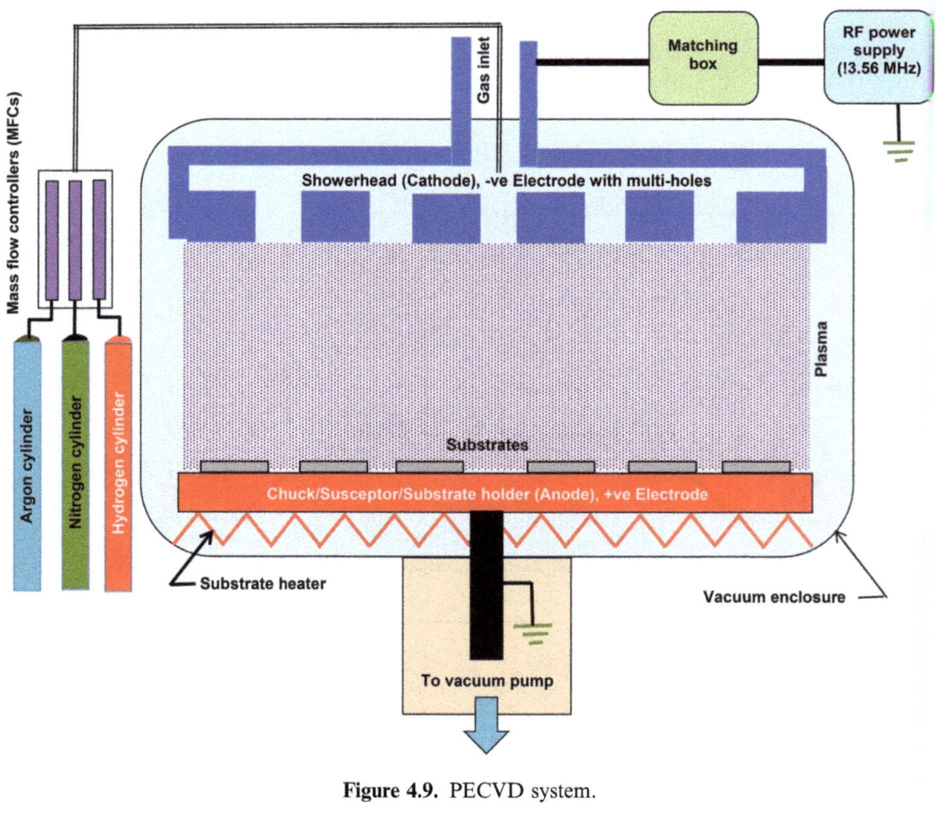

Figure 4.9. PECVD system.

having thicknesses in nanometers, are deposited in a precisely controlled manner (Ritala and Niinistö 2009). Consecutive atomic layers are formed by the reaction of, generally, two chemical species called precursors with the substrate in a sequential manner. The reaction is broken into two half-reactions and the precursors are always kept separate. The process is also self-limiting.

First the substrate surface is exposed to one precursor. Then any unreacted excess precursor is flushed away by introducing a cleaning gas in the reaction chamber. Precursor removal is followed by exposure of the substrate to the second precursor. Then purging of the reaction chamber is performed again to remove this second precursor. Several such cycles are repeated until the desired film thickness is achieved.

4.3.11 Ohmic (non-rectifying) contacts to Si

Metallization techniques are thermal or e-beam evaporation (figure 4.10) and sputtering (figure 4.11). Examples of materials used include: aluminum on p-type

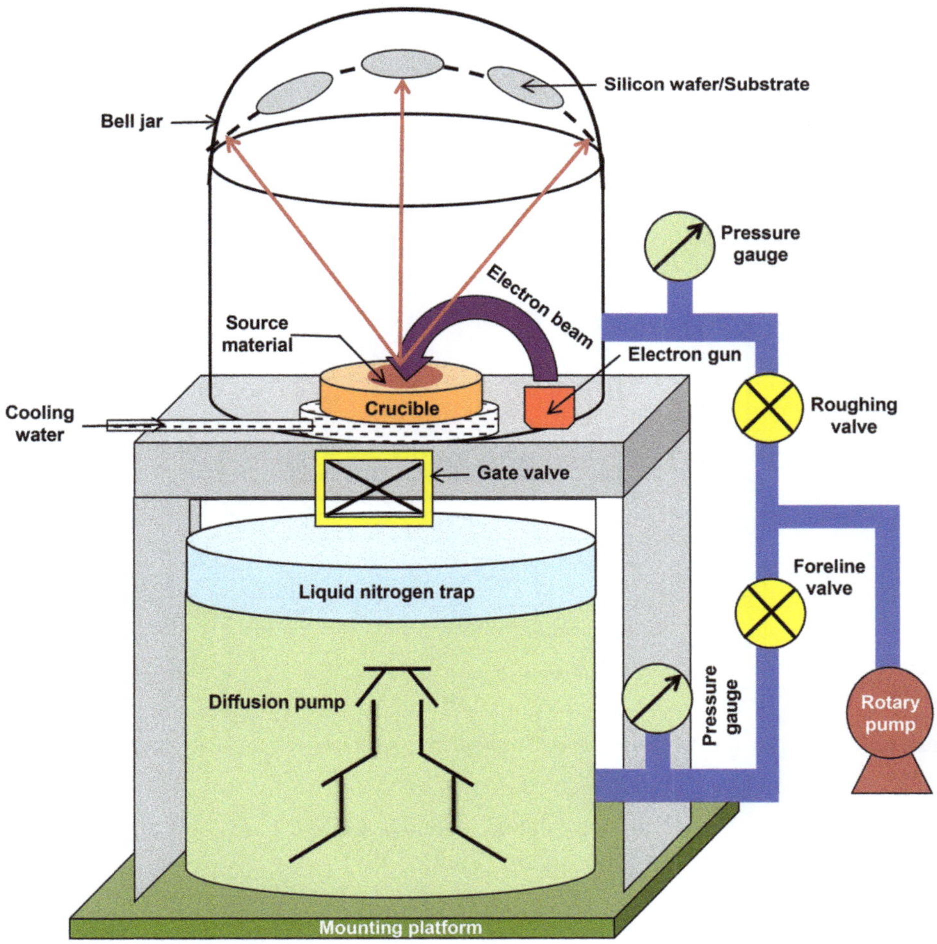

Figure 4.10. E-beam vacuum metallization system.

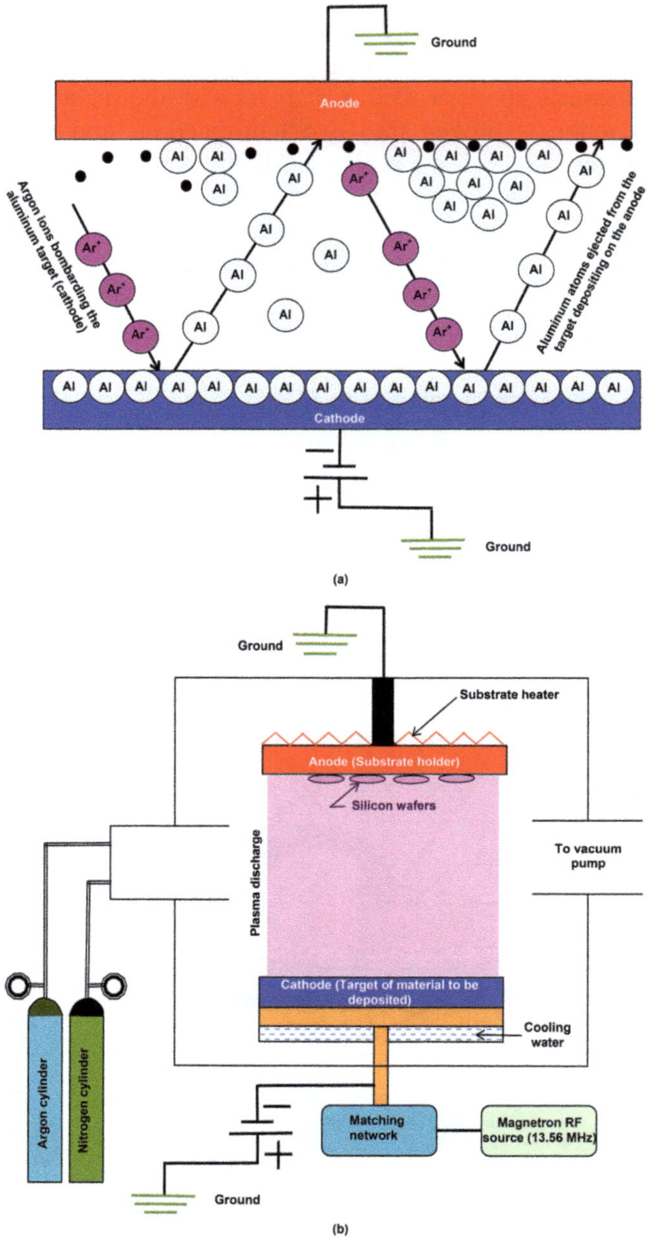

Figure 4.11. Sputtering. (a) Mechanism of metal deposition by sputtering and (b) an RF magnetron sputtering system.

silicon or heavily doped n-type silicon; silicides ($TiSi_2$, $CoSi_2$, WSi_2, $TaSi_2$, and $MoSi_2$), W, TiN, etc, on n-type silicon (Sinha 1981).

4.3.12 Schottky contacts to Si

An example is aluminum on moderately or low-doped n-type silicon.

4.3.13 p–n Junction and dielectric isolation in silicon integrated circuits

The wide variety of components fabricated on an IC chip is separated by two main types of silicon technology, namely junction and dielectric isolation. In junction isolation (figure 4.12), n-type islands are formed on a p-type starting substrate. In these islands, the different devices are fabricated. During operation of the circuit, the p-type substrate is maintained at the most negative potential with respect to the remaining chip. In this manner, the reverse-biased p–n diode serves as an electrically insulating material. The significant advantage gained with this method is its compatibility with IC processing. Disadvantages include the long diffusion times required to create isolation, the consumption of a significant proportion of chip area by lateral diffusion and the introduction of stray capacitances. At high temperatures, the junction isolation becomes excessively leaky, impairing the performance of the chip. This is further worsened by the electron–hole pair (EHP) generation due to photon irradiation.

Dielectric isolation (figure 4.13) solves both the stray capacitance and leakage current problems. The dielectric constant of silicon dioxide (3.9) is three times smaller than that of silicon (11.7), reducing the capacitive effect. Thermal SiO_2 has a DC resistivity of 10^{14}–10^{16} Ω cm. It need not be kept at a negative potential as required in junction isolation. SOI technology is a semiconductor fabrication technology which does not employ the traditional bulk silicon wafers as the starting material. Instead, the starting material is a composite material comprising a topmost thin layer of single-crystal silicon of several nanometers to micron thicknesses called the active silicon layer, formed on a thin thermally grown SiO_2 insulating layer, ranging in thickness from sub-microns up to 1 μm, known as the buried oxide (BOX) layer, and supported on a bottom silicon layer several hundred microns thick, termed the handle wafer. The transistors and other devices are formed in the active silicon layer. The underlying BOX layer provides the isolation.

4.4 Examining temperature effects on bipolar devices

The requirement of operation over a broad range of temperatures imposed by many applications makes it essential to take into account the variation of the performance parameters of semiconductor devices with temperature throughout device design, in order to guarantee that the device will work satisfactorily during thermal excursions. Interestingly, many device parameters change advantageously with decreasing temperature, while others vary favorably with increasing temperature. Therefore, the influence of temperature change in a given direction cannot always be said to be either beneficial or adverse for the application at hand, and each parameter must be examined separately.

4.4.1 The Shockley equation for the current–voltage characteristics of a p–n junction diode

The simplest semiconductor device is the p–n junction diode comprising the interface or boundary between a p-type and an n-type semiconductor (figure 4.14). It has the

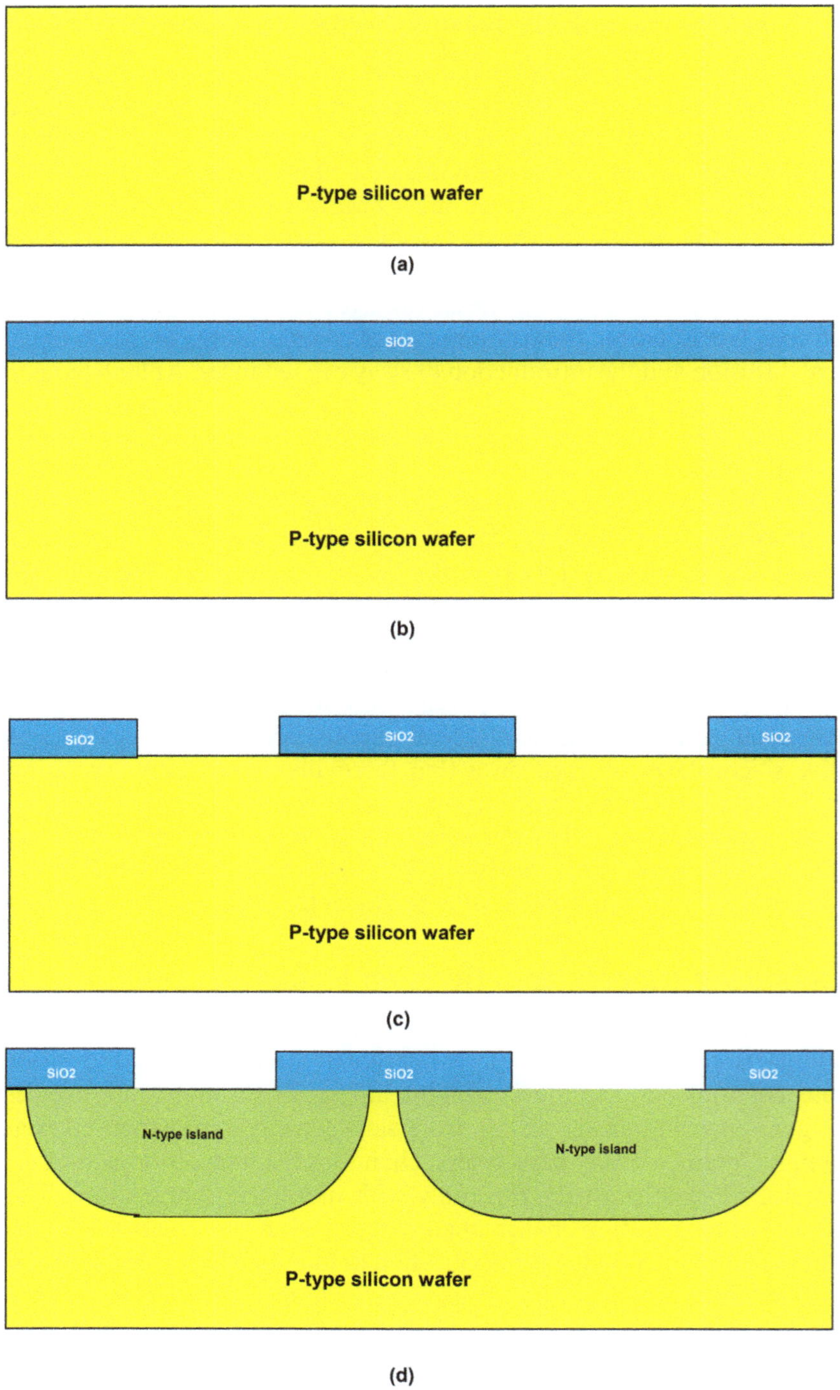

Figure 4.12. p–n junction isolation. (a) Initial p-type silicon wafer, (b) thermal oxidation, (c) photolithography and oxide etching, and (d) n-diffusion.

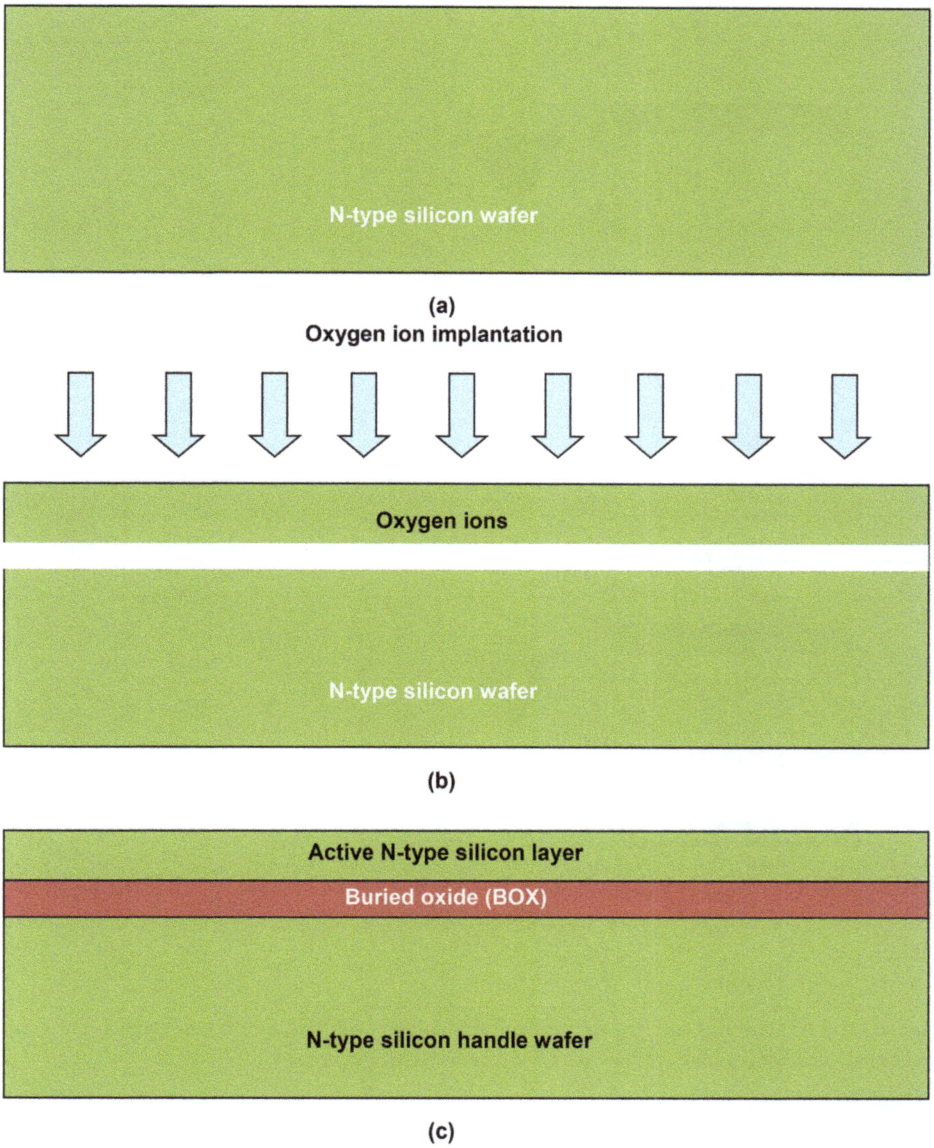

Figure 4.13. Dielectric isolation by SOI wafer production. (a) Initial n-type silicon wafer, (b) implantation of oxygen ions at 180 keV energy and 3×10^{17}–1.8×10^{18} ions cm^{-2}, and (c) thermal annealing at 1320 °C. The process is called separation by implantation of oxygen (SIMOX).

property of allowing current flow in one direction only and blocking it in the opposite direction.

The current I flowing through a parallel plane p–n junction diode as a function of voltage v applied across its terminals is given by the theoretically derived Shockley equation (Leach 2004):

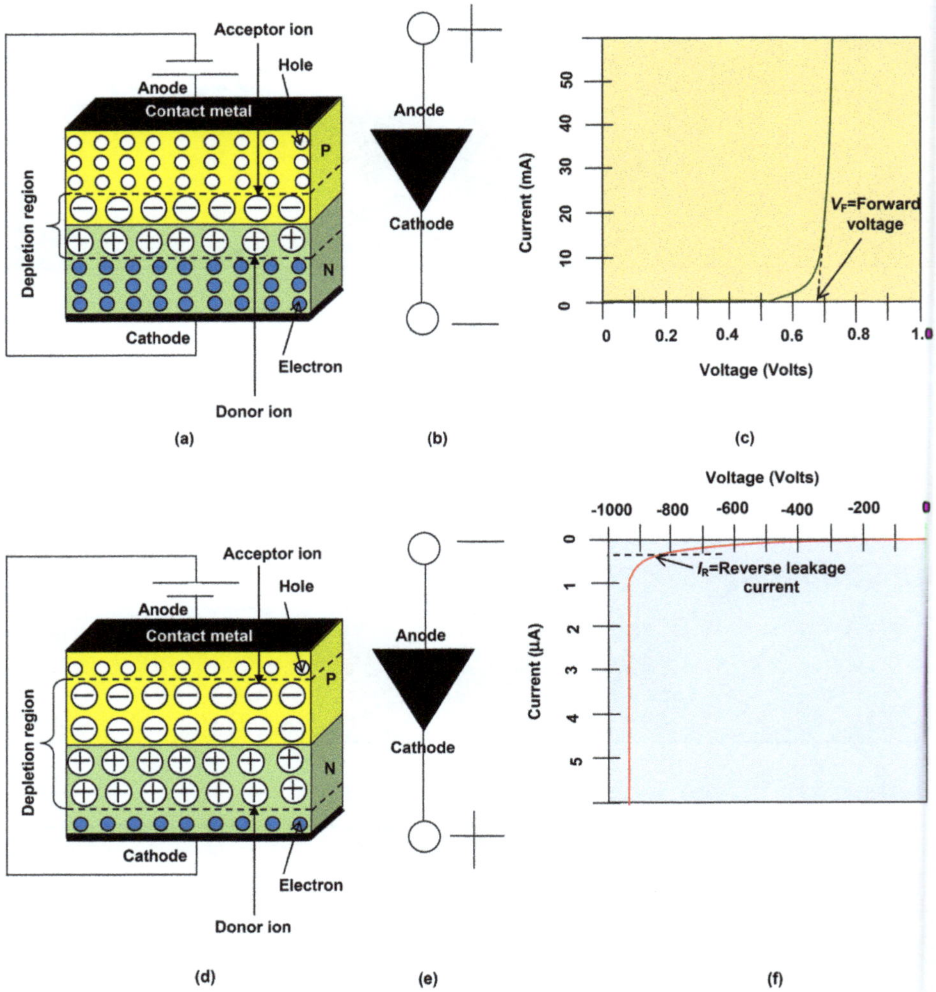

Figure 4.14. p–n junction diode. (a), (b), (c) forward-biased diode structure, circuit symbol and current–voltage characteristics; (d), (e), (f) the same for a reverse-biased diode.

$$i = I_S\left\{\exp\left(\frac{qv}{\eta k_B T}\right) - 1\right\}, \tag{4.31}$$

where I_S is the reverse saturation current in the dark (=1×10^{-12} A), η is the emission coefficient or ideality factor ($1 \leqslant \eta \leqslant 2$), q is the electronic charge (=1.6×10^{-19} C), k_B is the Boltzmann constant (= 8.62×10^{-5} eV K^{-1}) and T is the absolute temperature.

The emission coefficient η accounts for the decrease in current flowing through the diode by recombination of electrons and holes in the depletion region. When this recombination is small, $\eta = 1$. But if appreciable recombination takes place, $\eta = 2$. Therefore, the value of η depends upon two factors: (i) the semiconductor material and (ii) the doping concentrations of the p- and n-regions of the diode. Generally, $\eta = 1$ for indirect bandgap semiconductors, e.g., Si, Ge; $\eta = -2$ for direct bandgap semiconductors, e.g. GaAs, InP. Sometimes for Si diodes which form a sub-component of an IC $\eta = 1$, and for discrete silicon diodes $\eta = 2$.

The sub-expression $k_B T/q$ has the dimensions of voltage. It is called thermal voltage $V_{Thermal}$. At room temperature

$$V_{Thermal} = (k_B T)/q = 25.9 \text{ mV}. \tag{4.32}$$

In terms of the thermal voltage $V_{Thermal}$, the diode equation (4.31) can be written as

$$i = I_S\left\{\exp\left(\frac{v}{\eta V_{Thermal}}\right) - 1\right\}. \tag{4.33}$$

The saturation current varies with temperature according to the relation (Leach 2004)

$$I_S = KT^{3/\eta} \exp\left(-\frac{E_G}{\eta V_{Thermal}}\right), \tag{4.34}$$

where K is a constant which is directly proportional to the cross-sectional area of the junction and E_G is the energy bandgap of the semiconductor in which the diode is fabricated.

4.4.2 Forward voltage drop across a p–n junction diode

Figure 4.15 shows the trend of variation of the forward current–voltage characteristics of a p–n junction diode with temperature changes.

Equation (4.33) is solved for the diode voltage v as follows:

$$\frac{i}{I_S} = \exp\left(\frac{v}{\eta V_{Thermal}}\right) - 1$$

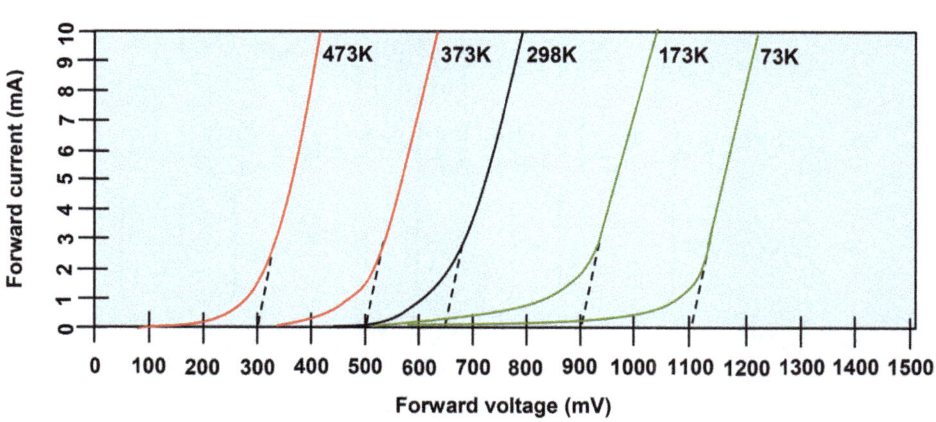

Figure 4.15. Forward current–voltage characteristics of a p–n junction diode as a function of temperature.

or

$$\frac{i}{I_S} + 1 = \exp\left(\frac{v}{\eta V_{\text{Thermal}}}\right)$$

or

$$\ln\left(\frac{i}{I_S} + 1\right) = \frac{v}{\eta V_{\text{Thermal}}} \tag{4.35}$$

$$\therefore v = \eta V_{\text{Thermal}} \ln\left(\frac{i}{I_S} + 1\right) \approx \eta V_{\text{Thermal}} \ln\left(\frac{i}{I_S}\right). \tag{4.36}$$

In this equation, V_{Thermal} and I_S are temperature-dependent. Keeping i constant and applying the chain and quotient rules for derivatives, we have

$$\begin{aligned}
\frac{\mathrm{d}v}{\mathrm{d}T} &= \eta \frac{\mathrm{d}V_{\text{Thermal}}}{\mathrm{d}T} \ln\left(\frac{i}{I_S}\right) + \eta V_{\text{Thermal}} \frac{\mathrm{d}}{\mathrm{d}T}\left\{\ln\left(\frac{i}{I_S}\right)\right\} \\
&= \eta \frac{\mathrm{d}V_{\text{Thermal}}}{\mathrm{d}T} \ln\left(\frac{i}{I_S}\right) + \eta V_{\text{Thermal}} \left\{1/\left(\frac{i}{I_S}\right)\right\}\left(\frac{-i}{I_S^2}\right)\left(\frac{\mathrm{d}I_S}{\mathrm{d}T}\right) \\
&= \eta \frac{\mathrm{d}V_{\text{Thermal}}}{\mathrm{d}T} \ln\left(\frac{i}{I_S}\right) - \eta V_{\text{Thermal}} \left(\frac{1}{I_S}\right)\left(\frac{\mathrm{d}I_S}{\mathrm{d}T}\right).
\end{aligned} \tag{4.37}$$

For determination of $\mathrm{d}v/\mathrm{d}T$, it is necessary to find $\mathrm{d}V_{\text{Thermal}}/\mathrm{d}T$ and $\mathrm{d}I_S/\mathrm{d}T$,

$$\frac{\mathrm{d}V_{\text{Thermal}}}{\mathrm{d}T} = \frac{\mathrm{d}}{\mathrm{d}T}\left(\frac{k_B T}{q}\right) = \left(\frac{k_B}{q}\right) \tag{4.38}$$

and

$$\begin{aligned}
\frac{\mathrm{d}I_S}{\mathrm{d}T} &= \frac{\mathrm{d}}{\mathrm{d}T}\left\{KT^{3/\eta} \exp\left(-\frac{E_G}{\eta V_{\text{Thermal}}}\right)\right\} = \frac{\mathrm{d}}{\mathrm{d}T}\left\{KT^{3/\eta} \exp\left(-\frac{qE_G}{\eta k_B T}\right)\right\} \\
&= \frac{3}{\eta} KT^{(3/\eta)-1} \exp\left(-\frac{qE_G}{\eta k_B T}\right) + KT^{3/\eta}\left(\frac{qE_G}{\eta k_B T^2}\right) \exp\left(-\frac{qE_G}{\eta k_B T}\right) \\
&= \frac{3}{\eta T} KT^{3/\eta} \exp\left(-\frac{qE_G}{\eta k_B T}\right) + \left(\frac{qE_G}{\eta k_B T}\right)\left(\frac{1}{T}\right) KT^{3/\eta} \exp\left(-\frac{qE_G}{\eta k_B T}\right) \\
&= \frac{3}{\eta T} KT^{3/\eta} \exp\left\{-\frac{E_G}{\eta\left(\frac{k_B T}{q}\right)}\right\} + \left\{\frac{E_G}{\eta\left(\frac{k_B T}{q}\right)}\right\}\left(\frac{1}{T}\right) KT^{3/\eta} \exp\left\{-\frac{E_G}{\eta\left(\frac{k_B T}{q}\right)}\right\} \\
&= \frac{3}{\eta T} KT^{3/\eta} \exp\left(-\frac{E_G}{\eta V_{\text{Thermal}}}\right) + \left(\frac{E_G}{\eta V_{\text{Thermal}}}\right)\left(\frac{1}{T}\right) KT^{3/\eta} \exp\left(-\frac{E_G}{\eta V_{\text{Thermal}}}\right) \\
&= \frac{3}{\eta T} I_S + \left(\frac{E_G}{\eta V_{\text{Thermal}}}\right)\frac{1}{T} I_S = I_S\left(\frac{3}{\eta T} + \frac{E_G}{\eta T V_{\text{Thermal}}}\right).
\end{aligned} \tag{4.39}$$

Substituting for $\mathrm{d}V_{\text{Thermal}}/\mathrm{d}T$ from equation (4.38) and $\mathrm{d}I_S/\mathrm{d}T$ from equation (4.39) into equation (4.37), we obtain

$$\frac{\mathrm{d}v}{\mathrm{d}T} = \eta\left(\frac{k_B}{q}\right)\ln\left(\frac{i}{I_S}\right) - \eta V_{\text{Thermal}}\left(\frac{1}{I_S}\right)I_S\left(\frac{3}{\eta T} + \frac{E_G}{\eta T V_{\text{Thermal}}}\right). \quad (4.40)$$

But from equation (4.36)

$$v = \eta V_{\text{Thermal}}\ln\left(\frac{i}{I_S}\right) \quad (4.41)$$

$$\begin{aligned} \therefore \frac{\mathrm{d}v}{\mathrm{d}T} &= \frac{1}{T}\eta\left(\frac{k_B T}{q}\right)\ln\left(\frac{i}{I_S}\right) - \eta V_{\text{Thermal}}\left(\frac{3}{\eta T} + \frac{E_G}{\eta T V_{\text{Thermal}}}\right) \\ &= \frac{v}{T} - \frac{3V_{\text{Thermal}}}{T} - \frac{E_G}{T} = \frac{v - (3V_{\text{Thermal}} + E_G)}{T}. \end{aligned} \quad (4.42)$$

For a silicon diode, $v = 0.7$ V, $E_G = 1.11$ eV. Since $V_{\text{Thermal}} = 0.0259$ V, $T = 300$ K at room temperature, we have

$$\begin{aligned} \frac{\mathrm{d}v}{\mathrm{d}T} &= \frac{0.7 - (3 \times 0.0259 + 1.11)}{300} = -\frac{0.4877}{300} \\ &= -0.0016\,\mathrm{VK}^{-1} = -1.6\,\mathrm{mVK}^{-1} \approx -2\,\mathrm{mVK}^{-1}. \end{aligned} \quad (4.43)$$

Thus, the forward voltage dropped across a silicon p–n junction diode decreases by ~2 mV K^{-1} or 2 mV $°\mathrm{C}^{-1}$.

The forward voltage drop across a diode increases by 2 mV for every 1 °C fall in temperature (Godse and Bakshi 2009). Hence, if the forward voltage of a Si diode is 0.71 V at 25 °C, and the temperature decreases to −55 °C, the forward voltage will increase by an amount = $2 \times 10^{-3} \times (25 + 55) = 0.16$ V to 0.71 + 0.16 V = 0.87 V.

4.4.3 Forward voltage of a Schottky diode

Figure 4.16 shows the schematic diagram of an Schottky barrier diode (SBD) and its circuit diagram symbol. The trend of variation of the forward voltage of a Schottky diode with temperature is depicted in figure 4.17.

From classical thermionic emission theory, the static current (I_d)–voltage (V_d) characteristic of a Schottky diode is described by the equation (Cory 2009)

$$I_d = I_{D0}[\exp\{(qV_d)/(nk_B T)\} - 1] \quad (4.44)$$

where I_{D0} is the reverse saturation current given by

$$I_{D0} = K_{SB}T^2 \exp\{(-q\phi_B)/(k_B T)\} \quad (4.45)$$

where $K_{SB} = A^*A$, A^* is the effective Richardson constant, A is the effective area of the diode and ϕ_B is the height of the Schottky barrier. The constant n is the ideality factor portraying the departure of the practical Schottky diode from theory; $n = 1$ for ideal diodes.

Taking $n = 1$, equation (4.44) is re-arranged as

$$I_d/I_{D0} = \exp\{(qV_d)/(k_B T)\} - 1. \quad (4.46)$$

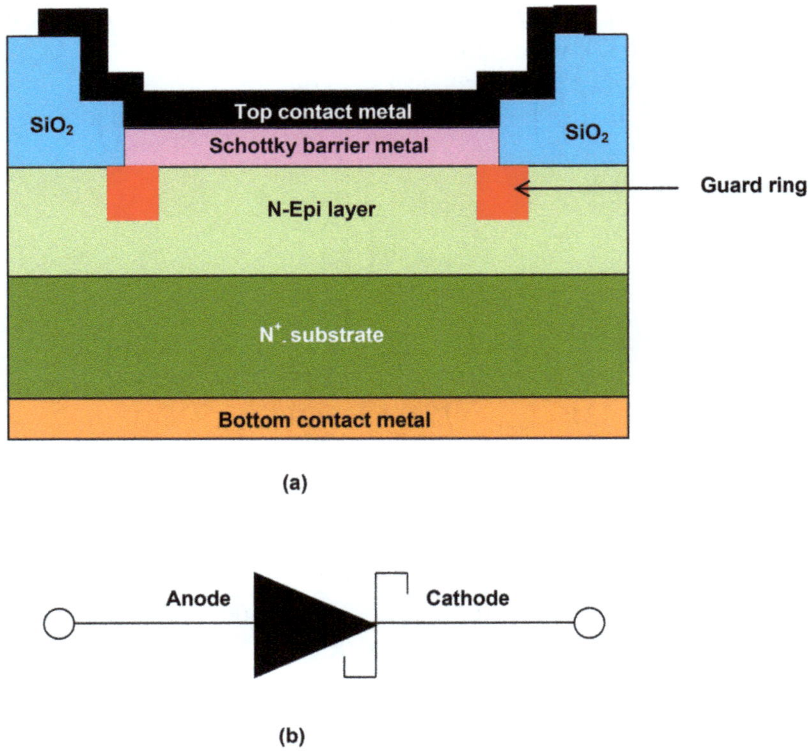

Figure 4.16. Schottky diode. (a) Cross-section and (b) circuit symbol.

Neglecting unity on the right-hand side, equation (4.46) reduces to

$$I_d/I_{D0} = \exp\{(qV_d)/(k_B T)\}. \tag{4.47}$$

Taking the natural logarithm of both sides

$$\ln(I_d/I_{D0}) = (qV_d)/(k_B T)$$

or

$$V_d = \{k_B T \ln(I_d/I_{D0})\}/q. \tag{4.48}$$

Differentiating both sides with respect to T, and keeping I_d constant,

$$\begin{aligned}\frac{dV_d}{dT} &= \left(\frac{k_B}{q}\right)\ln(I_d/I_{D0}) + \left(\frac{k_B T}{q}\right)\frac{1}{(I_d/I_{D0})} \times \frac{-I_d(dI_{D0}/dT)}{(I_{D0})^2}\\ &= \frac{1}{T}\left(\frac{k_B T}{q}\right)\ln(I_d/I_{D0}) - \left(\frac{k_B T}{q}\right)\frac{dI_{D0}/dT}{I_{D0}}\\ &= \frac{V_d}{T} - \left(\frac{k_B T}{q}\right)\frac{dI_{D0}/dT}{I_{D0}}.\end{aligned} \tag{4.49}$$

with the help of equation (4.48).

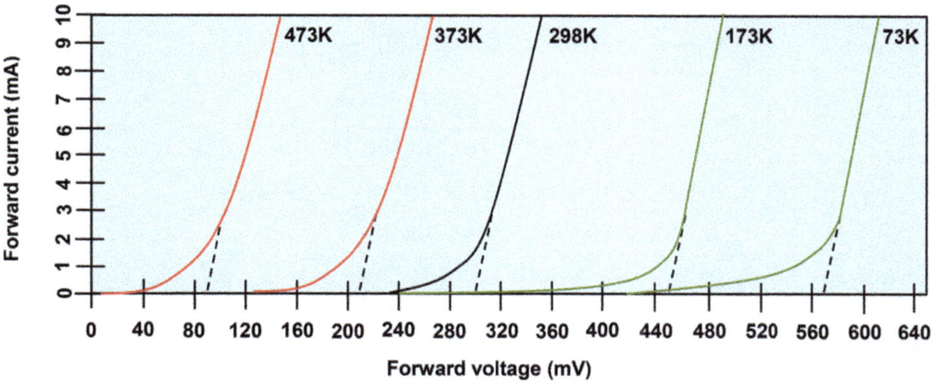

Figure 4.17. Variation of forward voltage of an SBD with temperature.

From equation (4.45)

$$\begin{aligned}\frac{\mathrm{d}I_{D0}}{\mathrm{d}T} &= \frac{\mathrm{d}\left[K_{SB}T^2\ \exp\{(-q\phi_B)/(k_BT)\}\right]}{\mathrm{d}T}\\ &= K_{SB}\times 2T\exp\{(-q\phi_B)/(k_BT)\}\\ &\quad + K_{SB}T^2\ \exp\{(-q\phi_B)/(k_BT)\}\times\left(\frac{-q}{k_B}\right)\times\frac{-\phi_B}{T^2}\\ &= \left(\frac{2}{T}\right)\left[K_{SB}\times T^2\ \exp\{(-q\phi_B)/(k_BT)\}\right]\\ &\quad + \left(\frac{1}{T}\right)K_{SB}T^2\ \exp\{(-q\phi_B)/(k_BT)\}\times\left(\frac{q}{k_B}\right)\times\frac{\phi_B}{T}\\ &= \left(\frac{2}{T}\right)I_{D0} + \left(\frac{1}{T}\right)I_{D0}\times\left(\frac{q\phi_B}{k_BT}\right)\end{aligned} \tag{4.50}$$

by using equation (4.45) again. Hence,

$$\frac{\mathrm{d}I_{D0}}{\mathrm{d}T} = \left(\frac{I_{D0}}{T}\right)\left\{2 + \left(\frac{q\phi_B}{k_BT}\right)\right\}. \tag{4.51}$$

Substituting for $\mathrm{d}I_{D0}/\mathrm{d}T$ from equation (4.51) into equation (4.49), we obtain

$$\frac{\mathrm{d}V_d}{\mathrm{d}T} = \frac{V_d}{T} - \left(\frac{k_BT}{q}\right)\frac{\left(\frac{I_{D0}}{T}\right)\left\{2 + \left(\frac{q\phi_B}{k_BT}\right)\right\}}{I_{D0}} = \frac{V_d}{T} - \left(\frac{k_BT}{q}\right)\left(\frac{1}{T}\right)\left\{2 + \left(\frac{q\phi_B}{k_BT}\right)\right\}. \tag{4.52}$$

For a Schottky diode fabricated using aluminum Schottky contact on silicon, $\phi_B = 0.7$ eV at $T = 300$ K. If $V_d = 0.4$ V,

$$\frac{\mathrm{d}V_{\mathrm{d}}}{\mathrm{d}T} = \frac{0.4}{300} - (8.62 \times 10^{-5} \times 300)\left(\frac{1}{300}\right)\left\{2 + \left(\frac{0.7}{8.62 \times 10^{-5} \times 300}\right)\right\}$$
$$= 1.33 \times 10^{-3} - (8.62 \times 10^{-5})(2 + 27.0688)$$
$$= 1.33 \times 10^{-3} - 2.505\,73 \times 10^{-3}$$
$$= -1.175\,73 \times 10^{-3}\ \mathrm{V\,K^{-1}}$$
$$= -1.175\,73\ \mathrm{mVK^{-1}} \approx -1.2\ \mathrm{mVK^{-1}}. \tag{4.53}$$

Thus, the temperature coefficient (TC) of the forward voltage drop for a Schottky diode ~ – 1.2 mV $\mathrm{K^{-1}}$ is smaller than that for a p–n junction diode ~ – 2 mV $\mathrm{K^{-1}}$.

4.4.4 Reverse leakage current of a p–n junction diode

When a diode is reverse biased, a small current flows across the junction. This current is due the diffusion of minority-carrier holes on the n-side and minority-carrier electrons on the p-side, across the junction. Hence, its magnitude depends on the diffusion coefficients of the minority carriers. Additionally, the carriers, which are generated in the depletion region, also contribute to this current flow. Within a certain limit, this current remains constant with increase of reverse voltage; hence it is called the reverse saturation current. But it is highly sensitive to temperature variations (figure 4.18). Its temperature sensitivity can be found as follows.

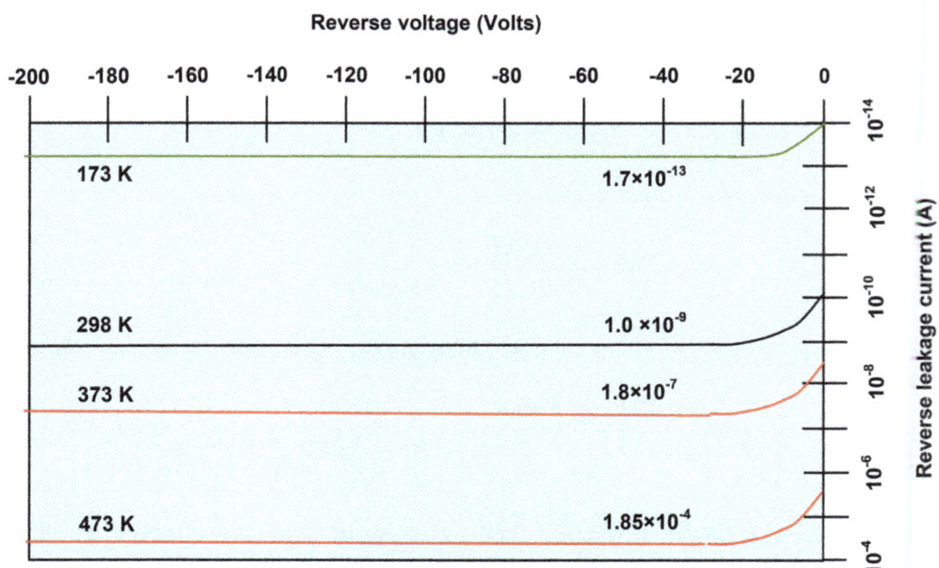

Figure 4.18. Reverse current–voltage characteristics of a p–n junction diode at different temperatures.

Since

$$\frac{\mathrm{d}\{\ln(I_S)\}}{\mathrm{d}T} = \left(\frac{1}{I_S}\right)\frac{\mathrm{d}I_S}{\mathrm{d}T} \tag{4.54}$$

putting the value of $\mathrm{d}I_S/\mathrm{d}T$ from equation (4.39)

$$\frac{\mathrm{d}\{\ln(I_S)\}}{\mathrm{d}T} = \left(\frac{1}{I_S}\right)\frac{\mathrm{d}I_S}{\mathrm{d}T} = \left(\frac{1}{I_S}\right)I_S\left(\frac{3}{\eta T} + \frac{E_G}{\eta T V_{\text{Thermal}}}\right)$$
$$= \frac{3}{\eta T} + \frac{E_G}{\eta T V_{\text{Thermal}}}$$

or

$$\frac{\Delta\{\ln(I_S)\}}{\Delta T} = \frac{3}{\eta T} + \frac{E_G}{\eta T V_{\text{Thermal}}} \tag{4.55}$$

$$\therefore\ \Delta\{\ln(I_S)\} = \left(\frac{3}{\eta T} + \frac{E_G}{\eta T V_{\text{Thermal}}}\right)\Delta T. \tag{4.56}$$

If $I_S = I_{S1}$ at temperature T and $I_S = I_{S2}$ at temperature $T + \Delta T$,

$$\ln\left(\frac{I_{S2}}{I_{S1}}\right) = \left(\frac{3}{\eta T} + \frac{E_G}{\eta T V_{\text{Thermal}}}\right)\Delta T$$

or

$$\frac{I_{S2}}{I_{S1}} = \exp\left\{\left(\frac{3}{\eta T} + \frac{E_G}{\eta T V_{\text{Thermal}}}\right)\Delta T\right\}. \tag{4.57}$$

For silicon diodes at room temperature $T = 300$ K, $E_G = 1.11$ eV, $V_{\text{Thermal}} = 0.0259$ V. Suppose $\Delta T = 10$ °C, taking $\eta = 1$,

$$\begin{aligned}\frac{I_{S2}}{I_{S1}} &= \exp\left\{\left(\frac{3}{1\times 300} + \frac{1.11}{1\times 300\times 0.0259}\right)\times 10\right\}\\ &= \exp\{(0.01 + 0.1429)\times 10\}\\ &= \exp(1.529) = 4.61.\end{aligned} \tag{4.58}$$

Taking $\eta = 2$,

$$\begin{aligned}\frac{I_{S2}}{I_{S1}} &= \exp\left\{\left(\frac{3}{2\times 300} + \frac{1.11}{2\times 300\times 0.0259}\right)\times 10\right\}\\ &= \exp\{(0.005 + 0.071\,43)\times 10\}\\ &= \exp(0.7643) = 2.15.\end{aligned} \tag{4.59}$$

Thus, for every 10 °C rise of temperature, the leakage current of a silicon diode quadruples if $\eta = 1$ and doubles if $\eta = 2$. For $\eta = 2$, the reverse leakage current of a Si diode is halved for every 10 °C fall in temperature. Hence, if the reverse current is 1 nA at 25 °C, and the temperature goes down to −55 °C, the reverse current will decrease to 1/2 nA at 15 °C, 1/4 nA at 5 °C, 1/8 nA at −5 °C, 1/16 nA at −15 °C,

1/32 nA at −25 °C, 1/64 nA at −35 °C, 1/128 nA at −45 °C and finally 1/256 nA = 0.0039 nA = 3.9 pA at −55 °C. Note that the temperature decreases by 80 °C = 8 × 10 °C. Hence, the reverse current = $(1/2^8) \times 1$ nA = (1/256) × 1 nA = 3.9 pA.

The doubling of reverse leakage current for each 10 °C increase in temperature is mathematically expressed as

$$I_S(T) = I_S(T_0)2^{(T-T_0)/10^{\circ}C}, \tag{4.60}$$

where $I_S(T)$ and $I_S(T_0)$ are the leakage currents at temperatures T, T_0, respectively.

4.4.5 Avalanche breakdown voltage of a p–n junction diode

The breakdown voltage of a p–n junction increases with an increase of temperature, as shown in figure 4.19.

At very high electric field called the critical field (≈3×10^5 V cm^{-1} for Si), electrons (minority carriers) gain energy very fast. The rate of energy acquisition by electrons is higher than the rate at which they can lose energy through the emission of optical phonons in a reverse-biased p–n junction. As a result, the high-energy electrons collide with bound electrons in the valence band. By collision, they excite the bonded electrons into the conduction band, creating an EPH. The phenomenon is termed impact ionization (Mukherjee 2011).

Avalanche breakdown is explained in terms of an ionization integral based on an important parameter, the ionization coefficient α. It is the average number of ionizing collisions accomplished by the carrier in traversing unit distance in the direction of electric field. The ionization integral becomes unity at the critical electric field

$$\int_0^W \alpha_p \exp\left\{\int_0^x (\alpha_n - \alpha_p)\mathrm{d}x\right\}\mathrm{d}x = 1, \tag{4.61}$$

where W is the width of the depletion region, and α_n, α_p are the ionization coefficients of electrons and holes, respectively. To explore the variation of breakdown voltage with temperature, attention is directed particularly towards the ionization coefficient.

Following the analysis by Chang *et al* (1971), the ionization rate α for an electron or a hole can be correlated with three material parameters:

(i) The ionization threshold energy W_I, i.e. the minimum energy required to cause an ionizing collision. It is approximately $1.5E_g$ for a parabolic band structure having the same effective masses for the carrier, if both energy and momentum conservation are taken into account (Maes *et al* 1990):

$$W_I = 1.5E_g. \tag{4.62}$$

(ii) The average energy loss due to optical phonon scattering (E_p):

$$\langle E_p \rangle = E_p \tanh\left(\frac{E_p}{2k_B T}\right), \tag{4.63}$$

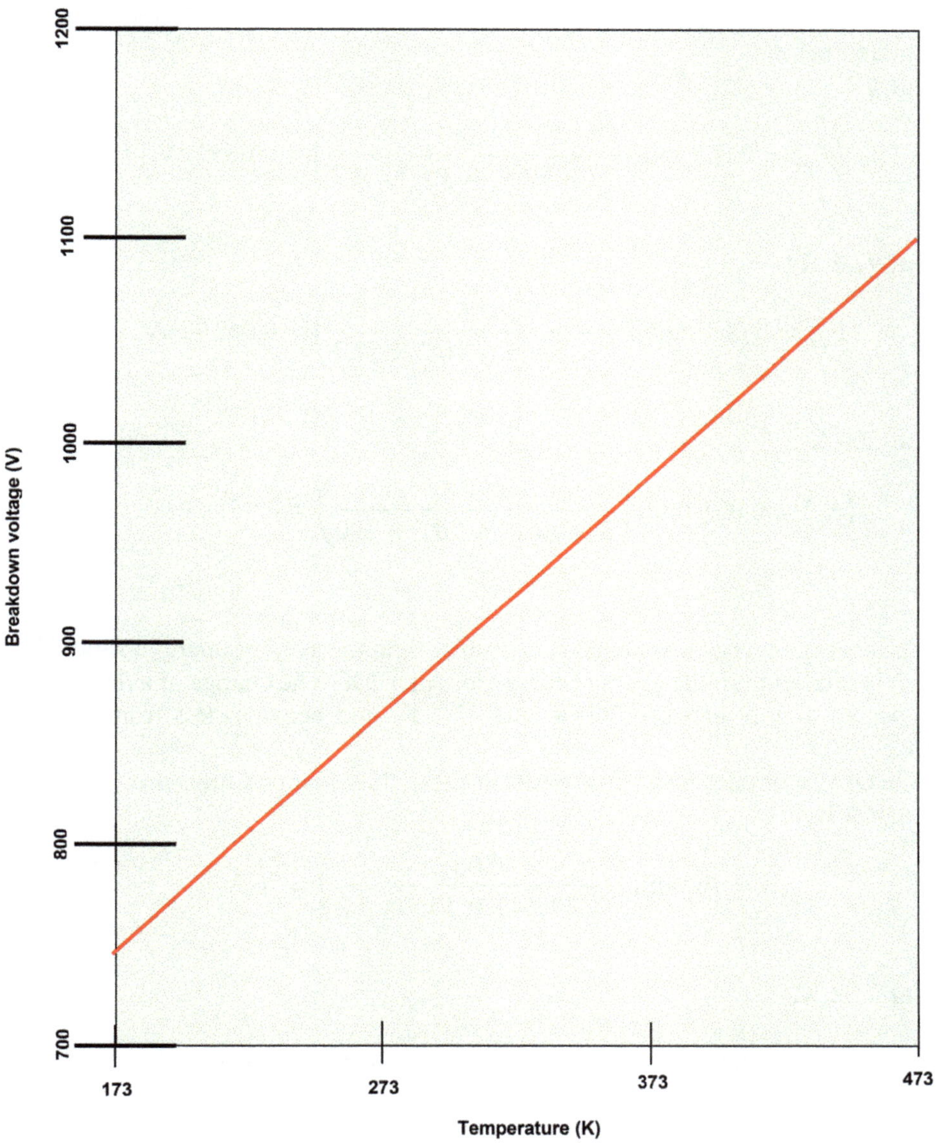

Figure 4.19. Change in the breakdown voltage of a diode with temperature.

where E_p is the optical phonon energy (=0.063 eV for Si).

(iii) The average carrier mean free path (λ) due to the scattering given by the Cromwell–Sze equation (Crowell and Sze 1966):

$$\lambda = \lambda_0 \tanh\left(\frac{E_p}{2k_B T}\right), \tag{4.64}$$

where λ_0 is the high-energy low-temperature asymptotic phonon mean free path (=76 Å for Si).

For Si, the average energy losses by scattering at 4.2, 77.2 and 300 K are calculated below:

at 4.2 K,

$$\langle E_\mathrm{p} \rangle = 0.063 \tanh\left(\frac{0.063}{2 \times 8.617 \times 10^{-5} \times 4.2}\right) = 0.0630\ \mathrm{eV}, \tag{4.65}$$

at 77.2 K,

$$\langle E_\mathrm{p} \rangle = 0.063 \tanh\left(\frac{0.063}{2 \times 8.617 \times 10^{-5} \times 77.2}\right) = 0.062\,99\ \mathrm{eV}, \tag{4.66}$$

at 300 K,

$$\langle E_\mathrm{p} \rangle = 0.063 \tanh\left(\frac{0.063}{2 \times 8.617 \times 10^{-5} \times 300}\right) = 0.052\,87\ \mathrm{eV}. \tag{4.67}$$

The average energy loss at 300 K is smaller than the average energy loss at 77.2 K, which is smaller than the average energy loss at 4.2 K. The change of average energy loss occurs mainly between 300 K and 77.2 K, and becomes less from 77.2 K to 4.2 K.

The corresponding mean free paths at these three temperatures are:

at 4.2 K,

$$\lambda = 76 \tanh\left(\frac{0.063}{2 \times 8.617 \times 10^{-5} \times 4.2}\right) = 76\ \text{Å}. \tag{4.68}$$

at 77.2 K,

$$\lambda = 76 \tanh\left(\frac{0.063}{2 \times 8.617 \times 10^{-5} \times 77.2}\right) = 75.9883\ \text{Å}, \tag{4.69}$$

at 300 K,

$$\lambda = 76 \tanh\left(\frac{0.063}{2 \times 8.617 \times 10^{-5} \times 300}\right) = 63.780\,53\ \text{Å}. \tag{4.70}$$

The mean free path at 4.2 K is larger than the mean free path at 77.2 K which is larger than the mean free path at 300 K. As before, the change occurs mainly between 300 K and 77.2 K, but is much smaller from 77.2 K to 4.2 K. Table 4.2 compiles the results of the above calculations.

Table 4.2. Average energy loss and mean free path at different temperatures.

Temperature (K)	Average energy loss (eV)	Mean free path (Å)
4.2	0.630	76.0
77.2	0.062 99	75.99
300	0.053	63.78

A carrier gains energy from the electric field and loses part of its energy by optical phonon scattering. At a low temperature, the following events take place:

(i) E_g and hence W_I increases as T decreases: as $T\downarrow$, $W_I\uparrow$. Hence, it is more difficult to cause ionization.

(ii) The mean free path λ of the carrier is longer. As $T\downarrow$, $\lambda\uparrow$. Therefore, a carrier can gain more energy from the field. Hence, ionization becomes easier.

(iii) Average energy loss by scattering is higher. As $T\downarrow$, $\langle E_p\rangle\uparrow$. So, a carrier becomes less effective for ionization.

(iv) The scattering incidents are less frequent. The number of scattering incidents per unit time is smaller. As a result, energy losses through scattering become negligible. The loss of energy by few scattering events, although larger for each event at lower temperature than at higher temperature, cannot prevent the carriers from acquiring high energy values. This considerably offsets the effect of (iii).

The predominance of factors (ii) and (iv) over (i) and (iii) leads to a large increase in energy by acceleration between two consecutive scattering incidents, whereby a greater number of carriers succeed in reaching the ionization threshold energy W_I necessary to dislodge electrons and promote them from the valence band edge to the conduction band. Hence, the breakdown voltage decreases at low temperatures.

Thus, the reason that the breakdown voltage decreases with a reduction in temperature is due to the increased mean free path of charge carriers from the reduced frequency of phonon scattering. As a result, higher energy can be gained in the electric field of the depletion region before the charge carriers collide with the lattice structure. The higher energy is the reason for impact ionization and device breakdown at lower temperatures.

4.4.6 Analytical model of temperature coefficient of avalanche breakdown voltage

A first-order model for impact ionization was formulated by Waldron *et al* (2005) starting from the Chynoweth formula (Chynoweth 1958) according to which the ionization coefficients of electrons and holes α_n, α_p are expressed as (Reisch 2003)

$$\alpha_n = a_n \exp(-b_n/E) \tag{4.71}$$

$$\alpha_p = a_p \exp(-b_p/E) \tag{4.72}$$

where a_n, b_n, a_p, b_p are fitting parameters dependent on the material and E is the intensity of the lateral electric field. In the presence of large electric fields perpendicular to the direction of motion of electrons, as in the channel of a MOSFET, the Chynoweth equations are modified as

$$\alpha_n = a_n \exp\{-(b_n \mid \mathbf{J}_n \mid)/(\mid \mathbf{E}\cdot\mathbf{J}_n \mid)\} \quad (4.73)$$

$$\alpha_p = a_p \exp\{-(b_p \mid \mathbf{J}_p \mid)/(\mid \mathbf{E}\cdot\mathbf{J}_p \mid)\} \quad (4.74)$$

where $\mathbf{J}_n$, $\mathbf{J}_p$ are the electron and hole current densities. In the present analysis, the same equation

$$\alpha = a \exp(-b/E) \quad (4.75)$$

will be used for both electrons and holes.

The calculation of ionization coefficients with local electric field dependence can be understood in terms of Shockley's lucky electron model of impact ionization. Shockley's model considers two principal mechanisms of energy loss by electrons, namely by optical phonon scattering and by generation of an EHP as soon as the energy of the electron exceeds the ionization threshold energy W_I. Shockley interpreted the parameter b in Chynoweth equations as:

$$b = \frac{W_I(T)}{q\lambda(T)} \quad (4.76)$$

for a mean free path λ of an electron of charge q. The mean free path is given by the Cromwell–Sze equation (4.64).

From modeling of threshold ionization energy by Ershov and Ryzhii (1995), based on the assumption of its proportionality to the bandgap energy of silicon, we can write

$$W_I(T) = C_1 + C_2T + C_3T^2, \quad (4.77)$$

where

$$C_1 = 1.1785 \text{ eV}, \quad (4.77a)$$

$$C_2 = -\,9.025 \times 10^{-5} \text{ eV K}^{-1}, \quad (4.77b)$$

and

$$C_3 = -\,3.05 \times 10^{-7} \text{ eV K}^{-2} \text{ for } T > 170 \text{ K}. \quad (4.77c)$$

The temperature dependence of the ionization coefficient is defined by the TC as

$$\mathrm{TC} = (1/\alpha)(\mathrm{d}\alpha/\mathrm{d}T). \quad (4.78)$$

Substituting for α from equation (4.75) into equation (4.78), we obtain

$$\begin{aligned}\mathrm{TC} &= \frac{1}{a\exp(-b/E)} \times \frac{\mathrm{d}}{\mathrm{d}T}\{a\exp(-b/E)\}\\ &= \frac{1}{a\exp(-b/E)} \times a\exp(-b/E) \times \frac{\mathrm{d}}{\mathrm{d}T}(-b/E)\\ &= \frac{\mathrm{d}}{\mathrm{d}T}(-b/E) = -\frac{1}{E}\left(\frac{\mathrm{d}b}{\mathrm{d}T}\right) = -\frac{1}{E}\frac{\mathrm{d}}{\mathrm{d}T}\left\{\frac{W_{\mathrm{I}}(T)}{q\lambda(T)}\right\}\\ &= -\frac{1}{qE}\frac{\mathrm{d}}{\mathrm{d}T}\left\{\frac{W_{\mathrm{I}}(T)}{\lambda(T)}\right\}\\ &= -\frac{1}{qE}\frac{(\mathrm{d}W_{\mathrm{I}}/\mathrm{d}T)\lambda - W_{\mathrm{I}}(\mathrm{d}\lambda/\mathrm{d}T)}{\lambda^2}\\ &= -\left(\frac{1}{qE}\right)\left\{\frac{(\mathrm{d}W_{\mathrm{I}}/\mathrm{d}T)\lambda}{\lambda^2}\right\} + \left(\frac{1}{qE}\right)\left\{\frac{W_{\mathrm{I}}(\mathrm{d}\lambda/\mathrm{d}T)}{\lambda^2}\right\}\\ &= \left(\frac{W_{\mathrm{I}}}{qE\lambda}\right)\left\{\frac{1}{\lambda}\left(\frac{\mathrm{d}\lambda}{\mathrm{d}T}\right) - \frac{1}{W_{\mathrm{I}}}\left(\frac{\mathrm{d}W_{\mathrm{I}}}{\mathrm{d}T}\right)\right\}.\end{aligned} \tag{4.79}$$

The first term inside the curly brackets in equation (4.79) is the TC of λ. From equation (4.64)

$$\begin{aligned}\frac{1}{\lambda}\left(\frac{\mathrm{d}\lambda}{\mathrm{d}T}\right) &= \left(\frac{1}{\lambda}\right)\frac{\mathrm{d}}{\mathrm{d}T}\left\{\lambda_0 \tanh\left(\frac{E_{\mathrm{p}}}{2k_{\mathrm{B}}T}\right)\right\}\\ &= \left(\frac{\lambda_0}{\lambda}\right)\frac{E_{\mathrm{p}}}{2k_{\mathrm{B}}}\left\{1 - \tanh^2\left(\frac{E_{\mathrm{p}}}{2k_{\mathrm{B}}T}\right)\right\} \times \frac{0-1}{T^2}\\ &= -\left(\frac{\lambda_0}{\lambda}\right)\frac{E_{\mathrm{p}}}{2k_{\mathrm{B}}T^2}\operatorname{sech}^2\left(\frac{E_{\mathrm{p}}}{2k_{\mathrm{B}}T}\right)\end{aligned} \tag{4.80}$$

since

$$\tanh^2 x + \operatorname{sech}^2 x = 1. \tag{4.81}$$

The second term inside curly brackets in equation (4.79) is the TC of W_{I}. From equation (4.77)

$$\begin{aligned}\frac{1}{W_{\mathrm{I}}}\left(\frac{\mathrm{d}W_{\mathrm{I}}}{\mathrm{d}T}\right) &= \left(\frac{1}{W_{\mathrm{I}}}\right)\frac{\mathrm{d}}{\mathrm{d}T}(C_1 + C_2T + C_3T^2)\\ &= \left(\frac{1}{W_{\mathrm{I}}}\right)(0 + C_2 \times 1 + C_3 \times 2T^{2-1}) = \left(\frac{1}{W_{\mathrm{I}}}\right)(C_2 + 2C_3T).\end{aligned} \tag{4.82}$$

Substituting the values of the TCs of λ and W_{I} from equations (4.80) and (4.82) into equation (4.79), we have

$$\begin{aligned}\mathrm{TC} &= -\left(\frac{W_\mathrm{I}}{qE\lambda}\right)\left\{\left(\frac{\lambda_0}{\lambda}\right)\left(\frac{E_\mathrm{p}}{2k_\mathrm{B}T^2}\right)\mathrm{sech}^2\left(\frac{E_\mathrm{p}}{2k_\mathrm{B}T}\right)\right.\\ &\quad\left.+\left(\frac{1}{W_\mathrm{I}}\right)\left(C_2+2C_3T\right)\right\} \\ &= -\left(\frac{1}{E}\right)\left[\left(\frac{W_\mathrm{I}\lambda_0E_\mathrm{p}}{2q\lambda^2k_\mathrm{B}T^2}\right)\mathrm{sech}^2\left(\frac{E_\mathrm{p}}{2k_\mathrm{B}T}\right)+\left(\frac{1}{q\lambda}\right)\left(C_2+2C_3T\right)\right]\end{aligned} \tag{4.83}$$

TC of λ is negative. Because both C_2 and C_3 are negative, the TC of W_I is also negative. Hence, the sign of TC of breakdown voltage is determined by the comparative magnitudes of the two terms in the above equation. Whether TC is positive or negative will be clear by performing an example calculation.

For Si at room temperature $T = 300$ K

$$\begin{aligned}W_\mathrm{I}(T) &= 1.1785 + (-9.025\times10^{-5})\times 300\\ &\quad + (-3.05\times10^{-7})\times(300)^2\\ &= 1.1785 - 2.7075\times10^{-2} - 2.745\times10^{-2} = 1.123\,975\,\mathrm{eV}\end{aligned} \tag{4.84}$$

$$\lambda_0 = 7.6\,\mathrm{nm},\ E_\mathrm{p} = 0.053\,\mathrm{eV},\ q = 1.6\times10^{-19}\,\mathrm{C},$$
$$\lambda = 6.378\,\mathrm{nm},\ k_\mathrm{B} = 1.381\times10^{-23}\ \mathrm{J\ K^{-1}}.$$

Putting the above values in equation (4.83)

$$\begin{aligned}\mathrm{TC} &= -\left(\frac{1}{E}\right)\left[\left\{\frac{1.123\,975\times1.6\times10^{-19}\times 7.6\times10^{-9}\times0.053\times1.6\times10^{-19}}{2\times1.6\times10^{-19}\times(6.378\times10^{-9})^2\times(1.381\times10^{-23})(300)^2}\right\}\right.\\ &\quad \mathrm{sech}^2\left\{\frac{0.053\times1.6\times10^{-19}}{2\times(1.381\times10^{-23})\times300}\right\}\\ &\quad +\left(\frac{1}{1.6\times10^{-19}\times6.378\times10^{-9}}\right)\{-9.025\times10^{-5}\times1.6\times10^{-19}\\ &\quad \left.+(2\times-3.05\times10^{-7}\times1.6\times10^{-19}\times300)\}\right]\\ &= -\left(\frac{1}{E}\right)\left[\left\{\frac{1.159\times10^{-47}}{1.6179\times10^{-53}}\right\}\mathrm{sech}^2\left\{\frac{8.48\times10^{-21}}{8.286\times10^{-21}}\right\}\right.\\ &\quad \left.+\left(\frac{1}{1.020\,48\times10^{-27}}\right)\{-1.444\times10^{-23}-2.928\times10^{-23}\}\right]\\ &= -\left(\frac{1}{E}\right)[(7.163\,607\,145\times10^5)\mathrm{sech}^2(1.023\,412\,985\,759)\\ &\quad +(9.799\,310\,128\times10^{26})(-4.372\times10^{-23})]\\ &= -\left(\frac{1}{E}\right)[(7.163\,607\,145\times10^5)(0.636\,5287)^2-(42\,842.583\,88)]\\ &= -\left(\frac{1}{E}\right)[290\,247.000\,977-42\,842.583\,88]\\ &= -\left(\frac{1}{E}\right)\left(\frac{247\,404.4171}{E}\right)\mathrm{Vm^{-1}\,K^{-1}} \approx -\left(\frac{1}{E}\right)2474\,\mathrm{Vcm^{-1}\,K^{-1}}.\end{aligned} \tag{4.85}$$

The calculation shows that:

(i) silicon has a negative TC of breakdown voltage, and

(ii) the small value of the TC reveals that the temperature dependence is a weak one.

4.4.7 Zener breakdown voltage of a diode

The reverse breakdown voltage of an avalanche diode decreases with decrease in temperature while that of a Zener diode shows the opposite behavior (table 4.3).

4.4.8 Storage time (t_s) of a p^+–n junction diode

The storage time is a principal figure of merit for characterizing the transient behavior of a diode. The storage time t_s of a p^+–n diode is expressed in terms of the forward current I_F, reverse current I_R and hole lifetime τ_p as (Dokić and Blanuša 2015)

Table 4.3. Temperature dependence of avalanche and Zener breakdown voltages.

Feature	Avalanche breakdown	Zener breakdown
Type of junction	Lightly doped	Heavily doped
Width of depletion region	Large	Small
Electric field across depletion region	Low	High
Range of breakdown voltage	Breakdown voltage > $6E_g/q$ = 6 × 1.11 = 6.66 V	Breakdown voltage < $4E_g/q$ = 4 × 1.11 = 4.44 V
TC of breakdown voltage	Positive and increases with an increase in breakdown voltage, e.g. from 3 to 6 mV K^{-1} for an 8 V diode to 13–18 mV K^{-1} for an 18 V diode.	Negative and independent of the breakdown voltage rating; typically −3 mV K^{-1}.
Thermal mechanism	As the temperature rises, the vibrational displacements of lattice atoms about their equilibrium positions increase. Hence, the carriers suffer more collisions from lattice atoms, and cannot be easily accelerated to high velocities to cause ionizations of lattice atoms. Greater voltage is required to trigger an avalanche.	With increasing temperature, the energies of valence electrons in silicon atoms increase. They can easily tear off from the covalent bonds. Less voltage is required to liberate these electrons from the bonds to participate in conduction. Therefore, the breakdown voltage decreases as the temperature increases because the bandgap decreases.

$$t_s = \tau_p \ln\left(1 + \frac{I_F}{I_R}\right). \quad (4.86)$$

The storage time is small if the hole lifetime is low; the forward current I_F is low because a smaller amount of charge is to be removed; and the reverse current I_R is high because the stored charge can be swept away at a faster rate. Hole lifetimes $\tau_p(T)$, $\tau_p(T_0)$ at two different temperatures T, T_0 are inter-related as (Dokić and Blanuša 2015)

$$\tau_p(T) = \tau_p(T_0)(T/T_0)^r \quad (4.87)$$

where r is a constant = 3.5 for Si and 2.2 for Ge diodes at low levels of carrier injection. For a temperature increase from −55 °C to +175 °C, the carrier lifetime

$$\tau_p(175 + 273\ \text{K}) = \tau_p(-55 + 273)\{(175 + 273)/(-55 + 273)\}^{3.5}$$

or

$$\tau_p(448\ \text{K}) = \tau_p(218)(448/218)^{3.5} = 12.44\ \tau_p(218). \quad (4.88)$$

Since, the carrier lifetime in silicon increases by a factor larger than 12, the storage time is considerably lengthened. This stretching of storage time is manifested as an increase in switching loss.

4.4.9 Current gain of a bipolar junction transistor

Figure 4.20 presents the schematic diagram and circuit symbols of a bipolar junction transistor (BJT).

The DC current gain–collector current curves of a BJT show a decline in current gain with decreasing temperature, as shown in figure 4.21.

Buhanan performed a comprehensive theoretical analysis of the current gain equation using the piece-by-piece methodology (Buhanan 1969). For examination of the temperature dependence of current gain, the common-base current gain α serves as a convenient platform. Since the current gain β of an n–p–n BJT in the common-emitter configuration,

$$\beta = \text{collector current } (I_C)/\text{base current } (I_B), \quad (4.89)$$

is related to its current gain α in the common-base connection

$$\alpha = \text{collector current } (I_C)/\text{emitter current } (I_E) \quad (4.90)$$

by the familiar equation

$$\beta = \alpha/(1 - \alpha), \quad (4.91)$$

the temperature dependence for β becomes evident from that of α. The common-base current gain α is a product of three factors:

(i) The emitter injection efficiency γ given by Kauffman and Bergh (1968)

$$\gamma = I_{nB}/(I_{nB} + I_{pE}), \quad (4.92)$$

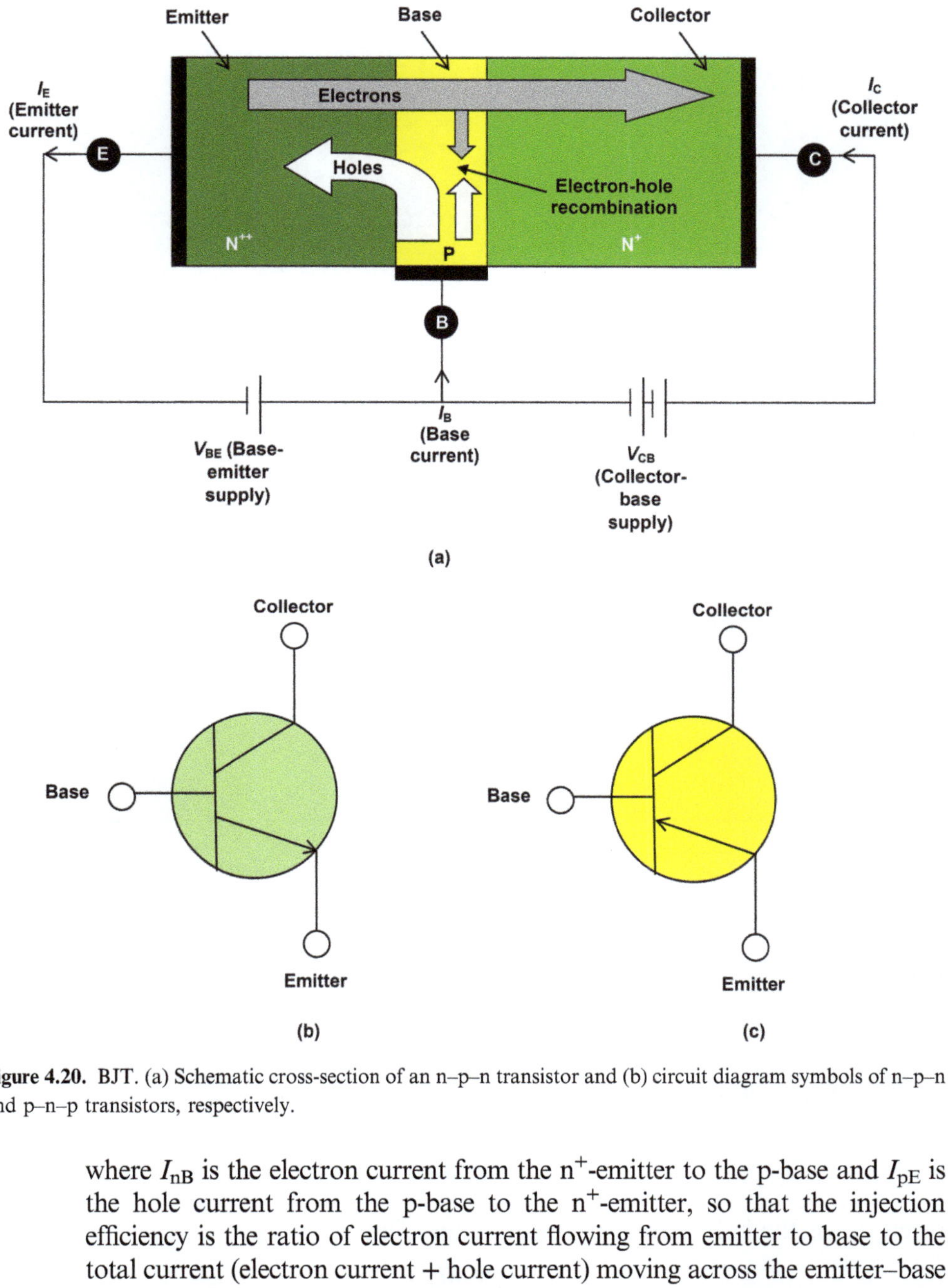

Figure 4.20. BJT. (a) Schematic cross-section of an n–p–n transistor and (b) circuit diagram symbols of n–p–n and p–n–p transistors, respectively.

where I_{nB} is the electron current from the n$^+$-emitter to the p-base and I_{pE} is the hole current from the p-base to the n$^+$-emitter, so that the injection efficiency is the ratio of electron current flowing from emitter to base to the total current (electron current + hole current) moving across the emitter–base junction. The notation used for currents designates the transistor terminal (emitter, base or collector) into which the electrons (n)/holes (p) are injected.

(ii) The base transport factor α_T written as

$$\alpha_T = I_{nC}/I_{nB} \tag{4.93}$$

where I_{nC} is the electron current in the collector.

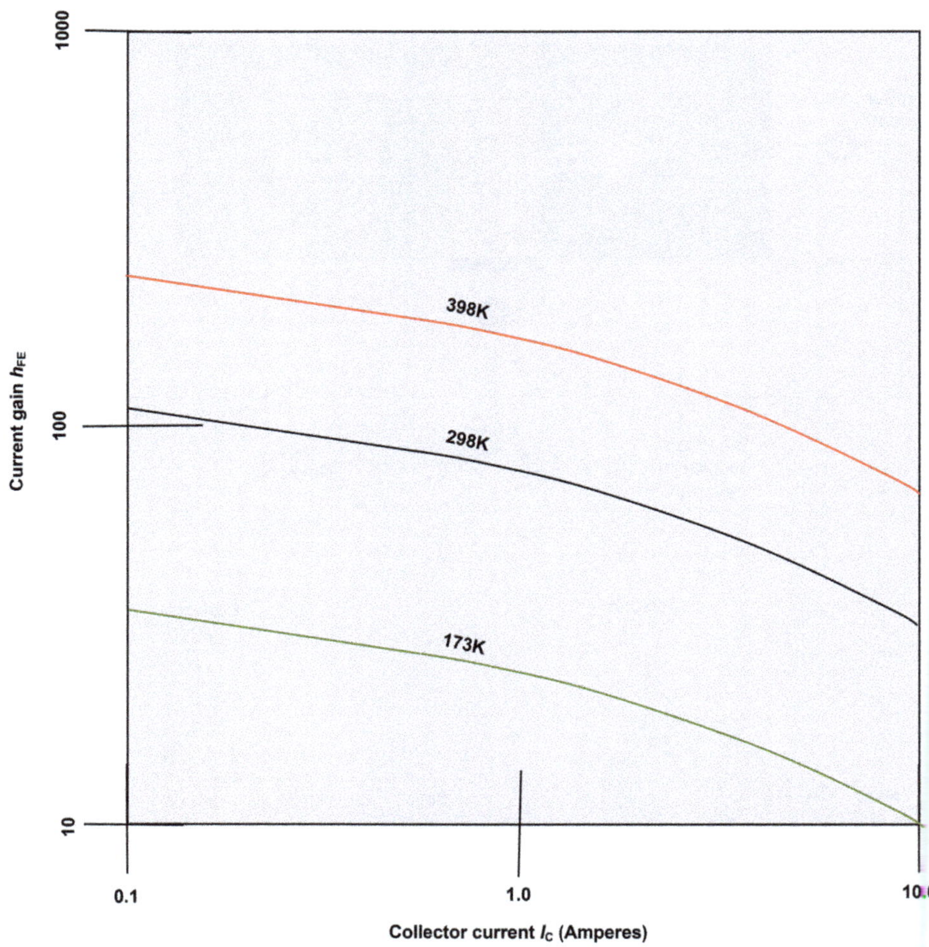

Figure 4.21. Shifts in the plots of DC current gain versus collector current of a Si bipolar transistor with temperature.

(iii) The collector multiplication ratio M expressed as

$$M = I_C/I_{nC} = (I_{nC} + I_{pC})/I_{nC}, \tag{4.94}$$

where I_{pC} is the hole current in the collector. Thus

$$\alpha = \gamma\alpha_T M. \tag{4.95}$$

At voltages $\ll BV_{CEO}$, $M = 1$; hence

$$\alpha = \gamma\alpha_T = \{1 + (D_{pE}/D_{nB})(p_{nE}/n_{pB})(W/L_{pE})\}^{-1} \times \{\text{sech}(W/L_{nB})\}, \tag{4.96}$$

where the first reciprocal term in curly brackets preceding the multiplication sign (×) represents γ while the second term after this sign gives α_T. The symbols have the following meanings: D_{pE} = the diffusion constant of minority-carrier holes in the n$^+$-emitter; D_{nB} = the diffusion constant of minority-carrier electrons in the p-base; p_{nE} = the number of holes in the n$^+$-emitter; n_{pB} = the number of electrons in the p-base; W = the base width; L_{pE} = the diffusion length of holes in the emitter and L_{nB} = the diffusion length of electrons in the base. Let us examine the behavior of each term and every factor in this equation individually to understand the effect of temperature on the current gain. Combining together the effects of temperature on each term separately will help us to visualize the influence of temperature on the current gain when considered as a whole.

(i) The ratio of minority-carrier diffusion constants (D_{pE}/D_{nB})

(a) D_{pE} is related to the mobility μ_{pE} of holes in the emitter through the Einstein equation

$$D_{pE} = \mu_{pE}\{(k_B T)/q\}. \tag{4.97}$$

The emitter of a transistor is a heavily doped region with electron concentration $>10^{19}$ cm^{-3}. In this heavily doped region, impurity scattering will dominate over lattice scattering. In lattice scattering, mobility decreases as temperature increases, but impurity scattering shows the reverse trend. Therefore, the predominance of impurity scattering will make the hole mobility increase with temperature. Thus, the diffusion constant increases with temperature.

(b) D_{nB} is connected with the mobility of electrons μ_{nB} in the base via the equation

$$D_{nB} = \mu_{nB}\{(k_B T)/q\}. \tag{4.98}$$

The base of a transistor is a moderately doped region. So, lattice scattering will foreshadow impurity scattering. Lattice scattering increases as temperature increases, and thereby decreases the mobility. So, the mobility μ_{nB} will decrease with temperature increase. As the D_{nB} equation contains the factor T, which is itself increasing, the decrease of μ_{nB} will be compensated by the increase of T. So, D_{nB} will remain constant.

In view of (a) and (b) above, the ratio (D_{pE}/D_{nB}) will increase with an increase in temperature.

(ii) The ratio of minority-carrier concentrations (p_{nE}/n_{pB})

A high impurity concentration in the emitter produces deformations in the lattice and other defects. These defects lower the silicon bandgap in the emitter by an amount $\Delta\xi_g$ from the initial value ξ_g to the final value $\xi_g - \Delta\xi_g$. Without bandgap narrowing, the minority-carrier concentrations are

$$p_{nE} = n_i^2/N_D, \tag{4.99}$$

where N_D is the donor concentration in the emitter, and

$$n_{pB} = n_i^2/N_A, \tag{4.100}$$

where N_A is the acceptor concentration in the base. Also, the intrinsic carrier concentration n_i is

$$n_i^2 = \text{constant} \times T^3 \exp\{-\xi_g/(2k_B T)\}. \tag{4.101}$$

After bandgap reduction, the intrinsic carrier concentration expression changes to

$$n_i^2 = \text{constant} \times T^3 \exp\{-(\xi_g - \Delta\xi_g)/(2k_B T)\}. \tag{4.102}$$

Hence, after bandgap narrowing

$$p_{nE} = [\text{constant}\, A \times T^3 \exp\{-(\xi_g - \Delta\xi_g)/(2k_B T)\}]/N_D. \tag{4.103}$$

But unlike the emitter, there is no bandgap narrowing in the moderately doped base region. Hence, the equation for n_{pB} contains the initial bandgap ξ_g:

$$n_{pB} = [\text{constant}\, B \times T^3 \exp\{-\xi_g/(2k_B T)\}]/N_A. \tag{4.104}$$

So, the minority-carrier concentration ratio is

$$\begin{aligned} p_{nE}/n_{pB} &= \frac{\left[\text{constant}\, A \times T^3 \exp\left\{-(\xi_g - \Delta\xi_g)/(2k_B T)\right\}\right]/N_D}{\left[\text{constant}\, B \times T^3 \exp\left\{-\xi_g/(2k_B T)\right\}\right]/N_A} \\ &= \frac{\text{constant}\, A}{\text{constant}\, B} \times \exp\left\{\left(-\xi_g + \Delta\xi_g + \xi_g\right)/(2k_B T)\right\} \\ &\times \frac{N_A}{N_D} = \frac{AN_A}{BN_D} \times \exp\ \left\{\Delta\xi_g/(2k_B T)\right\} \\ &= K \exp\ \left\{\Delta\xi_g/(2k_B T)\right\}, \end{aligned} \tag{4.105}$$

where $K = AN_A/(BN_D)$. This equation shows that as the temperature decreases, the minority-carrier concentration ratio increases exponentially, thereby causing a substantial reduction in current gain. As temperature increases, the current gain is enhanced. The exponential nature of variation lends this factor a prominent role in determining the temperature-induced variation of current gain.

(iii) Ratio of base width to diffusion length of minority carriers in the emitter (W/L_{pE})
The base width W is determined by the base profile. The base profile depends on the ionization of diffused impurity atoms in the base region. The acceptor level introduced by boron is very proximate to the edge of the valence band, around 0.045 eV above the valence band edge. Up to 100 K or −173 °C, all the impurity atoms may be considered to be ionized so that the free carrier concentration equals the concentration of diffused impurity atoms. So, at temperatures above 100 K, it may be safely assumed that all the impurity atoms are ionized and the base width is constant. For this temperature range, base width W has no role to play in the temperature dependence of current gain. The diffusion length L_{pE} is given by

$$L_{pE} = \sqrt{D_{pE}\tau_{pE}}, \tag{4.106}$$

where τ_{pE} is the minority-carrier lifetime of holes in the emitter. As already discussed, the diffusion constant D_{pE} varies directly with temperature, decreasing

Table 4.4. Effects of decreasing temperature on a transistor's parameters and its current gain.

Parameter	Effect of decreasing temperature	Effect on current gain
D_{pE}/D_{nB}	Decreases	Enhancement
p_{nE}/n_{pB}	Increases exponentially	Strong degradation
W/L_{pE}	Increases	Degradation
$\text{sech}(W/L_{nB})$	Decreases	Degradation

with a fall in temperature. Further, τ_{pE} also decreases with temperature. Therefore, the overall effect is that the factor (W/L_{pE}) increases as temperature decreases. Hence, the emitter injection efficiency and thereby the current gain are degraded.

(iv) Hyperbolic secant of the ratio of base width to diffusion length of minority carriers in the base $\{\text{sech}(W/L_{nB})\}$

The diffusion length L_{nB} is

$$L_{nB} = \sqrt{D_{nB}\tau_{nB}}. \tag{4.107}$$

As already pointed out above, the base is a lightly doped region in a transistor structure. Hence, lattice scattering is preponderant and impurity scattering is less pronounced. Lattice scattering decreases as temperature falls. Therefore, the mobility μ_{nB} increases with decrease of temperature. But the temperature T itself decreases. Accordingly, the diffusion constant D_{nB} given by equation (4.98) remains roughly constant. Since, τ_{nB} decreases when temperature is decreased, the parameter L_{nB} also declines. Hence, (W/L_{nB}) increases. Since we know that hyperbolic secant of a function decreases as the value of the function increases, the current gain will decrease.

The results of the above analysis (i)–(iv) are summarized in table 4.4.

Thus, on the whole, decreasing the temperature has a deprecating influence on the current gain of a bipolar transistor. In practice, the current gain falls by 0.3% K^{-1}–0.6% K^{-1}. It was also inferred by Buhanan (1969) that as transistor manufacturing was confined within a few orders of magnitude of doping concentrations, all the effects, whether advantageous or disadvantageous, acquire a secondary role in comparison to that of p_{nE}/n_{pB}. Therefore, the current gain degradation produced by heavy doping-induced bandgap narrowing (BGN) in the emitter by way of modifying the intrinsic carrier concentration is worthy of further attention.

4.4.10 Approximate analysis

Admittedly, the reduction in current gain with falling temperature is a result of the smaller bandgap in the emitter than in the base. The base current is typically limited by hole injection into the emitter, and its temperature dependence is largely determined by the effective bandgap of the emitter, which is reduced because of a combination of heavy doping effects. The collector current, on the other hand, consists of electrons traversing the base and its temperature dependence is therefore determined by the bandgap in the base. The difference in bandgap between emitter and base causes the ideal gain to decrease exponentially with temperature by 2–3 orders of magnitude between room (~300 K) and liquid-nitrogen (=80 K) temperatures.

Neglecting the space charge recombination current (non-ideal base current) and assuming unity for the base transport factor, we can write the following equations:

$$I_{\mathrm{nB}} = I_{\mathrm{C}} \tag{4.108}$$

$$I_{\mathrm{pE}} = I_{\mathrm{B}} \tag{4.109}$$

$$h_{\mathrm{fE}} = \frac{I_{\mathrm{C}}}{I_{\mathrm{B}}} = \frac{I_{\mathrm{nB}}}{I_{\mathrm{pE}}}. \tag{4.110}$$

The temperature dependence of both I_{nB} and I_{pE} is dominated by the temperature dependence of the intrinsic carrier concentration n_{i} in the emitter and base regions, respectively.

But

$$n_{\mathrm{i}}^2 \propto \exp\left(-\frac{\xi_{\mathrm{g}}}{k_{\mathrm{B}}T}\right), \tag{4.111}$$

where ξ_{g} is the bandgap, k_{B} is Boltzmann's constant ($= 8.617 \times 10^{-5}$ eV K^{-1}), and T is absolute temperature.

Therefore, we can write

$$I_{\mathrm{C}}(T) = I_{\mathrm{nB}}(T) \propto \exp\left(-\frac{\xi_{\mathrm{gB}}}{k_{\mathrm{B}}T}\right), \tag{4.112}$$

$$I_{\mathrm{B}}(T) = I_{\mathrm{pE}}(T) \propto \exp\left(-\frac{\xi_{\mathrm{gE}}}{k_{\mathrm{B}}T}\right), \tag{4.113}$$

where ξ_{gB}, ξ_{gE} denote the bandgaps of the base and emitter materials, respectively. Hence, the common-emitter current gain

$$\begin{aligned} h_{\mathrm{fE}} &= \frac{I_{\mathrm{C}}(T)}{I_{\mathrm{B}}(T)} = \frac{I_{\mathrm{nB}}(T)}{I_{\mathrm{pE}}(T)} \\ &= \frac{\exp\left(-\frac{\xi_{\mathrm{gB}}}{k_{\mathrm{B}}T}\right)}{\exp\left(-\frac{\xi_{\mathrm{gE}}}{k_{\mathrm{B}}T}\right)} = \exp\left\{-\left(\frac{\xi_{\mathrm{gB}} - \xi_{\mathrm{gE}}}{k_{\mathrm{B}}T}\right)\right\} = \exp\left(-\frac{\Delta\xi_{\mathrm{gBE}}}{k_{\mathrm{B}}T}\right). \end{aligned} \tag{4.114}$$

If $\Delta\xi_{\mathrm{gBE}} = 0$, then $h_{\mathrm{FE}} = 1 = h_{\mathrm{fE0}}$ (assume). Then for any value of $\Delta\xi_{\mathrm{gBE}}$

$$h_{\mathrm{fE}} = h_{\mathrm{fE0}} \exp\left(-\frac{\Delta\xi_{\mathrm{gBE}}}{k_{\mathrm{B}}T}\right). \tag{4.115}$$

For a typical value of $\Delta\xi_{gBE} = 0.05$ eV, the current gain at 300 K is $(h_{fE})_{300\ K} = 0.1445$, while that at 77.2 K is $(h_{fE})_{77.2\ K} = 5.4421 \times 10^{-4}$; decreasing by a factor of 265.5225. This shows that for a finite value of $\Delta\xi_{gBE}$, as the temperature decreases h_{fE} also decreases.

For a higher value of $\Delta\xi_{gBE} = 0.1$ eV, the current gain at 300 K is $(h_{fE})_{300\ K} = 0.020\,89$ while that at 77.2 K is $(h_{fE})_{77.2\ K} = 2.9617 \times 10^{-7}$; decreasing by a factor of 9.7579×10^{4}.

Table 4.5 presents the values of current gains at specified temperatures and also the decreasing factors, i.e. the ratios of current gains at two selected temperatures. It is observed that the greater ΔE_{gBE} is, the greater is the reduction in h_{FE} with a decrease in temperature.

From the above considerations, homojunction bipolar transistors are generally not considered for operation at liquid-nitrogen temperatures, because of insufficient DC current gain (Stork *et al* 1987).

4.4.11 Saturation voltage of a bipolar junction transistor

Figure 4.22 illustrates the translation of the curve drawn between the collector current and the collector–emitter voltage of a bipolar transistor with rising temperature.

The collector–emitter voltage v_{CES} during operation of a bipolar transistor in the saturation mode is given by the difference of base–emitter voltage v_{BE} and base–collector voltage v_{BC}

$$v_{CES} = v_{BE} - v_{BC}. \tag{4.116}$$

Differentiating both sides of equation (4.116) with respect to temperature, we find that

$$\frac{dv_{CES}}{dT} = \frac{dv_{BE}}{dT} - \frac{dv_{BC}}{dT}. \tag{4.117}$$

This equation shows that the TC of the collector–emitter voltage is equal to the TC of the base–emitter voltage minus the TC of the base–collector voltage. Applying equation (4.42) for the TC of the forward voltage of a p–n junction diode successively to each of the two diodes formed by the base–emitter and base–collector junctions, we obtain

Table 4.5. Current gains and related decreasing factors.

Current gain (h_{fE})/decreasing factors	$\Delta\xi_{gBE} = 0.05$ eV	$\Delta\xi_{gBE} = 0.1$ eV
$(h_{fE})_{300\ K}$	0.1445	0.020 89
$(h_{fE})_{77.2\ K}$	5.4421×10^{-4}	2.9617×10^{-7}
$(h_{fE})_{4.2\ K}$	1.007×10^{-60}	1.0015×10^{-120}
$(h_{fE})_{300\ K}/(h_{fE})_{77.2\ K}$	265.5225	9.7579×10^{4}
$(h_{fE})_{300\ K}/(h_{fE})_{4.2\ K}$	1.435×10^{59}	2.0859×10^{118}

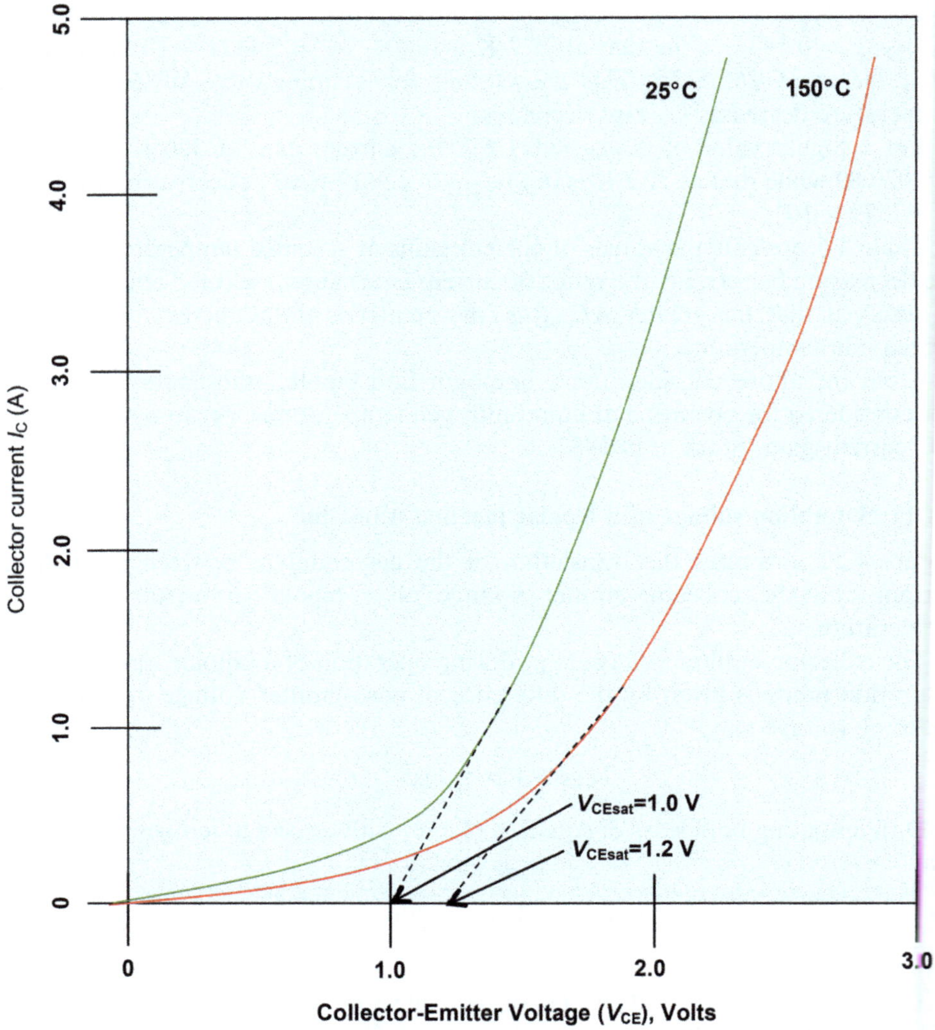

Figure 4.22. Effect of temperature on the saturation voltage of a bipolar transistor.

$$\begin{aligned}\frac{\mathrm{d}v_{\mathrm{CES}}}{\mathrm{d}T} &= \frac{\mathrm{d}v_{\mathrm{BE}}}{\mathrm{d}T} - \frac{\mathrm{d}v_{\mathrm{BC}}}{\mathrm{d}T} \\ &= \frac{v_{\mathrm{BE}} - (3V_{\mathrm{Thermal}} + E_{\mathrm{G}})}{T} - \frac{v_{\mathrm{BC}} - (3v_{\mathrm{Thermal}} + E_{\mathrm{G}})}{T} \\ &= \frac{v_{\mathrm{BE}}}{T} - \frac{v_{\mathrm{BC}}}{T}.\end{aligned} \tag{4.118}$$

The variation of built-in potential φ of a p–n junction diode with acceptor doping concentration N_{A}, donor doping concentration N_{D} is expressed as

$$\varphi_{\mathrm{bi}} = V_{\mathrm{Thermal}} \ln\left(\frac{N_{\mathrm{A}}N_{\mathrm{D}}}{n_{\mathrm{i}}^{2}}\right), \tag{4.119}$$

where V_{Thermal} is the thermal voltage = 25.9 mV at room temperature and n_i is the intrinsic carrier concentration. Using this equation to express the built-in potentials v_{BE} and v_{BC} of the base–emitter and base–collector diodes, we obtain for an n–p–n transistor

$$\begin{aligned}\frac{dv_{CES}}{dT} &= \frac{V_{\text{Thermal}}}{T}\ln\frac{(N_D)_{\text{Emitter}}(N_A)_{\text{Base}}}{n_i^2} \\ &\quad - \frac{V_{\text{Thermal}}}{T}\ln\frac{(N_A)_{\text{Base}}(N_D)_{\text{Collector}}}{n_i^2} \\ &= \frac{V_{\text{Thermal}}}{T}\ln\left[\left\{\frac{(N_D)_{\text{Emitter}}(N_A)_{\text{Base}}}{n_i^2}\right\}\Big/\left\{\frac{(N_A)_{\text{Base}}(N_D)_{\text{Collector}}}{n_i^2}\right\}\right] \\ &= \frac{V_{\text{Thermal}}}{T}\ln\left\{\frac{(N_D)_{\text{Emitter}}}{(N_D)_{\text{Collector}}}\right\}.\end{aligned} \tag{4.120}$$

At $T = 300$ K, for a transistor with $(N_D)_{\text{Emitter}} = 1 \times 10^{20}\ \text{cm}^{-3}$ and $(N_D)_{\text{Collector}} = 1 \times 10^{15}\ \text{cm}^{-3}$,

$$\frac{dv_{CES}}{dT} = \frac{25.9\,\text{mV}}{300}\ln\left(\frac{1 \times 10^{20}}{1 \times 10^{15}}\right) = 0.086\,333\ln(1 \times 10^{5}) = 0.994\,\text{mVK}^{-1}. \tag{4.121}$$

For a high-voltage power transistor with $(N_D)_{\text{Emitter}} = 1 \times 10^{20}\ \text{cm}^{-3}$ and $(N_D)_{\text{Collector}} = 1 \times 10^{14}\ \text{cm}^{-3}$

$$\frac{dv_{CES}}{dT} = \frac{25.9\,\text{mV}}{300}\ln\left(\frac{1 \times 10^{20}}{1 \times 10^{14}}\right) = 0.086\,333\ln(1 \times 10^{6}) = 1.1927\,\text{mVK}^{-1}. \tag{4.122}$$

Comparing the numerical values of TCs of bipolar junction transistor and p–n junction diode, it is found that the TC of the saturation voltage of a bipolar transistor is lower than the TC of the forward voltage of a diode. Looking at the signs of the TC, it is evident that the TC of the saturation voltage of a bipolar transistor is positive in sign, in contrast to that of the forward voltage of a diode, which was negative in sign. Practically, V_{CEsat} varies at a rate of +0.2% °C^{-1}–0.4% °C^{-1}.

4.4.12 Reverse base and emitter currents of a bipolar junction transistor (I_{CBO} and I_{CEO})

I_{CBO} and I_{CEO} denote collector currents with the collector junction reverse-biased, but in I_{CBO}, the emitter is open-circuited whereas in I_{CEO}, the base is open-circuited. They are inter-related as

$$I_{CEO} = I_{CBO}/(1 - \alpha) = (\beta + 1)I_{CBO} \tag{4.123}$$

where α, β are the current gains of the transistor in the common-base and common-emitter configurations, respectively. Both I_{CBO} and I_{CEO} are temperature-dependent.

I_{CBO} doubles for every 10 °C increment of temperature. For I_{CBO}, the following equation holds

$$I_{CBO}(T) = I_{CBO}(T_0)2^{(T-T_0)/10\,°C}. \tag{4.124}$$

Regarding I_{CEO}, equation (4.123) can be rewritten as

$$I_{CEO} = (\beta + 1)I_{CBO} \approx \beta I_{CBO}. \tag{4.125}$$

A growing exponential function is used to represent the variation of current gain β with temperature. However, within a limited range of temperatures (250–330 K), the nature of the graph is almost linear,

$$\beta(T) = \beta(T_0) + \varsigma\{1 + \beta(T_0)\}(T - T_0), \tag{4.126}$$

where ς is the TC:

$$\varsigma = \frac{1}{\beta(T_0)}\frac{\mathrm{d}\beta}{\mathrm{d}T} = 0.1 - 1.0 \text{ for Si transistors.} \tag{4.127}$$

Hence,

$$I_{CEO}(T) = [\beta(T_0) + \varsigma\{1 + \beta(T_0)\}(T - T_0)]\{I_{CBO}(T_0)2^{(T-T_0)/10\,°C}\}. \tag{4.128}$$

As an example, for a transistor of $\beta(300\text{ K}) = 50$ and $I_{CBO}(300\text{ K}) = 1$ nA, taking $\zeta = 0.5$, I_{CEO} values at $T = 300$ K and $T = 320$ K are:

$$\begin{aligned} I_{CEO}(300) &= [50 + 0.5\{1 + 50\}(300 - 300)]\{1 \times 2^{(300-300)/10\,°C}\} \\ &= 50 \times 1 = 50\,\text{nA} \end{aligned} \tag{4.129}$$

$$\begin{aligned} I_{CEO}(320) &= [50 + 0.5\{1 + 50\}(320 - 300)]\{1 \times 2^{(320-300)/10\,°C}\} \\ &= 560 \times 2^2 = 2240\,\text{nA}. \end{aligned} \tag{4.130}$$

But

$$I_{CBO}(320) = 1 \times 2^{(320-300)/10\,°C} = 1 \times 2^2 = 4\,\text{nA}. \tag{4.131}$$

Thus, I_{CBO} increases only four times but I_{CEO} is multiplied by 2240/50 = 44.8 times on raising the temperature from 300 K to 320 K.

4.4.13 Dynamic response of a bipolar transistor

The switching behavior of a bipolar transistor is determined by the charge stored in the base region. After reversal of the base current, the collector current continues flowing as long as sufficient charge persists in the base region. This charge decays by carrier recombination. Only when the excess charge in the base has been removed, is the base–emitter junction discharged so that the transistor is turned off. Thus, the minority-carrier lifetime in the base plays a vital role in transistor switching, as in a p–n junction diode. An increase in temperature is associated with a longer turn-off time and hence slower turn-off, which degrades transistor performance (table 4.6).

Table 4.6. Enhancing and degrading effects of temperature on device parameters.

Sl. No.	Device	Parameter	Effect of decreasing temperature	Remarks	Effect of increasing temperature	Remarks
1	Diode	(a) Forward voltage drop	Increases	Detrimental	Decreases	Beneficial
		(b) Reverse leakage current	Decreases	Beneficial	Increases	Detrimental
		(c) Breakdown voltage	Decreases	Detrimental	Increases	Beneficial
		(d) Storage time	Decreases	Beneficial	Increases	Detrimental
2	Bipolar transistor	(a) Current gain	Decreases	Detrimental	Increases	Beneficial
		(b) Saturation voltage	Decreases	Beneficial	Increases	Detrimental
		(c) Reverse breakdown voltages of E–B and C–B junctions	Decrease	Detrimental	Increase	Beneficial
		(d) Leakage currents	Decrease	Beneficial	Increase	Detrimental
		(e) Turn-off time	Decreases	Beneficial	Increases	Detrimental

4.5 Bipolar analog circuits in the 25 °C–300 °C range

The process and design considerations of a quad operational amplifier (OP-AMP) working in the 25 °C–300 °C range were described by Beasom and Patterson (1982). The fabrication process employed dielectric isolation in place of conventional junction isolation. A vertical p–n–p transistor structure complemented the high-performance lateral n–p–n device for operational stability over a broad temperature range. Vulnerability to interconnection pitting was avoided by using deeper (3.5 μm n-emitter and 4.5 μm p-emitter) structures.

Before embarking on the actual experiment, the researchers performed prior characterization of circuit components from 25 °C to 300 °C to ascertain the key issues and to decide what suitable corrective measures could be applied. Obviously, the parameter playing the pivotal role in a high-temperature circuit design is the leakage current. To this end, the characteristics of n–p–n and p–n–p transistors were measured at 25 °C, 200 °C and 300 °C. The characteristics at 300 °C were found to be offset by $I_B = 2$ μA for p–n–p and $I_B = 1.5$ μA for n–p–n transistors. The offset occurred because the collector–base leakage current I_{CBO} flows in the opposite direction to the base current I_B. As temperature increases from 200 °C to 300 °C, I_{CBO} increases to the extent that it exceeds the intrinsic base current I_B. At the same time, the incremental current gain h_{FE} is found to increase. Another noteworthy effect is the reduction in base–emitter voltage V_{BE} with increasing temperature. The V_{BE} change followed the rate −2 mV °C^{-1} up to 300 °C, at which V_{BE} was typically 50–100 mV. The diffused resistor values increased linearly with positive TC of resistance (TCR). For both p–n–p and n–p–n transistors, the unity-gain frequency f_T decreased to one-half at 300 °C from its value at 25 °C.

Particular emphasis was placed on the performance of aluminum metallization during heating from room temperature to 300 °C. The following facts were noted:

(i) No failures were observed in Al test structures up to 500 h at 325 °C and 3.3×10^4 A cm^{-2}.

(ii) No serious changes took place in contact resistances in p^+- and n^+-doped Si test structures up to 500 h at 325 °C.

(iii) No deterioration was noticed in Al-metallized transistors when they were kept at 325 °C for 500 h after applying $V_{CB} = 30$ V at $I_C = 1$ mA.

Learning from the above observations, circuit design adopted the following rules: (i) diode-connected transistors were considered unviable due to the low voltage magnitudes of forward-biased junctions; (ii) because of the reversal of the base current, the base voltage nodes for chains of current sources were provided with capabilities of current sourcing and sinking; and (iii) the values of diffused resistors at 300 °C were chosen as double their values at 25 °C to offset changes in forward junction voltages with temperature.

A combination of linear dependences of a Zener diode, a diffused resistor and several base–emitter voltages was exploited to obtain a stable current from the bias network over a wide temperature range. The circuit, consisting of a Zener diode in parallel with four temperature compensation diodes and a resistor, supplied an output current varying by < ±0.2 fraction in the −55 °C to +300 °C temperature range.

The currents produced by the bias network are coupled to the positive and negative supply rails through transistors, which are repeated in each of the four amplifiers of QUAD OP-AMP. The emitter resistors of these transistors provide negative feedback to supply the source and sink currents of each amplifier to counter the temperature gradients caused by the inequality of the power dissipations of the output stages. Biasing resistors of correct values are connected to transistors which are likely to saturate at high temperatures when the collector resistance increases and V_{BE} decreases. Provision is made for a sourcing/sinking current to/from the positive/negative bias lines upon reversal of direction of the base current. Proper scaling of the geometries of leakage compensation diodes is performed to match the sum of areas of sourcing/sinking devices.

In the input and gain stages, p–n–p transistors are favored as the input pair because their lower I_{CBO} results in smaller input bias and offset currents. Input currents at high temperatures are brought down by as much as 5–10 times by I_{CBO} compensation transistors. The collectors of the input pair of p–n–p transistors convey the signal to the grounded base of the n–p–n transistor pair, which translates the signal to the current mirror and to the high impedance point where voltage gain is produced. Protection against reversal of base current is provided by two p–n–p transistors. A Zener diode reduces the effect of output conductance on the offset voltage. This reduction has a major impact at high temperatures when the collector–base leakage current increases.

In the output stage, the voltage generated at the high impedance point is transmitted to the output through pairs of emitter followers. I_{CBO} is compensated through two transistors. This compensation reduces the leakage current which is reflected to the input as offset voltage.

The main performance parameters of the IC OP-AMP were presented by Beasom and Patterson (1982) for the temperature range −55 °C to +300 °C. The power supply current variation was <10% except at 300 °C. The influence of the leakage current at 300 °C was manifested in the form of an increase in the offset voltage and the input offset current. The decrease in the open-loop gain at 300 °C was caused by degradation of the output conductance. The gain bandwidth product decreased with increasing temperature but still remained higher at 300 °C than that of several OP-AMPs at 25 °C. This is ascribed to the use of vertical p–n–p transistors with $f_T = 100$ MHz in the circuit. A very low value of input noise (=4.2 nV $\mathrm{Hz}^{-1/2}$) was achieved.

4.6 Bipolar digital circuits in the 25 °C–340 °C range

Prince *et al* (1980) carried out extensive studies of the DC and dynamic characteristics of bipolar digital ICs in the range of temperatures from 25 °C to 340 °C. Transistor–transistor logic (TTL) is a type of saturating logic. In this logic, gold doping is used to decrease the storage time through reduction of the minority-carrier lifetime. As an alternative, a Schottky barrier clamping diode is connected between the base and collector terminals. Saturation does not take place because a Schottky diode has a lower forward voltage than that of the collector–base junction. Excess base current is routed through the Schottky diode. As the Schottky diode is a majority carrier device, it has negligible storage of minority carriers. Using Schottky diode-clamped transistors, storage times of 1–2 ns are obtained. These values are lower than 5–10 ns for gold-doped devices. Further, the gain of non-gold-doped transistors is higher than that of gold-doped ones.

Prince *et al* (1980) noted a difference between gold-doped TTL and Schottky-clamped TTL. The former remained functional up to 250 °C, whereas the latter worked up to 325 °C. The ICs tested consisted of two-input and four-input NAND gates. Apart from the lower temperature limits of gold-doped devices and the scaling behaviors of some parameters, both gold-doped and Schottky-clamped TTL devices performed equally well at high temperatures. The failure was observed with respect to noise margins. In particular, the output high-voltage V_{OH} decreased as the temperature was increased. By analyzing the voltage transfer characteristics of the ICs as a function of temperature, it was found that the primary reason for diminution in V_{OH} was the leakage current across the collector–base junction of the phase splitter transistor. This current flows through the phase splitter collector resistor. Failure mechanisms at high temperature were revealed by inspecting the behaviors of the power supply current I_{CCH} for output high conditions and output high voltage versus output high sourcing current for the IC, and correlating these data with the behavior of input low current I_{IL} at high temperatures. It was noticed that a large portion of the high-temperature increase in I_{CCH} could be explained by the increase in I_{IL} at zero input voltage. This increase in current is produced in the phase splitter transistor. It appears as an additional component of I_{IL}. This current flows through the collector resistor of the phase splitter transistor. The voltage drop resulting from this leakage current flow decreases V_{OH} because at medium output and supply currents, V_{OH} tracks the voltage drop across the collector resistor. Thus the leakage current and resistor values combine to yield an upper bound on functionality for Schottky-clamped TTL devices in the neighborhood of 325 °C.

The ICs showed a fan-out capability >1 at temperatures close to maximum operating temperatures. The current sourcing capability decreased owing to the increase in circuit resistance at high temperatures. The current sinking capability increased through augmentation of the current sink transistor gain at these temperatures.

These experiments led to the view that bipolar TTL logic gates can function properly at temperatures near 325 °C.

4.7 Discussion and conclusions

Data sheets of semiconductor components such as diode and bipolar transistors provide the specifications of the devices at a particular temperature. These values of the components listed at a given temperature are liable to vary with changes in temperature. Since the components rarely operate at the extreme temperature mentioned in the data sheets, it is important to know how their values vary at the actual working temperature in an application. In several cases, it is difficult to obtain the correct values as the procedure may require computer simulations. However, in many cases, the trends or directions of changes can be guessed from simple analytical models. The intent of this chapter was to give an overview of the available models in the literature which could help the reader in making a first-cut speculation.

The changes in the electrical parameters of discrete devices impact the performance of circuits fabricated using them as components. Examples showing the thermal effects of component parameter changes on circuit behavior were presented. Often, the temperature-dependent behavior can be pre-considered at the design stage and the necessary correction applied to ensure proper functioning at high temperatures.

Review exercises

4.1. What is the upper temperature bound on silicon devices? Does this bound imply that a silicon device stops functioning at this temperature? How far is this bound stretchable by suitable techniques?

4.2. Is silicon a metal or a non-metal or does it belong to some other class of materials? What is the resistivity of silicon in its intrinsic state? What is its intrinsic carrier concentration? What is, typically, the ratio of electron mobility/hole mobility for silicon?

4.3. Calculate the intrinsic temperature of silicon for a doping concentration of 1×10^{15} cm^{-3}. Assume that the effective masses of electrons and holes as well as the bandgap of silicon do not vary with temperature.

4.4. How is EG polysilicon obtained from MG polysilicon? Name two techniques used for growth of single-crystal silicon from EG polysilicon. Discuss their main features.

4.5. How is a silicon wafer doped as n- or p-type using the thermal diffusion process? Describe with relevant chemical equations.

4.6. How does ion implantation differ from thermal diffusion of impurities? Name the dopant gases commonly used. What are the typical ranges of ion energies and doses used in ion implantation for fabrication of semiconductor devices? Why is the ion implantation process followed by a thermal annealing step?

4.7. What are the main techniques used for depositing contact metals on semiconductor devices? What material is commonly used to form an ohmic contact on heavily doped n-type silicon? What type of contact will be formed if n-type silicon is lightly doped?

4.8. How are the different components in an IC separated by the p–n junction isolation technique? What are the advantages and disadvantages of this technique?

4.9. How are components of an IC separated by the dielectric isolation technique? What type of wafer is used for IC fabrication if this technique is adopted? What are the special features of these wafers? How do these wafers provide dielectric isolation?

4.10. Write down the Shockley equation for the current–voltage characteristics of a p–n diode, and explain the meanings of the symbols used. What is the significance of the emission coefficient η? What are the typical values of η? What are the factors on which its value depends?

4.11. Why is the expression $(k_B T/q)$ referred to as thermal voltage? What is the value of thermal voltage at room temperature?

4.12. Does the reverse saturation current I_S of a diode vary with temperature T? If so, how? Write the relevant equation. Show that

$$\frac{\mathrm{d}I_S}{\mathrm{d}T} = I_S\left(\frac{3}{\eta T} + \frac{E_G}{\eta T V_{\mathrm{Thermal}}}\right), \tag{4.132}$$

where η is the emission coefficient, E_G is the bandgap and V_{Thermal} is the thermal voltage.

4.13. If E_G denotes the bandgap and V_{Thermal} the thermal voltage, show that the TC $\mathrm{d}v/\mathrm{d}T$ of forward voltage drop across a p–n diode is given by

$$\frac{\mathrm{d}v}{\mathrm{d}T} = \frac{v - (3V_{\mathrm{Thermal}} + E_G)}{T}. \tag{4.133}$$

4.14. Prove that the TC $\mathrm{d}V_d/\mathrm{d}T$ of a Schottky diode is expressed as

$$\frac{\mathrm{d}V_d}{\mathrm{d}T} = \frac{V_d}{T} - \left(\frac{k_B T}{q}\right)\left(\frac{1}{T}\right)\left\{2 + \left(\frac{q\phi_B}{k_B T}\right)\right\}, \tag{4.134}$$

where ϕ_B is the Schottky barrier height.

4.15. If I_{S1}, I_{S2} are the reverse leakage currents of a p–n diode at temperature T, $T + \Delta T$, η is the emission coefficient, E_G is the bandgap and V_{Thermal} is thermal voltage, show that

$$\frac{I_{S2}}{I_{S1}} = \exp\left\{\left(\frac{3}{\eta T} + \frac{E_G}{\eta T V_{\text{Thermal}}}\right)\Delta T\right\}. \tag{4.135}$$

Taking $\eta = 2$, prove that for silicon diodes the leakage current doubles for every 10 °C rise in temperature.

4.16. What is the critical electric field of silicon? What happens at this field? Explain the phenomenon of impact ionization.

4.17. What is meant by the ionization coefficients of electrons and holes? Write the equation for the ionization integral. What is its value at the critical electric field?

4.18. What is the ionization threshold energy? How is it related to the energy bandgap? Write the Cromwell–Sze equation for the average carrier mean free path.

4.19. Explain why the breakdown voltage of a p–n junction decreases with temperature on the basis of the influence of temperature on ionization threshold energy, energy loss by scattering, the mean free path of carriers and the frequency of scattering events.

4.20. Write the Chynoweth equations for ionization coefficients α_n, α_p of electrons and holes. How did Shockley interpret the fitting parameter b in these equations?

4.21. Derive the following equation for TC of ionization coefficient

$$\text{TC} = \left(\frac{W_I}{qE\lambda}\right)\left\{\frac{1}{\lambda}\left(\frac{d\lambda}{dT}\right) - \frac{1}{W_I}\left(\frac{dW_I}{dT}\right)\right\}, \tag{4.136}$$

where W_I is the ionization threshold energy, E is the intensity of the lateral electric field, λ is the mean free path of an electron, q is the electronic charge and T is the temperature on the Kelvin scale. From the TCs of λ and W_I express the above equation in the form

$$\text{TC} = -\left(\frac{1}{E}\right)\left[\left(\frac{W_I\lambda_0 E_p}{2q\lambda^2 k_B T^2}\right)\text{sech}^2\left(\frac{E_p}{2k_B T}\right) + \left(\frac{1}{q\lambda}\right)(C_2 + 2C_3 T)\right], \tag{4.137}$$

where λ_0 is the high-energy low-temperature asymptotic phonon mean free path, E_p is the average energy loss due to optical phonon scattering, k_B is Boltzmann's constant; and C_2, C_3 are constants in the Ershov–Ryzhii model.

4.22. Distinguish between the avalanche and Zener breakdown mechanisms of a p–n junction diode with regard to the TCs. Elaborate the reasons for the opposite behaviors of the two mechanisms.

4.23. Why does the storage time of a p^+–n diode increase with an increase of temperature? What is the effect on the switching loss?

4.24. How is the current gain of a bipolar transistor in the common-emitter configuration (β) related to that in the common-base connection (α)?

4.25. Define the following terms for a bipolar transistor: (a) the emitter injection efficiency γ, (b) the base transport factor α_T, and (c) the collector multiplication ratio M. How is the common-base current gain α of a bipolar transistor related to γ, α_T and M?

4.26. Explain with reasons the effect of temperature on the term $\{1 + (D_{pE}/D_{nB})\}$ in the equation for common-base current gain α. Here, D_{pE} denotes the diffusion constant of holes in the n$^+$-emitter and D_{nB} represents the diffusion constant of electrons in the p-base.

4.27. Show that for a bipolar transistor, the ratio of the number of holes in the n$^+$-emitter (p_{nE}) to the number of electrons in the p-base (n_{pB}) is given by

$$p_{nE}/n_{pB} = K\exp\{\Delta\xi_g/(2k_BT)\}, \tag{4.138}$$

where K is a constant, $\Delta\xi_g$ is the bandgap reduction of the emitter caused by heavy doping-induced defects, k_B is Boltzmann's constant and T is temperature. Using this equation, explain the significant reduction in current gain with a decrease of temperature due to the effect of temperature on the ratio of minority-carrier concentrations.

4.28. Discuss the effect of temperature on the ratio of the base width to the diffusion length of minority carriers in the emitter (W/L_{pE}). Hence explain the decrease in current gain with a decrease in temperature.

4.29. If h_{fE} is the common-emitter current gain of a BJT at a temperature T, h_{fE0} is the common-emitter current gain of a BJT at a temperature $T = 0$ K, and $\Delta\xi_{gBE} = \xi_{gB} - \xi_{gE}$ = the difference in bandgaps between the emitter and base, show that

$$h_{fE} = h_{fE0}\exp\left(-\frac{\Delta\xi_{gBE}}{k_BT}\right). \tag{4.139}$$

Hence argue that the reduction in bandgap of emitter with respect to that of base ($\Delta\xi_{gBE}$) is the main reason responsible for the pronounced deterioration in current gain of a bipolar transistor with falling temperature.

4.30. Show that the TC of the saturation voltage of a bipolar transistor is given by

$$\frac{dv_{CES}}{dT} = \frac{V_{Thermal}}{T}\ln\left\{\frac{(N_D)_{Emitter}}{(N_D)_{Collector}}\right\}, \tag{4.140}$$

where $V_{Thermal}$ is the thermal voltage, T is the temperature, and $(N_D)_{Emitter}$ and $(N_D)_{Collector}$ are the donor doping concentrations of emitter and collector, respectively.

4.31. What do I_{CBO} and I_{CEO} of a bipolar transistor stand for? Which is more sensitive to changes of ambient temperature?

4.32. Describe, after Beasom and Patterson, the process and design considerations of a quad operational amplifier (OP-AMP) working in the 25 °C–300 °C range.

4.33. What experiments led to the inference that bipolar TTL logic gates can function properly at temperatures near 325 °C?

References

Beasom J D and Patterson R B 1982 Process characteristics and design methods for a 300 °C quad operational amplifier *IEEE Trans. Ind. Electron.* **29** 112–7

Buhanan D 1969 Investigation of current-gain temperature dependence in silicon transistors *IEEE Trans. Electron Devices* **16** 117–24

Chang C Y, Chiu S S and Hsu L P 1971 Temperature dependence of breakdown voltage in silicon abrupt p–n junctions *IEEE Trans. Electron Devices* **18** 391–3

Chynoweth A G 1958 Ionization rates for electron and holes in silicon *Phys. Rev.* **109** 1537

Cory R 2009 Schottky diodes *Skyworks Solutions* February, pp 1–5 www.skyworksinc.com/downloads/press_room/published_articles/MPD_022009.pdf

Crowell C R and Sze S M 1966 Temperature dependence of avalanche multiplication in semiconductors *Appl. Phys. Lett.* **92** 242–4

Dokić B L and Blanuša B 2015 Diodes and transistors *Power Electronics: Converters and Regulators* (Cham: Springer) ch 4 pp 141–3

Ershov M and Ryzhii V 1995 Temperature dependence of the electron impact ionization coefficient in silicon *Semicond. Sci. Technol.* **10** 138–42

Fisher G, Seacrist M R and Standley R W 2012 Silicon crystal growth and wafer technologies *Proc. IEEE* **100** 1454–74

Godse A P and Bakshi U A 2009 *Basic Electronics* vol 1 (Pune: Technical Publications), pp 3–30

Friedrich J, von Ammon W and Müller G 2015 Czochralski growth of silicon crystals *Handbook of Crystal Growth* ed P Rudolph 2nd edn (Amsterdam: Elsevier) ch 2 pp 45–104

Griffin P B, Deal M D and Plummer J D 2014 *Silicon VLSI Technology* (London: Pearson College Division)

Hu X 2021 Research on silicon wafer manufacturing process and physical properties testing using high-purity polysilicon *J. Phys.: Conf. Ser.* **2083** 022050

Jones S W 2008 *Diffusion in Silicon* (Georgetown, MA: IC Knowledge LLC), pp 1–68 http://www-eng.lbl.gov/~shuman/NEXT/MATERIALS&COMPONENTS/Xe_damage/Diffusionin%20siliconpdf.pdf

Kauffman W L and Bergh A A 1968 The temperature dependence of ideal gain in double diffused silicon transistors *IEEE Trans. Electron Devices* **15** 732–5

Leach W M 2004 The junction diode *Lecture Notes* ch 2 http://users.ece.gatech.edu/mleach/ece3040/notes/chap02.pdf

Madou M J 2002 *Fundamentals of Microfabrication: The Science of Miniaturization* (Boca Raton, FL: CRC Press)

Maes W, De Meyer K and Van Overstraeten R 1990 Impact ionization in silicon: a review and update *Solid-State Electron.* **33** 705–18

Muiznieks A, Virbulis J, Lüdge A, Riemann H and Werner N 2015 Floating zone growth of silicon *Handbook of Crystal Growth* ed P Rudolph 2nd edn (Amsterdam: Elsevier) ch 7 pp 241–79

Mukherjee M 2011 SiC devices on different polytypes: prospects and challenges *Silicon Carbide—Materials, Processing and Applications in Electronic Devices* ed M Mukherjee (Rijeka: InTech) pp 337–68

Nastasi M and Mayer J W 2006 Doping, diffusion and defects in ion-implanted Si *Ion Implantation and Synthesis of Materials* (Berlin: Springer) pp 107–26

Pandey K, Sharma A and Singh A K 2022 Silicon wafers; its manufacturing processes and finishing techniques: an overview *Silicon* **14** 12031–47

Prince J L, Draper B L, Rapp E A, Kronberg J N and Fitch L T 1980 Performance of digital integrated circuit technologies at very high temperatures *IEEE Trans. Compon. Hybrids Manuf. Technol.* **3** 571–9

Reisch M 2003 *High-Frequency Bipolar Transistors: Physics, Modeling, Applications* (Berlin: Springer) pp 149–50

Ritala M and Niinistö J 2009 Atomic layer deposition *Chemical Vapor Deposition: Precursors, Processes and Applications* ed A C Jones and M L Hitchman (London: Royal Society of Chemistry) ch 4 pp 158–206

Sinha A K 1981 Refractory metal silicides for VLSI applications *J. Vac. Sci. Technol.* **19** 778

Stork J M C, Harame D L, Meyerson B S and Nguyen T N 1987 High performance operation of silicon bipolar transistors at liquid nitrogen temperature *IEEE–IEDM Tech. Dig.* pp 405–8

Tedrow P K and Reif R 1994 Plasma-enhanced chemical vapor deposition *ASM Handbook, Volume 5: Surface Engineering* ed D M Mattox (Novelty, OH: ASM International) pp 532–7

Vergara-Irigaray N, Riesen M, Piazza G, Bronk L, Driessen W and Edwards J *et al* 2012 Low-pressure chemical vapor deposition (LPCVD) *Encyclopedia of Nanotechnology* ed B Bhushan (Dordrecht: Springer Science+Business Media B.V.)

Waldron N S, Pitera A J, Lee M L, Fitzgerald E A and del Alamo J A 2005 Positive temperature coefficient of impact ionization in strained-Si *IEEE Trans. Electron Devices* **52** 1627–33

Zant P V 2014 *Microchip Fabrication: A Practical Guide to Semiconductor Processing* 6th edn (New York: McGraw-Hill)

Zhang L 2014 Silicon process and manufacturing technology evolution: an overview of advancements in chip making *IEEE Consum. Electron. Mag.* **3** 44–8

Chapter 5

Temperature dependence of electrical characteristics of silicon MOS devices and circuits

Depending on the gate voltage, a MOSFET can be operated as a negative, zero or positive TC (temperature coefficient) device. This kind of MOSFET behavior owes its origin to the presence of opposing factors in its conduction mechanism, mainly the increase in carrier concentration with temperature being counterbalanced by the decrease in carrier mobility. At large gate bias values, when the drain current increases to excessively high magnitudes, it is the decrease in carrier mobility which helps in thwarting thermal runaway, and thus serves as a built-in protective mechanism. In this chapter, the effects of temperature on the critical electrical parameters of a MOSFET, e.g., threshold voltage, on-resistance, transconductance and breakdown voltages are examined. The effects of temperature on the dynamical response of the MOSFET are also explored. As the temperature increases or descends below room temperature, some parameters are adversely affected while changes in others are encouraging. Therefore, the variation of MOSFET parameters with temperature impacts the performance of MOS analog and digital circuits in both directions. The decrease in the latching susceptibility of CMOS circuits at low temperatures is a significant advantage derived by MOS circuit operation in a cool environment.

doi:10.1088/978-0-7503-5072-3ch5

5.1 Introduction

We begin by considering how temperature impacts the operation of a MOSFET in comparison to a bipolar transistor (Arora 2012, Siu 2022). In a bipolar transistor of common-emitter current gain β, the collector current I_C is expressed in terms of base current I_B and collector–base leakage current I_{CBO} with the emitter open as

$$I_C = \beta I_B + (\beta + 1)I_{CBO}. \tag{5.1}$$

The leakage current I_{CBO} is strongly temperature-dependent, doubling for every 10 °C increase of temperature. As the temperature increases, I_{CBO} increases and collector current I_C climbs appreciably by a factor of $(\beta + 1)\, I_{CBO}$. Consequent upon this steep rise in collector current I_C, the transistor is heated, increasing I_C still further through the heating effect. This regenerative or reinforcing cycle progresses cumulatively until I_C increases dangerously and the transistor burns out. This self-destruction of the transistor is known as thermal runaway.

Temperature affects the performance of a MOSFET quite differently from a bipolar transistor because the MOSFET is inherently thermally stable and not prone to the uncontrolled positive feedback mechanism, i.e., thermal runaway, which takes place as the temperature or drain current increases (Tsividis and Colin 2012, Suzuki 2016). Figure 5.1 shows the cross-section and circuit symbols of a MOS transistor.

5.2 Threshold voltage of an n-channel enhancement-mode MOSFET

The threshold voltages of n-channel and p-channel MOSFETs change in opposite directions with an increase of temperature, as shown in figure 5.2.

For an n-channel enhancement-mode MOS transistor, the formula for threshold voltage is (Wang *et al* 1971)

$$V_{Th} = V_{FB} + 2\phi_F + \frac{\sqrt{2\varepsilon_0\varepsilon_s q N_A(2\phi_F + V_{SB})}}{C_{ox}}, \tag{5.2}$$

where V_{FB} = flat-band voltage, ϕ_F = bulk potential, ε_0 = free-space permittivity, ε_s = dielectric constant of silicon, q = electronic charge, N_A = acceptor doping concentration, V_{SB} = substrate bias and C_{ox} = oxide capacitance per unit area. Also,

$$V_{FB} = \phi_{ms} - \frac{Q_f}{C_{ox}} - \frac{1}{\varepsilon_0\varepsilon_{ox}}\int_0^{t_{ox}} \rho_{ox}(x)x\,dx, \tag{5.3}$$

where ϕ_{ms} is the metal–semiconductor work function, Q_f is the fixed charge in the oxide, ε_{ox} is the relative permittivity of silicon dioxide, $\rho_{ox}(x)$ is the charge density in an oxide of thickness t_{ox} varying with distance x in the oxide. For an n-type polysilicon gate (STMicroelectronics 2006)

$$\phi_{ms} = -\left(\frac{k_B T}{q}\right)\ln\left(\frac{N_{D(g)}N_A}{n_i^2}\right), \tag{5.4}$$

where k_B is the Boltzmann constant, T is the temperature (K), q is electronic charge, $N_{D(g)}$ is the doping concentration of the polysilicon gate, N_A is the acceptor

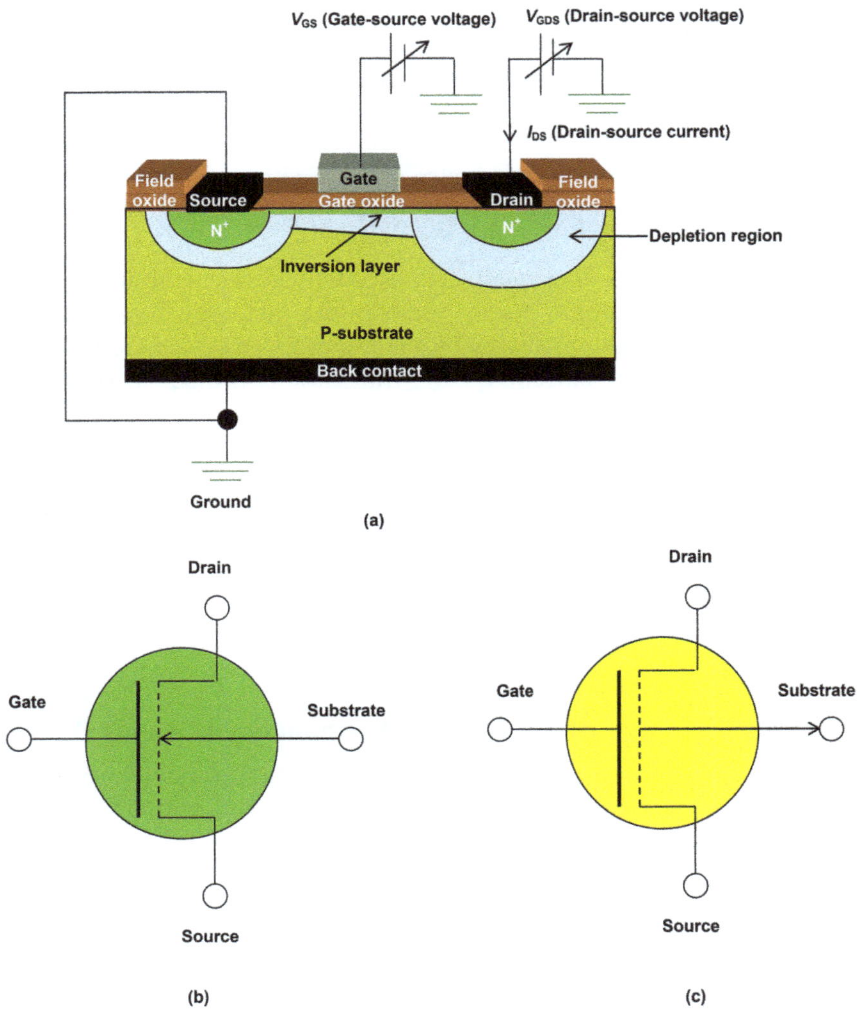

Figure 5.1. MOSFET. (a) Cross-sectional diagram of an n-channel enhancement-mode MOSFET, (b) a circuit diagram symbol of an n-channel device and (c) the symbol for a p-channel device.

concentration of the semiconductor and n_i is the intrinsic carrier concentration. For the p-substrate,

$$\phi_F = \left(\frac{k_B T}{q}\right) \ln \left(\frac{N_A}{n_i}\right). \tag{5.5}$$

From equations (5.2) and (5.3), the threshold voltage can be written in approximate form as

$$V_{Th} \approx \phi_{ms} + 2\phi_F + \frac{\sqrt{2\varepsilon_0\varepsilon_s q N_A(2\phi_F + 0)}}{C_{ox}} = \phi_{ms} + 2\phi_F + \frac{2\sqrt{\varepsilon_0\varepsilon_s q N_A \phi_F}}{C_{ox}}, \tag{5.6}$$

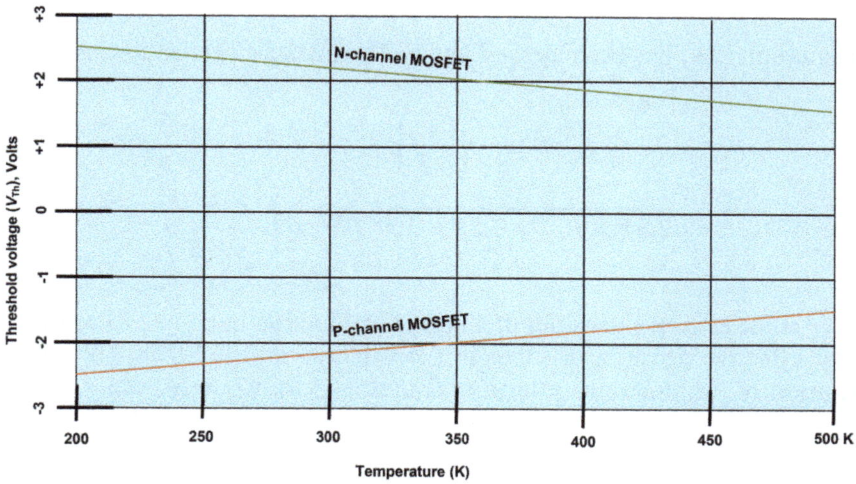

Figure 5.2. Changes in the threshold voltages of n-channel and p-channel MOSFETs with temperature.

where it has been assumed that oxide charges are negligible and V_{SB} has been taken as zero.

Differentiating both sides of equation (5.6) with respect to temperature,

$$\begin{aligned}\frac{\partial V_{Th}}{\partial T} &= \frac{\partial \phi_{ms}}{\partial T} + 2\left(\frac{\partial \phi_F}{\partial T}\right) + 2 \times \frac{1}{2}\frac{(\varepsilon_0\varepsilon_s q N_A \phi_F)^{1/2-1}}{C_{ox}} \times (\varepsilon_0\varepsilon_s q N_A) \times \frac{\partial \phi_F}{\partial T} \\ &= \frac{\partial \phi_{ms}}{\partial T} + 2\left(\frac{\partial \phi_F}{\partial T}\right) + \frac{(\varepsilon_0\varepsilon_s q N_A \phi_F)^{-1/2}}{C_{ox}} \times (\varepsilon_0\varepsilon_s q N_A) \times \frac{\partial \phi_F}{\partial T} \\ &= \frac{\partial \phi_{ms}}{\partial T} + 2\left(\frac{\partial \phi_F}{\partial T}\right) + \left(\frac{1}{C_{ox}}\right)\sqrt{\frac{\varepsilon_0\varepsilon_s q N_A}{\phi_F}}\left(\frac{\partial \phi_F}{\partial T}\right) \\ &= \frac{\partial \phi_{ms}}{\partial T} + \left(\frac{\partial \phi_F}{\partial T}\right)\left\{2 + \left(\frac{1}{C_{ox}}\right)\sqrt{\frac{\varepsilon_0\varepsilon_s q N_A}{\phi_F}}\right\}.\end{aligned} \tag{5.7}$$

Thus the TC of threshold voltage V_{Th} depends on those of the metal–semiconductor work function ϕ_{ms} and bulk potential ϕ_F. Now from equation (5.4),

$$\begin{aligned}\frac{\partial \phi_{ms}}{\partial T} &= -\frac{\partial}{\partial T}\left[\left(\frac{k_B T}{q}\right)\ln\left(\frac{N_{D(g)}N_A}{n_i^2}\right)\right] \\ &= -\left(\frac{k_B}{q}\right)\ln\left(\frac{N_{D(g)}N_A}{n_i^2}\right) - \left[\left(\frac{k_B T}{q}\right)\frac{1}{\left(\frac{N_{D(g)}N_A}{n_i^2}\right)}\left(N_{D(g)}N_A\right) \times -2n_i^{-2-1} \times \frac{\partial n_i}{\partial T}\right] \\ &= -\{(k_B)/q\}\ln\left(\frac{N_{D(g)}N_A}{n_i^2}\right) - \left[\left(\frac{k_B T}{q}\right) \times n_i^2 \times -2n_i^{-3} \times \frac{\partial n_i}{\partial T}\right] \\ &= -\{(k_B)/q\}\ln\left(\frac{N_{D(g)}N_A}{n_i^2}\right) + 2\left(\frac{k_B T}{q n_i}\right)\left(\frac{\partial n_i}{\partial T}\right) \\ &= \frac{\phi_{ms}}{T} + \left(\frac{2k_B T}{q n_i}\right)\left(\frac{\partial n_i}{\partial T}\right),\end{aligned} \tag{5.8}$$

where equation (5.4) has been applied again. The intrinsic carrier concentration n_i is

$$\begin{aligned} n_i &= \sqrt{N_C N_V}\, T^{3/2} \exp\left(-\frac{E_g}{2k_B T}\right) \\ &= \sqrt{2.81 \times 10^{19} \times 1.83 \times 10^{19}}\, T^{3/2} \exp\left(-\frac{E_g}{2k_B T}\right), \end{aligned} \tag{5.9}$$

where N_C is the effective densities of states in conduction band ($=2.81 \times 10^{19}$ cm^{-3}), N_V is the effective densities of states in the valence band ($=1.83 \times 10^{19}$ cm^{-3}). On computation of the square root term in equation (5.9), we have

$$n_i = 2.26766 \times 10^{19} T^{3/2} \exp\left(-\frac{E_g}{2k_B T}\right) = KT^{3/2} \exp\left(-\frac{E_g}{2k_B T}\right), \tag{5.10}$$

where $K = 2.267\,66 \times 10^{19}$ cm^{-3}. From equation (5.10),

$$\begin{aligned} \frac{\partial n_i}{\partial T} &= \frac{\partial}{\partial T}\left\{KT^{3/2} \exp\left(-\frac{E_g}{2k_B T}\right)\right\} \\ &= K \times \frac{3}{2} \times T^{3/2-1} \exp\left(-\frac{E_g}{2k_B T}\right) \\ &\quad + KT^{3/2} \exp\left(-\frac{E_g}{2k_B T}\right) \times \left(-\frac{E_g}{2k_B}\right) \times -T^{-1-1} \\ &= K \times \frac{3}{2} \times T^{1/2} \exp\left(-\frac{E_g}{2k_B T}\right) + n_i \times \frac{E_g}{2k_B} \times T^{-2} \\ &= \frac{3}{2T} KT^{3/2} \exp\left(-\frac{E_g}{2k_B T}\right) + \frac{n_i E_g}{2k_B T^2} \\ &= \frac{3n_i}{2T} + \frac{n_i E_g}{2k_B T^2} = \left(\frac{n_i}{2T}\right)\left(3 + \frac{E_g}{k_B T}\right), \end{aligned} \tag{5.11}$$

where equation (5.10) for n_i has been re-applied. Substituting for $\partial n_i/\partial T$ from equation (5.11) into equation (5.8),

$$\frac{\partial \phi_{ms}}{\partial T} = \frac{\phi_{ms}}{T} + \left(\frac{2k_B T}{qn_i}\right)\left(\frac{n_i}{2T}\right)\left(3 + \frac{E_g}{k_B T}\right) = \frac{\phi_{ms}}{T} + \left(\frac{k_B}{q}\right)\left(3 + \frac{E_g}{k_B T}\right). \tag{5.12}$$

Looking at $\partial \phi_F/\partial T$, we find from equation (5.5)

$$
\begin{aligned}
\frac{\partial\phi_F}{\partial T} &= \frac{\partial}{\partial T}\left\{\left(\frac{k_B T}{q}\right)\ln\left(\frac{N_A}{n_i}\right)\right\} = \left(\frac{k_B}{q}\right)\ln\left(\frac{N_A}{n_i}\right) + \left(\frac{k_B T}{q}\right) \\
&\quad \times \frac{1}{\left(\frac{N_A}{n_i}\right)} \times N_A \times -n_i^{-1-1} \times \frac{\partial n_i}{\partial T} \\
&= \left(\frac{k_B}{q}\right)\ln\left(\frac{N_A}{n_i}\right) - \left(\frac{k_B T}{q n_i}\right) \times \frac{\partial n_i}{\partial T} \\
&= \left(\frac{k_B}{q}\right)\ln\left(\frac{N_A}{n_i}\right) - \left(\frac{k_B T}{q n_i}\right) \times \left(\frac{n_i}{2T}\right)\left(3 + \frac{E_g}{k_B T}\right) \\
&= \left(\frac{k_B}{q}\right)\ln\left(\frac{N_A}{n_i}\right) - \left(\frac{k_B}{q}\right) \times \left(\frac{1}{2}\right)\left(3 + \frac{E_g}{k_B T}\right) \\
&= \frac{1}{T}\left\{\left(\frac{k_B T}{q}\right)\ln\left(\frac{N_A}{n_i}\right)\right\} - \frac{3}{2}\left(\frac{k_B}{q}\right) - \frac{E_g}{2qT} \\
&= \frac{\phi_F}{T} - \frac{3}{2}\left(\frac{k_B}{q}\right) - \frac{E_g}{2qT},
\end{aligned} \tag{5.13}
$$

where equations (5.11) and (5.5) for $\partial n_i/\partial T$ and ϕ_F have been used.

The temperature dependence of threshold voltage is obtained by substituting the expressions for $\partial\phi_{ms}/\partial T$ and $\partial\phi_F/\partial T$ from equations (5.12) and (5.13), respectively, into equation (5.7) for $\partial V_{Th}/\partial T$

$$
\begin{aligned}
\frac{\partial V_{Th}}{\partial T} &= \frac{\phi_{ms}}{T} + \left(\frac{k_B}{q}\right)\left(3 + \frac{E_g}{k_B T}\right) + \left\{\frac{\phi_F}{T} - \frac{3}{2}\left(\frac{k_B}{q}\right) - \frac{E_g}{2qT}\right\} \\
&\quad \left\{2 + \left(\frac{t_{ox}}{\varepsilon_0\varepsilon_{ox}}\right)\sqrt{\frac{\varepsilon_0\varepsilon_s q N_A}{\phi_F}}\right\},
\end{aligned} \tag{5.14}
$$

where we have put

$$
C_{ox} = \varepsilon_0\varepsilon_{ox}/t_{ox} \tag{5.15}
$$

where ε_{ox} is the dielectric constant of silicon dioxide and t_{ox} is oxide thickness.

Taking a typical example of a MOSFET,

$$
\begin{aligned}
&N_{D(g)} = 1 \times 10^{20}\ \text{cm}^{-3},\ N_A = 2 \times 10^{17}\ \text{cm}^{-3},\ T = 300\ \text{K}, \\
&E_g = 1.12\ \text{eV},\ \varepsilon_0 = 8.854 \times 10^{-14}\ \text{F cm}^{-1},\ \varepsilon_s = 11.9 \\
&\varepsilon_{ox} = 3.9,\ t_{ox} = 50\ \text{nm} = 50 \times 10^{-9}\ \text{m} = 50 \times 10^{-9} \times 100\ \text{cm} = 5 \times 10^{-6}\ \text{cm}
\end{aligned}
$$

with values

$$k_B = 8.62 \times 10^{-5}\ \text{eV K}^{-1},\ n_i = 1.45 \times 10^{10}\ \text{cm}^{-3},$$

$$\phi_{ms} = -\left(8.62 \times 10^{-5} \times 300\right) \ln \left\{\frac{1 \times 10^{20} \times 2 \times 10^{17}}{(1.45 \times 10^{10})^2}\right\} \tag{5.15}$$

$$= -0.02586 \ln (9.512485 \times 10^{16}) = -1.011\ \text{V}$$

$$\phi_F = \left(8.62 \times 10^{-5} \times 300\right) \ln \left(\frac{2 \times 10^{17}}{1.45 \times 10^{10}}\right) = 0.02586$$

$$\ln \left(1.37931 \times 10^{7}\right) \tag{5.17}$$

$$= 0.42513\ \text{V}$$

$$\frac{\partial V_{Th}}{\partial T} = \left[-\frac{1.011}{300} + (8.62 \times 10^{-5})\left(3 + \frac{1.12}{8.62 \times 10^{-5} \times 300}\right)\right.$$
$$+ \left\{\frac{0.42513}{300} - \frac{3}{2}(8.62 \times 10^{-5}) - \frac{1.12}{2 \times 300}\right\}$$
$$\times \left\{2 + \left(\frac{5 \times 10^{-6}}{8.854 \times 10^{-14} \times 3.9}\right)\right.$$
$$\left.\left.\times \sqrt{\frac{8.854 \times 10^{-14} \times 11.9 \times 1.6 \times 10^{-19} \times 2 \times 10^{17}}{0.42513}}\right\}\right]\text{V K}^{-1} \tag{5.18}$$

$$= \{-0.00337 + (8.62 \times 10^{-5})(3 + 43.31)$$
$$+ (1.4171 \times 10^{-3} - 1.293 \times 10^{-4} - 1.867 \times 10^{-3})$$
$$\times (2 + 1.44799 \times 10^{7} \times 2.816 \times 10^{-7})\}\ \text{V K}^{-1}$$
$$= \left\{-0.00337 + (8.62 \times 10^{-5})(46.31) + (-5.792 \times 10^{-4}) \times 6.07754\right\}\ \text{V K}^{-1}$$
$$= (-0.00337 + 3.9919 \times 10^{-3} - 3.5201 \times 10^{-3})\ \text{V K}^{-1}$$
$$= -2.8982 \times 10^{-3}\ \text{VK}^{-1} = -2.9\ \text{mVK}^{-1}.$$

Figure 5.3 shows how the plot of the subthreshold current curve of a MOSFET with respect to gate voltage is affected by temperature.

5.3 On-resistance ($R_{DS(ON)}$) of a double-diffused vertical MOSFET

The on-resistance of a MOSFET increases when its temperature rises. This variation is shown in figure 5.4.

A vertical MOSFET structure is shown in figure 5.5. Looking at the diagram, we find that the on-resistance $R_{DS(ON)}$ of this vertical MOSFET is the combination of many resistive elements. As these resistors are connected in series, the on-resistance is obtained by summation of values of the individual resistors. It can be described by the simple equation

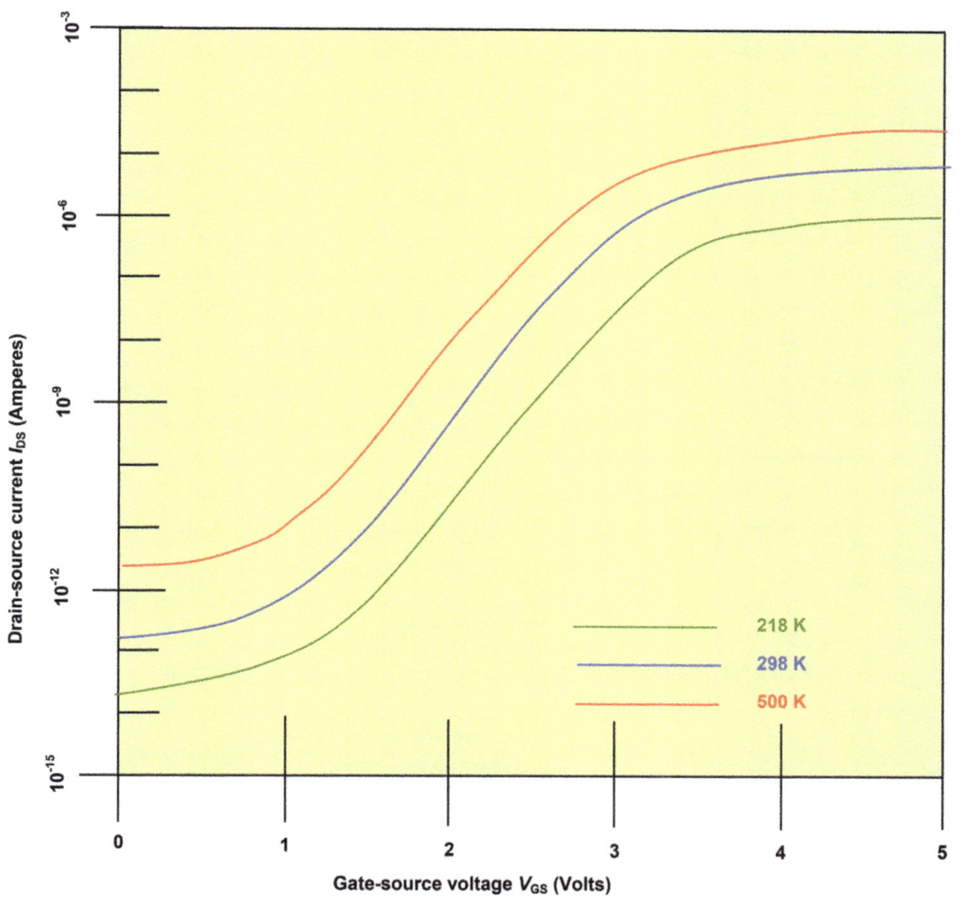

Figure 5.3. Variation in subthreshold current–gate voltage characteristics of a MOSFET with temperature.

$$R_{\mathrm{DS(ON)}} = R_{\mathrm{N^{+}SOURCE}} + R_{\mathrm{CHANNEL}} + R_{\mathrm{ACCUMULATION}} + R_{\mathrm{JFET}} + R_{\mathrm{DRIFT}} + R_{\mathrm{SUBSTRATE}} + R_{\mathrm{METALLIZATION+WIRE+LEADFRAME}} \tag{5.19}$$

where $R_{\mathrm{N^{+}SOURCE}}$ is the resistance of the heavily doped source region (usually negligible); R_{CHANNEL} is the resistance of the channel region dependent on the ratio of channel width to length, gate oxide thickness and gate drive voltage; $R_{\mathrm{ACCUMULATION}}$ is the resistance of the accumulation layer formed in the n^{-}-epi layer and joining the channel with the junction field-effect transistor (JFET) region (n^{-}-epi region between the p-bodies); R_{JFET} is the resistance of the JFET region with the p-body acting as the gate of the JFET; R_{DRIFT} is the resistance of the drift region; $R_{\mathrm{SUBSTRATE}}$ is the resistance of the substrate and $R_{\mathrm{METALLIZATION+WIRE+LEADFRAME}}$ is the contact resistance of source/drain metal layer, the resistance of the connecting wires and the resistance of the lead frames, the metal layers inside the package.

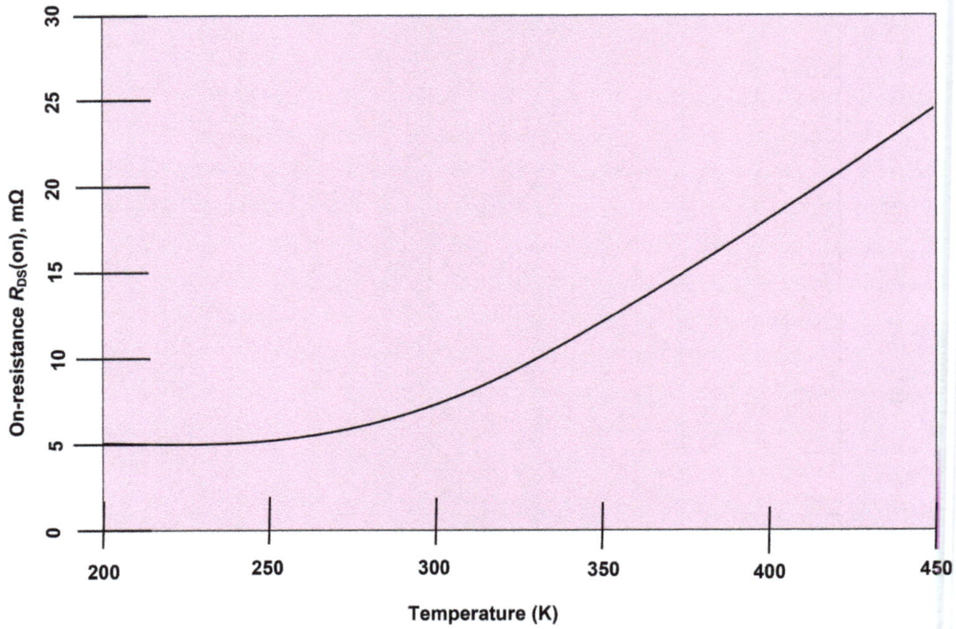

Figure 5.4. Increase in the on-resistance of a MOSFET with temperature.

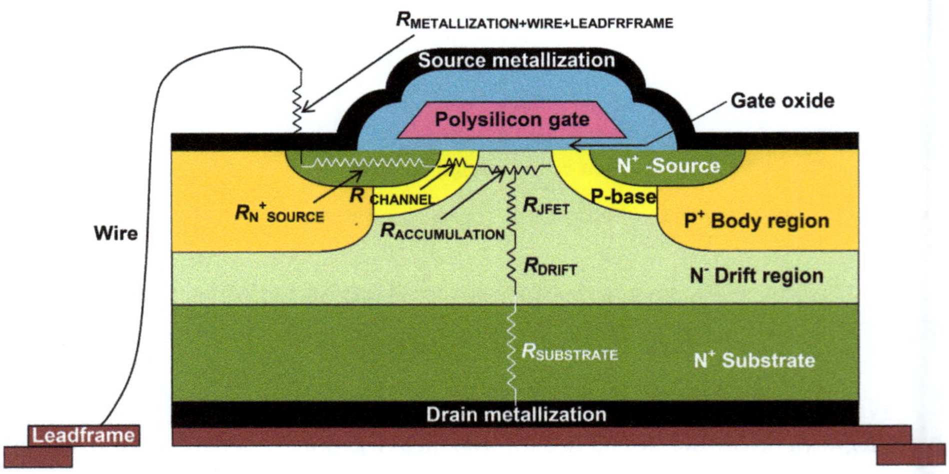

Figure 5.5. Components of the on-resistance of a MOSFET.

The components R_{CHANNEL} and R_{DRIFT} dominate in contribution over other components, and so the equation for on-resistance reduces to the form

$$R_{\mathrm{DS(ON)}} = R_{\mathrm{CHANNEL}} + R_{\mathrm{DRIFT}}. \tag{5.20}$$

To find R_{CHANNEL}, let us recall the equation for the static characteristic of a conventional long-channel MOSFET of channel width W, channel length L, electron mobility μ_{n} in the channel and oxide capacitance C_{ox} per unit area,

$$I_{DS} = (1/2)\mu_n C_{ox}(W/L)\left\{2\left(V_{GS} - V_{Th}\right)V_{DS} - V_{DS}^2\right\} \text{ for } V_{DS} < \left(V_{GS} - V_{Th}\right). \quad (5.21)$$

The on-resistance is a parameter characterizing the on-state operation of a MOSFET switch. In this state, the voltage V_{DS} is $\ll (V_{GS} - V_{Th})$. Hence, V_{DS}^2 term can be neglected, obtaining

$$I_{DS} = (1/2)\mu_n C_{ox}(W/L)\left\{2\left(V_{GS} - V_{Th}\right)V_{DS}\right\} \text{ for } V_{DS} \ll \left(V_{GS} - V_{Th}\right). \quad (5.22)$$

The channel resistance is

$$\begin{aligned} R_{CHANNEL} &= \frac{dV_{DS}}{dI_{DS}} = \frac{1}{\frac{dI_{DS}}{dV_{DS}}} \\ &= \frac{1}{\frac{d}{dV_{DS}}\left[(1/2)\mu_n C_{ox}(W/L)\left\{2\left(V_{GS} - V_{Th}\right)V_{DS}\right\}\right]} \\ &= \frac{2}{\mu_n C_{ox}(W/L)\left\{2\left(V_{GS} - V_{Th}\right)\right\}} \\ &= \frac{1}{\mu_n C_{ox}(W/L)\left(V_{GS} - V_{Th}\right)}. \end{aligned} \quad (5.23)$$

In this equation, the electron mobility and threshold voltage are temperature-dependent parameters. The mobility $\mu_n(T)$ at temperature T is related to mobility $\mu_n(T_0)$ at temperature T_0 by the equation

$$\mu_n(T) = \mu_n(T_0)(T/T_0)^{-n} \quad (5.24)$$

where $1.5 < n < 2.5$.

The threshold voltage has a TC $\kappa = -2$–4 mV K^{-1}, and the variation of threshold voltage with temperature is written as

$$V_{Th}(T) = V_{Th}(T_0) - \kappa(T - T_0) \quad (5.25)$$

$$\begin{aligned} \therefore R_{CHANNEL}(T) &= \frac{1}{\mu_n(T_0)(T/T_0)^{-n}C_{ox}(W/L)\{V_{GS} - V_{Th}(T_0) + \kappa(T - T_0)\}} \\ &= \frac{(T/T_0)^n}{\mu_n(T_0)C_{ox}(W/L)\left\{V_{GS} - V_{Th}(T_0)\right\}\left\{1 + \frac{\kappa(T-T_0)}{V_{GS} - V_{Th}(T_0)}\right\}} \\ &= \frac{1}{\mu_n(T_0)C_{ox}(W/L)\{V_{GS} - V_{Th}(T_0)\}} \\ &\quad \times \frac{(T/T_0)^n}{1 + \frac{\kappa(T-T_0)}{V_{GS} - V_{Th}(T_0)}} \\ &= R_{CHANNEL}(T_0) \times \frac{(T/T_0)^n}{1 + \frac{\kappa(T-T_0)}{V_{GS} - V_{Th}(T_0)}}. \end{aligned} \quad (5.26)$$

The resistance

$$\begin{aligned} R_{\text{DRIFT}} &= \text{Resistivity}(\rho) \times \frac{\text{Length}(d)}{\text{Cross sectional area}(S)} \\ &= \frac{1}{\text{Electronic charge}(q) \times \text{Electron mobility}(\mu_{\text{n}}) \times \text{Donor concentration}(N_{\text{D}})} \times \frac{d}{S} \\ &= \frac{d}{q\mu_{\text{n}} N_{\text{D}} S}, \end{aligned} \tag{5.27}$$

where d is the thickness of the drift region, N_{D} is the donor concentration in this region and S is the surface area of the MOSFET. Incorporating the temperature dependence of mobility, we have

$$\begin{aligned} R_{\text{DRIFT}}(T) &= \frac{d}{q\mu_{\text{n}}(T_0)(T/T_0)^{-n} N_{\text{D}} S} \\ &= \frac{d}{q\mu_{\text{n}}(T_0) S} \times (T/T_0)^n = R_{\text{DRIFT}}(T_0) \times (T/T_0)^n. \end{aligned} \tag{5.28}$$

Thus

$$R_{\text{DS(ON)}}(T) = R_{\text{CHANNEL}}(T_0) \times \frac{(T/T_0)^n}{1 + \frac{\kappa(T - T_0)}{V_{\text{GS}} - V_{\text{Th}}(T_0)}} + R_{\text{DRIFT}}(T_0) \times (T/T_0)^n. \tag{5.29}$$

This equation shows that on raising the temperature, the on-resistance of the MOSFET is augmented because the channel as well as drift region components of on-resistance increase. Since

$$\frac{\kappa(T - T_0)}{V_{\text{GS}} - V_{\text{Th}}(T_0)} \ll 1 \tag{5.30}$$

$$\begin{aligned} R_{\text{DS(ON)}}(T) &= R_{\text{CHANNEL}}(T_0) \times (T/T_0)^n + R_{\text{DRIFT}}(T_0) \times (T/T_0)^n \\ &= (R_{\text{CHANNEL}} + R_{\text{DRIFT}})(T/T_0)^n = R_{\text{DS(ON)}}(T_0)(T/T_0)^n. \end{aligned} \tag{5.31}$$

The $R_{\text{DS(ON)}}(T)$ of an n- or p-channel power MOSFET at a given temperature T can be calculated from its value $R_{\text{DS(ON)}}(T_0)$ at room temperature $T_0 = 27$ °C by using the approximate equation (Fairchild Semiconductor Corporation 2000)

$$R_{\text{DS(ON)}}(T) = R_{\text{DS(ON)}}(300)(T/300)^{2.3}. \tag{5.32}$$

In table 5.1, the $R_{\text{DS(ON)}}$ values of a power MOSFET at different operating temperatures T are estimated from the formula (5.32) in terms of its $R_{\text{DS(ON)}}$ value at 300 K.

An increase in the on-resistance of a MOSFET with temperature is a boon when paralleling several MOSFETs. A MOSFET passing more current is heated up. Heating of the MOSFET elevates its on-resistance. The increased on-resistance of this particular MOSFET automatically lowers the current being carried by it, and prevents it from being damaged by overdriving. Thus a MOSFET has a positive TC of on-resistance. The positive coefficient limits the current conducted by individual

Table 5.1. Calculated factors by which the on-resistance of a MOSFET changes with respect to its room temperature value (T = 27 °C).

Temperature T (K)	4.2	77.2	150	300	600
$R_{DS(ON)}(T)/R_{DS(ON)}(T = 300\text{ K})$	5.4462×10^{-5}	$4.406\,978 \times 10^{-2}$	0.203 0631	1	4.924 577 65

MOSFETs connected in a parallel arrangement and ensures proper sharing of current amongst the MOSFETs.

5.4 Transconductance (g_m) of a MOSFET

In many circuit applications of MOSFETs, the input signal is the gate–source voltage V_{GS} and the output signal is the drain–source current I_{DS}. For an applied drain–source voltage V_{DS}, the capability of a MOSFET to amplify the input signal is measured by its transconductance g_m defined as the ratio

$$g_m = \left(\frac{\partial I_{DS}}{\partial V_{GS}}\right)_{V_{DS}}. \tag{5.33}$$

If the value of the applied drain–source voltage provides MOSFET operation in the saturation region, the transconductance is said to be saturated transconductance. The transconductance–drain current curves of a MOSFET vary with temperature as shown in figure 5.6.

From equations (5.22) and (5.33), keeping V_{DS} constant,

$$\begin{aligned} g_m &= \frac{\partial}{\partial V_{GS}}[(1/2)\mu_n C_{ox}(W/L)\{2(V_{GS} - V_{Th})V_{DS}\}] \\ &= (1/2)\mu_n C_{ox}(W/L)\{2(1-0)V_{DS}\} = \mu_n C_{ox}(W/L)V_{DS}. \end{aligned} \tag{5.34}$$

Making mobility temperature-dependent

$$\begin{aligned} g_m(T) &= \mu_n(T_0)(T/T_0)^{-n}C_{ox}(W/L)V_{DS} \\ &= \mu_n(T_0)C_{ox}(W/L)V_{DS} \times (T/T_0)^{-n} \\ &= g_m(T_0) \times (T/T_0)^{-n}. \end{aligned} \tag{5.35}$$

When the temperature increases, the mobility of carriers decreases. Hence, the transconductance of a MOSFET falls. Like the on-resistance of a MOSFET, a crude relation representing this deterioration of transconductance $g_m(T_0)$ at temperature T_0 to the new value $g_m(T)$ at temperature T is (Fairchild Semiconductor Corporation 2000)

$$g_m(T) = g_m(300)(T/300)^{-2.3}. \tag{5.36}$$

This equation predicts the changes in transconductance by different factors relative to the value of transconductance at 300 K. A rough estimate of the trends can be seen from table 5.2.

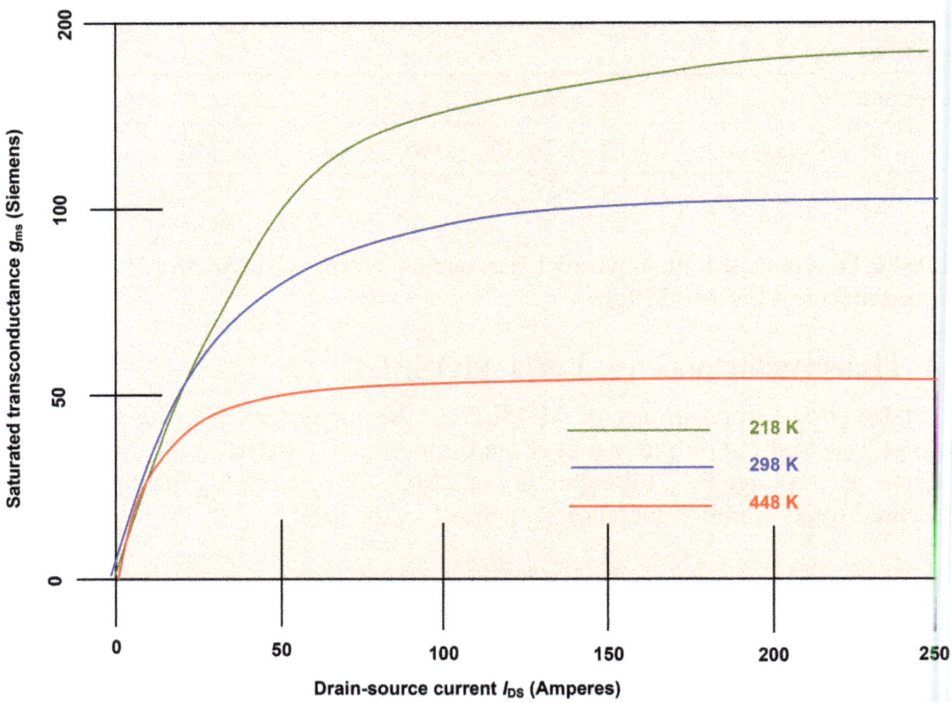

Figure 5.6. Variation in the transconductance versus drain current characteristics of a MOSFET with temperature.

Table 5.2. Effect of temperature on the transconductance of a MOSFET.

Temperature T (K)	4.2	77.2	150	300	600
$g_m(T)/g_m$ (300)	1.8136×10^4	22.69	4.924 58	1	0.203 063

5.5 BV_{DSS} and I_{DSS} of a MOSFET

BV_{DSS} stands for the drain–source breakdown voltage of a MOSFET with the gate shorted to the source. It is the maximum drain–source voltage which the device can withstand without the body–drain diode reaching the avalanche breakdown condition. I_{DSS} is the corresponding drain–source leakage current (Fairchild Semiconductor Corporation 2000). The astute reader may analyze the breakdown voltage and leakage current variations in MOSFET from the understanding of the corresponding parameters for p–n junction diodes and bipolar transistors.

5.6 Zero temperature coefficient biasing point of MOSFET

One possibility to operate a MOSFET-based circuit at a high temperature utilizes the property that the drain–source current exhibits zero or very small changes with temperature if the device is operated at a particular gate–source voltage, V_{GSO}. This value of gate–source voltage is called the zero TC biasing point (ZTC); see figure 5.7. Thus the ZTC of a MOSFET is the gate–source voltage at which its drain–source current is almost invariant with respect to temperature.

To find the ZTC of a MOSFET, we perform a simplified analysis in which the equation for the drain–source current I_{DS} is differentiated with respect to temperature, keeping V_{GS} = constant and V_{DS} = constant. We rewrite the MOSFET current equation (5.22) in saturation mode operation by substituting the temperature-dependent expressions for mobility and threshold voltage in equations (5.24) and (5.25):

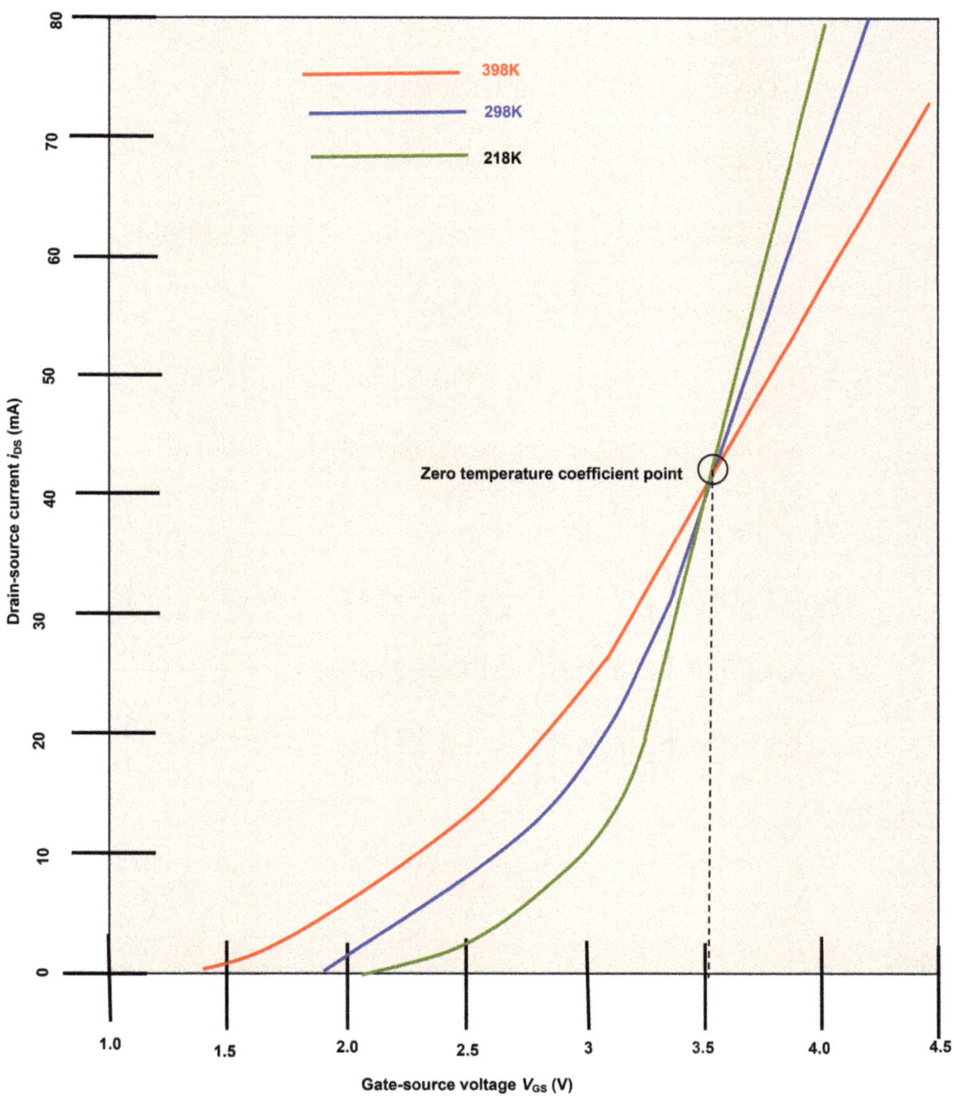

Figure 5.7. Transfer characteristics of a MOSFET at three temperatures.

$$
\begin{aligned}
I_{DS}(T) &= (1/2)\mu_n(T_0)(T/T_0)^{-n}C_{ox}(W/L) \\
&\quad \left[2\left\{V_{GS} - V_{Th}(T_0) + \kappa\left(T - T_0\right)\right\}V_{DS}\right] \\
&= \mu_n(T_0)(T/T_0)^{-n}C_{ox}(W/L)\left\{V_{GS} - V_{Th}(T_0)\right\}V_{DS} \\
&\quad + \mu_n(T_0)(T/T_0)^{-n}C_{ox}(W/L)\kappa T V_{DS} \\
&\quad - \mu_n(T_0)(T/T_0)^{-n}C_{ox}(W/L)\kappa T_0 V_{DS}
\end{aligned}
\tag{5.37}
$$

$$
\begin{aligned}
\therefore \frac{dI_{DS}}{dT} &= \mu_n(T_0)\left(\frac{1}{T_0}\right)^{-n}(-nT^{-n-1})C_{ox}(W/L)\{V_{GS} - V_{Th}(T_0)\}V_{DS} \\
&\quad + \mu_n(T_0)\left(\frac{1}{T_0}\right)^{-n}(-nT^{-n-1})C_{ox}(W/L)\kappa T V_{DS} + \mu_n(T_0)(T/T_0)^{-n}C_{ox}(W/L)\kappa V_{DS} \\
&\quad - \mu_n(T_0)\left(\frac{1}{T_0}\right)^{-n}(-nT^{-n-1})C_{ox}(W/L)\kappa T_0 V_{DS} \\
&= -n\mu_n(T_0)C_{ox}(W/L)\left(\frac{T}{T_0}\right)^{-n}(T^{-1})\left\{V_{GS} - V_{Th}(T_0)\right\}V_{DS} \\
&\quad - n\mu_n(T_0)C_{ox}(W/L)\left(\frac{T}{T_0}\right)^{-n}(T^{-1})\kappa T V_{DS} + \kappa\mu_n(T_0)(T/T_0)^{-n}C_{ox}(W/L)V_{DS} \\
&\quad + n\mu_n(T_0)C_{ox}(W/L)\left(\frac{T}{T_0}\right)^{-n}(T^{-1})\kappa T_0 V_{DS} \\
&= -\frac{n}{T}\mu_n(T_0)C_{ox}(W/L)\left(\frac{T}{T_0}\right)^{-n}\{V_{GS} - V_{Th}(T_0)\}V_{DS} \\
&\quad - n\mu_n(T_0)C_{ox}(W/L)\left(\frac{T}{T_0}\right)^{-n}\kappa V_{DS} + \kappa\mu_n(T_0)(T/T_0)^{-n}C_{ox}(W/L)V_{DS} \\
&\quad + \left(\frac{nT_0}{T}\right)\mu_n(T_0)C_{ox}(W/L)\left(\frac{T}{T_0}\right)^{-n}\kappa V_{DS} \\
&= \mu_n(T_0)C_{ox}(W/L)\left(\frac{T}{T_0}\right)^{-n}V_{DS}\left[\left(-\frac{n}{T}\right)\{V_{GS} - V_{Th}(T_0)\} - n\kappa + \kappa + \left(\frac{nT_0}{T}\right)\kappa\right] \\
&= \mu_n(T_0)C_{ox}(W/L)\left(\frac{T}{T_0}\right)^{-n}V_{DS}\left[\left(-\frac{n}{T}\right)\{V_{GS} - V_{Th}(T_0)\} + \left\{-n + 1 + \left(\frac{nT_0}{T}\right)\right\}\kappa\right] \\
&= \mu_n(T_0)C_{ox}(W/L)\left(\frac{T}{T_0}\right)^{-n}V_{DS}\left[\left\{1 - n + \left(\frac{nT_0}{T}\right)\right\}\kappa - \frac{n}{T}\{V_{GS} - V_{Th}(T_0)\}\right].
\end{aligned}
\tag{5.38}
$$

For the ZTC

$$
\frac{dI_{DS}}{dT} = 0 \tag{5.39}
$$

Hence,

$$
\left\{1 - n + \left(\frac{nT_0}{T}\right)\right\}\kappa - \frac{n}{T}\left\{V_{GS} - V_{Th}(T_0)\right\} = 0 \tag{5.40}
$$

or

$$
\begin{aligned}
\frac{n}{T}\{V_{GS} - V_{Th}(T_0)\} &= \left\{1 - n + \left(\frac{nT_0}{T}\right)\right\}\kappa \\
V_{GS} - V_{Th}(T_0) &= \frac{T}{n}\left\{1 - n + \left(\frac{nT_0}{T}\right)\right\}\kappa = \frac{T\kappa}{n} - T\kappa + T_0\kappa \\
&= \frac{T\kappa}{n} - \kappa(T - T_0) \qquad (5.41) \\
\text{or, } V_{GS} &= V_{Th}(T_0) - \kappa(T - T_0) + \frac{T\kappa}{n} \\
V_{GS} &= V_{Th}(T) + \frac{T\kappa}{n},
\end{aligned}
$$

using equation (5.25). Since this V_{GS} value corresponds to the ZTC, it will be designated by $V_{GS}(\text{ZTC})$. Hence,

$$V_{GS}(\text{ZTC}) = V_{Th}(T) + \frac{T\kappa}{n}. \tag{5.42}$$

Taking $n = 2$, $\kappa = -4$ mV K^{-1}, for a $V_{Th} = 1$ V MOSFET at room temperature $T = 300$ K,

$$V_{GS}(\text{ZTC}) = 1 + \frac{300 \times (4 \times 10^{-3})}{2} = 1.60 \text{ V}. \tag{5.43}$$

5.7 Dynamic response of a MOSFET

MOSFETs exhibit faster switching speeds than bipolar devices. The reason is that there is no delay in switching caused by storage of minority carriers, as experienced in p–n junction diodes and bipolar transistors. Primarily the intrinsic capacitance C_g and intrinsic resistance R_g of a MOSFET determine its switching characteristics (Sattar and Tsukanov 2007). The main components of intrinsic capacitance C_g are:

(i) input capacitance C_{iss} consisting of capacitance C_{gs} between the gate and source, and capacitance C_{gd} between the gate and drain (figure 5.8),

$$C_{iss} = C_{gs} + C_{gd}; \tag{5.44}$$

(ii) output capacitance comprising drain–source capacitance C_{ds} and gate–drain capacitance C_{gd},

$$C_{oss} = C_{ds} + C_{gd}; \tag{5.45}$$

(iii) reverse-transfer capacitance

$$C_{rss} = C_{gd}. \tag{5.46}$$

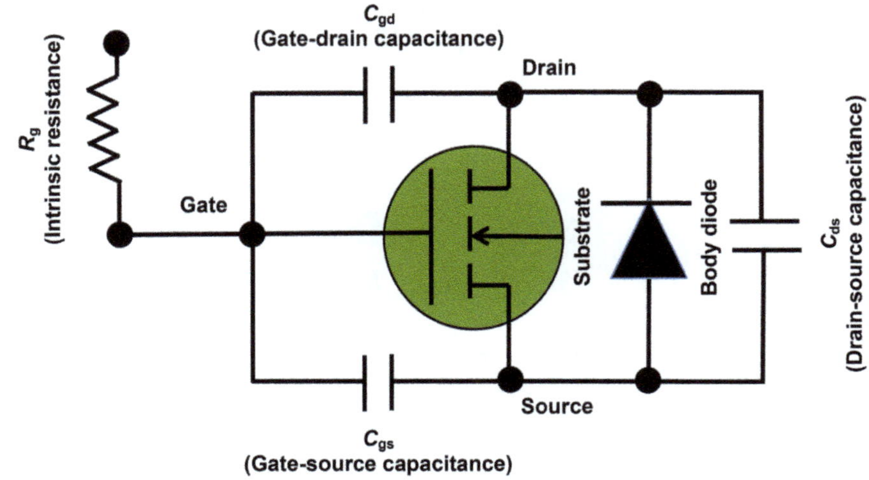

Figure 5.8. Parasitic components of a power MOSFET.

Intrinsic resistance R_g is a component of the total gate resistance, which includes externally connected resistances to the gate including the driver resistance. Due to its small magnitude, it exerts an imperceptible effect on the switching time.

The resistance–capacitance network formed between R_g and the input capacitance C_{iss} of the MOSFET has a time constant τ given by

$$\tau = R_g C_{iss}. \tag{5.47}$$

The charging and discharging times of the capacitance C_{iss}, i.e. the supply or removal of charges on these capacitances, through the resistance R_g decide how fast the voltage is established on the gate insulator to cause a current flow between the source and drain terminals. As the input capacitance is not affected by temperature variations, the switching speed of the MOSFET remains practically unaltered when the ambient temperature changes. Any subtle changes in C_{iss} or R_g may give rise to slight changes in switching parameters, but these are generally negligible. Naturally, the switching loss of the MOSFET is also invariant with temperature.

5.8 MOS analog circuits in the 25 °C to 300 °C range

Shoucair (1986) showed that analog CMOS circuits can be designed for operation up to 250 °C using commercial CMOS technologies, and considering design parameters such as transconductance, output conductance and gain. Below we explain the detailed analysis of the thermal behavior of a MOSFET, after Shoucair. In this analysis, the symbol for the TC of threshold voltage is p_0 in place of the κ used previously. To a good approximation, the variation of threshold voltage with temperature is described by a first-order polynomial of the form

$$V_{Th}(T) = p_0 T + q_0 \tag{5.48}$$

where

$$p_0 = \mathrm{d}V_{\mathrm{Th}}/\mathrm{d}T = -2.4\ \mathrm{mV\ K^{-1}},\ q_0 = +1.72\ \mathrm{V}\ \text{for an n-channel MOSFET} \quad (5.49)$$

$$p_0 = \mathrm{d}V_{\mathrm{Th}}/\mathrm{d}T = +2.4\ \mathrm{mV\ K^{-1}},\ q_0 = -1.72\mathrm{V}\ \text{for a p-channel MOSFET.} \quad (5.50)$$

Compared to the room temperature (25 °C) value, V_{Th} changes by ~0.55 V at 250 °C for either polarity of MOSFET. The change in threshold voltage affects the analog circuits through a shift in biasing voltages with the exception of the ZTC biasing points described below. For comparison, it may be recalled that in digital circuits, there is a decrease in the noise margins. The average carrier mobility follows the $T^{-1.5}$ relationship:

$$\mu(T) = \mu(T_0)\left(\frac{T}{T_0}\right)^{-1.5}. \quad (5.51)$$

At a ZTC point, the drain current remains constant in spite of temperature changes. Hence, the normalized derivative of the drain–source current I_{DS} with respect to temperature T can be equated to zero. For the linear region, this derivative is

$$\begin{aligned}\frac{\mathrm{d}I_{\mathrm{DS}}}{\mathrm{d}T} &= \frac{\mathrm{d}}{\mathrm{d}T}\left[\mu(T)C_{\mathrm{ox}}\left(\frac{W}{L}\right)\left\{V_{\mathrm{GS(ZTC)}} - V_{\mathrm{Th}}(T)\right\}V_{\mathrm{DS}}\right] \\ &= \frac{\mathrm{d}\mu(T)}{\mathrm{d}t}C_{\mathrm{ox}}\left(\frac{W}{L}\right)\left\{V_{\mathrm{GS(ZTC)}} - V_{\mathrm{Th}}(T)\right\}V_{\mathrm{DS}} \\ &\quad + \mu C_{\mathrm{ox}}\left(\frac{W}{L}\right)\left(0 - \frac{\mathrm{d}V_{\mathrm{Th}}(T)}{\mathrm{d}T}\right)V_{\mathrm{DS}} \\ &= \frac{\mathrm{d}\mu}{\mathrm{d}t}C_{\mathrm{ox}}\left(\frac{W}{L}\right)\left\{V_{\mathrm{GS(ZTC)}} - V_{\mathrm{Th}}(T)\right\}V_{\mathrm{DS}} \\ &\quad - \mu C_{\mathrm{ox}}\left(\frac{W}{L}\right)\left\{\frac{\mathrm{d}V_{\mathrm{Th}}(T)}{\mathrm{d}T}\right\}V_{\mathrm{DS}}.\end{aligned} \quad (5.52)$$

Normalizing this derivative, we have

$$\frac{1}{I_{\mathrm{DS}}}\left(\frac{\mathrm{d}I_{\mathrm{DS}}}{\mathrm{d}T}\right) = \frac{1}{\mu}\left(\frac{\mathrm{d}\mu}{\mathrm{d}t}\right) - \frac{1}{V_{\mathrm{GS(ZTC)}} - V_{\mathrm{Th}}(T)}\left\{\frac{\mathrm{d}V_{\mathrm{Th}}(T)}{\mathrm{d}T}\right\}. \quad (5.53)$$

Here,

$$\frac{1}{\mu}\left(\frac{\mathrm{d}\mu}{\mathrm{d}t}\right) = \frac{1}{\mu(T_0)\left(\frac{T}{T_0}\right)^{-1.5}} \times \mu(T_0) \times -1.5\left(\frac{T}{T_0}\right)^{-1.5-1} \times \frac{T_0 - 0}{T_0^2} = -\frac{1.5}{T_0}. \quad (5.54)$$

Putting

$$\frac{\mathrm{d}V_{\mathrm{Th}}}{\mathrm{d}T} = p_0 \quad (5.55)$$

$$\frac{1}{I_{DS}}\left(\frac{dI_{DS}}{dT}\right) = -\frac{1.5}{T_0} - \frac{p_0}{V_{GS(ZTC)} - V_{Th}(T)} \quad (5.56)$$

$$\therefore -\frac{1.5}{T_0} - \frac{p_0}{V_{GS(ZTC)} - V_{Th}(T)} = 0, \quad (5.57)$$

or

$$\frac{p_0}{V_{GS(ZTC)} - V_{Th}} = -\frac{1.5}{T_0} \quad (5.58)$$

$$\therefore V_{GS(ZTC)} - V_{Th} = -\frac{T_0}{1.5}p_0 \quad (5.59)$$

$$\therefore I_{DS} = \mu(T)C_{ox}\left(\frac{W}{L}\right)\left\{-\frac{T_0}{1.5}p_0\right\}V_{DS}. \quad (5.60)$$

In the saturation region, the temperature derivative of drain current is

$$\frac{dI_{DS}}{dT} = \frac{d}{dT}\left[(1/2)\mu(T)C_{ox}\left(\frac{W}{L}\right)\left\{V_{GS(ZTC)} - V_{Th}(T)\right\}^m\right] \quad (5.61)$$

where $m = 2$,

$$\begin{aligned}\frac{dI_{DS}}{dT} &= \frac{d\mu(T)}{dt}(1/2)C_{ox}\left(\frac{W}{L}\right)\left\{V_{GS(ZTC)} - V_{Th}(T)\right\}^m \\ &+ (1/2)\mu(T)C_{ox}\left(\frac{W}{L}\right)m\left\{V_{GS(ZTC)} - V_{Th}(T)\right\}^{m-1} \times -\frac{dV_{Th}(T)}{dT}\end{aligned} \quad (5.62)$$

$$\begin{aligned}\therefore \frac{1}{I_{DS}}\frac{dI_{DS}}{dT} &= \frac{1}{\mu(T)}\left\{\frac{d\mu(T)}{dt}\right\} - \frac{m}{V_{GS(ZTC)} - V_{Th}(T)}\frac{dV_{Th}(T)}{dT} \\ &= -\frac{1.5}{T_0} - \frac{m}{V_{GS(ZTC)} - V_{Th}(T)} \times p_0\end{aligned} \quad (5.63)$$

using equation (5.54). Setting

$$\frac{1}{I_{DS}}\frac{dI_{DS}}{dT} = 0 \quad (5.64)$$

$$-\frac{1.5}{T_0} - \frac{m}{V_{GS(ZTC)} - V_{Th}(T)} \times p_0 = 0, \quad (5.65)$$

or

$$\frac{mp_0}{V_{GS(ZTC)} - V_{Th}(T)} = -\frac{1.5}{T_0}$$
$$V_{GS(ZTC)} - V_{Th}(T) = -\frac{mT_0}{1.5}p_0 \tag{5.66}$$

$$\therefore I_{DS} = (1/2)\mu(T)C_{ox}\left(\frac{W}{L}\right)\left\{-\frac{mT_0}{1.5}p_0\right\}^m. \tag{5.67}$$

For a MOSFET with known parameters (p_0, q_0) over the temperature range (T_1, T_2), there exist two separate ZTC gate–source biasing voltages V_{GS}(ZTC), one located in the linear region and other in the saturation region, at which the drain current of the MOSFET exhibits the least sensitivity to temperature. These biasing voltages obey the following analytical equation derived by least squares minimization over (T_1, T_2) (Shoucair 1986)

$$V_{GS}(\text{ZTC}) = -(1/6)p_0(T_1 + T_2)(2m - 3) + q_0, \tag{5.68}$$

where m is the index in the approximate equation for drain–source current I_{DS}

$$I_{DS} \sim (V_{GS} - V_{Th})^m. \tag{5.69}$$

The index $m = 1$ in the linear region and $m = 2$ in the saturation region of MOSFET operation under ideal square law conditions. Equation (5.68) for V_{GS}(ZTC) has two forms corresponding to the two values of $m = 1, 2$:

$$\begin{aligned} V_{GS}(\text{ZTC})_{|\text{Linear}(m=1)} &\approx -(1/6)p_0(T_1 + T_2)(2 \times 1 - 3) + q_0 \\ &= +(1/6)p_0(T_1 + T_2) + q_0 \end{aligned} \tag{5.70}$$

$$\begin{aligned} V_{GS}(\text{ZTC})_{|\text{Saturation}(m=2)} &\approx -(1/6)p_0(T_1 + T_2)(2 \times 2 - 3) + q_0 \\ &= -(1/6)p_0(T_1 + T_2) + q_0. \end{aligned} \tag{5.71}$$

As an example, for an n-channel MOSFET operated in the saturation region in the temperature range 273–523 K,

$$\begin{aligned} V_{GS}(\text{ZTC})_{|\text{Saturation}(m=2)} &= -(1/6) \times -2.4 \times 10^{-3} \times (273 + 523) \\ &\quad + 1.72 \\ &= -0.4 \times 10^{-3} \times 796 + 1.72 \\ &= -0.3184 + 1.72 = 1.4016\ \text{V}. \end{aligned} \tag{5.72}$$

Drain currents for V_{GS} (ZTC) values are obtained from the drain current equations in the linear and saturation regions. Measurements of transfer characteristics of MOSFETs showed that there was a pronounced upswing in drain current above 250 °C. This originated from the marked increase in drain–body leakage current to the extent that it became comparable in magnitude with the forward MOSFET current. Leakage currents of source or drain junctions comprise contributions from generation and diffusion components, depending upon temperatures. For CMOS processes, a transition temperature $T_{trans} \sim 130$ °C–150 °C was found. Below the transition temperature, generation current dominated:

$$I_{gen}(T) = (qAw/\tau)n_i(T) \text{ for } T = 25°\text{C to } 150°\text{C}, \tag{5.73}$$

where A is the effective area of the p–n junction, w is the depletion region width and τ is the carrier lifetime.

Above the transition temperature, the diffusion component was predominant. For an n-channel MOSFET

$$I_{diff}(T) = (qAD_n/L_n)\{n_i^2(T)/N_A\} \text{ for } T = 150°\text{C to } 300°\text{C}, \tag{5.74}$$

where D_n, L_n are the diffusion coefficient and diffusion length of electrons, and N_A is the acceptor concentration. An identical equation applies to the p-channel device. The leakage currents and hence leakage conductances increase by five orders of magnitude in the temperature range 25 °C–300 °C. Consequently, gain is degraded. In analog circuits, the biasing points are shifted, whereas in digital circuits the noise margins decrease due to leakage current upswing. One possible reason for the higher leakage currents shown by n-channel MOSFETs is the additional processing required during their fabrication, namely, p-well formation, which introduces more defects, thereby decreasing carrier lifetime. Guard rings surrounding the source and drain diffusions help in preventing the minority carriers from reaching the crucial regions, and consequently avert damaging CMOS latchups.

Transconductance in the saturation region halves in the range 25 °C to 300 °C. It is given by

$$g_m(T) = 2I_{DS}(T)/\{V_{GS} - V_{Th}\}. \tag{5.75}$$

The body-effect conductance g_{mb} is expressed in terms of transconductance g_m as

$$g_{mb}(T) \approx \{\partial V_{Th}(T)/\partial V_{sub}\}g_m(T), \tag{5.76}$$

where V_{sub} is the substrate voltage. Like transconductance, the body-effect transconductance is also halved in the above temperature range.

Also halved is the channel output conductance (Shoucair and Early 1984)

$$g_d(T) = I_D(T)/V_a, \tag{5.77}$$

where V_a is the Early voltage. The overall impact is that the gain–bandwidth product is halved.

Shoucair (1986) laid out the guidelines for designing a two-stage topology of a CMOS OP-AMP. For stabilizing the circuit operation against temperature variations, the two gain stages are biased at their ZTC drain current values in the saturation region by applying a voltage $V_{GS}(\text{ZTC})|_{sat}$ to the current source biasing each stage, either by employing a voltage dividing string of MOSFETs or by using passive devices like polysilicon resistors. In the differential input stage of the CMOS OP-AMP, a diode is added to compensate for leakage currents. The low-frequency small-signal gain depends on the output conductances besides the transconductance and small-signal leakage-induced conductances, which become appreciably high at temperatures above 200 °C. In the output stage, provision is made to allow that the two drain–body junctions leak equal amounts of current. Leakage areas are matched if the leakage current densities are equal.

The MOSFET capacitance consists of overlap and junction capacitances. The overlap capacitances arise from overlap of metallization over the gate–source and gate–drain regions, whereas junction capacitances originate from the source–body and drain–body junctions. The overlap capacitances have a small TC ~25 ppm $°C^{-1}$ causing a smaller than 5% capacitance increase in the temperature range 25 °C–300 °C while junction capacitances have much larger TCs ~100–1500 ppm $°C^{-1}$ producing a 5%–50% increase in this range of temperatures. The former can be described as relatively weak temperature effect and the latter as a weak temperature effect. The influence of the increase in capacitance on analog circuits is observed as a slowing down of the circuit speed.

Considering that the source is shorted to the body, $V_{\text{Body-Source}} = 0$. Then the capacitance C_{DS} is the drain–body junction capacitance. If this junction is assumed to be an abrupt one,

$$C_{DS} = \sqrt{\frac{q\varepsilon_0\varepsilon_{Si}(N_A + N_D)}{2(V_{bi} + V_R)}}, \tag{5.78}$$

where ε_{Si} is the dielectric constant of Si, N_A is the acceptor concentration, N_D is the donor concentration, V_{bi} is the built-in potential and V_R is the applied reverse bias. The potential V_{bi} is given by

$$V_{bi} = \left(\frac{k_B T}{q}\right) \ln \left\{\frac{N_A N_D}{n_i^2(T)}\right\}. \tag{5.79}$$

Differentiating both sides of the equation (5.78) for C_{DS}, we obtain (Shoucair *et al* 1984)

$$\begin{aligned} \frac{dC_{DS}(T)}{dT} &= \left(\frac{1}{2}\right)\left\{\frac{q\varepsilon_0\varepsilon_{Si}(N_A + N_D)}{2(V_{bi} + V_R)}\right\}^{1/2-1} \\ &\quad \times \frac{0 - 2(dV_{bi}/dT)\{q\varepsilon_0\varepsilon_{Si}(N_A + N_D)\}}{\{2(V_{bi} + V_R)\}^2} \\ &= \left(\frac{1}{2}\right)\left\{\frac{2(V_{bi} + V_R)}{q\varepsilon_0\varepsilon_{Si}(N_A + N_D)}\right\}^{1/2} \\ &\quad \times \frac{-2(dV_{bi}/dT)\{q\varepsilon_0\varepsilon_{Si}(N_A + N_D)\}}{\{2(V_{bi} + V_R)\}^2} \\ &= -(dV_{bi}/dT)\frac{\{q\varepsilon_0\varepsilon_{Si}(N_A + N_D)\}^{1/2}}{\{2(V_{bi} + V_R)\}^{1.5}} \end{aligned} \tag{5.80}$$

$$\therefore \left(\frac{1}{C_{\rm DS}}\right)\frac{{\rm d}C_{\rm DS}(T)}{{\rm d}T} = -\frac{1}{2(V_{\rm bi}+V_{\rm R})}({\rm d}V_{\rm bi}/{\rm d}T), \tag{5.81}$$

but

$$\begin{aligned}\frac{{\rm d}V_{\rm bi}}{{\rm d}T} &= \left(\frac{k_{\rm B}}{q}\right)\ln\left\{\frac{N_{\rm A}N_{\rm D}}{n_{\rm i}^2(T)}\right\} + \left(\frac{k_{\rm B}T}{q}\right)\left\{\frac{n_{\rm i}^2(T)}{N_{\rm A}N_{\rm D}}\right\} \\ &\quad\times \frac{0 - 2n_{\rm i}(T)({\rm d}n_i/{\rm d}T)N_{\rm A}N_{\rm D}}{\{n_{\rm i}^2(T)\}^2} \\ &= \left(\frac{k_{\rm B}}{q}\right)\ln\left\{\frac{N_{\rm A}N_{\rm D}}{n_{\rm i}^2(T)}\right\} - \left(\frac{k_{\rm B}T}{q}\right)\frac{2({\rm d}n_i/{\rm d}T)}{n_{\rm i}(T)}.\end{aligned} \tag{5.82}$$

Since,

$$n_{\rm i}(T) = 4.68\times 10^{15}T^{3/2}\exp\left(-\frac{E_{\rm g}}{2k_{\rm B}T}\right) \tag{5.83}$$

$$\begin{aligned}\therefore \frac{{\rm d}n_{\rm i}(T)}{{\rm d}T} &= 4.68\times 10^{15}\left(\frac{3}{2}\right)T^{3/2-1}\exp\left(-\frac{E_{\rm g}}{2k_{\rm B}T}\right) + 4.68 \\ &\quad\times 10^{15}T^{3/2}\left\{\exp\left(-\frac{E_{\rm g}}{2k_{\rm B}T}\right)\right\}\times\frac{+2k_{\rm B}E_{\rm g}}{(2k_{\rm B}T)^2} \\ &= 4.68\times 10^{15}\left(\frac{3}{2}\right)T^{1/2}\exp\left(-\frac{E_{\rm g}}{2k_{\rm B}T}\right) + 4.68 \\ &\quad\times 10^{15}\left\{\exp\left(-\frac{E_{\rm g}}{2k_{\rm B}T}\right)\right\}\times\frac{E_{\rm g}}{2k_{\rm B}T^{1/2}} \\ &= 4.68\times 10^{15}\exp\left(-\frac{E_{\rm g}}{2k_{\rm B}T}\right)\left\{\left(\frac{3}{2}\right)T^{1/2} + \frac{E_{\rm g}}{2k_{\rm B}T^{1/2}}\right\}\end{aligned} \tag{5.84}$$

$$\begin{aligned}\frac{2({\rm d}n_i/{\rm d}T)}{n_{\rm i}(T)} &= \left[2\times 4.68\times 10^{15}\exp\left(-\frac{E_{\rm g}}{2k_{\rm B}T}\right)\left\{\left(\frac{3}{2}\right)T^{\frac{1}{2}} + \frac{E_{\rm g}}{2k_{\rm B}T^{\frac{1}{2}}}\right\}\right]/ \\ &\quad\left\{4.68\times 10^{15}T^{\frac{3}{2}}\exp\left(-\frac{E_{\rm g}}{2k_{\rm B}T}\right)\right\} \\ &= 2T^{-3/2}\left\{\left(\frac{3}{2}\right)T^{1/2} + \frac{E_{\rm g}}{2k_{\rm B}T^{1/2}}\right\} = T^{-3/2}\left(3T^{1/2} + \frac{E_{\rm g}}{k_{\rm B}T^{1/2}}\right) \\ &= \left(\frac{3}{T} + \frac{E_{\rm g}}{k_{\rm B}T^2}\right) = \frac{1}{T}\left(3 + \frac{E_{\rm g}}{k_{\rm B}T}\right)\end{aligned} \tag{5.85}$$

Combining equations (5.81), (5.82) and (5.85) we obtain

$$\left(\frac{1}{C_{DS}}\right)\frac{\mathrm{d}C_{DS}(T)}{\mathrm{d}T} = -\frac{1}{2(V_{bi}+V_R)}\left[\left(\frac{k_B}{q}\right)\ln\left\{\frac{N_A N_D}{n_i^2(T)}\right\} - \left(\frac{k_B T}{q}\right)\frac{2(\mathrm{d}n_i/\mathrm{d}T)}{n_i(T)}\right]$$
$$= -\frac{1}{2(V_{bi}+V_R)}\left(\frac{k_B}{q}\right)\left[\ln\left\{\frac{N_A N_D}{n_i^2(T)}\right\} - T\times\frac{1}{T}\left(3+\frac{E_g}{k_B T}\right)\right] \quad (5.86)$$
$$= -\frac{1}{2(V_{bi}+V_R)}\left(\frac{k_B}{q}\right)\left[\ln\left\{\frac{N_A N_D}{n_i^2(T)}\right\} - \left(3+\frac{E_g}{k_B T}\right)\right].$$

At $T = 300$ K, using $V_{bi} = 0.7$ V, $V_R = 0$ V, $k_B/q = 8.62\times10^{-5}$ eV K^{-1}, $N_A = 1\times10^{16}$ cm^{-3}, $N_D = 2\times10^{15}$ cm^{-3}, $n_i = 1.45\times10^{10}$ cm^{-3}, $E_g = 1.12$ eV, we have

$$\left(\frac{1}{C_{DS}}\right)\frac{\mathrm{d}C_{DS}(T)}{\mathrm{d}T} = -\frac{1}{2(0.7+0)}\left(8.62\times10^{-5}\right)$$
$$\times\left[\ln\left\{\frac{1\times10^{16}\times2\times10^{15}}{(1.45\times10^{10})^2}\right\} - \left\{3+\frac{1.12}{8.62\times10^{-5}\times300}\right\}\right] \quad (5.87)$$
$$= -6.157\times10^{-5}[\ln(9.51\times10^{10}) - 3 - 43.31]$$
$$= -6.157\times10^{-5}[25.28 - 46.31]$$
$$= 1.295\times10^{-4}/°\mathrm{C}.$$

At $V_R = 5$ V,

$$\left(\frac{1}{C_{DS}}\right)\frac{\mathrm{d}C_{DS}(T)}{\mathrm{d}T} = -\frac{1}{2(0.7+5)}\left(8.62\times10^{-5}\right)\left[25.28 - 46.31\right] \quad (5.88)$$
$$= 1.59\times10^{-4}/°\mathrm{C}.$$

At $V_R = 0$ V, when the temperature rises by 300 °C, the junction capacitance increases by 39%, whereas at $V_R = 5$ V, it does so by 4.8%.

5.9 Digital CMOS circuits in −196 °C to 270 °C range

Investigations by Prince and co-workers showed that satisfactory static and dynamic performance characteristics are obtainable from CMOS ICs, both unbuffered and buffered types, such as NAND gates, up to 270 °C (Prince *et al* 1980). Above this temperature, the high magnitude of the p-well substrate leakage current was the main hurdle. The cause of failure was the large value of the output low voltage V_{OL} at high temperatures arising from the inability of n-channel transistors to sink the leakage currents.

Enormous benefits are derived by operating MOSFET digital ICs below room temperature. Some of the advantages are:

(i) an increase in speed, which is made possible by enhanced mobility and conductance;
(ii) slow-down of degradation mechanisms, such as chemical reactions and electromigration, which are promoted by thermal energy;
(iii) noise reduction at low temperatures; and
(iv) tighter packing density due to easy heat withdrawal (Gaensslen *et al* 1977).

Particular interest in CMOS circuits is focused towards the increase in opposition to latchup phenomena as the temperature decreases (Estreich and Dutton 1982). Latchup is suppressed by using special structures such as guard rings, deep-trench isolations or epitaxial layers.

Latchup triggering starts by one of the following conditions: overvoltage stress, capacitive coupling or transitory irradiation. Latchup sustenance requires that a minimum current level called the holding current be maintained by the supply. On bringing the temperature down from 300 K to 77 K, it was found that the sustaining current increased by a factor of 2–4 (Dooley and Jaeger 1984). However, temperatures lower than 77 K may be necessary if complete disappearance of latchup is desired. Many n- and p-type epitaxial twin tub CMOS structures examined in the temperature range 77 K–400 K were deemed as latchup-free between 100 K and 200 K (Sangiorgi *et al* 1986).

The vulnerability of a CMOS circuit to latchup is decided by two important parameters: (i) the current gains of the parasitic bipolar transistors and (ii) related distributed base–emitter shunting resistances. Both these parameters decrease as the temperature falls (Shoucair 1988).

5.10 Discussion and conclusions

An increase in temperature affects the MOS device behavior favorably in some respects and unfavorably in others. The circuit designer must keep these variational trends in mind, along with their physical origins. In many situations, these trends can be turned to one's advantage and gainfully utilized in establishing proper circuit operation. A proper empathy of thermal limitations of MOSFET paves the way towards successful circuit design.

Review exercises

5.1. Write down the equation for the threshold voltage of a MOSFET and explain the meanings of the symbols used. Applying this equation show that the TC of threshold voltage V_{Th} of a MOSFET depends on the TC of the metal–semiconductor work function ϕ_{ms} and that of the bulk potential ϕ_F of the semiconductor.

5.2. Show that the intrinsic carrier concentration n_i in a semiconductor of bandgap E_g varies with temperature T according to the equation

$$\frac{\partial n_i}{\partial T} = \left(\frac{n_i}{2T}\right)\left(3 + \frac{E_g}{k_B T}\right), \tag{5.89}$$

where k_B is Boltzmann's constant.

5.3. Prove that the variation of the metal–semiconductor work function ϕ_{ms} with temperature T can be described by the equation

$$\frac{\partial \phi_{ms}}{\partial T} = \frac{\phi_{ms}}{T} + \left(\frac{k_B}{q}\right)\left(3 + \frac{E_g}{k_B T}\right), \tag{5.90}$$

where E_g is the bandgap of the semiconductor and k_B is Boltzmann's constant.

5.4. Show that the bulk potential ϕ_F of a semiconductor varies with temperature T according to the equation

$$\frac{\partial \phi_F}{\partial T} = \frac{\phi_F}{T} - \frac{3}{2}\left(\frac{k_B}{q}\right) - \frac{E_g}{2qT}, \tag{5.91}$$

where q is the electronic charge and E_g is the bandgap of the semiconductor.

5.5. Prove that the temperature dependence of the threshold voltage of an n-channel enhancement-mode MOSFET can be expressed by the equation

$$\frac{\partial V_{Th}}{\partial T} = \frac{\phi_{ms}}{T} + \left(\frac{k_B}{q}\right)\left(3 + \frac{E_g}{k_B T}\right) + \left\{2 + \left(\frac{t_{ox}}{\varepsilon_0 \varepsilon_{ox}}\right)\sqrt{\frac{\varepsilon_0 \varepsilon_s q N_A}{\phi_F}}\right\}. \tag{5.92}$$

The symbols are explained as follows: T denotes the temperature in absolute scale, ϕ_{ms} is the metal–semiconductor work function, k_B is the Boltzmann's constant, q is the electronic charge, E_g is the bandgap of the semiconductor, t_{ox} is the thickness of the gate oxide, ϕ_F is the bulk potential of the semiconductor and N_A is the doping concentration of p-substrate. The three epsilons ε_0, ε_{ox}, ε_s denote the permittivity of free space, the relative permittivity of silicon dioxide (SiO_2) and the relative permittivity of silicon (Si), respectively. Taking a representative example of a MOSFET with typical structural parameters, show that $\partial V_{Th}/\partial T = -2.9$ mV K^{-1}.

5.6. What are the seven components of the on-resistance of a MOSFET? Which two of these seven components are predominant?

5.7. Show that the channel resistance $R_{CHANNEL}(T)$ of a MOSFET at a temperature T is given by

$$R_{CHANNEL}(T) = R_{CHANNEL}(T_0) \times \frac{(T/T_0)^n}{1 + \frac{\kappa(T - T_0)}{V_{GS} - V_{Th}(T_0)}} \tag{5.93}$$

where κ is the TC of threshold voltage V_{Th}.

5.8. Taking into account the temperature dependence of the resistances of the channel and drift regions of a MOSFET, show that the temperature dependence of the on-resistance $R_{DS(ON)}$ of a MOSFET can be expressed as

$$R_{DS(ON)}(T) = R_{DS(ON)}(T_0)(T/T_0)^n. \quad (5.94)$$

5.9. Prove that the transconductance g_m of a MOSFET decreases with temperature T according to the relation

$$g_m(T) = g_m(T_0) \times (T/T_0)^{-n}. \quad (5.95)$$

5.10. Comment on the variation of BV_{DSS} and I_{DSS} of a MOSFET with temperature.

5.11. What is meant by the ZTC of a MOSFET? Using the equation for drain–source current I_{DS} of a MOSFET in the saturation region, and applying the ZTC condition $dI_{DS}/dT = 0$, derive the following equation for the voltage bias V_{GS}(ZTC)

$$V_{GS}(\text{ZTC}) = V_{Th}(T) + \frac{T\kappa}{n}, \quad (5.96)$$

where κ is the TC of threshold voltage V_{Th}, and n is the index in the equation for the temperature dependence of mobility.

5.12. Why is a MOSFET faster in speed than a bipolar transistor? Explain why a MOSFET experiences much smaller excursions in switching parameters with temperature.

5.13. Following Shoucair's analysis, show that there are two ZTC points for a MOSFET. One ZTC point lies in the linear region and the other ZTC point is in the saturation region of current–voltage characteristics. Write the analytical equation derived by him for ZTC gate–source biasing voltages V_{GS}(ZTC) applied to a MOSFET with known parameters (p_0,q_0) over the temperature range (T_1, T_2).

5.14. Prove, following Shoucair, that in the linear and saturation regions of MOSFET operation, the equations for drain–source current I_{DS} are, respectively,

$$I_{DS} = \mu(T)C_{ox}\left(\frac{W}{L}\right)\left\{-\frac{T_0}{1.5}p_0\right\}V_{DS}, \quad (5.97)$$

$$I_{DS} = (1/2)\mu(T)C_{ox}\left(\frac{W}{L}\right)\left\{-\frac{mT_0}{1.5}p_0\right\}^m, \quad (5.98)$$

where p_0 denotes the TC of the threshold voltage and index $m = 2$. The remaining symbols have their usual connotations.

5.15. Explain the term 'transition temperature' in the context of changeover of leakage current of a MOSFET from the generation to diffusion component. Write the equations for the generation and diffusion components of leakage current. What is the typical transition temperature for a CMOS device? Why does an n-channel MOSFET show a higher leakage current than a p-channel MOSFET?

5.16. What are the guidelines laid out by Shoucair for designing a two-stage topology of CMOS OP-AMP?

5.17. What are the overlap and junction capacitances in a MOSFET? Which of the two capacitances is more temperature sensitive?

5.18. Show that the drain–source capacitance C_{DS} of a MOSFET changes with temperature T according to the following equation

$$\left(\frac{1}{C_{DS}}\right)\frac{\mathrm{d}C_{DS}(T)}{\mathrm{d}T} = -\frac{1}{2(V_{bi}+V_R)}\left(\frac{k_B}{q}\right)\left[\ln\left\{\frac{N_A N_D}{n_i^2(T)}\right\} - \left(3 + \frac{E_g}{k_B T}\right)\right], \tag{5.99}$$

where V_{bi} is the built-in potential, V_R is the applied reverse voltage, k_B is Boltzmann's constant, q is the electronic charge, n_i is the intrinsic carrier concentration, E_g is the energy gap, and N_A, N_D are the acceptor and donor concentrations, respectively.

5.19. Mention three advantages of operating MOSFET digital ICs below room temperature. Discuss the effect of temperature on latchup in a CMOS circuit.

References

Arora N 2012 *MOSFET Models for VLSI Circuit Simulation: Theory and Practice* (Vienna: Springer)

Dooley J G and Jaeger R C 1984 Temperature dependence of latchup in CMOS circuits *IEEE Electron Device Lett.* **5** 41–3

Estreich D B and Dutton R W 1982 Modeling latch-up in CMOS integrated circuits *IEEE Trans. Comput.-Aided Des. Integr. Circuits Syst.* **4** 157–62

Fairchild Semiconductor Corporation 2000 MOSFET basics *Fairchild Semiconductor Corporation* www.fairchildsemi.com/application-notes/AN/AN-9010.pdf

Gaensslen F H, Rideout V L, Walker E J and Walker J J 1977 Very small MOSFETs for low-temperature operation *IEEE Trans. Electron Devices* **24** 218–29

Prince J L, Draper B L, Rapp E A, Kronberg J N and Fitch L T 1980 Performance of digital integrated circuits at very high temperatures *IEEE Trans. Compon. Hybrids Manuf. Technol.* **3** 571–9

Sangiorgi E, Johnston R L, Pinto M R, Bechtold P F and Fichtner W 1986 Temperature dependence of latch-up phenomena in scaled CMOS structures *IEEE Electron Device Lett.* **7** 28–31

Sattar A and Tsukanov V 2007 MOSFETs withstand stress of linear-mode operation *Power Electron. Technol.* April 2007 34–9

Shoucair F S and Early J M 1984 High-temperature diffusion leakage current-dependent MOSFET small signal conductance *IEEE Trans. Electron Devices* **31** 1866–72

Shoucair F S, Hwang W and Jain P 1984 Electrical characteristics of large scale integration (LSI) MOSFETs at very high temperatures: parts I and II *Microelectron. Rel.* **24** 465–85 and 487–510

Shoucair F S 1986 Design considerations in high temperature analog CMOS integrated circuits *IEEE Trans. Compon. Hybrids Manuf. Technol.* **9** 242–51

Shoucair F S 1988 High-temperature latchup characteristics in VLSI CMOS circuits *IEEE Trans. Electron Devices* **35** 2424–6

Siu C 2022 *Electronic Devices, Circuits, and Applications* (Cham: Springer Nature)

STMicroelectronics 2006 How to achieve the threshold voltage thermal coefficient of the MOSFET acting on design parameter *STMicroelectronics Application Note* www.bdtic.com/DownLoad/ST/Application_Note/AN2386.pdf

Suzuki K 2016 *Bipolar Transistor and MOSFET Device Models* (Sharjah: Bentham Science Publishers)

Tsividis Y and Colin M 2012 *Operation and Modeling of the MOS Transistor* 3rd edn (Cary, NC: Oxford University Press) International Edition

Wang R, Dunkley J, DeMassa T A and Jelsma L F 1971 Threshold voltage variations with temperature in MOS transistors *IEEE Trans. Electron Devices* **18** 386–8

Chapter 6

The influence of temperature on the performance of silicon–germanium heterojunction bipolar transistors

The constraints faced in designing homojunction bipolar junction transistors (BJTs) of the desired current gain and switching frequency necessitate trade-offs and compromise solutions. In this chapter, the design flexibility offered by the heterojunction structure is explained, and the heterojunction bipolar transistor (HBT) fabrication process is described. A comparative analysis between silicon BJTs and Si/SiGe HBTs is presented. The superior cryogenic performance of HBTs compared to BJTs in terms of current gain and frequency response characteristics is elaborated.

6.1 Introduction

A conventional BJT, which is made of the same indistinguishable silicon material throughout, is often referred to as a homojunction BJT because of its homogeneous composition. Two important parameters of a BJT are its current gain and switching speed. In order to maximize these parameters, the available device structural parameters to be varied are the emitter and base doping concentrations and the base width. To increase the current gain, the ratio of these doping concentrations must be large. Only then will there be a large concentration gradient from the emitter to base, so that a high emitter injection efficiency will be achieved. For a large ratio between the emitter and base doping concentrations, the base doping density must be low. This means that the depletion region extends to a larger depth

doi:10.1088/978-0-7503-5072-3ch6

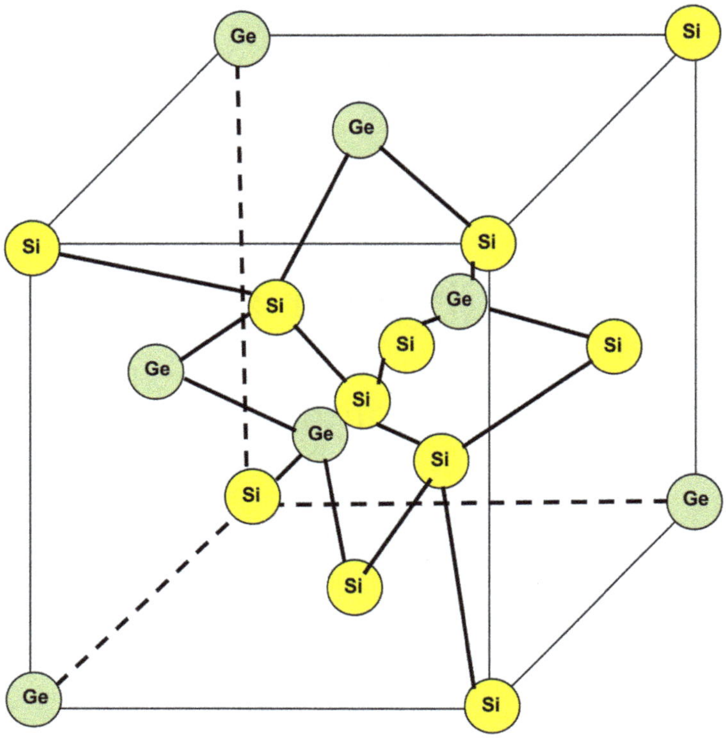

Figure 6.1. The structure of a silicon–germanium crystal.

on the base side at a low voltage when the collector–base junction is reverse biased. An obvious drawback is that this junction becomes susceptible to punch-through breakdown at a low voltage. To increase the punch-through breakdown voltage, a larger base width is necessary. But in a transistor with a large base width, the carriers injected from the emitter into the broader base take a longer time to cross the base and enter the collector, i.e. the transit time of carriers through the base region is lengthened. As a result, the transistor becomes slower in speed. Thus we must either sacrifice the punch-through breakdown voltage capability or the switching speed if we insist on keeping the current gain high.

Considering an alternative route, to increase the speed of the device the base must be thin. If the base is thin, its doping concentration must be high to prevent spreading of the depletion region across the base to the emitter edge causing punch-through breakdown. But a high doping concentration of the base lowers the emitter injection efficiency, and hence the current gain of the transistor. In effect, the transistor design involves a trade-off between different parameters. In view of these constraints, a compromise must be made between the electrical parameters to achieve the best solution suitable for a given application.

As the name suggests, an HBT contains a heterojunction. This heterojunction is a junction between two different materials. One material comprising this heterojunction is silicon and the other material is silicon–germanium, an alloy of silicon and germanium with the chemical formula $Si_{1-x}Ge_x$ where x is the mole fraction of germanium in the alloy with a value from 0 to 1 (Ioffe Institute 2015). Figure 6.1 illustrates the crystalline structure of silicon–germanium.

The silicon–germanium alloy is an indirect bandgap semiconductor. Its bandgap is given by the equations (Virginia Semiconductor 2002)

$$E_{\mathrm{g(indirect)}} = 1.155 - 0.43x + 0.206x^2 \text{ eV at } T = 300 \text{ K}, \quad \text{for } x < 0.85 \tag{6.1}$$

$$E_{\mathrm{g(indirect)}} = 2.010 - 1.27x \text{ eV at } T = 300 \text{ K}, \quad \text{for } x > 0.85. \tag{6.2}$$

The bandgap of SiGe is smaller than that of Si. The amount by which the SiGe bandgap is smaller than the silicon bandgap depends on the value of x. In an HBT, two regions, the emitter and collector, are made of silicon. The third region, which is the HBT base, is made of the lower bandgap material SiGe. The fact that the material used for base (SiGe) has a smaller bandgap than the emitter material (Si) has a favorable outcome. If an n–p–n transistor is considered, the injection of holes from the p-base into the n-emitter is prevented. In this way, the emitter injection efficiency is enhanced. The HBT structure with near-lattice matched or pseudomorphic epitaxial layers is well suited for n–p–n HBT fabrication because the band offset takes place exclusively in the valence band. No barrier is introduced opposing the injection of electrons from the emitter into the base (Shiraki and Usami 2011). The difference in operation between a Si BJT and a SiGe HBT can be understood by reference to figure 6.2.

In consequence of the above efforts, one does not need to rely on the ratio of doping concentrations of the emitter and base for achieving the required emitter injection efficiency for a high current gain value. Instead, the difference of bandgaps between the emitter and base regions takes care of this efficiency. This means that the limitation imposed on the base carrier concentration from the viewpoint of current gain is removed. A high carrier concentration can be used for this region. Obviously, the base region can be made thin without losing the capability for punch-through breakdown. Hence, the high current gain and punch-through breakdown capabilities can be provided in a device along with a high switching speed. By using a graded profile for germanium incorporation in the base, with the germanium concentration decreasing from the emitter edge towards the collector edge, an accelerating electric field is created across the base layer. This electric field helps the carriers injected from the emitter into the base to be swept fast into the collector by a field-assisted transport mechanism. Thus the speed of the transistor is further increased. In this way, the HBT structure provides an all-in-one solution for transistor designers for complying with the specifications of high current gain, large breakdown voltage and ultrafast speed.

6.2 HBT fabrication

Figure 6.3 shows the schematic cross-section of SiGe HBT. The high-quality p-type SiGe base layer is grown epitaxially. The epitaxy is carried out either using reduced pressure chemical vapor deposition (RPCVD) or ultra-high vacuum CVD (UHVCVD). There are two modalities of epitaxial growth: selective and

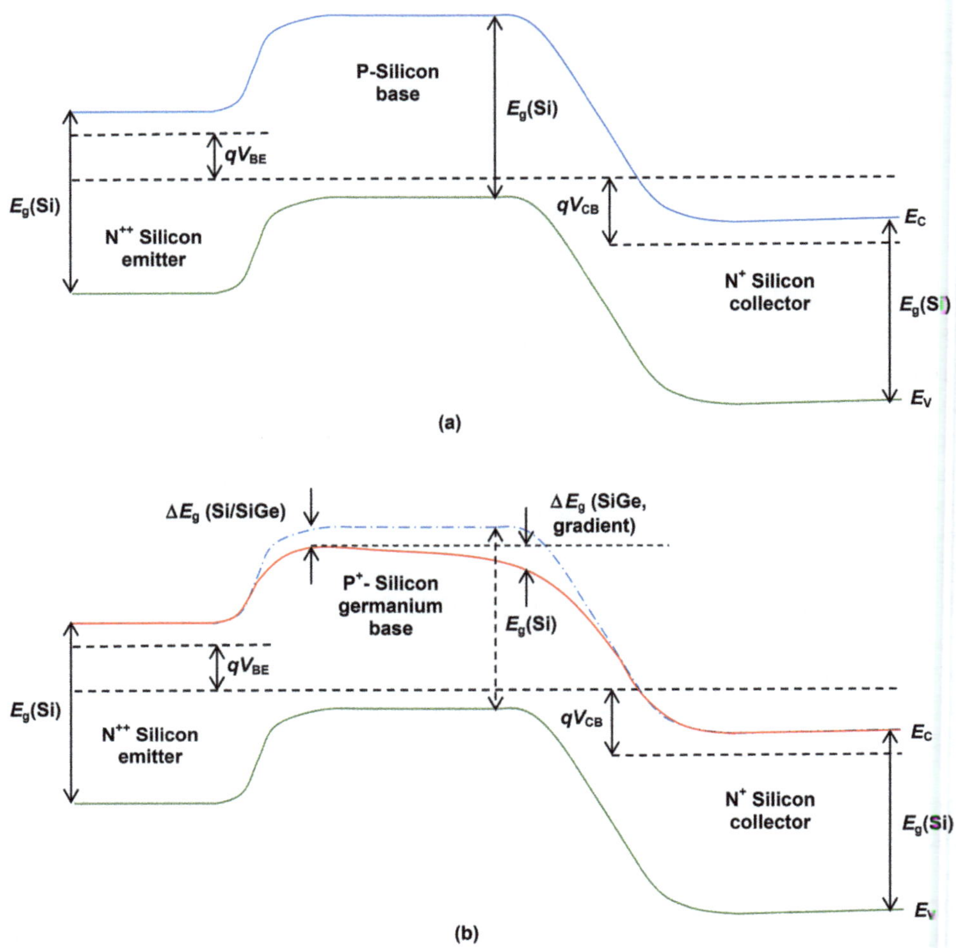

Figure 6.2. Energy band diagrams. (a) Silicon bipolar transistor and (b) silicon–germanium HBT compared to a silicon bipolar transistor. ΔE_g(Si, SiGe) is the bandgap difference between Si and SiGe at the emitter edge of the base to be denoted in terms of position x in the base as $\Delta E_{gB,Ge}(x = 0)$, and ΔE_g(SiGe, gradient) is the bandgap difference in the SiGe base from the emitter edge to collector edge to be represented by $\Delta E_{gB,Ge}(x = W_B)$ where W_B is base width. Then $\Delta E_{gB,Ge}(x = W_B) - \Delta E_{gB,Ge}(x = 0) = \Delta E_{g,Ge}(\text{grade})$.

non-selective epitaxy. Selective epitaxy enables the fabrication of self-aligned structures with small overlaps, minimizing parasitic effects. Non-selective epitaxy is a simpler process and provides better control over the thickness uniformity of the SiGe layer. Boron doping is performed *in situ* during epitaxial layer growth. Incorporating a small concentration of carbon $\leqslant 1\%$ in the SiGe layer is found to suppress boron out-diffusion appreciably during subsequent thermal processing (Hållstedt 2004). It thus prevents the base layer from enlarging and preserves the narrow base width necessary for a small transit time (Mitrovic *et al* 2005). After the base layer deposition, the polysilicon emitter layer is formed.

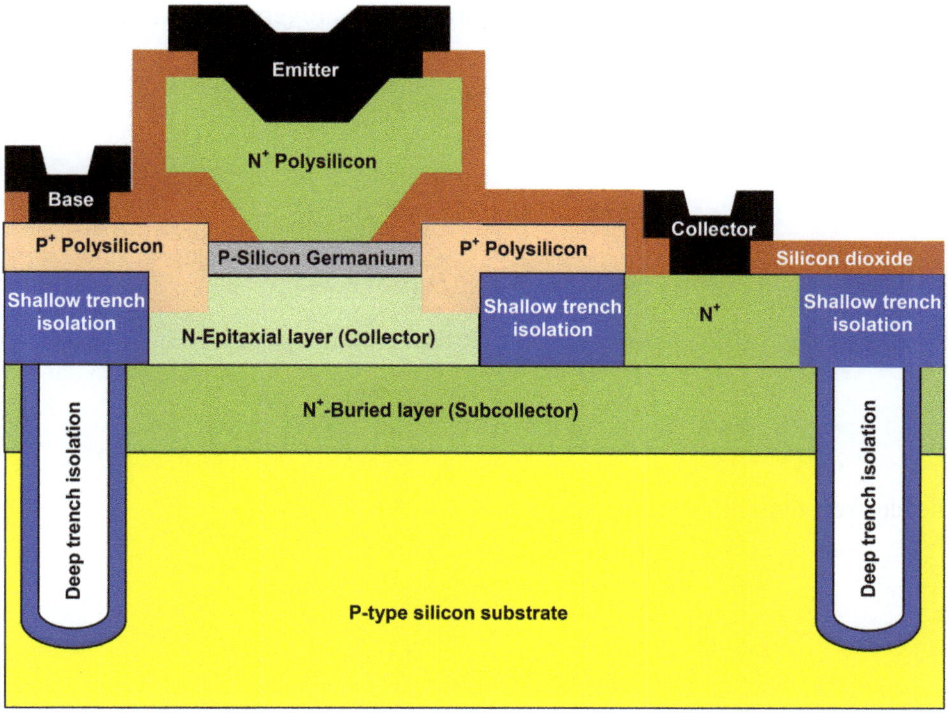

Figure 6.3. Schematic cross-sectional diagram of the silicon/silicon–germanium HBT.

6.3 Current gain and forward transit time of $Si/Si_{1-x}Ge_x$ HBT

Consider an n–p–n HBT with uniformly doped emitter and base regions. Let N_{DE} be the doping concentration in the n-emitter layer of width W_E and N_{AB} be the acceptor concentration in the p-base layer of width W_B. Let $D_n(T)$ be the diffusion coefficient of electrons at temperature T; electrons are minority carriers in the base. Let D_p be the diffusion coefficient of holes at temperature T; holes are minority carriers in the emitter. Let ΔE_g be the energy bandgap difference between the emitter and base semiconductor materials. Then the common-emitter current gain β is (Basu and Sarkar 2011)

$$\beta = \left(\frac{N_{DE}}{N_{AB}}\right)\left(\frac{W_E}{W_B}\right)\left\{\frac{D_n(T)}{D_p(T)}\right\}\exp\left(\frac{\Delta E_g}{k_B T}\right), \tag{6.3}$$

where, in accordance with Einstein's equation, the electron and hole diffusion coefficients $D_n(T)$, $D_p(T)$ are related to corresponding mobilities $\mu_n(T)$, $\mu_p(T)$ as

$$D_n(T) = \left(\frac{k_B T}{q}\right)\mu_n(T) \tag{6.4}$$

$$D_p(T) = \left(\frac{k_B T}{q}\right)\mu_p(T). \tag{6.5}$$

The forward transit time τ_F through the HBT is equal to (emitter transit time (τ_E) + base transit time (τ_B)), written as

$$\tau_F = \tau_E + \tau_B, \tag{6.5}$$

where

$$\tau_E(T) = W_E^2/\{2\beta D_p(T)\} \tag{6.7}$$

$$\tau_B(T) = W_B^2/\{2D_n(T)\}. \tag{6.8}$$

For $Si/Si_{1-x}Ge_x$ HBT, the equation for current gain $(\beta)_{Si/SiGe}$ is

$$(\beta)_{Si/SiGe} = \left(\frac{N_{DE}}{N_{AB}}\right)\left(\frac{W_E}{W_B}\right)\left\{\frac{D_{n,\,SiGe}(T)}{D_{p,\,Si}(T)}\right\}\exp\left(\frac{\Delta E_g}{k_B T}\right). \tag{6.9}$$

The electron mobility $\mu_{n,Si}$ in Si is

$$\mu_{n,\,Si}(T) \propto T^{-2.42}$$

or

$$\mu_{n,\,Si}(T) = K_1 T^{-2.42} \tag{6.10}$$

$$\therefore K_1 = \mu_{n,\,Si}(T)/T^{-2.42} = \mu_{n,\,Si}(T) \times T^{2.42}. \tag{6.11}$$

Since $\mu_{n,Si} = 1350\ \text{cm}^2\ \text{V}^{-1}\ \text{s}^{-1}$ at $T = 300$ K,

$$\therefore K_1 = 1350 \times T^{2.42} = 1350 \times (300)^{2.42} = 9.8772 \times 10^5\ \text{cm}^2\ \text{V}^{-1}\ \text{s}^{-1}\ \text{K}^{2.42} \tag{6.12}$$

$$\therefore \mu_{n,\,Si}(T) = 9.88 \times 10^5 T^{-2.42}. \tag{6.13}$$

The electron mobility in Ge is

$$\mu_{n,\,Ge}(T) \propto T^{-1.66}$$

or

$$\mu_{n,\,Ge}(T) = K_2 T^{-1.66} \tag{6.14}$$

$$\therefore K_2 = \mu_{n,\,Ge}(T)/T^{-1.66} = \mu_{n,\,Ge}(T) \times T^{1.66}. \tag{6.15}$$

Since $\mu_{n,Ge} = 3900\ \text{cm}^2\ \text{V}^{-1}\ \text{s}^{-1}$ at $T = 300$ K,

$$\therefore K_2 = 3900 \times T^{1.66} = 3900 \times (300)^{1.66} = 1.294 \times 10^4\ \text{cm}^2\ \text{V}^{-1}\ \text{s}^{-1}\ \text{K}^{1.66} \tag{6.16}$$

$$\therefore \mu_{n,\,Ge}(T) = 1.29 \times 10^4 T^{-1.66}. \tag{6.17}$$

Now,

$$\mu_{n,\,SiGe}(T) = (1 - x)\mu_{n,\,Si}(T) + x\mu_{n,\,Ge}(T) = 9.88 \times 10^5(1 - x)T^{-2.42} + 1.29 \times 10^4 xT^{-1.66} \tag{6.18}$$

$$\therefore D_{n,\,SiGe}(T) = \left(\frac{k_B T}{q}\right)\left\{9.88 \times 10^5(1 - x)T^{-2.42} + 1.29 \times 10^4 xT^{-1.66}\right\}. \tag{6.19}$$

The hole mobility in Si is

$$\mu_{p,\,Si}(T) \propto T^{-2.2}$$

or

$$\mu_{p,\,Si}(T) = K_3 T^{-2.2}. \tag{6.20}$$

Since $\mu_{p,Si} = 500$ cm^2 V^{-1} s^{-1} at $T = 300$ K,

$$\therefore K_3 = \mu_{p,\,Si}(T)/T^{-2.2} = \mu_{p,\,Si}(T) \times T^{2.2} = 500 \times (300)^{2.2} = 1.408 \times 10^8 \text{ cm}^2\,\text{V}^{-1}\,\text{s}^{-1}\,\text{K}^{2.2} \tag{6.21}$$

$$\therefore \mu_{p,\,Si}(T) = 1.408 \times 10^8 T^{-2.2} \tag{6.22}$$

$$\therefore D_{p,\,Si}(T) = \left(\frac{k_B T}{q}\right)1.408 \times 10^8 T^{-2.2}. \tag{6.23}$$

The difference between energy bandgaps of the Si emitter and Si_xGe_{1-x} base is (Basu and Sarkar 2011, Shur 1995)

$$(\Delta E_g)_{Si/SiGe} = 0.43x - 0.0206x^2. \tag{6.24}$$

Further, the emitter and base transit times are written as

$$\begin{aligned}\{\tau_E(T)\}_{Si/SiGe} &= W_E^2/[2\beta D_{p,\,Si}(T)] = W_E^2/\left[2(\beta)_{\frac{Si}{SiGe}} D_{p,\,Si}(T)\right] \\ &= W_E^2/\left[2\left(\frac{N_{DE}}{N_{AB}}\right)\left(\frac{W_E}{W_B}\right)\left\{\frac{D_{n,\,SiGe}(T)}{D_{p,\,Si}(T)}\right\}\exp\left(\frac{\Delta E_g}{k_B T}\right)D_{p,\,Si}(T)\right] \\ &= W_E^2/\left[2\left(\frac{N_{DE}}{N_{AB}}\right)\left(\frac{W_E}{W_B}\right)D_{n,\,SiGe}(T)\exp\left(\frac{\Delta E_g}{k_B T}\right)\right]\end{aligned} \tag{6.25}$$

$$\{\tau_B(T)\}_{Si/SiGe} = W_B^2/\{2D_{n,\,SiGe}(T)\}. \tag{6.26}$$

The model allows the prediction of HBT device performance with respect to current gain and forward transit time as a function of different emitter/base compositions and over a wide range of operating temperatures.

6.4 Comparison between Si BJT and Si/SiGe HBT

Figures 6.4 and 6.5 show the shifts in current–voltage characteristics and current gain–collector current curves of a SiGe HBT with temperature. In a silicon BJT, the temperature dependence of current gain is mainly controlled by the doping-induced bandgap narrowing of the emitter and base regions. To achieve high emitter injection efficiency, the doping level of the emitter region is kept much higher than that of the base region. Hence, the bandgap narrowing in the emitter region is much more pronounced than in the base region. If ΔE_{gE} is the bandgap narrowing of the emitter and ΔE_{gB} is the bandgap narrowing of the base, the difference between the bandgap narrowing of emitter and base

$$= \Delta E_{gE} - \Delta E_{gB} \Delta E_{gEB}. \tag{6.27}$$

The current gain of a silicon BJT is

$$\beta_{Si}(T) \propto \exp\{-(\Delta E_{gE} - \Delta E_{gB})/(k_B T)\} \propto \exp\{-(\Delta E_{gEB})/(k_B T)\}. \tag{6.28}$$

Consequently, the current gain of a silicon BJT exponentially decreases with temperature, resulting in a substantial reduction of gain when a temperature $T = 77$ K is reached.

In a Si/SiGe HBT, the inclusion of germanium in the silicon of the base layer to form the SiGe base layer increases the bandgap narrowing ΔE_{gB} of base. Thus the

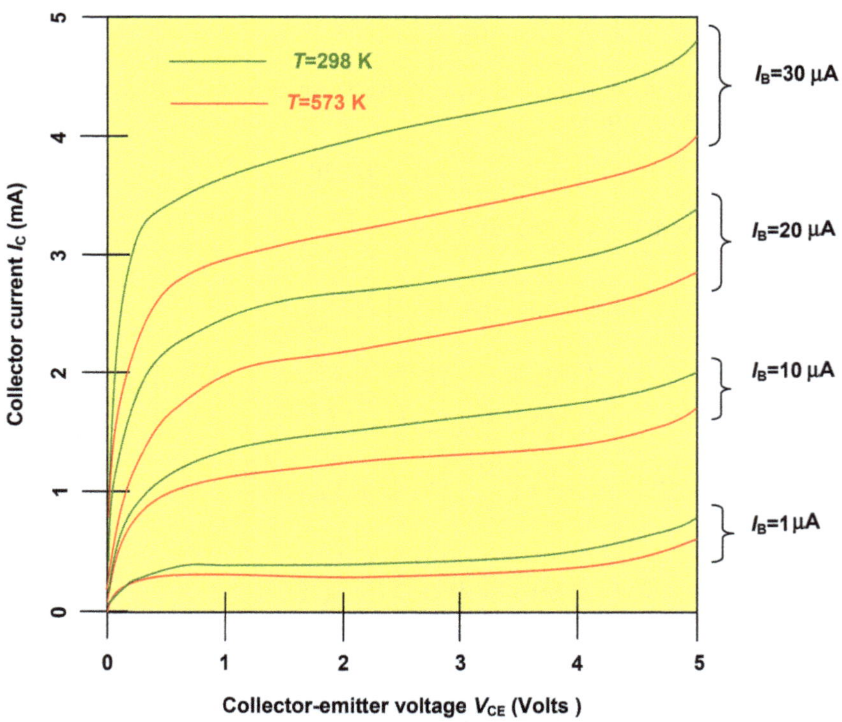

Figure 6.4. Change in current–voltage characteristics of a Si/SiGe HBT with temperature.

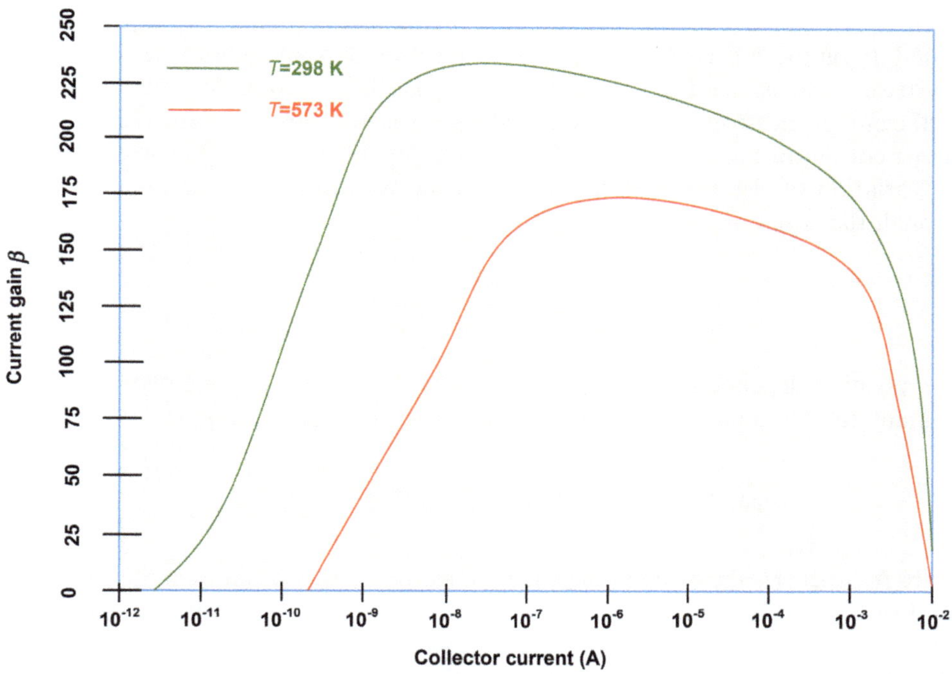

Figure 6.5. Shift in current gain–collector current curves of a silicon/silicon–germanium HBT with temperature.

bandgap narrowing effects in the emitter ΔE_{gE} and base ΔE_{gB} become comparable. The difference $\Delta E_{gE} - \Delta E_{gB} = \Delta E_{gEB}$ becomes smaller. Therefore, the degradation of current gain $\beta_{Si/SiGe}(T)$ of Si/SiGe HBT with a decrease in temperature becomes slower compared to $\beta_{Si}(T)$ of Si BJT. Stated in other words, the bandgap reduction occurring due to the presence of germanium in the SiGe base layer of Si/SiGe HBT serves to compensate the bandgap narrowing effect of heavy doping in the emitter on the bandgap of the emitter, whereby the current gain of the Si/SiGe HBT at a given temperature is higher than that of Si BJT.

For a deeper introspection into the operation of HBT, let us recall the Moll–Ross relationship, generalized by Kroemer. This relationship, valid for low injection conditions, is an important relationship because it shows that the collector current density is not affected by the nature of doping in the base but by the total quantity of charges present in the base. It relates the collector current density J_C of a bipolar transistor having a uniformly or non-uniformly doped base with the applied base–emitter voltage V_{BE} at a given temperature T as (Cressler and Niu 2003)

$$J_C = \frac{q\left\{\exp\left(\frac{qV_{BE}}{k_B T}\right) - 1\right\}}{\int_{x=0}^{x=W_B} \frac{p_B(x)dx}{D_{nB}(x)n_{iB}^2(x)}} \tag{6.29}$$

where $x = 0$ is the boundary of the base on the emitter side, $x = W_B$ is the base boundary on the collector side, $p_B(x)$ is the position-dependent hole concentration in the base varying with position x, $D_{nB}(x)$ is the position-dependent diffusion coefficient of electrons in the base and $n_{iB}(x)$ is the position-dependent intrinsic carrier concentration in the base. The position-dependence of $D_{nB}(x)$ originates from the variation of electron mobility $\mu_{nB}(x)$ in the base due to the germanium profile through the equation

$$D_{nB}(x) = \left(\frac{k_B T}{q}\right)\mu_{nB}(x). \tag{6.30}$$

The position-dependence of $n_{iB}(x)$ arises from the bandgap changes caused by germanium. The square of intrinsic carrier concentration in SiGe is

$$n_{iB}^2(x) = N_{C,\,SiGe}(x)N_{V,\,SiGe}(x)\exp\left\{-\frac{E_{gB}(x)}{k_B T}\right\}, \tag{6.31}$$

where $N_{C,SiGe}(x)$ is the position-dependent effective density of states in the conduction band of SiGe, $N_{V,SiGe}(x)$ is the position-dependent effective density of states in the valence band of SiGe and $E_{gB}(x)$ is the position-dependent energy bandgap of the SiGe base layer. In an HBT fabricated with a linearly graded base layer, suppose E_{gB0} = silicon bandgap at zero doping = 1.12 eV, $\Delta E_{gB,A}$ is the bandgap narrowing of the base layer due to the acceptor impurity doping effect, $\Delta E_{gB,Ge}(x = 0)$ is the bandgap offset of the base layer at $x = 0$ and $\Delta E_{gB,Ge}(x = W_B)$ is the bandgap offset of the base layer at $x = W_B$. Using these symbols, we can write

$$\begin{aligned} E_{gB}(x) = E_{gB0} - \Delta E_{gB,\,A} + \{\Delta E_{gB,\,Ge}(x = 0) - \Delta E_{gB,\,Ge}(x = W_B)\} \\ \times (x/W_B) - \Delta E_{gB,\,Ge}(x = 0) \end{aligned} \tag{6.32}$$

$$\begin{aligned} \therefore n_{ib}^2(x) &= N_{C,\,SiGe}(x)N_{V,\,SiGe}(x) \\ &\quad\times \exp\left[-\frac{E_{gB0} - \Delta E_{gB,\,A} + \{\Delta E_{gB,Ge}(x=0) - \Delta E_{gB,Ge}(x=W_B)\} \times (x/W_B) - \Delta E_{gB,Ge}(x=0)}{k_B T}\right] \\ &= N_{C,\,SiGe}(x)N_{V,\,SiGe}(x) \\ &\quad\times \exp\left[\frac{-E_{gB0} + \Delta E_{gB,\,A} - \{\Delta E_{gB,Ge}(x=0) - \Delta E_{gB,Ge}(x=W_B)\} \times (x/W_B) + \Delta E_{gB,Ge}(x=0)}{k_B T}\right] \\ &= N_{C,\,SiGe}(x)N_{V,\,SiGe}(x)\exp\left(-\frac{E_{gB0}}{k_B T}\right)\exp\left(\frac{\Delta E_{gB,\,A}}{k_B T}\right) \\ &\quad\times \exp\left[\left\{\frac{\Delta E_{gB,Ge}(x=W_B) - \Delta E_{gB,Ge}(x=0)}{k_B T}\right\} \times (x/W_B)\right]\exp\left\{\frac{\Delta E_{gB,Ge}(x=0)}{k_B T}\right\} \\ &= \frac{N_{C,\,SiGe}(x)N_{V,\,SiGe}(x)}{N_{C,\,Si}(x)N_{V,\,Si}(x)}N_{C,\,Si}(x)N_{V,\,Si}(x)\exp\left(-\frac{E_{gB0}}{k_B T}\right)\exp\left(\frac{\Delta E_{gB,\,A}}{k_B T}\right) \\ &\quad\times \exp\left[\left\{\frac{\Delta E_{gB,Ge}(x=W_B) - \Delta E_{gB,Ge}(x=0)}{k_B T}\right\} \times (x/W_B)\right]\exp\left\{\frac{\Delta E_{gB,Ge}(x=0)}{k_B T}\right\} \end{aligned} \tag{6.33}$$

Let us define

$$\frac{N_{C,\,SiGe}(x)N_{V,\,SiGe}(x)}{N_{C,\,Si}(x)N_{V,\,Si}(x)} = N \tag{6.34}$$

$$\Delta E_{gB,\,Ge}(x = W_B) - \Delta E_{gB,\,Ge}(x = 0) = \Delta E_{g,\,Ge}(\text{grade}). \tag{6.35}$$

Obviously,

$$N_{C,\,Si}(x)N_{V,\,Si}(x)\exp\left(-\frac{E_{gB0}}{k_B T}\right) = n_{i,\,Si}^2. \tag{6.36}$$

Hence,

$$\begin{aligned} n_{ib}^2(x) = Nn_{i,\,Si}^2 \exp\left(\frac{\Delta E_{gB,A}}{k_B T}\right)\exp\left[\left\{\frac{\Delta E_{g,Ge}(\text{grade})}{k_B T}\right\}(x/W_B)\right] \\ \times \exp\left\{\frac{\Delta E_{gB,Ge}(x=0)}{k_B T}\right\}. \end{aligned} \tag{6.37}$$

Substituting the expression for $n_{ib}^2(x)$ from (6.37) into (6.29) for J_C, we obtain

$$\begin{aligned} J_{C,\,SiGe} &= \frac{q\left\{\exp\left(\frac{qV_{BE}}{k_B T}\right) - 1\right\}}{\int_{x=0}^{x=W_B} \frac{p_B(x)\mathrm{d}x}{D_{nB}(x)Nn_{i,\,Si}^2 \exp\left(\frac{\Delta E_{gB,A}}{k_B T}\right)\exp\left[\left\{\frac{\Delta E_{g,Ge}(\text{grade})}{k_B T}\right\}(x/W_B)\right]\exp\left\{\frac{\Delta E_{gB,Ge}(x=0)}{k_B T}\right\}}} \\ &= \frac{q\left\{\exp\left(\frac{qV_{BE}}{k_B T}\right) - 1\right\}}{\left(\frac{1}{\widetilde{D_{nB}}}\right)\left(\frac{1}{\widetilde{N}}\right)N_{A,\,B}\left\{\frac{1}{n_{i,\,Si}^2\exp\left(\frac{\Delta E_{gB,A}}{k_B T}\right)}\right\}\left\{\frac{1}{\exp\left\{\frac{\Delta E_{gB,Ge}(x=0)}{k_B T}\right\}}\right\}\int_{x=0}^{x=W_B}\frac{\mathrm{d}x}{\exp\left[\left\{\frac{\Delta E_{g,Ge}(\text{grade})}{k_B T}\right\}(x/W_B)\right]}}, \end{aligned} \tag{6.38}$$

where $\widetilde{D_{nB}}$ is defined as the position-averaged diffusion coefficient across the base profile. $\widetilde{N}$ is defined as the position-averaged ratio of the effective densities of states in SiGe and Si, across the base profile. $N_{A,B}$ is the acceptor dopant concentration in the base:

$$\begin{aligned} \therefore J_{C,\,SiGe} &= q\frac{\widetilde{D_{nB}}\widetilde{N}}{N_{A,\,B}}\frac{n_{i,\,Si}^2\exp\left(\frac{\Delta E_{gB,A}}{k_B T}\right)\exp\left\{\frac{\Delta E_{gB,Ge}(x=0)}{k_B T}\right\}\left\{\exp\left(\frac{qV_{BE}}{k_B T}\right) - 1\right\}}{\int_{x=0}^{x=W_B}\exp\left[\{-\Delta E_{g,Ge}(\text{grade})\}\{x/(W_B k_B T)\}\right]\mathrm{d}x} \\ &= q\frac{\widetilde{D_{nB}}\widetilde{N}}{N_{A,\,B}}\frac{n_{i,\,Si}^2\exp\left(\frac{\Delta E_{gB,A}}{k_B T}\right)\exp\left\{\frac{\Delta E_{gB,Ge}(x=0)}{k_B T}\right\}\left\{\exp\left(\frac{qV_{BE}}{k_B T}\right) - 1\right\}}{\left[-\frac{1}{\{\Delta E_{g,Ge}(\text{grade})/(W_B k_B T)\}}\exp\left[\left\{-\Delta E_{g,Ge}(\text{grade})\right\}\{x/(W_B k_B T)\}\right]\right]_{x=0}^{x=W_B}} \end{aligned} \tag{6.39}$$

because

$$\int \exp(-ax)\mathrm{d}x = -\,(1/a)\exp(-ax). \tag{6.39a}$$

Therefore,

$$\begin{aligned}
J_{\mathrm{C,\,SiGe}} &= q\frac{\widetilde{D}_{\mathrm{nB}}\widetilde{N}}{N_{\mathrm{A,\,B}}} \times \frac{n_{\mathrm{i,\,Si}}^{2}\exp\left(\frac{\Delta E_{\mathrm{gB,A}}}{k_{\mathrm{B}}T}\right)\exp\left\{\frac{\Delta E_{\mathrm{gB,Ge}}(x=0)}{k_{\mathrm{B}}T}\right\}\left\{\exp\left(\frac{qV_{\mathrm{BE}}}{k_{\mathrm{B}}T}\right)-1\right\}}{\left[-\frac{1}{\left\{\Delta E_{\mathrm{g,Ge}}(\mathrm{grade})/\left(W_{\mathrm{B}}k_{\mathrm{B}}T\right)\right\}}\right]\left[\exp[\{-\Delta E_{\mathrm{g,Ge}}(\mathrm{grade})\}\{W_{\mathrm{B}}/(W_{\mathrm{B}}k_{\mathrm{B}}T)\}] - \exp[\{-\Delta E_{\mathrm{g,Ge}}(\mathrm{grade})\}\{0/(W_{\mathrm{B}}k_{\mathrm{B}}T)\}]\right]} \\
&= q\frac{\widetilde{D}_{\mathrm{nB}}\widetilde{N}}{N_{\mathrm{A,B}}}\frac{n_{\mathrm{i,\,Si}}^{2}\exp\left(\frac{\Delta E_{\mathrm{gB,\,A}}}{k_{\mathrm{B}}T}\right)\exp\left\{\frac{\Delta E_{\mathrm{gB,Ge}}(x=0)}{k_{\mathrm{B}}T}\right\}\left\{\exp\left(\frac{qV_{\mathrm{BE}}}{k_{\mathrm{B}}T}\right)-1\right\}}{\{-|(W_{\mathrm{B}}k_{\mathrm{B}}T)/\Delta E_{\mathrm{g,Ge}}(\mathrm{grade})\}\left[\exp[\{-\Delta E_{\mathrm{g,Ge}}(\mathrm{grade})\}\{1/(k_{\mathrm{B}}T)\}] - \exp(0)\right]} \\
&= q\frac{\widetilde{D}_{\mathrm{nB}}\widetilde{N}}{N_{\mathrm{A,B}}W_{\mathrm{B}}}\frac{n_{\mathrm{i,\,Si}}^{2}\exp\left(\frac{\Delta E_{\mathrm{gB,A}}}{k_{\mathrm{B}}T}\right)\{-|\Delta E_{\mathrm{g,Ge}}(\mathrm{grade})/(k_{\mathrm{B}}T)\}\exp\left\{\frac{\Delta E_{\mathrm{gB,Ge}}(x=0)}{k_{\mathrm{B}}T}\right\}\left\{\exp\left(\frac{qV_{\mathrm{BE}}}{k_{\mathrm{B}}T}\right)-1\right\}}{\left[\exp\{-\Delta E_{\mathrm{g,Ge}}(\mathrm{grade})/(k_{\mathrm{B}}T)\}-1\right]} \\
&= q\frac{\widetilde{D}_{\mathrm{nB}}\widetilde{N}}{N_{\mathrm{A,B}}W_{\mathrm{B}}}n_{\mathrm{i,\,Si}}^{2}\exp\left(\frac{\Delta E_{\mathrm{gB,A}}}{k_{\mathrm{B}}T}\right)\left\{\exp\left(\frac{qV_{\mathrm{BE}}}{k_{\mathrm{B}}T}\right)-1\right\} \\
&\quad\times\frac{\{\Delta E_{\mathrm{g,Ge}}(\mathrm{grade})/(k_{\mathrm{B}}T)\}\exp\left\{\frac{\Delta E_{\mathrm{gB,Ge}}(x=0)}{k_{\mathrm{B}}T}\right\}}{\left[1-\exp\{-\Delta E_{\mathrm{g,Ge}}(\mathrm{grade})/(k_{\mathrm{B}}T)\}\right]}.
\end{aligned} \tag{6.40}$$

In this equation, looking at the second factor after the multiplication sign (×), it is evident that the collector current density $J_{\mathrm{C,SiGe}}$ is directly proportional to the degree of bandgap grading $\Delta E_{\mathrm{g,Ge}}(\mathrm{grade})$ and exponentially dependent on the band offset $\Delta E_{\mathrm{gB,Ge}}(x = 0)$ caused by Ge at the edge of the base towards the emitter side. These two dependences empower the HBT designer with the capability to easily achieve the desired gain at any temperature.

If we focus attention on a Si BJT device which is fabricated using similar process parameters as the SiGe HBT, and assume that the germanium profile on the emitter side of the neutral base region in the SiGe HBT does not stretch far enough into the emitter to significantly alter the base current density, it is logical to infer that the base current density J_{B} of the Si BJT will be identical to that for SiGe HBT. Since, $\beta = J_{\mathrm{C}}/J_{\mathrm{B}}$, one can write for similarly constructed SiGe HBT and Si BJT (Cressler 1998)

$$\begin{aligned}
\frac{\beta_{\mathrm{SiGe}}}{\beta_{\mathrm{Si}}} = \frac{J_{\mathrm{C,\,SiGe}}}{J_{\mathrm{C,\,Si}}} &= \left\{\frac{\widetilde{N}_{\mathrm{C,\,SiGe}}(x)\widetilde{N}_{\mathrm{V,\,SiGe}}(x)}{\widetilde{N}_{\mathrm{C,\,Si}}(x)\widetilde{N}_{\mathrm{V,\,Si}}(x)}\right\}\left(\frac{\widetilde{D}_{\mathrm{nB,\,SiGe}}}{\widetilde{D}_{\mathrm{nB,\,Si}}}\right) \\
&\quad\times\frac{\{\Delta E_{\mathrm{g,\,Ge}}(\mathrm{grade})/(k_{\mathrm{B}}T)\}\exp\left\{\frac{\Delta E_{\mathrm{gB,\,Ge}}(x=0)}{k_{\mathrm{B}}T}\right\}}{[1-\exp\{-\Delta E_{\mathrm{g,\,Ge}}(\mathrm{grade})/(k_{\mathrm{B}}T)\}]},
\end{aligned} \tag{6.41}$$

where the superscript ~ over a variable means that position averaging has been performed over its value. The ratio $\beta_{\mathrm{SiGe}}/\beta_{\mathrm{Si}}$ represents the enhancement in current gain by use of SiGe over a silicon transistor.

Assuming constant doping in the base region, the forward base transit time of a SiGe HBT is given by Krömer's formula (Harame *et al* 1995)

$$
\begin{aligned}
\tau_{\mathrm{b,\,SiGe}} &= \int_{x=0}^{x=W_{\mathrm{B}}} \frac{n_{\mathrm{iB}}^2(x)}{N_{\mathrm{B}}(x)} \left[\int_{y=x}^{y=W_{\mathrm{B}}} \left\{ \frac{N_{\mathrm{B}}(y)\mathrm{d}y}{D_{\mathrm{nb}}(y) n_{\mathrm{iB}}^2(y)} \right\} \right] \mathrm{d}x \\
&= \int_{x=0}^{x=W_{\mathrm{B}}} \frac{n_{\mathrm{iB}}^2(x)}{N_{\mathrm{B}}(x)} \left[\frac{N_{\mathrm{B}}(y)}{D_{\mathrm{nb}}(y)} \int_{y=x}^{y=W_{\mathrm{B}}} \left\{ \frac{\mathrm{d}y}{n_{\mathrm{iB}}^2(y)} \right\} \right] \mathrm{d}x \\
&= \int_{x=0}^{x=W_{\mathrm{B}}} \frac{n_{\mathrm{iB}}^2(x)}{N_{\mathrm{B}}(x)} \left[\frac{N_{\mathrm{B}}(y)}{D_{\mathrm{nb}}(y)} \int_{y=x}^{y=W_{\mathrm{B}}} \left\{ \frac{\mathrm{d}y}{N n_{\mathrm{i,\,Si}}^2 \mathrm{e}^{\frac{\Delta E_{\mathrm{gB,\,A}}}{k_{\mathrm{B}}T}} \mathrm{e}^{\frac{\Delta E_{\mathrm{g,Ge}}(\mathrm{grade})y}{W_{\mathrm{B}}k_{\mathrm{B}}T}} \mathrm{e}^{\frac{\Delta E_{\mathrm{gB,Ge}}(x=0)}{k_{\mathrm{B}}T}}} \right\} \right] \mathrm{d}x \\
&= \int_{x=0}^{x=W_{\mathrm{B}}} \frac{n_{\mathrm{iB}}^2(x)}{N_{\mathrm{B}}(x)} \left[\frac{N_{\mathrm{B}}(y)}{D_{\mathrm{nb}}(y)} \int_{y=x}^{y=W_{\mathrm{B}}} \left\{ \frac{1}{N n_{\mathrm{i,\,Si}}^2} \mathrm{e}^{-\frac{\Delta E_{\mathrm{gB,\,A}}}{k_{\mathrm{B}}T}} \mathrm{e}^{-\frac{\Delta E_{\mathrm{g,Ge}}(\mathrm{grade})y}{W_{\mathrm{B}}k_{\mathrm{B}}T}} \mathrm{e}^{-\frac{\Delta E_{\mathrm{gB,Ge}}(x=0)}{k_{\mathrm{B}}T}} \mathrm{d}y \right\} \right] \mathrm{d}x \\
&= \int_{x=0}^{x=W_{\mathrm{B}}} \frac{n_{\mathrm{iB}}^2(x)}{N_{\mathrm{B}}(x)} \left[\frac{N_{\mathrm{B}}(y)}{D_{\mathrm{nb}}(y)} \times \frac{\mathrm{e}^{-\frac{\Delta E_{\mathrm{gB,\,A}}}{k_{\mathrm{B}}T}} \mathrm{e}^{-\frac{\Delta E_{\mathrm{gB,Ge}}(x=0)}{k_{\mathrm{B}}T}}}{N n_{\mathrm{i,\,Si}}^2} \int_{y=x}^{y=W_{\mathrm{B}}} \left\{ \mathrm{e}^{-\frac{\Delta E_{\mathrm{g,Ge}}(\mathrm{grade})y}{W_{\mathrm{B}}k_{\mathrm{B}}T}} \mathrm{d}y \right\} \right] \mathrm{d}x \\
&= \int_{x=0}^{x=W_{\mathrm{B}}} \frac{n_{\mathrm{iB}}^2(x) N_{\mathrm{B}}(y) \mathrm{e}^{-\frac{\Delta E_{\mathrm{gB,\,A}}}{k_{\mathrm{B}}T}} \mathrm{e}^{-\frac{\Delta E_{\mathrm{gB,Ge}}(x=0)}{k_{\mathrm{B}}T}}}{N_{\mathrm{B}}(x) D_{\mathrm{nb}}(y) N n_{\mathrm{i,\,Si}}^2} \left[-\frac{1}{\frac{\Delta E_{\mathrm{g,Ge}}(\mathrm{grade})}{W_{\mathrm{B}}k_{\mathrm{B}}T}} \mathrm{e}^{-\frac{\Delta E_{\mathrm{g,Ge}}(\mathrm{grade})y}{W_{\mathrm{B}}k_{\mathrm{B}}T}} \right]_{y=x}^{y=W_{\mathrm{B}}} \mathrm{d}x \qquad (6.42) \\
&= \int_{x=0}^{x=W_{\mathrm{B}}} \frac{n_{\mathrm{iB}}^2(x) N_{\mathrm{B}}(y) \mathrm{e}^{-\frac{\Delta E_{\mathrm{gB,\,A}}}{k_{\mathrm{B}}T}} \mathrm{e}^{-\frac{\Delta E_{\mathrm{gB,Ge}}(x=0)}{k_{\mathrm{B}}T}}}{N_{\mathrm{B}}(x) D_{\mathrm{nb}}(y) N n_{\mathrm{i,\,Si}}^2} \left[-\frac{W_{\mathrm{B}}k_{\mathrm{B}}T}{\Delta E_{\mathrm{g,Ge}}(\mathrm{grade})} \mathrm{e}^{-\frac{\Delta E_{\mathrm{g,Ge}}(\mathrm{grade})y}{W_{\mathrm{B}}k_{\mathrm{B}}T}} \right]_{y=x}^{y=W_{\mathrm{B}}} \mathrm{d}x \\
&= \int_{x=0}^{x=W_{\mathrm{B}}} \frac{n_{\mathrm{iB}}^2(x) N_{\mathrm{B}}(y) \mathrm{e}^{-\frac{\Delta E_{\mathrm{gB,\,A}}}{k_{\mathrm{B}}T}} \mathrm{e}^{-\frac{\Delta E_{\mathrm{gB,Ge}}(x=0)}{k_{\mathrm{B}}T}}}{N_{\mathrm{B}}(x) D_{\mathrm{nb}}(y) N n_{\mathrm{i,\,Si}}^2} \left[-\frac{W_{\mathrm{B}}k_{\mathrm{B}}T}{\Delta E_{\mathrm{g,Ge}}(\mathrm{grade})} \mathrm{e}^{-\frac{\Delta E_{\mathrm{g,Ge}}(\mathrm{grade})W_{\mathrm{B}}}{W_{\mathrm{B}}k_{\mathrm{B}}T}} \right. \\
&\qquad \left. + \frac{W_{\mathrm{B}}k_{\mathrm{B}}T}{\Delta E_{\mathrm{g,Ge}}(\mathrm{grade})} \mathrm{e}^{-\frac{\Delta E_{\mathrm{g,Ge}}(\mathrm{grade})x}{W_{\mathrm{B}}k_{\mathrm{B}}T}} \right] \mathrm{d}x \\
&= \int_{x=0}^{x=W_{\mathrm{B}}} \frac{n_{\mathrm{iB}}^2(x) N_{\mathrm{B}}(y) W_{\mathrm{B}}k_{\mathrm{B}}T \mathrm{e}^{-\frac{\Delta E_{\mathrm{gB,\,A}}}{k_{\mathrm{B}}T}} \mathrm{e}^{-\frac{\Delta E_{\mathrm{gB,Ge}}(x=0)}{k_{\mathrm{B}}T}}}{N_{\mathrm{B}}(x) D_{\mathrm{nb}}(y) N n_{\mathrm{i,\,Si}}^2 \Delta E_{\mathrm{g,Ge}}(\mathrm{grade})} \left[\mathrm{e}^{-\frac{\Delta E_{\mathrm{g,Ge}}(\mathrm{grade})x}{W_{\mathrm{B}}k_{\mathrm{B}}T}} \right. \\
&\qquad \left. - \mathrm{e}^{-\frac{\Delta E_{\mathrm{g,Ge}}(\mathrm{grade})W_{\mathrm{B}}}{W_{\mathrm{B}}k_{\mathrm{B}}T}} \right] \mathrm{d}x.
\end{aligned}
$$

Substituting for $n_{\mathrm{iB}}^2(x)$ from (6.37) into (6.42), we obtain

$$\tau_{\mathrm{b,\,SiGe}} = \int_{x=0}^{x=W_{\mathrm{B}}} \frac{Nn_{\mathrm{i,\,Si}}^2 \mathrm{e}^{\frac{\Delta E_{\mathrm{gB,A}}}{k_{\mathrm{B}}T}} \mathrm{e}^{\left\{\frac{\Delta E_{\mathrm{g,Ge}}(\mathrm{grade})}{k_{\mathrm{B}}T}\right\}(x/W_{\mathrm{B}})} \mathrm{e}^{\frac{\Delta E_{\mathrm{gB,Ge}}(x=0)}{k_{\mathrm{B}}T}}}{N_{\mathrm{B}}(x)D_{\mathrm{nb}}(y)Nn_{\mathrm{i,\,Si}}^2 \Delta E_{\mathrm{g,Ge}}(\mathrm{grade})} \tag{6.43}$$

$$N_{\mathrm{B}}(y)W_{\mathrm{B}}k_{\mathrm{B}}T\mathrm{e}^{-\frac{\Delta E_{\mathrm{gB,\,A}}}{k_{\mathrm{B}}T}} \mathrm{e}^{-\frac{\Delta E_{\mathrm{gB,Ge}}(x=0)}{k_{\mathrm{B}}T}} \times \left[\mathrm{e}^{-\frac{\Delta E_{\mathrm{g,Ge}}(\mathrm{grade})x}{W_{\mathrm{B}}k_{\mathrm{B}}T}} - \mathrm{e}^{-\frac{\Delta E_{\mathrm{g,Ge}}(\mathrm{grade})W_{\mathrm{B}}}{W_{\mathrm{B}}k_{\mathrm{B}}T}}\right]\mathrm{d}x$$

$$= \int_{x=0}^{x=W_{\mathrm{B}}} \frac{Nn_{\mathrm{i,\,Si}}^2 \mathrm{e}^{\frac{\Delta E_{\mathrm{gB,\,A}}}{k_{\mathrm{B}}T}} \mathrm{e}^{-\frac{\Delta E_{\mathrm{gB,\,A}}}{k_{\mathrm{B}}T}} \mathrm{e}^{\frac{\Delta E_{\mathrm{gB,Ge}}(x=0)}{k_{\mathrm{B}}T}} \mathrm{e}^{-\frac{\Delta E_{\mathrm{gB,Ge}}(x=0)}{k_{\mathrm{B}}T}}}{N_{\mathrm{B}}(x)D_{\mathrm{nb}}(y)Nn_{\mathrm{i,\,Si}}^2 \Delta E_{\mathrm{g,Ge}}(\mathrm{grade})} N_{\mathrm{B}}(y)W_{\mathrm{B}}k_{\mathrm{B}}T$$

$$\mathrm{e}^{\left\{\frac{\Delta E_{\mathrm{g,Ge}}(\mathrm{grade})}{k_{\mathrm{B}}T}\right\}(x/W_{\mathrm{B}})}$$

$$\times \left[\mathrm{e}^{-\frac{\Delta E_{\mathrm{g,Ge}}(\mathrm{grade})x}{W_{\mathrm{B}}k_{\mathrm{B}}T}} - \mathrm{e}^{-\frac{\Delta E_{\mathrm{g,Ge}}(\mathrm{grade})W_{\mathrm{B}}}{W_{\mathrm{B}}k_{\mathrm{B}}T}}\right]\mathrm{d}x$$

$$= \int_{x=0}^{x=W_{\mathrm{B}}} \frac{N_{\mathrm{B}}(y)W_{\mathrm{B}}k_{\mathrm{B}}T}{N_{\mathrm{B}}(x)D_{\mathrm{nb}}(y)\Delta E_{\mathrm{g,Ge}}(\mathrm{grade})} \mathrm{e}^{\left\{\frac{\Delta E_{\mathrm{g,Ge}}(\mathrm{grade})}{k_{\mathrm{B}}T}\right\}(x/W_{\mathrm{B}})} \tag{6.44}$$

$$\times \left[\mathrm{e}^{-\frac{\Delta E_{\mathrm{g,Ge}}(\mathrm{grade})x}{W_{\mathrm{B}}k_{\mathrm{B}}T}} - \mathrm{e}^{-\frac{\Delta E_{\mathrm{g,Ge}}(\mathrm{grade})W_{\mathrm{B}}}{W_{\mathrm{B}}k_{\mathrm{B}}T}}\right]\mathrm{d}x$$

$$= \int_{x=0}^{x=W_{\mathrm{B}}} \frac{N_{\mathrm{B}}(y)W_{\mathrm{B}}k_{\mathrm{B}}T}{N_{\mathrm{B}}(x)D_{\mathrm{nb}}(y)\Delta E_{\mathrm{g,Ge}}(\mathrm{grade})} \left[1 - \mathrm{e}^{\left\{\frac{\Delta E_{\mathrm{g,Ge}}(\mathrm{grade})}{k_{\mathrm{B}}T}\right\}(x/W_{\mathrm{B}})} \mathrm{e}^{-\frac{\Delta E_{\mathrm{g,Ge}}(\mathrm{grade})}{k_{\mathrm{B}}T}}\right]\mathrm{d}x$$

$$= \frac{W_{\mathrm{B}}}{D_{\mathrm{nb}}(y)} \frac{k_{\mathrm{B}}T}{\Delta E_{\mathrm{g,Ge}}(\mathrm{grade})} \int_{x=0}^{x=W_{\mathrm{B}}} \left[1 - \mathrm{e}^{\left\{\frac{\Delta E_{\mathrm{g,Ge}}(\mathrm{grade})}{k_{\mathrm{B}}T}\right\}(x/W_{\mathrm{B}})} \mathrm{e}^{-\frac{\Delta E_{\mathrm{g,Ge}}(\mathrm{grade})}{k_{\mathrm{B}}T}}\right]\mathrm{d}x$$

taking

$$N_{\mathrm{B}}(y) = N_{\mathrm{B}}(x).$$

Now,

$$
\begin{aligned}
\tau_{\mathrm{b,SiGe}} &= \frac{W_B}{D_{nb}(y)}\frac{k_BT}{\Delta E_{gGe}(\mathrm{grade})}\left[\int_{x=0}^{x=W_B}\mathrm{d}x \right.\\
&\quad \left. - \int_{x=0}^{x=W_B} e^{\left\{\frac{\Delta E_{gGe}(\mathrm{grade})}{k_BT}\right\}(x/W_B)} e^{-\frac{\Delta E_{gGe}(\mathrm{grade})}{k_BT}}\mathrm{d}x\right]\\
&= \frac{W_B}{D_{nb}(y)}\frac{k_BT}{\Delta E_{gGe}(\mathrm{grade})}[x]_{x=0}^{x=W_B} - \frac{W_B}{D_{nb}(y)}\\
&\quad \times \frac{k_BT}{\Delta E_{gGe}(\mathrm{grade})}\int_{x=0}^{x=W_B} e^{\left\{\frac{\Delta E_{gGe}(\mathrm{grade})}{k_BT}\right\}(x/W_B)} e^{-\frac{\Delta E_{gGe}(\mathrm{grade})}{k_BT}}\mathrm{d}x\\
&= \frac{W_B}{D_{nb}(y)}\frac{k_BT}{\Delta E_{gGe}(\mathrm{grade})}[W_B] - \frac{W_B}{D_{nb}(y)}\frac{k_BT e^{-\frac{\Delta E_{gGe}(\mathrm{grade})}{k_BT}}}{\Delta E_{gGe}(\mathrm{grade})}\\
&\quad \times \int_{x=W_B}^{x=0} e^{\left\{\frac{\Delta E_{gGe}(\mathrm{grade})}{k_BT}\right\}(x/W_B)}\mathrm{d}x\\
&= \frac{W_B^2}{D_{nb}(y)}\frac{k_BT}{\Delta E_{gGe}(\mathrm{grade})} - \frac{W_B}{D_{nb}(y)}\frac{k_BT e^{-\frac{\Delta E_{gGe}(\mathrm{grade})}{k_BT}}}{\Delta E_{gGe}(\mathrm{grade})} \qquad (6.45)\\
&\quad \frac{1}{\left\{\frac{\Delta E_{gGe}(\mathrm{grade})}{k_BT}\right\}(1/W_B)} \times \left[e^{\left\{\frac{\Delta E_{gGe}(\mathrm{grade})}{k_BT}\right\}(x/W_B)}\right]_{x=0}^{x=W_B}\\
&= \frac{W_B^2}{D_{nb}(y)}\frac{k_BT}{\Delta E_{gGe}(\mathrm{grade})} - \frac{W_B^2}{D_{nb}(y)}\frac{k_BT}{\Delta E_{gGe}(\mathrm{grade})}\\
&\quad \times \frac{k_BT}{\Delta E_{gGe}(\mathrm{grade})}e^{-\frac{\Delta E_{gGe}(\mathrm{grade})}{k_BT}}\left[e^{\frac{\Delta E_{gGe}(\mathrm{grade})}{k_BT}} - e^0\right]\\
&= \frac{W_B^2}{D_{nb}(y)}\frac{k_BT}{\Delta E_{gGe}(\mathrm{grade})} - \frac{W_B^2}{D_{nb}(y)}\frac{k_BT}{\Delta E_{gGe}(\mathrm{grade})}\\
&\quad \times \frac{k_BT}{\Delta E_{g,Ge}(\mathrm{grade})}\left\{1 - e^{-\frac{\Delta E_{g,Ge}(\mathrm{grade})}{k_BT}}\right\}\\
&= \frac{W_B^2}{D_{nb}(y)}\frac{k_BT}{\Delta E_{g,Ge}(\mathrm{grade})}\left[1 - \frac{k_BT}{\Delta E_{g,Ge}(\mathrm{grade})}\left\{1 - e^{-\frac{\Delta E_{g,Ge}(\mathrm{grade})}{k_BT}}\right\}\right].
\end{aligned}
$$

But

$$
\tau_{\mathrm{b,\ Si}} = \frac{W_B^2}{2D_{nb}(y)} \qquad (6.46)
$$

$$\therefore \tau_{\text{b, SiGe}} = 2 \times \frac{W_B^2}{2D_{\text{nb}}(y)} \times \frac{k_B T}{\Delta E_{\text{g, Ge}}(\text{grade})} \times \left[1 - \frac{k_B T}{\Delta E_{\text{g,Ge}}(\text{grade})}\left\{1 - e^{-\frac{\Delta E_{\text{g,Ge}}(\text{grade})}{k_B T}}\right\}\right]$$
$$= 2\tau_{\text{b,Si}} \times \frac{k_B T}{\Delta E_{\text{g,Ge}}(\text{grade})}\left[1 - \frac{k_B T}{\Delta E_{\text{g,Ge}}(\text{grade})}\left\{1 - e^{-\frac{\Delta E_{\text{g,Ge}}(\text{grade})}{k_B T}}\right\}\right] \quad (6.47)$$

or

$$\frac{\tau_{\text{b,SiGe}}}{\tau_{\text{b,Si}}} = \frac{2k_B T}{\Delta E_{\text{g,Ge}}(\text{grade})}\left[1 - \frac{k_B T}{\Delta E_{\text{g,Ge}}(\text{grade})}\left\{1 - e^{-\frac{\Delta E_{\text{g,Ge}}(\text{grade})}{k_B T}}\right\}\right]. \quad (6.48)$$

Taking $\Delta E_{\text{g,Ge}}(\text{grade}) = 75\text{ meV} = 75 \times 10^{-3}\text{ eV} = 0.075\text{ eV}$, $k_B = 8.62 \times 10^{-5}\text{ eV K}^{-1}$, at $T = 300$ K

$$\left(\frac{\tau_{\text{b, SiGe}}}{\tau_{\text{b, Si}}}\right)_{300K} = \frac{2 \times 8.62 \times 10^{-5} \times 300}{0.075} \times \left[1 - \frac{8.62 \times 10^{-5} \times 300}{0.075}\left\{1 - e^{-\frac{0.075}{8.62 \times 10^{-5} \times 300}}\right\}\right]$$
$$= 0.6896(1 - 0.3448 \times 0.9449895) = 0.4649. \quad (6.49)$$

At $T = 4.2$ K,

$$\left(\frac{\tau_{\text{b,SiGe}}}{\tau_{\text{b,Si}}}\right)_{4.2K} = \frac{2k_B \times 4.2}{\Delta E_{\text{g,Ge}}(\text{grade})}\left[1 - \frac{k_B \times 4.2}{\Delta E_{\text{g,Ge}}(\text{grade})}\left\{1 - e^{-\frac{\Delta E_{\text{g,Ge}}(\text{grade})}{k_B \times 4.2}}\right\}\right]$$
$$= \frac{2 \times 8.62 \times 10^{-5} \times 4.2}{0.075}\left[1 - \frac{8.62 \times 10^{-5} \times 4.2}{0.075}\left\{1 - e^{-\frac{0.075}{8.62 \times 10^{-5} \times 4.2}}\right\}\right] \quad (6.50)$$
$$= 0.0096544(1 - 0.0048272 \times 1) = 0.009608$$

$$\left(\frac{\tau_{\text{b, SiGe}}}{\tau_{\text{b, Si}}}\right)_{300K} \Big/ \left(\frac{\tau_{\text{b, SiGe}}}{\tau_{\text{b, Si}}}\right)_{4.2K} = \frac{0.4649}{0.009608} = 48.38676. \quad (6.51)$$

The above calculations show that at 300 K, $\tau_{\text{b,SiGe}} = 0.5 \times \tau_{\text{b,Si}}$. At 4.2 K, $\tau_{\text{b, SiGe}} = 0.01 \times \tau_{\text{b,Si}}$ becoming considerably shorter. The device becomes faster. Hence, the decrease in temperature favors the SiGe HBT, making it much faster than the Si HBT.

6.5 Discussion and conclusions

The outperformance of HBT compared to homojunction BJT at cryogenic temperatures provides a welcome relief to design engineers working on analog, digital and mixed-signal circuits for low-temperature operation. SiGe HBT technologies are key enablers for communication, radar and imaging systems in millimeter wave and

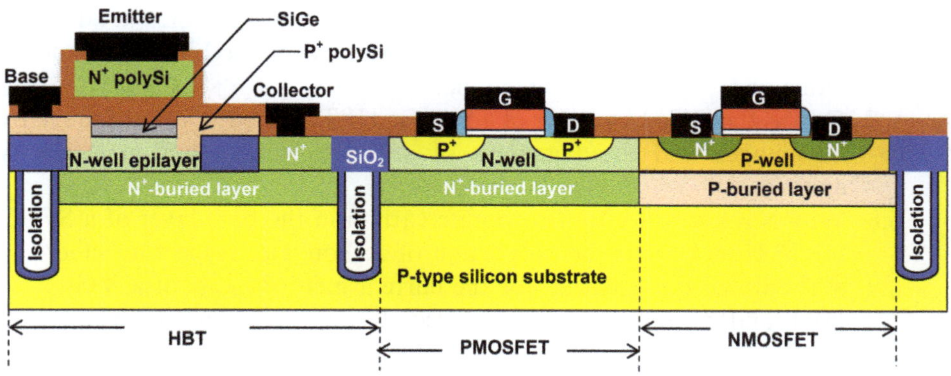

Figure 6.6. BiCMOS integration on a single platform.

terahertz ranges (Schroter *et al* 2018). Radiation effects and compact model of HBT are described by Sun (2018).

The SiGe HBT is compatible and easily integrable with Si CMOS structures, yielding bipolar CMOS (BiCMOS) circuits for wired/wireless RF communication networks and computing applications (figure 6.6). The SiGe BiCMOS is a flexible technological platform for mixed-signal RF and microwave circuits. By merging the high gain and speed provided by bipolar technology with the low power consumption logic gates of CMOS technology, it is possible to build high-performance circuits combining the best features of both technologies (Harame *et al* 2001). Fabrication steps of SiGe HBT BiCMOS critically impact the high-frequency performance demanding careful technology optimization (Cressler 2018, Rücker and Heinemann 2018a, 2018b).

Review exercises

6.1. How does a homojunction BJT differ from a heterojunction BJT? Argue why you cannot design a homojunction BJT having a high current gain together with fast switching capability. What structural parameters of the BJT are to be traded off in this design?

6.2. How does the use of a silicon–germanium heterojunction structure help in resolving the conflict between current gain and switching speed in a homojunction BJT? Explain how the heterojunction BJT achieves a high gain and low base transit time simultaneously.

6.3. How does the lower bandgap of the silicon–germanium base in comparison to the silicon emitter of a heterojunction BJT help in increasing the emitter injection efficiency of the heterostructure?

6.4. How does a graded profile of the silicon–germanium base in an HBT increase its switching speed?

6.5. Which has a smaller bandgap: silicon or germanium? Write the formula for the bandgap of the silicon–germanium alloy as a function of germanium percentage.

6.6. What is meant by BiCMOS technology? What is the principal advantage offered by this technology in comparison to either bipolar technology alone or CMOS technology alone?

6.7. How is the SiGe base layer of a Si/SiGe HBT formed? Discuss the relative advantages and disadvantages of selective epitaxy and non-selective epitaxy.

6.8. Why is it necessary to incorporate carbon in the base layer of a Si/SiGe HBT? How large is the proportion of carbon with respect to SiGe?

6.9. Write down the equation for the current gain $(\beta)_{\mathrm{Si/SiGe}}$ of a $\mathrm{Si/Si_{1-x}Ge_x}$ HBT and explain the symbols used. Using this equation for $(\beta)_{\mathrm{Si/SiGe}}$, write the equation for the forward transit time through the emitter of a $\mathrm{Si/Si_{1-x}Ge_x}$ HBT.

6.10. Given that electron mobility $\mu_{\mathrm{n,Si}}$ in Si is

$$\mu_{\mathrm{n,\,Si}}(T) \propto T^{-2.42} \tag{6.52}$$

and electron mobility in Ge is

$$\mu_{\mathrm{n,\,Ge}}(T) \propto T^{-1.66} \tag{6.53}$$

derive the following equation for diffusion coefficient of electrons in $\mathrm{Si_{1-x}Ge_x}$

$$D_{\mathrm{n,SiGe}}(T) = \left(\frac{k_{\mathrm{B}}T}{q}\right)\left\{9.88 \times 10^{5}(1 - x)T^{-2.42} + 1.29 \times 10^{4}xT^{-1.66}\right\}. \tag{6.54}$$

if $\mu_{\mathrm{n,Si}} = 1350\ \mathrm{cm^2\ V^{-1}\ s^{-1}}$ at $T = 300$ K, $\mu_{\mathrm{n,Ge}} = 3900\ \mathrm{cm^2\ V^{-1}\ s^{-1}}$ at $T = 300$ K.

6.11. When the temperature of a Si BJT is lowered down from room temperature towards absolute zero, its current gain decreases rapidly with temperature. However, when a Si/SiGe heterojunction BJT is subjected to a similar fall of temperature, its current gain degradation with temperature is much less severe. Explain, giving reasons, this difference in the current gain behaviors of a BJT and an HBT at low temperatures.

6.12. Starting from the Moll–Ross relationship connecting the collector current density J_{C} of a bipolar transistor with the applied base–emitter voltage V_{BE} at a given temperature T, derive the equation for the ratio of current gains $\beta_{\mathrm{SiGe}}/\beta_{\mathrm{Si}}$ of the silicon/silicon–germanium HBT and silicon BJT in terms of the degree of bandgap grading $\Delta E_{\mathrm{g,Ge}}(\mathrm{grade})$ and the band offset $\Delta E_{\mathrm{g,Ge}}(x = 0)$ caused by Ge at the edge of the base towards the emitter side. What is the importance of this equation to an HBT designer trying to design an HBT for a specific operating temperature?

6.13. Explain, with the help of the equation for the ratio of current gains $\beta_{\mathrm{SiGe}}/\beta_{\mathrm{Si}}$ of the silicon/silicon–germanium HBT and silicon BJT, why it is possible to design a silicon/silicon–germanium HBT with the required current gain β_{SiGe} at a given temperature T by choosing the appropriate degree of bandgap grading $\Delta E_{g,Ge}$(grade) and the band offset $\Delta E_{g,Ge}(x = 0)$ caused by Ge at the edge of the base towards the emitter side. Hence, elaborate the greater design flexibility afforded by an HBT over a BJT for low-temperature operation.

6.14. Derive the equation for the ratio of forward base transit times ($\tau_{b,SiGe}/\tau_{b,Si}$) of a silicon–germanium HBT and a Si BJT in terms of the degree of bandgap grading $\Delta E_{g,Ge}$(grade). Hence, show that a SiGe HBT becomes much faster than a Si BJT when the temperature is decreased from 300 K to 4.2 K.

6.15. Explain, with reference to the current gain and base transit time parameters, the superiority of a SiGe HBT over a Si BJT for operation at low temperatures.

References

Basu S and Sarkar P 2011 Analytical modeling of AlGaAs/GaAs and Si/SiGe HBTs including the effect of temperature *J. Electron Devices* **9** 325–9

Cressler J D 2018 *Fabrication of SiGe HBT BiCMOS Technology* (Boca Raton, FL: CRC Press)

Cressler J D 1998 SiGe HBT technology: a new contender for Si-based RF and microwave circuit applications *IEEE Trans. Microw. Theory Tech.* **46** 572–89

Cressler J D and Niu G 2003 *Silicon–Germanium Heterojunction Bipolar Transistors* (Boston, MA: Artech House) pp 98–104

Hållstedt J 2004 Epitaxy and characterization of SiGeC layers grown by reduced pressure chemical vapor deposition *Licentiate Thesis* (Stockholm: Royal Institute of Technology (KTH), Stockholm)

Harame D L, Ahlgren D C, Coolbaugh D D, Dunn J S and Freeman G G 2001 Current status and future trends of SiGe BiCMOS technology *IEEE Trans. Electron Devices* **48** 2575–94

Harame D L, Comfort J H, Cressler J D, Crabbé E F, Sun J Y-C, Meyerson B S and Tice T 1995 *IEEE Trans. Electron Devices* **42** 455–68

Ioffe Institute 2015 SiGe—silicon germanium: band structure and carrier concentration *Ioffe Institute* www.ioffe.ru/SVA/NSM/Semicond/SiGe/bandstr.html

Mitrovic I Z, Buiu O, Hall S, Bagnall D M and Ashburn P 2005 Review of SiGe HBTs on SOI *Solid-State Electron.* **49** 1556–67

Rücker H and Heinemann B 2018a High-performance SiGe HBTs for next generation BiCMOS technology *Semicond. Sci. Technol.* **33** 114003

Rücker H and Heinemann B 2018b SiGe HBT technology *Silicon–Germanium Heterojunction Bipolar Transistors for mm-Wave Systems: Technology, Modeling and Circuit Applications* ed N Rinaldi and M Schröter (Gistrup: River Publishers) ch 1 pp 11–54

Schroter M, Pfeiffer U and Jain R 2018 Future of SiGe HBT technology and its applications *Silicon–Germanium Heterojunction Bipolar Transistors for mm-Wave Systems: Technology, Modeling and Circuit Applications* ed N Rinaldi and M Schröter (Gistrup: River Publishers) ch 7 pp 309–24

Shiraki Y and Usami N 2011 *Silicon–Germanium (Si–Ge) Nanostructures: Production, Properties and Applications in Electronics* (Cambridge: Woodhead)

Shur M 1995 *Physics of Semiconductor Devices* (New Delhi: Prentice-Hall India)
Sun Y 2018 *Research on the Radiation Effects and Compact Model of SiGe HBT* (Singapore: Springer Nature Singapore Pte Ltd)
Virginia Semiconductor 2002 The general properties of Si, Ge, SiGe, SiO_2 and Si_3N_4 *Virginia Semiconductor* www.virginiasemi.com/pdf/generalpropertiesSi62002.pdf

Chapter 7

The temperature-sustaining capability of gallium arsenide electronics

Gallium arsenide ranks second to silicon with regard to two primary aspects; it is a wider bandgap semiconductor suitable for high-temperature operation and with technological maturity. The high electron mobility in GaAs together with its direct bandgap make this material a strong competitor to silicon for the fabrication of high-temperature microwave and optoelectronic devices and circuits. GaAs circuits fabricated with existing state-of-the-art technologies were found to succumb prematurely to elevated temperatures due to reasons for which GaAs itself was not directly responsible. For realization of the full thermal capability of GaAs prescribed by its fundamental physical limits, it is necessary to introduce thermally stable contact metallization, modify many processes, and formulate innovative device structures and designs so that failure from any cause unassociated with GaAs is completely eliminated.

7.1 Introduction

Attractive properties of gallium arsenide have fostered its use in semiconductor device fabrication (Einspruch and Wisseman 1985, Kirkpatrick 1988, Papež *et al* 2021). The wider bandgap of GaAs, by 1.424 eV − 1.12 eV = 0.304 eV, in comparison to silicon pushes the operating temperature range for GaAs electronics from between 200 °C and 300 °C for silicon to 400 °C for GaAs. Although, theoretically, the permissible range for GaAs extends up to approximately 500 °C, practical limitations have restricted the operation to lower temperatures. Table 7.1

doi:10.1088/978-0-7503-5072-3ch7

Table 7.1. Properties of gallium arsenide.

Property	Value	Property	Value	Property	Value
Chemical formula	GaAs	Melting point (°C)	1238	Hole mobility (cm^2 V^{-1} s^{-1})	400
Molecular mass	144.65	Dielectric constant	12.9	Electron diffusion coefficient (cm^2 s^{-1})	200
Classification	III–V compound semiconductor	Thermal conductivity (W cmK^{-1})	0.46	Hole diffusion coefficient (cm^2 s^{-1})	10
Crystal structure	Zinc blende	Energy bandgap E_g (eV) at 300 K	1.424	Electron saturated velocity (cm s^{-1})	4.4×10^7
Color	Dark red	Electrical breakdown field (V cm^{-1})	4×10^5	Hole saturated velocity (cm s^{-1})	1.8×10^7
Density at 300 K (g cm^{-3})	5.32	Intrinsic carrier concentration (cm^{-3})	2.1×10^6	Minority-carrier lifetime (s)	10^{-8}
Number of atoms cm^{-3}	4.42×10^{22}	Intrinsic resistivity (Ω-cm)	3.3×10^8		
Lattice constant (Å)	5.65	Electron mobility (cm^2 V^{-1} s^{-1})	8500		

compiles the properties of GaAs of interest to device and circuit designers. In figure 7.1, the crystal structure of gallium arsenide is displayed.

Some striking fundamental dissimilarities between gallium arsenide and silicon or germanium need attention (Ghandhi 2008):

(i) Both germanium and silicon are elemental semiconductors. A cursory glance at the periodic classification of chemical elements shows that group IV is the abode of these elements. Unlike elemental Ge and Si, gallium arsenide is a III–V compound semiconductor, formed by the union of elements which are members of group III and group V. It is so called because it comprises two distinct elements, gallium and arsenic. Gallium resides in group III of the periodic table while arsenic is housed in group V of the periodic table.

(ii) Another contrasting feature of gallium arsenide with germanium and silicon pertains to the crystal structure. It is well known that Ge and Si crystallize in the familiar diamond structure. But the crystal configuration of gallium arsenide is not like diamond, it has a zinc blende structure. A gallium arsenide crystal is made up of two face-centered cubic (fcc) sublattices. These fcc sublattices are displaced with respect to each other by a distance equal to half the length of the diagonal of the fcc sublattice.

(iii) In contrast to the indirect bandgap property of silicon, gallium arsenide is a direct bandgap semiconductor, enabling its utilization for fabrication of

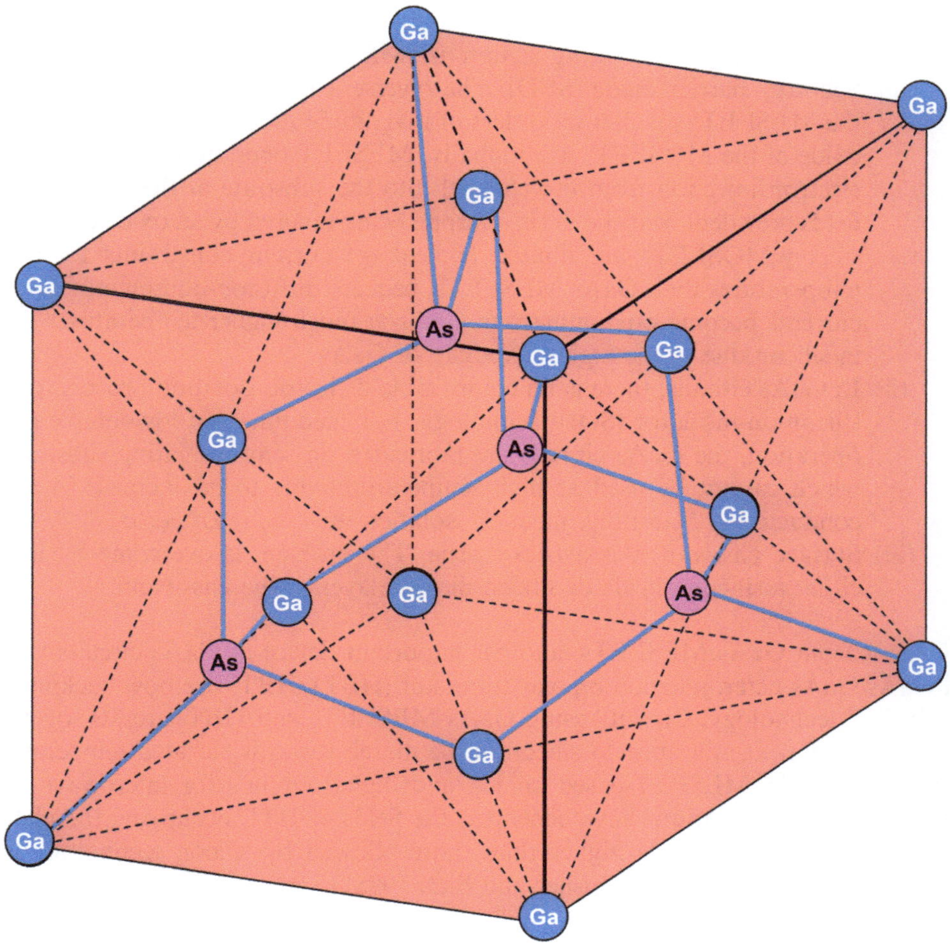

Figure 7.1. Structure of gallium arsenide crystal.

light-emitting diodes and lasers. In addition, the high electron mobility in GaAs makes it possible to use it for microwave and high-frequency devices.

(iv) A thin silicon dioxide film of thickness 1 nm spontaneously grows on a silicon surface due to the oxygen present in the atmosphere. This oxide layer, called the native oxide, is beneficial for microelectronics. On a GaAs surface, the available oxides Ga_2O_3, As_2O_3 and As_2O_5 are more problematic than useful.

Other noteworthy differences between silicon and GaAs ICs are as follows:

(i) In GaAs circuits, the primary device is a metal–semiconductor field-effect transistor (MESFET). It plays the same role in GaAs circuits as the metal–oxide–semiconductor field-effect transistor (MOSFET) in silicon technology. The absence of MOSFETs in GaAs technology is due to the non-availability of a tough adherent oxide layer comparable to the thermal

SiO_2, which is easily grown in the fabrication of silicon circuits. Consequently, the gating action in MESFETs is provided through a Schottky barrier diode (SBD). The reverse leakage current of an SBD in the MESFET is higher by orders of magnitude than that through the gate oxide of the MOSFET. Additionally, MESFET operation is influenced by electron injection from the channel into the substrate at the high electric fields prevalent near the drain. Despite being favored by an oxide dielectric, silicon MOSFETs are limited by leakage current constraints to lower temperatures than GaAs MESFETs, because diffusion-dominated leakage currents become predominant in GaAs at much higher temperatures than in silicon due to the higher bandgap of GaAs.

(ii) In GaAs circuits, there is no comparable dielectric isolation process to the silicon-on-insulator (SOI) technology. For medium- and low-temperature operation, the GaAs circuits are fabricated on semi-insulating substrates, which cannot be used at high temperatures due to an increase in their conductivity. Naturally, junction isolation must be resorted to.

(iii) Surface passivation is a tricky issue. The problem is overcome by using either a silicon nitride or silicon dioxide layer as the passivant.

Apart from GaAs MESFETs, another important device in GaAs circuits is the AlGaAs/GaAs heterojunction bipolar transistor (HBT). HBTs are now backing up MESFET technology. Circuits using GaAs MESFETs and HBT circuits together constitute the next milestone to silicon in the march towards elevated temperature operation. GaAs MESFET-based analog and digital circuits for microwave and optoelectronic applications have benefitted the fields of HTE occupying the niches where temperatures a little higher than those allowed by silicon technology are required (Bhattacharya 1990, Fisher and Bahl 1995).

7.2 The intrinsic temperature of GaAs

In analogy to calculations performed for silicon, the calculations for the intrinsic temperature of GaAs are carried out herein. For GaAs, we have

$$m_n^* = 0.85m_0,\ m_p^* = 0.53m_0,\ E_g = 1.424\ \text{eV}.$$
$$T\ln\left\{0.207/\left(m^* T^{1.5}\right)\right\} = -5.8025\times10^3 E_g \tag{7.1}$$

Equation (7.1) is to be recast for GaAs (refer to equation (4.6)).

Since, for GaAs

$$m^* = \left(\frac{m_n^* m_p^*}{m_0^2}\right)^{0.75} = \left(\frac{0.85m_0\times0.53m_0}{m_0^2}\right)^{0.75} = 0.5499 \tag{7.2}$$

we have

$$T\ \ln\ \{0.207/(0.5499T^{1.5})\} = -5.8025\times10^3\times1.424 \tag{7.3}$$

or

$$T \ln (0.3764/T^{1.5}) = -8262.76$$

or

$$T \ln 0.3764 - T \ln T^{1.5} = -8262.76 \tag{7.4}$$

$$\therefore -0.9771T - T \ln T^{1.5} = -8262.76. \tag{7.5}$$

When

$$\begin{aligned} T &= 756.5 \text{ K} \\ \text{lhs} &= -0.9771 \times 756.5 - 756.5 \ln 756.5^{1.5} = -739.176 - 7521.92 \\ &= -8261.096. \end{aligned} \tag{7.6}$$

If

$$\begin{aligned} T &= 756.6 \text{ K} \\ \text{lhs} &= -0.9771 \times 756.6 - 756.6 \ln 756.6^{1.5} \\ &= -739.274 - 7523.0645 \\ &= -8262.3385. \end{aligned} \tag{7.7}$$

If

$$\begin{aligned} T &= 756.7 \text{ K} \\ \text{lhs} &= -0.9771 \times 756.7 - 756.7 \ln 756.7^{1.5} \\ &= -739.3716 - 7524.2088 \\ &= -8263.58. \end{aligned} \tag{7.8}$$

$$\therefore T = 756.6 \text{ K}. \tag{7.9}$$

The intrinsic temperature of GaAs (756.6 K) is higher by 756.6 − 588.63 = 167.97 K than that for Si (588.63 K).

7.3 Growth of single-crystal gallium arsenide

For microelectronic device fabrication, single-crystal wafers of GaAs are required (Chang and Kai 1994, Baca and Ashby 2005). Techniques such as horizontal Bridgman, vertical gradient freeze or liquid encapsulated Czochralski (LEC) are applied for growing GaAs single crystals from the melt (Tiku and Biswas 2016, Cheng 2020). In the LEC method (figure 7.2), elemental Ga and As are placed inside a pyrolytic boron nitride (pBN) crucible. A pellet of boron trioxide (B_2O_3) is also added. The crucible is heated inside a high-pressure crystal puller. When the temperature reaches 460 °C, the boron trioxide melts. The molten boron trioxide (B_2O_3) is a thick liquid with high viscosity. The B_2O_3 layer surrounds the melt from all sides building an enclosure all around it. B_2O_3 lies between the crucible and the melt and thus is separated from the crucible. The top surface of the melt is fully

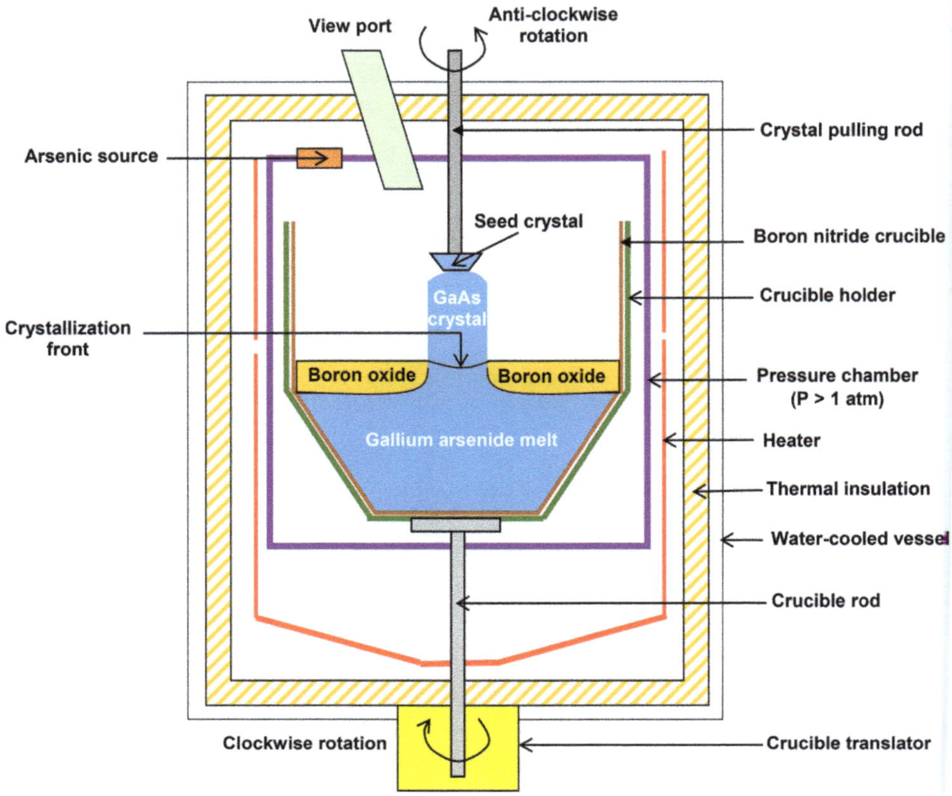

Figure 7.2. The LEC process for the growth of single-crystal gallium arsenide.

covered by B_2O_3, forming a protective cover. Thus, the GaAs melt is completely enclosed by liquid B_2O_3. Hence, this method is called LEC. Encapsulation of the GaAs melt by liquid B_2O_3 is necessary to prevent sublimation of the volatile As. If As sublimes, the melt will become gallium-rich and its stoichiometry will be disturbed. For crystal growth, a seed crystal is lowered into the melt, passing through the B_2O_3 layer and touching the GaAs surface. The crystal is slowly lifted upwards. During withdrawal, the crystal is also continuously rotated. As a result of this motion, a single crystal of GaAs propagates from the melt.

7.4 Doping of GaAs

Figure 7.3 presents the energy band diagram of GaAs showing the energy levels of impurities used for doping. For n-type doping of GaAs, group IV or group VI elements are used as the impurities (Shur 1987). Common group IV impurities are silicon (Si) and tin (Sn). These elements must occupy Ga sites to act as donors. Since the covalent radius of gallium (1.26 Å) is larger than that of arsenic (1.14 Å), group IV impurities (silicon, 1.11 Å and tin, 1.45 Å) tend to occupy gallium sites. Group VI impurities include sulfur (S), tellurium (Te) and selenium (Se). These

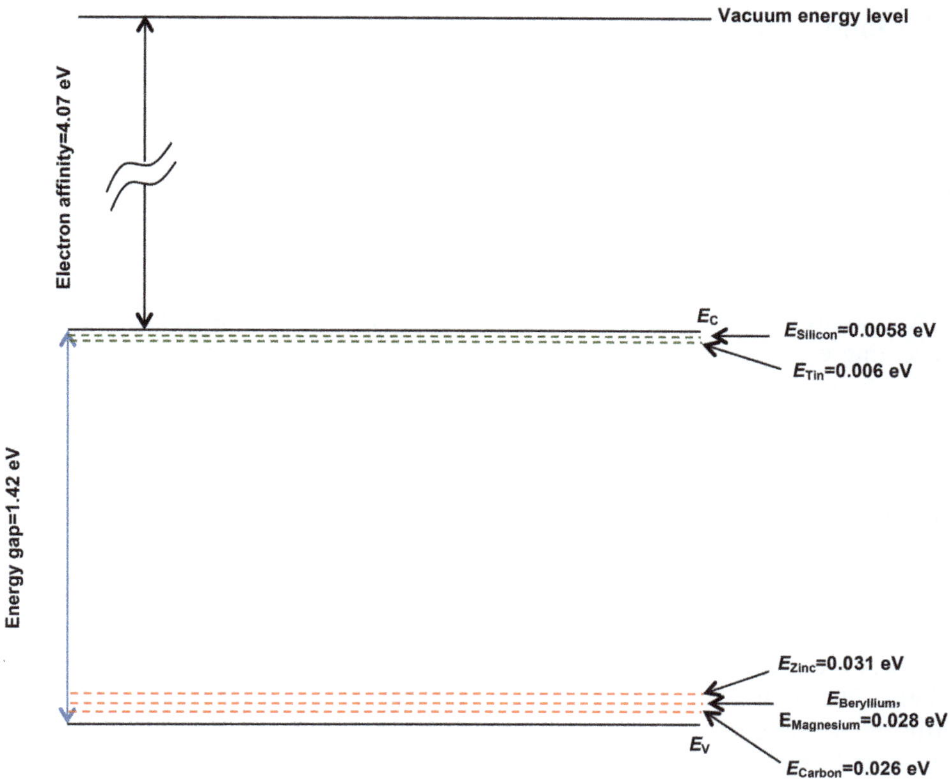

Figure 7.3. Energy band diagram of gallium arsenide showing the energy levels of commonly used impurities for N- (Si, Sn) and P-doping (Zn, Be, Mg, C).

elements must occupy As sites to act as acceptors. The doping efficiency is defined as the ratio of doping density to the implanted ion density. It is enhanced if the sample is heated during ion implantation. A lower dose gives a higher doping efficiency owing to comparatively smaller damage to the crystal lattice. The implantation step is followed by an annealing step between 850 °C and 950 °C. Plasma-enhanced chemical vapor depostion (PECVD) silicon nitride is used to passivate the GaAs surface, otherwise it loses As when heated at temperatures higher than 600 °C.

For p-type doping, beryllium (Be), zinc (Zn), carbon (C) or magnesium (Mg) ions are implanted into GaAs. The doping efficiency varies with the value of dose. It is 100% at low values of dose up to 10^{14} cm^{-2}, but decreases at higher doses. This decrease is due to the constraint of the solid solubility limit. The annealing temperature also affects the doping efficiency.

Chromium introduces an acceptor energy level near the middle of the bandgap. Cr-doping is performed to produce semi-insulating GaAs substrates (SI-GaAs: Cr) with conductivity in the dark of 3×10^{-9} S cm^{-1} at 300 K, close to the intrinsic conductivity of GaAs.

7.5 Ohmic contacts to GaAs

7.5.1 Au–Ge/Ni/Ti contact to n-type GaAs for room temperature operation

Au is a metal to which wire can be bonded. The Au–Ge eutectic temperature is 356 °C. Gold reacts with GaAs, producing the vacancies for Ge to move into GaAs. During alloying, Ga out-diffuses into the contact metal and Ge diffuses into the lattice sites. It is a dopant element and dopes GaAs degeneratively, forming an n^+ semiconductor layer. It also initiates the melting of the contact metal during alloying: Au + GaAs → Au–Ga + As. Nickel is a metal catalyzing the alloying process by driving Ge into GaAs and promoting uniformity: Ni + As → Ni–As. Titanium is a diffusion barrier, preventing excess diffusion of Au. The inter-diffusion of metals and formation of intermetallic compounds depends on the alloying conditions. Rapid thermal annealing (RTA) is best suited for alloying because undesired compounds may be formed if heating and cooling are prolonged. Primarily due to the metal inter-diffusion effects, the thermal stability of this metallization system is poor. However, it yields low contact resistance and is good for room temperature operation.

7.5.2 High-temperature ohmic contacts to n-type GaAs

Unusual problems associated with these contacts are discussed by Eun and Cooper (1993). A high-temperature metallization scheme has a layered structure (Fricke *et al* 1989), a 200 Å thick Ge film/25 Å thick Au film/100 Å thick Ni film/1000 Å thick diffusion barrier comprising nine alternating layers of e-beam evaporated W and Si with a 50 Å thick Ti film interposed between the fourth Si and fifth W layers/500 Å thick Au film (figure 7.4). The total W–Si layer thickness above the Ti film is 500 Å.

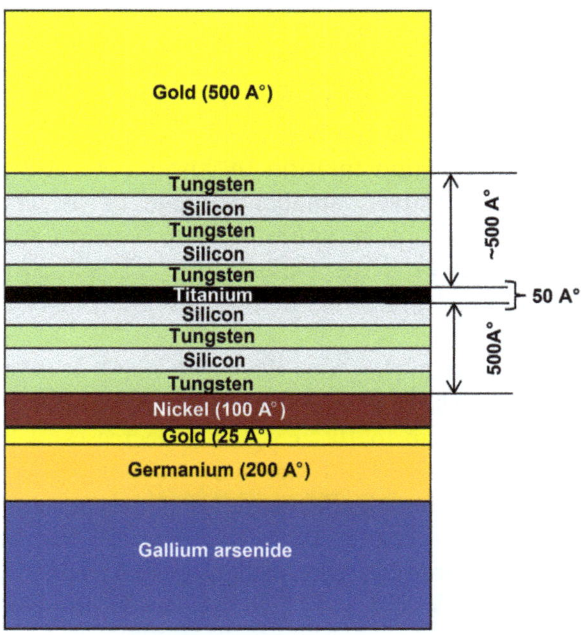

Figure 7.4. Metallization scheme for n-type GaAs.

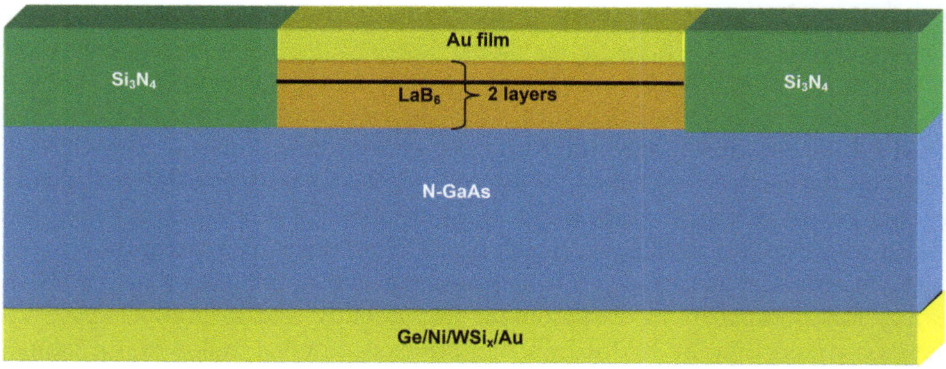

Figure 7.5. A LaB_6 diode.

The layer thickness below Ti is the same. As before, Ge dopes GaAs n-type, the reaction between Au and Ga leads to the formation of vacancies for the diffusion of Ge into GaAs, and Ni also drives Ge into GaAs during the annealing step. Due to the addition of Au, the annealing temperature is lowered to 590 °C. All Au must be consumed during annealing. If not consumed, any residual Au may cause degradation later because inter-diffusion between Au and GaAs is the main cause of degradation of contacts at high temperatures. After RTA at 640 °C, the contact resistance was $5 \times 10^{-6}\ \Omega\ cm^2$ on GaAs with a donor concentration of $1 \times 10^{17}\ cm^{-3}$. The contact showed no deterioration after storage at 300 °C for over 1000 h. MESFETs with this contact could be operated at 400 °C after the above high-temperature storage. The lifetime at 400 °C was over 100 h.

7.6 Schottky contacts to GaAs

A common multilayer contact scheme is Ti–Pt–Au (Fricke *et al* 1989), where Ti is the adhesion layer, Pt is the diffusion barrier and Au is the high-conductivity layer. The Ti–Pt–Au gate MESFETs did not degrade after storage at 300 °C for over 1000 h. They have a barrier height of 0.78 eV and an ideality factor of 1.1. Sputtering of WSi_2 on the GaAs followed by Ti–Pt–Au metal deposition improved the thermal stability of the contact. However, the barrier height decreases to 0.7 eV and the ideality factor becomes 1.4.

Au–LaB_6 Schottky contacts to GaAs (figure 7.5) yield a larger barrier height, ~0.9 eV, assuring a low leakage current during operation at high temperatures (Würfi *et al* 1990). The reliability of this contact depends on the processing and annealing conditions. The best reliability was achieved on moderately annealed samples (20 h for 20 min). The Schottky contacts were found to be stable after thermal stressing at 400 °C for several hundred hours.

7.7 Commercial GaAs device evaluation in the 25 °C–400 °C temperature range

As an initial step, the performance of the then state-of-the-art commercial GaAs devices was studied in the temperature range 25 °C–400 °C (Shoucair and

Ojala 1992). The tested MESFET devices had Au–Ge/Ni/Au ohmic contacts and tungsten nitride (WN_x) Schottky contacts. Fabricated using LEC semi-insulating substrates, these MESFETs contained a p-type buried layer underneath the n-channels for isolation from the substrate. The electrical behavior of these GaAs MESFETs at elevated temperatures was also compared with that of Si MOSFETs. The study brought to light that, in general, the GaAs devices exhibited identical degradation of electrical parameters to Si MOSFETs. Furthermore, they were riddled with complex difficulties, which necessitated immediate attention.

The temperature dependence of the transconductance parameter β of MESFET devices was dominated by variations in the mobility and saturation velocity of charge carriers. Like Si MOSFETs, the electron mobility followed the familiar power law dependence

$$\mu = \mu_0(T/T_0)^{-n} \tag{7.10}$$

on temperature with the exponent n lying between 1.6 and 2. The saturation velocity

$$v_{\text{sat}} \propto T^{-1}. \tag{7.11}$$

Both for enhancement- and depletion-mode MESFETs, the threshold voltage variation with temperature is modeled by a similar equation to Si MOSFETs:

$$V_{\text{Th}}(T) = p_0 T + q_0 \text{ with } p_0 = -1.2 \text{ mV K}^{-1} \text{ for } 25 \leqq T \leqq 250\ ^\circ\text{C}, \tag{7.12}$$

where q_0 is the value of V_{Th} at 0 K, as found by extrapolation. In analogy to Si MOSFETs, the existence of zero temperature coefficient (ZTC) bias points, one in the linear region and another in the saturation region, was confirmed.

While the above characteristics displayed similarity with Si MOSFETs, in opposition to Si MOSFETs, there was a significant leakage current flow from the gate. This was because the MESFET gate relied on the Schottky barrier junction in the reverse-biased mode, whereas the Si MOSFET gate depended on the SiO_2 dielectric layer, which allowed imperceptible leakage current flow. To the gate leakage current was added the substrate leakage current arising from the injection of electrons from the channel into the substrate under the influence of the high electric field prevailing near the drain of the MESFET. The substrate leakage current accounted for as much as 50% of the drain current. It was also affected by side-gating and back-gating. All throughout the temperature range 25 °C–400 °C, the predominant component of leakage current was the generation–recombination current. It showed a direct proportionality relationship with intrinsic carrier concentration. In GaAs, there is transition from the generation–recombination component dominated leakage current behavior at a much higher temperature because of the larger bandgap of this material. In silicon and therefore in Si MOSFETs, this transition occurs at a temperature between 125 °C and 150 °C. The ratio of on- and off-state currents was inferior by 2–3 orders of magnitude for GaAs MESFETs than Si MOSFETs. Transconductance decreased monotonically with temperature and drain output resistance increased as the temperature was raised over the temperature range 25 °C–250 °C and frequency range 1–10^6 Hz. The gate input resistance for a depletion-mode MESFET decreased by 4–6 orders of magnitude from 25 °C to 400 °C. The Schottky barrier height fell from 0.65 eV at 25 °C to 0.55 at 400 °C. On the whole, the study showed that performance-wise, GaAs MESFETs were in many respects worse than Si MOSFETs for high-temperature operation. Despite the larger bandgap of GaAs than Si, GaAs devices were succumbing to damage at lower temperatures than equivalent Si devices. This premature failure of GaAs devices fabricated by existing technology showed that considerable research and development efforts were necessary for the improvement of characteristics (Shoucair and Ojala 1992).

7.8 Structural innovations for restricting the leakage current of GaAs MESFET up to 300 °C

Computer simulations of the electrical characteristics of n-channel MESFETs show that an increase in temperature was accompanied by decrease in threshold voltage and transconductance together with an increase in leakage current (Kacprzak and Materka 1983, Wilson and O'Neill 1995). Among these three parameters, the leakage current increase was the issue of the most serious concern as it leads to a loss of transistor action. In GaAs, the crossover from the generation–recombination component of leakage current to the diffusion-related component takes place around 250 °C instead of the 150 °C for Si devices. Two types of leakage current were identified as crucial to device current:

(i) Gate leakage: this leakage current is determined by the Schottky barrier height, which is generally ~0.7 eV and decreases as temperature increases, leading to enhanced injection of electrons from the metal gate into the channel.
(ii) Drain leakage: in the off-state, the channel is depleted of carriers. With an increase in temperature, the background carrier concentration increases exponentially. Drift current flows through the conducting path between the drain, gate and source. Another path is formed towards the substrate contributing to additional leakage. The substrate leakage part constitutes a substantial chunk of the drain leakage.

Principally due to the above two factors, which increase the leakage current to alarming proportions, originating from the change in device structure from MOSFET to MESFET, the GaAs device lost favor to the Si device. These leakage currents in GaAs MESFETs must be subjugated or stopped from upsetting device operation at high temperatures by modifications in device design. Isolating the gate will prevent the thermionic emission of carriers from the gate into the channel. The conduction path to the substrate must be blocked to reduce substrate leakage. By introducing heterojunction barriers, one between the gate and channel and another between the channel and substrate, both leakages are suppressible. From these considerations, wider bandgap AlGaAs barriers are incorporated both above and below the MESFET channel, sealing the conduction paths from the gate and

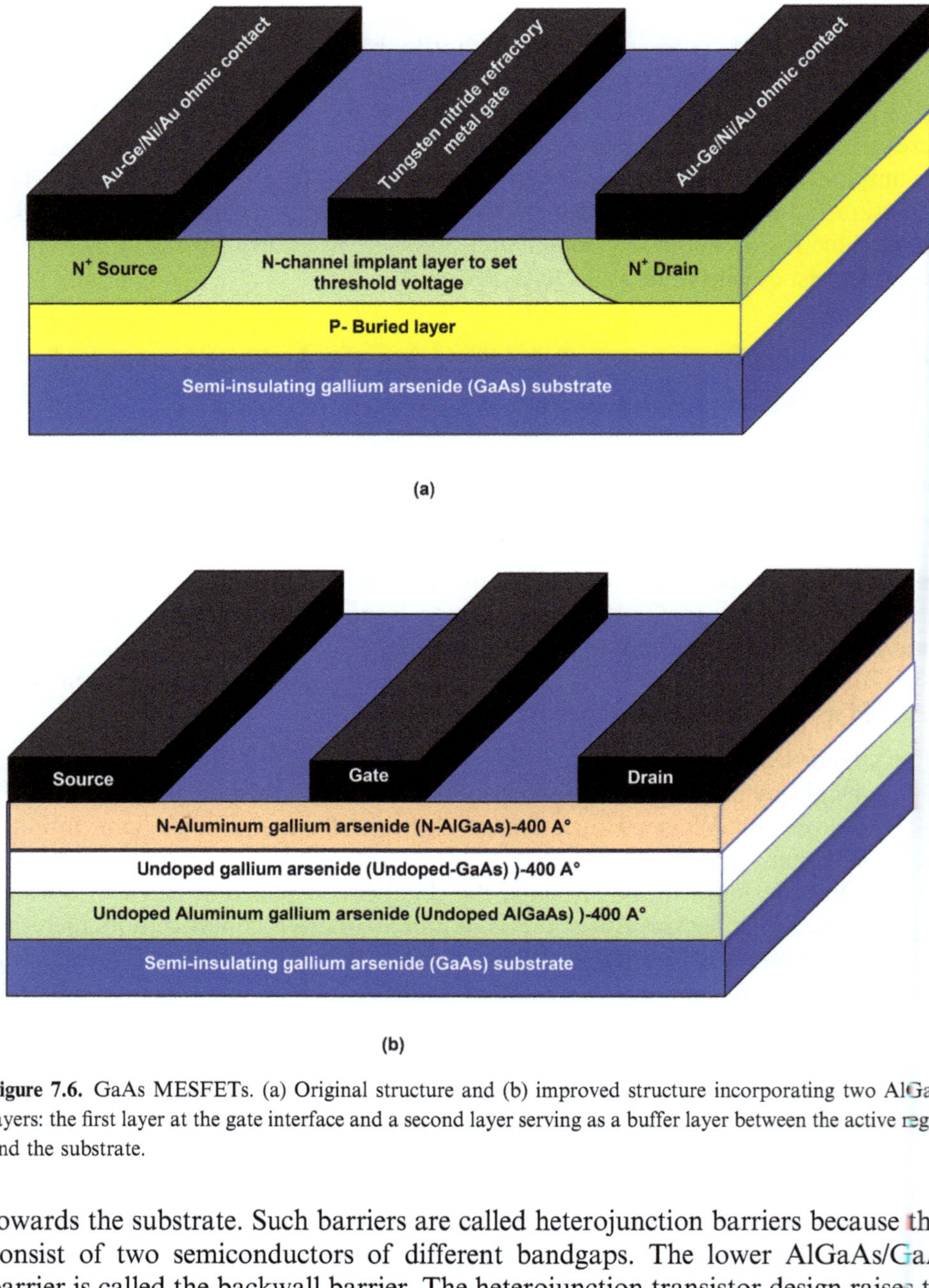

Figure 7.6. GaAs MESFETs. (a) Original structure and (b) improved structure incorporating two AlGaAs layers: the first layer at the gate interface and a second layer serving as a buffer layer between the active region and the substrate.

towards the substrate. Such barriers are called heterojunction barriers because they consist of two semiconductors of different bandgaps. The lower AlGaAs/GaAs barrier is called the backwall barrier. The heterojunction transistor design raises the upper limit of reliable operation of GaAs MESFETs to 300 °C. Figure 7.6 shows the original and improved GaAs MESFET structures.

In a separate experimental study, a stable MESFET technology was developed for operation up to 300 °C by inserting WSi_2 diffusion barriers in the ohmic contacts, supported by optimization of the surface passivation technique using PECVD Si_3N_4 to prevent out-diffusion of Au (Fricke *et al* 1989).

7.9 Won *et al* threshold voltage model for a GaAs MESFET

The drain current of a MESFET (figure 7.7) of channel width W, channel length L and threshold voltage V_{Th}, working under a drain–source voltage V_{DS} and a gate–source voltage V_{GS}, is given by (Won *et al* 1999)

$$I_{DS} = \beta\{2(V_{GS} - V_{Th})V_{DS} - V_{DS}^2\}(1 + \lambda V_{DS}) \text{ for } V_{DS} \leqslant (V_{GS} - V_{Th}) \tag{7.13}$$

$$I_{DS} = \beta(V_{GS} - V_{Th})^2(1 + \lambda V_{DS}) \text{ for } V_{DS} \geqslant (V_{GS} - V_{Th}) \tag{7.14}$$

where β is the transconductance parameter containing the electron mobility μ_n and λ is the channel length modulation parameter.

The drain current becomes temperature dependent through the decrease in both mobility and threshold voltage with temperature. A linear relation is used for the degradation of mobility with temperature. The decrease in threshold voltage with increase of temperature needs cautious inspection. Considering a depletion-mode MESFET, the threshold voltage is the voltage applied on the gate electrode to

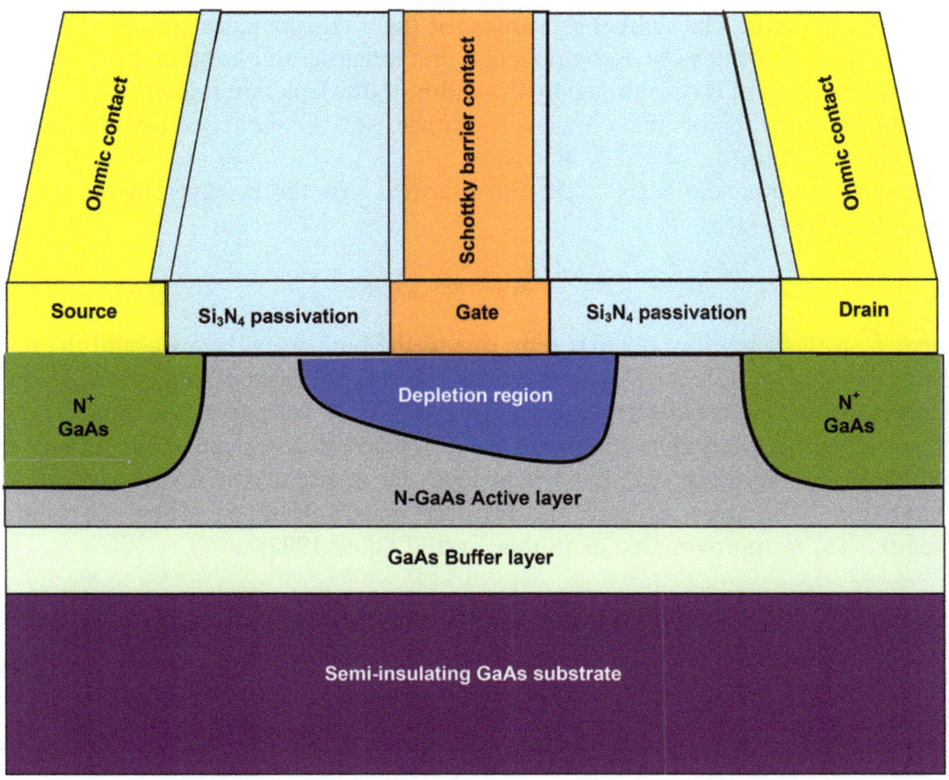

Figure 7.7. A gallium arsenide MESFET.

produce a depletion layer having a thickness equaling the depth a of the active region. Hence, threshold voltage expression consists of three terms contributed by different voltages present in the device as stated below:

$$\begin{aligned} V_{\mathrm{Th}} = &\text{ built-in potential of the Schottky barrier } (V_{\mathrm{bi}}) \\ &- \text{pinch-off voltage } (V_{\mathrm{po}}) \\ &- \text{voltage drop due to leakage current } (V_{\mathrm{l}}) \end{aligned} \tag{7.15}$$

$$= \left\{\Phi_{\mathrm{bn}} - \left(\frac{k_{\mathrm{B}}T}{q}\right) \ln\left(\frac{N_{\mathrm{C}}}{N_{\mathrm{D}}}\right)\right\} - \left(\frac{qN_{\mathrm{D}}a^2}{2\varepsilon_0\varepsilon_{\mathrm{s}}}\right) - \left\{R_{\mathrm{g}}WLA^*T^2 \exp\left(\frac{q\Phi_{\mathrm{bn}}}{k_{\mathrm{B}}T}\right)\right\}. \tag{7.16}$$

The symbols in this equation need explanation. Φ_{bn} is the Schottky barrier height of the metal–semiconductor interface, i.e. the intimate boundary of the metal layer with the n-type semiconductor, here GaAs. In magnitude, it equals the energy difference between two quantities, one on the metal side and the other on the semiconductor side. On the metal side, the Fermi level E_{F} of the metal is taken. On the semiconductor side, the energy of the conduction band edge E_{C} of the semiconductor is the relevant quantity. As usual, N_{C} is the effective density of states in the conduction band. It is related to temperature and is equal to 4.7×10^{17} $(T/300)^{1.5}$. Also, N_{D} is the donor concentration, and ε_{s} is the relative permittivity of the semiconductor. The symbol R_{g} stands for the total resistance confronted by the gate current. It includes two components, one intrinsic and another extrinsic. The intrinsic component is determined by the width of the depletion region. The extrinsic component consists of the contact resistance. A^* is the effective Richardson constant, equal to 14.7×10^4 A m^{-2} K^{-2}.

In this equation, the basis of the third term is that the leakage current flowing through an SBD (area A) is

$$I_{\mathrm{D0}} = A^*AT^2 \exp\left\{(-q\phi_{\mathrm{bn}})/(k_{\mathrm{B}}T)\right\}. \tag{7.17}$$

Further, in this equation, the first term due to the built-in voltage V_{bi} and the third term arising from voltage V_{l} generated by the leakage current depend on temperature, whereas the middle term due to the pinch-off voltage V_{p0} is temperature-independent. An alternative equation for V_{bi} in place of that given in equation (7.16) is derived by assuming that the Fermi level E_{F} is situated midway between the conduction band edge E_{C} and the intrinsic energy level E_{i}. Then the built-in potential V_{bi} is approximated as (Shoucair and Ojala 1992)

$$\begin{aligned} V_{\mathrm{bi}} &= \left(\frac{k_{\mathrm{B}}T}{q}\right) \ln\left(\frac{N_{\mathrm{D}}}{n_{\mathrm{i}}}\right) - \frac{E_{\mathrm{C}} - E_{\mathrm{F}}}{q} = \left(\frac{k_{\mathrm{B}}T}{q}\right) \ln\left(\frac{N_{\mathrm{D}}}{n_{\mathrm{i}}}\right) - \left(\frac{k_{\mathrm{B}}T}{q}\right) \ln\left(\frac{N_{\mathrm{C}}}{N_{\mathrm{D}}}\right) \\ &= \left(\frac{k_{\mathrm{B}}T}{q}\right) \ln\left(\frac{N_{\mathrm{D}}}{n_{\mathrm{i}}} \times \frac{N_{\mathrm{D}}}{N_{\mathrm{C}}}\right) = \left(\frac{k_{\mathrm{B}}T}{q}\right) \ln\left(\frac{N_{\mathrm{D}}^2}{n_{\mathrm{i}}N_{\mathrm{C}}}\right) \end{aligned} \tag{7.18}$$

$$\therefore\ V_{\mathrm{Th}} = \left(\frac{k_{\mathrm{B}}T}{q}\right)\ln\left(\frac{N_{\mathrm{D}}^2}{n_{\mathrm{i}}N_{\mathrm{C}}}\right) - \left(\frac{qN_{\mathrm{D}}a^2}{2\varepsilon_0\varepsilon_{\mathrm{s}}}\right) - \left\{R_{\mathrm{g}}WLA^{*}T^2\exp\left(\frac{q\Phi_{\mathrm{bn}}}{k_{\mathrm{B}}T}\right)\right\} \tag{7.19}$$

where

$$N_{\mathrm{C}} = 4.7\times10^{17}\left(\frac{T}{300}\right)^{1.5} \tag{7.20}$$

$$n_{\mathrm{i}} = 2.51\times10^{19}\left\{\left(\frac{m_{\mathrm{n}}^{*}}{m_0}\right)\left(\frac{m_{\mathrm{p}}^{*}}{m_0}\right)\right\}^{0.75}\left(\frac{T}{300}\right)^{1.5}\exp\left(-\frac{E_{\mathrm{g}}}{2k_{\mathrm{B}}T}\right) \tag{7.21}$$

$$\begin{aligned} n_{\mathrm{i}} = 2.51\times10^{19}[&\{1.028 + (6.11\times10^{-4})T - (3.09\times10^{-7})T^2\} \\ &\times\{0.61 + (7.83\times10^{-4})T - (4.46\times10^{-7})T^2\}]^{0.75} \\ &\times\left(\frac{T}{300}\right)^{1.5}\exp\left(-\frac{E_{\mathrm{g}}}{2k_{\mathrm{B}}T}\right) \end{aligned} \tag{7.22}$$

obtained by substituting the temperature-dependent expressions for m_{n}^{*}/m_0 and m_{p}^{*}/m_0. The model is applied to the calculation of the drain current in the subthreshold and saturation operational regimes of a MESFET in the temperature range 273–673 K (Won *et al* 1999). It may be mentioned here that this model takes into consideration the changes in electron and hole effective masses with temperature but overlooks the thermally induced energy bandgap variation.

7.10 The high-temperature electronic technique for enhancing the performance of MESFETs up to 300 °C

From the discussion in preceding sections, it appears as if the problems associated with operation at high temperatures can only be solved by cleverly improvised device process techniques. Process modification is a complex approach. For simplification, one must look for low-cost, process-free alternative solutions. One such technique, the high-temperature electronic technique, is based on electrical biasing of the substrate with a correct polarity voltage of the required magnitude (Narasimhan *et al* 1999). The I_{DS}–V_{DS} characteristics of the MESFET at room temperature (25 °C) and 300 °C showed that:

(i) the transconductance g_{m} decreased by a factor of 56% from 160 mS mm^{-1} at 25 °C to 70 mS mm^{-1} at 300 °C;
(ii) the leakage current increased by over five orders of magnitude from 25 °C to 275 °C. A careful inspection revealed that these temperature-induced variations in I_{DS}–V_{DS} characteristics were caused by an increase in the thermally generated leakage current. This leakage current altered the behavior of the GaAs substrate, which was semi-insulating at 25 °C. It made the substrate semiconducting at 300 °C. Consequently, the output

resistance in the saturation region had a lower finite value at 300 °C instead of the desired infinite value.

The device characteristics at 300 °C can be restored to the 25 °C values by applying a 6 V bias to the substrate (Narasimhan *et al* 1999). Upon re-measuring and re-plotting the I_{DS}–V_{DS} characteristics at 300 °C, it was found that:

(i) the transconductance g_m became comparable with the values at 25 °C;
(ii) the leakage current decreased to a value which was lower than at 25 °C.

Overall, the drain current at 300 °C was higher than its value at 25 °C and gate-controllable too. Thus stable operation of the device at high temperature was achieved without introducing any process complications.

7.11 The operation of GaAs complementary heterojunction FETs from 25 °C to 500 °C

A complementary heterostructure FET (CHFET; Wilson *et al* 1996) in GaAs parlance is the analogue of CMOS phraseology in silicon. As the CMOS structure consists of n- and p-channel MOSFETs, the CHFET structure comprises n- and p-channel GaAs FETs (figure 7.8). Looking at the layers in this diagram, the heterostructure is: $Al_{0.75}Ga_{0.25}As/In_{0.25}Ga_{0.75}As/GaAs$. In this heterostructure,

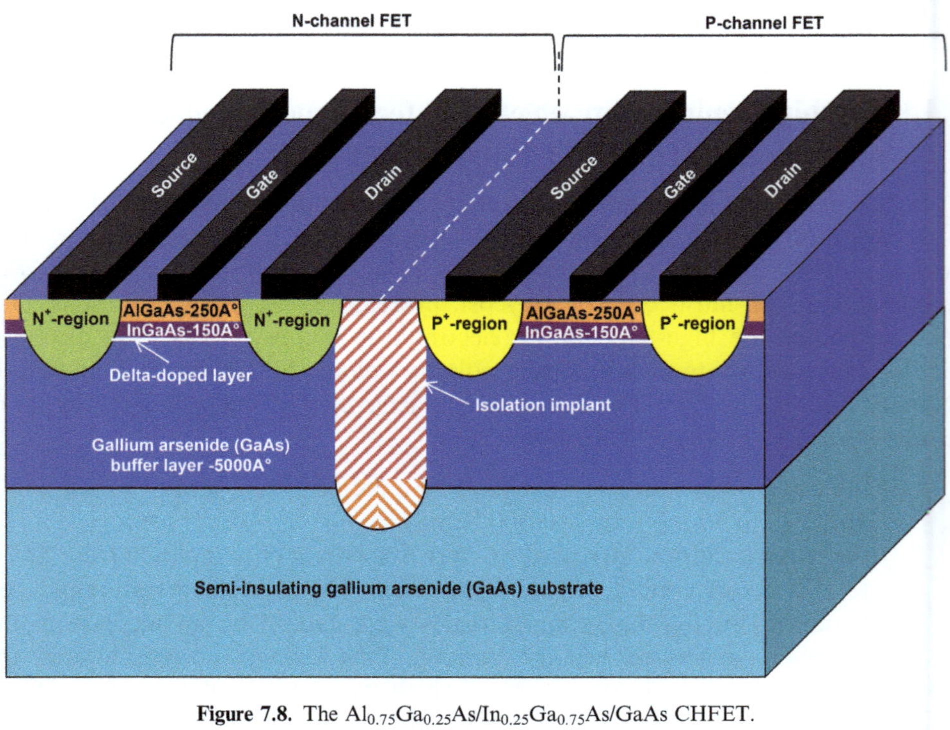

Figure 7.8. The $Al_{0.75}Ga_{0.25}As/In_{0.25}Ga_{0.75}As/GaAs$ CHFET.

the $Al_{0.75}Ga_{0.25}As$ layer with a bandgap 2 eV plays the same role in the MESFET device as the silicon dioxide in a MOSFET. Due to its insulating nature, the gate leakage current of a CHFET is substantially reduced in comparison to that of bulk GaAs MESFETs. The $In_{0.25}Ga_{0.75}As$ layer constitutes the channel of the CHFET. The layer offers the advantage that the mobility of carriers, both electrons and holes, is 30% higher in this layer than in bulk GaAs. Underneath the $In_{0.25}Ga_{0.75}As$ layer is a delta-doped layer with silicon as the dopant. It is meant to control the threshold voltage. Ohmic contacts are made of InGe alloy with stability established up to 400 °C. At 500 °C, the contact resistance began to increase but stayed below 2 Ω-mm. Gate contacts are based on refractory metallization (WSi). The n-channel CHFETs performed sufficiently well digitally at 400 °C. The same is not true for the p-channel device. For both n- and p-channel CHFETs, the leakage current was an issue with the p-channel device being inferior to the n-channel type. Nonetheless, the forward gate leakage current of an n-channel CHFET was less than that of an equivalent MESFET. Unmodified ring oscillators based on CHFETs were recorded as working up to 420 °C. The power consumption of the ring oscillator leaped to 10.9 mW at 350 °C from its room temperature (25 °C) value of 4.3 mW (Wilson *et al* 1996).

The CHFET provides a fast digital technology for high temperatures. Simulations showed that the CHFETs displayed a drain leakage current, which was limited by the gate due to effective Schottky barrier lowering under reverse bias conditions.

7.12 GaAs bipolar transistor operation up to 400 °C

For specific applications, bipolar transistors are favored in comparison to MESFETs. They have a lower $1/f$ noise. The intermodulation distortion introduced by them is also less troublesome. Moreover, they provide intrinsic current gain.

Bipolar transistors (working with the n^+ epitaxial layer as the emitter and the surface n-layer as the collector) were fabricated by ion implantation into bulk n^+ GaAs through the following sequence of steps (Doerbeck *et al* 1982): Se implantation (shallow) for the n-layer → annealing → Be implantation (deep) for p-base layer formation → annealing → localized Be implantation for p^+ region formation to make contacts → annealing → localized Be implantation for isolation region formation. An n–p–n structure was thus obtained. The masking for ion implantation and the surface passivation layer is done through silicon nitride. For n-type regions, the contact material is alloyed Au–Ge–Ni. For p-type regions, it is alloyed Au–Zn. For interconnections, it is Ti–Au.

The current gain of the bipolar transistor exhibited a small increase with increasing temperatures up to 300 °C. Above this temperature, it started to decrease. However, the device showed current gain even beyond 400 °C. The collector–emitter leakage current with the base open (I_{CEO}) is temperature-independent up to 200 °C but increases above this temperature due to an increase in the leakage current of the collector–base diode. The device failed cataclysmically at around 400 °C due to melting and flowing of the gold layer on the contacts.

A 15-stage ring oscillator circuit, mounted in a ceramic package, was tested in the temperature range 25 °C–390 °C. At a bias voltage of 1.75 V, the input current was 5 mA at 25 °C but increased to 7 mA at 385 °C. In the same temperature interval, the gate delay time stretched from 2.5 ns to 4.8 ns. The output signal increased three-fold. Circuit failure occurred at 390 °C due to metallization rupture (Doerbeck *et al* 1982).

7.13 A GaAs-based HBT for applications up to 350 °C

The performance characteristics of n–p–n HBTs with AlGaAs/GaAs heterojunctions were measured up to 350 °C (Fricke *et al* 1992). The Al mole fraction in the wide bandgap AlGaAs emitter layer was fixed at 0.45 to achieve a reasonable current gain value at the high operating temperature. The arguments for arriving at this value are as follows: if N_D and N_A are the doping concentrations of the emitter and base, respectively, v_{nB} is the velocity of electrons at the emitter end of the base, v_{pE} is the velocity of holes at the base end of the emitter and k_B stands for the Boltzmann constant, the maximum current gain β_{max} of an HBT at temperature T is expressed in terms of the valence band offset ΔE_V as

$$\beta_{max} = \left(\frac{N_D}{N_A}\right)\left(\frac{v_{nB}}{v_{pE}}\right)\exp\left(\frac{\Delta E_V}{k_B T}\right). \tag{7.23}$$

Since $\Delta E_V = 0.2$ eV for Al mole fraction 0.45, we have

$$\left[\exp\left(\frac{\Delta E_V}{k_B T}\right)\right]_{298\ K} = \exp\left(\frac{0.2}{8.617 \times 10^{-5} \times 298}\right) = \exp(7.7886) = 2412.94 \tag{7.24}$$

$$\left[\exp\left(\frac{\Delta E_V}{k_B T}\right)\right]_{623\ K} = \exp\left(\frac{0.2}{8.617 \times 10^{-5} \times 623}\right) = \exp(3.7255) = 41.49. \tag{7.25}$$

At room temperature, $T = 298$ K, the exponential term is 2413. At $T = 623$ K it is 42. The ratio of exponential terms in the two cases gives the factor by which the hole injection into the emitter is suppressed by the heterojunction. This factor is found to be 57.45. Considering this value as adequate for repressing the injection of holes, the above Al fraction was set at 0.45.

The emitter and collector contacts were made of Ni/Au/Ge/Ni (figure 7.9). The base contact was Ti–Pt–Au. The contact metallizations were subjected to RTA. $Si_xN_yO_z$ was used as surface passivation to prevent As out-diffusion from GaAs. The DC characteristics of the HBTs were measured at 300 K, 423 K, 573 K and 623 K. The HBTs were functional up to 623 K. However, the characteristics were degraded as the temperature increased. Two distinct temperature ranges were identified:

(i) Room temperature to 573 K: the hole current across the emitter–base heterojunction increases. Consequently, the ideality factor n_B of the base

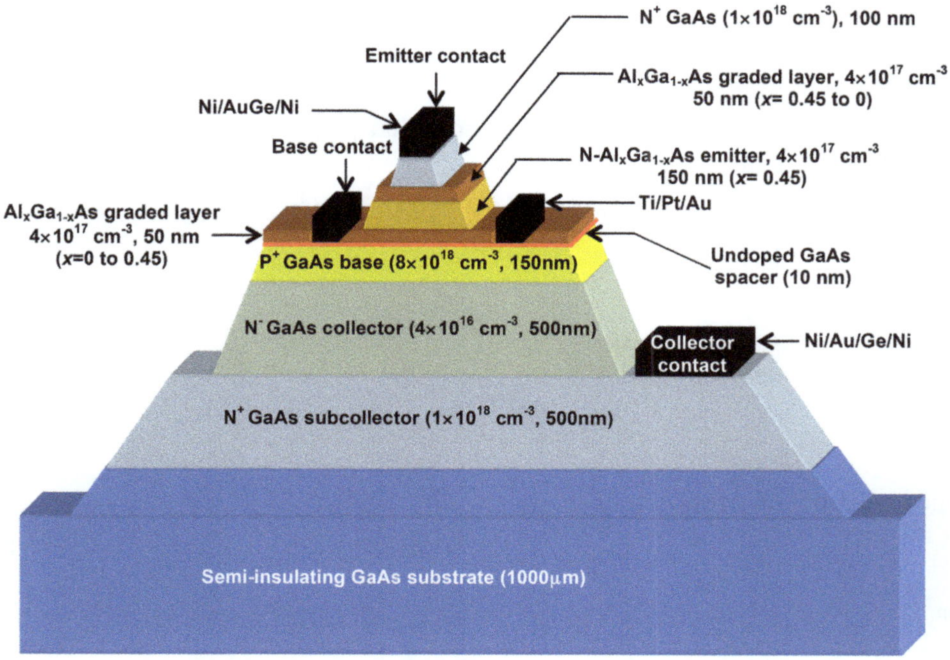

Figure 7.9. Layers in the AlGaAs/GaAs HBT.

current decreases. The current gain β also diminishes as does the small signal current gain h_{FE}.

(ii) 573 K to 623 K: the leakage current across the collector–base diode increases. As a result, n_{B}, β and h_{FE} decrease.

All throughout the temperature range from room temperature to 673 K, a common-emitter small-signal current gain of 35 was maintained. Furthermore, the stability of DC characteristics was confirmed after performing a large number of heating/cooling cycles as well as storage at 573 K for 24 h (Fricke *et al* 1992).

7.14 $\mathrm{Al}_x\mathrm{GaAs}_{1-x}$/GaAs HBT

Basu and Sarkar (2011) proposed an analytical model on similar lines to the SiGe HBT model presented by them and described in chapter 6. The main points of this model are given below: for $\mathrm{Al}_x\mathrm{GaAs}_{1-x}$/GaAs HBT, the equation for current gain $(\beta)_{\mathrm{AlGaAs/GaAs}}$ is

$$(\beta)_{\mathrm{AlGaAs/GaAs}} = \left(\frac{N_{\mathrm{DE}}}{N_{\mathrm{AB}}}\right)\left(\frac{W_{\mathrm{E}}}{W_{\mathrm{B}}}\right)\left\{\frac{D_{\mathrm{n,\,GaAs}}(T)}{D_{\mathrm{p,\,AlGaAs}}(T)}\right\}\exp\left(\frac{\Delta E_{\mathrm{g}}}{k_{\mathrm{B}}T}\right). \tag{7.26}$$

The electron mobility $\mu_{\mathrm{n,GaAs}}$ in GaAs is

$$\mu_{\mathrm{n,\,GaAs}}(T) \propto T^{-2.1} \tag{7.27}$$

or

$$\mu_{\text{n, GaAs}}(T) = K_1 T^{-2.1}, \tag{7.28}$$

where K_1 is a constant. To find K_1, we note that $\mu_{\text{n, GaAs}} = 8500\ \text{cm}^2\ \text{V}^{-1}\ \text{s}^{-1}$ at $T = 300$ K, so that

$$\begin{aligned} K_1 &= \mu_{\text{n, GaAs}}(T)/T^{-2.1} = \mu_{\text{n, GaAS}}(T) \times T^{2.1} \\ &= 8500 \times (300)^{2.1} = 1.353 \times 10^9\ \text{cm}^2\ \text{V}^{-1}\ \text{s}^{-1}\ \text{K}^{2.1} \end{aligned} \tag{7.29}$$

$$\therefore \mu_{\text{n, GaAs}}(T) = 1.353 \times 10^9 T^{-2.1} \tag{7.30}$$

$$\therefore D_{\text{n, GaAs}}(T) = \left(\frac{k_B T}{q}\right) 1.353 \times 10^9 T^{-2.1}. \tag{7.31}$$

The hole mobility in $\text{Al}_x\text{Ga}_{1-x}\text{As}$ is

$$\mu_{\text{p, AlGaAs}}(T) \propto T^{-1} \tag{7.32}$$

or

$$\mu_{\text{p, AlGaAs}}(T) = K_2 T^{-1}, \tag{7.33}$$

where K_2 is a constant determined from the condition that at $T = 300$ K, for AlGaAs (Shur 1995, Basu and Sarkar 2011)

$$\mu_{\text{p, AlGaAs}}(300\ \text{K}) = 370 - 970x + 740x^2, \tag{7.34}$$

where x is the mole fraction of aluminum in the alloy $\text{Al}_x\text{Ga}_{1-x}\text{As}$

$$\therefore K_2 = \mu_{\text{p, AlGaAs}}(T)/T^{-1} = \mu_{\text{p, AlGaAS}}(T) \times T \tag{7.35}$$

$$= \mu_{\text{p, AlGaAs}}(300) \times 300 = 300\left(370 - 970x + 740x^2\right)\ \text{cm}^2\ \text{V}^{-1}\ \text{s}^{-1}\ \text{K} \tag{7.36}$$

$$\therefore \mu_{\text{p, AlGaAs}}(T) = 300\left(370 - 970x + 740x^2\right)T^{-1} \tag{7.37}$$

$$\therefore D_{\text{p, AlGaAs}}(T) = \left(\frac{k_B T}{q}\right) 300\left(370 - 970x + 740x^2\right)T^{-1}. \tag{7.38}$$

The difference between the energy bandgaps of the $\text{Al}_x\text{Ga}_{1-x}\text{As}$ emitter and GaAs base is (Shur 1995, Basu and Sarkar 2011)

$$\left(\Delta E_g\right)_{\text{AlGaAs/GaAs}} = 1.25x \ \text{ for } \ x < 0.4. \tag{7.39}$$

The emitter and base transit times $\{\tau_E(T)\}_{\text{AlGaAs/GaAs}}$ and $\{\tau_B(T)\}_{\text{AlGaAs/GaAs}}$ are:

$$\begin{aligned}\{\tau_E(T)\}_{AlGaAs/GaAs} &= W_E^2/\left[2\beta D_{p,\,AlGaAs}(T)\right] \\ &= W_E^2/\left[2(\beta)_{AlGaAs/GaAs} D_{p,\,AlGaAs}(T)\right]\end{aligned} \tag{7.40}$$

$$\begin{aligned}&= W_E^2\Bigg/\left[2\left(\frac{N_{DE}}{N_{AB}}\right)\left(\frac{W_E}{W_B}\right)\left\{\frac{D_{n,\,GaAs}(T)}{D_{p,\,AlGaAs}(T)}\right\}\exp\left(\frac{\Delta E_g}{k_B T}\right)D_{p,\,AlGaAs}(T)\right] \\ &= W_E^2\Bigg/\left[2\left(\frac{N_{DE}}{N_{AB}}\right)\left(\frac{W_E}{W_B}\right)D_{n,\,GaAs}(T)\exp\left(\frac{\Delta E_g}{k_B T}\right)\right]\end{aligned} \tag{7.41}$$

$$\{\tau_B(T)\}_{AlGaAs/GaAs} = W_B^2/\{2D_{n,\,GaAs}(T)\}. \tag{7.42}$$

From the above analytical model, Basu and Sarkar (2011) found that high gain along with low forward transit time were achievable when a small fraction of Al was introduced in the emitter region of an AlGaAs/GaAs HBT.

7.15 GaAs x-ray and beta particle detectors

Radiation detection can be done using gallium arsenide in high temperature and intense radiation environments (Nava *et al* 1997). It exhibits better radiation resistance to gamma rays and electrons than silicon. It therefore provides improved radiation hardness. Cooling and shielding are not necessary. So, enormous saving in cost and volume of the detector is achievable making it suitable for space missions. Moreover, absorption efficiency of silicon falls rapidly at energies >20 keV because of its low atomic number (Z = 14). GaAs with atomic numbers Z_{Ga} = 31 and Z_{As} = 33 of its constitutive elements is a worthwhile alternative (Greiffenberg *et al* 2021).

Semi-insulating GaAs resistor structures fabricated by Cr diffusion in N-type GaAs show good sensitivity to α, β and γ radiation (Ayzenshtat *et al* 2002). Chromium compensated GaAs material is very useful for x-ray detection in the range 15–60 keV (Budnitsky *et al* 2013).

GaAs mesa photodiodes have been investigated in the temperature range −20 °C to +100 °C for x-ray photon counting spectroscopic detection (Lioliou *et al* 2017). The mesa photodiodes have the structure p^+–i–n^+, and the intrinsic layer is 10 μm thick. ^{55}Fe radioisotope x-ray spectra are characterized with these diodes by connecting to a charge sensitive preamplifier, followed by shaping, amplification and digitization of the output signal. For a diode having a diameter of 200 μm, the best energy resolution is recorded (full width at half maximum at 5.9 keV) at 625 eV. The energy resolution is 0.66 keV at −20 °C. At 100 °C, it rises to 2 keV. β-particle counting spectroscopy has also been done using GaAs photodiodes. The ^{63}Ni β spectra shows the use of these diodes for electron detection (Lioliou and Barnett 2016).

7.16 Discussion and conclusions

Gallium arsenide is fundamentally different from silicon in terms of it being a compound formed from two distinct elements, its crystal structure, the nature of its bandgap and the absence of a good quality insulating oxide layer. As an upshot of these basic differences, the devices which have generated the greatest interest in GaAs are MESFETs and HBTs. To prevent sublimation of the volatile arsenic, the LEC technique is used for growing single crystals of GaAs. Common ohmic and Schottky contact metallization schemes for GaAs are modified for high-temperature operation. Despite the higher intrinsic temperature of GaAs, initial experiments on commercial devices developed for room temperature operation showed that these devices were even inferior to silicon MOSFETs at high temperatures. They exhibited high leakage currents, which originated from the gate and drain leakage components. Therefore, blocking these leakage current paths by using heterojunctions helped in raising the upper temperature limit at which GaAs MESFETs operated well. The high-temperature performance of GaAs MESFETs can be restored to be equivalent to room temperature operation without any process iteration by taking advantage of high-temperature electronic techniques. The CHFET structure in GaAs is similar to the CMOS structure in silicon technology. The n-channel devices in the CHFET configuration show better performance than the p-channel devices. The GaAs bipolar transistor also functions well at high temperatures. AlGaAs/GaAs HBTs worked, with degraded performance, up to 623 K.

Review exercises

7.1. Complete the following sentences for gallium arsenide:

- a Silicon is an elemental semiconductor but gallium arsenide is …………
- b Silicon crystallizes in a diamond structure but gallium arsenide crystallizes in ………………
- c Silicon is an indirect bandgap semiconductor but gallium arsenide is an ……………
- d The oxide grown on silicon is advantageous for microelectronics but the oxides on gallium arsenide are ………………
- e The most popular device in silicon ICs is the MOSFET whereas in gallium arsenide ICs, it is the ………………
- f In a silicon MOSFET, the gating action is provided through the silicon dioxide dielectric, while in a gallium arsenide MESFET, the gating action is provided by …………

7.2. (a) Which allows a high leakage current: an SBD in a GaAs MESFET or an insulating oxide film in a Si MOSFET? (b) Which has a high diffusion-dominated leakage current at high temperatures: a Si MOSFET or GaAs MESFET?

7.3. Is there any analogue of SOI technology in GaAs ICs? If not, how are the different components isolated (a) at low temperatures and (b) at high temperatures?

7.4. Can you name another important device used in gallium arsenide ICs besides the GaAs MESFET?
7.5. What is the intrinsic temperature for gallium arsenide? How high is it compared to that of silicon?
7.6. Why is it necessary to surround the GaAs melt with liquid boron trioxide during single-crystal growth? What is this technique called? How are single crystals of GaAs grown?
7.7. What elements are used for p-type doping of gallium arsenide? What are the common n-type dopants in gallium arsenide? What is the doping technique used?
7.8. What happens if the sample is heated during ion implantation? How is damage introduced by implantation removed?
7.9. Discuss the contributions of the constituent layers in the Au/Ge/Ni/Ti metallization scheme for n-type GaAs? Why is RTA suitable for this scheme? Is this scheme thermally stable?
7.10. Describe a suitable metallization scheme for n-type GaAs which can sustain high-temperature operation. What is the maximum allowed temperature for this scheme? What happens if any residual gold is left unconsumed during annealing?
7.11. What is the barrier height for a Ti–Pt–Au Schottky contact to GaAs? Explain the roles of the three constituent layers. How can the thermal stability of this contact scheme be improved? What is the maximum temperature limit for the scheme?
7.12. What is the barrier height of a Au–LaB_6 Schottky contact to GaAs? How high a temperature can be allowed for this contact without damage?
7.13. Describe experiments carried out on commercial GaAs devices in the 25 ° C–400 °C temperature range. What are the ohmic contacts made of? What material is used for Schottky contacts? Write the equations describing the variation of mobility, saturation velocity and threshold voltage. Why did the GaAs MESFETs perform worse than Si MOSFETs despite the larger bandgap of gallium arsenide than silicon?
7.14. What are the two components of leakage current in a GaAs MESFET at high temperatures? What are the paths of flow for these components? Explain how it is possible to raise the operating temperature limit of GaAs MESFETs by taking advantage of heterojunction barriers to stop leakage current flows?
7.15. Write the equations for the drain–source current of a MESFET in the linear and saturation regions. Explain the symbols used. What parameters in these equations make the drain current temperature dependent?
7.16. Write the equation for the threshold voltage of a MESFET and explain the symbols in this equation. What is the physical explanation for the three terms in this equation? Which of the terms in this equation depend on temperature?

7.17. Is it possible to improve the performance of a GaAs MESFET up to 300 °C without process modification? If yes, what is the technique used for this purpose called?
7.18. Describe the high-temperature electronic technique for restoring the 300 °C values of electrical parameters of a GaAs MESFET to room temperature values.
7.19. What is the analog of CMOS structure of silicon technology in GaAs technology? What are the equivalents of NMOS and PMOS transistors in this analog?
7.20. Which layer in the $Al_{0.75}Ga_{0.25}As/In_{0.25}Ga_{0.75}As/GaAs$ heterostructure plays the same role in a GaAs MESFET as silicon dioxide in a MOSFET? What property of this layer helps in reducing the gate leakage?
7.21. Which layer in the $Al_{0.75}Ga_{0.25}As/In_{0.25}Ga_{0.75}As/GaAs$ heterostructure plays the role of a channel in a GaAs MESFET? What advantage is offered by this layer as compared to bulk GaAs?
7.22. How are the ohmic and Schottky contacts made for the $Al_{0.75}Ga_{0.25}As/In_{0.25}Ga_{0.75}As/GaAs$ heterostructure? How do the n- and p-channel CHFETs perform at a high temperature? How does the power consumption of ring oscillators based on CHFETs vary with temperature?
7.23. Describe the process sequence for fabrication of a GaAs bipolar transistor. How does this transistor work at 300 °C? What are the effects of temperature on current gain and leakage current? How does the ring oscillator circuit using this device function at high temperature? At what temperature does it fail?
7.24. Explain, with calculations, why the Al mole fraction in the AlGaAs emitter of an AlGaAs/GaAs HBT is fixed at 0.45? What are the emitter and collector contacts made of? What is the base contact made of? What is the passivation material used? Up to what temperature is the HBT device functional? How does the device function in the temperature ranges: room temperature to 573 K and 573 K to 623 K? How much is the performance degraded?
7.25. In what respects GaAs is better than silicon for radiation detection? Elaborate the reasons justifying its superiority.

References

Ayzenshtat G I, Budnitsky D L, Koretskaya O B, Novikov V A, Okaevich L S, Potapov A I and Tolbanov O P *et al* 2002 GaAs resistor structures for X-ray imaging detectors *Nucl. Instrum. Methods Phys. Res.* A **487** 96–101

Baca A G and Ashby C I H 2005 *Fabrication of GaAs Devices* (London: Institution of Electrical Engineers)

Basu S and Sarkar P 2011 Analytical modeling of AlGaAs/GaAs and Si/SiGe HBTs including the effect of temperature *J. Electron Devices* **9** 325–9

Bhattacharya D 1990 Gallium arsenide digital integrated circuits *Bull. Mater. Sci.* **13** 135–50

Budnitsky D, Tyazhev A, Novikov V, Zarubin A, Tolbanov O, Skakunov M and Hamann E *et al* 2013 Chromium-compensated GaAs detector material and sensors *15th Int. Workshop on Radiation Imaging Detectors (Paris, 23–27 June)* 1–7

Chang C Y and Kai F 1994 *GaAs High-Speed Devices: Physics, Technology, and Circuit Applications* (New York: Wiley) pp 7–54

Cheng K Y 2020 Material technologies *III–V Compound Semiconductors and Devices: An Introduction to Fundamentals* (Cham: Springer Nature Switzerland AG) pp 161–202

Doerbeck F H, Duncan W M, Mclevige W V and Yuan H-T 1982 Fabrication and high-temperature characteristics of ion-implanted GaAs bipolar transistors and ring-oscillators *IEEE Trans. Indust. Electron.* **29** 136–9

Einspruch N G and Wisseman W R (ed) 1985 VLSI electronics microstructure science *GaAs Microelectronics* vol 11 (Orlando, FL: Academic)

Eun J and Cooper J A Jr 1993 High-temperature ohmic contact technology to n-type GaAs *ECE Technical Reports, Purdue University TR-EE 93-7* pp 14–8

Fisher D and Bahl I (ed) 1995 *Gallium Arsenide IC Applications Handbook* vol 1 (San Diego, CA: Academic) pp 1–78

Fricke K, Hartnagel H L, Lee W-Y and Würfl J 1992 AlGaAs/GaAs HBT for high-temperature applications *IEEE Trans. Electron Devices* **39** 1977–81

Fricke K, Hartnagel H L, Schütz R, Schweeger G and Würfl J 1989 A new GaAs technology for stable FETs at 300 °C *IEEE Electron Device Lett.* **10** 577–9

Ghandhi S K 2008 *VLSI Fabrication Principles: Silicon and Gallium Arsenide* (New York: Wiley) pp 1–68

Greiffenberg D, Andrä M, Barten R, Bergamaschi A, Brückner M, Busca P, Chiriotti S and Chsherbakov I *et al* 2021 Characterization of chromium compensated GaAs sensors with the charge-integrating JUNGFRAU readout chip by means of a highly collimated pencil beam *Sensors* **21** 1550

Kacprzak T and Materka A 1983 Compact dc model of GaAs FETs for large-signal computer calculations *IEEE J. Solid State Circuits* **18** 211

Kirkpatrick C G 1988 Making GaAs integrated circuits *Proc. IEEE* **76** 792–815

Lioliou G and Barnett A M 2016 Gallium Arsenide detectors for X-ray and electron (beta particle) spectroscopy *Nucl. Instrum. Methods Phys. Res.* A **836** 37–45

Lioliou G, Whitaker M D C and Barnett A M 2017 High temperature GaAs X-ray detectors *J. Appl. Phys.* **122** 244506

Narasimhan R, Sadwick L P and Hwu R J 1999 Enhancement of high-temperature high-frequency performance of GaAs-based FETs by the high-temperature electronic technique *IEEE Trans. Electron Devices* **46** 24–31

Nava F, Bertuccio G, Vanni P, Fantacci M E, Canali C, Cavallini A and Peroni M 1997 Improved performance of GaAs radiation detectors with low ohmic contacts *IEEE Trans. Nucl. Sci.* **44** 943–9

Papež N, Dallaev P, Ţălu S and Kaštyl J 2021 Overview of the current state of gallium arsenide-based solar cells *Materials (Basel).* **14** 3075

Shoucair F S and Ojala P K 1992 High-temperature electrical characteristics of GaAs MESFETs (25–400 °C) *IEEE Trans. Electron Devices* **39** 1551–7

Shur M 1987 *GaAs Devices and Circuits* (New York: Springer) pp 161–2

Shur M 1995 *Physics of Semiconductor Devices* (New Delhi: Prentice-Hall India)

Tiku S and Biswas D 2016 *III–V Integrated Circuit Fabrication Technology* (Boca Raton, FL: Pan Stanford Publishing, CRC Press) pp 79–96

Wilson C D and O'Neill A G 1995 High temperature operation of GaAs based FETs *Solid-State Electron.* **38** 339–43

Wilson C D, O'Neill A G, Baier S M and Nohava J C 1996 High temperature performance and operation of HFETs *IEEE Trans. Electron Devices* **43** 201–6

Won C-S, Ahn H K, Han D-Y and El Nokali M A 1999 DC characteristic of MESFETs at high temperatures *Solid-State Electron.* **43** 537–42

Würfi J, Singh J K and Hartnagel H L 1990 Reliability aspects of thermally stable LaB_6–Au Schottky contacts to GaAs *Reliability Physics Symp. (New Orleans, LA, 26–29 March)* 87–93

Chapter 8

Silicon carbide electronics for hot environments

Silicon carbide is endowed with unique material properties that can be utilized for the fabrication of high-temperature, high-power and fast switching devices, promising a greater impact on high-voltage devices than Si and GaAs due to its higher breakdown field. The electron mobility in the 4H-SiC polytype is double the mobility in the 6H-SiC polytype perpendicular to the *c*-axis, while parallel to the *c*-axis the ratio of mobility in 4H-SiC to that in 6H-SiC is 10. This difference makes 4H-SiC more attractive. Important SiC devices are p–n junction diodes, Schottky barrier diodes (SBDs) and junction field-effect transistors (JFETs). MOSFET development has slowed down due to the low channel mobilities achieved. Bipolar junction transistors (BJTs) are receiving increasing attention due to their low on-resistance, made possible by electron injection and resulting conductivity modulation. SiC technology is progressing from the stage of research in semiconductor laboratories to commercial-scale production. The supply of low-cost, defect-free large-area substrates will considerably boost this transition. Availability of economical SiC wafers is a matter of the greatest concern in moving towards technological maturity.

8.1 Impact of silicon carbide devices on power electronics and its superiority over silicon

The third generation semiconductor SiC holds great promise in complementing silicon-based technologies (Yi *et al* 2022). Energy efficiency of converters and inverters depends on the power semiconductor devices used in these circuits. Therefore, the progress of power electronics relies on the high-performance power devices (Kimoto 2022). Keeping in view that GaN is still used in a small number of

doi:10.1088/978-0-7503-5072-3ch8

applications, the present trends suggest that SiC is the most viable substitute for silicon in high-power and high-temperature electronics (Alves *et al* 2017).

Silicon carbide electronics has progressed from the research phase to commercial manufacturing, mounted on the most appropriate polytype 4H-SiC for high-temperature circuits. Planar and trench MOSFETs and JFETs are widely commercially available, either as discrete devices or in power modules in the 650–3300 V range, replacing silicon in automotive and rail transport (She *et al* 2017, Veliadis 2022).

Silicon carbide occurs in a large number of polytypes, around 150–250. The packing sequences of the close-crowded bi-atomic strata in these polytypes constitute the criteria for distinguishing them from each other. All these polytypes do not allow easy growth. Amongst those found, the 4H-SiC and 6H-SiC polytypes are available as substrates for device fabrication. Between 4H-SiC and 6H-SiC, the former is superior for electronic device fabrication due to the higher mobility of charge carriers in this polytype and its wider bandgap (Langpoklakpam *et al* 2022, Papanasam *et al* 2022).

The bandgap of silicon carbide (3.23 eV for 4H-SiC) is larger than that for silicon (1.12 eV) by about 2.9 times. So, operation above 873 K is feasible. Its electrical breakdown field (3 MV cm^{-1}) is higher than that of silicon (0.3 MV cm^{-1}) by a factor of ten. Taking advantage of the higher field, the thickness of the uniformly doped conduction region is drastically reduced, enabling a sharp fall of the on-resistance of devices. Its thermal conductivity (4.9 W cm K^{-1}) is 3.27 times higher than that of silicon (1.5 W cm K^{-1}). Hence, a greater power density is obtainable from SiC, giving more power per unit area of the chip. On the opposite side, the lower electron mobility in SiC (800–900 $cm^2\ V^{-1}\ s^{-1}$) than silicon (1400 $cm^2\ V^{-1}\ s^{-1}$) is a major disadvantage of SiC. Similar remarks apply to the hole mobility. The low carrier mobilities are sufficient to provide RF performance in the frequency range 8×10^9–12×10^9 Hz but are inadequate beyond this limit. At microwave frequencies ($<10^{10}$ Hz), SiC devices compete well with Si and GaAs components. p–n junction diodes, JFETs and thyristors are fabricated using SiC, but MOSFETs have lagged behind, mainly due to the lower carrier mobilities. In table 8.1, the salient properties of silicon carbide related to device and circuit design are listed. Figures 8.1 and 8.2 show the crystalline structures of polytypes of silicon carbide.

8.2 Intrinsic temperature of silicon carbide

It is interesting to perform the intrinsic temperature calculations for silicon carbide in line with those for silicon, and compare the values. For 4H-SiC,

$$m_n^* = 0.39m_0,\ m_p^* = 0.82m_0,\ E_g = 3.23\ \text{eV}. \tag{8.1}$$

Hence,

$$m^* = \left(\frac{m_n^* m_p^*}{m_0^2}\right)^{0.75} = \left(\frac{0.39m_0 \times 0.82m_0}{m_0^2}\right)^{0.75} = 0.4253. \tag{8.2}$$

For 4H-SiC, the equation

Table 8.1. Properties of silicon carbide (4H-SiC polytype).

Property	Value	Property	Value	Property	Value
Chemical formula	4H-SiC	Dielectric constant	9.7	Electron saturated velocity (cm s^{-1})	2×10^7
Classification	IV–VI compound semiconductor	Thermal conductivity (W cm K^{-1})	4.9	Donors	Nitrogen phosphorous
Crystal structure	Wurtzite (hexagonal unit cell)	Energy bandgap E_g (eV) at 300 K	3.23	Donor ionization energies ΔE_D (meV)	50, 92 (nitrogen); 54, 93 (phosphorous)
Color	Light brown (low n-doping)	Electrical breakdown field (V cm^{-1})	3×10^6 parallel to *c*-axis	Acceptor impurities	Aluminum, boron
Density at 300 K (g cm^{-3})	3.211	Intrinsic carrier concentration (cm^{-3})	5×10^{-9}	Acceptor ionization energies ΔE_A (meV)	200 (aluminum), 285 (boron)
Lattice constant (Å)	$a = 3.073$, $b = 10.053$	Electron mobility (cm^2 V^{-1} s^{-1})	900 parallel to *c*-axis, 800 perpendicular to *c*-axis		
Melting point (°C)	3103 ± 40 K at 35 atm	Hole mobility (cm^2 V^{-1} s^{-1})	115		

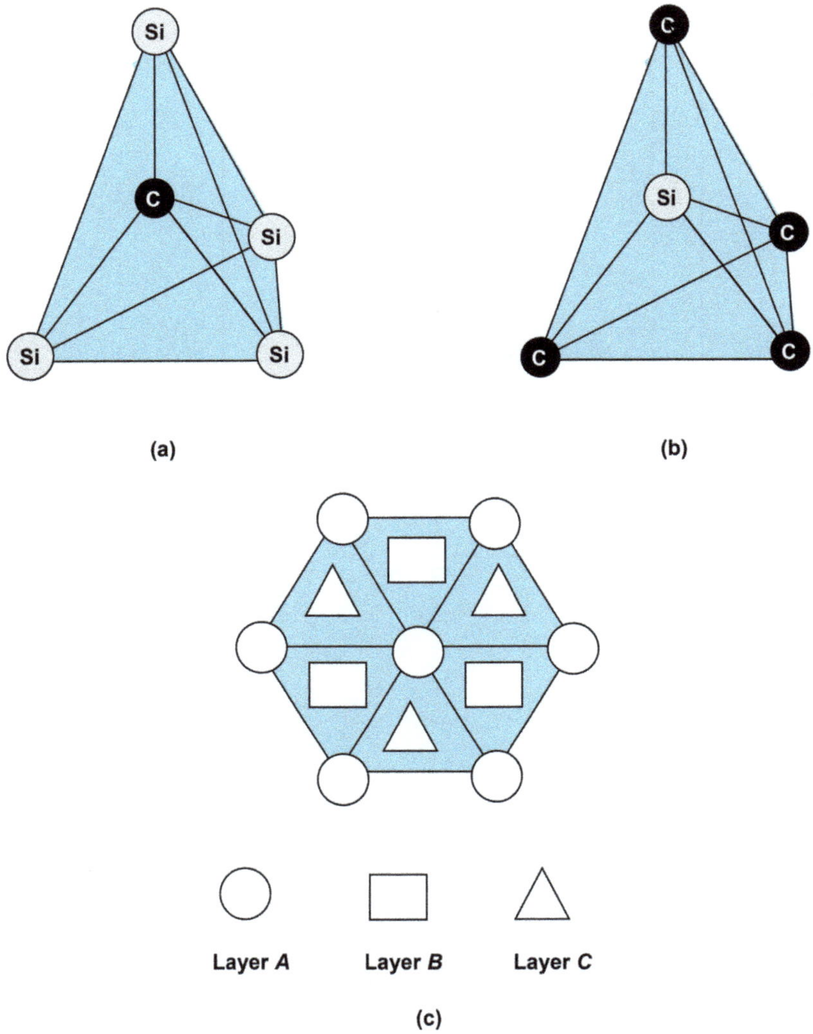

Figure 8.1. Silicon carbide crystal structure. (a) Arrangement of carbon and silicon atoms in a tetrahedron with one carbon atom surrounded by four silicon atoms, (b) a similar arrangement with one silicon atom surrounded by four carbon atoms and (c) naming of the layers as A, B and C according to the positions occupied by the atoms in the tetrahedron.

$$\ln\{0.207/(m^*T^{1.5})\} = -5.8025 \times 10^3 E_g \tag{8.3}$$

takes the form

$$T\ln\{0.207/(0.4253T^{1.5})\} = -5.8025 \times 10^3 \times 3.23 = -18\ 742.075 \tag{8.4}$$

or

$$\begin{aligned} T\ln\ 0.4867 - T\ln T^{1.5} &= -\ 18\ 742.075 \\ \therefore -\ 0.720\ 11T - T\ln T^{1.5} &= -\ 18\ 742.075. \end{aligned} \tag{8.5}$$

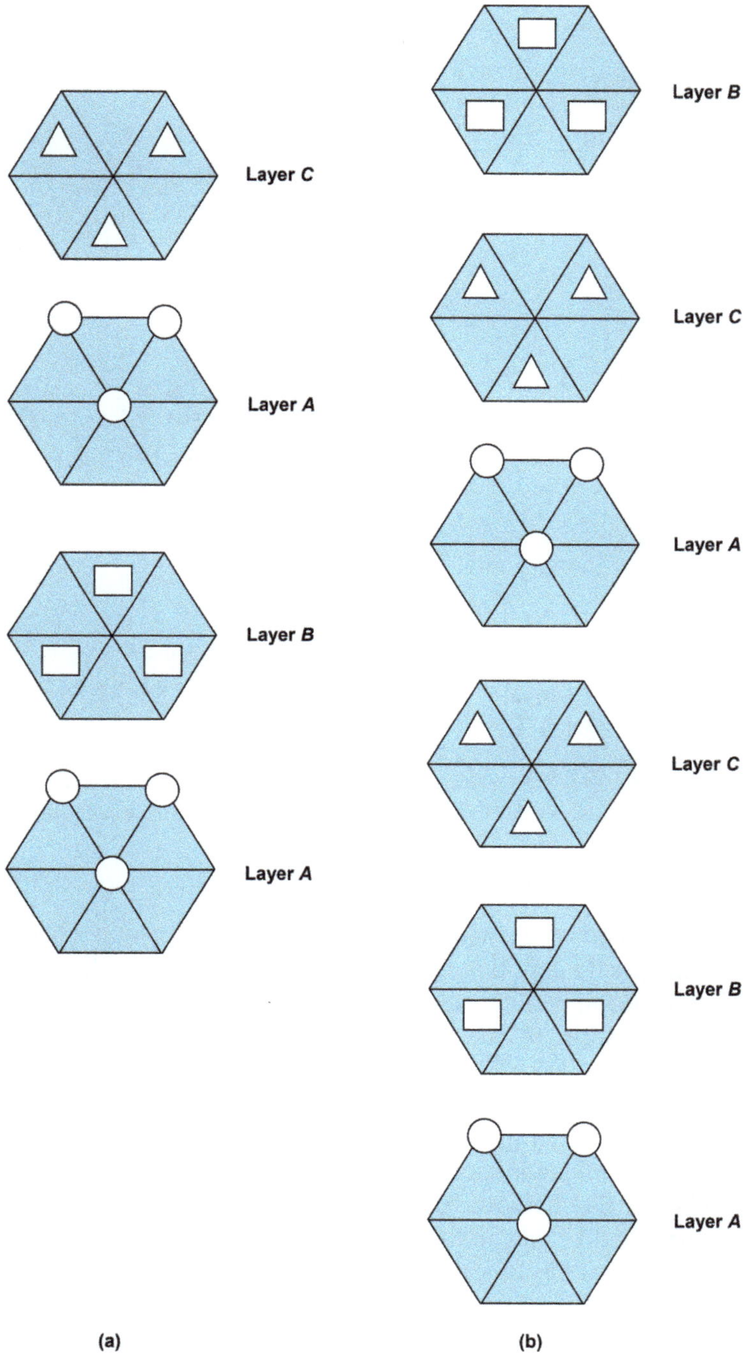

Figure 8.2. Stacking sequence of the double layers of silicon and carbon to produce the polytypes. (a) 4H-SiC (ABAC) and (b) 6H-SiC (ABCACB); H stands for hexagonal.

Suppose,

$$T = 1590 \text{ K} \tag{8.6}$$

$$\begin{aligned} \text{lhs} &= -0.720\ 11T - T\ln T^{1.5} \\ &= -0.720\ 11 \times 1590 - 1590\ln 1590^{1.5} \\ &= -1144.9749 - 17\ 581.001\ 969 = -18\ 725.976\ 869. \end{aligned} \tag{8.7}$$

If

$$T = 1591.5 \text{ K}$$

$$\begin{aligned} \text{lhs} &= -0.720\ 11T - T\ln T^{1.5} \\ &= -0.720\ 11 \times 1591.5 - 1591.5\ln 1591.5^{1.5} \\ &= -1146.055 - 17\ 599.838\ 88 = -18\ 745.893\ 88. \end{aligned} \tag{8.8}$$

When

$$T = 1592 \text{ K}$$

$$\begin{aligned} \text{lhs} &= -0.720\ 11T - T\ln T^{1.5} \\ &= -0.720\ 11 \times 1592 - 1592\ln 1592^{1.5} \\ &= -1146.415 - 17\ 606.1183 = -18\ 752.5333. \\ &\therefore T \approx 1591.5 \text{ K}. \end{aligned} \tag{8.9}$$

This temperature is 1591.5 K − 588.63 K = 1002.87 K higher than that for silicon (588.63 K) and 1591.5 K − 756.6 K = 834.9 K higher than that for GaAs (756.6 K).

8.3 Silicon carbide single-crystal growth

The traditional method of growing large diameter single-crystal silicon by crystal pulling from the melt, e.g. the CZ process, supplies the bulk of crystals for semiconductor device manufacturing. Unfortunately, this method is not applicable to silicon carbide owing to the fact that instead of melting, it sublimes at a temperature lower than 2000 °C.

Silicon carbide single crystals are grown by a sublimation method referred to as physical vapor deposition (figure 8.3) in which the SiC formed by the reaction between molecular species containing silicon and carbon is directly deposited on the seed crystal (Cooper and Kimoto 2014). These molecular species are obtained from a subliming source of SiC lying in the vicinity of the seed crystal. For crystal growth, a seed of {0001} substrate is generally used. The SiC single crystals are grown along the direction of the [0001] *c*-axis. This method of growth is called *c*-face growth.

Of the well-known defects in SiC crystals the following can be mentioned:

(i) Open core screw dislocations termed micropipes, which are responsible for the failure of SiC diodes below their avalanche breakdown limit.
(ii) Domain walls or grain boundaries leading to leakage current flow and producing cracks in the wafer during epitaxy.

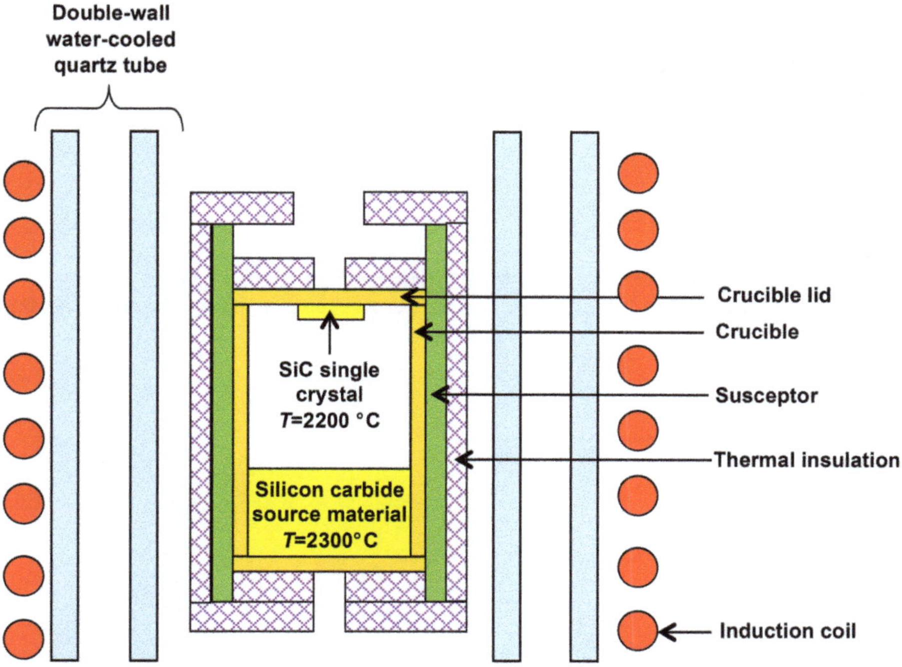

Figure 8.3. Physical vapor deposition chamber for growing single-crystal silicon carbide.

(iii) Threading screw dislocations affecting the leakage currents in Schottky diodes.
(iv) Basal plane dislocations, whose presence in the active layer of PIN devices degrades their forward voltage characteristics.
(v) Remnant nitrogen and boron contamination which is detrimental to highly pure undoped semi-insulating SiC substrates.

8.4 Doping of silicon carbide

An accepted n-type dopant for SiC is nitrogen. p-type doping is performed using aluminum. Even at very high temperatures, ~2000 °C, the diffusion coefficients of these dopants in SiC are extremely low. The energy levels of these impurities are shown in figure 8.4. Therefore, the popular technique of thermal diffusion of dopants into semiconductors is precluded outright. A practical option is ion implantation. During ion implantation, the SiC wafer may be kept at a temperature between normal room temperature to as high as 900 °C. The implantation step is invariably followed by annealing at a temperature higher than 1700 °C. The temperature of the SiC wafer during ion implantation as well as in the annealing cycle is a critical determinant of the resultant doping profile.

The above techniques are applied when doping needs to be done selectively in defined areas of the SiC wafer for device fabrication. Otherwise, *in situ* doping is carried out during epitaxial growth of SiC.

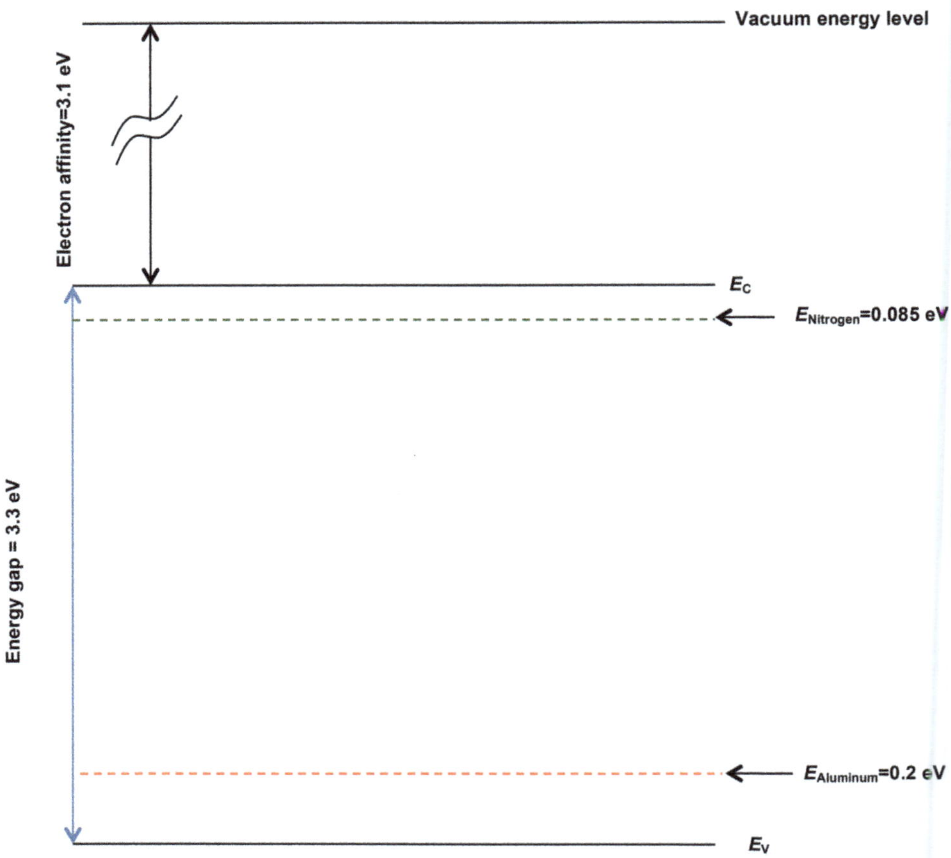

Figure 8.4. Energy band diagram of silicon carbide showing the energy levels of commonly used impurities for n- (nitrogen) and p-doping (Al).

8.5 Surface oxidation of silicon dioxide

An encouraging aspect of SiC wafer processing is that oxide films can be grown over SiC surfaces much like silicon, although the oxidation rates are comparatively slower than Si and also depend upon whether the silicon- or carbon-terminated face is oriented towards the growing oxide. Nevertheless, the growth of good-quality oxide films with low interface state and trap concentrations has been a major roadblock. The reliability of the oxide films, particularly at high electric fields and elevated temperatures, is questionable.

8.6 Schottky and ohmic contacts to silicon carbide

For Schottky contacts with n- and p-type SiC, various combinations of transition metals have been attempted. These metals have barrier heights in the range 0.9–1.7 eV. However, the Schottky behavior is either transformed to ohmic or serious deterioration of performance takes place at temperatures exceeding 873 K. An ohmic contact to n-type SiC is commonly established with Ni_2Si obtained by deposition of Ni followed by its silicidation by high-temperature annealing at >900 °C, providing a thermally stable contact. For an ohmic contact to p-type SiC, aluminum is still the preferred choice but its operation at high temperatures is not recommended because it has a low melting point. Low-resistance contact materials with high work functions are necessary to equipoise the large bandgap and electron affinity of SiC.

8.7 SiC p–n diodes

Using the edge-termination technique, planar p–n diodes with a blocking capability of 1400 V showed a forward drop of 6.2 V at 4000 A cm^{-2}. The differential on-resistance was <1 mΩ cm^{-2} (Peters *et al* 1997). 4.5 kV 4H-SiC diodes could switch waveforms up to 770 A. The diodes used a two-fold implanted junction termination extension (JTE) consisting of two p-rings surrounding the p^+-anode (Braun *et al* 2002).

8.7.1 SiC diode testing up to 498 K

A breakdown voltage of 12.9 kV was obtained on 100 μm thick epitaxial layers. These layers were grown on n^+ 4H-SiC substrates (Sundaresan *et al* 2012). Carrier lifetimes were measured using the open circuit voltage decay (OCVD) technique. They ranged from 2 to 4 μs at room temperature and increased to 14 μs at 498 K. To test the switching behavior at high currents, a module consisting of five 10 kV PIN rectifiers was connected as a free-wheeling diode with silicon insulated gate bipolar transistor (IGBT). 1100 V, 100 A switching was carried out through the IGBT-PIN rectifier module up to 498 K. The peak recovery current rose from −48 A at 298 K to −120 A at 498 K. The reverse recovery time increased from 168 ns at 298 K to 528 ns at 498 K. These effects were produced by the positive TC of the lifetime of carriers injected into the n-base.

8.7.2 SiC diode testing up to 873 K

A primary need to guarantee high-temperature operation of SiC diodes is provision of thermally stable ohmic contacts. This calls for proper selection of metals which can withstand the high-temperature conditions. It was found (Kakanakov *et al* 2002) that Ti/Al/p-SiC and Ni/n-SiC contacts were not degraded upon ageing at 773 K to 783 K in N_2 ambient for 100 h. On injecting high current density (10^3 A cm^{-2}) through these contacts and raising the temperature to 723 K, the contacts exhibited stability. In a related study (Chand *et al* 2014), Ni or Pt was sputtered on Ni/Si ohmic contacts on SiC diodes (figure 8.5). Ageing experiments were conducted in air at different temperatures: 673 K, 773 K and 873 K. Forward current–voltage characteristics of the diodes were recorded. The motive was to examine the changes in junction and series resistances of the diodes. Platinum metallized diodes showed stability of the current–voltage characteristics as well as invariability of series resistance at all the above temperatures. The finding that the diode ideality factor was 1.02 at 873 K evidenced that the diode was functioning properly at this temperature. However, the series resistance of the nickel-metallized diodes

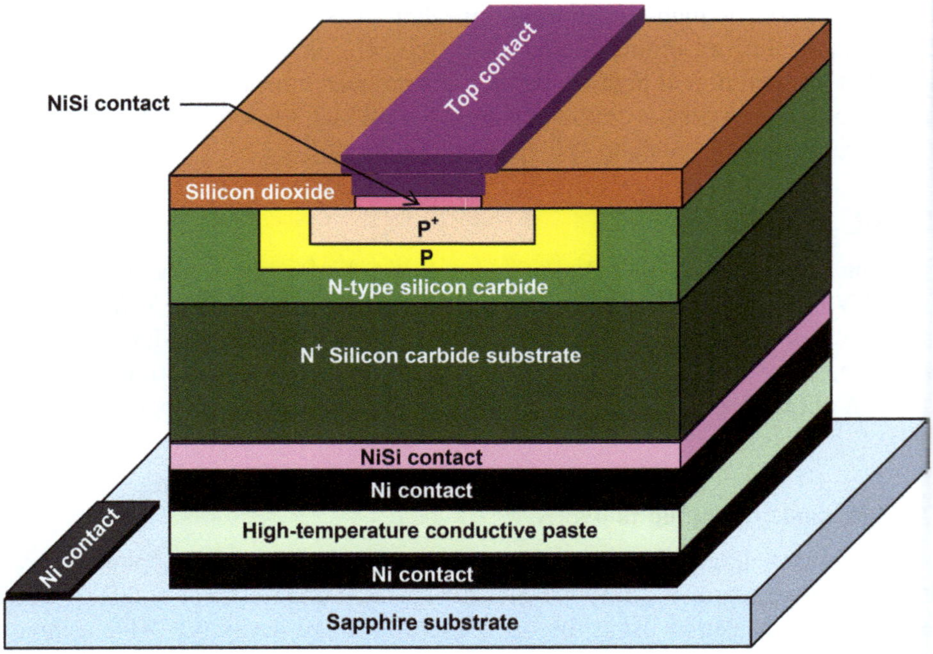

Figure 8.5. SiC p–n junction diode with NiSi ohmic contact.

increased, although only slightly, due to oxidation of the nickel surface. This happened after 7 h at 673 K.

8.7.3 Operation of SiC integrated bridge rectifier up to 773 K

Stable operation of an integrated bridge rectifier circuit was demonstrated up to 773 K (Shao *et al* 2014). The turn-on voltage of the diode decreased from 2.6 V at room temperature (290 K) to 1.4 V at 773 K. The average shift rate was 2.48 mV K^{-1}. The on-resistance decreased from 2.3 kΩ to 0.9 kΩ in the same temperature range with an average rate of 2.9 Ω K^{-1}. From room temperature to 773 K, the DC output voltage of the rectifier changed from 5.4 V to 6.4 V. An important figure-of-merit of the rectifier is the voltage conversion efficiency (V_{CE}) = output DC voltage/effective DC voltage of the input sine wave (0.707 V_{peak}). The V_{CE} was estimated as 76.4% at room temperature and 90.5% at 773 K. The success of the SiC rectifier circuit promises applications in extreme environments.

8.8 SiC Schottky barrier diodes

Schottky diodes are closer to ideal switches than p–n diodes with reference to reverse recovery charge and recovery softness. SiC SBDs essentially resemble Si SBDs in construction (figure 8.6). They are majority carrier devices. They do not show any reverse recovery behavior like p–n diodes. However, some reverse recovery effects are observed due to the parasitic capacitances and inductances in the circuit, and

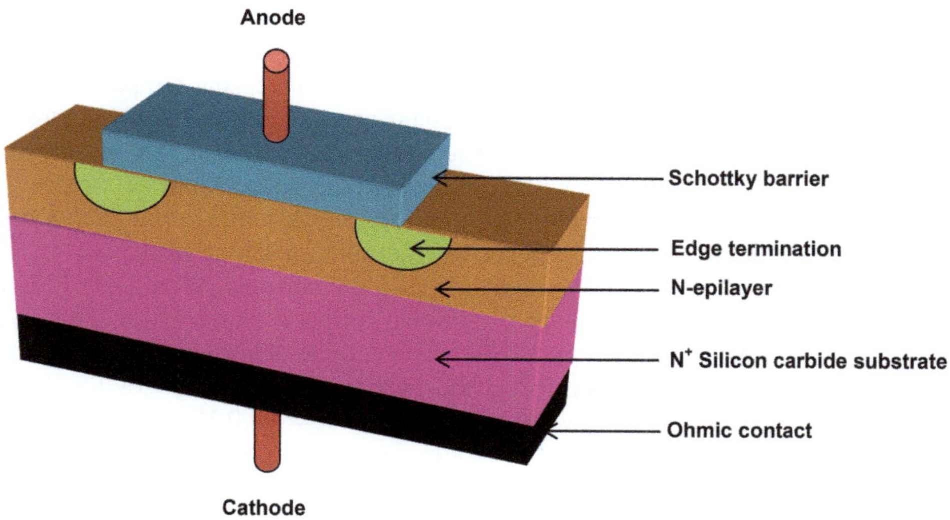

Figure 8.6. Silicon carbide SBD.

also due to the p-rings, which are incorporated in the chip to decrease leakage current.

SBDs rank as the first commercialized SiC device. SiC SBDs expanded the blocking voltage range from 200 V for Si SBDs to over 1000 V for SiC SBDs. But the forward voltage of SiC SBDs, ~1–2 V, is greater than that of Si SBDs. The larger forward voltage arises from the wider bandgap of SiC than Si.

8.8.1 Temperature effects on Si and SiC Schottky diodes

The forward voltage of both Si and SiC diodes decreases with temperature. In one comparative study, through the characterization of commercial Si and SiC diodes it was found that the forward voltage V_D of Si and SiC diodes can be represented as a function of temperature by the equations (Ozpineci and Tolbert 2003)

$$V_D^{Si} = 0.3306\ \exp(-0.0103t\ ^\circ C) + 0.5724 \tag{8.10}$$

$$V_D^{SiC} = 0.2785\ \exp(-0.0046t\ ^\circ C) + 0.7042. \tag{8.11}$$

As can be seen, the forward voltages for both materials decrease as temperature increases. However, the same remarks are not true in both cases for the on-resistances R_D of Si and SiC diodes. The equations for Si and SiC on-resistance-temperature variation are (Ozpineci and Tolbert 2003)

$$R_D^{Si} = 0.2136\ \exp(-0.293t\ ^\circ C) + 0.0529 \tag{8.12}$$

$$R_D^{SiC} = -0.1108\ \exp(-0.0072t\ ^\circ C) + 0.2023. \tag{8.13}$$

In opposition to the negative TC of the on-resistance of Si SBDs, the TC of the on-resistance of SiC SBDs is positive, enabling easy paralleling of SiC SBDs.

The reverse recovery time (t_{rr}) of SiC SBDs (~20 ns) is shorter than that of Si SBDs (~40 ns). Therefore, they can be operated at higher switching frequencies. Moreover, the t_{rr} of SiC SBDs remains constant with increasing temperature. We know that power losses consist of two components: conduction and switching losses. Due to temperature-independent reverse recovery characteristics, SiC diodes can work at elevated temperatures with lower switching losses than Si diodes. In contrast, the t_{rr} of Si SBDs increases with an increase in temperature. The longer t_{rr} of Si SBDs at high temperatures leads to enormous switching losses.

6H-SiC SBDs with a forward voltage drop of 1.1 V between 298 K and 473 K at 100 A cm^{-2} current density and sharp breakdown voltage characteristics of 400 V at 298 K were reported (Bhatnagar *et al* 1992). These diodes showed improved reverse recovery behavior over fast silicon PIN diodes. Merged PIN/Schottky (MPS) diodes made from 4H-SiC with reverse voltage capability V_B = 4300 V showed a specific on-resistance = 20.9 mΩ cm (Wu *et al* 2006).

8.8.2 Schottky diode testing up to 623 K

The thermal stability of the ohmic contacts in two commercially available SiC Schottky diode dies was examined under 1000 h vacuum storage conditions at 623 K without applying any bias (O'Mahony *et al* 2011). These diodes had Al metallization as the Schottky anode for wiring. They had Ni/Ag metallization on the cathode bond pad for affixing the die. After high-temperature storage, the Al anode remained stable and showed <5% change in Schottky barrier height or diode ideality factor. But the die attach metal was seriously affected, resulting in an appreciable increase in series resistance to the extent of a hundred-fold increase in one type of diode.

8.8.3 Schottky diode testing up to 523 K

High-temperature (523 K) reverse-bias endurance tests were carried out on 600 V, 6 A 4H-SiC Schottky diodes (Testa *et al* 2011). The forward voltages at 100 μA to 6 A and the reverse currents of the diodes at 600 V changed insignificantly after 1000 h. This experiment confirmed the stability of the electrical parameters of these devices.

8.9 SiC JFETs

The late 1980s and early 1990s witnessed efforts towards the development of SiC JFETs, hindered by the low mobilities, transconductance and SiC material quality issues. But a SiC JFET was easier to implement than the MOSFET because of the absence of oxide–semiconductor interface quality problems (Baliga 2006). The fruition of these efforts led to the emergence of two main designs (figure 8.7): the lateral channel JFET (LCJFET) and vertical trench JFET (VTJFET) structures. The structural differences between LCJFET and VTJFET are observed as differences in their performance characteristics. Again, two versions of LCJFET, both n-type and with a buried p-layer, were put forward to be used for applications in either high-speed switching or low on-state losses:

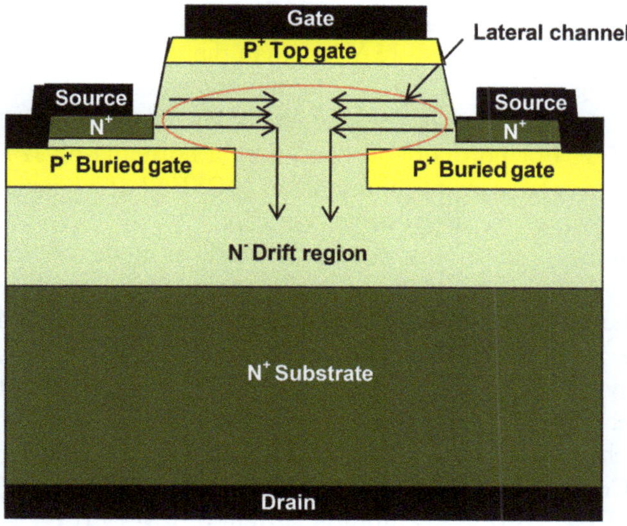

(a)

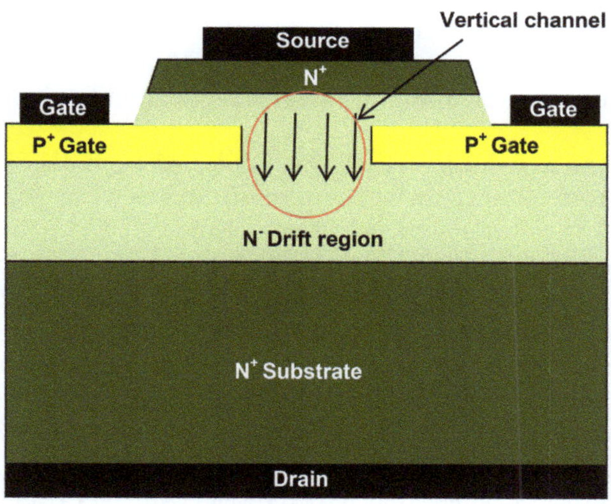

(b)

Figure 8.7. SiC JFET structures: (a) lateral and (b) vertical.

(i) The fast switching version has a smaller gate–drain area producing lower gate–drain capacitance. The pre-channel region between the source junction and the channel is responsible for the high on-resistance.
(ii) The low on-state resistance version has a structure such that the direct flow of current from the source through the channel to the drift region reduces the on-resistance, but the high gate resistance due to the high resistivity p-type SiC material at the gate junction slows down the switching rate.

Both versions (i) and (ii) are normally-on or depletion-mode devices which require a negative gate–source voltage ∼−15 V less than the pinch-off voltage to be applied between the p^+ gate and n^+-source terminals to constrict the channel and keep the JFET turned off. The reverse breakdown voltage of the gate–source junction is >−35 V. For normally-off operation, a normally-on JFET is combined with a low-voltage enhancement-mode silicon MOSFET (Siemieniec and Kirchner 2011). This is known as the direct-driven concept.

Power electronic applications need enhancement-mode devices. Applying a voltage of ∼2–3 V can turn them on but these devices have larger specific on-resistance. In the VTJFET, both enhancement-mode VTJFET (EMVTJFET) and depletion-mode VTJFET (DMVTJFET) types are realized (Casady *et al* 2010). Cross-sectionally, EMVTJFET and DMVTJFET are identical. The thickness of the vertical channel of EMVTJFET is 10% narrower than that of DMVTJFET. Further, the channel doping of EMVTJFET is 10% lower than that of DMVTJFET. Consequently, the on-resistance of EMVTJFET is 15% higher than that of DMVTJFET. Also, the saturation current of EMVTJFET is 50% of that of DMVTJFET. These JFETs were made from 4H-SiC. They had equivalent die areas (4.5 mm^2) and were rated at 1200 V.

In a VTJFET, the anti-parallel body diode is absent. The presence of such a diode in LCJFET is a distinct advantage of this structure. An external anti-parallel diode must therefore be connected to the VTJFET. It is utilized during switching transitions.

An enhancement type 5.3 kV JFET was fabricated from 4H-SiC. It has a low specific on-resistance (69 mΩ cm^2). The turn-off time is 47 ns (Asano *et al* 2001, 2002). This JFET is called a static expansion channel JFET (SEJFET). An area-efficient cell structure is used and a suitable JTE technique is employed.

A buried-grid JFET (BGJFET) (Malhan *et al* 2006, Tanaka *et al* 2006, Malhan *et al* 2009, Lim *et al* 2010) consists of several cells at a small pitch. As a result, the on-resistance is low and the saturation current is high. Fabrication is more difficult than for a LCJFET. As decided by the channel width and channel doping, this type of JFET is configured as either a normally-on or a normally-off device. The on-resistance of the normally-off BGJFET is 30% lower than that of the VTJFET (Lim *et al* 2010).

A dual-gate vertical channel trench JFET (DGVTJFET) (Malhan *et al* 2006, 2009) combines the attributes of a LCJFET and BGJFET. Its low gate–drain capacitance leads to fast switching while the decrease in cell pitch together with gate control minimizes the on-resistance. It is available as both types, normally-on as well as normally-off, with the latter type providing a much larger saturation current. However, the fabrication process of these JFETs is complex.

8.9.1 Characterization of SiC JFETs from 25 °C to 450 °C

Several 2.5 A, 1200 V SiC JFETs (figure 8.8) have been comprehensively characterized (Funaki *et al* 2006) from room temperature to 450 °C by packaging the devices in dedicated TO-258 packages, that were able to withstand high temperatures, and by using specially fabricated fixtures. These studies were conducted for

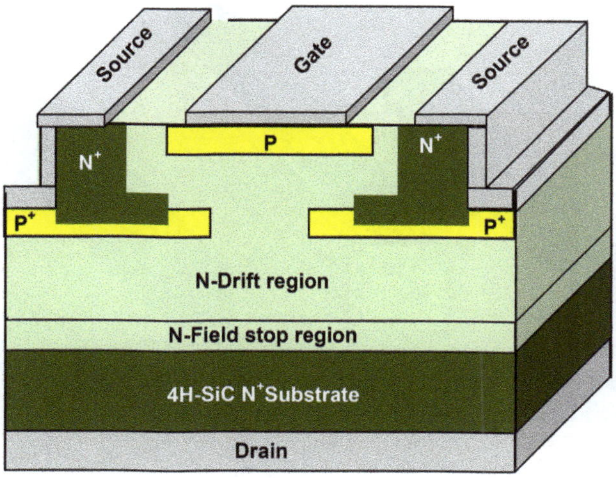

Figure 8.8. Vertical topology SiC JFET.

modeling and the extraction of parameters which can be used for circuit simulations. DC current–voltage characteristics were measured on a curve tracer while AC capacitance–voltage and impedance–voltage measurements were performed on an impedance analyzer.

From the forward conduction characteristics, for $V_{GS} = 0$ V, the saturated drain current is ~4 A at 25 °C. It decreases to ~0.75 A at 450 °C. The threshold voltage is −15.8 V at 25 °C and becomes more negative at 450 °C; the new value is −18.2 V. The transconductance parameter decreases by a factor of 9.6 from 25 °C to 450 °C. From the reverse conduction characteristics, the saturation current increases from 10^{-10} A at 25 °C to 10^{-6} A at 450 °C. From the AC characteristics, the capacitance C_{GS} at $V_{GS} = 0$ V increases from 450 pF at 25 °C to 575 pF at 450 °C. The capacitance C_{GD} shows an identical trend. The capacitance C_{DS} decreases as V_{DS} increases but remains practically unaffected by temperature.

8.9.2 500 °C operational test of 6H-SiC JFETs and ICs

The vital electrical parameters of depletion-mode epitaxial JFETs with gate lengths of 10 μm showed changes smaller than 10% over 3007 h of operation at 500 °C in air ambient (Neudeck *et al* 2008). In these JFETs (figure 8.9), an n-type source and drain were formed by nitrogen implantation. The p-type epitaxial gate layer was formed by Al implantation. Multilayer Ti/$TaSi_2$/Pt metallization was used to make the three contacts.

The differential amplifier IC consists of three resistors and two JFETs. The JFETs were source-coupled and operated with a 40 V source.

The NOR gate IC consists of three JFETs and the resistors. +20 V and −24 V power supplies were used for the NOR gate.

The test run was carried out by Neudeck *et al* by keeping the JFETs and ICs in an oven containing ordinary room air. Measurement data were periodically recorded

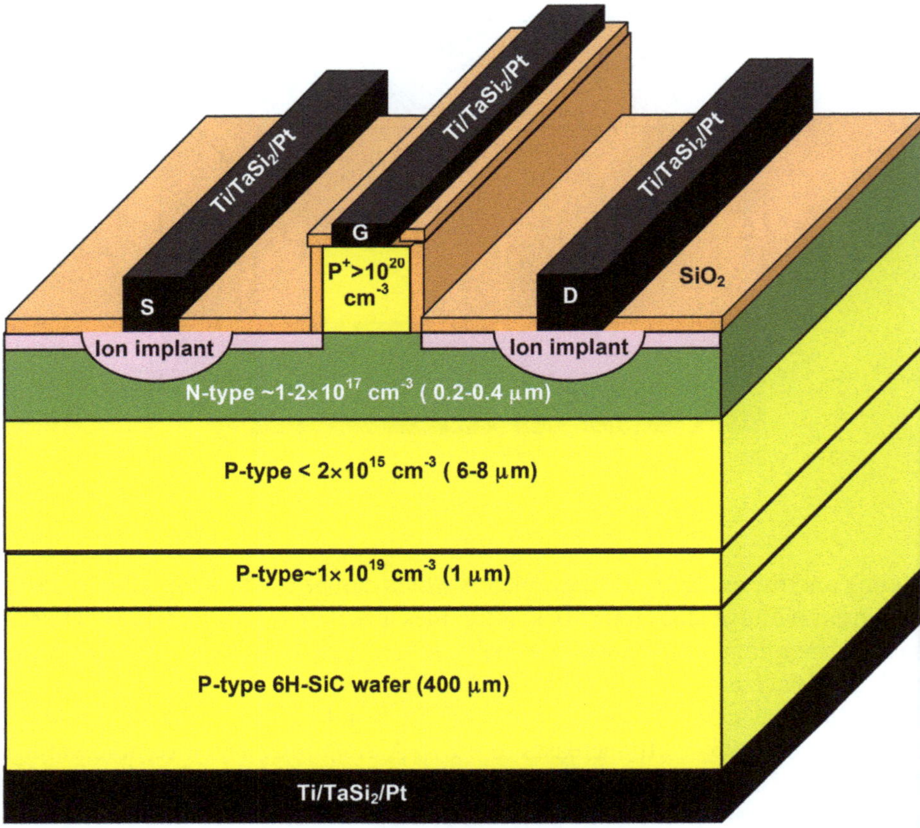

Figure 8.9. A 6H-SiC JFET without Si_3N_4 surface passivation.

during continuous operation under electrical bias at 500 °C. Typical magnitudes of changes in JFET parameters at 500 °C storage can be visualized from the following readings (Neudeck *et al* 2008):

(i) the capacitor leakage current density J_R under 50 V stress decreased from 604 μA cm^{-2} after 100 h to 104 μA cm^{-2} after 3007 h;
(ii) the threshold voltage remained at −11.8 V throughout;
(iii) the drain current I_{DSS} with $V_G = 0$ V, $V_D = 20$ V decreased from 1.36 mA after 100 h to 1.31 mA after 3007 h;
(iv) the transconductance g_{m0} at $V_D = 20$ V decreased from 214 μS after 100 h to 205 μS after 3007 h;
(v) the drain–source resistance R_{DS} for $V_G = 0$ V increased from 4.64 kΩ after 100 h to 4.83 kΩ after 3007 h;
(vi) the drain leakage current I_{OFF} at $V_D = 50$ V, $V_G = -15$ V decreased from 37.1 μA after 100 h to 0.19 μA after 3007 h.

For the differential amplifier IC:

(i) the voltage gain A_V at 100 Hz was 2.99 and 2.91 after 100 h and 3007 h, respectively;
(ii) voltage gain A_V at 10 kHz was 2.18 after 100 h and 2.19 after 3007 h;
(iii) the unity voltage gain frequency f_T was 32 kHz after 100 h and rose to 33 kHz after 3007 h.

For the NOR gate IC:

(i) the output high-voltage V_{OH} with both inputs at low voltage (= −7.5 V) was −1.47 V after 100 h and −1.67 V after 3007 h;
(ii) the output low voltage V_{OL} keeping one input at high voltage (= −2.5 V) and the other input at low voltage (= −7.5 V) was −8.12 V after 100 h and −8.14 V after 3007 h.

The above data of Neudeck *et al* amply demonstrate the feasibility of JFETs and ICs for long-duration functioning capability at 500 °C.

8.9.3 6H-SiC JFET-based logic circuits for the 25 °C–550 °C range

Depletion-mode JFETs were used to fabricate inverter, NAND gate and NOR gate circuits to provide a DC transfer characteristic with a broad noise margin at 550 °C (Soong *et al* 2012). In the JFET structure, the top p^+ layer serving as the gate, controls the depletion of the underlying n-type epitaxial layer acting as the channel layer. Below this n-type epi-layer is a p^+ epi-layer on a p-type SiC substrate. Nitrogen ion implantation is used to form the n^+ source and n^+ drain. Following deposition of a 20 nm SiO_2 layer, a stack of Ti, $TaSi_2$ and Pt is sputtered and protected by a Si_3N_4 layer. JFETs and logic circuits are completed, and interconnections are formed. The circuits are encapsulated in dual-in-line packages using gold bonding wires.

The performance of the inverter is evaluated from its voltage transfer characteristic (VTC). Figure 8.10 shows the typical shape of the VTC of the inverter. For the temperature interval 25 °C–550 °C, Soong *et al* noticed:

(i) a gradient of VTC, falling sharply in an almost perpendicular manner;
(ii) up to a temperature of 550 °C, the voltage gain exceeds −20; and
(iii) the logic threshold is located in the middle of the logic swing.

To get an idea of the influence of temperature on a SiC inverter:

(i) V_{OL} = low-level output voltage = −8.00 V at 25 °C and −8.69 V at 550 °C;
(ii) V_{OH} = high-level output voltage = −0.50 V at 25 °C and −3.97 V at 550 °C;
(iii) V_{NML} = low-level noise margin = 1.80 V at 25 °C and 1.52 V at 550 °C;
(iv) V_{NMH} = high-level noise margin = 4.91 V at 25 °C and 2.17 V at 550 °C; and
(v) voltage gain = −26 at 25 °C, −20 at 500 °C and −12 at 550 °C.

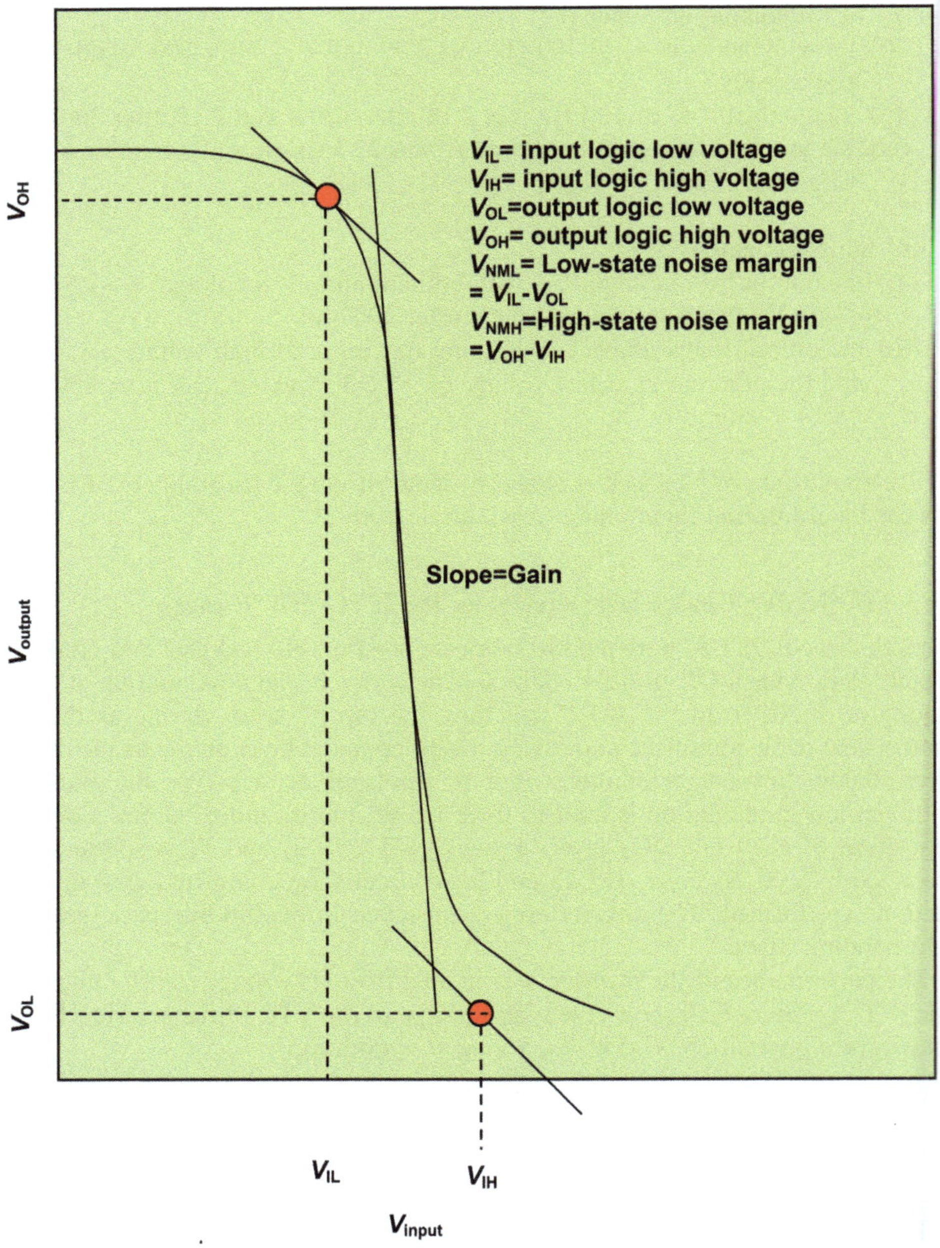

Figure 8.10. Voltage transfer characteristic of the inverter circuit.

The DC characteristics of NAND and NOR gates were similar to those of the inverter circuit. Dynamic testing of NAND and NOR gates showed satisfactory performance up to 550 °C. Thus a technological platform for the fabrication of reliable digital circuits with operational capability up to 550 °C was upheld.

8.9.4 Long operational lifetime (10 000 h), 500 °C, 6H-SiC analog and digital ICs

The JFETs were fabricated on a 6H-SiC wafer over which a p^-SiC epi-layer (2×10^{15} cm^{-3}) was grown followed by an n-SiC epi-layer (1×10^{17} cm^{-3}) (Neudeck *et al* 2009). The devices employed a mesa-etched heavily Al-doped p^+ (2×10^{19} cm^{-3}) epi-gate structure with nitrogen-doped n-type implantations for the source and drain. The thickness and doping concentration of the sub-channel p-type layer were carefully controlled to reduce parasitic substrate-to-channel capacitance. A self-aligned nitrogen implant minimized the parasitic resistance between the gate and source/drain. Contacts were made of Ti/$TaSi_2$/Pt.

The diced chips were encapsulated by Neudeck *et al* in custom ceramic packages with 96% ceramic substrate and Au metallization. Wire bonding was performed with 1 mil diameter Au wires. The packages were fixed to ceramic printed circuit boards (PCBs). Device and circuit testing was carried out under electrical bias at 500 °C in an oven filled with room air of uncontrolled humidity. The JFET with W/L = 100 μm/10 μm was tested under drain voltage V_D = 50 V sweep. Gate bias step V_G was stepped from 0 to −16 V in −2 V intervals. The 200 μm/10 μm JFET was operated under V_D = 50 V and V_G = −5 V. After the first, 100th, 1000th and 10 000th hours at 500 °C, the I_D–V_D characteristics were recorded. The drain saturation current at zero gate bias (I_{DSS}), transconductance g_m and threshold voltage V_{Th} of the JFETs were normalized with respect to their corresponding values I_{DSS0}, g_{m0} and V_{Th0} at the 100th hour of 500 °C testing. The current I_{DSS} changed by less than 10% over 10 000 h at 500 °C. The transconductance g_m showed a similar trend. V_{Th} varied by less than 1% over this time span (Neudeck *et al* 2009).

8.9.5 Characterization of 6H-SiC JFETs and differential amplifiers up to 450 °C

JFETs were fabricated on Al-doped 6H-SiC wafers by growing an Al-doped p^- epi-SiC layer, followed by an n^- epi-SiC layer and a p^+ epi-SiC layer (Patil *et al* 2007, 2009), see figure 8.11. Nitrogen ion implantation at a high dose was performed for the source and drain. The contact metal for the source, drain and gate was Al deposited by sputtering. Up to 300 °C, the chips were probed by mounting on a hot chuck. For electrical characterization at temperatures above 300 °C, the diced JFET chips were encapsulated in ceramic dual-in-line packages (DIPs). Measurement of the I_{DS}–V_{DS} and I_{DS}–V_{GS} characteristics of the W/L = 100 μm/100 μm JFETs from 25 °C to 450 °C showed that the characteristics were well-behaved throughout, i.e. the nature of characteristics was preserved in the whole range. However, Al contacts created some instability near the upper limit of temperature. The experimental characteristics fully conformed to the 3/2 power JFET model. This allowed the extraction and computation of many JFET parameters. The pinch-off current (I_p) is 0.48 mA at 25 °C. It decreases to one-half at 450 °C. The pinch-off voltage (V_{po}) is 11.90 V at 25 °C, changing almost negligibly to 11.13 V at 450 °C. The channel length modulation parameter (λ) also varies insignificantly. Up to 300 °C, the threshold voltage V_{Th} follows the −2.3 mV °C^{-1} rule. The source/drain resistance $R_{S,D}$ increases from 4.54 kΩ at 25 °C to 13.39 kΩ at 450 °C.

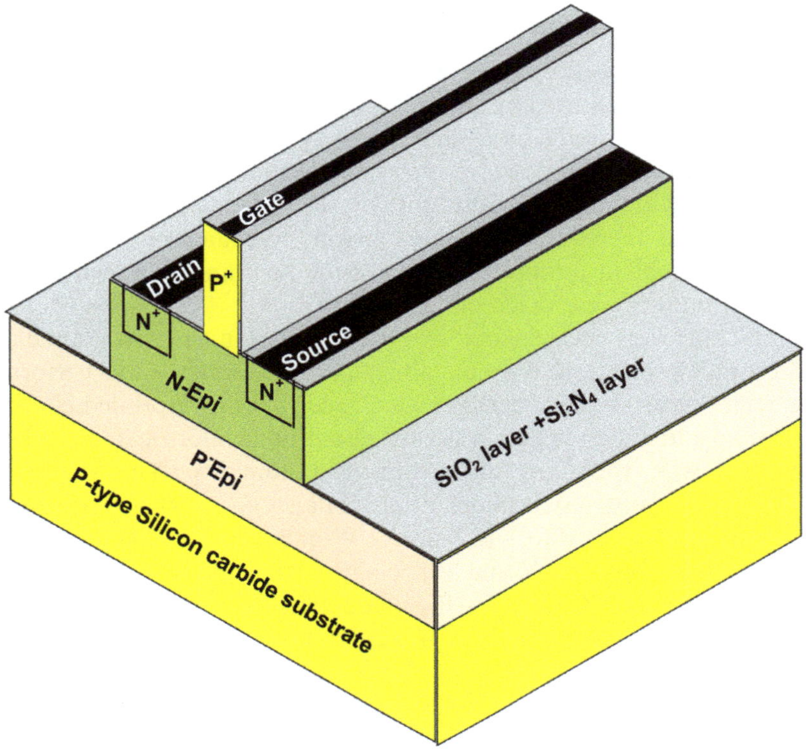

Figure 8.11. The 6H-SiC JFET structure.

Differential amplifiers fabricated with SiC JFETs were operated at high temperature with off-chip components at room temperature (Patil *et al* 2009). They worked successfully up to 450 °C albeit with altered parametric values. The DC transfer characteristics of a three-stage differential amplifier with external biasing and passive loads were measured at room temperature and 450 °C. These 6H-SiC JFETs had W/L = 110 μm/10 μm. The JFET-based differential amplifiers gave a low-frequency voltage gain of 87 dB at 25 °C, falling to 50 dB at 450 °C. The unity gain bandwidth product is 335 kHz at 25 °C, decreasing to 201 kHz at 450 °C. Further, the input offset voltage at 450 °C is larger than the room temperature value. For comparison, an externally biased SiC multistage amplifier with active loads gave a low-frequency voltage gain of 70 dB and a gain bandwidth product of 1300 kHz at 450 °C.

8.10 SiC bipolar junction transistors

SiC BJTs (figure 8.12) can provide a low on-resistance feature along with a high breakdown voltage. Freedom from the gate oxide problem permits their operation at a high temperature, unlike SiC MOSFETs.

4H-SiC BJTs with a voltage V_{CEO} above 1000 V and a DC current gain of 32 at collector current density $J_C = 319$ A cm^{-2} showed specific on-resistance 17 mΩ cm^2

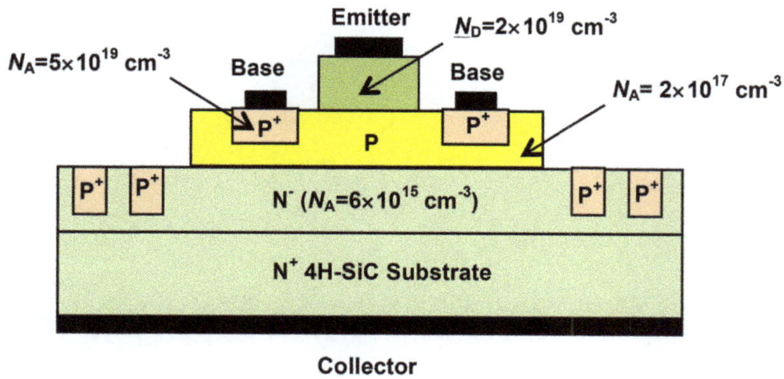

Figure 8.12. Cross-section of a SiC BJT.

at $J_C = 289$ A cm^{-2} (Luo *et al* 2003). These BJTs used an Al-free ohmic contact to the base.

4H-SiC BJTs blocked more than 480 V and conducted at $J_C = 239$ A cm^{-2} with specific on-resistance of 14 mΩ cm^2. A current gain of ⩾35 was obtained at $J_C = 40$–239 A cm^{-2} (Zhang *et al* 2003). The peak gain value was 38 at $J_C = 114$ A cm^{-2}.

4H-SiC BJTs with a blocking voltage of 9.2 kV at a leakage current of 4.6 μA showed a DC common-emitter current gain of 7 and a specific on-resistance of 33 mΩ cm^2 without considering current spreading (Zhang *et al* 2004).

4H-SiC BJTs (blocking voltage 1836 V) were tested for power switching (300 V, >7 A) and found to switch at 20–80 ns at par with Si unipolar devices. The switching speed remained unaffected up to a high temperature of 275 °C. The high-frequency, high-temperature capability of these BJTs was proved (Sheng *et al* 2005).

4H-SiC BJTs having a collector–emitter voltage with the base open ($V_{CEO} = 757$ V) and a current gain of 18.8, conducted up to 5.24 A at a forward voltage $V_{CE} = 2.5$ with a specific on-resistance of 2.9 mΩ cm^2 up to $J_C = 859$ A cm^{-2} (Zhang *et al* 2006).

4H-SiC BJTs were demonstrated using a graded base profile (Zhang *et al* 2009). They had a current gain of ∼33 and a collector–emitter breakdown voltage (V_{CEO}) > 1000 V. The specific on-resistance was 2.9 mΩ cm^2.

8.10.1 Characterization of SiC BJTs from 140 K to 460 K

4H-SiC BJTs were characterized for temperature dependence in the range 140–460 K (Asada *et al* 2015). Surprisingly, the temperature behavior of SiC BJTs is strikingly different from that of Si BJTs. Whereas the current gain of Si BJTs decreases monotonically as temperature falls, SiC BJTs show two types of behavior, namely the reverse behavior of an increase in current gain up to a certain temperature followed by similar behavior of a decrease in current gain with further decrease of temperature, as shown in figure 8.13.

(i) When the temperature descended from 460 to 200 K, the current gain rose from 110 to 1200. This rise of current gain with decreasing temperature was

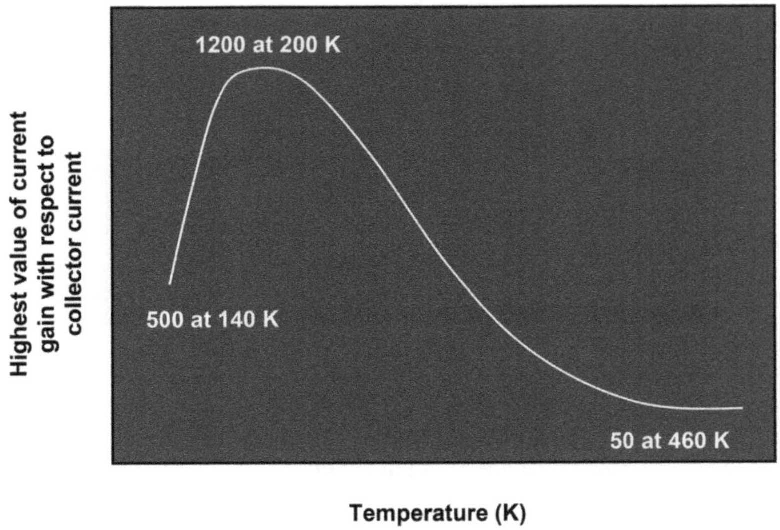

Figure 8.13. Current gain of silicon carbide bipolar junction transistor versus temperature.

interpreted in terms of an uplift of emitter injection efficiency. The efficiency was augmented because the aluminum acceptors in the base region of the transistor were only partially ionized at the low temperature; a higher temperature was necessary for the completion of ionization. So, the free carrier concentration in the base was low.

(ii) Subsequently, when the temperature was brought down from 200 to 140 K, the trend was opposite. The current gain started to decrease. It dropped down from 1200 at 200 K to 515 at 140 K. Interpretation for this change of trend is provided on the premise that the injected carrier concentration from the emitter into the base surpasses the low hole concentration in the base leading to the creation of a high injection condition at low collector currents. This high injection condition is held responsible for the decline of current gain.

8.10.2 Performance assessment of SiC BJT from −86 °C to 550 °C

In another investigation (Nawaz *et al* 2009), the current gain of SiC n–p–n BJTs was reported to decrease from 50 at room temperature to half the value (25) at 548 K. On decreasing the temperature below room temperature, the current gain attained a peak value of 111 at 187 K. Below 187 K, the current gain fell abruptly through the carrier freeze-out effect. The on-resistance measurements for the SiC BJT were also performed. Starting from a room temperature value of 7 mΩ cm^2, the on-resistance increased to 28 mΩ cm^2 at 548 K. However, the on-resistance remained practically constant when the temperature was decreased to 187 K. At temperatures lower than 187 K, the on-resistance climbed up suddenly, which is caused by carrier freeze-out. Packaged BJTs displayed satisfactory performance up to 823 K.

8.11 SiC MOSFETs

The voltage-controlled, normally-off devices are favorites of power electronic circuit designers. In the vertical trench U-shaped MOSFET (UMOSFET), figure 8.14(a), the channel is formed on the sidewalls of trenches excavated by reactive ion etching. The structure suffers from the problem of oxide breakdown at the corners of the trench (Cooper *et al* 2002). When close to avalanche breakdown, the peak electric field in SiC is 3×10^6 V cm^{-1}. Then the electric field in the oxide is higher by a factor equal to the ratio of the dielectric constant of SiC against the dielectric constant of

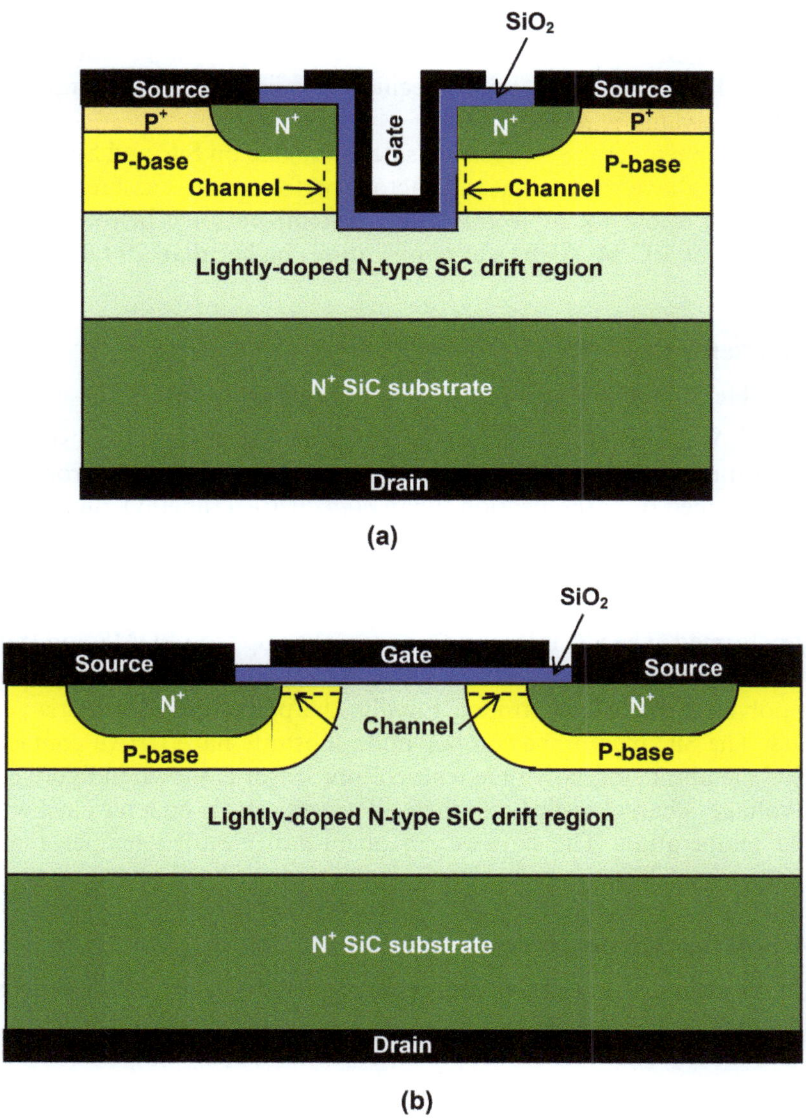

Figure 8.14. SiC MOSFETs: (a) UMOSFET and (b) DMOSFET.

SiO_2 (= 9.7/3.9 = 2.49). Hence, this electric field = 2.49 × 3 × 10^6 V cm^{-1} = 7.46 × 10^6 V cm^{-1}. This high field value is very near to the dielectric strength of SiO_2 (10^7 V cm^{-1}). Current crowding at the trench corner further intensifies the field. At high temperatures or during prolonged operation, the situation is even more serious. SiC UMOSFETs with a blocking capability ~260 V and on-resistance equal to 10–50 mΩ cm^2 were fabricated in the 1990s. The channel mobility of these devices is low and the gate oxide quality needs improvement. In 1996, a planar DMOSFET structure (figure 8.14(b)) was advanced as a step to circumvent some of the oxide problems faced with UMOSFET. It was fabricated by successive implantation of aluminum or boron ions for the p-type base and nitrogen ions for the n^+ source. A three-fold improvement in blocking voltage up to 760 V could be achieved. Several variants of DMOSFET design were developed. A 6.1 kV static induction injected accumulated FET (SIAFET) with specific on-resistance of 732 mΩ cm^2 was developed (Takayama *et al* 2001).

Regarding the temperature effects on silicon dioxide on SiC and SiC MOSFETs, reliability studies have shown that satisfactory performance is assured if the electric field is restricted below 4 × 10^6 V cm^{-1}, and the temperature is below 150 °C. Long-term operation of SiC MOSFETs at temperatures higher than 200°C–250°C seems to be unviable.

8.12 SiC sensors

8.12.1 Flexible 3C-SiC temperature sensors working up to 450 °C

130 nm thick Al-doped 3C-SiC film is deposited on silicon substrate using chemical vapor deposition at 1250 °C by alternately supplying silane and propane over a buffer layer formed by carbonization (Phan *et al* 2020). Trimethyl aluminum is used for Al doping to achieve a concentration ~10^{18} cm^{-3}. The SiC films are transferred from hosting silicon substrate on soft polymide platforms by etching silicon to form free-standing SiC membranes. From these SiC membranes, suspended SiC micro-bridges are formed. The SiC microbridges are picked up with PDMS and transferred on polymide/glass. Cr/Au metallization is done followed by annealing. An encapsulating polymide layer is deposited. Finally, the polymide substrate is peeled off from glass. The SiC membranes on polymide substrate have Cr/Au contacts.

The SiC membrane-on-polymide temperature sensor is shown in figure 8.15. The current–voltage characteristics of these devices rotate counter-clockwise with increasing temperature. The devices can operate at elevated temperatures up to 450 °C (Phan *et al* 2020).

8.12.2 4H-SiC gas sensors operating up to 500 °C

Accidents in chemical industries can be prevented by using efficient sensors for flammable and explosive gases.

8.12.2.1 Hydrogen sensor

A hydrogen gas sensor with metal–insulator semiconductor structure has been fabricated on SiC in the configuration Pd/Ta_2O_5/4H-SiC/Ni (Kim *et al* 2013). The

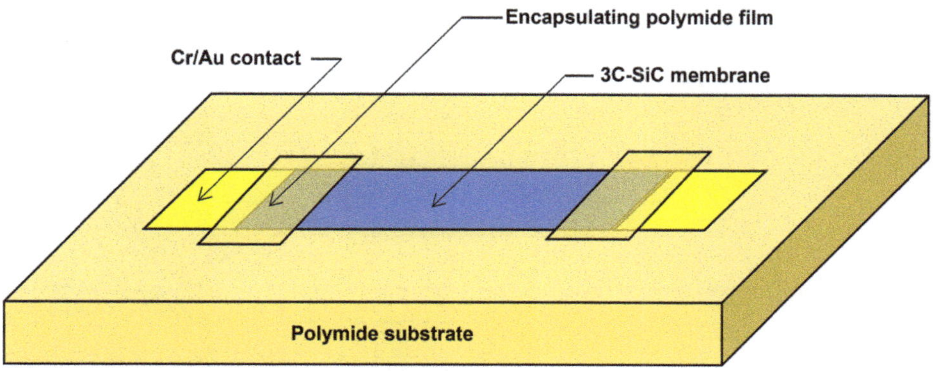

Figure 8.15. Temperature sensor formed with SiC nanomembrane on flexible polymide.

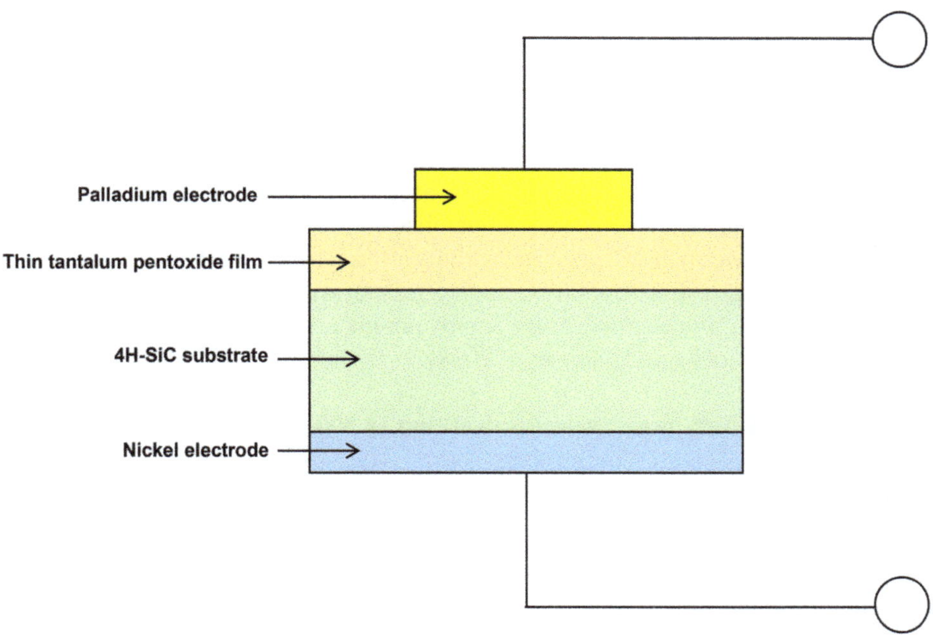

Figure 8.16. Cross-sectional diagram of hydrogen sensor with the metal–insulator–semiconductor (MIS) structure comprising the layers Pd/Ta_2O_5/4H-SiC/Ni.

sensor is shown in figure 8.16. Tantalum is deposited on 4H-SiC substrate by sputtering and oxidized by rapid thermal processing to form tantalum pentoxide. The nickel electrode is formed by sputtering, and palladium film is deposited using a shadow mask. The sensor can work up to 500 °C, as revealed by current–voltage and capacitance–voltage characteristics. Its capacitance varies with hydrogen concentration (Kim *et al* 2013).

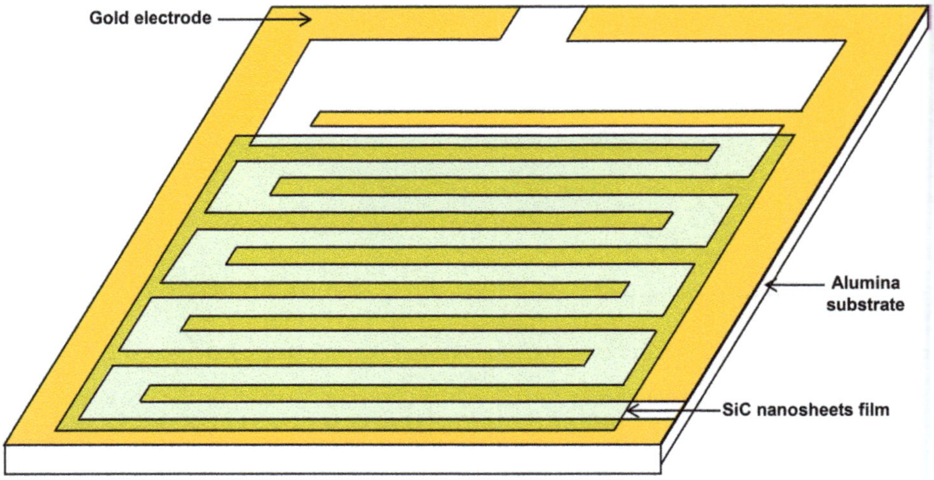

Figure 8.17. Silicon carbide nanosheets gas sensor.

8.12.2.2 Acetone, methanol, ethanol and ammonia gas sensor

Silicon carbide nanosheets prepared via a carbothermal reduction reaction between graphene oxide and Si powder are coated in the form of a paste over gold electrodes on alumina substrate and air-dried for several hours (figure 8.17). The resistance of the device is measured after injecting controlled amounts of target gas into the test chamber. The device shows good sensitivity to acetone, methanol, ethanol and ammonia gases under 400 °C–500 °C testing temperature confirming its application for gas sensing in high-temperature environments. Fast response and recovery towards ethanol is observed (Sun *et al* 2018).

8.12.3 3C-SiC MEMS pressure sensor working at 500 °C

A capacitive pressure sensor is fabricated with 1 μm thick 3C-SiC diaphragm using bulk micromachining (figure 8.18). The fabrication process is divided into three parts: making the 3C-SiC diaphragm, making the silicon substrate and bonding together the diaphragm and substrate. The substrate is the backplate. The sensor is encapsulated in a stainless-steel package using O-rings and screws. The package provides the structural support to the device. The sensor can operate up to 500 °C in the pressure range 1–5 MPa (Marsi *et al* 2015).

8.13 Discussion and conclusions

Amongst the two popular polytypes of SiC, namely 4H-SiC and 6H-SiC, the former reigns supreme by virtue of its mobility and bandgap advantages. Owing to the inapplicability of the conventional CZ (Czochralski) process for single-crystal growth of silicon carbide, physical vapor deposition is widely used. Nitrogen and aluminum are used as n- and p-type dopants, respectively, and dopants are introduced by ion implantation followed by thermal annealing. Silicon carbide allows surface oxidation. Nonetheless, the oxide quality needs appreciable

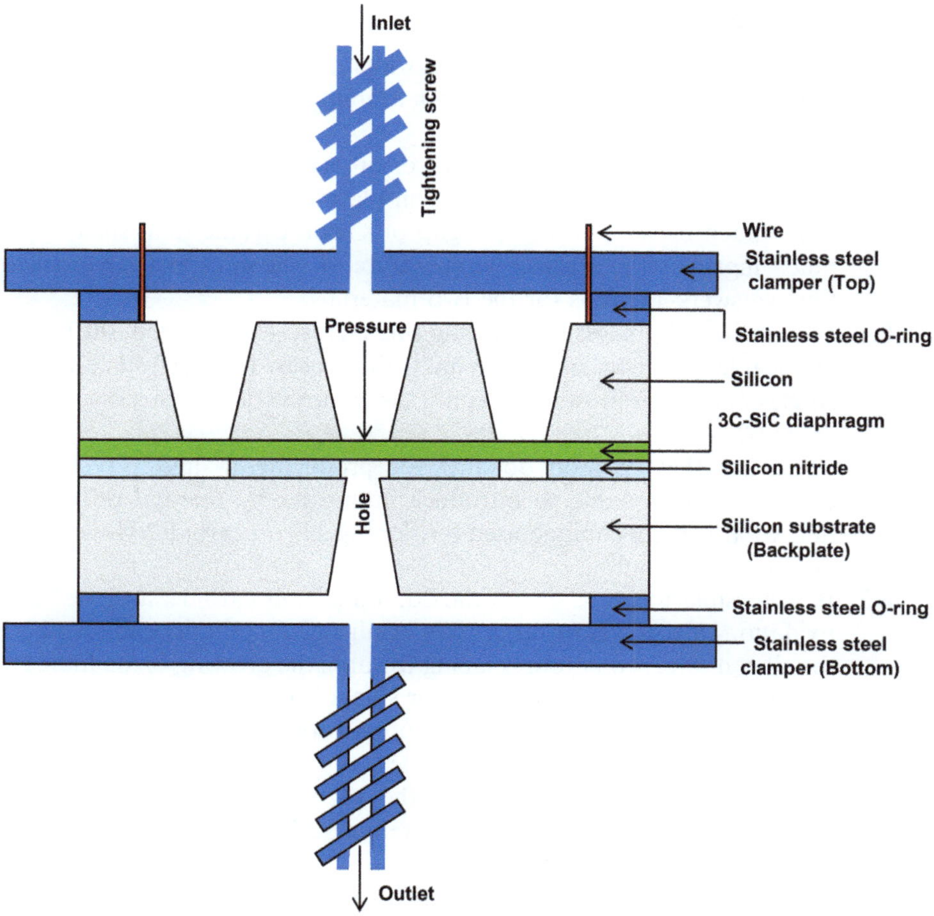

Figure 8.18. Packaged MEMS capacitive pressure sensor using 3C-SiC diaphragm.

improvement. In SiC technology, different contact metallizations are available for making reliable ohmic and Schottky contacts. In several experimental studies conducted at temperatures up to 783 K, SiC p–n diodes as well as Schottky diodes have been found to function in a satisfactory manner, establishing their credentials for high-temperature survivability. Concerning SiC JFETs and MOSFETs, the former could easily be implemented in two versions, namely lateral channel and VTJFETs. JFETs and JFET-based analog and digital circuits were subjected to several trials at temperatures up to 550 °C. Their performance was encouraging. Silicon carbide BJTs have been fabricated providing low on-resistance with high breakdown voltage, but they behave differently from Si BJTs with regard to the variation of current gain with temperature, namely the opposite behavior in a certain temperature range and similar behavior in another range. SiC MOSFETs have hitherto shown relatively poor performance compared to Si MOSFETs because of inferior oxide quality and the low mobility of carriers.

Review exercises

8.1. Mention two properties of 4H-SiC which make it superior to 6H-SiC.

8.2. Up to what frequencies are carrier mobilities in silicon carbide able to provide RF performance?

8.3. Why is it possible to provide a lower on-resistance with a higher breakdown voltage using silicon carbide in place of silicon?

8.4. How much broader is the bandgap of silicon carbide than silicon? How much higher is the intrinsic temperature of silicon carbide than silicon? Give answers as ratios for the two materials.

8.1. Why is the CZ process for growing single-crystal silicon from the melt not applicable to silicon carbide? What is the process by which silicon carbide single crystals are grown? What are the common defects in silicon carbide crystals?

8.6. What are the common dopants for producing n- and p-type silicon carbide? Is it possible to introduce impurities by thermal diffusion? If not, what is the technique used for doping silicon carbide? How is *in situ* doping carried out?

8.7. Is it possible to grow silicon dioxide on silicon carbide surfaces? Is the oxidation rate slower or faster than in silicon? Does it depend on whether the silicon or carbon atom is facing towards the growing oxide film? How reliable are these oxides grown over silicon carbide?

8.8. What metals are used for making ohmic contacts to: (a) n-type silicon carbide and (b) p-type silicon carbide?

8.9. What was the carrier lifetime for the 12.9 kV SiC p–n diode at room temperature and at 498 K? What were the peak recovery current values at the two temperatures?

8.10. Mention two types of contacts, which were not degraded upon ageing at 773–783 K. Discuss the relative performance of p–n diodes in which Ni or Pt was sputtered on Ni/Si ohmic contacts on SiC diodes, and the diodes were kept at 873 K.

8.11. Describe the performance of a SiC integrated bridge rectifier up to 773 K with reference to the effect of temperature on the following parameters as compared to their values at room temperature: (i) the turn-on voltage and the on-resistance of a diode and (ii) the DC output voltage and voltage conversion efficiency of the rectifier.

8.12. In what respects is an SBD better than a p–n diode? How does a silicon carbide Schottky diode outperform a silicon p–n diode? In what way is it inferior?

8.13. How does temperature affect the forward voltages across SBDs made of silicon carbide and silicon? What is the effect on on-resistances of the two types of diodes? Which type of diodes can be easily paralleled without risk of thermal runaway and why?

8.14. Differentiate between silicon carbide and silicon SBD regarding their reverse recovery times. How does this difference impact the performances of the two types of diodes at high temperatures?
8.15. What were the results of 623 K testing on commercial Schottky diodes with Al metallization as the Schottky anode and Ni/Ag metallization on the cathode bond pad? How did the Schottky diodes perform in high-temperature (523 K) reverse-bias endurance tests?
8.16. Why did SiC JFET allow easier implementation than SiC MOSFET? What are the two main JFET designs pursued in SiC?
8.17. What are the main features of the two types of lateral channel JFET? Are these designs of normally-on or normally-off type? Can a normally-on JFET be operated as a normally-off type? If so, how?
8.18. What types of switching devices are required in power electronics: enhancement or depletion type? Does the VTJFET enable realization of both types? Which type of structure has lower saturation current? Which type has lower on-resistance?
8.19. An anti-parallel diode is present in which type: LCJFET or VTJFET? How is the deficiency of the diode overcome in the structure in which this diode is not present?
8.20. What are full forms of the acronyms of the following types of JFETs: (a) SEJFET, (b) BGJFET and (c) DGVTJFET? Describe the construction and salient features of each type of JFET.
8.21. Describe the performance parameter variation of SiC JFETs from 25 °C to 450 °C in respect of: (a) the saturated drain current, (b) threshold voltage, (c) transconductance, (d) reverse current and (f) gate–source and drain–source capacitances.
8.22. Describe the variations in vital parameters over the extended time period for the 500 °C operational test of: (a) 6H-SiC JFETs, (b) differential amplifier IC and (c) NOR gate IC in respect of drain current I_{DSS} and transconductance g_{m0} for JFET; voltage gain A_V at 10 kHz and unity voltage gain frequency f_T for a differential amplifier IC; and output high-voltage V_{OH} and output low voltage V_{OL} for a NOR-gate IC.
8.23. Describe the characteristics of a SiC JFET-based inverter with respect to V_{OH}, V_{OL} and V_{NML} variation in the 25 °C–550 °C range. How did the NAND and NOR gates work in this temperature range?
8.24. How did the current I_{DSS}, the transconductance g_m, and threshold voltage V_{Th} of JFETs vary during the 10 000 h, 500 °C test on 6H-SiC analog and digital ICs?
8.25. What is the low-frequency voltage gain of the JFET-based differential amplifier at 25 °C and at 450 °C? What are the unity gain bandwidth products at 25 °C and at 450 °C? How do the pinch-off current, pinch-off voltage and source/drain resistance of JFET change from 25 °C to 450 °C?
8.26. Give two examples of silicon carbide bipolar transistors mentioning the breakdown voltage and specific on-resistance in each case.

8.27. How is the trend of variation in current gain with temperature (140–460 K) different for Si BJT and SiC BJT? How does one account for this difference of behavior between the BJTs made from these two materials?
8.28. Explain the increase in current gain of a SiC BJT as the temperature is lowered from 460 K to 200 K.
8.29. Explain the decrease in current gain of a silicon carbide bipolar transistor when the temperature is decreased from 200 K to 140 K.
8.30. How does the on-resistance of a SiC BJT vary on decreasing the temperature from 548 K to 187 K and below 187 K? Why?
8.31. What is the problem of oxide breakdown in a vertical trench UMOSFET? Does the problem become more serious at high temperatures? Does SiC MOSFET operation at temperatures above 200 °C–250 °C seem to be possible?
8.32. Give examples of SiC sensors which can function up to 500 °C. Describe in detail the working of two of these sensors.

References

Alves L F S, Gomes R C M, Lefranc P, Pegado R D A, Jeanin P-O, Luciano B A and Rocha F V 2017 SIC power devices in power electronics: an overview *2017 Brazilian Power Electronics Conf. (COBEP)* (Juiz de Fora, Brazil, 19–22 November) 1–8

Asada S, Okuda T, Kimoto T and Suda J 2015 Temperature dependence of current gain in 4H-SiC bipolar junction transistors *Jpn. J. Appl. Phys.* **54** 04DP13

Asano K, Sugawara Y, Hayashi T, Ryu S, Singh R, Palmour J and Takayama D 2002 5 kV 4H-SiC SEJFET with low RonS of 69 mΩ cm^2 *Proc. 14th Int. Symp. on Power Semiconductor Devices and ICs* (Piscataway, NJ: IEEE) pp 61–4

Asano K, Sugawara Y, Ryu S, Singh R, Palmour J, Hayashi T and Takayama D 2001 5.5 kV normally-off low RonS 4H-SiC SEJFET *Proc. 13th Int. Symp. Power Semiconductor Devices and ICs (Osaka 4–7 June)* (Piscataway, NJ: IEEE) pp 23–6

Baliga B J 2006 *Silicon Carbide Power Devices Devices* (Singapore: World Scientific)

Bhatnagar M, McLarty P K and Baliga B J 1992 Silicon-carbide high-voltage (400 V) Schottky barrier diodes *IEEE Electron Device Lett.* **13** 501–3

Braun M, Weis B, Bartsch W and Mitlehner H 2002 4.5 kV SiC pn-diodes with high current capability *10th Int. Conf. Power Electron. Motion Control (Cavtat and Dubrovnik)* 1–8

Casady J B, Sheridan D C, Kelley R L, Bondarenko V and Ritenour A 2010 A comparison of 1200 V normally-off and normally-on vertical trench SiC power JFET devices *Mater. Sci. Forum* **679–680** 641–4

Chand R, Esashi M and Tanaka S 2014 P–N junction and metal contact reliability of SiC diode in high temperature (873 K) environment *Solid-State Electron.* **94** 82–5

Cooper J A and Kimoto T 2014 *Fundamentals of Silicon Carbide Technology: Growth, Characterization, Devices and Applications* (Singapore: Wiley)

Cooper J A Jr, Melloch M R, Singh R, Agarwal A and Palmour J W 2002 Status and prospects for SiC power MOSFETs *IEEE Trans. Electron Devices* **49** 658–64

Funaki T, Kashyap A S, Mantooth H A, Balda J C, Barlow F D, Kimoto T and Hikihara T 2006 Characterization of SiC JFET for temperature dependent device modeling *37th IEEE Power Electronics Specialists Conf.*

Kakanakov R, Kassamakova-Kolaklieva L, Hristeva N, Lepoeva G and Zekentes K 2002 Thermally stable low resistivity ohmic contacts for high power and high temperature SiC device applications *Proc. 23rd Int. Conf. Microelectronics (Niš, 12–15 May)* vol 1 (Piscataway, NJ: IEEE) pp 205–8

Kim S, Choi J, Jung M, Joo S and Kim S 2013 Silicon carbide-based hydrogen gas sensors for high-temperature applications *Sensors* **13** 13575–83

Kimoto T 2022 High-voltage SiC power devices for improved energy efficiency *Proc. Japan Acad.* B **98** 161–89

Langpoklakpam C, Liu A-C, Chu K-H, Hsu L-H, Lee W-C, Chen S-C and Sun C W *et al* 2022 Review of silicon carbide processing for power MOSFET *Crystals* **12** 245

Lim J K, Bakowski M and Nee H P 2010 Design and gate drive considerations for epitaxial 1.2 kV buried grid N-on and N-off JFETs for operation at 250 °C *Mater. Sci. Forum* **645–8** 961–4

Luo Y, Zhang J and Alexandrov P 2003 High voltage (>1 kV) and high current gain (32) 4H-SiC power BJTs using Al-free ohmic contact to the base *IEEE Electron Device Lett.* **24** 695–7

Malhan R K, Bakowski M, Takeuchi Y, Sugiyama N and Schöner A 2009 Design, process, and performance of all-epitaxial normally-off SiC JFETs *Phys. Status Solidi* A **206** 2308–28

Malhan R K, Takeuchi Y, Kataoka M, Mihaila A P, Rashid S J, Udrea F and Amaratunga G A J 2006 Normally-off trench JFET technology in 4 H silicon carbide *Microelectron. Eng.* **83** 107–11

Marsi N, Majlis B Y, Hamzah A A and Mohd-Yasin F 2015 Development of high temperature resistant of 500 °C employing silicon carbide (3C-SiC) based MEMS pressure sensor *Microsyst. Technol.* **21** 319–30

Nawaz M, Zaing C, Bource J, Schupbach M, Domeij M, Lee H-S and Östling M 2009 Assessment of high and low temperature performance of SiC BJTs *Mater. Sci. Forum* **615–7** 825–8

Neudeck P G, Garverick S L, Spry D J, Chen L-Y, Beheim G M, Krasowski M J and Mehregany M 2009 Extreme temperature 6H-SiC JFET integrated circuit technology *Phys. Status Solidi* A **206** 2329–45

Neudeck P G, Spry D J, Chen L-Y, Beheim G M and Okojie R S 2008 Stable electrical operation of 6H-SiC JFETs and ICs for thousands of hours at 500 °C *IEEE Electron Device Lett.* **29** 456–9

O'Mahony D, Duane R, Campagno T, Lewis L, Cordero N, Maaskant P, Waldron F and Corbett B 2011 Thermal stability of SiC Schottky diode anode and cathode metalisations after 1000 h at 350 °C *Microelectron. Reliab.* **51** 904–8

Ozpineci B and Tolbert L M 2003 Characterization of SiC Schottky diodes at different temperatures *IEEE Power Electron. Lett.* **1** 54–7

Papanasam E, Prashanth K B, Chanthini B, Manikandan E and Agarwal L 2022 A comprehensive review of recent progress, prospect and challenges of silicon carbide and its applications *Silicon* **14** 12887–900

Patil A C, Fu X-A, Anupongongarch C, Mehregany M and Garverick S L 2007 Characterization of silicon carbide differential amplifiers at high temperature CSIC 2007 *IEEE Compound Semiconductor Integrated Circuit Symp. (Portland, OR, 14–17 October)* 1–4

Patil A C, Fu X-A, Anupongongarch C, Mehregany M and Garverick S L 2009 6H-SiC JFETs for 450 °C differential sensing applications *J. Microelectromech. Syst.* **18** 950–61

Peters D, Schörner R, Hölzlein K-H and Friedrichs P 1997 Planar aluminum-implanted 1400 V 4 H silicon carbide p–n diodes with low on resistance *Appl. Phys. Lett.* **71** 2996–7

Phan H-P, Dinh T, Nguyen T-K, Qamar A, Nguyen T, Dau V T, Han J, Dao D V and Nguyen N-T 2020 High temperature silicon-carbide-based flexible electronics for monitoring hazardous environments *J. Hazard. Mater.* **394** 122486

Shao S, Lien W-C, Maralani A and Pisano A P 2014 Integrated 4H-silicon carbide diode bridge rectifier for high temperature (773 K) environment *44th European Solid State Device Research Conf. (Venice, 22–26 September)* 138–41

She X, Huang A Q, Lucía Ó and Ozpineci B 2017 Review of silicon carbide power devices and their applications *IEEE Trans. Ind. Electron.* **64** 8193–205

Sheng K, Yu L C, Zhang J and Zhao J H 2005 High temperature characterization of SiC BJTs for power switching applications *Int. Semiconductor Device Research Symp. (Bethesda, MD, 7–9 December)* 168–9

Siemieniec R and Kirchner U 2011 The 1200 V direct-driven SiC JFET power switch *EPE 2011 (Birmingham)* 1–10

Soong C-W, Patil A C, Garverick S L, Fu X and Mehregany M 2012 550 °C integrated logic circuits using 6H-SiC JFETs *IEEE Electron Device Lett.* **33** 1369–71

Sun L, Han C, Wu N, Wang B and Wang Y 2018 High temperature gas sensing performances of silicon carbide nanosheets with an n–p conductivity transition *RSC Adv.* **8** 13697–707

Sundaresan S G, Sturdevant C, Marripelly M, Lieser E and Singh R 2012 12.9 kV SiC PiN diodes with low on-state drops and high carrier lifetimes *Mater. Sci. Forum* **717–20** 949–52

Takayama D, Sugawara Y, Hayashi T, Singh R, Palmour J, Ryu S and Asano K 2001 Static and dynamic characteristics of 4–6 kV 4H-SiC SIAFETs *Proc. 2001 Int. Symp. on Power Semiconductor Devices and ICs (Osaka)* 41–4

Tanaka Y, Okamoto M, Takatsuka A, Arai K, Yatsuo T, Yano K and Kasuga M 2006 700-V 1.0-mΩ-cm^2 buried gate SiC-SIT (SiC-BGSIT) *IEEE Electron Device Lett.* **27** 908–10

Testa A, De Caro S, Russo S, Patti D and Torrisi L 2011 High temperature long term stability of SiC Schottky diodes *Microelectron. Reliab.* **51** 1778–82

Veliadis V 2022 SiC power device mass commercialization *ESSDERC 2022 – IEEE 52nd European Solid-State Device Research Conf. (ESSDERC) (Milan, 19–22 September)* 31–6

Wu J, Fursin L, Li Y, Alexandrov P, Weiner M and Zhao J H 2006 4.3 kV 4H-SiC merged PiN/Schottky diodes *Semicond. Sci. Technol.* **21** 987

Yi A, Wang C, Zhou L, Zhu Y, Zhang S, You T, Zhang J and Ou X 2022 Silicon carbide for integrated photonics *Appl. Phys. Rev.* **9** 031302

Zhang J, Alexandrov P, Burke T and Zhao J H 2006 4H-SiC power bipolar junction transistor with a very low specific on-resistance of 2.9 mΩ cm^2 *IEEE Electron Device Lett.* **27** 368–70

Zhang J, Luo Y, Alexandrov P, Fursin L and Zhao J H 2003 A high current gain 4H-SiC NPN power bipolar junction transistor *IEEE Electron Device Lett.* **24** 327–9

Zhang J H, Fursin L, Li X Q, Wang X H, Zhao J H, VanMil B L, Myers-Ward R L, Eddy C R and Gaskill D K 2009 4H-SiC bipolar junction transistors with graded base doping profile *Mater. Sci. Forum* **615–7** 829–32

Zhang J, Zhao J H, Alexandrov P and Burke T 2004 Demonstration of first 9.2 kV 4H-SiC bipolar junction transistor *Electron. Lett.* **40** 1381–82

Chapter 9

Gallium nitride electronics for very hot environments

Gallium nitride is of interest not only for optical devices covering the full visible spectrum and stretching far into the ultraviolet (UV) region, but also for microwave power devices capable of operating at much higher temperatures compared to traditional silicon and gallium arsenide components. In GaN electronics, the device that has drawn the most attention is the high electron mobility transistor (HEMT). Difficulty in the p-doping of GaN impedes bipolar device development. Therefore, investigations have primarily focused on metal–semiconductor field-effect transistors (MESFETs), metal–insulator–semiconductor FETs (MISFETs), heterojunction bipolar transistors (HBTs) and HEMTs. Amongst these, the HEMTs offer better carrier transport properties than MESFETs, and have consequently received more consideration. While AlGaN/GaN HEMTs could not withstand temperatures ~500 °C, the use of lattice-matched InAlN/GaN heterojunctions was found to extend the capability up to 1000 °C owing to nonexistence of mechanical strain. Polarization could be preserved in the heterostructures up to 1000 °C. Thus, ferroelectric polarization instabilities were avoided, and thermal and chemical stabilities comparable to ceramic materials could be attained.

9.1 Introduction

Silicon-based electronics has almost reached its performance pinnacle primarily owing to limitations of material properties. A stage is reached where device capabilities can be pushed further only marginally through innovations in device principles, architectures and fabrication processes. Besides SiC, another example of third generation

doi:10.1088/978-0-7503-5072-3ch9

semiconductor is GaN (Yu and Duan 2017). Thanks to the inimitable properties of gallium nitride. It has emerged as a material of choice for fabrication of power semiconductor devices. Improvement in efficiency of motor drives is achieved (Ding *et al* 2019). GaN power devices are expected to find increasing applications in future power converters (Meneghini *et al* 2021, Zhong *et al* 2022).

A side-by-side comparison of the properties of gallium nitride with silicon and silicon carbide reveals that the bandgap of gallium nitride (3.39 eV) is 3 times that of silicon (1.12 eV) and 1.05 times that of silicon carbide (3.23 eV) (Baliga 2013, 2017). Insofar as high-temperature operation is concerned, gallium nitride compares well with silicon carbide but is far ahead of silicon because gallium nitride has the ability to work at temperatures above 600 °C. From the viewpoint of thermal conductivity, gallium nitride (1.3 W cm K^{-1}) is slightly inferior to silicon (1.5 W cm K^{-1}) but much inferior to silicon carbide (4.9 W cm K^{-1}). The electron mobility (1000 cm^2 V^{-1} s^{-1}) of GaN is of the same order as that of SiC (800–900 cm^2 V^{-1} s^{-1}) as is its hole mobility (200 cm^2 V^{-1} s^{-1} for GaN) against 115 cm^2 V^{-1} s^{-1} for SiC. High-speed ICs for high-temperature applications use AlGaN/GaN heterojunction FETs (HFETs), also called HEMTs. In table 9.1, the physical properties of wurtzite gallium nitride structure routinely required for design computations of GaN devices and circuits are compiled; the other structure of GaN is the zinc blende variety. The wurtzite crystalline structure is depicted in figure 9.1(a) and the zinc blende structure in figure 9.1(b).

Table 9.1. Properties of gallium nitride (wurtzite).

Property	Value	Property	Value	Property	Value
Chemical formula	GaN	Lattice constant (Å)	A = 3.186 Å, c = 5.186 Å		
Molecular mass (g mol^{-1})	83.73	Melting point (°C)	2500	Electron mobility (cm^2 V^{-1} s^{-1})	1000
Classification	III–V compound semiconductor	Dielectric constant	9.5 (static), 5.35 (high frequency)	Hole mobility (cm^2 V^{-1} s^{-1})	200
Crystal structure	Wurtzite	Thermal conductivity (W cm K^{-1})	1.3	Electron diffusion coefficient (cm^2 s^{-1})	25
Color	Yellow	Energy bandgap E_g (eV) at 300 K	3.39 (direct)	Hole diffusion coefficient (cm^2 s^{-1})	5
Density at 300 K (g cm^{-3})	6.15	Electrical breakdown field (V cm^{-1})	5×10^6	Electron saturated velocity (cm s^{-1})	2×10^7
		Intrinsic carrier concentration (cm^{-3})	1.9×10^{-10}	Minority-carrier lifetime (s)	10^{-8}

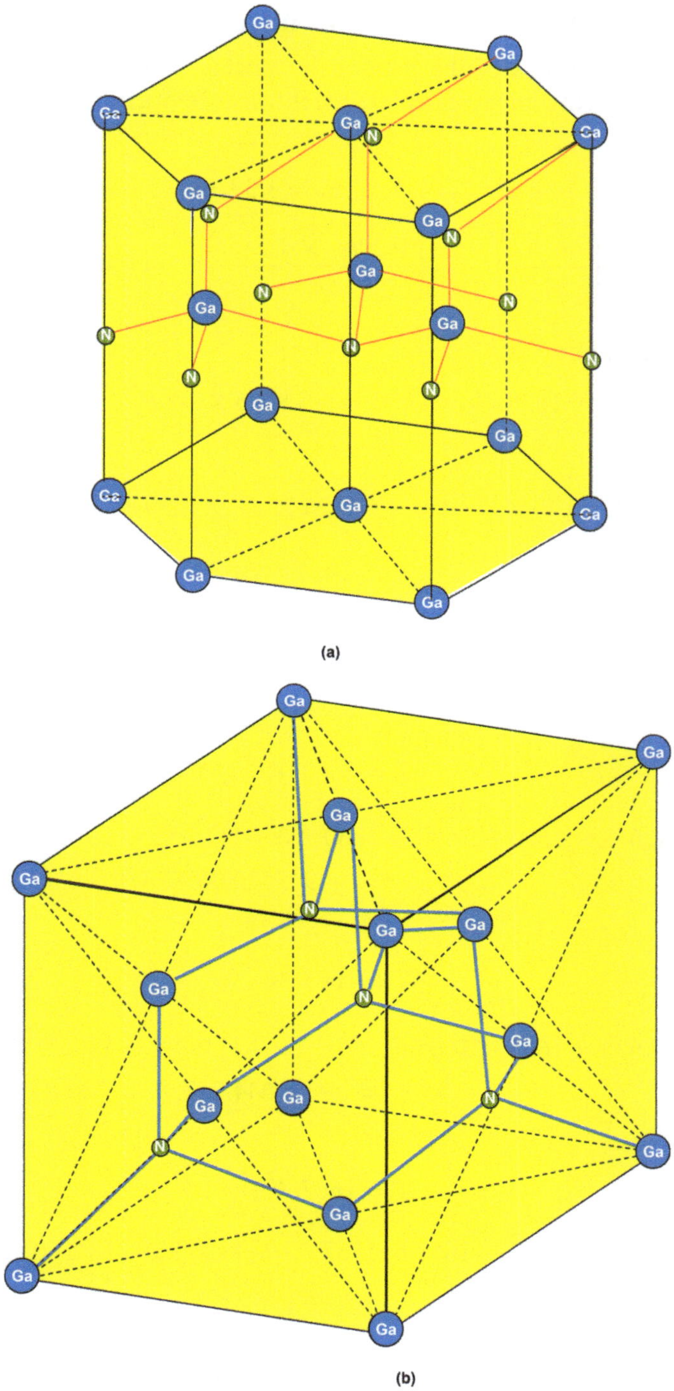

Figure 9.1. The structure of gallium nitride crystal: (a) wurtzite structure and (b) zinc blende.

9.2 Intrinsic temperature of gallium nitride

Following in the footsteps of the previously considered semiconductors Si, GaAs and 4H-SiC, it is worthwhile to calculate the intrinsic temperature of gallium nitride. For GaN,

$$m_n^* = 0.20m_0,\ m_p^* = 1.5m_0,\ E_g = 3.39\ \text{eV}$$

$$m^* = \left(\frac{m_n^* m_p^*}{m_0^2}\right)^{0.75} = \left(\frac{0.20m_0 \times 1.5m_0}{m_0^2}\right)^{0.75} = 0.4054. \tag{9.1}$$

For GaN, the equation

$$\ln\{0.207/(m^* T^{1.5})\} = -5.8025 \times 10^3 E_g \tag{9.2}$$

is recast as

$$T\ln\{0.207/(0.4054T^{1.5})\} = -5.8025 \times 10^3 \times 3.39 = -19\ 670.475$$

or

$$T\ln 0.510\ 61 - T\ln\ T^{1.5} = -19\ 670.475 \tag{9.3}$$

$$\therefore -0.672T - T\ln T^{1.5} = -19\ 670.475. \tag{9.4}$$

Suppose,

$$T = 1660\ \text{K}$$

$$\begin{aligned}\text{lhs} &= 0.672 \times 1660 - 1660\ln\ 1660^{1.5} = -1115.52 - 18\ 462.286\\ &= -19\ 577.806.\end{aligned} \tag{9.5}$$

If

$$T = 1665\ \text{K}$$

$$\begin{aligned}\text{lhs} &= 0.672 \times 1665 - 1665\ln\ 1665^{1.5} = -1118.88 - 18\ 525.407\\ &= -19\ 644.287.\end{aligned} \tag{9.6}$$

When

$$T = 1667\ \text{K}$$

$$\begin{aligned}\text{lhs} &= 0.672 \times 1667 - 1667\ln\ 1667^{1.5} = -1120.22 - 18\ 550.662\\ &= -19\ 670.882.\end{aligned} \tag{9.7}$$

$$\therefore T \approx 1667\ \text{K} \tag{9.8}$$

The intrinsic temperature for gallium nitride (1667 K) is slightly higher, by 1667 − 1591.5 = 75.5 K, than SiC (1591.5 K) but much higher, by 1667 − 588.63 = 1078.37 K, for Si (588.63 K).

9.3 Growth of the GaN epitaxial layer

GaN, AlN, AlGaN/GaN and InGaN heterostructures are grown epitaxially by the metal–organic chemical vapor deposition (MOCVD) technique (figure 9.2) (Parikh and Adomaitis 2006, Denis *et al* 2006, Kucharski *et al* 2020). Substrates used include silicon carbide, sapphire, silicon, etc. Silicon carbide is preferred due to its high thermal conductivity. Sapphire and silicon are low-cost materials. A resistive AlN nucleation layer is used to isolate the devices from silicon or silicon carbide. The non-uniformities on 100 mm diameter SiC substrates are <2%. The precursors for Ga, Al, In and N are trimethylgallium, trimethylaluminum, trimethylindium and ammonia, respectively. Growth is carried out at a typical pressure of 76 Torr. The growth temperature is ~1000 °C. The carrier gas for metal–organic compounds is hydrogen. Flow rate of trimethylgallium is generally kept at 1–10 mmoles min^{-1}; that of trimethylaluminum is taken as 0.6–1.5 mmoles min^{-1}.

9.4 Doping of GaN

The carrier concentration in the deposited films is $<10^{15}$ cm^{-3}. The resistive film can be doped as n-type with silicon as the donor impurity using disilane. Concentrations up to 10^{19} cm^{-3} are easily attainable. p-type doping is performed with magnesium as the acceptor impurity. As the energy level of magnesium is 160 meV above the valence band of GaN, only a small fraction <0.01 of the dopant is ionized at room

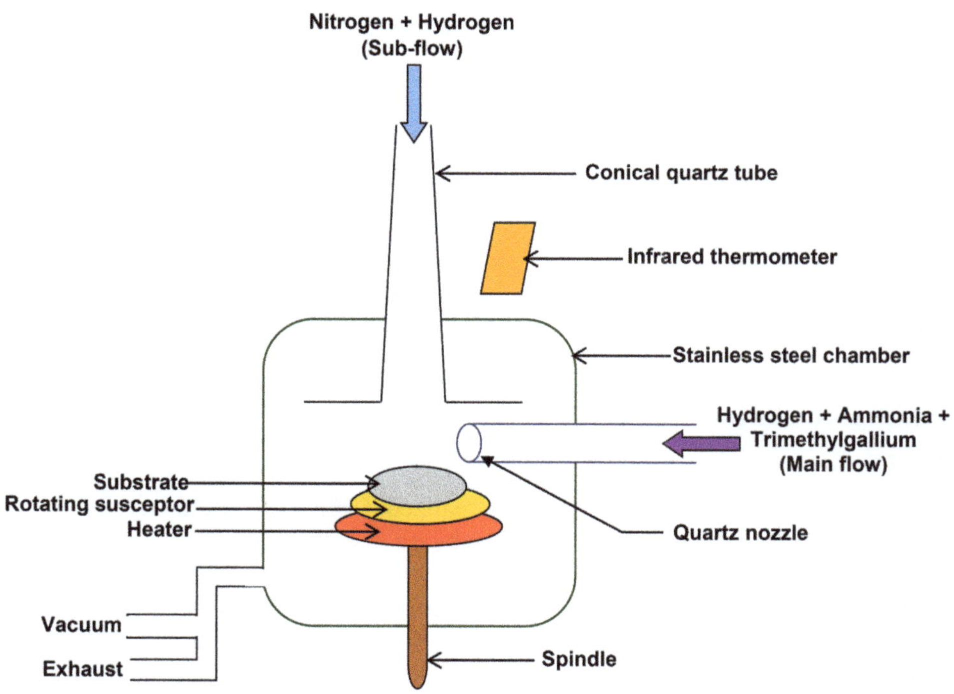

Figure 9.2. A two-flow MOCVD reactor for single-crystal growth of gallium nitride.

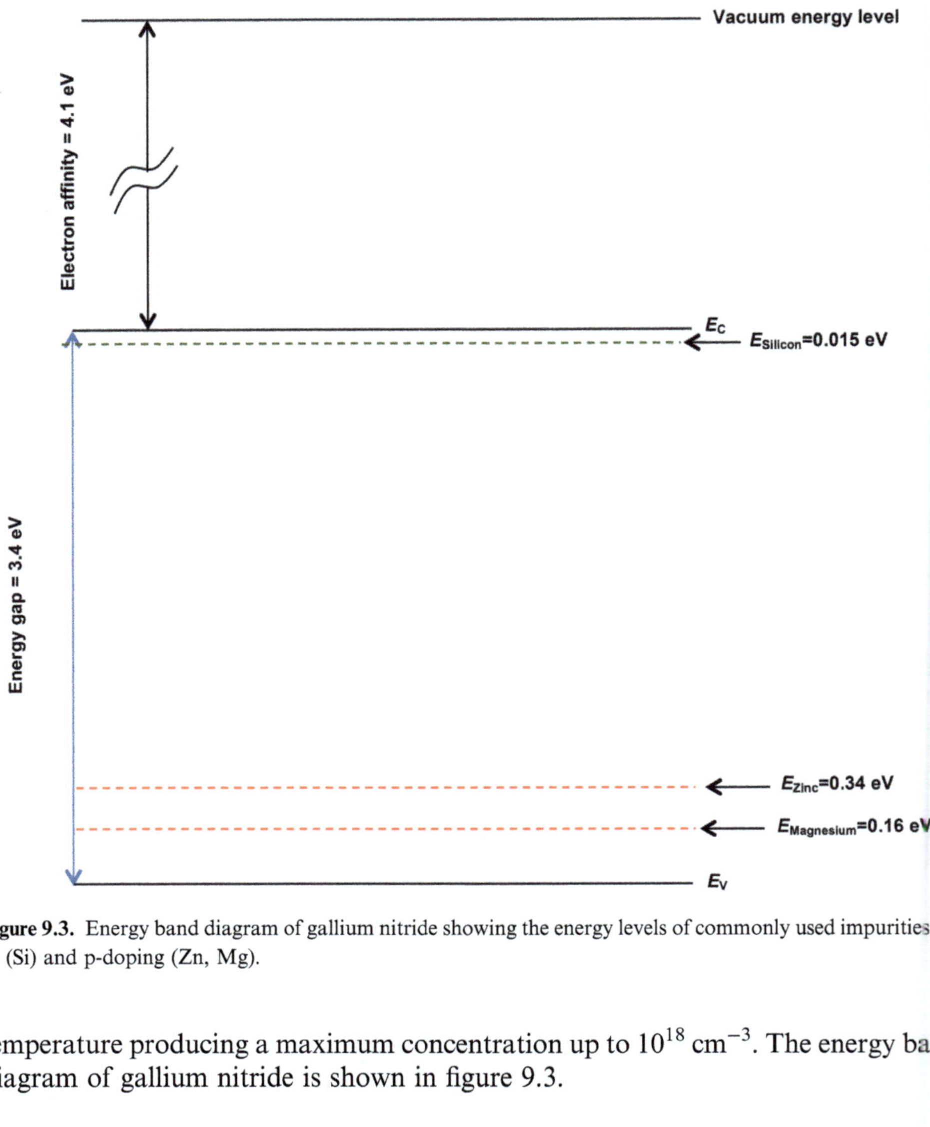

Figure 9.3. Energy band diagram of gallium nitride showing the energy levels of commonly used impurities for n- (Si) and p-doping (Zn, Mg).

temperature producing a maximum concentration up to 10^{18} cm^{-3}. The energy band diagram of gallium nitride is shown in figure 9.3.

9.5 Ohmic contacts to GaN

9.5.1 Ohmic contacts to n-type GaN

Ohmic contacts to n-type GaN are easily realized. Composite metal layer Ti/Al/Ni/Au (15 nm/220 nm/40 nm/50 nm) contact on reactive-ion etched n-type GaN (4×10^{17} cm^{-3}) surface, followed by RTA at 900 °C for 30 s in N_2 ambient, gives a low resistance (=8.9×10^{-8} Ω cm^2). Here, Ti and Ni are e-beam evaporated; Al and Au are thermally evaporated (Fan *et al* 1996, Ruvimov *et al* 1996). The reaction of GaN with Ti forms TiN, which produces a large number of n-vacancies acting as donors in GaN. The resultant increase of donor density decreases the contact resistance. The Ni layer between Al and Au serves as a diffusion barrier between these layers, otherwise Al will react with Au to form $AlAu_4$. It is viscous at low temperatures and its lateral flow may lead to shorting.

A low contact resistance ~10^{-4} Ω cm^2 was obtained on n$^+$ GaN with impurity concentration $\geqslant 10^{19}$ cm^{-3} by sputtering 500–1200 Å thick $WSi_{0.45}$ (Pearton *et al* 1998). The metallization was stable up to 1000 °C.

9.5.2 Ohmic contacts to p-type GaN

Low resistance ohmic contacts to p-type GaN are more difficult to accomplish, mainly due to the low acceptor concentrations available, the rarity of high work function metals, and the presence of compensating hydrogen interstitials and nitrogen vacancies. Commonly used metals are Pd, Pt, Ni or a high work function metal. This metal is covered with a gold layer to prevent oxidation. Pd/Ag/Au/Ti/Au (1 nm/50 nm/10 nm/30 nm/20 nm) contacts deposited by e-beam evaporation and subsequently annealed at 800 °C for 1 min give a resistance = 1×10^{-6} Ω cm^2 due to formation of an alloy between Ag, Au and p-type GaN (Adivarahan *et al* 2001) producing a p$^+$ region at the metal–semiconductor interface.

9.6 Schottky contacts to GaN

Multilayer Au/Pt/Ti (50 Å/300 Å/1500 Å) gave a Schottky barrier height of 0.84 eV in the as-deposited condition on an n-GaN layer (Macherzyński *et al* 2009). RTA in a hydrogen/nitrogen mixture in the ratio 1:10 at 300 °C for 20 s decreased the barrier height to 0.61 eV. The barrier height decreased further to 0.48 eV when the annealing was performed at a temperature from 400 °C to 700 °C.

9.7 GaN MESFET model with hyperbolic tangent function

In this model, the drain–source current I_{DS} is approximately represented for large-signal devices by a simpler version of the Shockley equation in terms of a hyperbolic tangent function (Kacprzak and Materka 1983, Kabra *et al* 2008, Shashikala and Nagabhushana 2010); see figure 9.4. The equation given by Kacprzak and Materka (1983) is

$$I_{DS} = I_{DSS}\left(1 - \frac{V_{GS}}{V_{Th} + \gamma V_{DS}}\right)^2 \tanh\left(\frac{\alpha V_{DS}}{V_{Th} - \gamma V_{DS}}\right). \tag{9.9}$$

This equation was modified by Kabra *et al* (2008) by replacing V_{GS} by ($V_{GS} - V_{bi}$), and for temperature dependence as follows:

$$I_{DS} = I_{DSS}\left(\frac{T_0}{T}\right)\left(1 - \frac{V_{GS} - V_{bi}}{V_{Th} + \gamma V_{DS}}\right)^2 \tanh\left(\frac{\alpha V_{DS}}{V_{Th} - \gamma V_{DS}}\right), \tag{9.10}$$

where I_{DSS} is the saturated drain current at $V_{GS} = 0$, T_0 is room temperature (300 K), T is the temperature at which the calculation is performed, V_{Th} is the threshold voltage, $\alpha = -0.56$ is an empirical constant determining the saturation

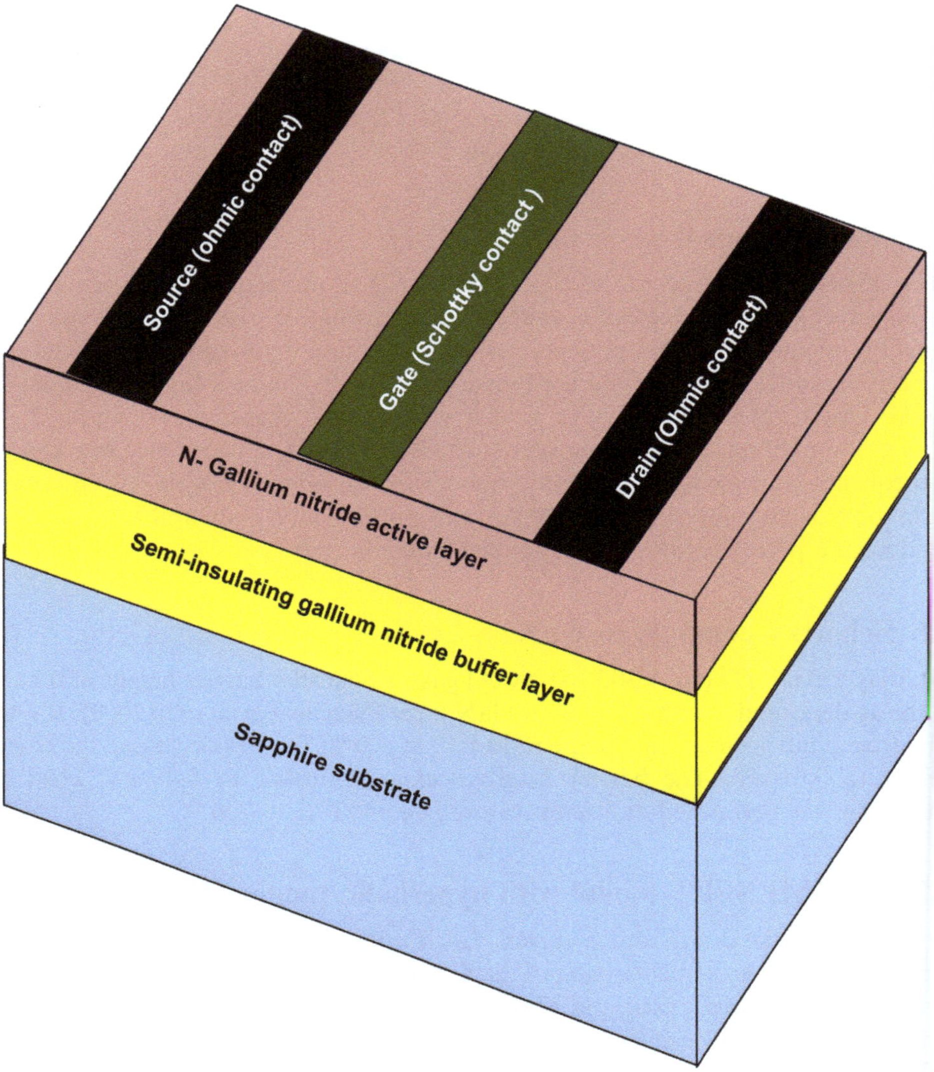

Figure 9.4. Gallium nitride MESFET.

voltage of the drain current and $\gamma = -0.0001$ represents the effective threshold voltage displacement with V_{DS} (Kabra *et al* 2008). The current I_{DSS} is given by

$$I_{DSS} = (I_{fc}/2)\left[\left(2 + \frac{R_s I_{fc}}{V_p}\right) - \sqrt{\left(2 + \frac{R_s I_{fc}}{V_p}\right)^2 - \left(\frac{4}{V_p}\right)\left\{V_p - \left(V_{bi} + V_{GS}\right)\right\}}\right], \quad (9.11)$$

where $R_s = 160\ \Omega$ is the series resistance of the source, I_{fc} is the full saturation current expressed in terms of channel width W and the temperature-dependent variables namely the saturation velocity $v_{sat}(T)$ and the conducting channel charge $Q_a(T)$ as

$$I_{\mathrm{fc}} = qWv_{\mathrm{sat}}(T)Q_{\mathrm{a}}(T) \tag{9.12}$$

with

$$v_{\mathrm{sat}}(T) = v_{\mathrm{sat}}(T_0) - \sigma_v T \tag{9.13}$$

$$Q_{\mathrm{a}}(T) = N_{\mathrm{D}}\sqrt{\frac{2\varepsilon_0\varepsilon_{\mathrm{s}}(T)V_{\mathrm{p}}}{qN_{\mathrm{D}}}}, \tag{9.14}$$

where σ_v is an empirical constant = 98 $\mathrm{ms}^{-1}\ \mathrm{K}^{-1}$ and $v_{\mathrm{sat}}(T_0)$ is the saturation velocity at a temperature T_0(= 300 K) = 2.8743×10^7 cm s^{-1}. The symbols V_{bi}, V_{p} and V_1 have the same meanings as for GaAs (see equations (7.15), (7.16), (7.18) and (7.19)) with proper modifications for the differences between GaN and GaAs, e,g., in the equation for V_{p}, the relative permittivity ε_{s} is replaced by a temperature-dependent dielectric constant defined by an identical equation as saturation velocity

$$\varepsilon_{\mathrm{s}}(T) = \varepsilon_{\mathrm{s}}(T_0) - \sigma_{\mathrm{s}}T, \tag{9.15}$$

where σ_{s} is an empirical constant = 10^{-4} K^{-1}. The dielectric constant $\varepsilon_{\mathrm{s}}(T_0)$ at T_0(=300 K) is 9. Furthermore, the intrinsic carrier concentration $n_{\mathrm{i}}(T)$ is written in terms of the effective density of states in the conduction band $N_{\mathrm{c}}(T)$, the effective density of states in the valence band N_v (T) and the energy gap $E_{\mathrm{g}}(T)$ of the semiconductor as

$$n_{\mathrm{i}}(T) = \sqrt{N_{\mathrm{c}}(T)N_{\mathrm{V}}(T)}\exp\left\{-\frac{qE_{\mathrm{g}}(T)}{2k_{\mathrm{B}}T}\right\} \tag{9.16}$$

$$N_{\mathrm{c}}(T) = N_{\mathrm{c}}(T_0)T^{1.5} = 4.3 \times 10^{14}T^{1.5}\ \mathrm{cm}^{-3}\ \mathrm{K}^{-3/2} \tag{9.17}$$

$$N_{\mathrm{V}}(T) = N_{\mathrm{V}}(T_0)T^{1.5} = 8.9 \times 10^{15}T^{1.5}\ \mathrm{cm}^{-3}\ \mathrm{K}^{-3/2} \tag{9.18}$$

$$E_{\mathrm{g}}(T) = E_{\mathrm{g}}(T_0) - \frac{\eta_{\mathrm{a}}T^2}{T + \eta_{\mathrm{b}}}\ \text{(Varshini expression)} \tag{9.19}$$

$N_{\mathrm{c}}(T_0)$ = effective density of states in the conduction band of GaN at T_0 (= 300 K)
$N_{\mathrm{V}}(T_0)$ = effective density of states in the valence band of GaN at T_0 (= 300 K)
$E_{\mathrm{g}}(T_0)$ = energy bandgap of GaN at T_0 (=300 K) = 3.427 eV
η_{a} = an an empirical constant = 9.39×10^{-4} eV K^{-1}
η_{b} = an empirical constant = 772 K.

Transconductance g_{m} is obtained by differentiating the I_{DS} expression with respect to V_{GS}

$$
\begin{aligned}
g_{\mathrm{m}} &= \left(\frac{\partial I_{\mathrm{DS}}}{\partial V_{\mathrm{GS}}}\right)_{V_{\mathrm{DS}}=\text{constant}} \\
&= \frac{\partial}{\partial V_{\mathrm{GS}}}\left\{I_{\mathrm{DSS}}\left(\frac{T_0}{T}\right)\left(1-\frac{V_{\mathrm{GS}}-V_{\mathrm{bi}}}{V_{\mathrm{Th}}+\gamma V_{\mathrm{DS}}}\right)^2 \tanh\left(\frac{\alpha V_{\mathrm{DS}}}{V_{\mathrm{Th}}-\gamma V_{\mathrm{DS}}}\right)\right\} \\
&= \left(\frac{T_0}{T}\right)\tanh\left(\frac{\alpha V_{\mathrm{DS}}}{V_{\mathrm{Th}}-\gamma V_{\mathrm{DS}}}\right)\times\frac{\partial}{\partial V_{\mathrm{GS}}}\left\{I_{\mathrm{DSS}}\left(1-\frac{V_{\mathrm{GS}}-V_{\mathrm{bi}}}{V_{\mathrm{Th}}+\gamma V_{\mathrm{DS}}}\right)^2\right\} \\
&= \left(\frac{T_0}{T}\right)\tanh\left(\frac{\alpha V_{\mathrm{DS}}}{V_{\mathrm{Th}}-\gamma V_{\mathrm{DS}}}\right)\left\{\frac{\mathrm{d}I_{\mathrm{DSS}}}{\mathrm{d}V_{\mathrm{GS}}}\times\left(1-\frac{V_{\mathrm{GS}}-V_{\mathrm{bi}}}{V_{\mathrm{Th}}+\gamma V_{\mathrm{DS}}}\right)^2+I_{\mathrm{DSS}}\right. \\
&\quad \left.\times\,2\left(1-\frac{V_{\mathrm{GS}}-V_{\mathrm{bi}}}{V_{\mathrm{Th}}+\gamma V_{\mathrm{DS}}}\right)\left(0-\frac{1-0}{V_{\mathrm{Th}}+\gamma V_{\mathrm{DS}}}\right)\right\} \\
&= \left(\frac{T_0}{T}\right)\tanh\left(\frac{\alpha V_{\mathrm{DS}}}{V_{\mathrm{Th}}-\gamma V_{\mathrm{DS}}}\right)\left(1-\frac{V_{\mathrm{GS}}-V_{\mathrm{bi}}}{V_{\mathrm{Th}}+\gamma V_{\mathrm{DS}}}\right) \\
&\quad \times\left\{\frac{\mathrm{d}I_{\mathrm{DSS}}}{\mathrm{d}V_{\mathrm{GS}}}\times\left(1-\frac{V_{\mathrm{GS}}-V_{\mathrm{bi}}}{V_{\mathrm{Th}}+\gamma V_{\mathrm{DS}}}\right)+I_{\mathrm{DSS}}\times 2\left(-\frac{1}{V_{\mathrm{Th}}+\gamma V_{\mathrm{DS}}}\right)\right\} \\
&= \left(\frac{T_0}{T}\right)\tanh\left(\frac{\alpha V_{\mathrm{DS}}}{V_{\mathrm{Th}}-\gamma V_{\mathrm{DS}}}\right)\left(1-\frac{V_{\mathrm{GS}}-V_{\mathrm{bi}}}{V_{\mathrm{Th}}+\gamma V_{\mathrm{DS}}}\right) \\
&\quad \times\left\{\frac{\mathrm{d}I_{\mathrm{DSS}}}{\mathrm{d}V_{\mathrm{GS}}}\left(1-\frac{V_{\mathrm{GS}}-V_{\mathrm{bi}}}{V_{\mathrm{Th}}+\gamma V_{\mathrm{DS}}}\right)-\frac{2I_{\mathrm{DSS}}}{V_{\mathrm{Th}}+\gamma V_{\mathrm{DS}}}\right\},
\end{aligned}
\tag{9.20}
$$

where from equation (9.11),

$$
\begin{aligned}
\frac{\mathrm{d}I_{\mathrm{DSS}}}{\mathrm{d}V_{\mathrm{GS}}} &= (I_{\mathrm{fc}}/2)\times\frac{\mathrm{d}}{\mathrm{d}V_{\mathrm{GS}}}\left[\left(2+\frac{R_{\mathrm{s}}I_{\mathrm{fc}}}{V_{\mathrm{p}}}\right)-\sqrt{\left(2+\frac{R_{\mathrm{s}}I_{\mathrm{fc}}}{V_{\mathrm{p}}}\right)^2-\left(\frac{4}{V_{\mathrm{p}}}\right)\{V_{\mathrm{p}}-(V_{\mathrm{bi}}+V_{\mathrm{GS}})\}}\right] \\
&= (I_{\mathrm{fc}}/2)\times\frac{\mathrm{d}}{\mathrm{d}V_{\mathrm{GS}}}\left(2+\frac{R_{\mathrm{s}}I_{\mathrm{fc}}}{V_{\mathrm{p}}}\right) \\
&\quad -(I_{\mathrm{fc}}/2)\times\frac{\mathrm{d}}{\mathrm{d}V_{\mathrm{GS}}}\left[\sqrt{\left(2+\frac{R_{\mathrm{s}}I_{\mathrm{fc}}}{V_{\mathrm{p}}}\right)^2-\left(\frac{4}{V_{\mathrm{p}}}\right)\{V_{\mathrm{p}}-(V_{\mathrm{bi}}+V_{\mathrm{GS}})\}}\right] \\
&= 0-(I_{\mathrm{fc}}/2)\times(1/2)\left[\left(2+\frac{R_{\mathrm{s}}I_{\mathrm{fc}}}{V_{\mathrm{p}}}\right)^2-\left(\frac{4}{V_{\mathrm{p}}}\right)\{V_{\mathrm{p}}-(V_{\mathrm{bi}}+V_{\mathrm{GS}})\}\right]^{1/2-1} \\
&\quad \times -\left(\frac{4}{V_{\mathrm{p}}}\right)\times\frac{\mathrm{d}}{\mathrm{d}V_{\mathrm{GS}}}\{V_{\mathrm{p}}-(V_{\mathrm{bi}}+V_{\mathrm{GS}})\} \\
&= -(I_{\mathrm{fc}}/2)\times(1/2)\left[\left(2+\frac{R_{\mathrm{s}}I_{\mathrm{fc}}}{V_{\mathrm{p}}}\right)^2-\left(\frac{4}{V_{\mathrm{p}}}\right)\{V_{\mathrm{p}}-(V_{\mathrm{bi}}+V_{\mathrm{GS}})\}\right]^{1/2-1}\times-\left(\frac{4}{V_{\mathrm{p}}}\right)\times -1 \\
&= -\frac{I_{\mathrm{fc}}/V_{\mathrm{p}}}{\sqrt{\left(2+\frac{R_{\mathrm{s}}I_{\mathrm{fc}}}{V_{\mathrm{p}}}\right)^2-\left(\frac{4}{V_{\mathrm{p}}}\right)\{V_{\mathrm{p}}-(V_{\mathrm{bi}}+V_{\mathrm{GS}})\}}}.
\end{aligned}
\tag{9.21}
$$

Output conductance g_d is found by differentiation of the expression for I_{DS} with respect to V_{DS}

$$
\begin{aligned}
\left(\frac{\partial I_{DS}}{\partial V_{DS}}\right)_{V_{GS}=\text{constant}} &= \frac{\partial}{\partial V_{DS}}\left\{I_{DSS}\left(\frac{T_0}{T}\right)\left(1-\frac{V_{GS}-V_{bi}}{V_{Th}+\gamma V_{DS}}\right)^2 \tanh\left(\frac{\alpha V_{DS}}{V_{Th}-\gamma V_{DS}}\right)\right\} \\
&= I_{DSS}\left(\frac{T_0}{T}\right)\times\frac{d}{dV_{DS}}\left\{\left(1-\frac{V_{GS}-V_{bi}}{V_{Th}+\gamma V_{DS}}\right)^2 \tanh\left(\frac{\alpha V_{DS}}{V_{Th}-\gamma V_{DS}}\right)\right\} \\
&= I_{DSS}\left(\frac{T_0}{T}\right)\left[2\left(1-\frac{V_{GS}-V_{bi}}{V_{Th}+\gamma V_{DS}}\right)^{2-1}\left\{0-\frac{0-\gamma(V_{GS}-V_{bi})}{(V_{Th}+\gamma V_{DS})^2}\right\}\right. \\
&\quad\times\tanh\left(\frac{\alpha V_{DS}}{V_{Th}-\gamma V_{DS}}\right)+\left(1-\frac{V_{GS}-V_{bi}}{V_{Th}+\gamma V_{DS}}\right)^2 \operatorname{sech}^2\left(\frac{\alpha V_{DS}}{V_{Th}-\gamma V_{DS}}\right) \\
&\quad\left.\times\frac{\alpha(V_{Th}-\gamma V_{DS})+\gamma\alpha V_{DS}}{(V_{Th}-\gamma V_{DS})^2}\right] \qquad (9.22) \\
&= I_{DSS}\left(\frac{T_0}{T}\right)\left(1-\frac{V_{GS}-V_{bi}}{V_{Th}+\gamma V_{DS}}\right)\left[2\tanh\left(\frac{\alpha V_{DS}}{V_{Th}-\gamma V_{DS}}\right)\left\{\frac{\gamma(V_{GS}-V_{bi})}{(V_{Th}+\gamma V_{DS})^2}\right\}\right. \\
&\quad\left.+\left(1-\frac{V_{GS}-V_{bi}}{V_{Th}+\gamma V_{DS}}\right)\operatorname{sech}^2\left(\frac{\alpha V_{DS}}{V_{Th}-\gamma V_{DS}}\right)\times\frac{\alpha V_{Th}}{(V_{Th}-\gamma V_{DS})^2}\right] \\
&= I_{DSS}\left(\frac{T_0}{T}\right)\left(1-\frac{V_{GS}-V_{bi}}{V_{Th}+\gamma V_{DS}}\right)\left[2\tanh\left(\frac{\alpha V_{DS}}{V_{Th}-\gamma V_{DS}}\right)\left\{\frac{\gamma(V_{GS}-V_{bi})}{(V_{Th}+\gamma V_{DS})^2}\right\}\right. \\
&\quad\left.+\left(1-\frac{V_{GS}-V_{bi}}{V_{Th}+\gamma V_{DS}}\right)\left\{1-\tanh^2\left(\frac{\alpha V_{DS}}{V_{Th}-\gamma V_{DS}}\right)\right\}\frac{\alpha V_{Th}}{(V_{Th}-\gamma V_{DS})^2}\right].
\end{aligned}
$$

The cut-off frequency f_T of the MESFET is

$$f_T = g_m/\{2\pi(C_{gs}+C_{gd})\}, \qquad (9.23)$$

where C_{gs} is the gate–source capacitance

$$C_{gs} = \{\varepsilon_0\varepsilon_s(T)WL\}/[a - I_{DS}/\{qN_Dv_{sat}(T)W\}], \qquad (9.24)$$

where a is the thickness of the active n-GaN layer, and W and L stand for channel width and length, respectively. C_{gd} is the gate–drain capacitance

$$C_{gd} = \{\varepsilon_0\varepsilon_s(T)\pi W_S L\}/(2W_D), \qquad (9.25)$$

where W_S is the depletion region thickness at the source and W_D is the depletion region thickness at the drain. The variation of the relative permittivity of the semiconductor and the saturation velocity of the charge carriers with temperature are described by the aforementioned equations (9.15) and (9.13), respectively.

The maximum frequency f_{max} of oscillations is

$$f_{max} = (kf_T)/\sqrt{g_d(R_S+R_G)}, \qquad (9.26)$$

where $k = 0.34$ is a constant of proportionality, $R_S = 160\ \Omega$ is the source resistance and $R_G = 10\ \Omega$ is the gate resistance.

The I_{DS}–V_{GS} characteristics of the GaN MESFET simulated from this model, along with the various MESFET parameters such as transconductance, output conductance, cut-off frequency and maximum frequency of oscillation, were found to agree well with experimental results up to 200 °C (Kabra *et al* 2008).

9.8 AlGaN/GaN HEMTs

9.8.1 Operation of AlGaN/GaN HEMTs on 4H-SiC/sapphire substrates from 25 °C to 500 °C

A comparative experimental study was performed to understand the high-temperature performance of $Al_{0.26}Ga_{0.74}N$/GaN HEMTs fabricated on 4H-SiC and sapphire substrates using atmospheric pressure MOCVD (Arulkumaran *et al* 2002). HEMT fabricated on a SiC substrate is shown in figure 9.5. The devices comprised the following successive layers: an AlGaN barrier layer (3 nm thick and undoped), an AlGaN carrier supply layer (15 nm thick, Si-doped), an AlGaN spacer layer (7 nm thick, undoped), an i-GaN layer (3000 nm thick, insulating), a buffer layer (200 nm thick AlN for the SiC substrate/30 nm thick i-GaN for the sapphire

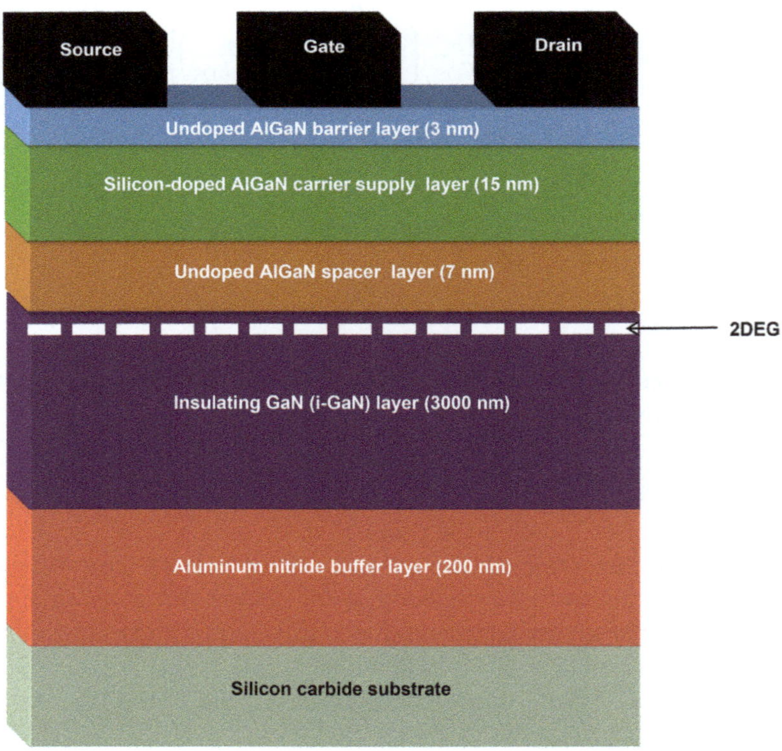

Figure 9.5. The structure of AlGaN/GaN HEMT on a SiC substrate showing the several constituent layers.

substrate) and the SiC/sapphire substrate. Gate metallization was Pd/Ti/Au (40 nm/40 nm/80 nm) of proven capability up to 500 °C. After being subjected to thermal stress at 500 °C, HEMTs fabricated on 4H-SiC substrates showed superior DC characteristics compared to those on sapphire substrates. For both types of HEMTs, the transconductance g_m and the drain current I_D decreased as the temperature was increased, mainly because of the reduction in two-dimensional electron gas (2DEG) mobility and carrier velocity. The general trends of variation of parameters for HEMTs on SiC substrates are as follows:

(i) The maximum transconductance $g_{m(max)}$ decreased from 210 mS mm^{-1} at 25 °C to 33 mS mm^{-1} at 500 °C, but was 201 mS mm^{-1} when re-measured at 25 °C.
(ii) The maximum drain current I_{Dmax} was 510 mA mm^{-1} at 25 °C and declined to 110 mA mm^{-1} at 500 °C but shot up to 550 mA mm^{-1} when measured after restoring the temperature to 25 °C (figure 9.6). In this graph, the I_{DS} values were measured from 0 to 550 mA mm^{-1}, V_{DS} from 0 V to 20 V and the steps were taken at V_{GS} = 0.5 V (Arulkumaran *et al* 2002).
(iii) The source resistance R_s was 2.6 Ω-mm and 12.8 Ω-mm at 25 °C and 500 °C, respectively; it recovered to 3.5 Ω-mm on returning to 25 °C.
(iv) The drain resistance R_d climbed from 8.4 Ω-mm at 25 °C to 35 Ω-mm at 500 °C, but became 6.5 Ω-mm upon returning to 25 °C.

Thus, the HEMTs operated with diminished capabilities at elevated temperatures. They resumed their normal functioning after cooling to 25 °C with practically the

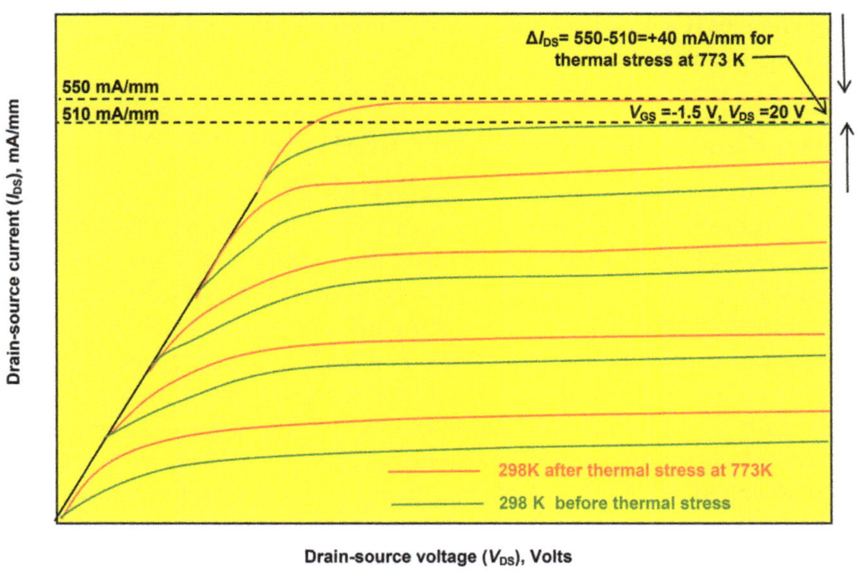

Figure 9.6. Increase in the room temperature drain–source current of AlGaN/GaN HEMTs after thermal stressing at 773 K.

same I_{DS}–V_{DS} characteristics. However, they showed an increase in their room temperature I_{DS} with respect to the values they had before thermal stress (Arulkumaran *et al* 2002).

9.8.2 Life testing of AlGaN/GaN HEMTs from 150 °C to 240 °C

MOCVD-grown 0.25 μm AlGaN/GaN HEMTs fabricated on SiC substrates and mounted on a dual-in-line package, were subjected to systematic life testing at high temperatures in nitrogen ambience (Chou 2004). In terms of their construction, these HEMTs contained the following layers: SiC substrate, AlN nucleation layer (100 nm), GaN buffer layer (1000 nm) and top $Al_{0.28}Ga_{0.72}N$ layer (25 nm). Ohmic contacts were made using Ti/Al/Pt/Au annealed at 860 °C in N_2. The Schottky contacts were made of Pt/Au metal stacks. The T-shaped gate was patterned and passivated with plasma-enhanced CVD (PECVD) nitride.

Commencing from a temperature of 150 °C, the temperature was increased in steps of 15 °C up to a maximum limit of 240 °C. The HEMTs were kept for 48 h at each temperature step. An electrical stress was applied with V_{DS} = 10 V and I_{DS} = 500 mA mm^{-1}. An important result of this experiment was that the degradation of I_{DS}–V_{DS} characteristics starts at a temperature of 195 °C. The degradation is observed as a fall in transconductance. The drain current decreases. The channel on-resistance registers an increase. Despite the variation of these electrical parameters of HEMTs, neither the gate area nor the ohmic contacts showed signs of any inter-diffusion of metal. There was no degradation of the gate diode as there was neither any modification of the ideality factor of the diode nor that of the Schottky barrier height of the gate. Gate leakage current also did not show any changes (Chou 2004). The quality of the AlGaN/GaN epitaxial layers was of utmost significance.

9.8.3 Power characteristics of AlGaN/GaN HEMTs up to 368 °C

The power characteristics of AlGaN/GaN HEMTs with a surface charge-controlled structure on SiC substrates, were examined at high temperatures (Adachi *et al* 2005). These HEMTs are provided with a capping n-GaN layer on the surface so that the constituent layers are: SiC, i-GaN, i-AlGaN, n-AlGaN and n-GaN; the 2DEG was formed between the i-GaN and i-AlGaN layers. They had Ti/Al ohmic contacts for source and drain connections, and Ni/Au gate electrodes. At a channel temperature of 269 °C, a frequency of 2.14 GHz and an operating voltage of 50 V, the linear gain was 12.3 dB and the maximum efficiency was 53.6%. When the channel temperature rose to 368 °C, the linear gain went down to 10.4 dB and the efficiency dropped to 43.9%. Further, from a channel temperature >100 °C onwards, the saturated output power begins to decrease. Its value at 368 °C is 2.1 dB less than at 59 °C. These changes were correlated with a decrease in drain current and hence with that in electron velocity. The drain current at 300 °C is half of its value at 30 °C. The runaway threat to AlGaN/GaN was very low due to the gate bias voltage remaining virtually constant (Adachi *et al* 2005).

9.8.4 Mechanisms of the failure of high-power AlGaN/GaN HEMTs at high temperatures

A review of the modes and mechanisms of the failure of high-power AlGaN/GaN HEMTs (Meneghesso *et al* 2008) showed that at frequencies <3.5 GHz, these HEMTs were highly reliable with a mean-time-to-failure (MTTF) $\geqslant 10^6$ h at a junction temperature of 180 °C. Good thermal stability was attained with Pt/Au, Ni/Au and Mo/Au Schottky contacts. The Mo/Au Schottky contacts along with Ti/Al/Ni/Au ohmic contacts, annealed at 900 °C for 30 s, showed a <2% change in ohmic contact resistivity after storage in nitrogen at 340 °C for 2000 h. The Schottky barrier height increased from 0.8 eV to 0.95 eV and became stable at this value (Sozza 2005).

9.9 InAlN/GaN HEMTs

9.9.1 AlGaN/GaN versus InAlN/GaN HEMTs for high-temperature applications

The commonly used AlGaN/GaN structure in HEMT devices has the drawback that it is a highly strained and polarized heterojunction (Maier *et al* 2010). The mechanical strain affects its thermal stability, thereby causing its performance to deteriorate at high temperatures. Therefore, AlGaN/GaN HEMTs have often failed in the temperature range 25 °C–500 °C. For reliable operation, it is advisable to keep away from the prescribed failure limit.

When comparing the InAlN/GaN heterostructure to the usual AlGaN/GaN heterostructure, it was found that:

(i) mechanical stress was absent in the lattice-matched InAlN/AlGaN structure;
(ii) the quantum well polarization-induced charge obtained from the InAlN/GaN structure was higher than from the AlGaN/GaN structure by a factor of 2–3 giving a density of 2DEG sheet charge = 2.8×10^{-13} cm^{-2}.This translates to a possibility of a 205% increase in drain current with the InAlN/GaN structure (Kuzmik 2001).

9.9.2 InAlN/GaN HEMT behavior up to 1000 °C

The above facts have encouraged the adoption of the InAlN/GaN structure for designing robust HEMT devices. The electrical performance of InAlN/GaN HEMTs was assessed to stake its claim as an alternative choice to conventional AlGaN/GaN HEMTs for high-power and high-temperature applications. For sapphire substrate HEMTs of gate length 0.25 μm, the maximum output current density was 3 A mm^{-1} at 77 K and 2 A mm^{-1} at room temperature (Medjdoub *et al* 2006). The transition frequency was 50 GHz and the maximum frequency was 60 GHz for a 0.15 μm gate length. The HEMT structure consists of a sapphire substrate, a GaN buffer layer (2 μm), an AlN spacer (1 nm) and an $Al_{0.81}In_{0.19}N$ layer (13 nm, undoped). The ohmic contact was a Ti/Al/Ni/Au multilayer metallization annealed at 890 °C for 1 min. The Schottky contact was Ni/Au. On-chip in-vacuum testing of devices was carried out by increasing the temperature by intervals of 100 °C and maintaining the device at every temperature for 10 min. At 800 °C, the maximum drain current density was 850 mA mm^{-1}. A marginal decrease in pinch-off voltage was noticed. At 1000 °C, the maximum current density dropped to 600 mA mm^{-1}, but pinch-off voltage suffered serious degradation. However, the device was not irreversibly degraded. It recovered its original characteristics after cooling, indicating the temporary nature of the departure in device behavior (Medjdoub *et al* 2006).

9.9.3 Thermal stability of barrier layer in InAlN/GaN HEMTs up to 1000 °C

Barrier-layer scaling of InAlN/GaN HEMTs was studied by varying the $In_{0.17}Al_{0.83}N$ barrier layer thickness from 33 nm to 3 nm (Medjdoub *et al* 2008). The HEMT structures studied included the GaN buffer layer (2 μm), an AlN spacer layer (1.0 nm) and different thicknesses of the barrier layer, with Ti/Al/Ni/Au ohmic contacts and Ni/Au Schottky gates. In the thermal stress experiment, the temperature was incremented by 100 °C, kept at 100 °C for half an hour, and then lowered to room temperature to examine the electrical characteristics and to ascertain any possible damage. This sequence was continued up to 1000 °C. Above 1000 °C, melting of Au did not permit any further experimentation. Despite the rigorous thermal stressing undergone by the HEMTs, no detrimental effects on the gate forward current capability came to light. Neither were there any changes in the open channel current nor in the pinch-off voltage. The properties of the metallurgical interfaces remained unaffected. The integrity and characteristics of the InAlN/GaN barrier layer up to 3 nm thickness were preserved. The reverse current of the SBD was constant. This is a major advantage of the InAlN/GaN structure over $Al_{0.3}Ga_{0.7}N$/GaN devices, which showed degeneration of Schottky diode characteristics to ohmic behavior after a thermal stress trial, even with 25 nm thick barrier layers (Medjdoub *et al* 2008). These observations could vastly benefit ultra-high frequency HEMT design and fabrication.

9.9.4 Feasibility demonstration of HEMT operation at gigahertz frequency up to 1000 °C

The ruggedness and reliability of a HEMT device with respect to temperature is determined by the thermal stability of the different constituent layers and metal stacks (Maier *et al* 2010). $In_{0.185}Al_{0.815}N$/GaN HEMTs on sapphire substrates contained a 12 nm thick InAlN layer with a 1 nm thick AlN spacer layer separating the buffer and barrier layers; the buffer layer thickness was 3 μm (Maier *et al* 2012). The ohmic contact was Ti/Al/Ni/Au annealed at 800 °C in nitrogen. The thickness ratio of the different metal films in this stack was optimized to achieve stability at 1000 °C. The refractory metal molybdenum (Mo) was used as the gate metal. The passivation coating was a 30 nm thick silicon nitride film formed by the PECVD process at 340 °C. This process was ammonia-free. The surface was specially prepared prior to Si_3N_4 deposition. These measures assured avoidance of blistering and cracking issues on raising the temperature to 1100 °C. Operated under a 1 MHz large-signal condition, the HEMT devices were ramped up to 1000 °C in 48 h with intermediate stops at 500 °C, 700 °C and 900 °C before the final destination of 1000 °C. As the temperature was increased, the gate leakage current became obvious. At 1000 °C, the gate leakage current showed an incessant increase as testing time elapsed. The drain current increased at 1000 °C on account of the extra leakage produced in the buffer layer around 800 °C. Throughout the temperature range from room temperature up to 1000 °C, and over the total time period of testing, the threshold voltage was stable. The HEMTs were operated for 25 h at 1000 °C. Thus a feasibility demonstration of HEMT operation was successfully executed in the gigahertz frequency range on a temperature scale hitherto considered the sole province of ceramic materials and refractory metals. The study also showed that the polarization in HEMT structure was preserved up to this temperature (Maier *et al* 2012).

However, it was evident that the full capability of the HEMT structure remained unutilized. Parasitic effects influenced the operation of the device at high temperatures. One conspicuous effect was the increase in leakage current of the buffer layer in this temperature regime. This predicament is avoided by resorting to mesa technology using an ultra-thin HEMT body (Herfurth *et al* 2013), see figure 9.7. In this improved structure, the active region of the device is restricted only to the channel area. An exceedingly thin (50 nm) AlN nucleation layer was chosen. The buffer layer thickness was also reduced to 50 nm. Besides these layers, the structure included a 1 nm thick AlN interfacial layer and a 5 nm thick InAlN barrier layer. The carrier concentration in 2DEG was 1.4×10^{13} cm^{-2}. The ohmic contact stack was Ti/Al/Ni without any Au capping layer, annealed at 800 °C for 30 s. The Schottky contact for the gate was Cu overlaid with Pt. The HEMT devices were tested using tungsten carbide needles. Testing at 600 °C was carried out at a pressure of 10^{-6} mbar. At room temperature, the maximum drain current was 0.4 A mm^{-1} and the threshold voltage was −1.4 V. Further, at room temperature, the off-state current I_{off} was ~1 pA for gate width W_G = 50 μm with a current on/off ratio of $I_{on}/I_{off} > 10^{10}$ and subthreshold swing SS = 73 mV dec^{-1}. At 600 °C, the maximum drain current and threshold voltage values remained nearly unaltered. But I_{off} decreased to ~1 μA mm^{-1} and I_{on}/I_{off} slumped to a low but tolerable level of 10^6; the SS was 166 mV dec^{-1}. The increase in off-state gate leakage current at 600 °C was a major shortcoming. Room temperature RF testing at 1 MHz gave a small-signal peak transconductance g_m = 110 mS mm^{-1}. The value of g_m was the same at 600 °C. No chemically or electrically induced material degradation was noticed during a 30 min trial experiment. The output power measurements in class A mode operation at 600 °C gave an RF output power = 109 mW mm^{-1} at V_{DS} = 8.75 V. The transition frequency was 6.6 GHz and the maximum frequency was 30 GHz (Herfurth *et al* 2013).

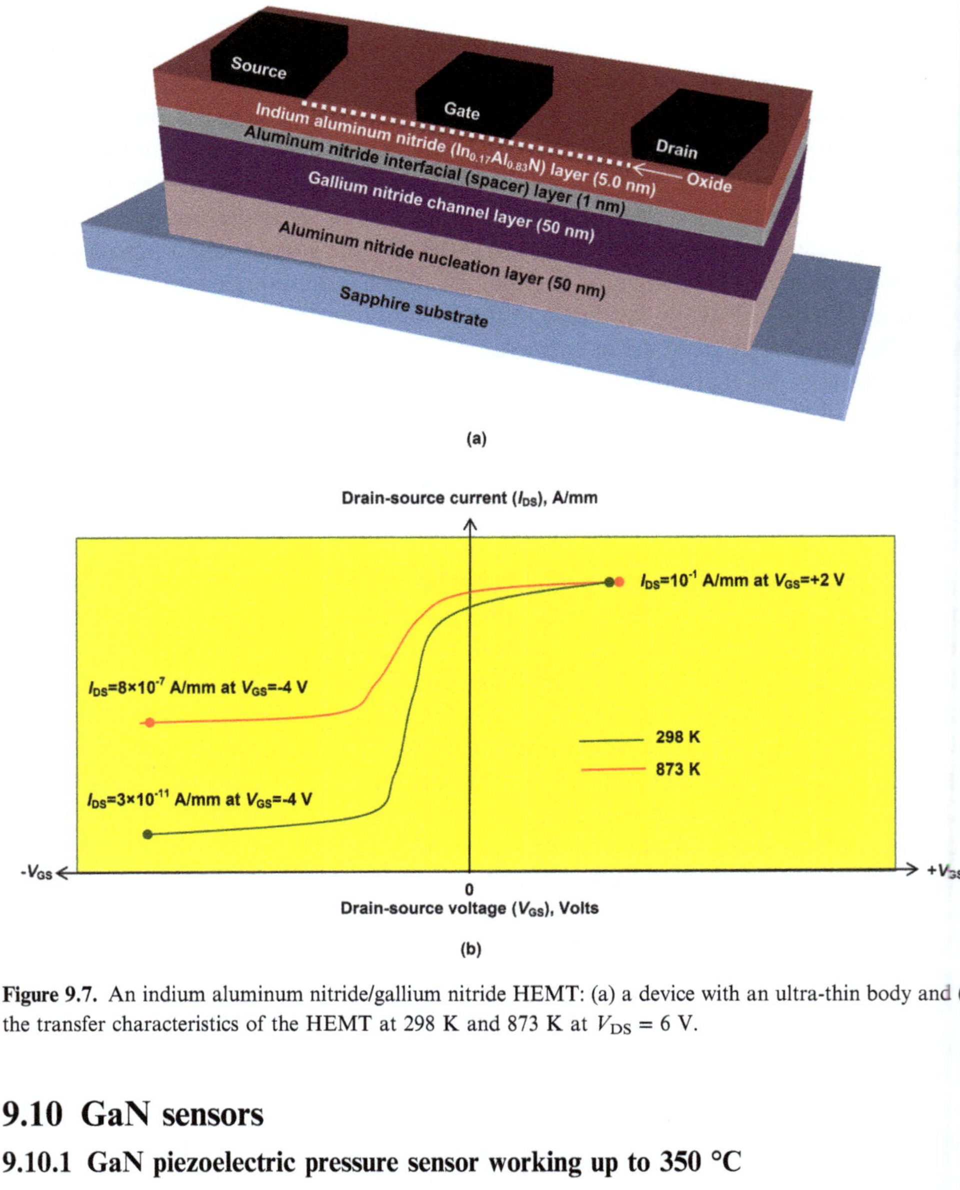

Figure 9.7. An indium aluminum nitride/gallium nitride HEMT: (a) a device with an ultra-thin body and (b) the transfer characteristics of the HEMT at 298 K and 873 K at V_{DS} = 6 V.

9.10 GaN sensors

9.10.1 GaN piezoelectric pressure sensor working up to 350 °C

A piezoelectric pressure sensor is fabricated using epitaxially grown GaN thin film (2 μm) by MOCVD on (111) silicon substrate (Kim *et al* 2020). Chromium (10 nm) and gold (100 nm) films are deposited over the GaN film by electron beam evaporation. 10 μm thick copper layer is electroplated over the gold film. The copper layer is covered with another gold film for protecting its surface. We now have the structure: Si substrate/GaN/Cr/Au/Cu/Au (from bottom upwards). A sapphire substrate is glued to the top Au film yielding: Si substrate/GaN/Cr/Au/Cu/Au/sapphire substrate. The silicon substrate is etched away in an acidic solution and Cr/Au is deposited over the exposed GaN film giving: Au/Cr/GaN/Cr/Au/Cu/

Au/sapphire substrate. The sapphire substrate is removed in acetone and isopropyl alcohol leaving the sensor structure shown in figure 9.8(a). The thick Cu layer serves as the support for this bendable sensor device. For measurements, the sensor is bonded to the stainless-steel foil of a housing with high-temperature glue and Ag paste is painted to reduce the contact resistance between the sensor and foil. Figure 9.8(b) shows the packaging arrangement.

The sensor shows an output potential value of 42.3 mV at 50 pounds per square inch (psi) pressure. The potential increases to 122.1 mV at 200 psi. The sensor yields reliable output potential up to 350 °C and 200 psi (Kim *et al* 2020).

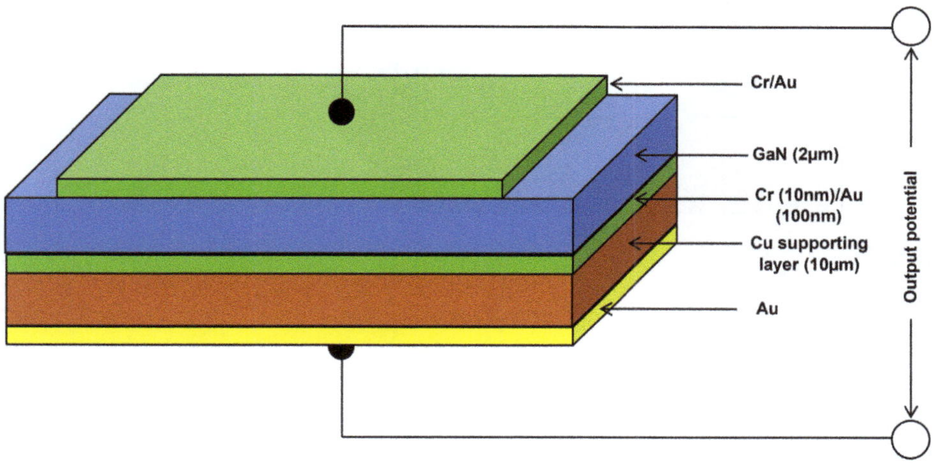

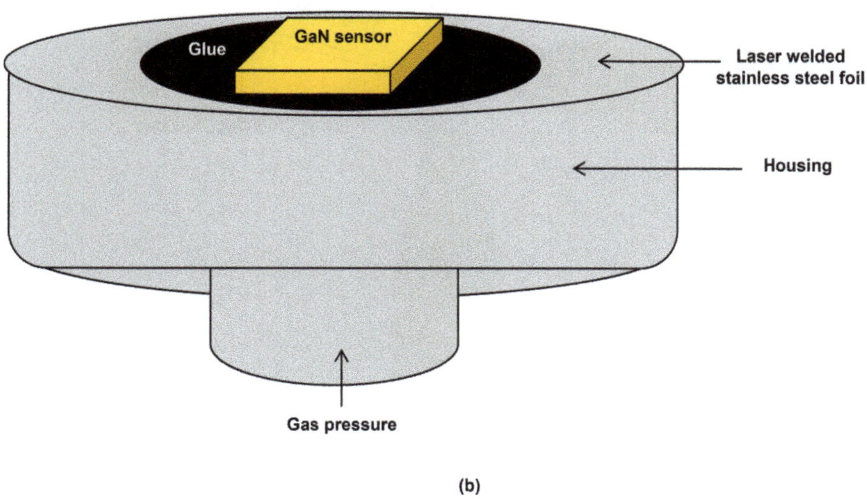

Figure 9.8. Piezoelectric pressure sensor made with GaN thin film.

9.10.2 GaN-based Hall-effect magnetic field sensors operating up to 400 °C

AlGaN/GaN heterostructure sensors are grown on sapphire substrate using MOCVD. The details of the epitaxial stack are shown in figure 9.9(a). The layers are arranged from the bottom upwards as: sapphire substrate, GaN

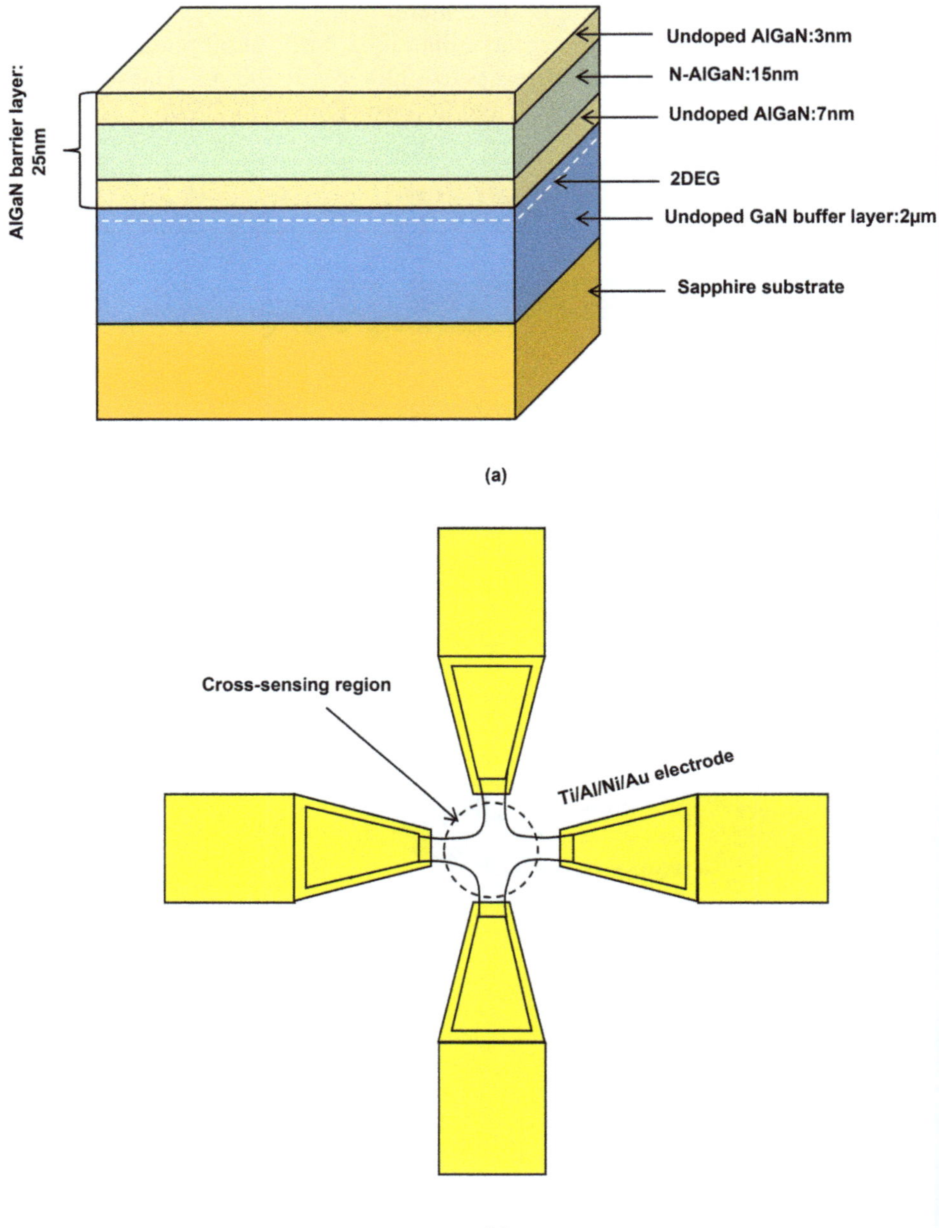

Figure 9.9. Hall-effect magnetic field sensor: (a) AlGaN/GaN heterostructure and (b) Hall cross pattern of the device.

buffer layer, undoped AlGaN layer, N-doped AlGaN layer with carrier concentration 4×10^{18} cm^{-3} using silicon dopant, and undoped AlGaN layer. The two-dimensional electron gas (2DEG) layer is located at the AlGaN/GaN interface. The AlGaN/GaN Hall sensors are fabricated by photolithographically defining the geometry. The mesa structure and cross patterns are formed by reactive-ion etching. Ti/Al/Ni/Au ohmic contacts are evaporated and annealed. Figure 9.9(b) illustrates the Hall cross.

The Hall-effect sensor shows stable operation between room temperature and 400 °C. The supply current-related magnetic field sensitivity is 77 V A^{-1} T^{-1} at 400 °C. This performance is much better than that of traditional Si, InAs or InSb-based sensors which become unstable at temperatures >125 °C due to increase in intrinsic carrier concentration (Koide *et al* 2012).

9.11 Discussion and conclusions

The merits and demerits of GaN over SiC and Si were pointed out. The MOCVD technique for synthesis of GaN was described. n- and p-type dopants for GaN were mentioned and the difficulties in achieving high carrier concentrations in p-type doping were indicated. Various ohmic and Schottky contact schemes on GaN were mentioned, and the fabrication and performance characteristics of AlGaN/GaN HEMTs were discussed. This structure was crippled by the stress problem. The lattice-matched InGaN/GaN HEMT structure greatly helped in overcoming this issue. HEMT operation is feasible at gigahertz frequencies up to 1000 °C using this structure.

Using high thermal conductivity dielectrics and substrates, adopting device designs with suitable electric field distribution to avoid hot spot creation and development of packaging and cooling techniques for efficient heat removal can significantly boost the performance of wide bandgap semiconductor devices (Spencer *et al* 2022).

Review exercises

9.1. Compare GAN with SiC and Si with regard to thermal conductivity.
9.2. Compare GaN with SiC with respect to electron and hole mobilities.
9.3. The energy bandgap of GaN is slightly larger than that of SiC. What is the difference between their intrinsic temperatures?
9.4. What is the technique used for growing GaN-based heterostructures called? What are the substrates used? What precursors are used for Ga, Al, In and N? What is the typical pressure of growth? What is the temperature used?
9.5. What are the commonly used n-type and p-type dopants in gallium nitride? What are the maximum concentrations achieved in n- and p-type doped GaN?
9.6. What is the position of the energy level for p-type doping in gallium nitride? What practical difficulty is experienced in p-type doping due to this energy level position?

9.7. Discuss the roles of titanium and nickel in Ti/Al/Ni/Au multilevel metallization on n-type GaN. Give a metallization scheme which is stable up to 1000 °C.
9.8. Mention one metallization scheme for making an ohmic contact on p-type GaN. Give one example of Schottky contact metallization on n-type GaN.
9.9. What are the different structural layers used in the fabrication of AlGaN/GaN HEMTs on 4H-SiC/sapphire substrates? What was the gate metallization used? Which type of HEMTs showed superior DC characteristics? What were the values of the following parameters at 25 °C and 500 °C: (a) maximum transconductance, (b) maximum drain current, (c) source resistance and (d) drain resistance?
9.10. Describe the systematic life testing of AlGaN/GaN HEMTs from 150 °C to 240 °C. At what temperature does the degradation of I_{DS}–V_{DS} characteristics begin? In what form is it observed? Does the gate diode degrade in any way? Describe the variation in power characteristics of AlGaN/GaN HEMTs up to 368 °C.
9.11. How do the following contacts on HEMTs perform at high temperatures: (a) Mo/Au Schottky contacts and (b) Ti/Al/Ni/Au ohmic contacts?
9.12. What is the main shortcoming of the frequently used AlGaN/GaN structure in HEMT devices? In what respects is the InAlN/GaN heterostructure superior to AlGaN/GaN heterostructure?
9.13. Describe the structure of InAlN/GaN HEMT. What is the ohmic contact metallization? What is the Schottky contact metallization used? What is the maximum drain current density at 800 °C? What is its value at 1000 °C? Is the device permanently degraded at 1000 °C?
9.14. Compare the thermal stability of the barrier layer in InAlN/GaN HEMTs with AlGaN/GaN HEMTs up to 1000 °C.
9.15. Describe the performance of $In_{0.185}Al_{0.815}N$/GaN HEMTs with refractory metallization up to 1000 °C in the gigahertz frequency range. What was the passivation coating used? Did the threshold voltage remain stable?
9.16. What were the Schottky and ohmic contacts used in the mesa InAlN/GaN HEMTs? What was the small-signal peak transconductance g_m at room temperature and at 600 °C? What was the RF output power in class A operation at 600 °C?
9.17. Elaborate the process steps in the fabrication of GaN piezoelectric pressure sensor. How is this sensor packaged? Up to what temperature can it be used?
9.18. Describe the construction and working of GaN magnetic field sensor. Is it useful for high-temperature measurements? Up to what temperature?

References

Adachi N, Tateno Y, Mizuno S, Kawano A, Nikaido J and Sano S 2005 High temperature operation of AlGaN/GaN HEMT *2005 IEEE MTT-S Int. Microwave Symp. Digest (12–17 June)* 507–10

Adivarahan V, Lunev A, Asif Khan M, Yang J, Simin G, Shur M S and Gaska R 2001 Very-low-specific-resistance Pd/Ag/Au/Ti/Au alloyed ohmic contact to p GaN for high-current devices *Appl. Phys. Lett.* **78** 2781–3

Arulkumaran S, Egawa T, Ishikawa H and Jimbo T 2002 High-temperature effects of AlGaN/GaN high-electron-mobility transistors on sapphire and semi-insulating SiC substrates *Appl. Phys. Lett.* **80** 2186–8

Baliga B J 2017 *Gallium Nitride and Silicon Carbide Power Devices* (Singapore: World Scientific)

Baliga B J 2013 Gallium nitride devices for power electronic applications *Semicond. Sci. Technol.* **28** 074011

Chou Y C 2004 Degradation of AlGaN/GaN HEMTs under elevated temperature life testing *Microelectron. Reliab.* **44** 1033–8

Denis A, Goglio G and Demazeau G 2006 Gallium nitride bulk crystal growth processes: a review *Mater. Sci. Eng. R Rep.* **50** 167–94

Ding X, Zhou Y and Cheng J 2019 A review of gallium nitride power device and its applications in motor drive *CES Trans. Electr. Mach. Syst.* **3** 54–64

Fan S Z, Mohammad S N, Kim W, Aktas O, Botchkarev A E and Morkoc H 1996 Very low resistance multilayer ohmic contact to n-GaN *Appl. Phys. Lett.* **68** 1672–4

Herfurth P, Maier D, Lugani L, Carlin J-F, Rösch R, Men Y, Grandjean N and Kohn E 2013 Ultrathin body InAlN/GaN HEMTs for high-temperature (600 °C) electronics *IEEE Electron Device Lett.* **34** 496–8

Kabra S, Kaur H, Haldar S, Gupta M and Gupta R S 2008 Temperature dependent analytical model of sub-micron GaN MESFETs for microwave frequency applications *Solid-State Electron.* **52** 25–30

Kacprzak T and Materka A 1983 Compact DC model of GaAs FETs for large-signal computer calculation *IEEE J. Solid State Circuits* **18** 211–3

Kim N-I, Chang Y-L, Chen J, Barbee T, Wang W, Kim J-Y and Kwon M-K *et al* 2020 Piezoelectric pressure sensor based on flexible gallium nitride thin film for harsh-environment and high-temperature applications *Sens. Actuators* A **305** 111940

Koide S, Takahashi H, Abderrahmane A, Shibasaki I and Sandhu A 2012 High temperature Hall sensors using AlGaN/GaN HEMT structures *Asia-Pacific Interdisciplinary Research Conf. 2011 (AP-IRC 2011) J. Phys.: Conf. Ser.* 352 012009

Kucharski R, Sochacki T, Lucznik B and Bockowski M 2020 Growth of bulk GaN crystals *J. Appl. Phys.* **128** 050902-1–13

Kuzmik J 2001 Power electronics on InAlN/(In)GaN: prospect for a record performance *IEEE Electron Device Lett.* **22** 510–2

Macherzyński W, Paszkiewicz B, Szyszka A, Paszkiewicz R and Tłaczała M 2009 Effect of annealing on electrical characteristics of platinum based Schottky contacts to n-GaN layers *J. Electr. Eng.* **60** 276–8

Maier D, Alomari M, Grandjean N, Carlin J-F and di Forte-Poisson M-A 2010 Testing the temperature limits of GaN-based HEMT devices *IEEE Trans. Device Mater. Reliab.* **10** 427–36

Maier D, Alomari M, Grandjean N, Carlin J-F, Diforte-Poisson M-A, Dua C, Delage S and Kohn E 2012 InAlN/GaN HEMTs for operation in the 1000 °C regime: a first experiment *IEEE Electron Device Lett.* **33** 985–7

Medjdoub F, Alomari M, Carlin J-F, Gonschorek M, Feltin E, Py M A, Grandjean N and Kohn E 2008 Barrier-layer scaling of InAlN/GaN HEMTs *IEEE Electron Device Lett.* **29** 422–5

Medjdoub F, Carlin J-F, Gonschorek M, Feltin E, Py M A, Ducatteau D, Gaquière C, Grandjean N and Kohn E 2006 Can InAlN/GaN be an alternative to high power/high temperature AlGaN/GaN devices? *IEDM'06 Int. Electron Devices Meeting (San Francisco, CA, 11–13 December)* 1–4

Meneghesso G, Verzellesi G, Danesin F, Rampazzo F, Zanon F, Tazzoli A, Meneghini M and Zanoni E 2008 Reliability of GaN high-electron-mobility transistors: state of the art and perspectives *IEEE Trans. Device Mater. Reliab.* **8** 332–43

Meneghini M, De Santi C, Abid I, Buffolo M, Cioni M, Khadar R A and Nela L *et al* 2021 GaN-based power devices: physics, reliability, and perspectives *J. Appl. Phys.* **130** 181101

Parikh R P and Adomaitis R A 2006 An overview of gallium nitride growth chemistry and its effect on reactor design: application to a planetary radial-flow CVD system *J. Cryst. Growth* **286** 259–78

Pearton S J, Donovan S M, Abernat C R, Ren F, Zolper J C, Cole M W and Shul R J 1998 High temperature stable WSi_x ohmic contacts on GaN *IEEE 4th Int. High Temperature Electronics Conf.* 296–300

Ruvimov S, Liliental-Weber Z, Washburn J, Duxstad K J, Haller E E, Fan Z-F, Mohammad S N, Kim W, Botchkarev A E and Morkoc H 1996 Microstructure of Ti/Al and Ti/Al/Ni/Au ohmic contacts for n-GaN *Appl. Phys. Lett.* **69** 1556–8

Shashikala B N and Nagabhushana B S 2010 Modeling of GaN MESFETs at high temperature *22nd Int. Conf. on Microelectronics (ICM 2010)* 4

Sozza A 2005 Evidence of traps creation in GaN/AlGaN/GaN HEMTs after a 3000 h on-state and off-state hot-electron stress *IEEE Electron Device Meeting* 590–3

Spencer J A, Mock A L and Zhang Y 2022 Heating issues in wide-bandgap semiconductor devices *Thermal Management of Gallium Nitride Electronics* ed M J Tadjer and T J Anderson (Cambridge: Woodhead Publishing) ch 1 pp 1–19

Yu H and Duan T (ed) 2017 *Gallium Nitride Power Devices* (Dubai: Jenny Stanford Publishing)

Zhong Y, Zhang J, Wu S, Jia L, Yang X, Liu Y, Zhang Y and Sun Q 2022 A review on the GaN-on-Si power electronic devices *Fundam. Res.* **2** 462–75

Chapter 10

Diamond electronics for ultra-hot environments

Ranked at the highest place for mechanical hardness and thermal conductivity among all known materials, diamond has the most interesting thermal properties of semiconductors, and is seen as the perfect material for power electronics, fulfilling the requirements of a niche market for robust high-power devices. From its calculated figures of merit based on intrinsic diamond of low defect density and with shallow donors/acceptors for doping, diamond may in future replace SiC and GaN by providing less lossy devices with a better trade-off between on-resistance and breakdown voltage. Practically, however, the achievements are too far from predicted values. Tremendous efforts will be necessary to overcome the technological difficulties in order to reach the goal of making this human ornamental gemstone the real jewel of microelectronics.

10.1 Introduction

As diamond has a high thermal conductivity, it is the correct material choice for fabrication of power semiconductor devices whose reliable operation crucially depends on heat dissipation. This capability is a premonition of great possibilities of diamond power electronics (Han *et al* 2022). Diamond semiconductors are expected to provide improved electronics in the power sector and at high-temperatures than conventional semiconductors (Koizumi *et al* 2018, Perez *et al* 2020, Shukla *et al* 2022).

Compared to the silicon bandgap of 1.12 eV, diamond has a bandgap of 5.5 eV, which is 4.9 times higher than that of silicon (Pan and Kania 1995). In fact, the bandgap of diamond approaches that of an insulator. Operating temperatures of >1000 °C can be realized by using diamond as the fabrication material for devices.

doi:10.1088/978-0-7503-5072-3ch10

The breakdown electric field for diamond (10 MV cm^{-1}) is 33 times larger than that of silicon (0.3 MV cm^{-1}). It is 3.33 times that for silicon carbide (3 MV cm^{-1}). A much lower specific on-resistance is achievable using the thin conducting regions in diamond devices than is possible with silicon or silicon carbide devices. The thermal conductivity of diamond (15 W cm K^{-1}) is 10 times that of silicon (1.5 W cm K^{-1}) and 3 times that of silicon carbide (4.9 W cm K^{-1}). Therefore, diamond devices can provide substantially higher power per unit area than either silicon or silicon carbide components. Unlike silicon carbide, the carrier mobilities in diamond, ~2000 $cm^2 V^{-1} s^{-1}$, are better than those available in silicon (electron mobility 1400 $cm^2 V^{-1} s^{-1}$). So, the mobility disadvantage upsetting silicon carbide does not impose any encumbrance on the good features of diamond. In addition to the above good features, diamond displays exceptional hardness and outstanding wear resistance. It offers high stability against the degradation effects due to chemical or thermal effects or upon irradiation. Thus, diamond displays a unique set of properties which are several-fold superior in every category when compared to any other wide bandgap semiconductor, making it the 'true dream material' for designers of microelectronic devices, sensors, microwave and acoustic wave filters, and microelectromechanical systems (MEMS) (Yang *et al* 2021). Table 10.1 lists the major properties of interest to engineers working on diamond device and circuit design. Diamond crystallizes in the structural form displayed in figure 10.1.

Diamond electronics is still in its preliminary phase. Many processing issues for the exploitation of diamond as a base material for semiconductor device fabrication are still technological challenges to be resolved. The present research outcomes trail far behind the idealistic prospects.

10.2 Intrinsic temperature of diamond

Before embarking on detailed discussions of diamond electronics, let us repeat the intrinsic temperature calculations for diamond and examine where diamond stands with respect to other semiconductors for high-temperature applications. This will help us in appreciating the significance of diamond in a proper perspective relative to competing materials. For diamond

$$m_n^* = 1.9m_0,\ m_p^* = 0.8m_0,\ E_g = 5.46 - 5.6\ \text{eV}$$

$$\therefore m^* = \left(\frac{m_n^* m_p^*}{m_0^2}\right)^{0.75} = \left(\frac{1.90m_0 \times 0.8m_0}{m_0^2}\right)^{0.75} = 1.3689 \tag{10.1}$$

The equation for intrinsic temperature calculation

$$\ln\{0.207/(m^* T^{1.5})\} = -5.8025 \times 10^3 E_g \tag{10.2}$$

is converted to the form

$$T \ln\{0.207/(1.37T^{1.5})\} = -5.8025 \times 10^3 \times 5.5 \tag{10.3}$$

or

$$T \ln 0.151 - T \ln T^{1.5} = -31\,913.75 \tag{10.4}$$

Table 10.1. Properties of diamond.

Property	Value	Property	Value	Property	Value
Chemical symbol	C (carbon)	Lattice constant (Å) at 300 K	3.5668	Intrinsic resistivity (Ω-cm)	1×10^{42}
Refractive index	2.424 at 546.1 atm.	Melting point (°C)	3773	Electron mobility ($cm^2\ V^{-1}\ s^{-1}$)	2400
Classification	Non-metal	Dielectric constant	5.7 (1–10 kHz)	Hole mobility ($cm^2\ V^{-1}\ s^{-1}$)	2100
Crystal structure	Diamond cubic	Thermal conductivity ($W\ cm\ K^{-1}$)	10–20	Electron saturated velocity ($cm\ s^{-1}$)	2.7×10^7
Color	Colorless to yellow/ brown	Energy bandgap E_g (eV) at 300 K	5.46–5.60	Donors	Phosphorous, arsenic
Density at 300 K ($g\ cm^{-3}$)	3.516–3.525	Electrical breakdown field ($V\ cm^{-1}$)	1×10^7	Donor ionization energies ΔE_D (meV)	590 (phosphorous), 410 (arsenic)
Number of atoms/ cm^3	1.763×10^{23}	Intrinsic carrier concentration (cm^{-3})	$\sim 1 \times 10^{-27}$	Acceptor and ionization energy ΔE_A (meV)	Boron, 370

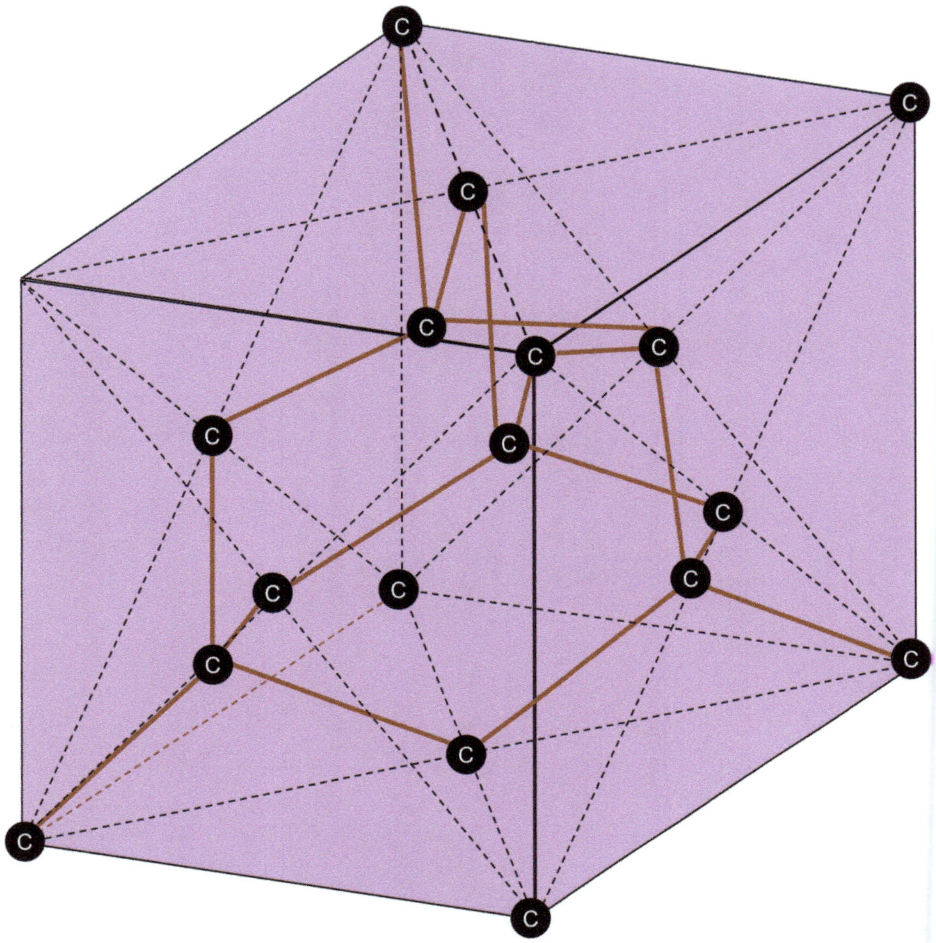

Figure 10.1. Structure of diamond crystal.

$$\therefore -1.89T - T\ \ln T^{1.5} = -31\ 913.75. \tag{10.5}$$

Suppose,

$$T = 2357.3\ \text{K} \tag{10.6}$$

$$\begin{aligned} \text{lhs} &= -1.89T - T\ \ln T^{1.5} = -1.89 \times 2357.3 - 2357.3\ \ln 2357.3^{1.5} \\ &= -4455.297 - 27\ 457.6141 = -31\ 912.9111. \end{aligned} \tag{10.7}$$

If

$$T = 2357.35\ \text{K} \tag{10.8}$$

$$\begin{aligned} \text{lhs} &= -1.89T - T\ \ln T^{1.5} = -1.89 \times 2357.35 - 2357.35\ \ln 2357.35^{1.5} \\ &= -4455.3915 - 27\ 458.2715 = -31\ 913.663. \end{aligned} \tag{10.9}$$

When

$$T = 2357.4\ \text{K} \tag{10.10}$$

$$\begin{aligned} \text{lhs} &= -\ 1.89T - T\ \ln\ T^{1.5} = -1.89 \times 2357.4 - 2357.4\ \ln 2357.4^{1.5} \\ &= -\ 4455.486 - 27\ 458.9289 = -31\ 914.4149. \end{aligned} \tag{10.11}$$

$$\therefore\ T = 2357.35\ \text{K}. \tag{10.12}$$

This temperature represents the highest of all the materials examined in the preceding chapters. It is much higher than for GaN (1667 K) or SiC (1591.5 K).

10.3 Synthesis of diamond

Diamond has been known to humankind for more than 200 years, but its identity as elemental carbon was recognized only in 1796. Various types of diamonds have been extensively studied and subdivided into four categories (Araujo *et al* 2021). This classification is based on the type of chemical impurities present and their concentration in the diamond. The categories of diamond are: type Ia, type Ib, type IIa and type IIb. The type Ia category comprises 98% of all diamonds obtained naturally. Their main impurity content is nitrogen, up to 0.1%. Type Ib diamonds represent about 0.1% of diamonds found naturally with nitrogen content up to 0.05%. Type IIa diamonds constitute 1%–2% of natural diamonds and are practically free from impurities. Type IIb diamonds are present in around 0.1% of natural diamonds with a very low impurity content like type IIa; they are the one of the rarest diamond varieties.

The artificial synthesis of diamond through a high-pressure and high-temperature (HPHT) process was reported in 1955 (Bundy *et al* 1955); hence it is called HPHT diamond. Diamond is synthesized by another method known as chemical vapor deposition (CVD), see figures 10.2 and 10.3. Hence, this diamond is termed CVD-grown diamond or CVD diamond (Arnault *et al* 2022). In general, man-made diamond is called synthetic diamond, cultured diamond or cultivated diamond.

CVD is an established commercial technique. It is an atomistic process yielding highly dense, good-quality, adherent films. In this technique, a solid film is slowly deposited on a hot substrate kept inside a vacuum chamber at low pressure, and serving as a reactor. This deposition takes place by chemical reaction between gaseous carbon-containing precursor molecules (usually a hydrocarbon such as methane and hydrogen) introduced into the reactor. In the growth chamber, the gases are ionized into chemically reactive radicals by any of the following means: a hot filament, a DC or microwave plasma, an oxyacetylene flame, a welding torch, a laser or an e-beam. Typical growth conditions are: 160 Torr, 1000 °C–1200 °C, 3% N_2/CH_4 and high methane proportion (12% CH_4/H_2) (Yan *et al* 2002). To produce gem-quality diamond, a seed crystal is inserted into the gas mixture. Then seeded growth of diamond occurs, replicating the structure of the seed crystal.

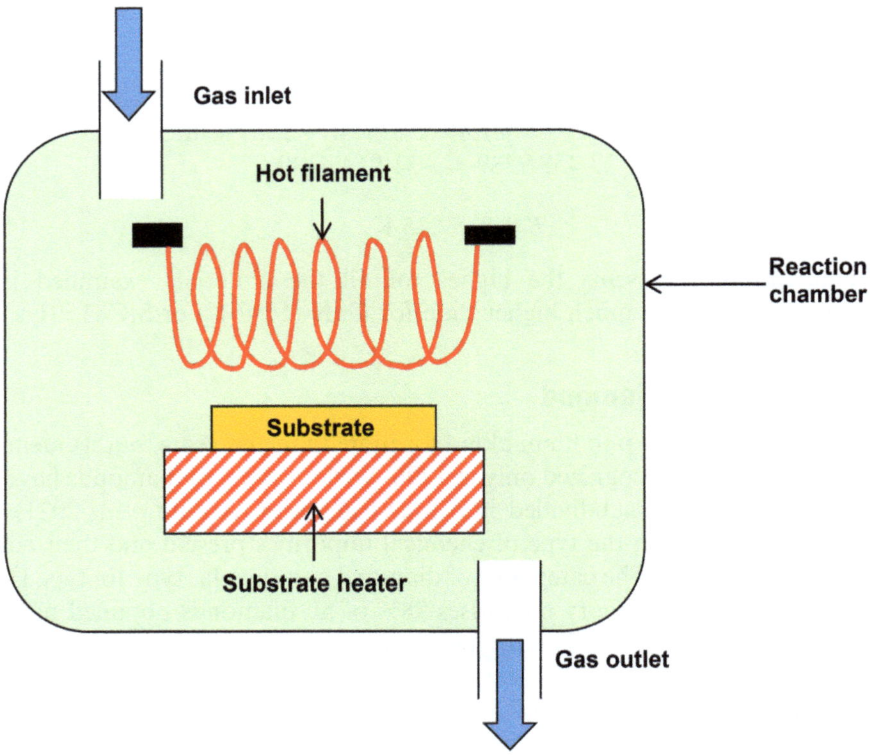

Figure 10.2. Hot-filament CVD of diamond.

10.4 Doping of diamond

The absence of donor and acceptor impurities with shallow energy levels, i.e. at small distances below the conduction band and above the valence band, is the main bottleneck confronting diamond electronics. Substitutional doping by conventional methods such as thermal diffusion and ion implantation has not yet been able to provide high carrier concentrations and mobilities. The energy levels of the available impurities are appreciably deep. Hence, the ionization energies of these impurities are much larger than the thermal energy at room temperature.

10.4.1 n-Type doping

Attempts to produce n-type diamond, either by diffusion, by impurity incorporation during CVD growth, or by ion implantation and high-temperature annealing, have so far met with limited success. Very often, nitrogen is used to dope diamond as n-type. But the energy level of the nitrogen impurity is very deep (1.7 eV below the conduction band), see figure 10.4. Hence, thermal energy is unable to cause ionization at room temperature. The fraction of impurities that are ionized at room temperature is very small. Nitrogen can be thermally activated only at temperatures >600 °C–700 °C. As a result, nitrogen doping cannot provide the required high electron concentrations. Nitrogen-doped diamond is practically an insulator at room temperature.

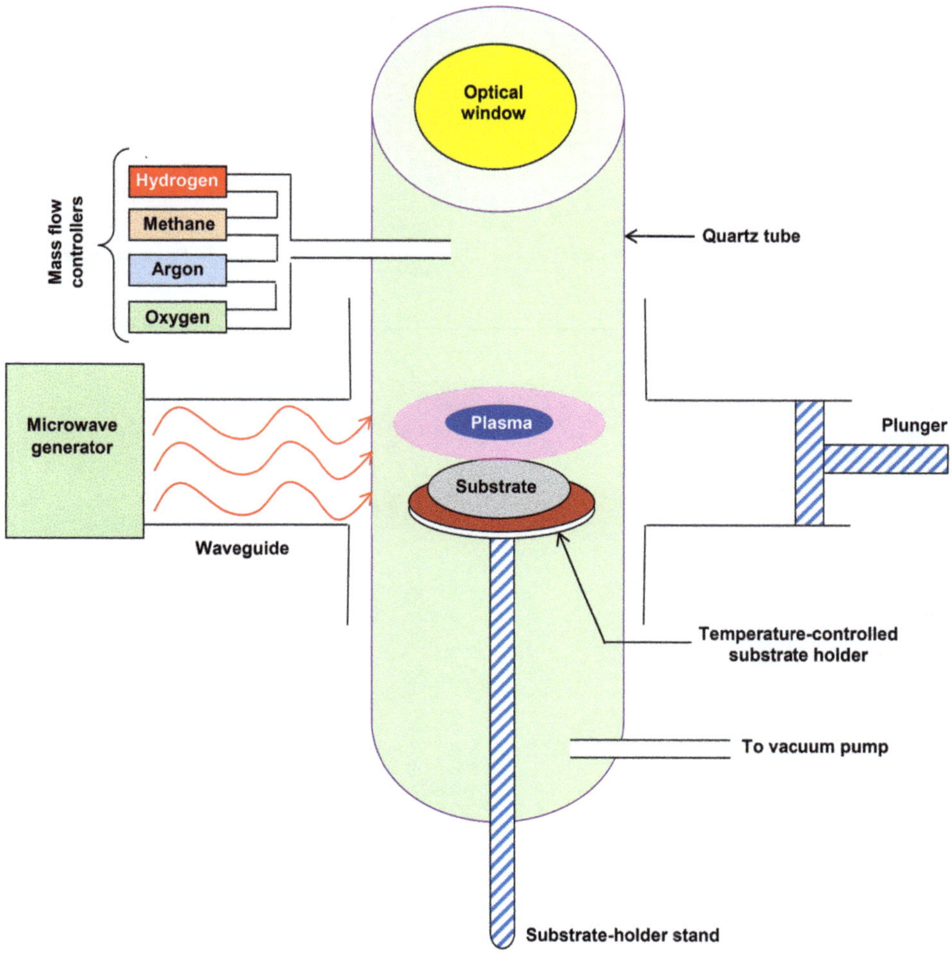

Figure 10.3. Microwave plasma reactor for diamond synthesis by CVD.

Besides nitrogen, phosphorous has been used for n-type doping of diamond (Koizumi 2006, Kato *et al* 2007). The energy level of phosphorous, located at 0.59 eV, is much shallower than that for nitrogen but its depth is still too large to meet the requirement.

Phosphorous and nitrogen co-doped diamond epitaxial films have been grown with a phosphorous concentration of 2×10^{16}–3.6×10^{17} atoms/cm^3 and a nitrogen concentration from 4×10^{17} to 3×10^{18} atoms/cm^3 (Cao *et al* 1995).

Sulfur (0.32 eV) and lithium (0.16 eV) have also been tried. Sulfur was implanted with energies up to 400 keV (Hasegawa *et al* 1999).

10.4.2 p-Type doping

p-Type doping of diamond is done by ion implantation of boron. The activation energy of boron decreases as its concentration increases. It lies in the range 0–0.43 eV above the valence band. It becomes zero at boron concentrations $>1.7 \times 10^{20}$ cm^{-3} and

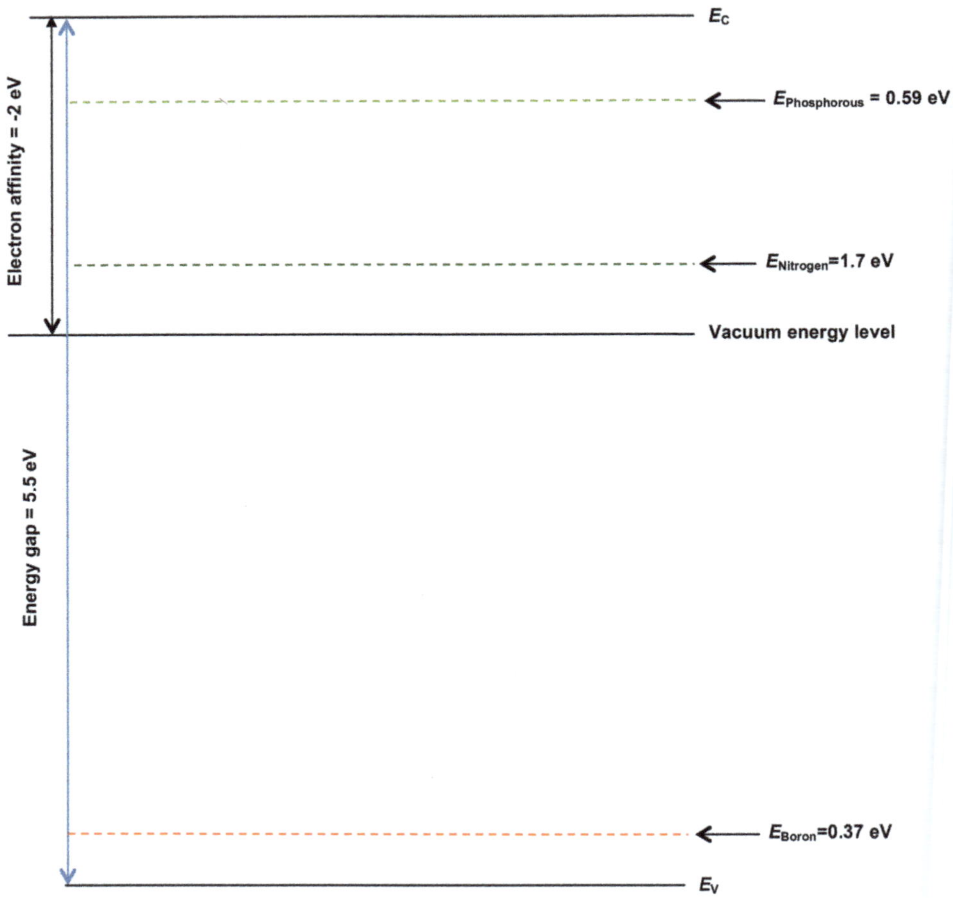

Figure 10.4. The energy band diagram of diamond showing the energy levels of commonly used impurities for n- (phosphorous, nitrogen) and p-doping (boron).

is 0.35 eV at a boron concentration of 1×10^{18} cm^{-3}. Boron can also be incorporated during CVD growth of diamond. However, the presence of any hydrogen in the plasma inhibits the control of boron doping to obtain p-type diamond. Efforts to achieve high hole mobility have been made by several workers, e.g., by a cold-implantation–rapid-annealing process (CIRA; Prins 1988), with a final annealing temperature of 1723 K (Fontaine *et al* 1996) giving a hole mobility of 400 cm^2 V^{-1} s^{-1}; by ion implantation at a high energy of ~MeV (Prawer *et al* 1997) yielding a mobility of 600 cm^2 V^{-1} s^{-1} (Uzan-Saguy *et al* 1998).

10.4.3 p-Doping by hydrogenation termination of the diamond surface

p-Type conduction in diamond can also be obtained by hydrogenation of the diamond surface in hydrogen plasma and cooling to room temperature in a hydrogen atmosphere (Kawarada 1996). More than 90% of charge carriers exist in a <10 nm thick conducting layer formed on the surface showing p-type conductivity ∼10^{-4}–10^{-5} S, with a hole concentration of ∼10^{13} cm^{-2} and hole mobility ∼100–150 cm^2 V^{-1} s^{-1}. Such a surface with a two-dimensional hole gas is said to be hydrogen-terminated. Thus, an intrinsic conducting channel containing holes as charge carriers is formed in a sub-surface layer without any impurity doping, in sharp contrast to the extrinsic boron-doped acceptor layer, which is only partially activated and hence only incompletely conducting at room temperature.

The p-type conduction is chemically and thermally unstable. It is vulnerable to environmental effects. By exposure to oxygen ambient, by heating in air above 200 °C or annealing in vacuum, the p-type conduction is destroyed because of dehydrogenation of the surface. For utilization of this p-type layer as a part of an electronic device, it needs proper protection. Passivation of the hydrogenated surface by hydrogenated carbon films formed by an RF plasma CVD process was found to improve its thermal stability up to 200 °C (Yamada *et al* 2004). Surface passivation is crucial for robust device development.

The low density of surface states <10^{11} cm^{-2} exhibited by the H-terminated surface makes it appropriate for a field-effect transistor (FET) channel. Moreover, an ohmic contact on the p-type conducting layer can be formed using metals of high electronegativity. The contact resistance for Au or Pd is <10^{-5} Ω cm^{-2}.

In opposition to the conducting nature of a H-terminated diamond surface, an oxygen-terminated diamond surface behaves as an insulator. Oxygen termination is accomplished by exposure of the diamond surface to oxygen plasma. A H-terminated surface can be rendered insulating by such treatment in which hydrogen is substituted by oxygen. This affords a method of making the active areas of a device conducting and the adjoining passive areas insulating by selective treatment with hydrogen and oxygen.

10.5 A diamond p–n junction diode

A UV light-emitting diode (LED) was fabricated on diamond (Koizumi *et al* 2001), see figure 10.5. The starting substrate was boron-doped single-crystal diamond. The diode was formed on its mechanically polished {111} surface. Upon the surface, the boron-doped layer (1–2 × 10^{17} cm^{-3}, 1 μm) was grown using trimethylboron (TMB; $(CH_3)_3B$) as the source of boron impurity. Then a phosphorous-doped layer (7–8 × 10^{18} cm^{-3}, 2 μm) was grown with phosphine (PH_3), as the source of phosphorous impurity. Separate microwave plasma-enhanced CVD (MPCVD) systems were used for boron and phosphorous doping. The activation energies for B and P were determined as 0.37 eV and 0.59 eV, respectively. The hole mobility was 60 cm^2 V^{-1} s^{-1} and the electron mobility was 300 cm^2 V^{-1} s^{-1}. For the p-layer, Ti–Au contacts were made. For the n-layer, graphitic dots formed by Ar implantation through a mask were capped with Ti–Au. The turn-on voltage was 6–7 V. The reverse leakage current was 10^{-8} A at −20 V bias. UV light emission was observed at a wavelength of 235 nm on applying a forward voltage of 20 V.

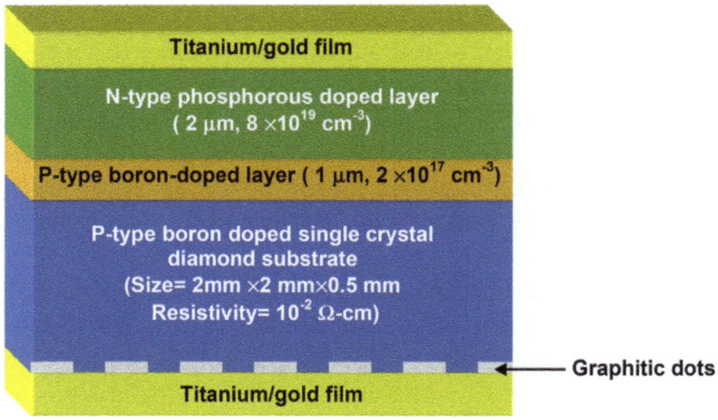

(a)

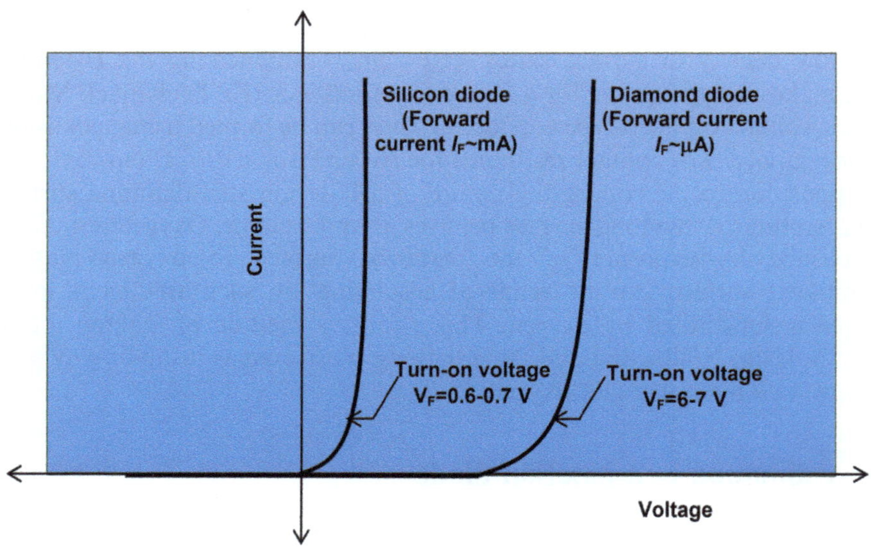

(b)

Figure 10.5. Diamond p–n junction diode: (a) construction and (b) comparison of its forward current–voltage characteristics with those of a silicon diode.

10.6 Diamond Schottky diode

10.6.1 Diamond Schottky diode operation up to 1000 °C

The vertical-structure Schottky diodes (figure 10.6) were fabricated on synthetic boron-doped IIb substrates. The boron concentration was $>10^{20}$ cm^{-3} (Vescan *et al* 1997a). In the MPCVD-grown active layer, the surface doping concentration decreased sharply to 10^{16} cm^{-3} to achieve a breakdown voltage >50 V. The surface was exposed to oxygen plasma. For the Schottky contact, a nitrogen-doped Si:W

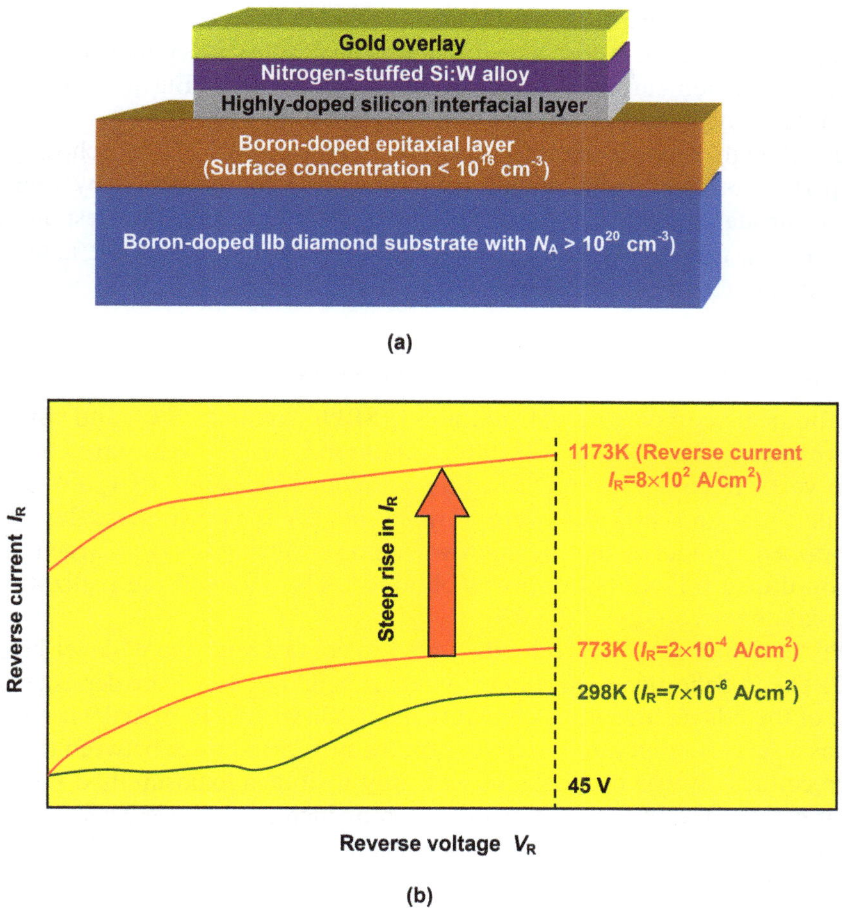

Figure 10.6. Diamond Schottky diode: (a) constituent layers, (b) satisfactory reverse operation up to 773 K and the sharp increase in leakage current above this temperature.

alloy was deposited by ion beam sputtering. It was covered with gold. Up to 773 K, temperature has little effect on the reverse leakage current. Above 773 K, the leakage current increases. It happened because the amorphous Si contact layer re-crystallizes at ~873 K. At still higher temperatures, the Schottky barrier diode (SBD) failed permanently as the W:Si diffusion barrier stopped working.

In order to operate SBDs at higher temperatures, the Au layer was not applied in some diodes. These diodes, fabricated without the Au film, were functional up to 1000 °C (Vescan *et al* 1997a). In these diodes as well, the reverse leakage current was boosted up above a temperature of 500 °C. Nonetheless, the slope of the reverse current was degraded along with decline of breakdown voltage to 30 V only when the diodes were stored at 1000 °C for 15 min before applying the voltage. The on-state current I_{on} was still 10 times the off-state current I_{off} at 1000 °C.

From the Richardson plot, the barrier height was estimated to be 1.9 eV. At low temperatures, the ideality factor departs from the unity value. This shows the presence of current generation mechanisms deviating from thermionic emission. As the temperature increases, the ideality factor approaches unity indicating that the thermionic emission current is the predominant component. Further, as the diodes were un-passivated, surface leakage was apparent above 800 °C.

Thus, at high temperatures, the forward characteristics of the Schottky diode show perfect behavior with an ideality factor of 1.01, governed by thermionic emission through a 1.9 eV high barrier. The reverse characteristics responsible for the low I_{on}/I_{off} ratio point towards the presence of defects, which are activated by temperature (Vescan *et al* 1997a).

10.6.2 Long-term operation of diamond Schottky barrier diodes up to 400 °C

A methodical comparison of diamond SBDs against SiC and Si diodes was presented (Tatsumi *et al* 2009). Diamond SBDs were fabricated in a pseudo-vertical configuration. Starting with an insulating Ib (100) diamond substrate, p^+- and p-layers were formed by CVD. Both these layers were boron-doped. By increasing the microwave power during CVD, a large thickness of boron-doped p-layer (14 μm) was obtained. This enabled the realization of a high breakdown voltage of 2.8 kV.

Ti/Pt/Au ohmic electrodes were deposited to contact with the p^+-layer below the p-layer. The Pt Schottky electrodes with barrier height of 2 eV were deposited on the surface of the p-layer after subjecting it to oxidation treatment to make it insulating. This was a surface preparation step to form the proper interface between the surface and the contact. It was necessary because any hydrogen exposure had the adverse effect of making the surface conducting, thereby increasing the leakage current and reducing the Schottky barrier height. A breakdown field of 3.1×10^6 V cm^{-1} was attained, which was $>2.4 \times 10^6$ V cm^{-1} for SiC p–n diode.

In comparison to Si or SiC, high temperature favorably affects the forward operation of a diamond Schottky diode. This difference originates from the position of impurity energy level in the bandgap. For Si, the energy level of the boron impurity is 44 meV above the valence band. The energy level for boron in SiC is 200 meV and in diamond it is 370 meV. Looking at these energy levels, it is evident that in silicon all impurities are ionized at room temperature and maximum conductivity is obtained at room temperature. As temperature increases, the mobility decreases and hence the conductivity decreases. In SiC, complete ionization of impurity atoms takes place at a higher temperature. Therefore, the temperature of maximum conductivity is higher than that for silicon. For the same reason, the temperature of the maximum conductivity of diamond is even higher than that for SiC. It is in the range of 100 °C–200 °C. For a diamond SBD, the forward current density at 8 V is 3000 A cm^{-2}, which is threefold that available with SiC SBD.

Regarding the reverse characteristics, the leakage current of the SiC Schottky diode was very high at room temperature. It rose to 100 mA cm^{-2} at 162 °C at 1.5×10^6 V cm^{-1}. The leakage current of the diamond Schottky diode was reported to be less than the limit of measurement at room temperature. At 142 °C it was 0.1 mA cm^{-2} (Tatsumi *et al* 2009).

The long-term thermal stability of diodes with Ru Schottky electrodes was judged by keeping the diodes at 400 °C, and measuring the forward and reverse current–voltage characteristics after 100 h, 250 h, 500 h, 1000 h and 1500 h. The characteristics did not show any perceptible changes during this period, confirming that the diamond SBDs provide energy saving and high-power capabilities at 400 °C (Tatsumi *et al* 2009).

10.7 Diamond bipolar junction transistor operating at < 200 °C

Prins (1982) demonstrated bipolar transistor action in natural p-type diamond by implanting carbon to produce n-type regions. A low current gain was achieved. The prospects of bipolar diamond devices were examined by Aleksov and fellow workers (Aleksov *et al* 2000). Heavily doped (1×10^{20} cm^{-3}) p-type boron-doped synthetic diamond crystals were used as the starting material. p^+-emitter (10^{20} cm^{-3}) and p^--collector (10^{17} cm^{-3}) regions were grown using a solid boron source. The base region was formed by nitrogen doping (1.5×10^{18} cm^{-3}). Emitter contacts were made with Ti/Au while base contacts were formed with sputtered phosphorous-doped silicon capped with W–Si/Au. Measurements of the forward and reverse current–voltage characteristics of the diodes from 20 °C to 400 °C showed that the ideality factor was close to unity, pointing towards the domination of the diffusion current component, whereas above 10–15 V, the high value of the current was due to leakage currents arising from the strong electric field, estimated as 5×10^5 V cm^{-1}, in the neutral base. The base has a high resistivity (10 Ω-cm at 20 °C). Further, the reverse-biased junction showed large leakage currents, which together with base resistance limited the operation of diamond bipolar junction transistor (BJT) to the nA range and temperatures <200 °C. The common-emitter current gain β was 1.1. The low gain was interpreted to be caused by high leakage current. It was shown theoretically that nitrogen-doped diamond BJTs can give a DC current gain of ~30 000 upon reduction of leakage current (Aleksov *et al* 2000).

10.8 Diamond metal–semiconductor FET

10.8.1 Hydrogen-terminated diamond metal–semiconductor FETs

Several research groups fabricated diamond metal–semiconductor FETs (MESFETs) on hydrogen-terminated diamond with successively improved maximum drain current, transconductance, switching parameters and other characteristics. Salient features of a few studies are outlined below.

An enhancement-mode MESFET was fabricated using an aluminum Schottky contact on a p-type hydrogen-terminated homoepitaxial layer, and gold for the source and drain contacts (Kawarada *et al* 1994). The aluminum gate length was 10–40 µm. A transconductance g_m = 20–200 µS mm^{-1} was obtained.

Enhancement-mode MESFETs were fabricated by Gluche *et al* 1997 by growing 100 nm thick homoepitaxial diamond films on 1b (nitrogen-doped) single-crystal diamond substrates by microwave plasma-enhanced CVD (MPCVD) at 595 °C using 1.5% CH_4 in H_2 (figure 10.7). As-grown surfaces were hydrogen-terminated

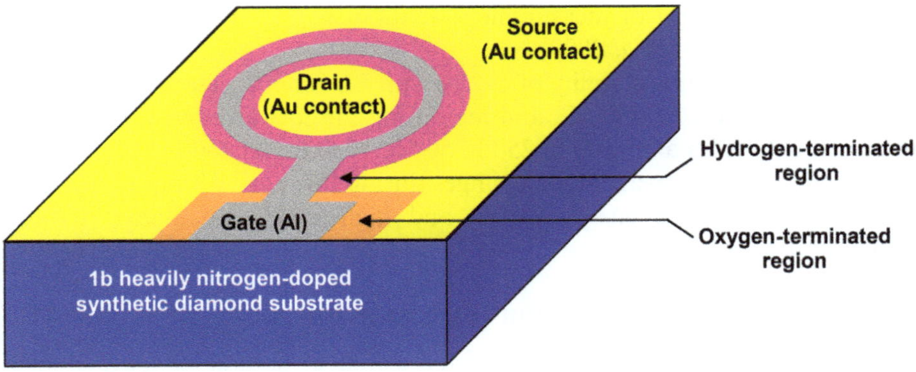

Figure 10.7. Circular diamond surface-channel MESFET structure.

showing p-type conductivity. Al-Schottky and Au ohmic contacts were formed. A channel length of 3 μm gave a maximum drain current of 90 mA mm^{-1} and a 200 V gate–drain breakdown voltage (Gluche *et al* 1997).

Cu-gate diamond MESFETs of gate length 2–3 μm, fabricated on hydrogen-terminated MPCVD-grown diamond, gave a maximum transconductance = 70 mS mm^{-1} at $V_{GS} = -1$ 5 V and $V_{DS} = 5$ V, with a cut-off frequency $f_T = 2 \times 10^9$ Hz and maximum frequency of oscillation $f_{max} = 7$ GHz (Taniuchi *et al* 2001).

Al-Schottky gated MESFETs with Au source/drain contacts and a channel length $L_G = 0.2$ μm gave a maximum drain current $I_{Dmax} = 275$ mA mm^{-1}, at gate voltage = −3.5 V, maximum transconductance $g_m = 100$ mS mm^{-1} and high cut-off frequencies: transition frequency $f_T = 24.6 \times 10^9$ Hz, maximum frequencies $f_{max(MAG)} = 63 \times 10^9$ Hz and $f_{max(U)} = 80 \times 10^9$ Hz, where MAG is the maximum available gain and U is unilateral gain (Kubovic *et al* 2004). The achieved saturated output power density was 0.35 W mm^{-1} at 1 GHz.

Kasu *et al* (2006) reported substantial improvement in RF output power. For a gate width of 1 mm and gate length of 0.4 μm, the maximum output power was 1.26 W. At an output power of 0.84 W, the device temperature rose by merely 0.6 °C. This was enabled by the extremely high thermal conductivity of diamond.

Al-Schottky gate diamond MESFETs of gate length 0.1 μm were fabricated on polycrystalline diamond films (Ueda *et al* 2006). This work deviated from the trend of using single-crystal diamond to explore the polycrystalline material as an alternative (see figure 10.8). This was necessary because the commercial HPHT diamond substrate was 4 mm in size, whereas for mass manufacturing of diamond devices, a minimum of 4″ diameter diamond wafers is required. A quasi-2D hole channel was formed by H-passivation in an MPCVD system. For this device, $f_T \sim 45 \times 10^9$ Hz, $f_{max} \sim 129 \times 10^9$ Hz and $g_m = 143$ mS mm^{-1} at $V_{DS} = -8$ V (Ueda *et al* 2006).

For production of hole carriers and formation of a hole channel on the hydrogen-terminated diamond surface, the presence of some adsorbed molecules is necessary. Significant improvement in the characteristics of diamond transistors was observed on exposure to nitrogen dioxide (NO_2) gas (Kubovic and Kasu 2009). By this

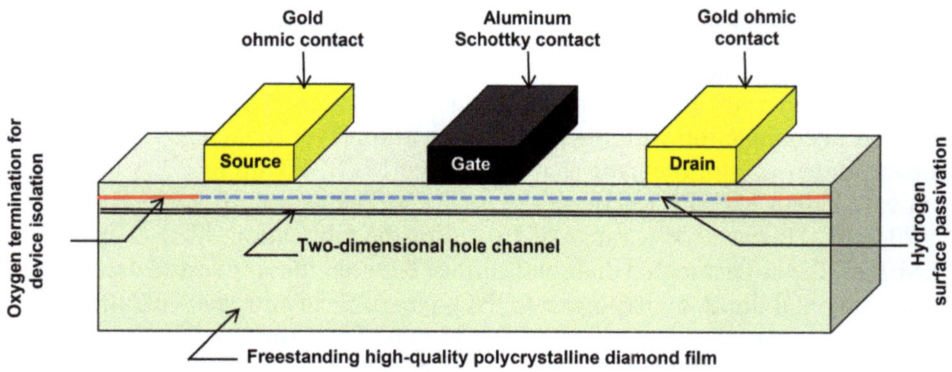

Figure 10.8. MESFET formed on polycrystalline diamond film.

exposure, the hole sheet charge concentration increased to 1.3×10^{14} cm^{-2}, which is much higher than in air. Maximum I_{DS} increased by a multiplication factor of 1.8, g_m by a factor of 1.5 and f_T by 1.6.

Hydrogen-terminated diamond MESFETs with gate lengths of 250, 120 and 50 nm were fabricated on a homoepitaxial diamond substrate (Russell *et al* 2015). For hydrogen termination, the substrate was exposed to hydrogen plasma for 30 min at 580 °C. The gate metal was Al/Au (25 nm/25 nm). The device had Au ohmic contacts on the source and drain. Extrinsic peak transconductance for 50 nm, 120 nm and 250 nm MESFETs were 78, 137 and 92 mS mm^{-1}, respectively, while the extrinsic f_T of respective devices was 53×10^9 Hz, 45×10^9 Hz and 19×10^9 Hz.

10.8.2 Electrical characteristics of diamond MESFETs in 20 °C–100 °C temperature range

The above efforts concentrated on studying diamond MESFETs as a microwave or power device. A microwave power device needs scrupulous attention to thermal stability. The heat produced during device operation may considerably degrade its performance. Therefore, an extensive investigation of temperature effects on DC and RF characteristics of diamond MESFETs was called for (Ye *et al* 2006). This study was performed between 20 °C and 100 °C. At 20 °C, the maximum I_{DS} is 160 mA mm^{-1} which decreased to 120 mA mm^{-1} at 100 °C. The on-resistance was virtually temperature-independent, but the threshold voltage moved towards the negative side. Its value was 0.4 V at 20 °C and 0.18 V at 100 °C. The maximum transconductance showed a decline from 55 mS mm^{-1} at 20 °C to 48 mS mm^{-1} at 100 °C. The transition frequency f_T was 9×10^9 Hz at 20 °C. It decreased marginally to 8.5×10^9 Hz at 100 °C. The sheet resistance of the hydrogen-terminated surface maintained a constant value of 5 kΩ per sq in the range 20 °C–150 °C, showing good thermal stability. On the whole, the device performance was satisfactory between 20 °C and 100 °C with the cut-off frequency nearly constant at $8–9 \times 10^9$ Hz (Ye *et al* 2006).

10.8.3 Hydrogen-terminated diamond MESFETs with a passivation layer

Passivation of the hole channel by an Al_2O_3 layer was found to provide a stable hole channel on the hydrogen-terminated surface up to 200 °C. NO_2 exposure was employed in conjunction with the Al_2O_3 passivation layer to increase the maximum drain current I_{DSmax} of the hydrogen-terminated MESFET up to 1.3 A mm^{-1} for a 1 μm gate length MESFET (Hirama *et al* 2012). The f_T was 10×10^9 Hz and f_{max} was 20×10^9 Hz over a V_{GS} range of 10 V. For the fabrication of Al_2O_3-passivated MESFETs, the H-terminated diamond surface between the source and drain contact regions was stabilized by exposure to NO_2 gas (2% in nitrogen) at atmospheric pressure. The exposure was performed for 35 min. Thereafter, the drain–source gap was covered with the 17 nm thick Al_2O_3 layer formed at 150 °C by atomic layer deposition (ALD).

In addition to Al_2O_3, several other dielectric materials have been used for passivation of diamond FETs, notable examples being AlN, HfO_2, $LaAlO_3$, Ta_2O_5, ZrO_2 and SiN_x (Wang *et al* 2015). The techniques used for this purpose include thermal evaporation, ALD and MOCVD.

10.8.4 Operation of pulse or delta boron-doped diamond MESFETs up to 350 °C

Boron dopant in diamond can be fully activated at room temperature if the peak doping concentration is in the range $\sim 10^{20}$ cm^{-3}. Based on this condition for full activation, it is possible to fabricate an FET working on full charge activation. To achieve full modulation of the channel charge by the gate diode without exceeding the breakdown limit of the material, $\sim 3 \times 10^6$ to 1×10^7 V cm^{-1}, the total channel sheet charge must be $<10^{13}$ cm^{-2}. The restriction applies when the thickness of the doped region is around 1–2 nm. This means that the doping profile required has a high peak concentration and a well-defined, narrow distribution. Such a distribution of dopant is defined by the delta (δ) function, and the semiconductor is said to be δ-doped or pulse-doped. The resultant spiky profile is a δ-doping profile or pulse profile.

For fabrication of a pulse-doped diamond MESFET (figure 10.9), a homoepitaxial diamond film was formed by MPCVD in CH_4 and H_2 plasma at 700 °C on a 1b insulating diamond substrate with nitrogen doping, followed by growth of an undoped 1 μm thick buffer layer (Vescan *et al* 1997b). A boron rod was inserted

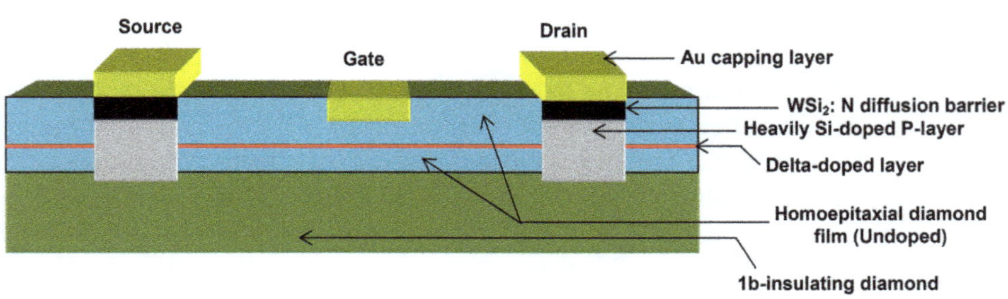

Figure 10.9. A delta-doped diamond MESFET with selectively grown ohmic contact zones.

into the plasma for 5 s to obtain the pulse doping profile with peak concentration 10^{19} cm^{-3}. Selectively grown p^{+}-regions were formed for source and drain. Silicon-based metallization, metallurgically stable up to 750 °C, was employed both for Schottky and ohmic contacts. It consisted of a 50 nm thick contact layer made of highly doped sputter-deposited silicon, a 30 nm thick WSi_2:N diffusion barrier, and finally a 250 nm thick gold capping layer. Operation of the pulse-doped diamond MESFET was demonstrated up to 350 °C. The maximum drain–source voltage was 70 V. The maximum drain current was 35 μA mm^{-1} at room temperature. At 350 °C, it was 5 mA mm^{-1}. The maximum channel conductance was 0.22 mS mm^{-1} at 350 °C (Vescan *et al* 1997b).

10.8.5 Alternative approach to boron δ-doping profile

In another investigation (Aleksov *et al* 1999), a boron δ-doping profile was formed by eliminating any parasitic boron doping tails by compensation of p-type boron doping with n-type nitrogen doping. This leads to the formation of a p–n junction. In this junction, the n-type nitrogen-doped portion is unactivated at room temperature. It behaves as a semi-insulating or lossy dielectric when the temperature is low and the frequency is high. But when the temperature is high and the frequency is low, the nitrogen dopant is activated, rendering the n-type portion conducting. Then this portion acts as a series resistance connected with the p–n junction. Utilizing this concept, two FET structures were studied. In the first structure, the nitrogen-doped diamond substrate served as a back gate. The second structure consisted of a nitrogen-doped gate layer formed on top of the δ-channel. These FETs showed drain currents up to 100 mA mm^{-1}. The channel was fully modulated at temperatures ~200 °C–250 °C.

10.9 Diamond JFET

10.9.1 Diamond JFETs with lateral p–n junctions

A boron-doped p-type channel is sandwiched between two selectively grown highly phosphorous-doped n^{+} sidewall gates (Iwasaki *et al* 2013), see figure 10.10. Control of the depletion region thickness in the channel region is accomplished by applying the same voltage on both gates. The contact metallization scheme is Ti (30 nm)/Pt (30 nm)/Au (100 nm). By controlling the depletion regions from two sides, the channel width and therefore the drain current can be modulated more effectively than from a single side.

10.9.2 Operation of diamond JEFTs up to 723 K

Measurement of the I_{DS}–V_{DS} characteristics at 300 K and 673 K showed that:

(i) clearly distinguishable linear and saturation regions are obtained at both temperatures;
(ii) the current density obtained at 673 K is 1300 A cm^{-2}. It is 50 times the current density at 300 K, which is 25 A cm^{-2}; and

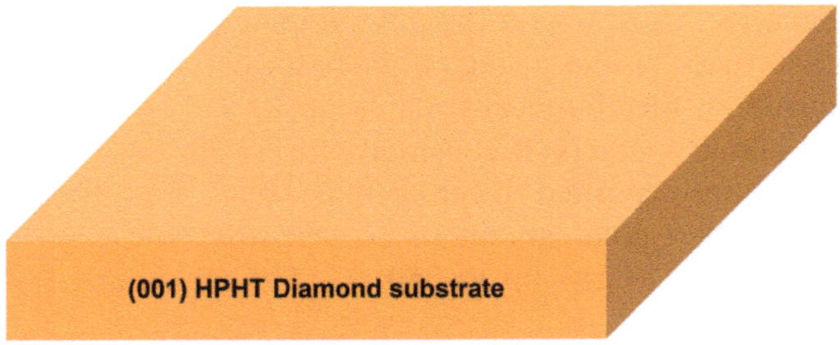

(a) Starting high-pressure high-temperature (HPHT) diamond substrate

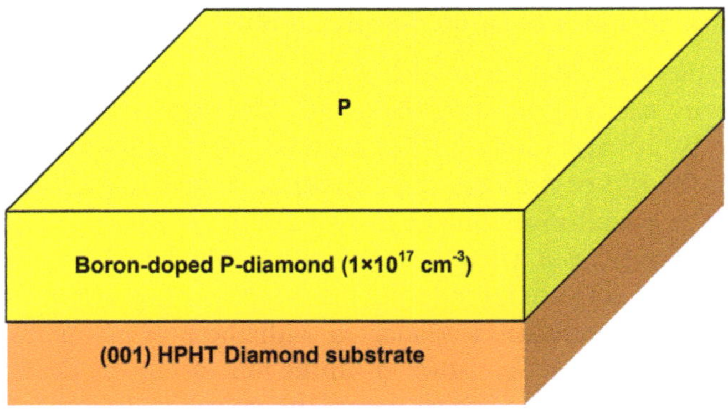

(b) P-layer formation by microwave plasma chemical vapor deposition (MPCVD).

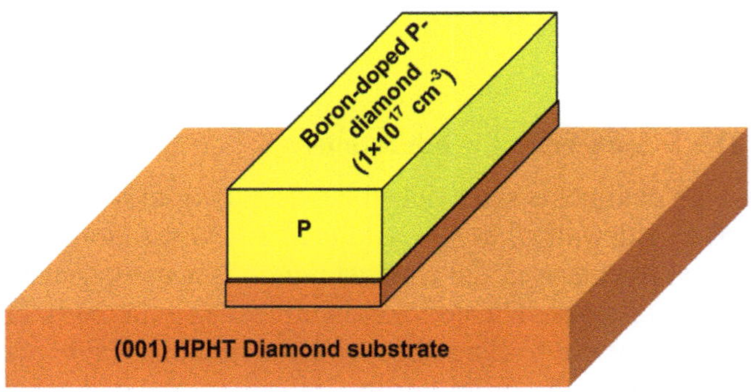

(c) Patterning of the P-layer by electron beam (e-beam) lithography followed by inductively-coupled plasma etching (ICP).

Figure 10.10. Fabrication process steps of diamond JFET with selectively deposited n^+ side gates.

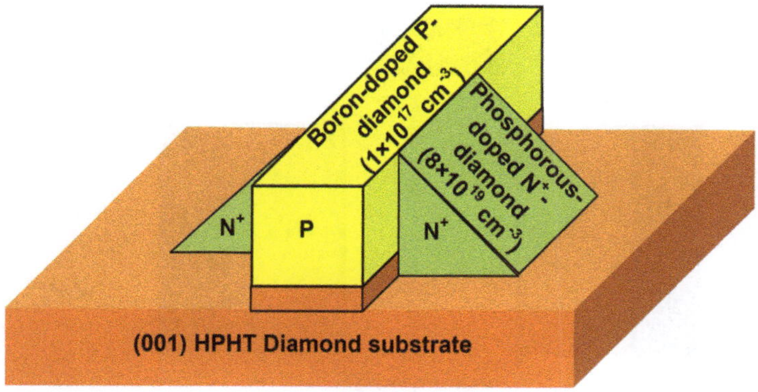

(d) Selectively-grown N⁺ diamond forming a lateral P-N junction with the P-layer. The P-channel is sandwiched between N⁺ side gates from its two sides.

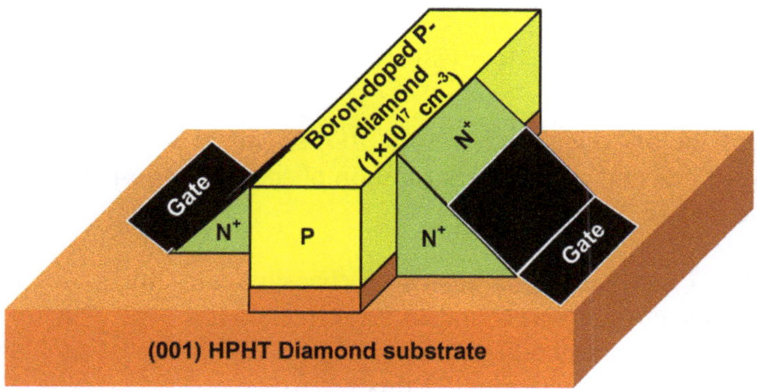

(e) Electrode deposition on the walls of side gates.

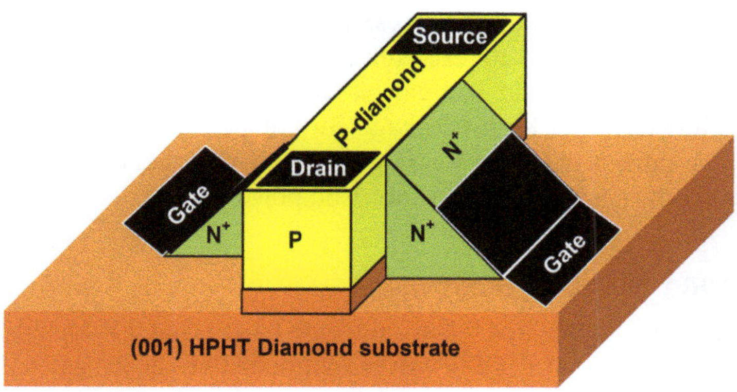

(f) Source and drain contact deposition

Figure 10.10. (Continued.)

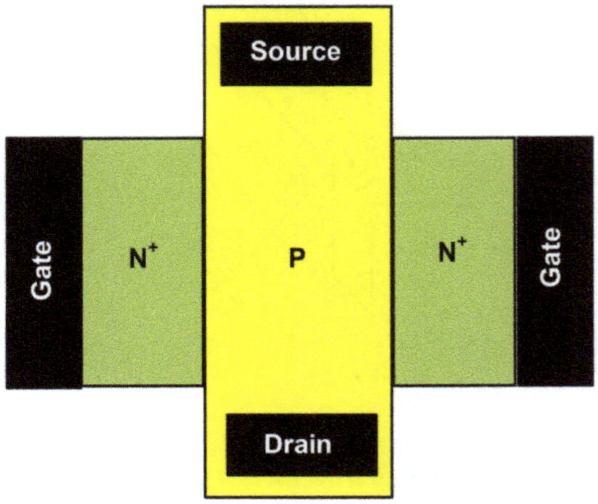

(g) Top view of the fabricated JFET structure showing the P-channel and N+ side gates with contact metallization.

Figure 10.10. (Continued.)

(iii) the specific on-resistance is 52.2×10^{-3} Ω cm^2 at 300 K and plummets to 1.8×10^{-3} Ω cm^2 at 723 K. Saturation of specific on-resistance was noticed above 500 K.

From the I_{DS}–V_{GS} transfer characteristics (Iwasaki *et al* 2013):

(i) the leakage current was found to be extremely low ~10^{-14} A up to 673 K;
(ii) the maximum transconductance at 300 K is 1.6 μS mm^{-1}. At 723 K, it rises to 34.6 μS mm^{-1}, increasing 22 times.

A high ratio between on-state and off-state currents $I_{on}/I_{off} > 10^6$ is maintained from 300 K to 673 K with a peak value of 4×10^7 at 623 K. A very steep subthreshold swing, SS, is observed. It is nearly ideal, and this is particularly true for high temperatures. An interesting achievement is the attainment of high-voltage operation at high temperature with exceedingly low leakage current from the lateral p–n junctions between the drain and gate. From the transfer curves of the JFET (figure 10.11) plotted at 423 K, the leakage current was found to be ~10^{-14} A at $V_{DS} = -10$ V; the same is true even for $V_{DS} = -100$ V (Iwasaki *et al* 2013). The low leakage current confirms its berth as a futuristic power device.

10.10 Diamond MISFET

Matsudaira *et al* (2004) fabricated 0.2–0.5 μm gate length Cu/CaF_2/diamond metal–insulator–semiconductor (MISFETs) with evaporated CAF_2 as the gate insulator on p-type hydrogen-terminated homoepitaxial diamond films deposited on synthetic 1b (001) substrates by MPCVD using H_2-diluted CH_4 source gas (4%). With a 0.2 μm gate length, the highest f_T was 23×10^9 Hz and f_{max} was 25×10^9 Hz.

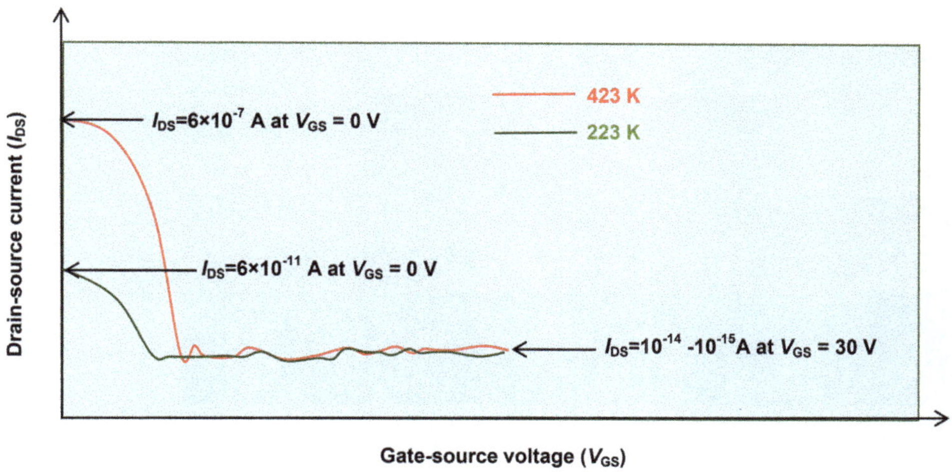

Figure 10.11. Transfer characteristics of diamond JFET at two temperatures at $V_{DS} = -10$ V.

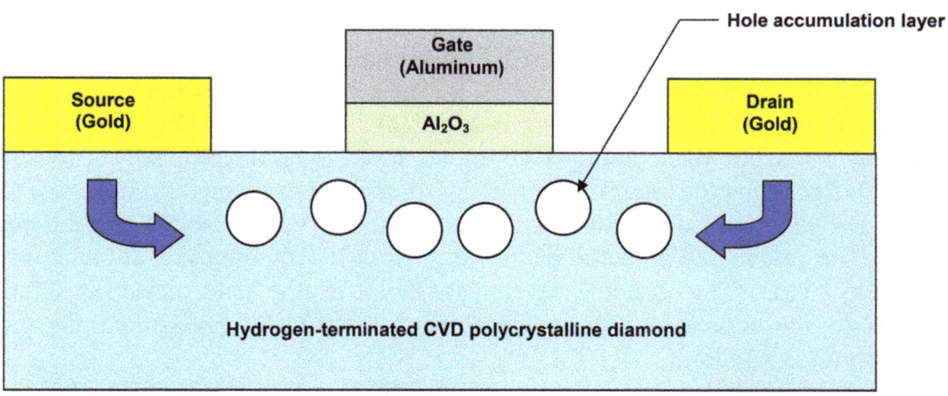

Figure 10.12. MISFET on CVD H-terminated diamond.

High-quality IIa-type polycrystalline CVD diamond has been used as a substrate for fabrication of diamond MISFETs (Hirama *et al* 2007a). A hole accumulation layer was used as the p-channel. Naturally oxidized evaporated Al formed the Al_2O_3 gate insulator. On a 0.1 μm gate length device, f_T was 42×10^9 Hz. The maximum drain current was 650 mA mm^{-1}.

p-channel MOSFETs were fabricated with an Al_2O_3 gate insulator on (001) homoepitaxial and (110) preferentially oriented CVD diamond films of large grain size (figure 10.12). The gate electrode was formed with Al_2O_3/Al by oxidation of 3 nm thick Al film. Ohmic contacts of the drain and source were formed with gold. The devices showed a channel mobility of 20 $cm^2\ V^{-1}\ s^{-1}$, a drain–source current $I_{DS} = -790$ mA mm^{-1}, $f_T = 45 \times 10^9$ Hz and a power density of 2.14 W mm^{-1} at 1 GHz (Hirama *et al* 2007b).

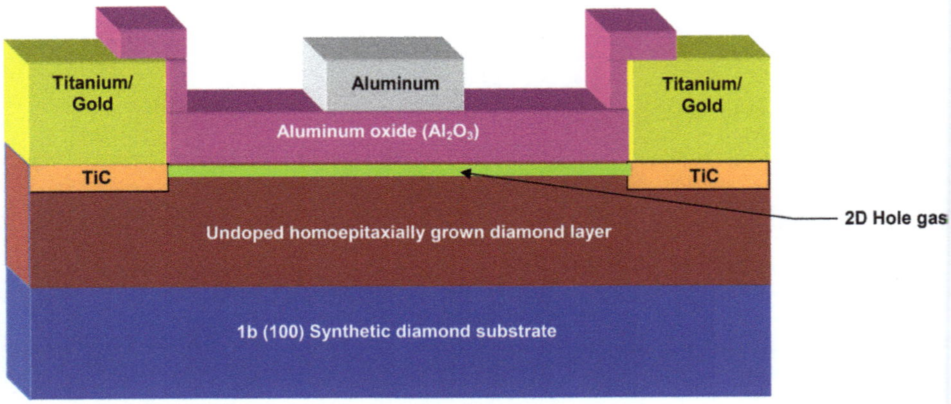

Figure 10.13. Diamond MOSFET with alumina gate dielectric and H-terminated channel.

Let us now look at the studies conducted regarding MISFET operation in the 10–673 K range. Thermally stable two-dimensional hole gas (2DHG) up to 800 K was obtained at Al_2O_3/H-terminated C–H diamond interface using Al_2O_3 formed by high-temperature ALD process. MISFET performance was preserved in the temperature range 10–673 K (Kawarada *et al* 2014). The substrate used is 1b (100) synthetic diamond (figure 10.13). It is nitrogen-doped at $\sim 10^{17}$ cm^{-3}. On this diamond substrate, a 0.5 μm thick undoped diamond layer is homoepitaxially grown by MPCVD. Excluding the regions occupied by source and drain metallization, the top surface is hydrogenated. Oxidation of the surface serves to isolate the different regions. The gate is insulated and passivated with a 10–200 nm thick alumina layer formed by ALD. This alumina layer also serves as the gate dielectric. Ohmic contacts to the source and drain are formed with Ti/Au layers and the gate metallization is Al film.

The I_{DS}–V_{GS} characteristics of a diamond MISFET with a gate insulator thickness of 32 nm were plotted at room temperature, 100 °C, 200 °C, 300 °C and 400 °C. For each case, the on-state current was ~30 A. At room temperature, the off-state current was 8×10^{-8} A, giving a ratio $I_{on}/I_{off} = 30/(8 \times 10^{-8}) = 3.75 \times 10^8$. At 300 °C, the off-state current increased to 6×10^{-4} A. Then the ratio $I_{on}/I_{off} = 30/(6 \times 10^{-4}) = 5 \times 10^4$. At 400 °C, $I_{off} = 2 \times 10^{-2}$ A, so that $I_{on}/I_{off} = 30/(2 \times 10^{-2}) = 1.5 \times 10^3$. For this MISFET, the maximum drain current I_{DSmax} was −23 mA mm^{-1} at 10 K. It was −38 mA mm^{-1} at room temperature and changed to −35 mA mm^{-1} at 400 °C. Hence, the I_{DSmax} at a particular bias varies by <50% from room temperature to 400 °C.

The breakdown voltage was 996 V without any field plate. The maximum breakdown field was 3.6×10^6 V cm^{-1}. In further improvement, the drain current was 100 mA mm^{-1} with a breakdown voltage >1600 V (Kawarada *et al* 2015). The values achieved are superior to those of diamond MESFETs and SiC planar MOSFETs.

10.11 Diamond radiation detectors

Diamond detectors are used in medical physics for accurate dosimetry during cancer radiotherapy, nuclear engineering in reactor containment vessels (requiring capability of withstanding 500 °C operating temperatures), nuclear fusion for plasma diagnostics, and high energy physics in accelerators and synchrotron radiation sources. Diamond radiation detectors can work satisfactorily at temperatures 150 °C–500 °C (Bäni *et al* 2018, Bossini and Minafra 2020, Shimaoka *et al* 2021).

10.11.1 Structural configuration

The most popular configuration of the detector is a two-terminal electronic device in the form of a planar metal–insulator–metal (MIM) structure. It consists of a CVD diamond film sandwiched between metal electrodes on both sides (figure 10.14(a)). A p–n diode or p–i–n diode structure is also used (figure 10.14(b)). Another variant is the stacked structure in which a heavily boron-doped diamond layer is used for contact and a high purity intrinsic diamond layer for drift layer (figure 10.14(c)). A membrane detector version is made by etching to form an extremely thin drift layer (figure 10.14(d)).

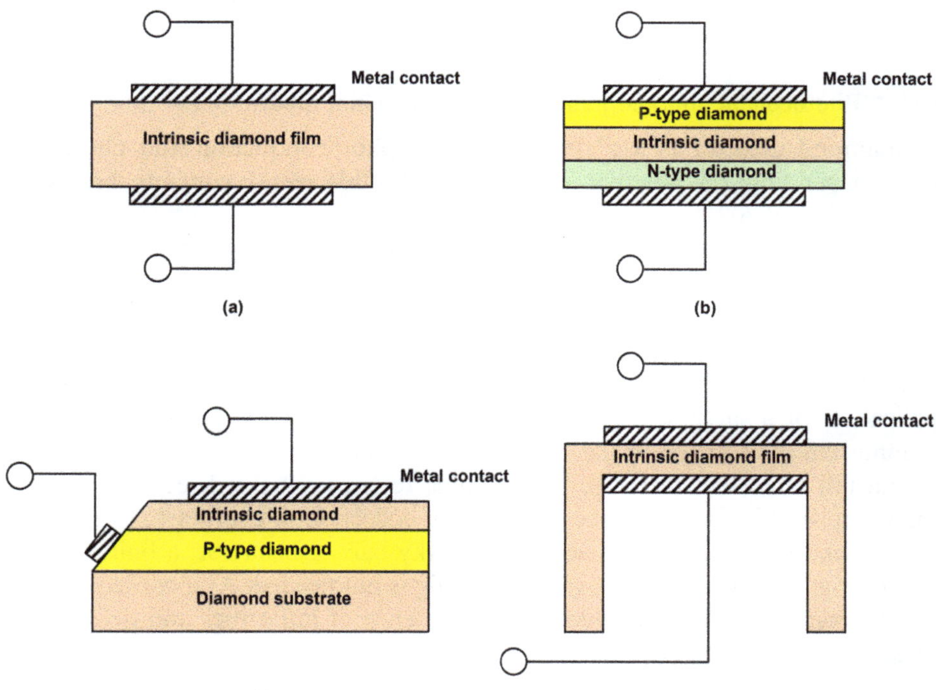

Figure 10.14. Different diamond detector configurations: (a) planar MIM structure, (b) p–i–n diode structure, (c) stacked structure and (d) membrane structure.

10.11.2 Radiation detection principles

Detection of different forms of radiation takes place by the following principles:

(i) The device detects high energy charged particle radiation (alpha particles, electrons, etc) through the ionization caused by them in the diamond film when they travel through it. It behaves as a solid-state ionization chamber.

(ii) It detects electromagnetic radiation with energy >5.5 eV (ultraviolet rays, x-rays and gamma rays). These radiations cause free carrier generation in the diamond (figure 10.15(a)).

(iii) It detects neutrons with the help of a neutron-to-charged particle converter (figure 10.15(b)). The converter has a high absorption cross-section for neutrons. When the neutrons are absorbed in the converter, charged particles are generated. Thermal neutrons are detected through the secondary alpha particles emitted during inelastic neutron reaction in the ^{6}Li absorber leading to ^{6}Li (n, α). Fast neutrons are detected via the inelastic neutron reaction ^{12}C (n, α) ^{9}Be. The released alpha particles give a measure of neutron dose (Angelone and Verona 2021).

(iv) Neutrons and photons also undergo absorption or scattering in diamond. From the absorption or scattering events, secondary charged particles such as electrons, protons or alpha particles are liberated. The charged particles resulting from these events can give an indirect indication of the incident neutron/photon flux.

10.11.3 Photoconduction and photovoltaic operational modes

The diamond detector is used in two modes: photoconduction and photovoltaic modes depending on whether ohmic or Schottky metal contacts are formed (Talamonti *et al* 2021).

In the photoconduction mode based on ohmic behavior, a high voltage in the range of 10–1000 V (~1 V μm^{-1}) is applied across the two faces of the diamond detector, and the detector is exposed to radiation. The electron–hole pairs produced by the impinging radiation in the diamond film drift under an applied electric field causing a flow of current, and the magnitude of current indicates the absorbed radiation dose. Application of a large voltage is necessary because the charges generated in the diamond film are trapped in the defects thereby influencing the output current. The rise and fall times and therefore the dynamic response of the detector is degraded. A high applied bias improves the efficiency of collection of charges.

In the photovoltaic mode based on Schottky barrier behavior, a built-in electric field exists in the structure. So external biasing is not needed. As a consequence, the dynamic response is less degraded in the presence of impurities and defects in the crystalline structure of diamond.

10.11.4 Current and pulse counting modes

A detector yielding a current signal proportional to the intensity of incident radiation is said to be working in the current mode.

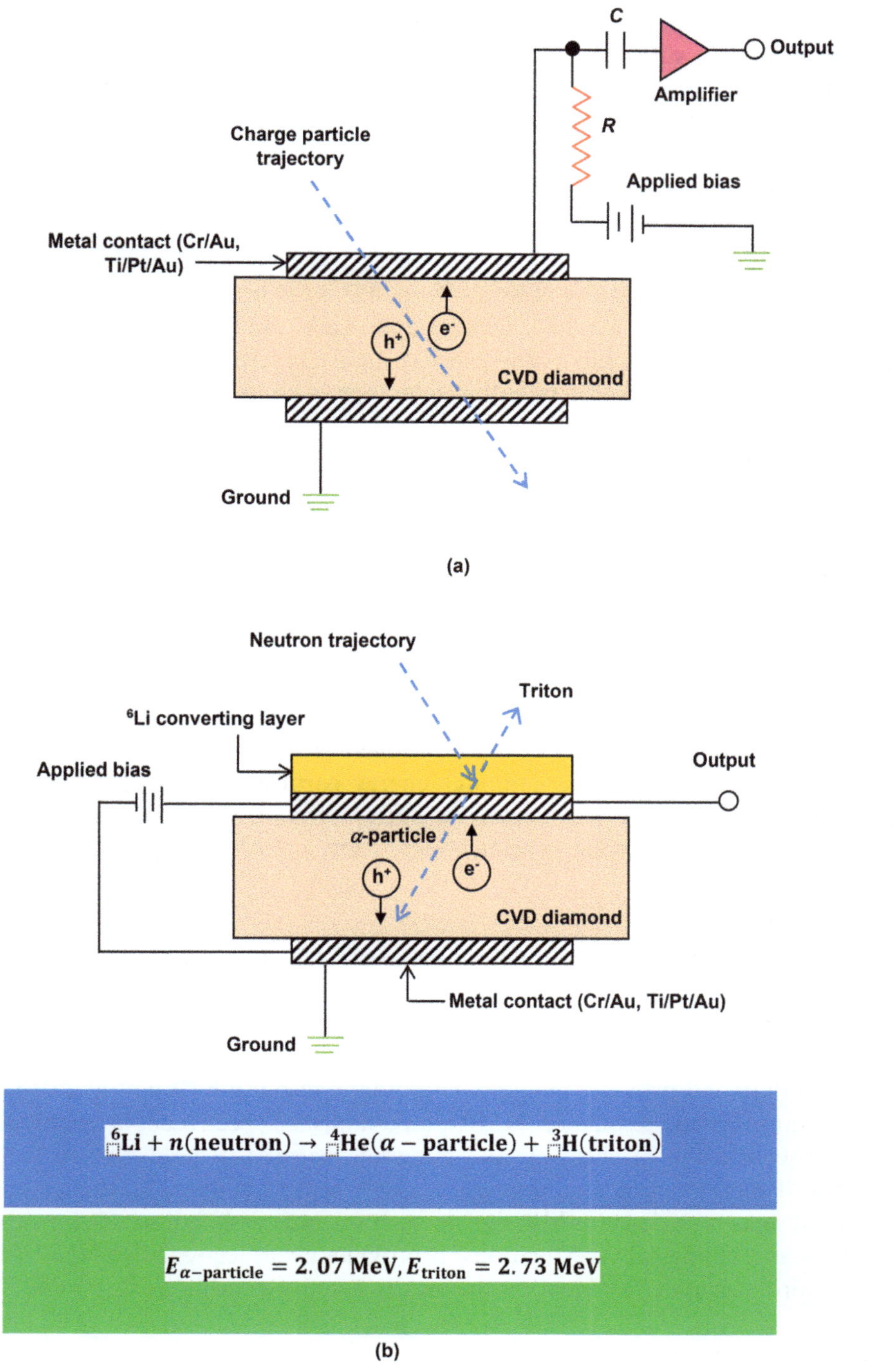

Figure 10.15. Diamond detectors operated in the photoconduction mode: (a) detector for photons with energies exceeding 5.5 eV and charged particles, and (b) detector for neutrons using the ^{6}Li converter.

Pulse counting pertains to individual bursts or quanta of radiation. When an incoming particle strikes the diamond, two kinds of pulses are observed. When a particle moves through the detector, a linear charge density is ionized. A triangular-shaped pulse is obtained. In case, a particle is stopped inside the diamond near an electrode, the pulse is rectangular in shape. So, different particles can be discriminated by performing an analysis of pulse shapes recorded.

10.11.5 Advantages

Diamond radiation detectors offer fast response times in the range of nanoseconds. Their dark current is very low, around a few pA. Due to the large bandgap of diamond, the dark current is low even at high temperatures. Additionally, these detectors exhibit remarkable radiation tolerance (Abbott *et al* 2022).

10.12 Diamond quantum sensors

10.12.1 N-V center in diamond

The nitrogen-vacancy (N-V) center is a point defect in diamond consisting of a nearest neighbor pair of a nitrogen (N) atom and a directly connected adjacent lattice vacancy (V); the nitrogen atom is the replacement for a carbon atom (figure 10.16).

10.12.2 N-V center creation in bulk diamond

Bulk diamond crystals are prepared by synthetic techniques. They contain substitutional nitrogen atoms. Sources of this nitrogen are the solvent, metal and carbon materials. Irradiation is done with high energy particles such as electrons, protons and neutrons to produce vacancies in the lattice. Around 800 °C, the nitrogen atoms create lattice strain capturing vacancies and forming N-V centers (Wu *et al* 2016).

10.12.3 Applications

The N-V center has its own quantum spin which is influenced by magnetic field allowing its utilization for sensor applications (Savage 2021). The N-V center is a quantum sensing tool for monitoring physical parameters such as magnetic field, temperature, strain, etc (Ho *et al* 2021). Spin dependent photoluminescence of N-V center in diamond provides imaging with atomic resolution (Wu *et al* 2016).

Nuclear magnetic resonance (NMR), an established spectroscopic technique used for characterization in chemical and materials sciences is insensitive to the small number of spins at surfaces and interfaces. N-V centers in diamond are exploited as quantum sensors to detect NMR signals from surfaces with femtomole sensitivity in catalysis and biological investigations. It provides a surface-sensitive NMR tool for surface probing under ambient or chemically reactive conditions (Liu *et al* 2022).

10.13 Discussion and conclusions

Diamond provides a multitude of useful physical and chemical properties of which only a small percentage have been utilized. Diamond can be synthesized by CVD. It can be doped as n-type by nitrogen and phosphorous. p-type doping is performed

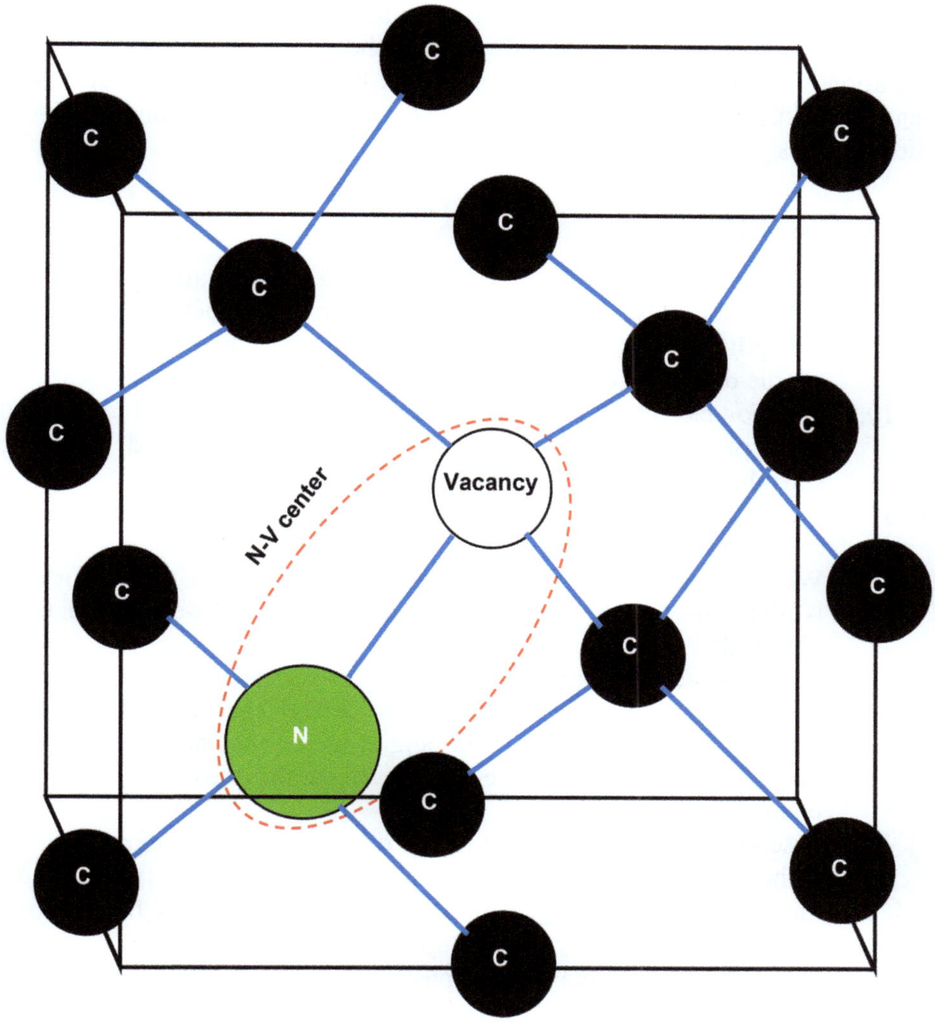

Figure 10.16. Nitrogen-vacancy center in the diamond lattice.

through boron and hydrogen termination. But there are many practical problems. The energy levels of these dopants lie deep inside the bandgap, so they are not activated at room temperature as in silicon. Hydrogen termination is unstable and needs to be properly stabilized and protected. Diamond Schottky diodes have performed well up to 1000 °C. The current gain of BJT must be increased. Boron delta-doped MESFETs have been demonstrated to function up to 350 °C while MESFETs fabricated by hydrogen termination with a passivation layer worked at lower temperatures. Functional JFETs up to 723 K and MISFETs in the temperature range 10 K to 673 K have been reported, asserting their capabilities for high-temperature applications. Some technological challenges have been successfully dealt with, while many others are waiting to be solved if the gigantic opportunities offered by diamond are to be tapped. Despite the tremendous potentialities, the expectations from diamond devices have not yet been met, mainly because of the absence of shallow donor and acceptor species, and a dearth of consistent device models and design strategies (Donato *et al* 2019).

Review exercises

10.1. Compare diamond with Si, GaAs, SiC and GaN regarding its bandgap, carrier mobility, breakdown field, thermal conductivity and fissure strength. Explain with proper justification why diamond is said to be a 'true dream material' of microelectronic device design engineers.

10.2. If silicon electronics is described as attaining maturity and touching the basic physical limits, how would you describe diamond electronics? Calculate the intrinsic temperature of diamond.

10.3. What are the four classes of diamond called? Mention the features of each class.

10.4. What do the terms HPHT diamond and CVD diamond stand for? How is diamond synthesized by CVD? How is gem-quality diamond obtained?

10.5. Why have conventional doping techniques such as thermal diffusion and ion implantation not been able to provide high carrier concentrations in diamond? Name two n-type doping elements that have been used for diamond. What are their energy level depths below the conduction band?

10.6. Discuss the use of boron as a p-type doping impurity in diamond. What is the effect of boron concentration on its activation energy? How does the presence of hydrogen in the plasma affect *in situ* boron doping during CVD?

10.7. How is p-type doping accomplished in diamond by surface hydrogenation? What is the typical depth below the diamond surface within which most of the charge carriers lie? How does this channel containing holes differ from a boron-doped layer?

10.8. Is the p-type diamond surface formed by hydrogenation stable from chemical and temperature viewpoints? How can it be destroyed? How can it be protected?

10.9. Can a hydrogen-terminated diamond surface be used as the channel of an FET?

10.10. How does an oxygen-terminated diamond surface behave? How is oxygen termination of a diamond surface achieved? Can a hydrogen-terminated diamond surface be modified using oxygen termination? What is the result of this modification?

10.11. Describe the fabrication of an UV emitting diamond p–n junction diode using boron and phosphorous as doping impurities. Discuss its optical characteristics.

10.12. Describe one method of fabrication of a diamond Schottky diode using a nitrogen-doped Si:W alloy as the Schottky metal. What is the Schottky barrier height? What is the ideality factor at low temperatures? What is its value at high temperatures? What do these values signify?

10.13. Why is oxygen treatment necessary before deposition of a Schottky contact? What happens if it this step is omitted? Up to what temperature does the Schottky diode perform satisfactorily?

10.14. The energy level for boron impurity in silicon is 44 meV above the valence band. In SiC the energy position for boron is at 200 meV and in diamond it is at 370 meV. Explain how high temperature affects the forward operation of a diamond Schottky diode more favorably than for a SiC diode. Do the same remarks hold true for a comparison between a SiC diode and a Si diode? If so, how?

10.15. Compare the relative magnitudes of reverse leakage current at room temperature and at high temperature in a SiC Schottky diode with a diamond Schottky diode.

10.16. How are emitter, base and collector regions of a diamond BJT formed? What contact materials are used for these regions? What is the common-emitter current gain achieved? Why is it so low? How can the current gain be improved? How much current gain is expected theoretically after these improvements are made?

10.17. Describe the structure and fabrication of a hydrogen-terminated diamond MESFET without any passivation layer. Give typical experimental results on its electrical characteristics. How does the MESFET perform in the 20 °C–100 °C temperature range?

10.18. What is the effect of the Al_2O_3 passivation layer deposition in a hydrogen-terminated MESFET on the hole carrier concentration? How are the RF small-signal characteristics of the MESFET improved by this passivation layer? What other materials are used for passivation apart from Al_2O_3? What deposition techniques are used for preparing these layers?

10.19. What is meant by a delta doping profile? How does it provide full charge activation for the boron dopant at room temperature without exceeding the critical breakdown field of boron?

10.20. How is a delta doping profile of boron achieved by inserting a boron rod into the plasma? How is a diamond MESFET fabricated by this method? What is the maximum drain current of this device at room temperature? What is the drain current value at 350 °C?

10.21. How is the delta doping profile formed by elimination of boron tails by compensating with opposite type of impurity? How is the MESFET fabricated based on this principle? At what temperatures is the channel fully modulated?

10.22. How is the channel doping performed in a diamond JFET? How are the sidewall gates doped? What is the contact metallization scheme used? How does this JFET perform electrically at 300 K and 673 K? What electrical features assure its bright prospects as a diamond microelectronic device for high-temperature operation?

10.23. Describe the fabrication of a diamond MISFET with thermally stable 2DHG up to 800 K. What is the ratio I_{on}/I_{off} at room temperature and at 400 °C? What is the maximum drain current at 10 K and at 400 °C? How does its performance compare with that of a MESFET?

10.24. Mention a few applications of diamond radiation detectors. What are their advantages?

10.25. In what structural configurations are diamond detectors made? Which is the most popular configuration?

10.26. Explain the mechanisms of detection of electromagnetic waves and neutrons with the help of diamond devices.

10.27. What is the difference between photoconduction and photovoltaic operational modes of diamond detectors. Differentiate between their current and pulse counting modes.

10.28. What is a nitrogen-vacancy center in diamond? How is it created? What features enable its utilization as a quantum sensor?

References

Abbott A, Sorensen M, Tobin Z, McIntosh A B, Hannaman A, Gauthier J and Hagel K *et al* 2022 Investigation of sCVD diamond detectors for low energy heavy-ion reactions *Nucl. Instrum. Methods Phys. Res., Sect.* A **1034** 166796

Aleksov A, Denisenko A and Kohn E 2000 Prospects of bipolar diamond devices *Solid-State Electron.* **44** 369–75

Aleksov A, Vescan A, Kunze M, Gluche P, Ebert W, Kohn E, Bergmeier A and Dollinger G 1999 Diamond junction FETs based on δ-doped channels *Diam. Relat. Mater.* **8** 941–5

Angelone M and Verona C 2021 Properties of diamond-based neutron detectors operated in harsh environments *J. Nucl. Eng.* **2** 422–70

Araujo D, Suzuki M, Lloret F, Alba G and Villar P 2021 Diamond for electronics: materials, processing and devices *Materials (Basel)* **14** 7081

Arnault J-C, Saada S and Ralchenko V 2022 Chemical vapor deposition single-crystal diamond: a review *Phys. Status Solidi RRL* **16** 2100354

Bäni L, Alexopoulos A, Artuso M, Bachmair F, Bartosik M, Beacham J and Beck H *et al* 2018 Diamond detectors for high energy physics experiments *J. Instrum.* **13** C01029 11th Int. Conf. on Position Sensitive Detectors (PSD11)

Bossini E and Minafra N 2020 Diamond detectors for timing measurements in high energy physics *Front. Phys.* **8** 248 Sec. Radiation Detectors and Imaging

Bundy F P, Hall H T, Strong H M and Wentorf Jun R H 1955 Man-made diamonds *Nature* **176** 51–5

Cao G Z, Giling L J and Alkemade P F A 1995 Growth of phosphorus and nitrogen co-doped diamond films *Diam. Relat. Mater.* **4** 775–9

Donato N, Rouger N, Pernot J, Longobardi G and Udrea F 2019 Diamond power devices: state of the art, modelling, figures of merit and future perspective *J. Phys. D: Appl. Phys.* **53** 093001

Fontaine F, Uzan-Saguy C, Philosoph B and Kalish R 1996 Boron implantation *in situ* annealing procedure for optimal p-type properties of diamond *Appl. Phys. Lett.* **68** 2264–6

Gluche P, Aleksov A, Vescan A, Ebert W and Kohn E 1997 Diamond surface-channel FET structure with 200 V breakdown voltage *IEEE Electron Device Lett.* **18** 547–9

Han Z, Lee H-P, Bayram B and Bayram C 2022 Diamond power electronics: From 1 kV towards 10 kV breakdown voltage *2022 Compound Semiconductor Week (CSW) (Ann Arbor, MI, 1–3 June 2022)* 1–2

Hasegawa M, Takeuchi D, Yamanaka S, Ogura M, Watanabe H, Kobayashi N, Okushi H and Kajimura K 1999 n-Type control by sulfur ion implantation in homoepitaxial diamond films grown by chemical vapor deposition *Jpn. J. Appl. Phys.* **38** L1519–22

Hirama K, Sato H, Harada Y, Yamamoto H and Kasu M 2012 Diamond field-effect transistors with 1.3 A/mm drain current density by Al_2O_3 passivation layer *Jpn. J. Appl. Phys.* **51** 090112

Hirama K, Takayanagi H, Yamauchi S, Jingu Y, Umezawa H and Kawarada H 2007a Diamond MISFETs fabricated on high quality polycrystalline CVD diamond *Proc. 19th Int. Symp. on Power Semiconductor Devices and ICs (Jeju, Korea, 27–30 May)* 269–72

Hirama K, Takayanagi H, Yamauchi S, Jingu Y, Umezawa H and Kawarada H 2007b High-performance p-channel diamond MOSFETs with alumina gate insulator *IEEE Int. Electron Devices Meeting (Washington, DC, 10–12 December)* 873–6

Ho K O, Shen Y, Pang Y Y, Leung W K, Zhao N and Yang S 2021 Diamond quantum sensors: from physics to applications on condensed matter research *Funct. Diamond* **1** 160–73

Iwasaki T, Hoshino Y, Tsuzuki K, Kato H, Makino T, Ogura M, Takeuchi D, Okushi H, Yamasaki S and Hatano M 2013 High-temperature operation of diamond junction field-effect transistors with lateral p–n junctions *IEEE Electron Device Lett.* **34** 1175–7

Kasu M, Ueda K, Ye H, Yamauchi Y, Sasaki S and Makimoto T 2006 High RF output power for H-terminated diamond FETs *Diam. Relat. Mater.* **15** 783–6

Kato H, Makino T, Yamasaki S and Okushi H 2007 n-Type diamond growth by phosphorus doping *MRS Proc.* **1039** 1039-P05-01

Kawarada H 1996 Hydrogen-terminated diamond surface and interface *Surf. Sci. Rep.* **26** 205–59

Kawarada H, Aoki M and Ito M 1994 Enhancement mode metal semiconductor field effect transistors using homoepitaxial diamonds *Appl. Phys. Lett.* **65** 1563–5

Kawarada H, Yamada T, Xu D, Tsuboi H, Saito T and Hiraiwa A 2014 Wide temperature (10 K–700 K) and high voltage (1000 V) operation of C–H diamond MOSFETs for power electronics application *IEEE Int. Electron Devices Meeting (San Francisco, CA, 15–17 December)* 11.2.1–4

Kawarada H, Yamada T, Xu D, Tsuboi H, Saito T, Kitabayashi Y and Hiraiwa A 2015 Diamond power MOSFETs using 2D hole gas with >1600 V breakdown *6th NIMS/MANA-Waseda University Int. Symp. (Tokyo, 29 July)* 1

Koizumi S 2006 n-Type doping of diamond *MRS Proc.* **956** 0956-J04-01

Koizumi S, Umezawa H, Pernot J and Suzuki M (ed) 2018 *Power Electronics Device Applications of Diamond Semiconductors* (Duxford: Woodhead Publishing, An Imprint of Elsevier)

Koizumi S, Watanabe K, Hasegawa M and Kanda H 2001 Ultraviolet emission from a diamond p–n junction *Science* **292** 18991–901

Kubovic M and Kasu M 2009 Improvement of hydrogen-terminated diamond field effect transistors in nitrogen dioxide atmosphere *Appl. Phys. Express* **2** 086502

Kubovic M, Kasu M, Kallfass I, Neuburger M, Aleksov A, Koley G, Spencer M G and Kohn E 2004 Microwave performance evaluation of diamond surface channel FETs *Diam. Relat. Mater.* **13** 802–7

Liu K S, Henning A, Heindl M W, Allert R D, Bartl J D, Sharp I D, Rizzato R and Bucher D B 2022 Surface NMR using quantum sensors in diamond *Proc. Natl Acad. Sci. USA* **119** e2111607119

Matsudaira H, Miyamoto S, Ishizaka H, Umezawa H and Kaw H 2004 Over 20-GHz cutoff frequency submicrometer-gate diamond MISFETs *IEEE Electron Device Lett.* **25** 480–2

Pan L S and Kania D R (ed) 1995 *Diamond: Electronic Properties and Applications* (New York: Springer Science+Business Media)

Perez G, Maréchal A, Chicot G, Lefranc P, Jeannin P-O, Eon D and Rouger N 2020 Diamond semiconductor performances in power electronics applications *Diam. Relat. Mater.* **110** 108154

Prawer S, Nugent K W and Jamieson D N 1997 The Raman spectrum of amorphous diamond *Diam. Relat. Mater.* **7** 106–10

Prins J F 1982 Bipolar transistor action in ion implanted diamond *Appl. Phys. Lett.* **41** 950–2

Prins J F 1988 Activation of boron-dopant atoms in ion-implanted diamonds *Phys. Rev.* B **38** 5576–84

Russell S, Sharabi S, Tallaire A and Moran D A J 2015 RF operation of hydrogen-terminated diamond field effect transistors: a comparative study *IEEE Trans. Electron Devices* **62** 751–6

Saada D 2000 *p-type Diamond* http://phycomp.technion.ac.il/~david/thesis/node16.html

Savage N 2021 Quantum diamond sensors, Nature Outlines *Nature* **591** S37

Shimaoka T, Koizumia S and Kaneko J H 2021 Recent progress in diamond radiation detectors *Funct. Diamond* **1** 205–20

Shukla H, Rajvanshi T, Gupta T, Kumar V, Gaurav K and Kumar D 2022 Diamond based power electronic devices for aerospace application *AIP Conf. Proc.* **2615** 030006

Talamonti C, Kanxheri K, Pallotta S and Servoli L 2021 Diamond detectors for radiotherapy X-ray small beam dosimetry *Front. Phys.* **9** 632299

Taniuchi H, Umezawa H, Arima T, Tachiki M and Kawarada H 2001 High-frequency performance of diamond field-effect transistor *IEEE Electron Device Lett.* **22** 390–2

Tatsumi N, Ikeda K, Umezawa H and Shikata S 2009 Development of diamond Schottky barrier diode *SEI Tech. Rev.* **68** 54–61

Ueda K, Kasu M, Yamauchi Y, Makimoto T, Schwitters M, Twitchen D J, Scarsbrook G A and Coe S E 2006 Diamond FET using high-quality polycrystalline diamond with f_T of 45 GHz and f_{max} of 120 GHz *IEEE Electron Device Lett.* **27** 570–2

Uzan-Saguy C, Kalish R, Walker R, Jamieson D N and Prawer S 1998 Formation of delta-doped, buried conducting layers in diamond, by high-energy, B-ion implantation *Diam. Relat. Mater.* **7** 1429–32

Vescan A, Daumiller I, Gluche P, Ebert W and Kohn E 1997a Very high temperature operation of diamond Schottky diode *IEEE Electron Device Lett.* **18** 556–8

Vescan A, Gluche P, Ebert W and Kohn E 1997b High-temperature, high-voltage operation of pulse-doped diamond MESFET *IEEE Electron Device Lett.* **18** 222–4

Wang W, Hu C, Li S Y, Li F N, Liu Z C, Wang F, Fu J and Wang H X 2015 Diamond based field-effect transistors of a Zr gate with SiN_x dielectric layers *J. Nanomater.* **2015** 124640

Wu Y, Jelezko F, Plenio M B and Weil T 2016 Diamond quantum devices in biology *Angew. Chem. Int. Ed.* **55** 6586–98

Yamada T, Kojima A, Sawabe A and Suzuki K 2004 Passivation of hydrogen terminated diamond surface conductive layer using hydrogenated amorphous carbon *Diam. Relat. Mater.* **13** 776–9

Yan C-S, Vohra Y K, Mao H-K and Hemley R J 2002 Very high growth rate chemical vapor deposition of single-crystal diamond *Proc. Natl Acad. Sci.* **99** 12523–5

Yang H, Ma Y and Dai Y 2021 Progress of structural and electronic properties of diamond: a mini review *Funct. Diamond* **1** 150–9

Ye H, Kasu M, Ueda K, Yamauchi Y, Maeda N, Sasaki S and Makimoto T 2006 Temperature dependent DC and RF performance of diamond MESFET *Diam. Relat. Mater.* **15** 787–91

Chapter 11

High-temperature passive components, interconnections and packaging

Progress in passive electronic devices for high-temperature electronics (HTE) must proceed hand-in-hand with the active devices using wide bandgap semiconductors. A passive component failure is no less catastrophic than that of an active one. Passive components cannot be ignored because their role is equally important for circuit operation as active components. However, they have hitherto received much less attention. This chapter surveys the passive component aspects of HTE, including resistors, capacitors and inductors. Several new metallization schemes proposed for interconnections are also examined. The packaging and housing needs of HTE are addressed. Most low-cost plastic packages are unable to cope with the thermal challenges, leaving the scenario in favor of metallic hermetic seals.

11.1 Introduction

To assemble an electronic circuit, active devices are the vital components. But together with active devices, several passive devices are required. These include resistors, capacitors, inductors and conductors/interconnections. This chapter will deal with these circuit components, and will also be concerned with packaging of the assembly.

11.2 High-temperature resistors

11.2.1 Metal foil resistors

A metal foil resistor (figure 11.1) consists of a metal alloy, e.g. nichrome with additives cemented to a ceramic substrate by an adhesive. The foil is photo-etched to form a

doi:10.1088/978-0-7503-5072-3ch11

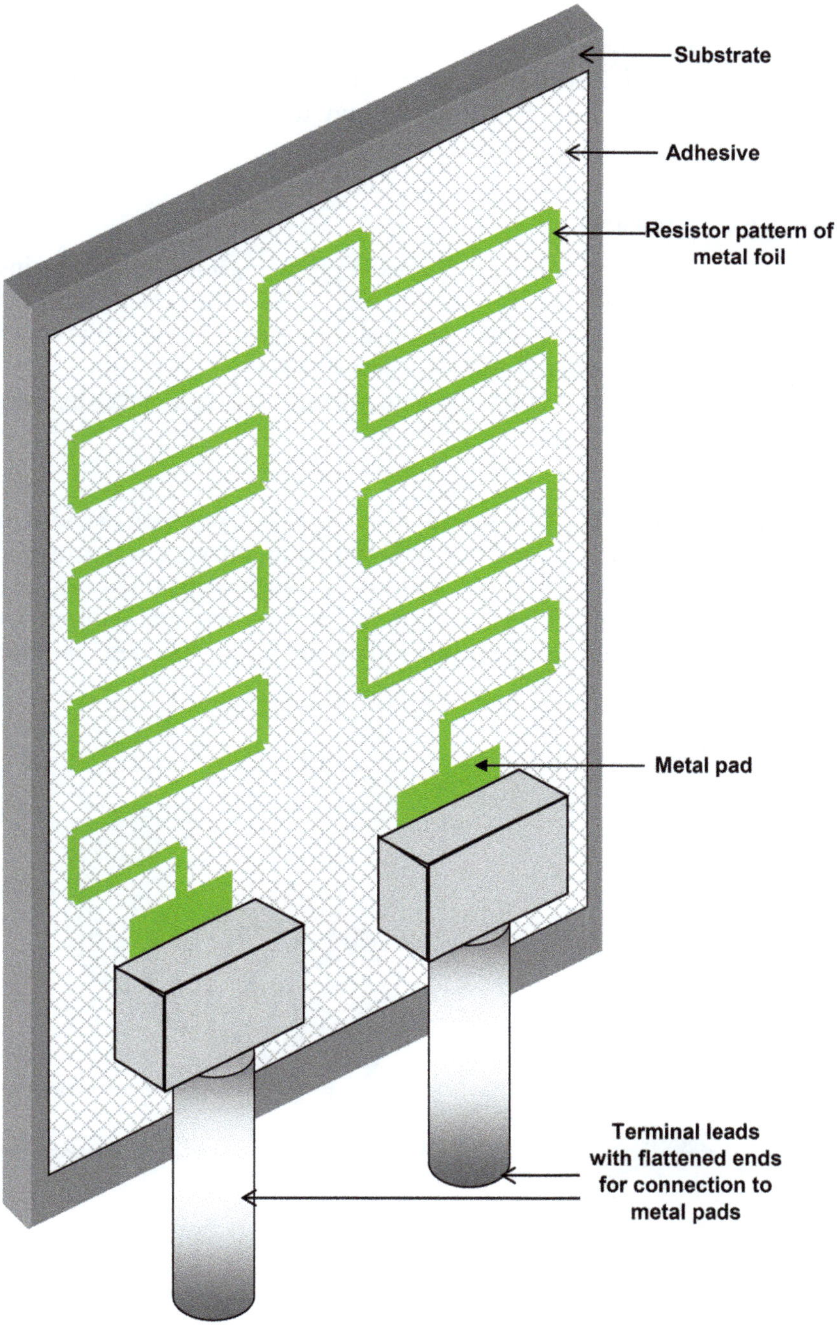

Figure 11.1. Metal foil resistor.

metal pattern. Foil resistors can survive temperatures up to 240 °C (Hernik 2012). A low temperature coefficient of resistance (TCR) is achieved by the balancing effect of the increase in resistance with rising temperature and the decrease in resistance by compressive forces generated in the foil bound to the substrate by its thermal expansion.

11.2.2 Wire wound resistors

A wire wound resistor (figure 11.2) is formed by winding long strands of insulated resistive wire, e.g., 20Cr–80Ni (nichrome) or 86Cu–2Ni–12Mn (manganin) or 72Fe–20Cr–5Al–3Co (kanthal A) or W (tungsten) wire around a non-conducting core such as a ceramic bobbin. Wire wound resistors have been reported to work from −55 °C to 275 °C (Ebbert 2014). Applying derating, they can operate up to still higher temperatures.

11.2.3 Thin-film resistors

A thin-film resistor (figure 11.3) is fabricated by thermal evaporation or sputtering of a metallic film 0.1 μm thick, e.g., NiCr or TaN (tantalum nitride), on a substrate, and then photolithographically etching the film into a resistive metal pattern. A typical example is as follows: film material: nichrome (NiCr); substrate: alumina (Al_2O_3); passivation: silicon nitride (Si_3N_4); protection: epoxy + silicone; terminations:

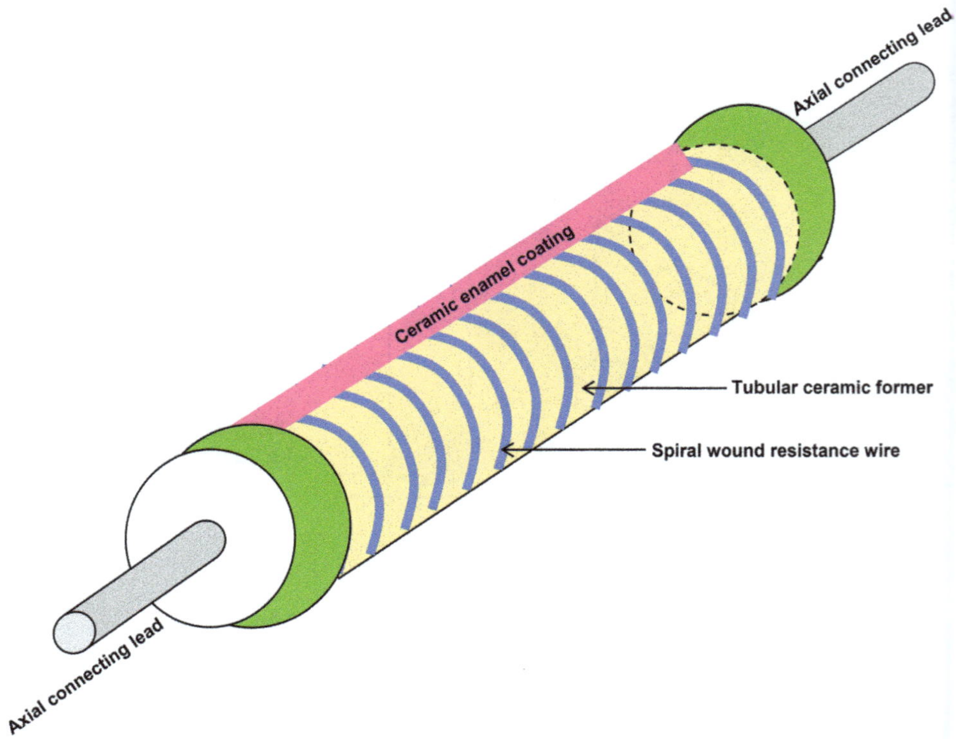

Figure 11.2. Wire wound resistor.

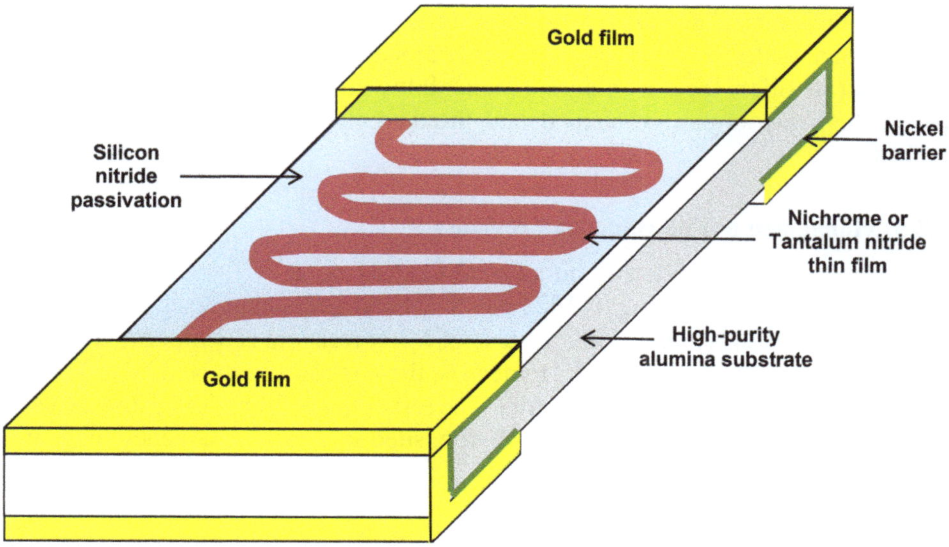

Figure 11.3. Thin-film wrap-around resistor.

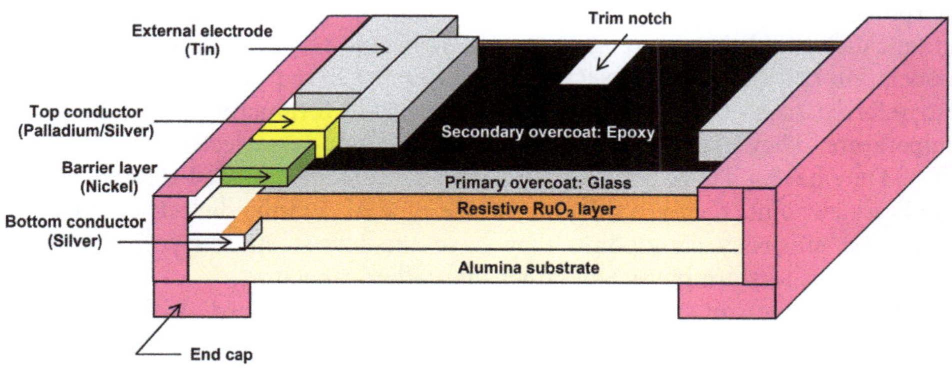

Figure 11.4. Anatomy of a thick-film chip resistor.

gold <1 μm over nickel; operating temperature range: −55 °C to 215 °C; TCR: 15 ppm; structure: wrap-around chip resistor (Vishay 2015).

11.2.4 Thick-film resistors

A thick-film resistor (figure 11.4) is made by screen and stencil printing of a mixture of a glassy frit binder with a carrier of organic solvents and plasticizers, and metal oxides of ruthenium, iridium and rhenium. The resistor film is ~100 μm thick. An example is as follows: film material: ruthenium oxide (RuO_2); substrate: alumina (Al_2O_3); protection: organic; terminations: gold <1 μm over nickel; operating temperature range: −55 °C to 215 °C; structure: chip resistor (Vishay 2014).

Thick-film-on-steel resistors for operation up to 400 °C are made by binding a thick ceramic dielectric glaze on a substrate of stainless steel (Morrison 2000). The steel substrate offers the advantages of bendability into desired shapes, and welding. Hence, the resistors can be shaped and fitted to the dimensions required by the application.

11.2.5 Manganese nitride compound resistors

TCR of manganese nitride compound resistors (54 ppm $°C^{-1}$) is 1/21 times lower than that of aluminum wirings (1162 ppm $°C^{-1}$) and polysilicon resistors (1188 ppm $°C^{-1}$) (Kino *et al* 2021). As the TCR value of MnN resistors is comparable with that of other near-zero TCR (NZ-TCR) materials, these resistors can be used to fabricate high thermal stability integrated circuits. The resistors are fabricated using a CMOS-compatible process. The MnN films are deposited in a radio-frequency magnetron sputtering system.

11.3 High-temperature capacitors

11.3.1 Ceramic capacitors

These capacitors (figure 11.5) comprise a ceramic material as the dielectric interposed between two conducting layers, e.g., silver-plated ceramic discs. The ceramic dielectric belongs to class 1 of the classification by the Electronics Industries Association (EIA). Class 1 dielectrics are of the negative positive zero (NP0) type, also referred to as C0G. They are the most stable formulations with respect to temperature. They are based on para-electric, i.e. non-ferroelectric, materials, e.g. TiO_2. They have a TC of 0 ± 30 ppm from −55 °C to +125 °C. From the thermal stability viewpoint, they are the most stable type of ceramic dielectrics. From the volumetric efficiency standpoint, they have the lowest efficiency due to which ceramic capacitors are physically larger than other capacitors.

For compactness of size, a high dielectric constant is required. A dielectric based on $(1-x)(0.6Bi_{1/2}Na_{1/2}TiO_3-0.4Bi_{1/2}K_{1/2}TiO_3)-xK_{0.5}Na_{0.5}NbO_3$ has been reported

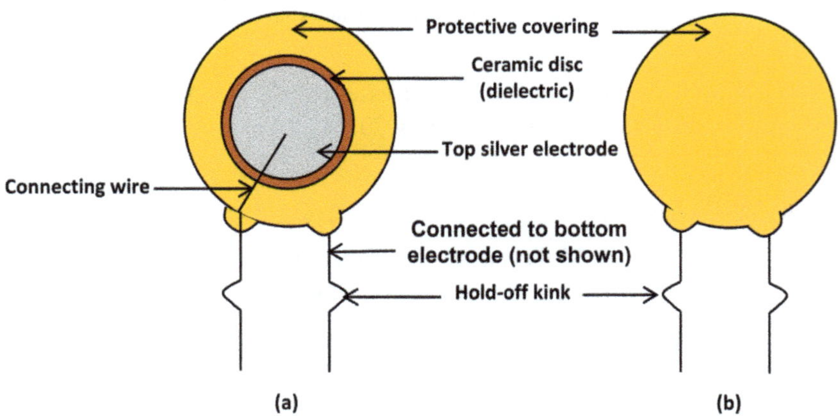

Figure 11.5. Ceramic capacitor: (a) without protective covering and (b) with protective covering.

(Dittmer *et al* 2012) to show a deviation of ±10% in relative permittivity of 2167 in the temperature range 54 °C to 400 °C for $x = 0.15$. The temperature insensitivity of this dielectric was ascribed to the existence of two types of polar nanoregions providing dissimilar relaxation mechanisms.

11.3.2 Solid and wet tantalum capacitors

Tantalum capacitors are of two types: solid and wet. These capacitors are electrolytic capacitors made with a tantalum anode.

In a solid tantalum capacitor (figure 11.6), the Ta anode is formed by compressing Ta powder around a Ta wire called the riser wire and sintering it at 1500 °C to 1800 °C for pellet formation. The tantalum pellet is oxidized by anodization to tantalum pentoxide:

$$2Ta \rightarrow 2Ta^{5+} + 10e^{-}; 2Ta^{5+} + 10OH^{-} \rightarrow Ta_2O_5 + 5H_2O. \quad (11.1)$$

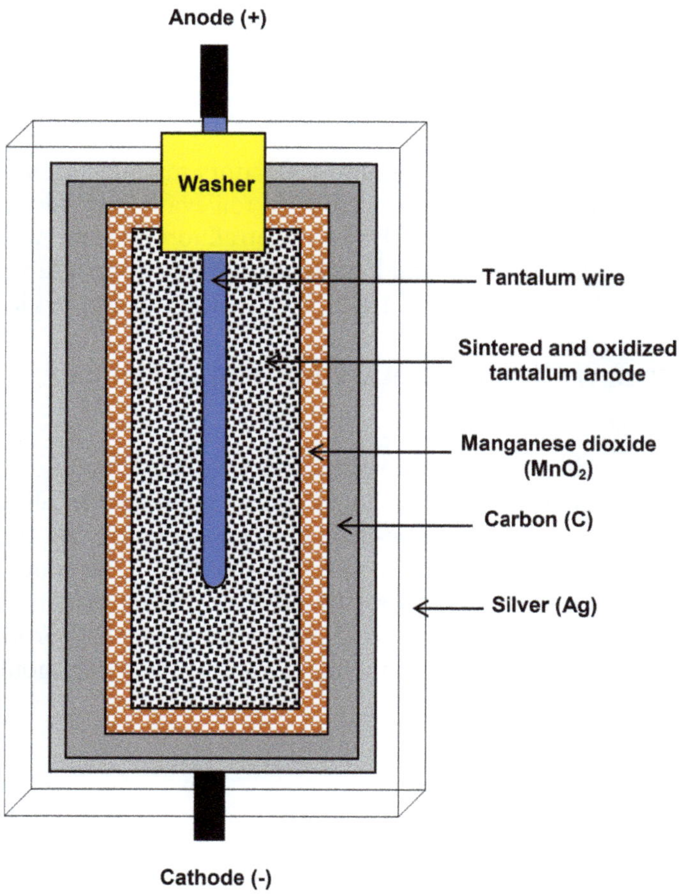

Figure 11.6. Solid tantalum capacitor.

The tantalum pentoxide, thus electrochemically formed, has a dielectric constant of 26, providing high capacitance per unit volume. By immersion of the Ta_2O_5 grown pellet into manganese nitrate aqueous solution, and baking it at 250 °C, it is coated with manganese dioxide electrolyte via a pyrolysis reaction:

$$Mn(NO_3)_2 \rightarrow MnO_2 + 2NO_2. \tag{11.2}$$

Dipping it successively into graphite and silver forms the silver cathode termination while a tinned nickel anode lead is attached to the Ta riser wire for the external anode termination. For high-temperature operation >200 °C, the silver in the top coating is substituted by electroplated nickel (Freeman *et al* 2013).

In a wet tantalum capacitor (figure 11.7), the electrolyte is non-solid. The advantage derived by keeping the electrolyte in the wet condition is that it becomes possible to supply oxygen in weaker portions of the dielectric film. Thus, an improved oxide layer can be formed in these regions. This facility helps in safely forming very thin tantalum pentoxide films, enabling the realization of very high capacitance values per unit volume with extremely low leakage current, permitting high-voltage operation. Additionally, the silver migration problems affecting the silver case led to the use of the sintered tantalum encasing. The electrolyte is gelled sulfuric acid. It is filled in a case of pure tantalum serving as the cathode.

11.3.3 Teflon capacitors

These capacitors (figure 11.8) are film/foil capacitors manufactured by covering a polytetrafluoroethylene (PTFE) film/foil with Al foil/evaporated film on both sides. PTFE is a chemically inert fluoropolymer of tetrafluoroethylene with a dielectric constant of 2.1, low dielectric loss and high insulation resistance, usable up to 250 °C (Bauchman 2011). Teflon capacitors are used in critical mission mode applications.

11.4 High-temperature magnetic cores and inductors

11.4.1 Magnetic cores

Powder cores for elevated temperature operation are made from alloys of Ni and Fe in various compositions (Magnetics 2016). After grinding the alloys into small particles, they are coated with a non-magnetic, non-organic, and non-conducting material. The powder thus obtained is annealed at 500 °C. The annealed powder is pressed to form toroids of the desired size (figure 11.9). The stresses in the toroids are relieved by annealing at temperatures >500 °C. The annealed toroids are coated with epoxy paint. During operation, the maximum temperature should not exceed 200 °C.

11.4.2 Inductors

11.4.2.1 Wire wound inductors

Surface mount device (SMD) power inductors in a wire wound configuration for operation from −40 °C to +150 °C are made by winding a coil around a ferrite drum core (TDK 2014); see figure 11.10. Magnetic coupling with neighboring electronic

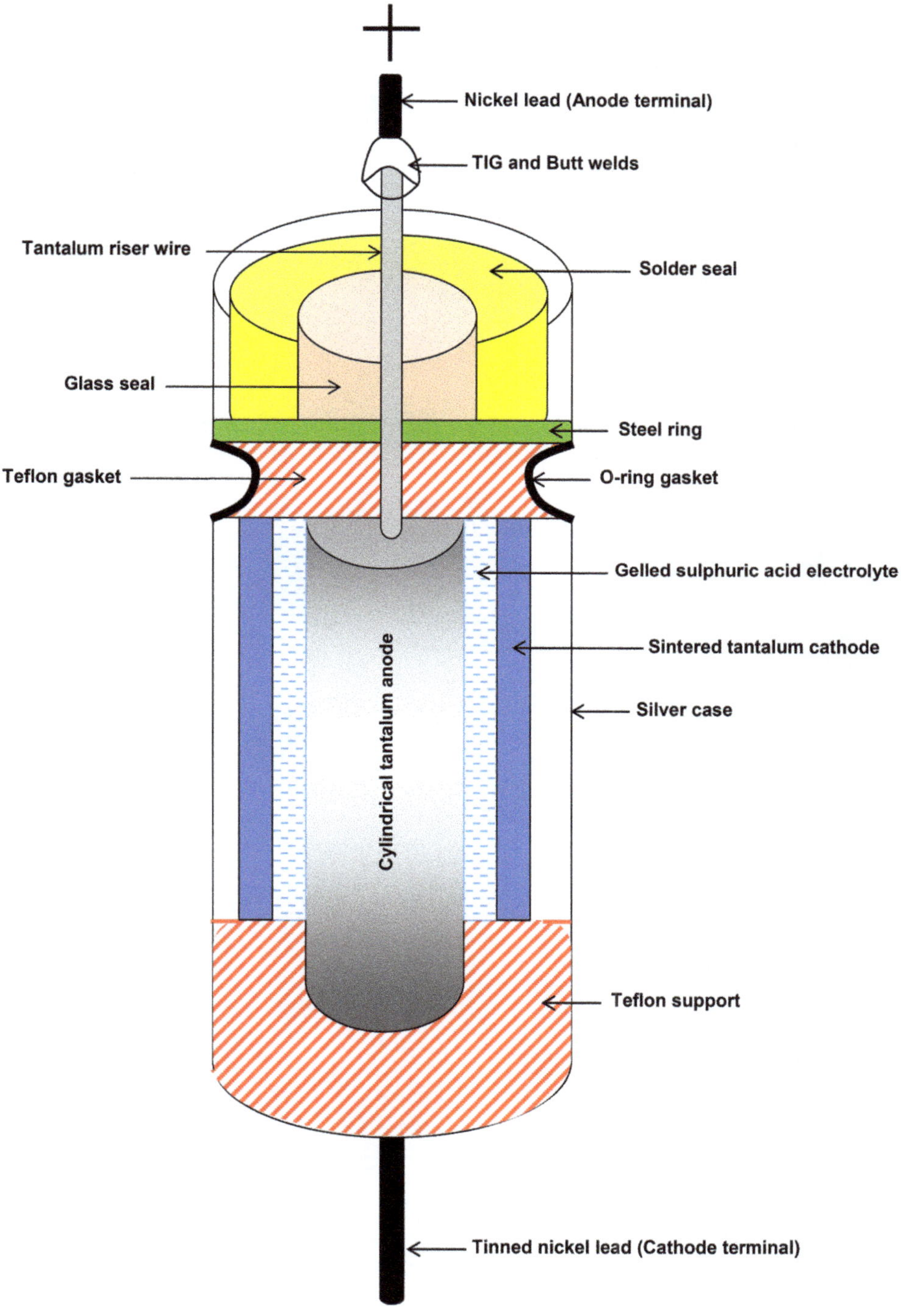

Figure 11.7. Wet tantalum capacitor.

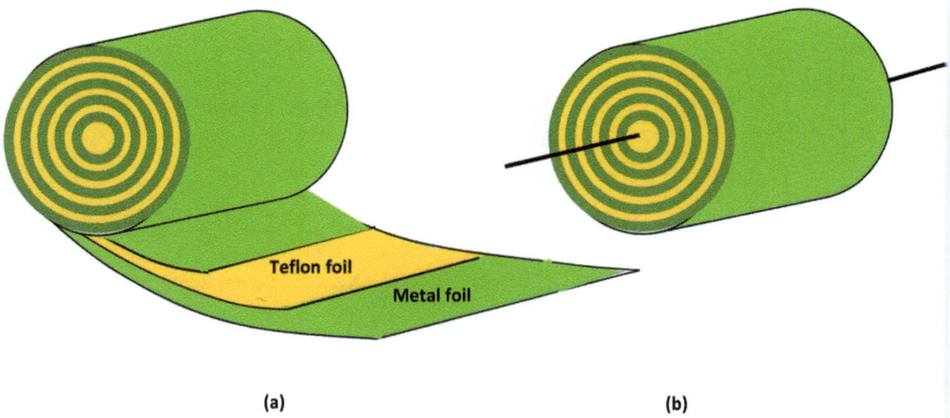

Figure 11.8. Teflon capacitor: (a) winding of the teflon/metal foils and (b) lead fixation.

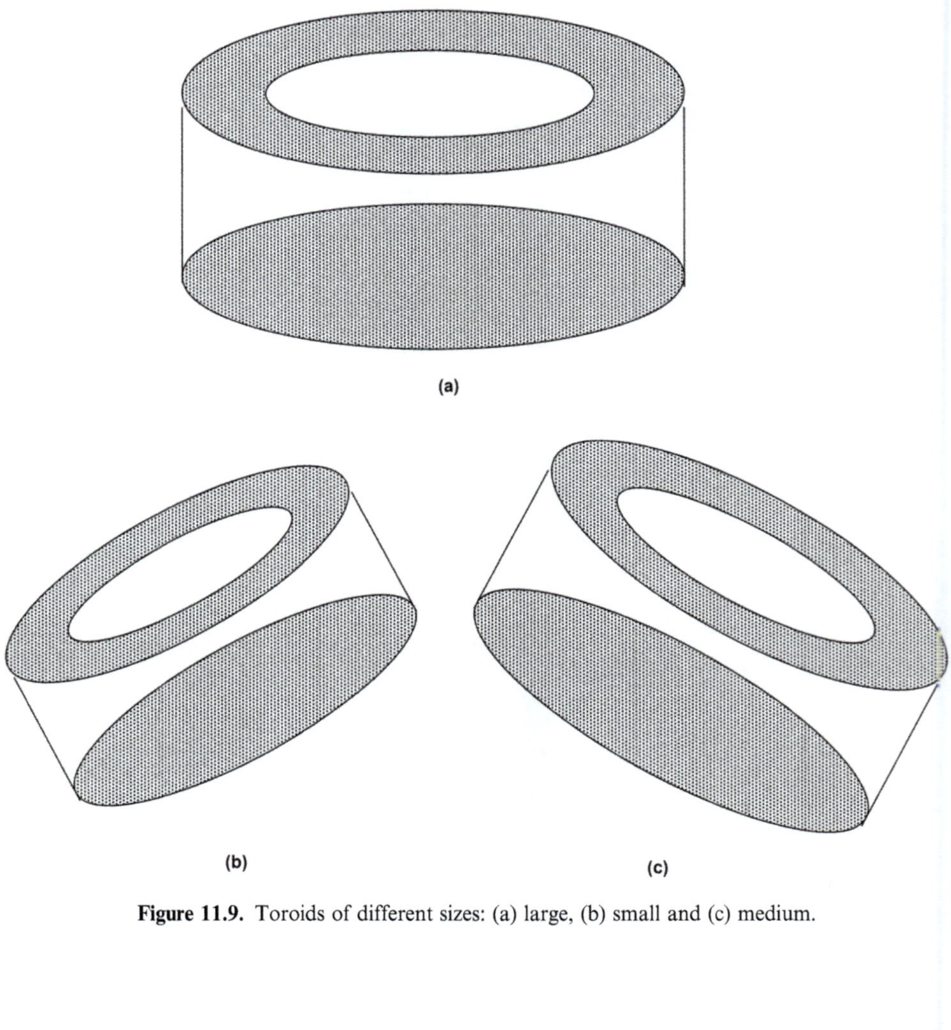

Figure 11.9. Toroids of different sizes: (a) large, (b) small and (c) medium.

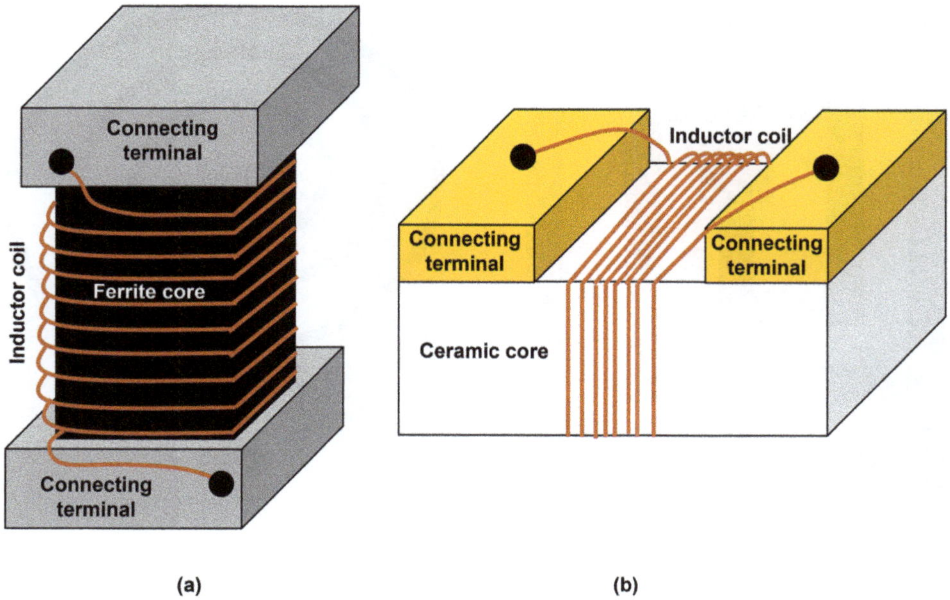

Figure 11.10. Wire wound SMD chip inductor with: (a) a ferrite core and (b) ceramic core.

components is secured by an octagonal ferrite core. Interference and power losses are thereby prevented.

11.4.2.2 Micro-inductors

Planar micro-inductors (figure 11.11) for operation up to 200 °C are fabricated using yttrium iron garnet (YIG) magnetic layers (Haddad *et al* 2013) by the following sequence of steps:

(i) Piercing 500 μm diameter holes in a 1 mm thick YIG layer in alignment with the terminals of the inductor.
(ii) Stuffing the holes with copper and removing the superfluous copper by polishing to achieve surface flatness.
(iii) Depositing a Ti/Cu seed layer on the substrate by e-beam evaporation.
(iv) Laminating and patterning a thick layer of dry film photoresist.
(v) Electroplating the copper windings using the photoresist as a mold.
(vi) Removing the dry film photoresist with NaOH and wet etching the Ti–Cu seed layer for separation of copper turns.
(vii) Polishing the copper windings and covering them with a thick bismaleimide (BMI) resin.
(viii) Adding another YIG magnetic layer and clasping the inductor to bond the two YIG layers with BMI resin.

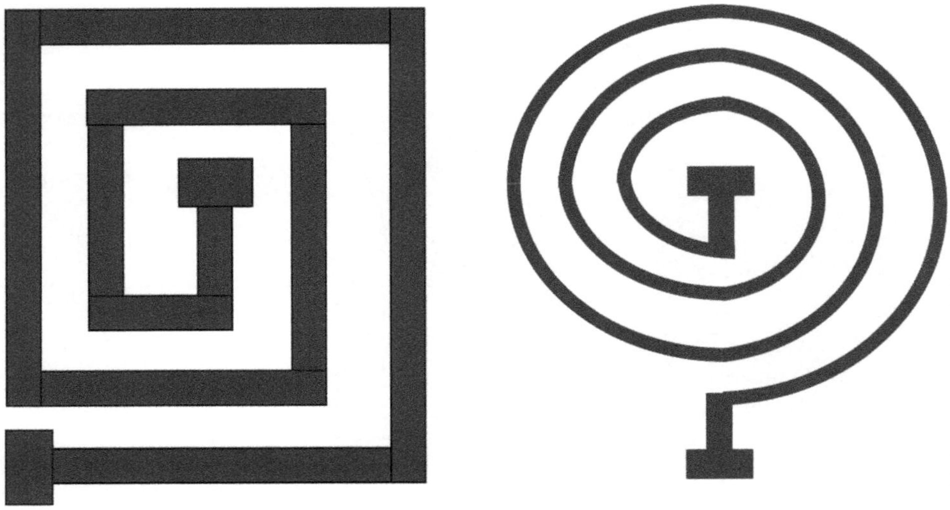

Figure 11.11. Planar micro-inductor layouts: (a) rectangular spiral and (b) circular spiral.

11.5 High-temperature metallization

The adjunct metallization schemes for high-temperature electronic devices should be on par with the thermal capabilities of the devices.

11.5.1 Tungsten metallization on silicon

To raise the high-temperature limit of metallization on silicon to temperatures beyond 450 °C, the refractory metal tungsten is a favorable choice because of its matching coefficient of thermal expansion (CTE) with silicon. The obstacle to be overcome is the formation of tungsten discilicide (WSi_2) by reaction between tungsten and silicon at 600 °C. The formation of WSi_2 layer can be prevented by providing a diffusion barrier on Si. A $TiSi_2$/TiN layer, obtained by rapid thermal annealing (RTA) in nitrogen, serves as an effective diffusion barrier and adhesion layer (Chen and Colinge 1995, 1996, Madou 2002). The $TiSi_2$ layer is formed at the Ti–Si interface and TiN layer on the exposed surface of the Ti film.

11.5.2 Tungsten: nickel metallization on nitrogen-doped homoepitaxial layers on p-type 4H- and 6H-SiC substrates

An n-type ohmic contact was formed on silicon carbide by sputtering a tungsten: nickel layer containing 75 atomic % tungsten and 25 atomic % nickel (Evans *et al* 2012). The thickness of the W:Ni layer was 100 nm. The composite layer was covered with 20 nm thick silicon film. The role of silicon film was to prevent premature oxidation of tungsten, which occurs during venting of the chamber and annealing. The W:Ni metallization performed well at high temperatures. It survived at 1000 °C in an argon atmosphere for 15 h. This metallization scheme is expected to provide reliable operation at 600 °C.

11.5.3 Nickel metallization on n-type 4H-SiC and Ni/Ti/Al metallization on p-type 4H-SiC

The metal deposition was performed by e-beam evaporation and sputtering (Smedfors *et al* 2014, Smedfors 2014). RTA of the contacts was performed in Ar or N_2 ambient. The metallizations were tested in the temperature range from −40 °C to +500 °C. The Ni/Ti/Al ohmic contacts on p-type material showed an increase in specific contact resistivity at −40 °C by a factor of 5 with respect to the room temperature value (6.75×10^{-4} Ω-cm^2 at 25 °C). They exhibited a tenfold decrease in the specific contact resistivity at 500 °C with reference to the 25 °C value. The Ni ohmic contacts on n-type material showed improvement with an increase in temperature. The variation in specific contact resistivity was comparatively smaller.

11.5.4 A thick-film Au interconnection system on alumina and aluminum nitride ceramic substrates

Au thick-film was screen printed on a ceramic substrate. The thick Au film proved to be a low resistance and stable substrate metallization through a 1500 h exposure to atmospheric oxygen at 500 °C (Chen *et al* 2001). The metal film was tested in both unbiased and biased electrical conditions (50 mA DC). The resistance varied by ~0.1% in 1000 h. The shear strength of Au film on an Al_2O_3 substrate at 500 °C was reported to decrease by a fraction 0.8 of its value at 350 °C. But the shear strength value at 350 °C was approximately the same as at room temperature. Thermal cycling tests were performed on 44 gold thick-film to gold wire bonds at a temperature rate of 32 °C min^{-1} for 120 cycles and at 53 °C min^{-1} for 100 more cycles in an oxidizing ambient under an electrical bias condition. The survival of the bonds indicated good bond strength.

11.6 High-temperature packaging

Electronic packaging is a functional link between two vital segments of an electronic system. On one side are the delicate electronic devices and on the other side is the remaining system incorporating a variety of operations such as die attachment, wire bonding, passivation, interconnections between the components and housing of the complete assembly inside a safe enclosure that protects it from mechanical disturbances and the surroundings. High-temperature electronic packaging differs from the conventional moderate temperature packaging catering to room temperature conditions with small excursions, primarily in the exclusion of materials and related processes, producing components that cannot withstand the high temperatures, and inclusion of any new materials with the demonstrated capability to tolerate the same.

Advantages offered by wide bandgap semiconductors for extreme temperature operation can be fully exploited by development of suitable packages, e.g., packaging challenges confronting reliable operation of SiC MOSFET devices demonstrated to be working up to 500 °C are overcome through a robust hermetic package design (Yin *et al* 2015).

From such considerations, it is necessary to re-examine the list of commonly used materials starting from substrate selection, and then moving to die attachment, wire bonding and further, and discard those which do not fulfill the expectations, and add any novel material that promises high-temperature forbearance.

11.6.1 Substrates

Ceramic substrates are ideal for HTE. Among the competing materials—alumina (Al_2O_3), aluminum nitride (AlN), boron nitride (BN) and silicon nitride (Si_3N_4)—aluminum nitride stands out supreme (Chasserio *et al* 2009). In addition to very high thermal conductivity (175 W mK^{-1} against 28.1 W mK^{-1} for alumina), it offers good chemical stability, dielectric strength and stability, flexural strength, a matching CTE with Si and SiC, and good thermal shock resistance. The matching of the thermal expansion coefficient assures less stress creation, both on the device and at the solder joints during thermal cycling. However, it suffers from a cost disadvantage, where alumina is distinctively better. Alumina is inferior to aluminum nitride from the point of view of thermal expansion coefficient mismatch, thermal shock resistance and thermal conductivity. Silicon nitride provides very high flexural strength and a good thermal coefficient matching with that of Si and SiC. Together with these qualities, it has good thermal shock resistance and chemical stability, but it suffers from being unacceptably costly. Boron nitride fails on many parameters, particularly on flexural strength, and is eliminated from the contest.

11.6.2 Die-attach materials

Examples of lead-based materials are (Manikam and Cheong 2011) Pb95–Sn5 and Pb97.5–Ag1.5–Sn1 with liquidus temperatures of 312 °C and 309 °C, respectively. Some lead-free gold-based solutions are Au100, AuNi18, Au thick-film paste and AuGe12, with maximum operating temperatures of 1063 °C, 950 °C, >600 °C and 356 °C. Silver nanoparticle paste was sintered at a temperature <350 °C (Guo *et al* 2015). The sintering joint showed good electrical and thermal characteristics with a melting point near that of pure silver (960 °C), and is useful for high-temperature operation >350 °C.

11.6.3 Wire bonding

Due to the intermetallic compound formation between Al–Au and Al–Cu with the resultant decrease in bond strength at temperatures <200 °C, aluminum wires bonded to nickel surfaces serve as an alternative to the above combinations (Barlow and Elshabini 2006). They do not degrade below 350 °C by intermetallic formation.

A comprehensive investigation of nickel wire (25 μm) bonding to nickel metallization pads (750 nm) on a 3C-SiC substrate is described by Burla *et al* (2009) using the ultrasonic technique. The wire bonds provided a pull strength of

13.1 gf (gram-force), about 4.4 times stronger than for Au wire bonds and 5.2 times stronger than for Al wire bonds. The wire bonds were mechanically and electrically tested up to 550 °C. Nickel wire bonds exhibited good performance on a chemical sensor (up to 280 °C) and a resonant device (up to 950 °C).

Problems of weak wire bonds and mismatching of the CTEs between the bond wires and devices are avoidable by attaching a CTE-matched cover as the top contact over the semiconductor device using spherical bumps of a high-temperature soft solder in place of the wires (Barlow and Elshabini 2006).

11.6.4 Hermetic packaging

The hermetic package provides an airtight connection between the semiconductor device(s) inside and the wiring outside. Hermetic packages are based on glass-to-metal and ceramic-to-metal seal technologies.

11.6.4.1 Glass-to-metal seal packages

Glass-to-metal seals are of two types:

(i) *Matched seal.* Here, the CTEs of the metal and glass are identical. The sealing takes place through a chemical reaction between the oxide film on the metal surface and the glass. This type of seal is weak.

(ii) *Compression seal.* Here the metal has a higher CTE than glass. The resulting seal is much stronger than the matched seal. To create feed-throughs for passing electric current from a part inside the package to the external circuit, a metal body with the required number of holes is formed. Through the holes pass conducting metal wires surrounded by glass. The CTE of the metal body is higher than that of glass and the conducting metal wires. The parts are placed inside a fixture and transferred to a furnace at a high temperature. On cooling, the metal body contracts around the solidifying glass. Due to the concentric compressive force exerted by the metal body on the solidifying glass, the conducting metal wires are strongly fixed in place, surrounded by and insulated from the metal body by the intervening glass layer.

Figure 11.12 illustrates the use of glass-to-metal sealing in the common three-pin headers used for semiconductor device packaging.

11.6.4.2 Ceramic-to-metal seal packages

A metal film is deposited on the ceramic. The ceramic is sealed with the metal body by brazing the metal film on ceramic with the metal body (figure 11.13). In the case that a metal with a high CTE is to be sealed with ceramic, a transition piece of metal with a low CTE is included between the metal piece with a high CTE and the ceramic to avoid direct interfacing between the ceramic and the metal piece with a high CTE.

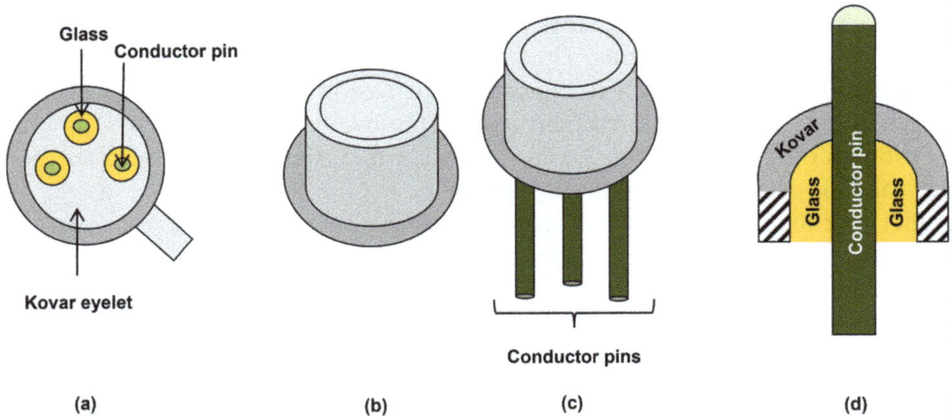

Figure 11.12. A three-pin header for semiconductor device packaging: (a) base of the header, (b) cap, (c) header fitted with cap and (d) diagram showing the glass-to-metal seal at the conductor pin and the hole in the kovar eyelet.

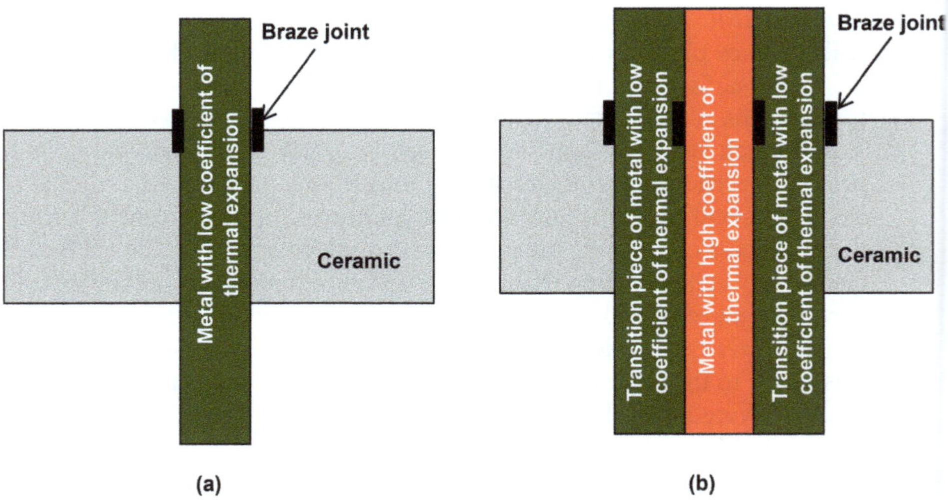

Figure 11.13. Ceramic-to-metal packaging for metal with (a) low CTE and (b) metal with high CTE.

11.6.5 Joining the two parts of hermetic packages

Hermetic packages consist of a base plate and an exactly fitting cap. The two parts are joined together to create a cavity in which the electronic die is placed by one of following processes (Schott 2009):

(i) *Resistance welding.* Electrodes are fixed on opposite sides to pass an electric current through the base plate and the cap which are mounted by aligning them in correct positions for welding. When the current flows, the heat produced by the electro-thermal process due to the electrical resistance of the base plate/cap melts them at the welding ridges to form a bond on cooling.

(ii) *Rolled seam welding.* This is a variant of the resistance welding in which the parts to be welded are pressed between rolling electrodes through which current is made to flow. Thus the parts are welded point by point using rollers in place of static electrodes. This process is applicable only when the parts to be welded together are made of conducting materials. In the case of ceramic packages, metallic frames are soldered to the ceramic parts. Electric current flowing between the rollers through the metallic frames joins the ceramic parts together.

(iii) *Cold welding.* Pressure applied to the surfaces of package parts causes ring welding when the surfaces are flat, parallel and clean, and the tooling is accurately designed.

(iv) *Laser welding.* The laser beam is focused at different points of the parts to be welded, fusing and connecting them together. However, the welding becomes difficult for surfaces coated with gold, silver or nickel due to reflection of the beam instead of it being absorbed.

(v) *Soldering.* Here a solder is used as the joining glue. The surfaces of the two parts must be wettable by the solder for soldering to take place.

11.7 Discussion and conclusions

Neither conventional materials nor the processes followed in the electronics industry can be used to fabricate systems capable of operating reliably at high temperatures. An all-round effort is necessary to choose the appropriate materials and processes at each step of system development to comply with temperature specifications. HTE involves mastering and leveraging the aforesaid technological advancements to fabricate rugged versions of previous medium-temperature designs or to evolve altogether new designs. To achieve this goal, diverse technologies capable of providing high-temperature sustaining components have to be interwoven with existing fabrication facilities to supply robust electronic systems for elevated temperature operation. This will necessitate the re-arrangement, rejuvenation and restructuring of the entire infrastructure.

Review exercises

11.1. What is a metal foil resistor? Describe the mechanism by which low TCR is achieved in this resistor structure.

11.2. What is a wire wound resistor? What is its typical upper temperature limit?

11.3. How are the following types of resistors fabricated: (a) thin-film resistor and (b) thick-film resistor? Give an idea of the resistive film thicknesses in the two types. How high a temperature can they sustain?

11.4. How do thick-film-on-steel resistors differ from conventional thick-film resistors? What is the advantage of using steel as a substrate material?

11.5. What is the composition of a class 1 dielectric used in ceramic capacitors? What are the virtues of this class? What is its limitation?
11.6. Give an example of a dielectric composition providing a high dielectric constant with low temperature sensitivity. What is the reason for the temperature insensitivity?
11.7. What are the materials used for the anode, cathode and dielectric layers in a wet tantalum capacitor? How is the dielectric film formed? What is the relative permittivity of the dielectric? How is the cathode film formed and how is the cathode metal termination deposited? What is the cathode metal termination used for high-temperature operation?
11.8. What is the dielectric constant of Teflon? How high are the temperatures that a teflon capacitor can withstand?
11.9. How is a magnetic core for high-temperature operation made? How is a high-temperature sustaining wire wound inductor assembled?
11.10. Name the magnetic layer material used in a micro-inductor. Describe the process sequence for micro-inductor fabrication.
11.11. What difficulty is faced in using tungsten metallization over silicon at high temperatures? How is the difficulty overcome?
11.12. Up to what operating temperatures can tungsten:nickel metallization be used over silicon carbide? What is the reason for covering it with a silicon layer?
11.13. Compare the performance of Ni/Ti/Al metallization on p-type SiC with that of Ni metallization on n-type SiC with respect to the variation of specific contact resistivity of the metal layer with temperature.
11.14. How is high-temperature electronic packaging fundamentally different from low- and moderate temperature packaging?
11.15. Prepare a comparative performance analysis of four substrate materials for HTE: aluminum nitride, aluminum oxide, silicon nitride and boron nitride.
11.16. Which has the higher thermal conductivity: aluminum nitride or aluminum oxide?
11.17. Which has higher flexural strength: silicon nitride or boron nitride?
11.18. Give one example each of lead-based, gold-based and silver-based die-attach materials and mention the temperature limit for each case.
11.19. Give two examples of the materials of bonding wire-metal pad combinations used in microelectronics, which are prone to failure by intermetallic compound formation.
11.20. Describe the high-temperature performance of nickel wire bonding on nickel pads.
11.21. For reliable operation at high temperatures, it is necessary to eliminate the wire bonding step from the process. Discuss.
11.22. What are the two types of glass-to-metal seals? How are they made? How do they differ in sealing strengths?
11.23. How is a ceramic-to-metal seal made?

11.24. What are the main techniques used for joining together the base plate and cap of a hermetic package?

11.25. Discuss with reference to the following, what steps you would take to fabricate an electronic system working at high temperatures: substrates, die-attach materials, wire bonding, and packages.

References

Barlow F D and Elshabini A 2006 High-temperature high-power packaging techniques for HEV traction applications *Technical Report ORNL/TM-2006/515* 1–20 http://info.ornl.gov/sites/publications/files/Pub2285.pdf

Bauchman M E 2011 A look at film capacitors *TTI* https://www.tti.com/content/ttiinc/en/resources/marketeye/categories/passives/me-tti-20120111.html

Burla R K, Chen L, Zorman C A and Mehregany M 2009 Development of nickel wire bonding for high-temperature packaging of SiC devices *IEEE Trans. Adv. Packag.* **32** 564–74

Chasserio N, Guillemet-Fritsch S, Lebey T and Dagdag S 2009 Ceramic substrates for high temperature electronic integration *J. Electron. Mater.* **38** 164–74

Chen J and Colinge J P 1995 Tungsten metallization for high-temperature SOI devices *Mater. Sci. Eng.* B **29** 18–20

Chen J and Colinge J P 1996 Tungsten metallization system with TiN/$TiSi_2$ contact structure for thin film SOI devices *Trans. 3rd Int. High Temperature Electronics Conf.* 1 *(Albuquerque, NM, June)* V.27–31

Chen L-Y, Okojie R S, Neudeck P G, Hunter G W and Lin S-T 2001 Material system for packaging 500 °C SiC microsystems, in microelectronic, optoelectronic and MEMS packaging *Materials Research Society Symp. Proc., MRS Spring Meeting* 682 *(April 16–20)* ed J C Boudreaux, R H Dauskardt, H R Last and F P McCluskey 79–90

Dittmer R, Anton E-M, Jo W, Simons H, Daniels J E, Hoffman M, Pokorny J, Reaney I M and Rödel J 2012 A high-temperature-capacitor dielectric based on $K_{0.5}Na_{0.5}NbO_3$-modified $Bi_{1/2}Na_{1/2}TiO_3$–$Bi_{1/2}K_{1/2}TiO_3$ *J. Am. Ceram. Soc.* **95** 3519–24

Ebbert P 2014 The wirebound resistor: the report of my death was an exaggeration *Riedon White Paper.* https://riedon.com/media/pdf-tech/Pulse_Wirewound_Resistors.pdf

Evans L J, Okojie R S and Lukco D 2012 Development of an extreme high temperature n-type ohmic contact to silicon carbide *Mater. Sci. Forum* **717–720** 841–4

Freeman Y, Chacko A, Lessner P, Hessey S, Marques R and Moncada J 2013 High-temperature Ta/MnO_2 capacitors *CARTS Int. Proc. (Houston, TX, 25–28 March)* 151–5

Guo W, Zeng Z, Zhang X, Peng P and Tang S 2015 Low-temperature sintering bonding using silver nanoparticle paste for electronics packaging *J. Nanomater.* **2015** 897142

Haddad E, Martin C, Buttay C, Joubert C and Allard Bergogne D B 2013 High temperature, high frequency micro-inductors for low power DC–DC converters *EPE'13-ECCE Europe (Lille, France, September)* Paper 390, hal-00874475

Hernik Y 2012 Precision resistors for energy, transportation, and high-temperature applications *Vishay Precision Groups* www.digikey.com/Web%20Export/Supplier%20Content/VishayPrecisionGroup_804/PDF/vishay-glance-precision-resistors-for-energy.pdf

Kino H, Fukushima T and Tanaka T 2021 High-thermal-stability resistor formed from manganese nitride compound that exhibits the saturation state of the mean free path *Appl. Phys. Express* **14** 091003-1–5

Madou M J 2002 *Fundamentals of Microfabrication: The Science of Miniaturization* (Boca Raton, FL: CRC Press)

Magnetics © 2016 Using magnetic cores at high temperatures *Technical Bulletin* No. CG-06 http://wenku.baidu.com/view/74a86e170b4e767f5acfcee2.html

Manikam V R and Cheong K Y 2011 Die attach materials for high temperature applications: a review *IEEE Trans. Compon. Packag. Manuf. Technol.* **1** 457–78

Morrison D G 2000 Thick-film-on-steel resistors prove worthy in high-temperature, high-power applications *Electronic Design* http://electronicdesign.com/energy/thick-film-steel-resistors-prove-worthy-high-temperature-high-power-applications

Schott 2009 *New Hermetic Packaging and Sealing Technology Handbook* http://schott.com/english/news/press.html?NID=com2516

Smedfors K 2014 Ohmic contacts for high temperature integrated circuits in silicon carbide *Licentiate Thesis* Stockhlom: Royal Institute of Technology pp 1–39

Smedfors K, Lanni L, Östling M and Zetterling C M 2014 Characterization of ohmic Ni/Ti/Al and Ni contacts to 4H-SiC from −40 °C to 500 °C *Mater. Sci. Forum* **778–780** 681–4

TDK 2014 Power inductors for advanced engine management: built tough for high temperatures *TDK* http://en.tdk.eu/tdk-en/373562/tech-library/articles/applications---cases/applications---cases/built-tough-for-high-temperatures/790022

Vishay 2014 High temperature (245 °C) thick film chip resistor *Vishay* www.vishay.com/docs/52032/chpht.pdf

Vishay 2015 High stability–high temperature (230 °C) thin film wraparound chip resistors, sulfur resistant *Vishay* www.vishay.com/docs/53050/pht.pdf

Yin L, Chen C-P, Kapusta C and Ghandi R 2015 Electronic packaging of SiC MOSFET-based devices for reliable high temperature operation *2015 IEEE Int. Symp. on Circuits and Systems (ISCAS) (Lisbon, 24–27 May)* 1170–3

Chapter 12

Superconductive electronics for ultra-cool environments

Superconductive electronics is an extensive field covering both low- and high-temperature superconductors (HTS), and their applications in quantum interference devices, power transmission, microwave filters and logic circuits. The microscopic origins of superconductivity are explained by Bardeen–Cooper–Schieffer (BCS) theory whereas a macroscopic treatment is provided by the Ginzburg–Landau theory. The London equations can explain the well-known Meissner effect of exclusion of magnetic induction by superconductors. A widely used device in superconductive electronics is the Josephson junction (JJ). The DC and AC Josephson effects as well as the inverse AC Josephson effects are described. The application of DC and AC superconducting quantum interference devices (SQUIDs) in magnetometry are discussed. Rapid single flux quantum (RSFQ) logic, touted as a competitive beyond-CMOS technology in nanoelectronics, is briefly surveyed.

12.1 Introduction

Superconductivity is at the forefront of research even after more than a century has elapsed since it was discovered in 1911 (Onnes 1911, Koblischka and Koblischka-Veneva 2022). As the name insinuates, superconductive electronics means electronics using superconducting materials. These are the materials which exhibit zero electrical resistance when cooled below a particular temperature known as the critical or transition temperature (T_C). Superconductive electronics has a broad scope covering (Superconductivite 2015):

doi:10.1088/978-0-7503-5072-3ch12

(i) *SQUID magnetometers.* These are instruments used for accurate measurement of the strength and direction of a magnetic field at a point in space.
(ii) *Current limiters.* These are safety devices used for protecting an electrical plant from accidental faults such as short circuits or voltage overshoots.
(iii) *Efficient electronic filters.* These high-efficiency filters are made of superconductors.
(iv) *RSFQ logic.* This is a logic system based on flux quantum.

The above topics will now be elaborated.

12.2 Superconductivity basics

12.2.1 Low-temperature superconductors

A low-temperature superconductor is a material, element or alloy which offers practically no resistance to the flow of electricity at a temperature near absolute zero (−273.15 °C). In such a material, the current can flow for an indefinitely long time, even for years without any reduction in value.

The resistivity of any normal conductor decreases with cooling but levels off at a certain value beyond which it does not decrease with temperature. Interestingly, the best conductors, such as copper, silver and gold, belong to this class. The resistivity of a superconductor also decreases up to a certain value, as does that of a normal conductor, but in contrast it suddenly drops to zero when a particular temperature is reached; this is the critical temperature, as indicated earlier. Many materials behave as superconductors below 30 K (−243.15 °C), e.g. Ga (1.1 K), Al (1.2 K), Hg (4.2 K), Nb (9.3 K) and La–Ba–Cu oxide (17.9 K). Superconductivity is the property by which these materials show extremely small electrical resistance/resistivity below their respective critical temperatures.

12.2.2 Meissner effect

A superconductor opposes any magnetic field applied by an external magnet (figure 12.1). This expulsion of the magnetic field by a superconductor from its interior is known as the Meissner effect (figures 12.2 and 12.3). Thus a superconductor is said to exclude any magnetic field, or superconductivity opposes magnetism.

The Meissner effect is not an abrupt transition at the superconductor surface. The distance up to which the magnetic field makes an incursion into the superconductor has a finite non-zero value. The characteristic depth ~10–100 nm of a superconductor up to which the external magnetic field can penetrate inside the material, as determined from the decay of the magnetic field to $1/e$ times its value at the surface, is referred to as its London penetration length/depth, denoted by λ or λ_L. Thus, the variation of magnetic induction $B(x)$ along the x-direction is given by

$$B(x) = B(0)\ \exp(-x/\lambda_L), \tag{12.1}$$

where $B(0)$ is magnetic induction at the surface. The penetration depth is

$$\lambda_L = \sqrt{\frac{m}{\mu_0 n q^2}}, \tag{12.2}$$

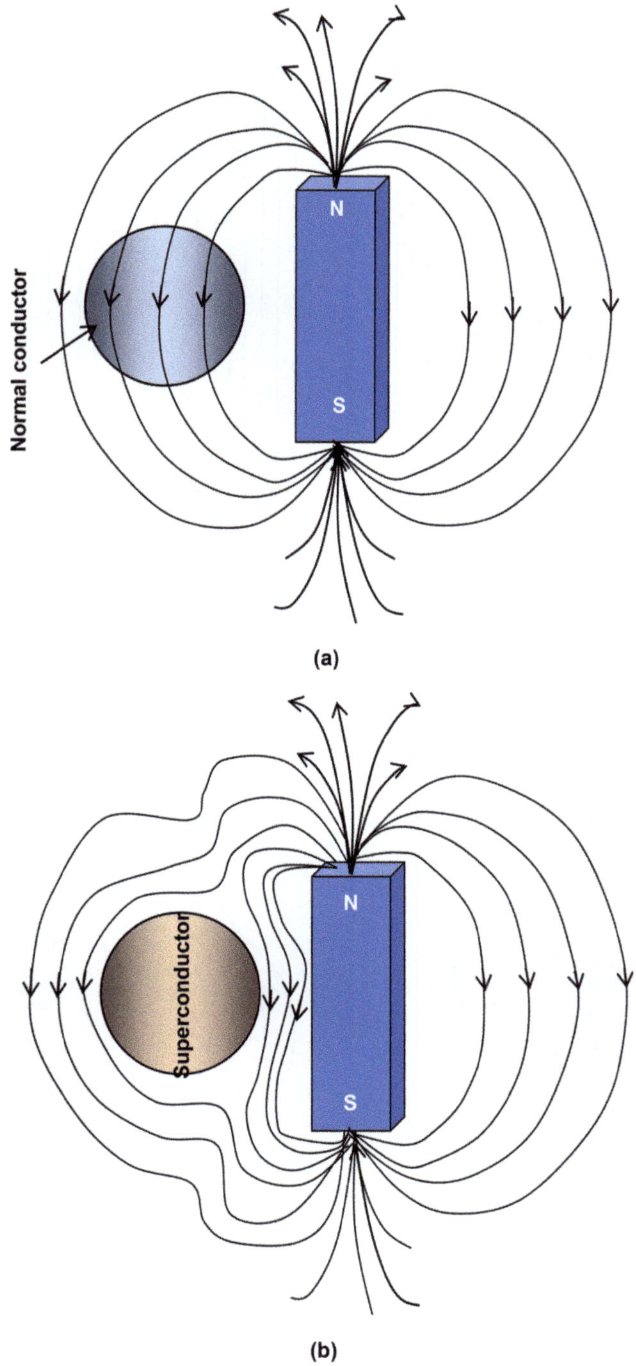

Figure 12.1. Difference in the behavior of superconductor from a normal electrical conductor when placed in an external magnetic field: (a) penetration of the magnetic lines of force through a normal conductor and (b) their exclusion by a superconductor.

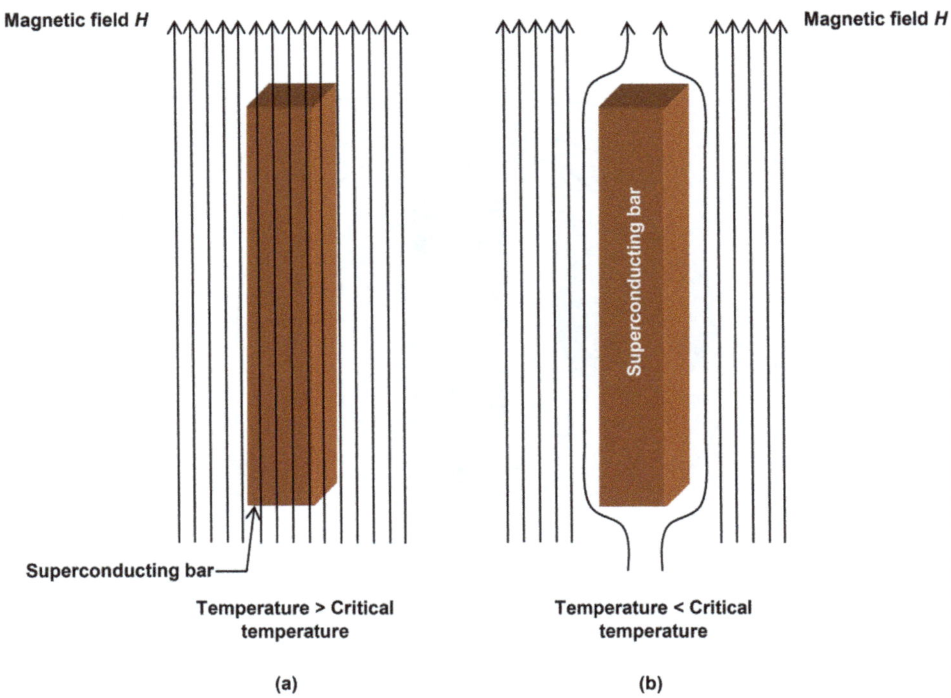

Figure 12.2. The Meissner effect for a bar-shaped superconductor: (a) at a temperature above the critical temperature and (b) at a temperature below the critical temperature.

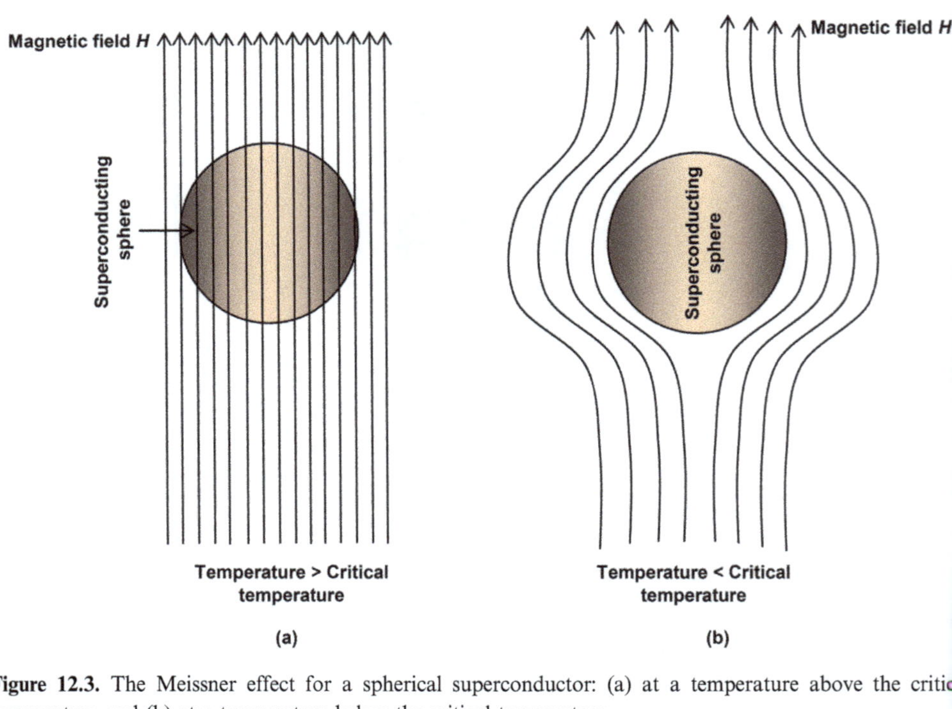

Figure 12.3. The Meissner effect for a spherical superconductor: (a) at a temperature above the critical temperature and (b) at a temperature below the critical temperature.

where m, n, q are the mass, density and charge of superconducting electrons, and μ_0 is the vacuum permeability (=1.257×10^{-6} V s A m^{-1}). Taking $n = 1 \times 10^{29}$ m^{-3}, $m = 9.11 \times 10^{-31}$ kg, $q = 1.6 \times 10^{-19}$ C,

$$\lambda_L = \sqrt{\frac{9.11 \times 10^{-31}}{1.257 \times 10^{-6} \times 1 \times 10^{29} \times (1.6 \times 10^{-19})^2}} = 1.683 \times 10^{-8}\ \text{m} = 16.83\ \text{nm} \quad (12.3)$$

12.2.3 Critical magnetic field (H_C) and critical current density (J_C)

When maintaining the temperature and the current flowing through a semiconductor as constant, if the external magnetic field applied on the superconductor becomes greater than a particular value known as the critical magnetic field H_C, the superconductivity behavior of the material is lost. Similarly, keeping the temperature and the magnetic field acting on a superconductor unchanged, when the current density J flowing through a superconductor exceeds a critical value J_C, superconductivity is disrupted. Thus, including the critical temperature T_C, a superconductor loses its superconducting property when any one of the group of three parameters is surpassed. These vital parameters are: critical temperature, critical magnetic field or critical current density value.

12.2.4 Superconductor classification: type I and type II

Superconductors are classified into two types (figure 12.4):

(i) *Type I or soft.* These are the materials which completely pass into the superconducting state through an abrupt transition when the temperature is brought down below the critical temperature, and vice versa. This implies that all the material under investigation is transformed into a superconductor. No chunk of the material is left untransformed. Examples of pure metals are aluminum (Al), lead (Pb), mercury (Hg), etc. The alloy $TaSi_2$ also behaves in this manner.

(ii) *Type II or hard.* The conversion of these materials into the superconducting state does not occur altogether at once in completeness. It is preceded by an intermediate stage in which some fraction of the material is superconducting while the remaining portion is conducting. An example of such materials is the alloy NbTi which is step-by-step, little-by-little changed into a superconductor. The mixed state comprising superconducting and normal conducting materials is known as the vortex state. It is so-called because in this state vortices (whirlwinds or eddies) of superconducting current encircle normal conductors. Slowly, the amount of superconducting material increases while that of conducting material decreases until all the material becomes a superconductor.

Type I superconductors retain their superconducting nature up to very low critical temperatures and critical magnetic fields. Hence, their practical utility is low. In contrast, type II superconductors are able to maintain their superconducting

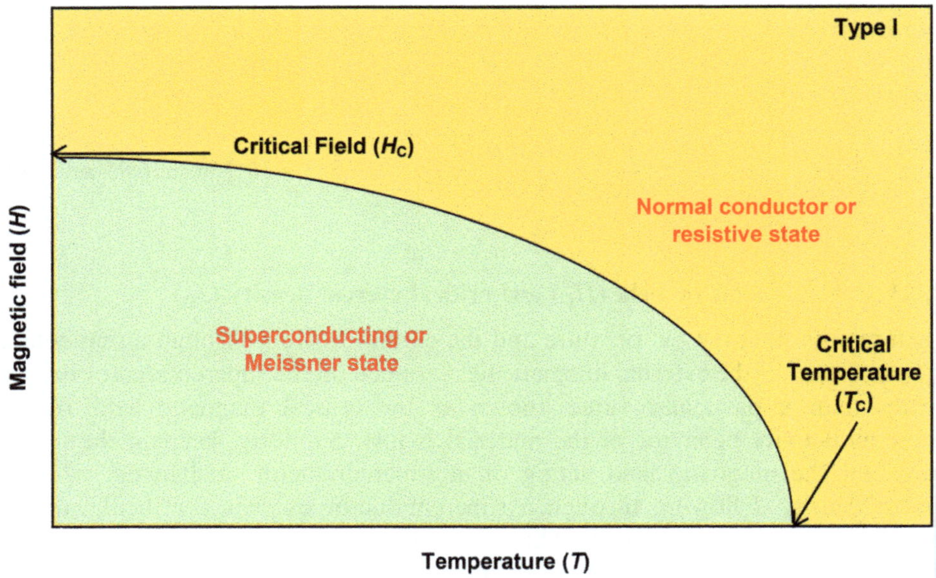

(a)

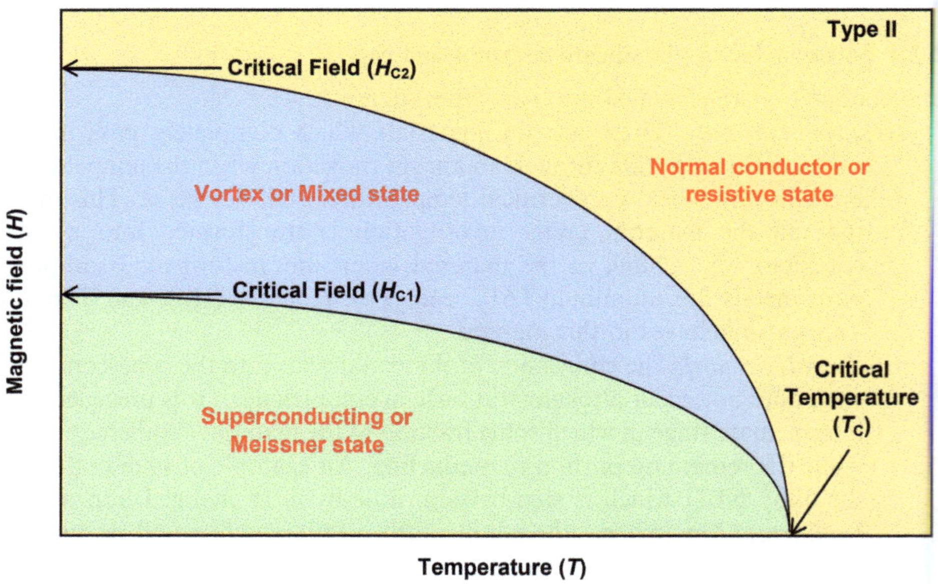

(b)

Figure 12.4. Critical magnetic field-temperature plots of: (a) a type I superconductor and (b) a type II superconductor.

characteristics up to very high critical temperatures and critical magnetic fields. This capability considerably enhances their usefulness.

Type I superconductors display a single critical magnetic field strength H_C below which superconductivity begins or above which it ends (figure 12.5(a)). Type II

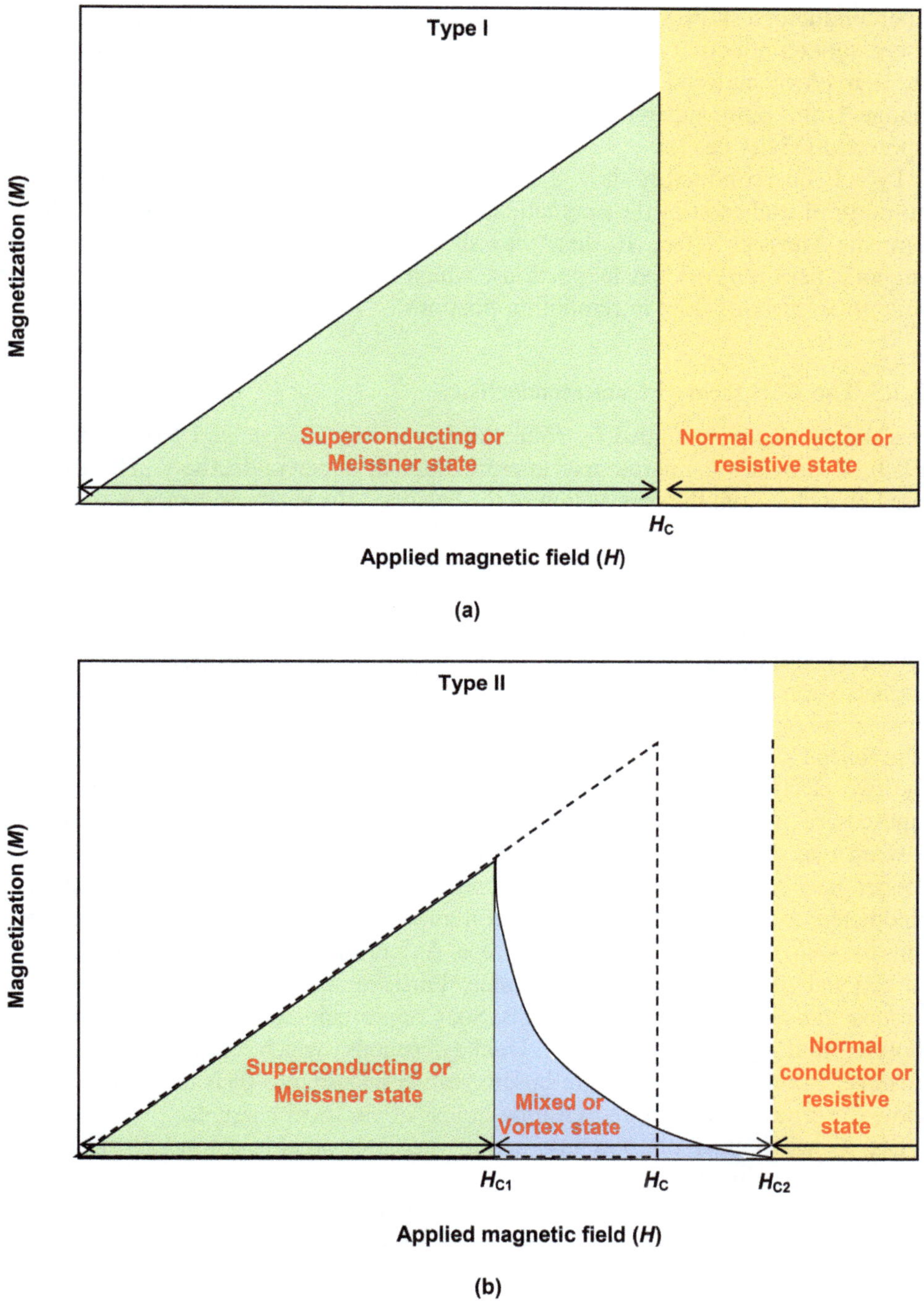

Figure 12.5. The magnetization of superconductors under the influence of an applied magnetic field: (a) type I and (b) type II.

superconductors show two critical magnetic field strengths (figure 12.5(b)): H_{C2} where superconductivity starts to produce a mixed material and H_{C1} where the whole mass of material becomes superconducting. On reversal, at H_{C1} there is a change from pure superconductor to mixed superconductor while at H_{C2} the superconductivity ceases.

Type I superconductors show a complete Meissner effect due to the thorough or out-and-out exclusion of the magnetic field. Type II superconductors do not show a complete Meissner effect. Rather, they show a mixed Meissner effect because the magnetic field is restricted to portions, which are in the normal conductor state, while supercurrents flow in remaining portions.

12.2.5 The BCS theory of superconductivity

This theory was propounded by John Bardeen, Leon Cooper and Robert Schrieffer (1959). It provides a microscopic interpretation of superconductivity. The theory is based on a clue from the observation of the behavior of a superconductor in a magnetic field. It is a well-known fact that a superconductor does not allow a magnetic field to enter inside its body. This affirmation implies that superconductivity has some relationship to the magnetic properties of materials. As we know, the magnetism results from the alignment of spins of electrons. Electrons are particles obeying Fermi–Dirac statistics; hence called fermions. They have two permissible spin values, ±½. During the magnetization of a material, the randomly oriented spins of electrons are re-arranged to form an aligned configuration. In a superconductor, the electrons are arranged into pairs called Cooper pairs. The electrons in a Cooper pair work in a coordinated fashion. The two electrons in a Cooper pair may be located far apart, up to a distance of hundreds of nanometers. Thus they can coordinate from a distance. The distance between two electrons comprising the Cooper pair or the distance up to which the Cooper pair extends is known as the coherence length ξ. This length ranges from nanometers to micrometers. Each pair contains electrons of oppositely directed spins. Thus the spin of a Cooper pair becomes zero. As a result, the electron spin changes from a fractional number to an integral value. Particles with integral spins obey Bose–Einstein statistics and are called bosons. So, Cooper pairs are bosons like phonons or photons, and no longer fermions. Unlike fermions, which obey Pauli's exclusion principle, Cooper pairs, which are bosons, can condense into their lowest energy level below the critical temperature, resulting in loss of electrical resistance.

Let us see how a phonon interaction mediates weakly bound electrons in a Cooper pair (figure 12.6):

(i) At a temperature below critical temperature, the lattice vibrations are minimum. So, the phonons are called virtual phonons.

(ii) Contemplate the motion of an electron in a material. As the electron moves past the lattice ions, it disturbs the lattice ions from their normal positions, not destructively by collision but constructively by interaction of the negative charge on the electron with the positive charge of the nuclear cores of the ions in the material. Creating this disturbance, the electron continues its motion and moves past the lattice ions.

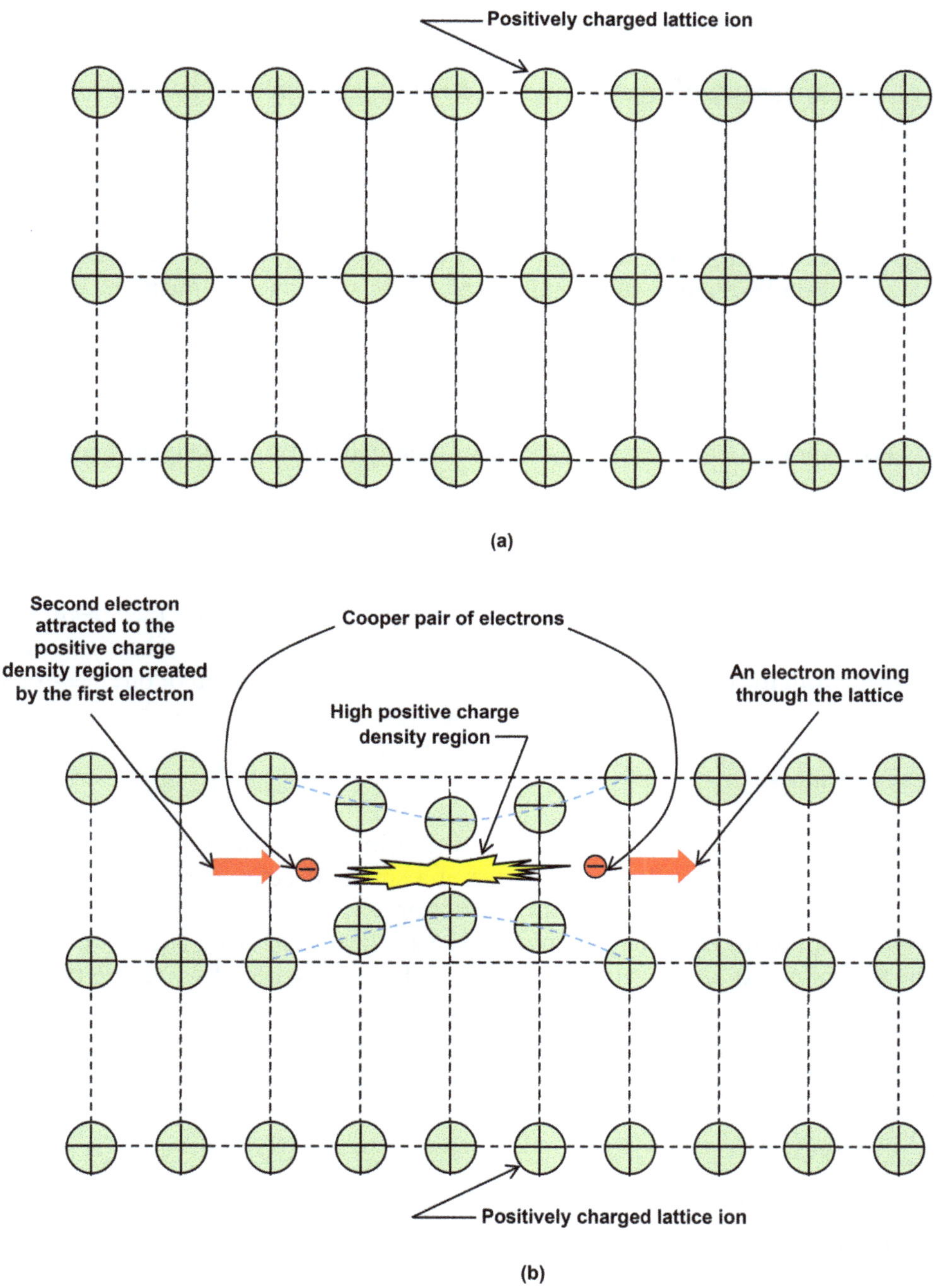

Figure 12.6. Illustration of BCS theory: (a) an undisturbed lattice of positive ions and (b) a distorted region of increased positive charge density created by a passing electron towards which another electron is attracted, resulting in the formation of a Cooper pair. This region disseminates through the crystal as a phonon.

(iii) The lattice ions momentarily shift from their normal positions in a delayed action after the electron has passed by. Due to this distortion, a region of net positive charge density is created in the vicinity of the electron.

(iv) A second electron having opposite momentum and spin to the first electron trails behind it. This second electron is attracted to the above region of positive charge created by the passage of the first electron. This trailing electron pairs up with the leading electron. Thus the attraction between the two electrons is mediated by the distortion of the lattice. Hence, it is an electron–phonon interaction because the phonon is a quantum of vibrational energy. The attractive force produced by lattice deformation is greater than the Coulomb repulsive force that will tend to push the electrons apart.

(v) After passage of the second electron, the lattice ions bounce back from their shifted positions to their original positions.

(vi) Another Cooper pair of electrons then undergoes an identical sequence of steps, causing electrical conduction.

(vii) The sequence is repeated for more Cooper pairs. Thus electric current constantly flows through the agency of Cooper pairs in the lattice. Cooper pairs serve as the carriers of current. The temperature-induced lattice vibrations cannot impede their movements in the manner they disturb the normal single electrons. This insusceptibility to lattice vibrations arises from the fact that the lattice itself mediates the attraction of one electron of the pair to its partner. Thus, the Cooper pairs are more stable than single electrons. This leads to decrease in the electrical resistance to negligible values.

To restate, at temperatures lower than the critical temperature, the electrons are kept in a paired state under the influence of a small attractive force between them. The attractive force originates from interaction of the electron with the lattice. In this interaction, the electrons do not undergo any scattering from lattice ions. There is no loss of energy as the supercurrent flows. In the absence of scattering from the ions, there is no electrical resistance. Above the critical temperature, the repulsive force between electrons causes them to become unpaired. These single electrons suffer scattering from ions as they traverse through the lattice. Due to this scattering, the material shows resistance.

12.2.6 Ginzburg–Landau theory

This is a macroscopic theory of superconductors based on physical intuition. It was proposed in 1950 before the microscopic BCS theory. It begins with proposition of a wavefunction Ψ. This wavefunction is a complex order parameter characterizing the superconducting state. The parameter $|\Psi|^2$ represents the density of superconducting electrons. Assuming that the free energy density of a superconductor can be expanded in powers of supercarrier density, the free energy density is minimized with respect to two parameters: the order parameter Ψ and the vector potential $\mathbf{A}$.

This is done for stabilizing the energy of the superconducting state. The calculations are complex and merely numerical.

The free energy function is expanded as

$$F = \alpha \mid \Psi \mid^2 + \frac{\beta}{2} \mid \Psi \mid^4 + \gamma \left| \left(\nabla + \frac{2iq}{\hbar c} \mathbf{A} \right) \Psi \right|^2 + \frac{B^2}{8\pi}, \tag{12.4}$$

where **A** is the magnetic vector potential, q is the electronic charge and c is the velocity of light. Ψ is a complex-valued order parameter. The symbols α and β represent phenomenological parameters dependent on temperature. The derivative of the free energy function with respect to Ψ is equated to zero to minimize free energy with respect to Ψ

$$\frac{dF}{d\Psi} = 0 = 2\alpha \mid \Psi \mid + \frac{4\beta}{2} \mid \Psi \mid^2 \Psi + 2\gamma \left(\nabla + \frac{2iq}{\hbar c} \mathbf{A} \right)^2 \Psi. \tag{12.5}$$

Normalizing ψ such that

$$\gamma = \hbar^2/(4m) \tag{12.6}$$

$$\begin{aligned}
\frac{dF}{d\Psi} &= 2\alpha\Psi + 2\beta \mid \Psi \mid^2 \Psi + 2\left(\frac{\hbar^2}{4m}\right)\left(\nabla + \frac{2iq}{\hbar c}\mathbf{A}\right)^2 \Psi \\
&= 2\alpha\Psi + 2\beta \mid \Psi \mid^2 \Psi + 2\left(\frac{1}{4m}\right)\left\{\hbar\left(\nabla + \frac{2iq}{\hbar c}\mathbf{A}\right)\right\}^2 \Psi \\
&= 2\alpha\Psi + 2\beta \mid \Psi \mid^2 \Psi \\
&\quad + \left(\frac{1}{2m}\right) \times \left\{\hbar^2\nabla^2 + 2 \times \nabla \times \left(\frac{2iq}{\hbar c}\mathbf{A}\right) \times \hbar^2 \right. \\
&\quad \left. + \left(\frac{2iq}{\hbar c}\mathbf{A}\right)^2 \hbar^2 \right\}\Psi \\
&= 2\alpha\Psi + 2\beta \mid \Psi \mid^2 \Psi + \left(\frac{1}{2m}\right)\left[-(-i\hbar\nabla)^2 - 2 \times (-i\hbar\nabla)\right. \\
&\quad \left. \times (-i\hbar)\left(\frac{2iq}{\hbar c}\mathbf{A}\right) - \left\{(-i\hbar)\left(\frac{2iq}{\hbar c}\mathbf{A}\right)\right\}^2\right] \Psi \\
&= 2\alpha\Psi + 2\beta \mid \Psi \mid^2 \Psi + \left(\frac{1}{2m}\right) \\
&\quad \left[-(-i\hbar\nabla)^2 - 2 \times (-i\hbar\nabla) \times \left(\frac{2q}{c}\mathbf{A}\right) - \left(\frac{2q}{c}\mathbf{A}\right)^2\right]\Psi \\
&= 2\alpha\Psi + 2\beta \mid \Psi \mid^2 \Psi + \left(\frac{1}{2m}\right)\left\{-i\hbar\nabla - \left(\frac{2q}{c}\mathbf{A}\right)\right\}^2 \Psi
\end{aligned} \tag{12.7}$$

$$\therefore \alpha\Psi + \beta \mid \Psi \mid^2 \Psi + \left(\frac{1}{4m}\right)\left\{-i\hbar\nabla - \left(\frac{2q}{c}\mathbf{A}\right)\right\}^2 \Psi = 0 \tag{12.8}$$

or

$$\left\{\alpha + \beta \mid \Psi \mid^2\right\}\Psi + \left(\frac{1}{4m}\right)\left\{-i\hbar\nabla - \left(\frac{2q}{c}\mathbf{A}\right)\right\}^2 \Psi = 0. \tag{12.9}$$

This is the first equation of Ginzburg–Landau.

The derivative of the free energy function with respect to **A** is equated to zero to minimize it with respect to **A**. Rewriting F by putting $\mathbf{B} = \nabla \times \mathbf{A}$

$$F = \alpha \mid \Psi \mid^2 + \frac{\beta}{2} \mid \Psi \mid^4 + \gamma\left\{\left(\nabla + \frac{2iq}{\hbar c}\mathbf{A}\right)\Psi \times \left(\nabla - \frac{2iq}{\hbar c}\mathbf{A}\right)\Psi^*\right\} + \frac{(\nabla \times \mathbf{A})^2}{8\pi}$$

$$\begin{aligned}\frac{\mathrm{d}F}{\mathrm{d}\mathbf{A}} &= 0 = 0 + 0 + \gamma\left(\frac{2iq}{\hbar c}\right)\left\{\Psi\left(\nabla - \frac{2iq}{\hbar c}\mathbf{A}\right)\Psi^* - \Psi^*\left(\nabla + \frac{2iq}{\hbar c}\mathbf{A}\right)\Psi\right\} \\ &\quad + \frac{2\{\nabla \times (\nabla \times \mathbf{A})\}}{8\pi} \\ &= 0 + 0 + \gamma\left(\frac{2iq}{\hbar c}\right)\left\{\Psi\nabla\Psi^* - \Psi\frac{2iq}{\hbar c}\mathbf{A}\Psi^* - \Psi^*\nabla\Psi - \Psi^*\frac{2iq}{\hbar c}\mathbf{A}\Psi\right\} \\ &\quad + \frac{\nabla \times \mathbf{B}}{4\pi} \\ &= \gamma\left(\frac{2iq}{\hbar c}\right)\left\{(\Psi\nabla\Psi^* - \Psi^*\nabla\Psi) - \frac{4iq}{\hbar c}\mathbf{A}\Psi^*\Psi\right\} + \frac{4\pi j}{4\pi c},\end{aligned} \tag{12.10}$$

since from Maxwell's fourth equation

$$\nabla \times \mathbf{B} = \frac{4\pi j}{c}, \tag{12.11}$$

where the displacement current has been neglected,

$$\therefore \left(\frac{2\gamma iq}{\hbar c}\right)\left\{(\Psi\nabla\Psi^* - \Psi^*\nabla\Psi) - \frac{4iq}{\hbar c}\mathbf{A}\Psi^*\Psi\right\} + \frac{4\pi j}{4\pi c} = 0 \tag{12.12}$$

or

$$\frac{j}{c} = -\left(\frac{2\hbar^2 iq}{4m\hbar c}\right)\left\{(\Psi\nabla\Psi^* - \Psi^*\nabla\Psi) - \frac{4iq}{\hbar c}\mathbf{A}\Psi^*\Psi\right\} \tag{12.13}$$

or

$$\frac{j}{c} = -\left(\frac{\hbar iq}{2mc}\right)(\Psi\nabla\Psi^* - \Psi^*\nabla\Psi) - \left(\frac{2q^2}{mc^2}\right)\mathbf{A}\Psi^*\Psi \tag{12.14}$$

$$\therefore j = -\left(\frac{\hbar iq}{2m}\right)(\Psi\nabla\Psi^{*} - \Psi^{*}\nabla\Psi) - \left(\frac{2q^2}{mc}\right)\mathbf{A}\Psi^{*}\Psi. \tag{12.15}$$

This is the second equation of Ginzburg–Landau.

The Ginzburg–Landau theory does not treat the mechanisms responsible for superconductivity. Instead, the existence of superconductivity is postulated, and equations are derived to model its properties.

12.2.7 London equations

These equations were proposed by two brothers, Fritz and Heinz London, in the year 1935 as restrictions on classical electromagnetism for interpretation of the Meissner effect.

In a normal conductor, the acceleration force F_1 acting on an electron of charge q moving in an electric field E is

$$F_1 = qE. \tag{12.16}$$

The electron is accelerated by this force to a velocity v. On its way, it suffers collisions with metal ions which retard its motion. If τ is the time in which the electron comes to rest, its deceleration is given by

$$a = v/\tau. \tag{12.17}$$

If m is the mass of the electron, the deceleration force F_2 is

$$F_2 = -(mv)/\tau. \tag{12.18}$$

The equation of motion of the electron is

$$m\frac{\mathrm{d}v}{\mathrm{d}t} = F_1 + F_2 = qE - (mv)/\tau. \tag{12.19}$$

In a superconductor, there are no collisions causing scattering of an electron. Therefore,

$$F_2 = -(mv)/\tau = 0 \tag{12.20}$$

and the equation of motion reduces to

$$m\frac{\mathrm{d}v}{\mathrm{d}t} = qE \tag{12.21}$$

or

$$\frac{\mathrm{d}v}{\mathrm{d}t} = \frac{qE}{m}. \tag{12.22}$$

The current I is

$$I = \frac{\text{total charge/cross} - \text{sectional area}}{\text{time}}. \tag{12.23}$$

The current density is

$$J = \frac{(\text{total charge/volume}) \times \text{path length traversed}}{\text{time}} = nqv, \qquad (12.24)$$

where nq = (number of electrons/volume) × charge of one electron, and v = path length traversed/time.

Taking derivative of both sides of equation (12.24) with respect to time

$$\frac{\mathrm{d}J}{\mathrm{d}t} = nq\frac{\mathrm{d}v}{\mathrm{d}t}. \qquad (12.25)$$

Substituting the value of dv/dt from equation (12.22) into equation (12.25)

$$\frac{\mathrm{d}J}{\mathrm{d}t} = nq\left(\frac{qE}{m}\right) = \frac{nq^2E}{m}. \qquad (12.25)$$

This is the first equation of London.

Taking the curl of London's first equation

$$\nabla \times \frac{\mathrm{d}\mathbf{J}}{\mathrm{d}t} = \nabla \times \left(\frac{nq^2\mathbf{E}}{m}\right) = \left(\frac{nq^2}{m}\right)(\nabla \times \mathbf{E}) = \left(\frac{nq^2}{m}\right)\left(-\frac{\mathrm{d}\mathbf{B}}{\mathrm{d}t}\right) \qquad (12.27)$$

since from Maxwell's third equation

$$\nabla \times \mathbf{E} = -\frac{\mathrm{d}\mathbf{B}}{\mathrm{d}t}. \qquad (12.28)$$

Integrating both sides of equation (12.27) with respect to time

$$\nabla \times \mathbf{J} = -\frac{nq^2\mathbf{B}}{m}. \qquad (12.29)$$

This is the second equation of London.

12.2.8 Explanation of Meissner's effect from London equations

Maxwell's fourth equation is differential form of Ampere's circuital law:

$$\nabla \times \mathbf{B} = \mu_0\mathbf{J} \qquad (12.30)$$

excluding the displacement current term.

Taking the curl of both sides,

$$\nabla \times \nabla \times \mathbf{B} = \mu_0(\nabla \times \mathbf{J}) \qquad (12.31)$$

or

$$\nabla \times \nabla \times \mathbf{B} = \mu_0\left(-\frac{nq^2\mathbf{B}}{m}\right), \qquad (12.32)$$

where London's second equation has been used.

Applying the formula

$$\nabla \times \nabla \times \mathbf{B} = \nabla(\nabla\cdot\mathbf{B}) - \nabla^2 B, \tag{12.33}$$

equation (12.32) can be expressed as

$$\nabla(\nabla\cdot\mathbf{B}) - \nabla^2 B = \mu_0\left(-\frac{nq^2\mathbf{B}}{m}\right). \tag{12.34}$$

From Maxwell's second equation (Gauss law of magnetostatics)

$$\nabla\cdot\mathbf{B} = 0 \tag{12.35}$$

equation (12.34) simplifies to

$$\nabla^2 B = \mu_0\left(\frac{nq^2\mathbf{B}}{m}\right) = \frac{B}{\left\{m/\left(\mu_0 nq^2\right)\right\}} = \frac{B}{\lambda_L^2}, \tag{12.36}$$

where

$$\lambda_L = \sqrt{\left\{m/\left(\mu_0 nq^2\right)\right\}}, \tag{12.37}$$

having the dimensions of length, is called the penetration depth; see equation (12.2). Equation (12.36) clearly shows that B is spatially uniform or

$$\nabla^2 B = 0 \tag{12.38}$$

only if $B = 0$. Hence, it is possible for B to be a uniform field inside the superconductor only if it is zero everywhere. So, a non-zero field cannot exist uniformly inside the superconductor. In other words, the field must be non-uniform or dependent on position.

Writing equation (12.36) in the x-direction,

$$\frac{d^2 B}{dx^2} = \left(\frac{1}{\lambda_L^2}\right)B. \tag{12.39}$$

Let

$$B = \exp(px). \tag{12.40}$$

Then

$$p^2 \exp(px) - \left(\frac{1}{\lambda_L^2}\right)\exp(px) = 0 \tag{12.41}$$

or

$$\exp(px)\left\{p^2 - \left(\frac{1}{\lambda_L^2}\right)\right\} = 0 \tag{12.42}$$

$$\therefore p^2 - \left(\frac{1}{\lambda_L^2}\right) = 0. \quad (12.43)$$

Hence,

$$p = \pm\left(\frac{1}{\lambda_L}\right). \quad (12.44)$$

The general solution of the differential equation (12.39) is

$$B = C_1 \exp\left(+\frac{1}{\lambda_L}\right)x + C_2 \exp\left(-\frac{1}{\lambda_L}\right)x, \quad (12.45)$$

$$\text{At } x = \infty, B = 0. \quad (12.46)$$

As the first term of equation (12.45) shows opposite behavior, the physically reasonable solution is

$$B = C_2 \exp\left(-\frac{1}{\lambda_L}\right)x. \quad (12.47)$$

$$\text{At } x = 0, B = B(0) \quad (12.48)$$

$$\therefore C_2 = B(0) \quad (12.49)$$

and

$$\therefore B = B(0)\exp\left(-\frac{1}{\lambda_L}\right)x. \quad (12.50)$$

Compare this to equation (12.1). Equations (12.1) and (12.50) are the same. The equation predicts an exponential decay of magnetic induction with distance from the external surface of the superconductor. So, λ_L is the distance over which the magnetic induction falls from the value $B(0)$ at the surface to e^{-1} times value inside the superconductor. It is the penetration length. This exclusion of magnetic induction from the superconductor is the Meissner effect. The expulsion of magnetic field is achieved by the shielding supercurrents flowing in a sheath of thickness λ_L at the surface without any ohmic losses.

12.2.9 Practical applications

As superconductivity phenomena take place at very low temperatures, a primary requirement for utilizing these phenomena practically is to reach these low temperatures and maintain the temperatures, both of which are uphill tasks. So, these phenomena have not been put to use in everyday life but have found specific applications. One such application is MRI equipment, in which large magnetic fields are generated by electromagnets. If common conductors are used, the high currents flowing through the electromagnets will cause enormous heating. Coils made of superconducting materials are used in these gigantic electromagnets to solve the heating problem. Another application area of superconducting magnets is in particle accelerators. These are large devices in which elementary particles are paced to high speeds to smash against other particles in order to produce new particles.

12.2.10 High-temperature superconductor

In the context of superconductivity, the term 'high temperature' has different connotation than its common meaning because superconductivity was first observed close to absolute zero. A HTS (high-T_C) simply means one which shows superconducting behavior farther from absolute zero than a traditional low-temperature superconductor. In this context, a temperature around 90–100 K being close to liquid-nitrogen temperature (77.2 K) is sufficiently high for a material to qualify as an HTS. The HTS is generally made of a ceramic material. It behaves in the same way as an older low-temperature superconductor but at somewhat elevated temperatures with respect to its predecessors. The ultimate dream is to obtain a superconductor which works at room temperature, so that no extensive cooling arrangements are required for its practical utilization.

12.3 Josephson junction

The JJ is an important active device in superconductive electronics (Scientific American 1997). A JJ, also called a superconducting tunnel junction (STJ), is a sandwich structure consisting of two superconducting electrodes separated by a weak link such as a normal metal or a thin insulating tunnel barrier (figure 12.7). A JJ is artificially fabricated by using the following structures:

(i) superconductor–non-superconducting-metal–superconductor; or
(ii) superconductor–insulator–superconductor.

In structure (i), the metal thickness can be several microns, but in structure (ii), the insulator thickness must be very small ~3 nm. Several other ways of forming weak links have also been used, such as narrow constrictions, grain boundaries, etc.

12.3.1 The DC Josephson effect

Even when no DC voltage is applied across the junction, a DC current flows through the junction as long as the current is below a critical level (figure 12.8(a)). This DC current symbolizes the maximum value of supercurrent that can pass through the junction. The Cooper pairs on the opposite sides of the junction can be represented by wavefunctions ψ_1 and ψ_2. The current spontaneously flows by quantum-mechanical tunneling of Cooper pairs. In this tunneling, the collective quantum wave of Cooper pairs spills over to the other side of the junction causing current conduction. Tunneling of Cooper pairs is dissimilar to the tunneling of electrons in a normal tunnel junction because:

(i) The tunneling of Cooper pairs does not need any excitation or voltage, whereas electrons require a finite voltage for tunneling.

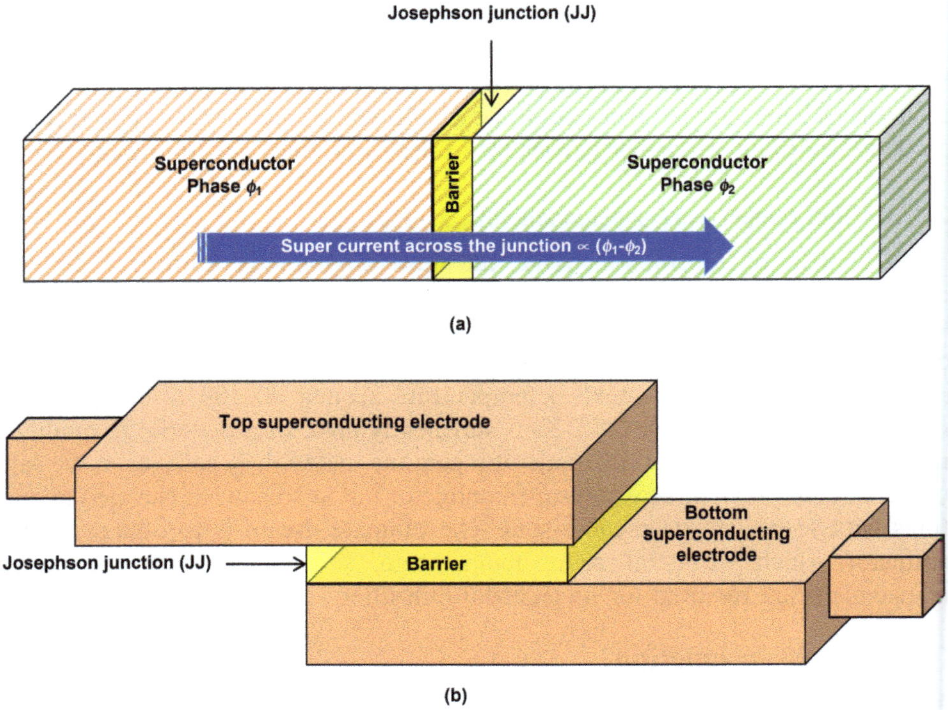

Figure 12.7. A JJ: (a) structure of the junction and (b) junction with electrodes.

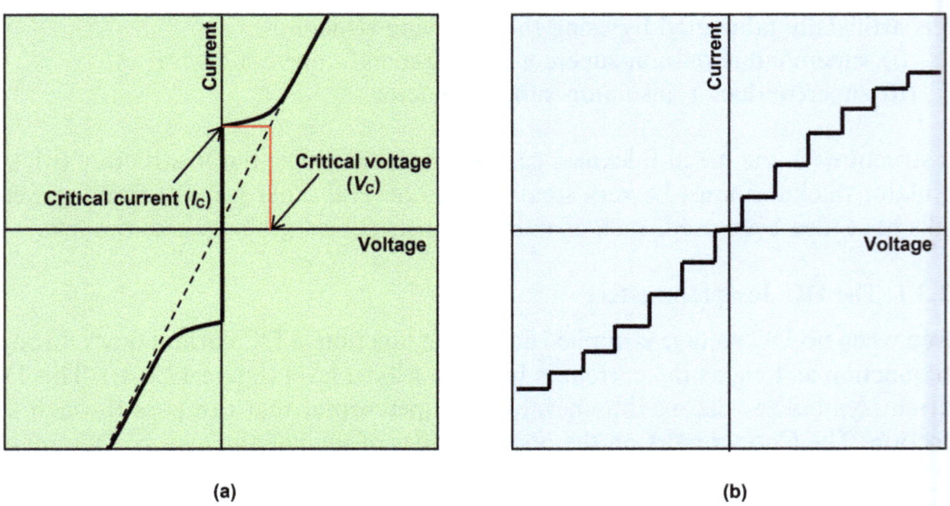

Figure 12.8. Current–voltage characteristics for: (a) the DC Josephson effect and (b) the AC Josephson effect.

(ii) Unlike a normal tunnel junction, the tunneling current of Cooper pairs does not encounter any resistance. So, if a current source is connected across the junction there is no voltage drop or power dissipation.

The current at zero DC voltage is called the Josephson current, and the phenomenon is known as the DC Josephson effect. The Josephson current is proportional to the trigonometric sine function of the phase difference between the wavefunctions ψ_1 and ψ_2. The phenomenon of DC Josephson effect was a surprise because it means that current can flow without any electric field. It is opposite to the commonly taught principle in physics that voltage = current × resistance. So, if voltage = 0, current = 0.

12.3.2 The AC Josephson effect

This effect is observed when a DC voltage is applied across the junction and the DC current exceeds the critical current. Then an oscillatory voltage is developed across the junction in reactionary response to phase variations (figure 12.8(b)). The characteristic frequency of this oscillation depends neither on the size of the superconductors on the two sides of the junction nor on their properties such as chemical composition or critical temperature. It depends only on the applied voltage and on the fundamental physical constants such as elementary electronic charge and Planck's constant. It is proportional to the voltage across the junction. Oscillations are generated at ~500 GHz mV^{-1} across the junction. This is the AC Josephson effect. It is also strange because it means that on applying a DC voltage, an AC current flows across the JJ.

12.3.3 Theory

For superconductivity, the system containing the JJ is assumed to be immersed inside a liquid-helium bath. The temperature of the bath is −269 °C. Further, the system is completely isolated from magnetic fields.

When the wavefunctions ψ_1 and ψ_2 of the Cooper pairs in the two superconducting electrodes of a JJ come closer together, they penetrate the barrier and try to couple with each other. As soon as the energy associated with coupling of wavefunctions becomes greater than the thermal fluctuation energy, the phases of the wavefunctions are locked. This enables the passage of Cooper pairs from one superconductor to another without incurring any energy loss. A simple derivation of the governing equations of the Josephson effect will be presented for a random location in the plane of the junction.

The temporal evolution of the wavefunctions ψ_1 and ψ_2 on the opposite sides of the tunnel barrier is described by the coupled time-dependent Schrödinger equations

$$i\hbar\frac{\partial\psi_1}{\partial t} = U_1\psi_1 + K\psi_2 \tag{12.51}$$

$$i\hbar\frac{\partial\psi_2}{\partial t} = U_2\psi_2 + K\psi_1, \tag{12.52}$$

where K is a constant representing the coupling through the tunnel barrier. It is a measure of the interaction between the wavefunctions. Its value depends on barrier width and composition. It also varies with temperature. It is called the barrier opacity constant. U_1, U_2 denote the states of lowest energy on the opposite sides of the barrier resulting from application of a voltage source $V = (V_1 - V_2)$ between the two sides. The difference of energies is

$$\Delta U = U_2 - U_1 = -2q(V_1 - V_2) = -2qV, \tag{12.53}$$

where $2q$ is the charge on the Cooper pair, q is the electronic charge and V is the potential difference across the junction.

The wavefunctions ψ_1 and ψ_2 can be conveniently expressed as

$$\psi_1 = \sqrt{\rho_1}\ \exp(\mathrm{i}\phi_1) \tag{12.54}$$

$$\psi_2 = \sqrt{\rho_2}\ \exp(\mathrm{i}\phi_2), \tag{12.55}$$

where ρ_1, ρ_2 are the densities of the Cooper pairs and ϕ_1, ϕ_2 are the phases of the wavefunctions.

Substituting for ψ_1, ψ_2 from equations (12.54) and (12.55) into equation (12.51), we obtain

$$\mathrm{i}\hbar\frac{\partial\{\sqrt{\rho_1}\ \exp(\mathrm{i}\phi_1)\}}{\partial t} = U_1\{\sqrt{\rho_1}\ \exp(\mathrm{i}\phi_1)\} + K\{\sqrt{\rho_2}\ \exp(\mathrm{i}\phi_2)\} \tag{12.56}$$

or

$$\begin{aligned} &\mathrm{i}\hbar \times (1/2)\rho_1^{-1/2}\left(\frac{\partial\rho_1}{\partial t}\right)\exp(\mathrm{i}\phi_1) + \left\{\mathrm{i}\hbar\sqrt{\rho_1} \times \exp(\mathrm{i}\phi_1) \times \mathrm{i}\left(\frac{\partial\phi_1}{\partial t}\right)\right\} \\ &\quad = U_1\sqrt{\rho_1}\ \exp(\mathrm{i}\phi_1) + K\sqrt{\rho_2}\ \exp(\mathrm{i}\phi_2). \end{aligned} \tag{12.57}$$

Multiplying both sides by $\sqrt{\rho_1}/\{\exp(\mathrm{i}\phi_1)\}$, we have

$$\begin{aligned} &\mathrm{i}\hbar \times (1/2)\left(\frac{\partial\rho_1}{\partial t}\right) - \left\{\hbar\rho_1\left(\frac{\partial\phi_1}{\partial t}\right) \times \exp(\mathrm{i}\phi_1)\right\}/\{\exp(\mathrm{i}\phi_1)\} \\ &\quad = U_1\rho_1 + K\sqrt{\rho_1\rho_2}\{\exp(\mathrm{i}\phi_2)/\exp(\mathrm{i}\phi_1)\} \end{aligned} \tag{12.58}$$

or

$$\begin{aligned} &\mathrm{i}\hbar\left(\frac{\partial\rho_1}{\partial t}\right) - 2\hbar\rho_1\left(\frac{\partial\phi_1}{\partial t}\right) = 2U_1\rho_1 + 2K\sqrt{\rho_1\rho_2}\left[\exp\{\mathrm{i}(\phi_2 - \phi_1)\}\right] \\ &\quad = 2U_1\rho_1 + 2K\sqrt{\rho_1\rho_2}\{\cos(\phi_2 - \phi_1) + \mathrm{i}\sin(\phi_2 - \phi_1)\}. \end{aligned} \tag{12.59}$$

Equating real and imaginary parts,

$$-2\hbar\rho_1\left(\frac{\partial\phi_1}{\partial t}\right) = 2U_1\rho_1 + 2K\sqrt{\rho_1\rho_2}\ \cos(\phi_2 - \phi_1) \tag{12.60}$$

or

$$\frac{\partial \phi_1}{\partial t} = -\frac{U_1}{\hbar} - \frac{K}{\hbar}\sqrt{\frac{\rho_2}{\rho_1}}\ \cos(\phi_2 - \phi_1) \tag{12.61}$$

and

$$\mathrm{i}\hbar\left(\frac{\partial \rho_1}{\partial t}\right) = 2K\sqrt{\rho_1 \rho_2} \times \mathrm{i}\sin(\phi_2 - \phi_1) \tag{12.62}$$

or

$$\frac{\partial \rho_1}{\partial t} = \frac{2K}{\hbar}\sqrt{\rho_1 \rho_2}\ \sin(\phi_2 - \phi_1). \tag{12.63}$$

Substituting for ψ_1, ψ_2 from equations (12.54) and (12.55) into equation (12.52), we obtain

$$\mathrm{i}\hbar\frac{\partial\left\{\sqrt{\rho_2}\ \exp(\mathrm{i}\phi_2)\right\}}{\partial t} = U_2\{\sqrt{\rho_2}\ \exp(\mathrm{i}\phi_2)\} + K\{\sqrt{\rho_1}\ \exp(\mathrm{i}\phi_1)\} \tag{12.64}$$

or

$$\begin{aligned} &\mathrm{i}\hbar \times (1/2)\rho_2^{-1/2}\left(\frac{\partial \rho_2}{\partial t}\right)\exp(\mathrm{i}\phi_2) + \left\{\mathrm{i}\hbar\sqrt{\rho_2} \times \exp(\mathrm{i}\phi_2) \times \mathrm{i}\left(\frac{\partial \phi_2}{\partial t}\right)\right\} \\ &= U_2\sqrt{\rho_2}\ \exp(\mathrm{i}\phi_2) + K\left\{\sqrt{\rho_1}\ \exp(\mathrm{i}\phi_1)\right\}. \end{aligned} \tag{12.65}$$

Multiplying both sides by $\sqrt{\rho_2}/\{\exp(\mathrm{i}\phi_2)\}$,

$$\begin{aligned} &\mathrm{i}\hbar \times (1/2)\left(\frac{\partial \rho_2}{\partial t}\right) - \left\{\hbar\rho_2\left(\frac{\partial \phi_2}{\partial t}\right) \times \exp(\mathrm{i}\phi_2)\right\}/\{\exp(\mathrm{i}\phi_2)\} \\ &= U_2\rho_2 + K\sqrt{\rho_1\rho_2}\left\{\exp(\mathrm{i}\phi_1)/\exp(\mathrm{i}\phi_2)\right\} \end{aligned} \tag{12.66}$$

or

$$\begin{aligned} \mathrm{i}\hbar\left(\frac{\partial \rho_2}{\partial t}\right) - 2\hbar\rho_2\left(\frac{\partial \phi_2}{\partial t}\right) &= 2U_2\rho_2 + 2K\sqrt{\rho_1\rho_2}[\exp\{\mathrm{i}(\phi_1 - \phi_2)\}] \\ &= 2U_2\rho_2 + 2K\sqrt{\rho_1\rho_2}\left\{\cos(\phi_1 - \phi_2) + \mathrm{i}\sin(\phi_1 - \phi_2)\right\}. \end{aligned} \tag{12.67}$$

Equating real and imaginary parts,

$$-2\hbar\rho_2\left(\frac{\partial \phi_2}{\partial t}\right) = 2U_2\rho_2 + 2K\sqrt{\rho_1\rho_2}\ \cos(\phi_1 - \phi_2) \tag{12.68}$$

or

$$-\frac{\partial\phi_2}{\partial t}=\frac{U_2}{\hbar}+\frac{K}{\hbar}\sqrt{\frac{\rho_1}{\rho_2}}\cos(\phi_1-\phi_2)$$

or

$$\frac{\partial\phi_2}{\partial t}=-\frac{U_2}{\hbar}-\frac{K}{\hbar}\sqrt{\frac{\rho_1}{\rho_2}}\cos(\phi_2-\phi_1) \tag{12.69}$$

and

$$\frac{\partial\rho_2}{\partial t}=\frac{2K}{\hbar}\sqrt{\rho_1\rho_2}\times\sin(\phi_1-\phi_2)=-\frac{2K}{\hbar}\sqrt{\rho_1\rho_2}\times\sin(\phi_2-\phi_1). \tag{12.70}$$

Comparison of equations (12.63) and (12.70) shows that

$$\begin{aligned}\frac{\partial\rho_1}{\partial t}&=\text{rate of decrease of Cooper pair density } \rho_1\\ &\quad\text{in superconductor 1}\\ &=-\frac{\partial\rho_2}{\partial t}=-\text{rate of decrease of Cooper pair density } \rho_2\\ &\quad\text{in superconductor 2.}\end{aligned} \tag{12.71}$$

As these changes may bring about a charge imbalance between the electrons and background ions, currents flowing in the circuit connected to the JJ oppose them. The current density flowing from one superconductor to another is given by the time derivative of the density of Cooper pairs. The sign of the current density is obtained by comparing the JJ with a bulk superconductor. The direction of current density is opposite to that of the gradient of phase. In the JJ, the positive gradient of phase from superconductor 1 to superconductor 2 is represented by $\phi_2-\phi_1>0$. For this polarity of current, transference of electrons in Cooper pairs must take place from superconductor 1 to superconductor 2. Hence, $\frac{\partial\rho_2}{\partial t}>0$ and K is negative in sign, so that from equation (12.63), the current density is

$$j=\frac{\partial\rho_1}{\partial t}=2\frac{K}{\hbar}\sqrt{\rho_1\rho_2}\,\sin(\phi_2-\phi_1), \tag{12.72}$$

which can be recast in the form

$$j=j_C\,\sin\phi, \tag{12.73}$$

where

$$j_C=2\frac{K}{\hbar}\sqrt{\rho_1\rho_2} \tag{12.74}$$

is the critical current density, and

$$\phi_2 - \phi_1 = \phi. \tag{12.75}$$

The critical current density j_C depends on K and the external magnetic field which is assumed to be absent.

For simplification, the superconducting metallic layers of the JJ are made of the same element niobium (Nb). Then $\rho_1 = \rho_2$, by taking identical superconductors and subtracting equation (12.61) from equation (12.69),

$$\frac{\partial\phi_2}{\partial t} - \frac{\partial\phi_1}{\partial t} = \frac{U_1}{\hbar} - \frac{U_2}{\hbar} \tag{12.76}$$

or

$$\frac{\partial(\phi_2 - \phi_1)}{\partial t} = \frac{U_1 - U_2}{\hbar}$$

or

$$\frac{\partial\phi}{\partial t} = \frac{2qV}{\hbar} \tag{12.77}$$

from equation (12.53).

Equations (12.71) and (12.77) constitute the basic governing equations of the Josephson effect. Let us see how they lead to the DC and AC Josephson effects.

12.3.3.1 The DC effect

Integrating equation (12.77) over time,

$$\int_{\phi_0}^{\phi(t)} \left(\frac{\partial\phi}{\partial t}\right) \mathrm{d}t = \frac{2q}{\hbar} \int_0^t V \mathrm{d}t \tag{12.78}$$

or

$$\phi(t) - \phi_0 = \frac{2q}{\hbar} \int_0^t V \mathrm{d}t. \tag{12.79}$$

If $V = 0$, $\phi(t) - \phi_0 = 0$ or, $\phi(t) = \phi_0$, hence

$$j = j_C \sin\phi_0, \tag{12.80}$$

which is a constant. Therefore, for $V = 0$, a constant DC current flows. The maximum value of this current is attained when $\sin\phi_0$ has peak value = 1. This maximum value is

$$j = j_C \sin\phi_0 = j_C \times 1 = j_C = 2\frac{K}{\hbar}\sqrt{\rho_1\rho_2} \tag{12.81}$$

from equation (12.74).

This is exactly the DC Josephson effect, namely the appearance of a constant current without applied voltage and sustenance of this current up to a peak value j_C.

12.3.3.2 The AC effect

Now, if V is non-zero but has a magnitude = V, we have

$$\phi(t) - \phi_0 = \phi = \frac{2qVt}{\hbar} \tag{12.82}$$

so that

$$j = j_C \sin \phi = j_C \sin\left(\frac{2qVt}{\hbar}\right), \tag{12.83}$$

which is an oscillating current with a frequency

$$f = \frac{1}{2\pi}\left(\frac{2qV}{\hbar}\right) = \frac{1}{2\pi}\left\{\frac{2qV}{h/(2\pi)}\right\} = \frac{2qV}{h}. \tag{12.84}$$

For $V = 1$ mV $= 10^{-3}$ V,

$$f = \frac{2q}{h} = \frac{2 \times 1.6 \times 10^{-19} \times 10^{-3}}{6.63 \times 10^{-34}} = 4.827 \times 10^{11} \text{ Hz}. \tag{12.85}$$

This is the AC Josephson effect in which application of a DC voltage V produces high-frequency AC with a frequency of 4.827×10^{11} Hz mV^{-1}.

12.3.4 Gauge-invariant phase difference

The foregoing treatment was based on the assumption that the JJ was not exposed to any magnetic field. The phase difference was taken as $\phi = \phi_2 - \phi_1$. This is not a gauge-invariant quantity. But it did not lead to any error because the effect of magnetic field on the JJ was not considered. Nevertheless, this ϕ value cannot represent a general situation wherein a magnetic field is present. Therefore, the current density J thus determined is not true in a generalized perspective. In order to remedy this non-inclusion, in the presence of a magnetic vector potential **A**, defined in terms of the magnetic field **B** as

$$\mathbf{B} = \nabla \times \mathbf{A}, \tag{12.86}$$

the phase difference ϕ will be replaced by the gauge-invariant phase difference θ defined as

$$\theta = \phi_2 - \phi_1 - (2\pi/\Phi_0) \oint_1^2 \mathbf{A}\cdot d\mathbf{s}, \tag{12.87}$$

where Φ_0 is the magnetic flux quantum given by

$$\Phi_0 = h/(2q) = 6.63 \times 10^{-34}/(2 \times 1.6 \times 10^{-19}) = 2.072 \times 10^{-15} \text{ Wb}. \tag{12.88}$$

It has the same value for all superconductors. The quantization arises from the requirement for the macroscopic wavefunction to be single-valued. Due to this restriction, the total magnetic flux trapped by a superconducting ring can acquire only quantized values, which are integral multiples of the flux quantum. The magnetic flux $\mathbf{\Phi}$ threading a loop is

$$\mathbf{\Phi} = \mathbf{B}\cdot\mathbf{s}, \tag{12.89}$$

where **s** is the area of the loop.

Let the Cooper pair concentration in the superconductor be $\rho(\mathbf{r})$ where **r** is the position vector, and let the wavefunction be $\psi(\mathbf{r})$. Then (Tsang 1997)

$$\rho(\mathbf{r}) = \Psi^*(\mathbf{r})\Psi(\mathbf{r}) = |\ \Psi(\mathbf{r})|^2 \tag{12.90}$$

$$\therefore\ \Psi(\mathbf{r}) = \sqrt{\rho(\mathbf{r})}\exp\left\{+\mathrm{i}\theta(\mathbf{r})\right\} \tag{12.91}$$

and

$$\Psi^*(\mathbf{r}) = \sqrt{\rho(\mathbf{r})}\exp\left\{-\mathrm{i}\theta(\mathbf{r})\right\}, \tag{12.92}$$

where $\theta(\mathbf{r})$ is the phase of the wavefunction.

The canonical momentum p of a classical particle of mass m moving with a velocity **v** is

$$\boldsymbol{p} = m v + (q/c)\boldsymbol{A}, \tag{12.93}$$

where q is the electronic charge, c is the velocity of light and **A** is the magnetic vector potential.

Replacing **p** by the quantum-mechanical momentum operator $-\mathrm{i}\hbar\nabla$, we can write

$$m v = -\mathrm{i}\hbar\nabla - (q/c)\boldsymbol{A} \tag{12.94}$$

$$v = (1/m)\{-\mathrm{i}\hbar\nabla \ -\ (q/c)\boldsymbol{A}\}. \tag{12.95}$$

Let us recall the derivation of the relationship between the current density J and velocity v of free electrons. Consider a conductor of cross-sectional area A and length L. Then the volume of the conductor is

$$V = AL. \tag{12.96}$$

If n is the number of electrons/volume of the conductor, then the total number of electrons in the conductor is

$$N = ALn. \tag{12.97}$$

If q is the charge of a single electron, the total charge on all the electrons in the conductor is

$$Q = qALn. \tag{12.98}$$

Time taken by charge carriers to traverse the length L of the conductor is

$$t = L/v. \tag{12.99}$$

Current I is defined as

$$I = Q/t. \tag{12.100}$$

Substituting for Q and t from equations (12.98) and (12.99) into equation (12.100), we obtain

$$I = Q/t = (qALn)/(L/v) = qAnv. \tag{12.101}$$

The current density is

$$J = I/A = qAnv/A = qnv. \tag{12.102}$$

In quantum mechanics

$$nv = \Psi^* v \Psi. \tag{12.103}$$

Hence,

$$\mathbf{J} = q\Psi^* \mathbf{v}\Psi \tag{12.104}$$

Putting the values of Ψ from equation (12.91), Ψ^* from equation (12.92) and v from equation (12.95) into (12.104) for $\mathbf{J}$, we have

$$\begin{aligned}
\mathbf{J} &= q\Psi^* \mathbf{v}\ \Psi = q\sqrt{\rho}\ \exp(-\mathrm{i}\theta)(1/m)\{-\mathrm{i}\hbar\nabla - (q/c)\boldsymbol{A}\}\sqrt{\rho}\ \exp(+\mathrm{i}\theta) \\
&= q\sqrt{\rho}\ \exp(-\mathrm{i}\theta)(1/m)\left[\{-\mathrm{i}\hbar\nabla - (q/c)\boldsymbol{A}\}\sqrt{\rho}\ \exp(+\mathrm{i}\theta)\right] \\
&= q\sqrt{\rho}\ \exp(-\mathrm{i}\theta)(1/m) \\
&\quad\times\left[\left\{-\ \mathrm{i}\hbar\sqrt{\rho}\ \exp(+\mathrm{i}\theta)\times(+\mathrm{i})\nabla\theta - (q/c)\boldsymbol{A}\sqrt{\rho}\ \exp(+\mathrm{i}\theta)\right\}\right] \\
&= q\sqrt{\rho}\ \exp(-\mathrm{i}\theta)(1/m)\sqrt{\rho}\ \exp(+\mathrm{i}\theta)[\{-\mathrm{i}\hbar\times(+\mathrm{i})\nabla\theta - (q/c)\boldsymbol{A}\}] \\
&= (q\rho/m)[\{-\mathrm{i}\hbar\times(+\mathrm{i}\theta)\nabla\theta - (q/c)\boldsymbol{A}\}] = (q\rho/m)\{\hbar\nabla\theta - (q/c)\boldsymbol{A}\}.
\end{aligned} \tag{12.105}$$

But from Maxwell's fourth equation

$$\mathbf{J} = (c/4\pi)(\nabla\times\mathbf{B}). \tag{12.106}$$

Inside the superconductor, $\mathbf{B} = 0$ by Meissner's effect. Hence, $\mathbf{J} = 0$, and therefore

$$(q\rho/m)\{\hbar\nabla\theta - (q/c)\boldsymbol{A}\} = 0 \tag{12.107}$$

or

$$\hbar\nabla\theta - (q/c)\boldsymbol{A} = 0$$

or

$$\hbar\nabla\theta = (q/c)\boldsymbol{A}. \tag{12.108}$$

Performing line integration of both sides along a closed path C inside the superconducting ring, we obtain

$$\oint \hbar\nabla\theta\cdot d\boldsymbol{l} = \oint (q/c)\boldsymbol{A}\cdot d\boldsymbol{l} \tag{12.109}$$

or

$$(\hbar c)\oint \nabla\theta\cdot d\boldsymbol{l} = q\oint \boldsymbol{A}\cdot d\boldsymbol{l} \tag{12.110}$$

On the left-hand side,

$$\oint \nabla\theta\cdot d\boldsymbol{l} = \int_{\theta_1}^{\theta_2} \nabla\theta\cdot d\boldsymbol{l} = \theta_2 - \theta_1. \tag{12.111}$$

This is the phase difference on travelling once around the loop. The uniqueness of the Cooper-pair wavefunction imposes the restriction that the integral of the phase difference taken once around a closed loop can take on values equal to integer multiples of 2π only. Hence,

$$\theta_2 - \theta_1 = 2\pi\Xi, \tag{12.112}$$

where Ξ is an integer. So,

$$(\hbar c)\oint \nabla\theta\cdot d\boldsymbol{l} = (\hbar c)2\pi\Xi. \tag{12.113}$$

On the right-hand side of equation (12.110), Stoke's theorem is applied to obtain

$$q\oint \boldsymbol{A}\cdot d\boldsymbol{l} = q\iint (\nabla\times\mathbf{A})\cdot d\boldsymbol{s}. \tag{12.114}$$

Note that $\oint \boldsymbol{A}.\ d\boldsymbol{l}$ is a line integral over the closed contour C; $d\boldsymbol{l}$ is a linear element whereas

$$\iint (\nabla\times\mathbf{A})\cdot d\boldsymbol{s} \tag{12.115}$$

is a surface integral over the surface covering the interior of the closed path C; $d\mathbf{s}$ is an areal element.

But

$$\nabla\times\mathbf{A} = \mathbf{B} \tag{12.116}$$

so that

$$q\oint \boldsymbol{A}\cdot d\boldsymbol{l} = q\iint \mathbf{B}\cdot d\boldsymbol{s} = q\mathbf{B}\cdot\mathbf{s} = q\boldsymbol{\Phi}, \tag{12.117}$$

where equation (12.89) has been used.

Combining equations (12.110), (12.113) and (12.117),

$$(\hbar c)2\pi\Xi = q\boldsymbol{\Phi} \tag{12.118}$$

$$\therefore \boldsymbol{\Phi} = \frac{(\hbar c)2\pi\Xi}{q} = \frac{hc}{2\pi q} \times 2\pi\Xi = \left(\frac{hc}{q}\right)\Xi = \Phi_0\Xi. \tag{12.119}$$

The line integral of **A** around a contour passing through the superconducting electrodes and the tunnel barrier yields the enclosed flux Φ. The path of integration of magnetic vector potential is taken from superconductor 2 to superconductor 1 across the tunnel barrier. Thus the modified supercurrent density is

$$J = J_C \sin\theta = J_C \sin\left\{\phi_2 - \phi_1 - (2\pi/\Phi_0)\oint_1^2 \mathbf{A}\cdot\mathrm{d}\boldsymbol{s}\right\} \tag{12.120}$$

from equation (12.87). J_C is the modified critical current density. Its relationship with previous critical current density j_C will be derived. In terms of θ, the modified form of equation (12.77) is

$$\frac{2q(V_1 - V_2)}{\hbar} = \frac{\partial\theta}{\partial t} = \frac{\partial}{\partial t}\left\{\phi_2 - \phi_1 - (2\pi/\Phi_0)\oint_1^2 \mathbf{A}\cdot\mathrm{d}\boldsymbol{s}\right\}. \tag{12.121}$$

If, however, no magnetic field is present, one can take $\mathbf{A} = 0$. Then the equality $\theta = \phi$ is valid.

After knowing about this gauge-invariant form of current, let us apply a magnetic induction **B** across the junction along the $-\hat{y}$-direction. Then

$$\mathbf{A}\cdot\mathrm{d}\boldsymbol{s} = -Bx\mathrm{d}z \tag{12.122}$$

and

$$\oint_1^2 \mathbf{A}\cdot\mathrm{d}\boldsymbol{s} = \oint_0^d -Bx\mathrm{d}z = -Bx[z]_0^d = -Bxd \tag{12.123}$$

where x, dz are the dimensions of the JJ element in the x, z directions respectively, and d is the thickness of the tunnel barrier.

Substituting for $\oint_1^2 \mathbf{A}\cdot\mathrm{d}\boldsymbol{s}$ in the equation for current density, i.e. equation (12.120), equation (12.73) is rewritten as

$$j = j_C \sin\{\phi - (2\pi/\Phi_0)(-Bxd)\} = j_C \sin\{\phi + (2\pi/\Phi_0)(Bxd)\}. \tag{12.124}$$

Integrating from 0 to L, the length over which JJ extends along the x-direction

$$\begin{aligned} J &= \int_0^L j(x)\mathrm{d}x = \int_0^L j_C \sin\{\phi + (2\pi/\Phi_0)(Bxd)\}\mathrm{d}x \\ &= j_C \int_0^L \sin\{\phi + (2\pi/\Phi_0)(Bxd)\}\mathrm{d}x. \end{aligned} \tag{12.125}$$

Putting

$$(2\pi/\Phi_0)(Bd) = \eta \tag{12.126}$$

$$J = j_C \int_0^L \sin(\phi + \eta x)\ \mathrm{d}x. \tag{12.127}$$

Let

$$u = \phi + \eta x. \tag{12.128}$$

Then

$$\mathrm{d}u = \eta \mathrm{d}x$$

or,

$$\mathrm{d}x = \mathrm{d}u/\eta. \tag{12.129}$$

Further,

$$\begin{aligned}
J &= j_C \int_0^L (\sin u\mathrm{d}u)/\eta = (j_C/\eta) \int_0^L \sin u\mathrm{d}u \\
&= \left[(j_C/\eta) \times (-\cos u)\right]_0^L \\
&= (j_C/\eta) \times [-\cos\{\phi + (2\pi/\Phi_0)(Bxd)\}]_0^L \\
&= [j_C/\{(2\pi/\Phi_0)(Bd)\}] \times [-\cos\{\phi + (2\pi/\Phi_0)(Bxd)\}]_0^L \\
&= [Lj_C/\{(2\pi/\Phi_0)(BdL)\}] \\
&\quad \times [-\cos\{\phi + (2\pi/\Phi_0)(BLd)\} + \\
&\qquad \cos\{\phi + (2\pi/\Phi_0)(B \times 0 \times d)\}] \\
&= [Lj_C/(2\pi\Phi/\Phi_0)] \times [-\cos\{\phi + (2\pi/\Phi_0)(\Phi)\} + \cos(\phi)] \\
&= [Lj_C/(2\pi\Phi/\Phi_0)][\cos\phi - \cos\{\phi + (2\pi\Phi/\Phi_0)\}].
\end{aligned} \tag{12.130}$$

To find the maximum current flowing through the junction for all possible values of ϕ, let us find

$$\begin{aligned}
\mathrm{d}J/\mathrm{d}\phi &= [Lj_C/(2\pi\Phi/\Phi_0)] \times (\mathrm{d}/\mathrm{d}\phi)\big[\cos\phi \\
&\qquad - \cos\{\phi + (2\pi\Phi/\Phi_0)\}\big] \\
&= [Lj_C/(2\pi\Phi/\Phi_0)] \times [-\sin\phi + \sin\{\phi + (2\pi\Phi/\Phi_0)\}].
\end{aligned} \tag{12.131}$$

Setting

$$\mathrm{d}J/\mathrm{d}\phi = 0 \tag{12.132}$$

$$-\sin\phi + \sin\{\phi + (2\pi\Phi/\Phi_0)\} = 0 \tag{12.133}$$

or,

$$
\begin{aligned}
\sin\{\phi + (2\pi\Phi/\Phi_0)\} - \sin\phi &= 0\\
2\sin[\{(\phi + 2\pi\Phi/\Phi_0) - \phi\}/2]\cos[\{(\phi + 2\pi\Phi/\Phi_0) + \phi\}/2] &= 0\\
2\sin(\pi\Phi/\Phi_0)\cos\{\phi + (\pi\Phi/\Phi_0)\} &= 0\\
2\sin(\pi\Phi/\Phi_0)\{\cos\phi\cos(\pi\Phi/\Phi_0) - \sin\phi\sin(\pi\Phi/\Phi_0)\} &= 0\\
\cos\phi\cos(\pi\Phi/\Phi_0) - \sin\phi\sin(\pi\Phi/\Phi_0) &= 0\\
\cos\phi\cos(\pi\Phi/\Phi_0) &= \sin\phi\sin(\pi\Phi/\Phi_0)\\
\cos(\pi\Phi/\Phi_0)/\sin(\pi\Phi/\Phi_0) &= \sin\phi/\cos\phi\\
\cot(\pi\Phi/\Phi_0) &= \tan\phi.
\end{aligned}
\tag{12.134}
$$

Therefore, from equations (12.130) and (12.134), the critical current density J_C corresponding to the gauge-invariant phase difference is obtained as follows

$$
\begin{aligned}
J_C &= \{Lj_C/(2\pi\Phi/\Phi_0)\}2\sin[\{2\phi + (2\pi\Phi/\Phi_0)\}/2]\\
&\quad\times\sin[\{(\phi + 2\pi\Phi/\Phi_0) - \phi\}/2]\\
&= \{Lj_C/(\pi\Phi/\Phi_0)\}\sin\{\phi + (\pi\Phi/\Phi_0)\}\sin(\pi\Phi/\Phi_0)\\
&= \{Lj_C/(\pi\Phi/\Phi_0)\}\{\sin\phi\cos(\pi\Phi/\Phi_0) + \cos\phi\sin(\pi\Phi/\Phi_0)\}\sin(\pi\Phi/\Phi_0)\\
&= [\{Lj_C/(\pi\Phi/\Phi_0)\}\sin(\pi\Phi/\Phi_0)]\\
&\quad\times[\cos\phi\sin(\pi\Phi/\Phi_0)\{\tan\phi\cot(\pi\Phi/\Phi_0) + 1\}]\\
&= [\{Lj_C/(\pi\Phi/\Phi_0)\}\sin(\pi\Phi/\Phi_0)]\{\cos\phi\sin(\pi\Phi/\Phi_0)(\tan^2\phi + 1)\}
\end{aligned}
\tag{12.135}
$$

by applying equation (12.134). Hence,

$$
\begin{aligned}
J_C &= \left[\left\{Lj_C/\left(\frac{\pi\Phi}{\Phi_0}\right)\right\}\sin(\pi\Phi/\Phi_0)\right]\{\cos\phi\ \sin(\pi\Phi/\Phi_0)(\sec^2\phi)\}\\
&= \left[\left\{Lj_C/\left(\frac{\pi\Phi}{\Phi_0}\right)\right\}\sin(\pi\Phi/\Phi_0)\right]\left\{\sin\left(\frac{\pi\Phi}{\Phi_0}\right)\cos\phi\right\}.
\end{aligned}
\tag{12.136}
$$

But

$$
\begin{aligned}
\cos\phi &= 1/\sqrt{1 + \cot^2\left(\frac{\pi\Phi}{\Phi_0}\right)} = 1/\sqrt{1 + 1/\left\{\tan^2\left(\frac{\pi\Phi}{\Phi_0}\right)\right\}}\\
&= \tan(\pi\Phi/\Phi_0)/\sqrt{\tan^2(\pi\Phi/\Phi_0) + 1}\\
&= \tan\left(\frac{\pi\Phi}{\Phi_0}\right)\Big/\sqrt{\sec^2\left(\frac{\pi\Phi}{\Phi_0}\right)} = \tan(\pi\Phi/\Phi_0)/\sec(\pi\Phi/\Phi_0)\\
&= \left\{\frac{\sin\left(\frac{\pi\Phi}{\Phi_0}\right)}{\cos\left(\frac{\pi\Phi}{\Phi_0}\right)}\right\}\left\{\frac{1}{\cos\left(\frac{\pi\Phi}{\Phi_0}\right)}\right\} = \sin(\pi\Phi/\Phi_0).
\end{aligned}
\tag{12.137}
$$

Therefore,

$$J_C = \left[\left\{Lj_C/\left(\frac{\pi\Phi}{\Phi_0}\right)\right\}\sin(\pi\Phi/\Phi_0)\right]\left\{\sin\left(\frac{\pi\Phi}{\Phi_0}\right)\Big/\sin\left(\frac{\pi\Phi}{\Phi_0}\right)\right\} \tag{12.138}$$

or

$$J_C = Lj_C\left|\sin\left(\frac{\pi\Phi}{\Phi_0}\right)\Big/\left(\frac{\pi\Phi}{\Phi_0}\right)\right|. \tag{12.139}$$

12.4 Inverse AC Josephson effect: Shapiro steps

Let us apply a combined (DC + AC) voltage $V(t)$. Let DC voltage = V_{DC}, AC voltage = $V_0 \cos \omega t$ where V_0 is the amplitude and ω is the angular frequency. Here, $V_0 \ll V_{DC}$ and ω is very high in the RF or microwave range. Thus a high-frequency (ω), small amplitude (V_0) AC voltage is superimposed on a large DC voltage (V_{DC}):

$$V(t) = V_{DC} + V_0 \cos \omega t. \tag{12.140}$$

Writing equation (12.77) for the AC Josephson effect

$$\frac{\partial\delta}{\partial t} = \frac{2q\{V(t)\}}{\hbar} = \frac{2q(V_{DC} + V_0 \cos \omega t)}{\hbar}, \tag{12.141}$$

where δ is the phase difference.

Integrating both sides with respect to time,

$$\int_0^t \left(\frac{\partial\delta}{\partial t}\right)\mathrm{d}t = \left(\frac{2q}{\hbar}\right)\int V_{DC}\mathrm{d}t + \left(\frac{2q}{\hbar}\right)\int V_0 \cos \omega t \ \mathrm{d}t \tag{12.142}$$

or

$$\int_0^t \partial\delta = \left(\frac{2q}{\hbar}\right)V_{DC}t + \left(\frac{2qV_0}{\hbar\omega}\right)\sin \omega t$$

or

$$\begin{aligned} \delta(t) - \delta(0) &= \left(\frac{2q}{\hbar}\right)V_{DC}t + \left(\frac{2qV_0}{\hbar\omega}\right)\sin \omega t \\ \therefore \delta(t) &= \delta(0) + \left(\frac{2q}{\hbar}\right)V_{DC}t + \left(\frac{2qV_0}{\hbar\omega}\right)\sin \omega t. \end{aligned} \tag{12.143}$$

From equations (12.143) and (12.73), the current density j through the junction is

$$j = j_C \ \sin\left\{\delta(0) + \left(\frac{2q}{\hbar}\right)V_{DC}t + \left(\frac{2qV_0}{\hbar\omega}\right)\sin \omega t\right\}, \tag{12.144}$$

where j is a frequency-modulated current that can be analyzed using the following approximation:

$$\sin(x + \delta x) \approx \sin x + \delta x \ \cos x \tag{12.145}$$

giving

$$j = j_{\mathrm{C}}\left[\sin\left\{\delta(0) + \left(\frac{2q}{\hbar}\right)V_{\mathrm{DC}}t\right\} + \left(\frac{2qV_0}{\hbar\omega}\right)\sin\omega t \ \cos\left\{\delta(0) + \left(\frac{2q}{\hbar}\right)V_{\mathrm{DC}}\right\}\right], \tag{12.146}$$

where we have put

$$x = \delta(0) + \left(\frac{2q}{\hbar}\right)V_{\mathrm{DC}}t \tag{12.147}$$

$$\delta x = \left(\frac{2qV_0}{\hbar\omega}\right)\sin\omega t. \tag{12.148}$$

In the first term within square brackets of equation (12.146),

$$\sin\left\{\delta(0) + \left(\frac{2q}{\hbar}\right)V_{\mathrm{DC}}t\right\} \tag{12.149}$$

$\hbar \to 0$ because $\hbar$ is very small, hence,

$$\left\{\delta(0) + \left(\frac{2q}{\hbar}\right)V_{\mathrm{DC}}t\right\} \to \infty. \tag{12.150}$$

$\sin(\infty)$ oscillates between -1 and $+1$; therefore, its average value over time is zero. So, this term is zero.

The second term within square brackets

$$\left(\frac{2qV_0}{\hbar\omega}\right)\sin\omega t \ \cos\left\{\delta(0) + \left(\frac{2q}{\hbar}\right)V_{\mathrm{DC}}t\right\} \tag{12.151}$$

will also time-average to zero since $\cos(\infty)$ behaves like $\sin(\infty)$. But it can be made non-zero by choosing the frequency ω of the applied AC field such that

$$\omega = 2qV_{\mathrm{DC}}/\hbar \tag{12.152}$$

because then this term becomes

$$\left(\frac{2qV_0}{\hbar\omega}\right)\sin\left\{\left(2qV_{\mathrm{DC}}/\hbar\right)t\right\}\cos\left\{\delta(0) + \left(\frac{2q}{\hbar}\right)V_{\mathrm{DC}}t\right\}. \tag{12.153}$$

Using the trigonometric identity

$$\sin a \ \cos b = (1/2)\{\sin(a + b) + \sin(a - b)\}, \tag{12.154}$$

equation (12.153) can be rewritten as

$$\begin{aligned}&(1/2)\left[\sin\left\{\left(2qV_{\mathrm{DC}}/\hbar\right)t + \delta(0) + \left(\frac{2q}{\hbar}\right)V_{\mathrm{DC}}t\right\}\right.\\ &\left.+\sin\left\{\left(2qV_{\mathrm{DC}}/\hbar\right)t - \delta(0) - \left(\frac{2q}{\hbar}\right)V_{\mathrm{DC}}t\right\}\right]\\ &= (1/2)\left[\sin\left\{\left(\delta(0) + \left(\frac{4q}{\hbar}\right)V_{\mathrm{DC}}t\right)\right\} + \sin\left\{-\delta(0)\right\}\right].\end{aligned} \tag{12.155}$$

As argued previously, the term

$$\sin\left\{\left(\delta(0) + \left(\frac{4q}{\hbar}\right)V_{\mathrm{DC}}t\right)\right\} \tag{12.156}$$

will time-average to zero because $\hbar \to 0$, so that

$$j = j_{\mathrm{C}}\left(\frac{2qV_0}{\hbar\omega}\right)(1/2)\sin\{-\delta(0)\} = -j_{\mathrm{C}}\left(\frac{qV_0}{\hbar\omega}\right)\sin\{\delta(0)\}. \tag{12.157}$$

This is a constant DC. This phenomenon of obtaining zero-frequency or DC supercurrent from combined (DC + AC) excitation is called the inverse AC effect.

Furthermore, the above analysis shows that at multiples of frequency ω, i.e. whenever

$$n\omega = 2qV_{\mathrm{DC}}/\hbar \tag{12.158}$$

or

$$2qV_{\mathrm{DC}} = n\hbar\omega, \tag{12.159}$$

where $n = 0, 1, 2, 3, \ldots$, the current density j has zero frequency. Hence, the DC current–voltage (I–V) characteristics of the junction contain a series of discrete steps having width $= 2q/\hbar$. The well-defined constant-voltage DC spikes generated in the I–V characteristics in response to distinct frequencies of external RF or microwave signals are known as Shapiro spikes or steps. For these distinct frequency values, the JJ acts as a frequency-to-voltage converter. A precise determination of the ratio $q/\hbar$ can be done from this experiment.

In general, the current through the JJ can be written as (Grosso and Parravicini 2014)

$$j = j_{\mathrm{C}}\sum_{+\infty}^{n=-\infty} J_{\mathrm{n}}\left(\frac{2qV_0}{\hbar\omega}\right)\sin\left\{\delta(0) + \left(\frac{2q}{\hbar}\right)V_{\mathrm{DC}}t + n\omega t\right\}, \tag{12.160}$$

where J_{n} are Bessel functions of the first kind and order n obtained from an infinite power series expansion:

$$J_n(x) = \sum\nolimits_{k=0}^{\infty}\frac{(-1)^k}{k!\Gamma(k+n+1)}\left(\frac{x}{2}\right)^{n+2k}. \tag{12.161}$$

12.5 Superconducting quantum interference devices

A SQUID works as an extremely sensitive magnetometer using one or more JJs and based on the principle of quantum interference. It is used for measuring very weak magnetic fields in microtesla or nanotesla (10^{-6}–10^{-9} Tesla) ranges (Mehta 2011), and up to 5×10^{-18} Tesla. A magnetometer is an instrument used to measure the magnitude and direction of a magnetic field at a point in space or the magnetization of a material. SQUIDs are of two types: (i) DC SQUID and (ii) AC or RF SQUID.

12.5.1 DC SQUID

A DC SQUID consists of two or more JJs connected in parallel (figure 12.9). Essentially, it is a dual-junction superconducting loop. It has a high sensitivity for magnetic field detection. Its fabrication is difficult and expensive.

Let us consider two weak links arranged in parallel connection as shown in figure 12.9. Let ϕ_{A1}, ϕ_{A2} be the phases adjacent to the junction A, and ϕ_{B1}, ϕ_{B2} be the phases adjacent to the junction B:

$$\phi_A = \phi_{A2} - \phi_{A1} \tag{12.162}$$

$$\phi_B = \phi_{B2} - \phi_{B1}. \tag{12.163}$$

Let ϕ_A, ϕ_B be the phase differences across the Josephson junctions A and B. Then the phase difference $\Delta\phi$ between junctions A, B is

$$\begin{aligned}\Delta\phi &= \phi_A - \phi_B = \phi_{A2} - \phi_{A1} - (\phi_{B2} - \phi_{B1}) = \phi_{A2} - \phi_{A1} - \phi_{B2} + \phi_{B1} \\ &= (\phi_{B1} - \phi_{A1}) + (\phi_{A2} - \phi_{B2}) = -\frac{2q}{\hbar}\int_{A1}^{B1} \mathbf{A}.\, ds - \frac{2q}{\hbar}\int_{B2}^{A2} \mathbf{A}.\, ds \\ &= -\frac{2q}{\hbar}\oint \mathbf{A}.\, ds = -\frac{2q}{\hbar}\Phi \\ &= -2\pi\frac{1}{h/(2q)}\Phi = -2\pi\frac{\Phi}{\Phi_0}\end{aligned} \tag{12.164}$$

since

$$h/(2q) = \Phi_0 \tag{12.165}$$

Φ is the magnetic flux penetrating the loop. Total current density flowing through the SQUID is

$$\begin{aligned}J &= J_C(\sin\phi_A + \sin\phi_B) \\ &= 2J_C\left[\sin\{(\phi_A + \phi_B)/2\}\cos\{(\phi_A - \phi_B)/2\}\right] \\ &= 2J_C\left[\sin\{(\phi_A + \phi_B)/2\}\cos\{(\Delta\Phi)/2\}\right]\end{aligned} \tag{12.166}$$

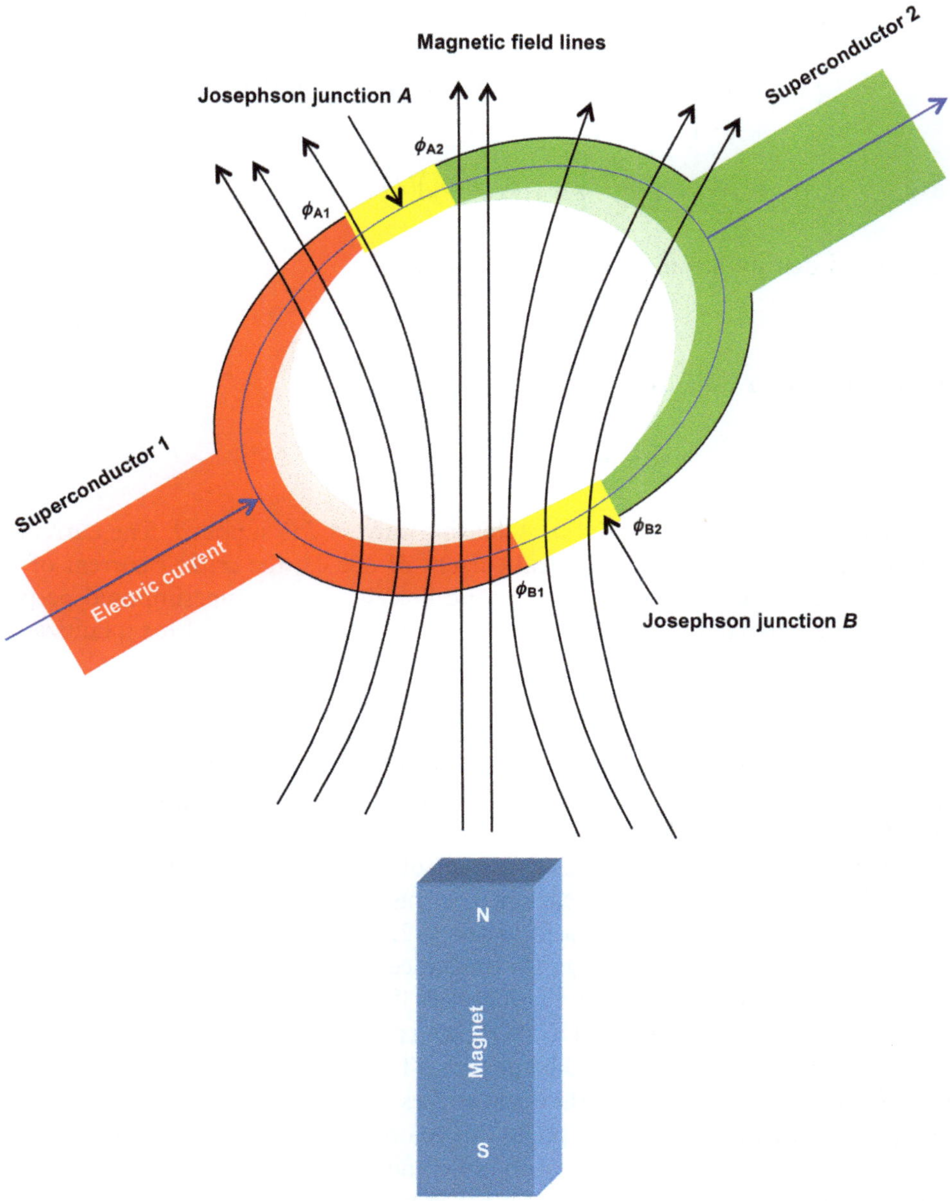

Figure 12.9. A DC SQUID.

where the formula

$$\sin a + \sin b = 2\sin\{(a + b)/2\}\cos\{(a - b)/2\} \tag{12.167}$$

has been applied.

Substituting for $\Delta\phi$ from equation (12.164) into equation (12.166)

$$\begin{aligned} J &= 2J_C\left[\cos\left\{\left(-2\pi\frac{\Phi}{\Phi_0}\right)\Big/2\right\}\sin\{(\phi_A + \phi_B)/2\}\right] \\ &= 2J_C\cos\left(\pi\frac{\Phi}{\Phi_0}\right)\sin\{(\phi_A + \phi_B)/2\} \end{aligned} \tag{12.168}$$

The current density has maximum value when

$$\sin\{(\phi_A + \phi_B)/2\} = 1 \tag{12.169}$$

and the maximum value is

$$J = 2J_C\ \cos\left(\pi\frac{\Phi}{\Phi_0}\right). \tag{12.170}$$

Consider a symmetric DC SQUID. A biasing current I is supplied. The current in each branch = $I/2$. If I_C is the critical current of one JJ, the critical current of the SQUID is $2I_C$. When an external magnetic flux $\Phi_{External}$ is superimposed perpendicular to the plane of the loop, a screening current I_S is produced. Flux quantization requires that

$$\Phi_{Total} = \Phi_{External} + LI_S = n\Phi_0, \tag{12.171}$$

where L is inductance of the loop, n is an integer and Φ_0 is a flux quantum. When $\Phi_{External} = n\Phi_0$, then $I_S = 0$. But when $\Phi_{External} = (n + ½)\Phi_0$, then $I_S = \pm(\Phi_0/2\ L)$. Thus I_S varies periodically with $\Phi_{External}$. Due to the flow of screening current I_S around the SQUID loop, the critical current of the SQUID decreases from $2I_C$ to $(2I_C - 2I_S)$. Hence, the critical current is a periodic function of $\Phi_{External}$. If the SQUID is biased with a current a little above $2I_C$, the output voltage is a periodic function of $\Phi_{External}$. Thus the SQUID transduces the magnetic flux variations into voltage changes. An electronic circuit linearizes the periodic response of the SQUID.

An optical analogy for the DC SQUID is provided by the famous two-slit interference experiment performed by physicist Thomas Young (figure 12.10). In this experiment, a monochromatic light beam strikes two slits. The light waves emerging from the slits are diffracted. The spread out beams meet in synchronization at some points. Here, they are aligned crest-to-crest and trough-to-trough. At these locations, they interfere constructively. At other points, they fall out of synchronization. At these points, they interfere destructively. Thus a pattern consisting of alternate bright and dark bands called interference fringes is formed on a screen placed behind the slits. Similar to the optical experiment, the critical current intensity of a DC SQUID is modulated in proportion to the magnetic flux. The critical current shows an ideal Fraunhoffer interference pattern. The two JJs behave as the two slits. The interfering waves are the supercurrents flowing through the opposite halves of the ring. The difference in the phases of the supercurrents is produced by the magnetic field.

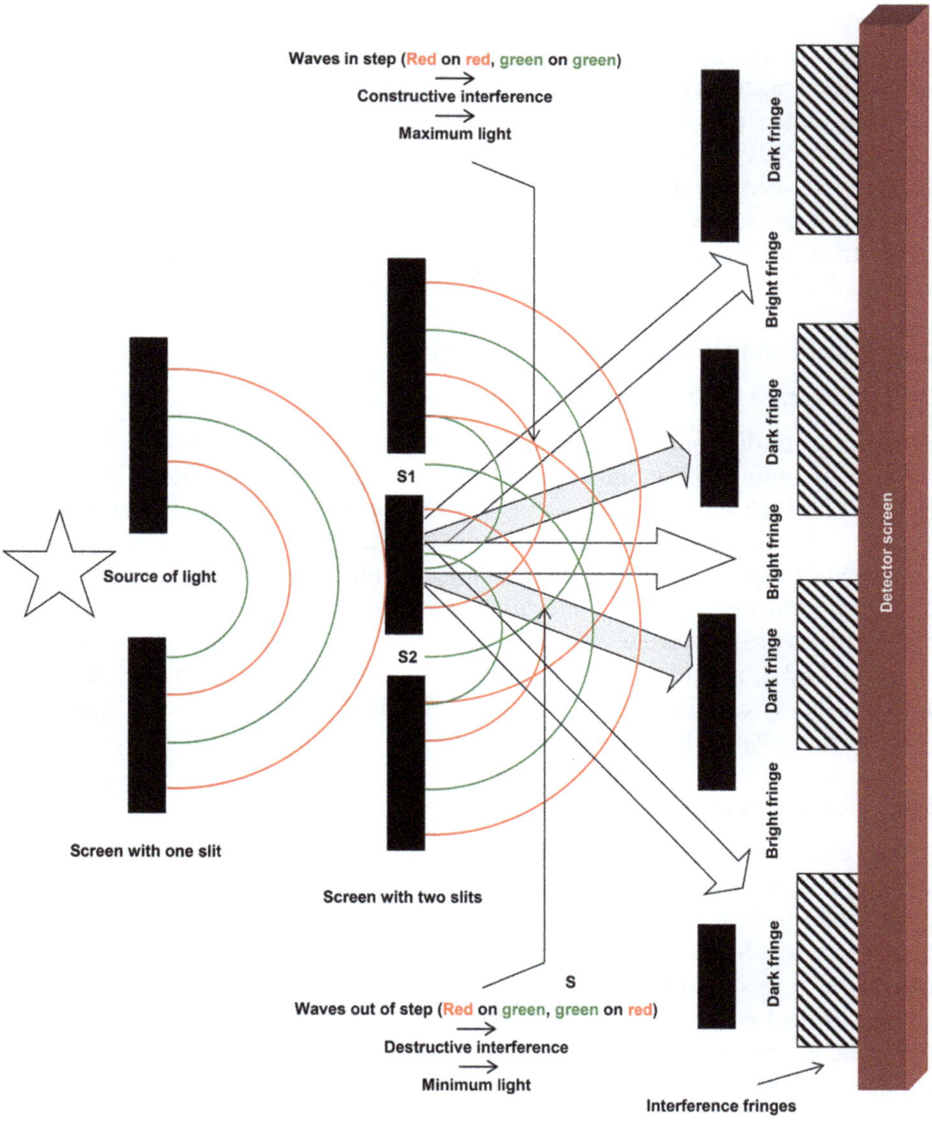

Figure 12.10. Young's double-slit experiment.

12.5.2 The AC or RF SQUID

Unlike the DC SQUID, an AC SQUID contains only one JJ. Its sensitivity is less, its fabrication is simpler and it is less costly. The sample is moved through the superconducting pickup coil. Its movement produces an alternating flux in the pickup coil. The pickup coil is part of a superconducting circuit used for transferring the magnetic flux from the sample to the RF SQUID placed inside a liquid-helium bath. The magnetic flux received by the SQUID is converted into a voltage. The voltage signal is amplified and read out by an electronic circuit.

The reason for the lower sensitivity of the single junction RF SQUID in comparison to the dual-junction DC SQUID is that in the RF SQUID, only the junction is engaged in gathering the flux, whereas in the DC SQUID, the total area of the loop takes part in picking up the magnetic flux, not the junctions alone.

The JJs in SQUIDs are fabricated from pure niobium or a lead alloy containing 10% Au or In, because pure Pb is thermally unstable. The tunnel barrier is formed by oxidation of the surface of the niobium base electrode. Over the tunnel barrier, the top electrode of the lead alloy is deposited, forming a JJ of structure: niobium–oxide–lead alloy.

12.6 Rapid single flux quantum logic

12.6.1 Difference from traditional logic

Operation of traditional logic circuits is based on two voltage levels, a high voltage level corresponding to the full power supply voltage and a low voltage level for zero voltage. The high voltage level represents the logic high or logic 1 state. The low voltage level indicates a logic low or logic 0 state. Rather than using two voltage levels, RSFQ logic works with two distinct conditions, one in which the voltage pulse is present and a second in which it is absent (Hutchby *et al* 2002). These voltage pulses are generated from the quantization of magnetic flux. As mentioned in section 12.3.4, magnetic flux threading a superconducting loop does not change continuously. It changes in discrete steps. Each step is known as a quantum of magnetic flux (Φ_0) or fluxon.

12.6.2 Generation of RSFQ voltage pulses

The device, which produces the voltage pulse corresponding to the change of magnetic flux, is the non-hysteric JJ consisting of a JJ shunted by an external resistor R_n. It is said to be non-hysteric in the sense that its current–voltage characteristics are non-hysteric. Suppose at a particular instant of time, the junction is biased in the superconducting state ($V = 0$) by supplying a biasing current of magnitude lower than the critical current I_C. Suppose at this instant, the flux supported by the loop equals one fluxon. An input signal current pulse of magnitude exceeding the critical current (typically 100 μA) switches the junction into the resistive state ($V \neq 0$). For this purpose, a short-duration DC pulse may be used, such as from a semiconductor device. Due to switching of the JJ, a very short voltage pulse is induced across the JJ. This voltage pulse has an area given by

$$\text{area} = \int V(t)\mathrm{d}t = 2.07 \ \text{mV ps}. \tag{12.172}$$

If the pulse is 1 ps wide, the amplitude will be 2 mV. If is <1 ps wide, the amplitude is proportionally higher. Generally, the pulse amplitude is given by the equation

$$\text{amplitude} \approx 2I_C R_n. \tag{12.173}$$

Consequent upon temporary closure of the JJ, a quantum of flux is ejected out of the loop comprising the JJ. After the input signal current pulse is over, the JJ again switches back to its original superconducting state ($V = 0$). Such restoration of the circuit to its initial state is analogous to the 2π rotation of a pendulum bob. The pendulum is said to be overdamped and does not continue its oscillations after reaching the starting point.

Logic circuits are constructed around the aforesaid voltage pulses that are produced each time that a fluxon is ejected out or enters a superconducting loop. The voltage pulses propagate along superconducting transmission lines energizing complex circuits implementing NOT, AND, OR, XOR, etc, logic functions.

12.6.3 RSFQ building blocks

By organization of three basic structural units (Brock *et al* 2000), one can perform various binary logic functions:

(i) *Active transmission stage*. This consists of a JJ and an inductor with small inductance.

(ii) *Storage loop*. This loop contains a JJ and an inductor with a large inductance through which a persistent current I_p is maintained around a loop.

(iii) *Decision-making pair*. This consists of two JJs of different sizes and hence different critical currents I_{C1} and I_{C2}.

12.6.4 RSFQ reset–set flip-flop

Looking at the circuit diagram of reset–set (R–S) flip-flop (figure 12.11), it is easily surmised that the circuit is essentially a DC two-JJ SQUID. The two JJs comprising the DC SQUID are the junctions J_3 and J_4. This flip-flop functions as follows (Brock

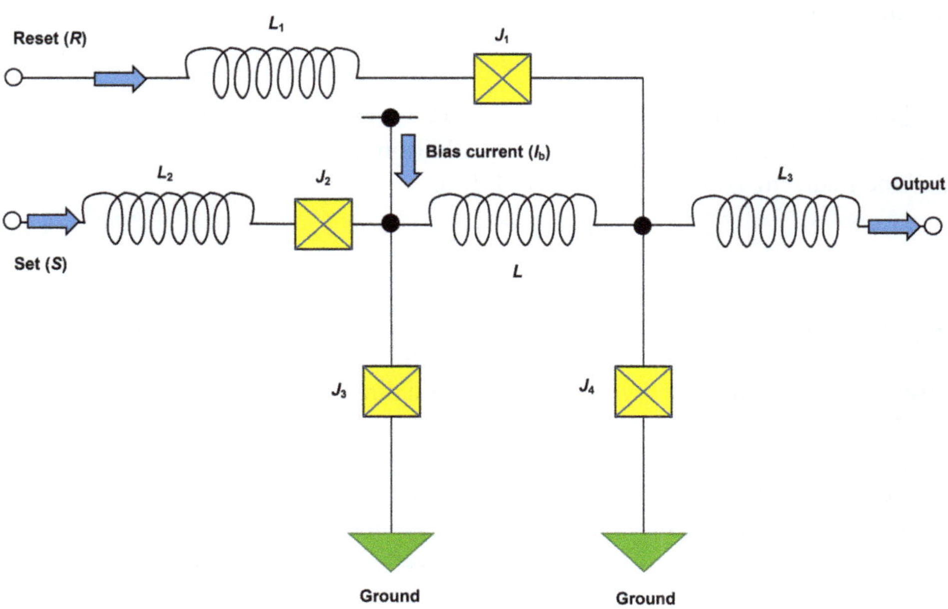

Figure 12.11. An RSFQ R–S flip-flop.

et al 2000): in the initial condition, the circuit is in the logic low state. In this logic low state, the persistent current flowing through JJ J_3 has a magnitude I_p and its direction is anticlockwise. From the circuit diagram, it is obvious that the direction of the persistent current I_p is the same as the direction of the biasing current $I_b/2$ in JJ J_3 (Likharev and Semenov 1991). The biasing current is halved because it is divided into two branches while entering JJ J_3. Hence, the currents $I_b/2$ and I_p add up to give the total current flowing through JJ J_3 as

$$I_3 = I_b/2 + I_p \tag{12.174}$$

For JJ J_4 the persistent current I_p and the biasing current I_b are flowing in opposite directions. Hence the total current I_4 flowing through J_4 is obtained by subtracting I_p from $I_b/2$, so that we can write

$$I_4 = I_b/2 - I_p. \tag{12.175}$$

Clearly $I_4 < I_3$.

On arrival of a set pulse, JJ J_1 transmits it to the storage loop. This set pulse can trigger a 2π jump in JJ J_3 carrying a larger current I_3 and not in JJ J_4, which carries a smaller current I_4. Consequent upon this switching of the JJ J_3 by 2π, the direction of current flowing through JJ J_3 reverses from the original anticlockwise direction to the clockwise direction. This current circulating around the storage inductance through JJ J_3 constitutes the logic high state of the flip-flop. Thus in the logic high state, the persistent current flowing through JJ J_3 has the same magnitude I_p as in the logic low state but its direction of flow is reversed from the anticlockwise to clockwise. Due to this reversal of current, the two currents flowing through JJ J_3 are I_p in the clockwise direction and $I_b/2$ in the anticlockwise direction, so that the total current I_3 flowing through JJ J_3 becomes

$$I_3 = I_b/2 - I_p. \tag{12.176}$$

Similarly, the two currents flowing through JJ J_4 are I_p and $I_b/2$. Both these currents are flowing in the clockwise direction. By addition of these two currents, we find the combined current I_4 flowing through JJ J_4 as

$$I_4 = I_b/2 + I_p. \tag{12.177}$$

As soon as a reset pulse arrives, a 2π jump is triggered in JJ J_4 carrying a larger current whereby the current flowing through the junction J_4 starts flowing in the clockwise direction. But current flow through JJ J_4 in the clockwise direction corresponds to current flow in the anticlockwise direction through JJ J_3, which is the logic low state, from which we started. At the same time, a single flux quantum pulse $V(t)$ is produced across the JJ J_2 which serves as an output signal F.

Now consider what happens if the flip-flop is in the logic high state with current I_p flowing in the clockwise direction through JJ J_3, and the set pulse arrives. Then the set pulse cannot cause switching of the JJ J_3 carrying a low current

$$I_3 = I_b/2 - I_p. \tag{12.178}$$

Instead, it switches the junction J_2 carrying a higher current $I_b/2$. Due to this switching the pulse falls out of the circuit, i.e. the fluxon escapes from the circuit. Hence, no pulse appears at the output. Similarly, the junction J_1 protects the source of the RSFQ pulse from the reverse reaction of the interferometer on receipt of an erroneous reset pulse.

Thus the flip-flop has two stable states, one in which the current I_p is flowing through JJ J_3 in the anticlockwise direction (logic low state) and the other in which the current I_p is flowing through JJ J_3 in the clockwise direction (logic high state).

12.6.5 RSFQ NOT gate or inverter

The RSFQ NOT gate (figure 12.12) contains one DC SQUID consisting of two parallel JJs J_2 and J_3 (Likharev and Semenov 1991). An additional JJ J_1 is connected

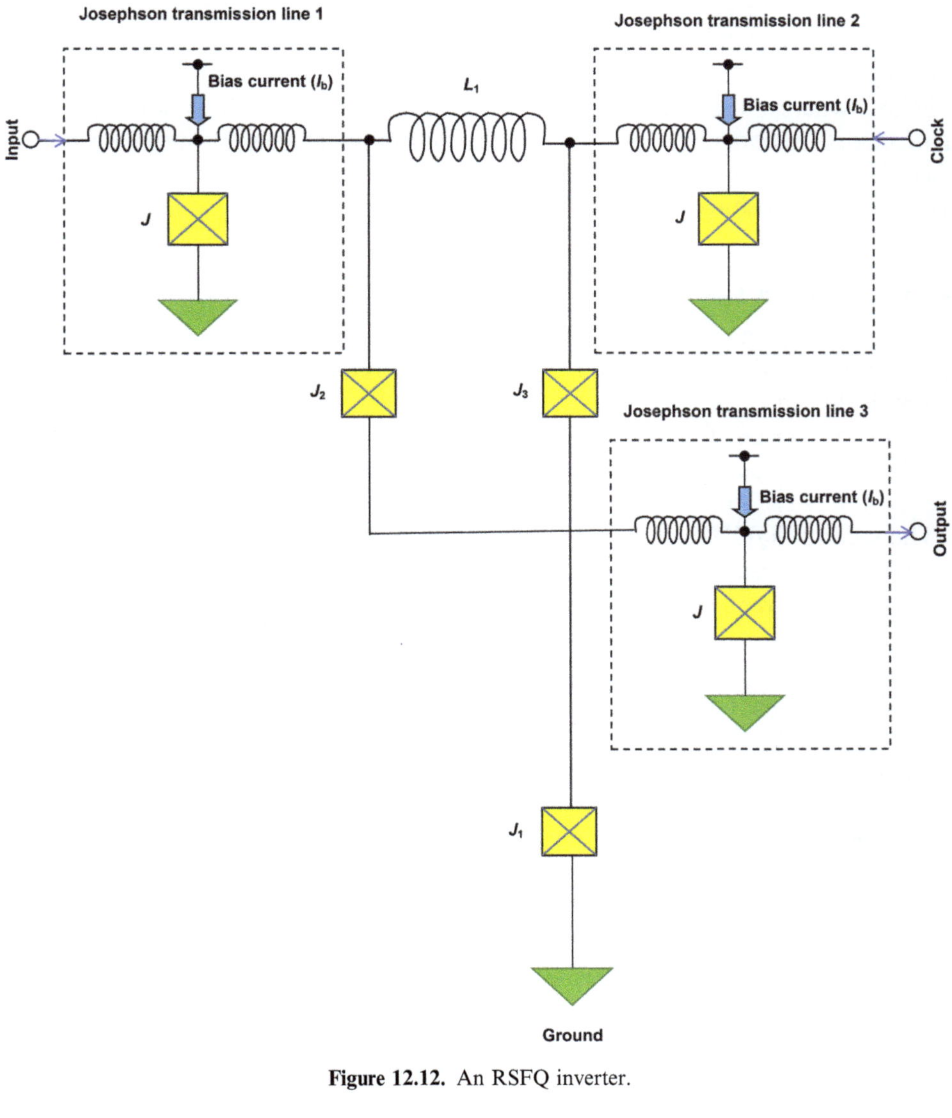

Figure 12.12. An RSFQ inverter.

between JJ J_3 and the ground terminal. The circuit contains two parallel branches for the flow of current: J_2–J_1–ground and L_1–J_3–J_1–ground. In the initial state of the circuit, suppose a high current is flowing through JJ J_2 while low current is flowing through JJ J_3. Thus a high current is flowing along the branch J_2–J_1–ground, whereas a low current is flowing along the branch L_1–J_3–J_1–ground. The input pulse is zero. Application of a clock pulse switches the JJ J_1 in the high-current branch J_2–J_1–ground. It does not trigger the junction J_3 in the low-current branch L_1–J_3–J_1–ground. The resultant RSFQ pulse produced in JJ J_1 appears at the output of the circuit. Thus in the absence of any input pulse, a pulse appears at the output. This is NOT operation.

Now if an input pulse arrives, the operation is similar to that of an R–S flip-flop with a pulse applied to the set terminal. The JJ J_2 switches so that JJ J_2 carries a low current and JJ J_3 carries a high current. At this stage, a high current is flowing along the branch L_1–J_3–J_1–ground, whereas a low current is flowing along the branch J_2–J_1–ground. On applying a clock pulse, the junction J_3 in the high-current branch L_1–J_3–J_1–ground is switched, whereas the junction J_1 in the low current branch J_2–J_1–ground remains unaffected. Consequently, there is no pulse at the output. Hence, the output pulse is zero when the input pulse is present. Once again, NOT operation is verified. Thus in both cases, the circuit behaves as a NOT gate or inverter.

12.6.6 RSFQ OR gate

In this circuit (figure 12.13), two DC SQUIDs can be identified (Likharev and Semenov 1991). The first SQUID consists of the JJs (J_1, J_2). The two stable states of this SQUID are:

(i) high current flowing through JJ J_1 with low current through JJ J_2; and
(ii) high current flowing through JJ J_2 with low current through JJ J_1.

The second SQUID comprises the JJs (J_5, J_6). Its two stable states are:

(i) high current flowing through J_5, low current in J_6; and
(ii) high current in J_6, low current in J_5.

The following cases arise:

(i) There are no input pulses during the clock period. The next clock pulse produces RSFQ pulses across J_4, J_8, which do not reach output.
(ii) Only input pulse IN1 arrives during the clock period. High-current-carrying junction J_2 switches due to the CLK pulse. The generated RSFQ pulse passes through L_3, triggering an RSFQ pulse across J_9, and passes to the output through J_{10} and L_5. At the same time, an RSFQ pulse is produced across J_{12}.
(iii) Only input pulse IN2 arrives during the clock period. A similar process to step (ii) occurs in the symmetrically located lower part of the circuit.
(iv) Both input pulses IN1 and IN2 arrive during the clock period. Both J_2 J_6 produce RSFQ pulses at the same time. RSFQ pulses are created across J_9 and J_{11} at the same time. Only one RSFQ pulse occurs at the output

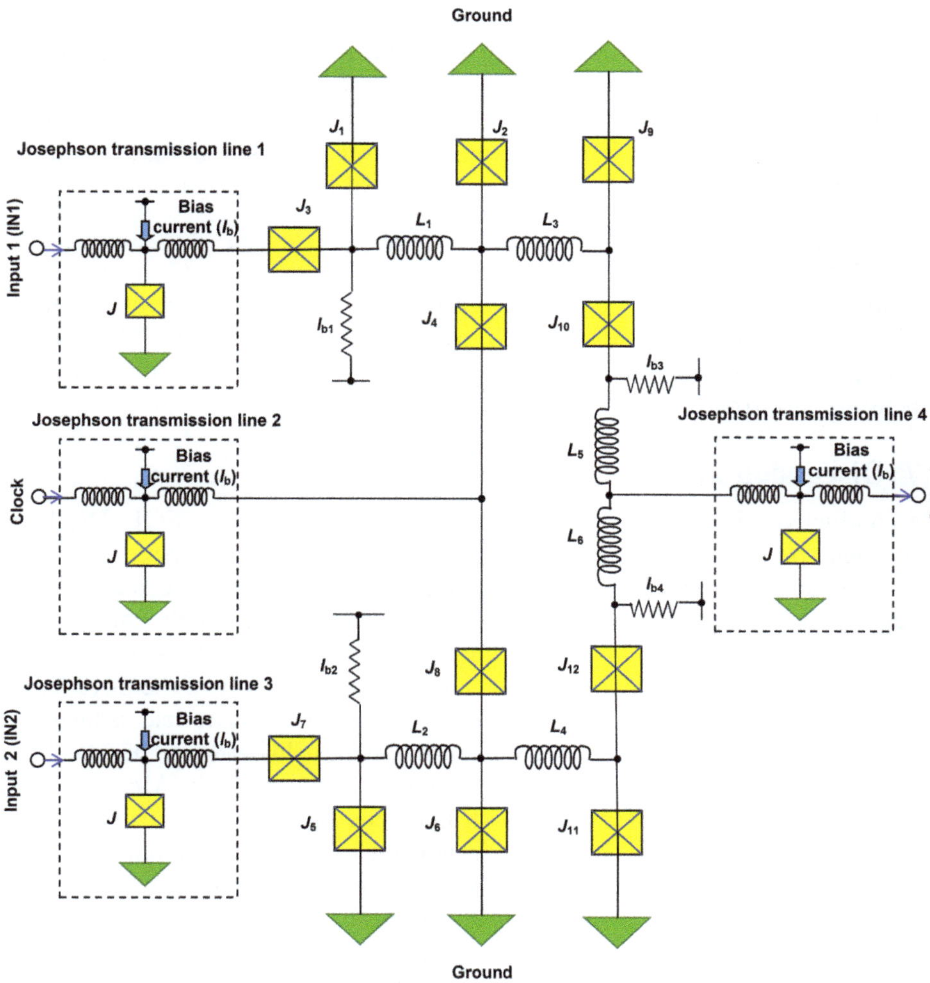

Figure 12.13. An RSFQ OR circuit.

because J_9, J_{11} are connected in parallel, whereas J_{10} and J_{12} remain in the superconducting state.

All the above operations are in conformity with that of an OR gate. Hence, the circuit works as an OR gate.

12.6.7 Advantages of RSFQ logic

The fabrication technology for RSFQ logic is simple. The commonly used superconductor is niobium or niobium nitride. The tunnel barrier in the JJs is aluminum oxide. The power consumption of RSFQ logic is extremely low. Each time an RSFQ pulse passes through the junction, the energy consumed = circulating current (100×10^{-6} A) $\times$ quantity of flux = 10^{-4} A $\times 2 \times 10^{-15}$ Wb = 2×10^{-19} J. The voltage pulses are conveyed to neighboring gates either along microstrip transmission lines or through active lines containing JJs to introduce timing delays necessary for synchronization. The superconducting interconnections are virtually loss-free. The losses are practically zero at DC and very low in comparison to metals up to frequencies around 750 GHz. Together with low power, the RSFQ logic circuits also provide superfast operation at frequencies of several hundred gigahertz.

12.6.8 Disadvantages of RSFQ logic

The main disadvantage is the cooling requirement for cryogenic operation, but the popularity of small-size closed-cycle refrigeration units has considerably eased the situation, making possible the integration of cryo-cooled RSFQ circuits with room temperature electronics.

12.7 Discussion and conclusions

Superconductive electronics encompasses various fields such as SQUIDs, filters, RSFQ logic, power transmission and current limiting devices, to name a few of its diverse applications. Superconductors belong to one of the two categories, low-temperature and high-temperature. An HTS does not work at a high temperature in reality; the term is used in a relative context to distinguish this class from the low-temperature class. Superconducting behavior vanishes when any of the three parameters, either critical temperature, critical current or critical magnetic field, is surpassed. A superconductor does not permit entry of any magnetic field inside it. The prohibition or expulsion of magnetic field by a superconductor is known as the Meissner effect. The correlation of superconductivity with magnetic phenomena leads to the BCS theory of superconductivity. The most important device in superconductive electronics is the JJ. A phenomenon of paramount significance is the Josephson effect, which is of two types: a DC Josephson effect characterized by current flow across a JJ in the absence of an applied voltage and an AC Josephson effect in which a DC voltage produces an alternating electric field. The wave functions of the Cooper pairs in the two superconducting electrodes of the JJ, are defined to develop the theory of the Josephson effect. The theory leads to two simple equations, one concerning the rate of change of Cooper pair density in the two superconducting electrodes, and the other dealing with the rate of change of phase difference between the wavefunctions. These equations are used to interpret the DC and AC Josephson effects. The theory is placed on a rigorous foothold by introducing a gauge-invariant phase difference.

Unlike the BCS theory of superconductivity, the Ginzburg–Landau theory is a macroscopic theory in which the two basic equations are derived by minimizing the free energy function with respect to an order parameter. The Ginzburg–Landau theory does not provide any explanation of the physical origin of superconductivity. On the contrary, it commences on the presumption that superconductivity exists. Then it proceeds towards modeling of the superconducting properties.

London's first equation is derived by writing the equation of motion of an electron in an electric field. For an electron in a superconductor, the retarding force due to electron scattering is set to zero. By taking the curl of London's first equation and applying Maxwell's third equation, London's second equation is obtained. Applying London's second equation and Maxwell's fourth equation, and with the help of Maxwell's second equation, it is shown after mathematical manipulation that a magnetic field can be uniform inside a superconductor only if it is zero everywhere. The differential equation for the spatial dependence of magnetic induction shows that the magnetic induction decays exponentially with distance from the exterior surface of the superconductor, explaining the Meissner effect.

The inverse AC Josephson effect involves the creation of a supercurrent by exciting a JJ with a combination of small amplitude, high-frequency AC signal on a large DC signal. The DC current contains a series of discrete steps called Shapiro steps at multiples of the frequency of the AC signal.

Magnetic fields of very feeble strengths can be accurately measured using quantum interference devices based on superconductivity. These devices are known as SQUIDs. A DC SQUID consists of two or more parallel-connected JJs. On applying a biasing current on the SQUID and superimposing an external magnetic flux on the loop, the supercurrents flowing through the two halves of the loop undergo interference, producing a periodic voltage waveform, which is processed by an electronic circuit. Thus the SQUID transduces the flux variations into voltage changes. An AC SQUID contains only a single JJ. The sample is moved through a pickup coil, which is part of the SQUID circuit. The resulting time-varying flux produced in the pickup coil influences the SQUID converting the flux changes into voltage signals.

The rudimentary principles of RSFQ logic circuits have been introduced. This logic is differentiated from routine voltage-based logic. The key switching elements in this logic are overdamped JJs. Operations of R–S flip-flop, NOT and OR gates are outlined.

Review exercises

12.1. What is a superconductive material? Define superconductive electronics and elaborate its scope mentioning the fields of study which fall under it.

12.2. What is the full form of: (a) SQUID and (b) RSFQ? What are they used for?

12.3. What is a low-temperature superconductor? Can the current flow through such a material for years without decreasing?

12.4. What happens when the temperature of a superconductor is lowered to a value close to absolute zero? How does a normal conductor behave near absolute zero? What is meant by the critical temperature of a superconductor? Give three examples of superconductors. Mention the critical temperature in each case.

12.5. What is the Meissner effect? What is meant by the London penetration depth of a superconductor? Write the equation for penetration depth and perform its calculation.

12.6. What are the three parameters whose value when exceeded leads to loss of the superconducting property? Define these parameters.

12.7. Define the following terms for a superconductor: (a) critical temperature, (b) critical magnetic field and (c) critical current density.
12.8. What are type I and type II superconductors? Give one example of each class. Which type is practically more useful? How do type I and type II superconductors differ with regard to the Meissner effect?
12.9. What behavior of a superconductor in a magnetic field leads to the correlation of superconductivity with the magnetic properties of materials? How are the electron spins arranged when a material is magnetized?
12.10. What is a Cooper pair? What is meant by the coherence length of a superconductor?
12.11. What type of particle is an electron: fermion or boson? What type of particle is a Cooper pair: fermion or boson? Do bosons obey Pauli's exclusion principle?
12.12. Explain the main idea of the BCS theory of superconductivity in terms of mediation of electrons in a Cooper pair via phonon interaction.
12.13. Why are the electrons paired in the form of Cooper pairs at low temperatures but become unpaired at high temperatures?
12.14. Describe the applications of superconductivity in MRI machines and particle accelerators.
12.15. How does an HTS differ from a low-temperature superconductor? Is the high temperature required for a superconductor really high or it is a relative comparison with temperatures near absolute zero required for conventional superconductors?
12.16. What is a JJ? What are the two types of structures used for fabricating JJs?
12.17. What is the DC Josephson effect? How is it interpreted on the basis of tunneling of Cooper pairs? How does the tunneling of Cooper pairs differ from the tunneling of electrons? How is the Josephson current related with the phase difference between wavefunctions of Cooper pairs on opposite sides of the JJ? How does the flow of Josephson current refute common physics laws?
12.18. What is the AC Josephson effect? Do the oscillations produced across a JJ depend on the size of the superconductors? Do they depend on the voltage applied across the junction?
12.19. What happens when the wavefunctions ψ_1 and ψ_2 of the Cooper pairs in the two superconducting electrodes of a JJ come closer together? Write the time-dependent Schrödinger equations for ψ_1 and ψ_2? What is the opacity constant? Derive the following governing equations for the Josephson effect:

$$\frac{\partial \rho_1}{\partial t} = - \frac{\partial \rho_2}{\partial t} \tag{12.179}$$

$$\frac{\partial \phi}{\partial t} = \frac{2qV}{\hbar}, \tag{12.180}$$

where ρ_1, ρ_2 are densities of Cooper pairs in superconductors 1 and 2, respectively, $\phi = \phi_2 - \phi_1$, where ϕ_1, ϕ_2 are the phases of ψ_1 and ψ_2, V is the applied voltage, q is the electronic charge and $\hbar = h/2\pi$; h is Planck's constant. Utilizing these basic equations, explain the phenomena of DC and AC Josephson effects.

12.20. Define gauge-invariant phase difference θ. What is the magnetic flux quantum Φ_0? Taking the gauge-invariant phase difference into account, derive the equation relating the modified critical current density J_C when a gauge-invariant phase difference θ is considered, with the critical current density j_C when a phase difference ϕ is taken in place of θ:

$$J_C = Lj_C\left|\sin\left(\frac{\pi\Phi}{\Phi_0}\right)\right| \Big/ \left|\left(\frac{\pi\Phi}{\Phi_0}\right)\right|, \tag{12.181}$$

where L is the length across which the JJ extends and Φ is the flux enclosed by a contour passing through the superconducting electrodes and the tunnel barrier.

12.21. How does the Ginzburg–Landau theory of superconductivity differ from the BCS theory? What is the wavefunction ψ proposed in this theory? Expand the free energy function with respect to ψ and derive the first equation of Ginzburg–Landau by equating to zero its derivative with respect to ψ. Equate to zero the derivative of free energy with respect to $\mathbf{A}$ and derive the second equation of Ginzburg–Landau.

12.22. Derive London's two equations starting from the equation of motion of an electron moving in an electric field. Provide an explanation for the Meissner effect by applying London's second equation.

12.23. What is the inverse AC Josephson effect? Show that application of a (DC + AC) field to a JJ produces a DC supercurrent containing a series of steps at multiples of the AC frequency. What are these steps called?

12.24. What is a SQUID? Mention one important application of this device.

12.25. How many JJs does a DC SQUID contain? How is the quantum interference produced in a DC SQUID? Draw the similarity between the operation of a DC SQUID and Young's two-slit experiment. To which parts of the DC SQUID do the two slits in Young's experiment correspond? What physical entities in a DC SQUID represent the interfering waves in Young's experiment?

12.26. How many JJs does an AC SQUID contain? How does an AC SQUID work? Why is the AC SQUID less sensitive than a DC SQUID?

12.27. Describe the use of flux quantization for performing logic operations. What is this logic called?

12.28. What are the main structural units of RSFQ logic? Why is a shunt resistor included along with the JJ?

12.29. Draw the circuit diagram of an R–S flip-flop in RSFQ logic. Explain its operation. How do you define the two stable states of the R–S flip-flop?

12.30. It is necessary to implement a NOT operation in RSFQ logic? How many SQUIDs are used for this implementation? Explain the operation of the NOT gate with its circuit diagram.

12.31. Explain with the help of a circuit diagram the operation of an OR gate in RSFQ logic.
12.32. Is RSFQ logic inter-operable with CMOS logic? What are its advantages and disadvantages relative to CMOS logic?

References

Brock D K, Track E K and Rowell J M 2000 Superconductor ICs: the 100-GHz second generation *IEEE Spectr.* **37** 40–6

Grosso G and Parravicini G P 2014 *Solid State Physics* (New York: Academic)

Hutchby A, Bourianoff G I, Zhirnov V V and Brewer J E 2002 Extending the road beyond CMOS *IEEE Circuits Devices Mag.* **18** 28–41

Koblischka M R and Koblischka-Veneva A 2022 Editorial: superconductivity 2022 *Metals* **12** 1–10

Likharev K and Semenov V 1991 RSFQ logic/memory family: a new Josephson-junction technology for sub-terahertz clock frequency digital systems *IEEE Trans. Appl. Supercond.* **1** 3–28

Onnes H K 1911 The resistance of pure mercury at helium temperatures *Communications from the Physical Laboratory of the University of Leiden* **12** 120

Mehta N 2011 *Applied Physics for Engineers* (New Delhi: PHI Learning)

Scientific American 1997 What are Josephson junctions? How do they work? *Sci. Am.* November www.scientificamerican.com/article/what-are-josephson-juncti/

Superconductivite 2015 March 2 Electronics with superconductors *Superconductivite* www.supraconductivite.fr/en/index.php?p=applications-electronique-more

Tsang T 1997 *Classical Electrodynamics* (Singapore: World Scientific)

IOP Publishing

Extreme-Temperature and Harsh-Environment Electronics
(Second Edition)
Physics, technology and applications
Vinod Kumar Khanna

Chapter 13

Superconductor-based microwave circuits operating at liquid-nitrogen temperatures

High-temperature superconductor (HTS)-based compact planar microwave circuits have been designed, fabricated and packaged, and their functioning has been corroborated. Prominent materials used for substrates include sapphire, magnesium oxide and lanthanum aluminate. Practical realizations of circuits largely exploit microstrip designs that are fabricated utilizing thin-film techniques. High-temperature superconducting films of YBCO and TBCCO have been used. These HTS thin films fulfil the critical insertion loss specifications of cellular communication networks, satellite broadcasting, international navigation and space flights. While filters fabricated by traditional technologies decrease in sensitivity if selectivity is increased, HTS filters can provide maximum sensitivity concomitantly with maximum selectivity. These filters exhibit characteristics closely approximating those of an ideal filter, namely 100% acceptance of frequencies in the passband and 100% rejection of frequencies in the stopband. They assure complete rejection of frequencies in proximity to the allowed band, thereby reducing interference. The combination of steep-skirted filters exhibiting rapid attenuation of transmission outside the desired band with related microwave components enables the realization of a low-noise front end of a receiver. These HTS microwave circuits provide the necessary weight and volume reduction, although the cryogenic cooling equipment for keeping temperatures in the 60–80 K range is an indispensable appendage to these circuits.

doi:10.1088/978-0-7503-5072-3ch13

13.1 Introduction

A superconductor does not offer any resistance to the flow of DC or zero-frequency current, i.e. the opposition to zero-frequency current flow is zero. However, similar remarks are not applicable to a current of any frequency. In the case of an AC, the superconductor acts differently from its DC behavior. In fact, the resistance of a superconductor must be mentioned with reference to frequency, because it changes with the frequency of the current. The higher the frequency of current, the larger is the resistance of a superconductor.

Although during operation at a frequency in the microwave range, the surface resistance of a superconductor might be quite large, it is still appreciably less than that of a conductor at that frequency, e.g. at 1 GHz the surface resistance of a superconductor at liquid-nitrogen temperature (77 K) is 3–4 orders of magnitude lower than that of copper at the same temperature. For this reason, it is advantageous to use a superconductor in a microwave circuit in place of a conductor. Although this has been known for a long time, the application of superconductors in microwave circuits did not gain popularity because of the associated expensive and cumbersome cooling gadgetry required for building such circuits. However, with the advent of HTSs requiring liquid-nitrogen temperatures instead of liquid-helium temperatures necessary for low-temperature superconductors, the circumstances improved and the stage was arranged for acceptance of superconductors in microwave circuits (Vendik *et al* 2000).

Substitution of a normal conductor by an HTS helps to decrease the insertion loss drastically. Additionally, a microwave component can be fabricated at an enormously reduced dimensional scale leading to a tremendously compact form factor. For microwave resonators, an extremely high quality factor is achievable by using HTS. The performance of microwave resonators, filters, phase shifters and antennas has been vastly improved by following this strategy (Willemsen 2001, Faisal 2012).

The high quality factor resonators enabled by low loss properties of superconductors can handle high current densities facilitating the development of superconducting particle accelerators and high performance analog electronics. Their macroscopic quantum properties have paved the way towards high-speed digital electronics working on the Josephson effect. Their microscopic quantum properties have led to quantum computers, which are fundamentally quantum engines fueled by microwaves (Anlage 2021).

13.2 Substrates for microwave circuits

The following are desirable properties for substrates to be used for microwave circuits. First, the loss tangent of the substrate material must be very low at the high frequencies. Second, the material must be mechanically strong. Third, the temperature coefficient of the substrate material must match closely with that of the HTS film to be deposited over it. If this matching is not very close, the thickness of the HTS film is restricted by temperature-induced stresses, which cause the film to crack and delaminate. Fourth, the substrate material should not undergo any chemical reaction with the HTS film, in order that its chemical integrity is preserved. Lastly, but most importantly, the material should be economically worthwhile. Only when it is available at an affordable cost will it be widely adopted.

Materials which fulfill these criteria, to some extent, are sapphire (aluminum oxide, Al_2O_3), lanthanum aluminate ($LaAlO_3$ or LAO) and magnesium oxide (MgO). Sapphire has an anisotropic dielectric constant $\varepsilon = 8.9\text{–}11$ and its thermal expansion properties match poorly with those of the HTS films. However, it is relatively inexpensive. Its mechanical robustness combined with the availability of large-size sapphire substrates may lead to its extensive adoption. LAO has a relative permittivity of 25. It is expensive and suffers from local permittivity variations when subjected to thermal cycling owing to structural transitions of phase. LAO single crystals are used for epitaxially growing cuprate superconductors. MgO has found widespread use as a substrate material for microwave filters (Ramesh *et al* 1990).

13.3 HTS thin-film materials

13.3.1 Yttrium barium copper oxide

YBCO, also called yttrium barium cuprate or Y123, and represented by the formula $YBa_2Cu_3O_{7-\delta}$ ($0 \leqslant \delta \leqslant 1$) is a black crystalline solid with perovskite structure. When $\delta = 0.07$, the material becomes superconducting at 92 K. The process known as the solid-state reaction is used to prepare YBCO. The process starts by mixing the precursor powders, yttrium oxide (Y_2O_3), barium carbonate ($BaCO_3$) and copper oxide (CuO), in the stoichiometric ratio 1:2:3 for 2–3 h in an agate mortar. The mixture is calcined at 840 °C for 14 h in an alumina crucible in a programmable furnace:

$$Y_2O_3 + 4BaCO_3 + 6CuO \rightarrow 2YBa_2Cu_3O_{7-\delta} + 4CO_2. \tag{13.1}$$

The temperature is increased to 950 °C to remove CO_2 gas. The calcined powder is pressed into pellets, followed by sintering and annealing in oxygen ambience.

YBCO thin films are deposited by techniques such as sputtering from a ceramic target or using pulsed laser deposition (PLD) by melting and evaporating the target material. Metal–organic chemical vapor deposition (MOCVD) is another method applied for HTS thin-film deposition.

13.3.2 Thallium barium calcium copper oxide

TBCCO films in the 2212 phase ($Tl_2Ba_2CaCu_2Oy$) have a transition temperature $T_C > 105$ K, so that a smaller cooling system with low input power suffices. To prepare a TBCCO film, a Ba–Ca–Cu–O precursor film free from thallium is first deposited epitaxially. This film is thallinated by annealing at 800 °C in the presence of thallium.

13.4 Fabrication processes for HTS microwave circuits

The process steps for the fabrication of these circuits are borrowed from standard semiconductor technology. Necessary modifications are performed wherever applicable. The steps followed are photolithography, ion milling, metallization for contact formation, and dicing. Ion milling is used in the absence of a reliable wet or dry etching chemistry. The aim is to provide reliable control of the line width of cuprate superconductors. The usual contact metals are used. Lift-off or chemical etching processes are employed for pattern definition. During dicing, care must be taken to avoid edge damage. The brittleness of MgO or ceramic substrates makes dicing difficult.

13.5 Design and tuning approaches for HTS filters

HTS filters are predominantly based on the popular microstrip technology. The microstrip is a planar transmission line. This transmission line is used for conveying signals at microwave frequencies. It consists of a conducting strip separated from a ground plane by a dielectric layer. This dielectric layer is the insulating substrate.

The filter design is based on four parameters:

(i) the size of the filter (and hence the chip area);
(ii) the quality factor;
(iii) the power handling capability; and
(iv) the insensitivity of filter response to manufacturing variations (Simon *et al* 2004).

For a distributed element resonator, maximization of Q and power handling requires a large line width. The disadvantage of this design is the large size necessary for a half-wavelength resonator at high frequencies. The constraints on wafer size and cost have led to two possibilities of reducing the die size. These are the lumped-element resonator or folded half-wavelength resonator. A lumped-element resonator yields a compact structure as a capacitively loaded inductance (figure 13.1). But the extremely small line widths required make it prone to variations in fabrication.

The folded half-wave type resonators include the clip resonator (figure 13.2) and spiral in/spiral out (SISO) resonator. In filter circuits, the commonly used geometry for distributed element resonators is the single spiral resonator (figure 13.3) because of the compactness in size achieved. The main issue with the single spiral structure is the inconvenience in accessing its inner terminal for which either a via hole or an air bridge is necessary. This difficulty is avoided by using the SISO geometry. For further size miniaturization, an on-chip capacitor is used, as shown in figure 13.4. The terminals of SISO geometry are easily accessible. Two feedline structures are used. These are: insert-tapped and insert-coupled (figure 13.5).

The clip resonator serves as a high-performance filter for wireless applications. Quasi-elliptic filters of bandpass and band-reject type are fabricated with the SISO resonator.

In a practical application, accurate adjustment and alignment of resonators constituting an HTS filter is essential. Generally, the frequency response deviates marginally from the designed value due to manufacturing tolerances. The frequency response is adjusted to the desired value by a process called tuning or trimming the filter. Filter tuning involves transferring the passband of the filter to the desired frequency as well as optimization of return loss of the filter. The conventional method using mechanical tuning elements such as metal screws introduces disappointingly high losses. Therefore, a low-loss device, e.g. a sapphire cylinder fixed to a threaded metal element, is employed. As the resonant frequency of an HTS

Figure 13.1. Schematic diagram of the capacitance loaded inductor filter.

resonator varies with temperature, it can be reset by a temperature control scheme. For electronically tuning a filter, either the capacitance or inductance of its resonator is changed.

Often during the design of an HTS thin-film microwave device, a ferroelectric thin-film such as single-crystal strontium titanate (STO) is included (Wooldridge *et al* 1999) for *in situ* tuning of the frequency response based on the electric field dependence of the ferroelectric material. A change in permittivity with electric field allows tunability of capacitance.

13.6 Cryogenic packaging

The package uses a cryo-cooler, a standalone cooler and a Dewar; the Dewar is a vacuum insulated vessel, built to provide long-lasting, reliable operation. The cryo-cooler must be small in size to decrease the overall size of the system and also reduce power consumption. Typically, the cryo-coolers must provide several watts of cooling capacity at filter operating temperatures ~60–80 K and surround temperatures up to 50 °C, with minimal maintenance and at affordable costs.

Figure 13.2. The clip resonator configuration.

Figure 13.3. The single spiral resonator.

A long-life Dewar may be built by reducing vacuum leaks and controlling the outgassing by materials. The Dewar is permanently welded after evacuation and the DC/RF feedthroughs are sealed for hermeticity at leakage rates $<1 \times 10^{-10}$ cm^3 s^{-1} helium. For outgassing, the Dewar must be maintained in a vacuum for several days to weeks at a temperature ~100 °C prior to sealing. A gettering agent may be incorporated to remove any leftover gas. Although a higher temperature quickens the outgassing, it may degrade the HTS material.

13.7 HTS bandpass filters for mobile telecommunications

Highly selective HTS filters improve the performance of mobile communication networks by increasing the data rates and redeeming frequency spectrum (Shen *et al* 1997).

Figure 13.4. On-chip capacitor-loaded SISO resonator.

13.7.1 Filter design methodology

A filter with a passband of 10 MHz is considered (figure 13.6). The response of the filter follows a ten-pole quasi-elliptic function. It is designed using a special topological layout. This geometry is a cascaded quadruplet trisection (CQT) coupling structure (Hong *et al* 2005). To design the filter, a set of design parameters is extracted from filter synthesis. The frequency response of the synthesized filter is analyzed using a circuit model, which has the required frequency offsets, coupling coefficients and external quality factors. The advantage of CQT topology is that the asymmetrical design leads to a smaller number of resonators and lower insertion loss in the passband. Furthermore, the efforts in tuning are minimized because the cross-couplings in the filter are mutually independent. The subsequent step in filter design

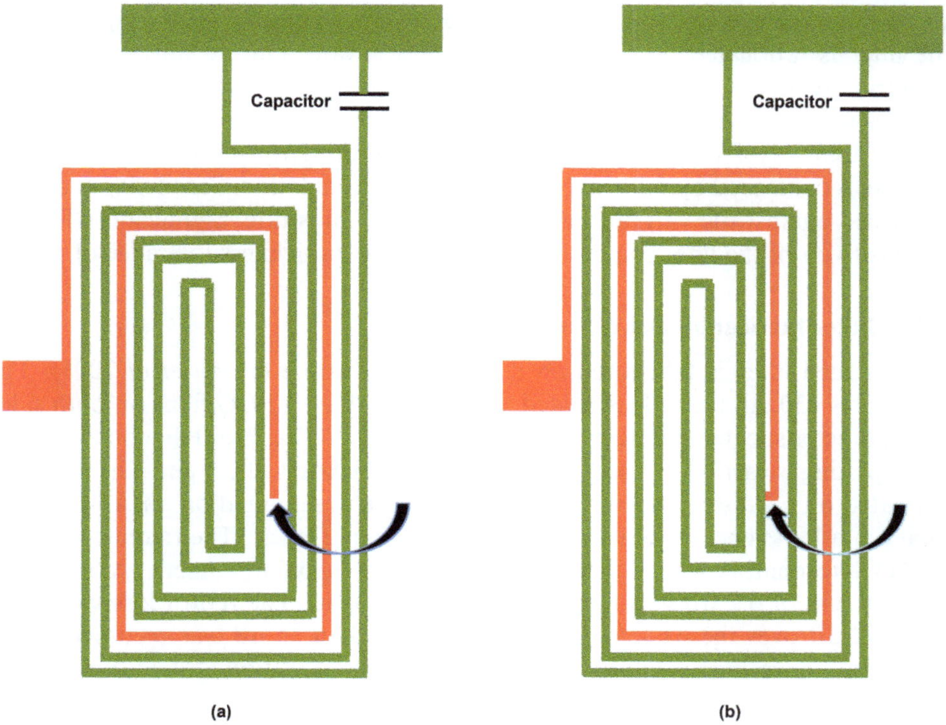

Figure 13.5. Feedline structures: (a) insert-tapped and (b) insert-coupled.

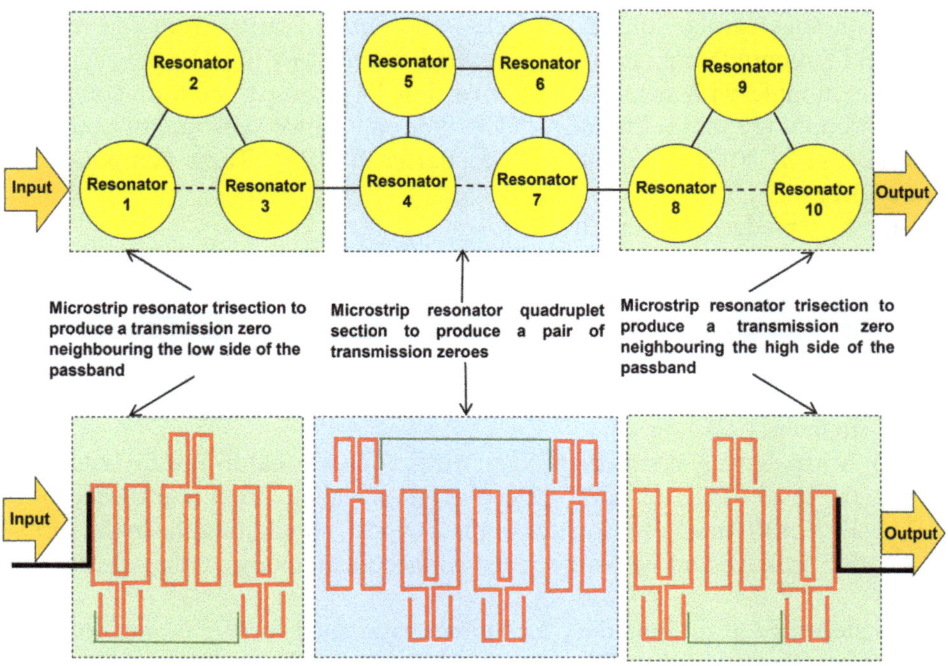

Figure 13.6. Execution plan and layout topology of the filter with a CQT coupling structure.

consists in implementation of the circuit model with a suitable microstrip structure. The filter is fabricated on an *r*-cut sapphire substrate. Due to anisotropy of the dielectric properties of sapphire, its relative permittivity is represented by a tensor. Arrangement of microstrip resonators on the anisotropic substrate is done in such a manner that all of them experience the same permittivity tensor. The frequency response is confirmed by full-wave electromagnetic simulation. Sensitivity analysis is performed against the variation in the thickness of substrate. Significant effects on the sensitivity of the filter response were noticed for a thickness tolerance of ±5 μm.

13.7.2 Filter fabrication and characterization

The designed filter is fabricated on a sapphire substrate (Hong *et al* 2005). This substrate is 430 μm thick. It is coated with 0.3 μm thick YBCO films on both sides. The characteristic temperature of YBCO films is 87 K. For RF contacts, gold electroplating is carried out on both sides up to a thickness of 0.3 μm. The filter thus fabricated on a substrate of size (47 mm × 17 mm) is mounted on an Au-plated Ti carrier in a brass enclosure placed inside a cryogenic Dewar. The frequencies of all the HTS resonators in the filter are tuned using sapphire tuners. A minimum insertion loss of 0.2 dB was measured in the passband. The return loss was better than −12 dB over the passband. Rejection better than 85 dB was recorded over a broad range covering the universal Mobile Telecommunication System transmission band extending from 2110 MHz to 2170 MHz.

13.8 HTS JJ-based frequency down-converter

HTS technology can be utilized in wireless telecommunication front-end receivers. An all-HTS monolithic microwave integrated circuit (MMIC) JJ frequency down-converter module with nearly zero conversion loss consists of four components integrated on a (10 mm × 20 mm) YBCO film on a single substrate of magnesium oxide (Du *et al* 2012, 2013, 2014, 2015; Bai *et al* 2013). These components are (figure 13.7):

(i) A step-edge JJ mixer circuit.
(ii) A 10–12 GHz bandpass filter (BPF) circuit for RF signal input. This eight-pole filter uses stepped-impedance hairpin resonators. Its best insertion loss is measured as 0.4 dB at 40 K and 0.8 dB at 77 K. The out-of-band rejection is better than 40 dB.
(iii) A low-pass filter (LPF) having a ~4 GHz cut-off frequency for intermediate frequency (IF) output.
(iv) A capacitively coupled HTS microstrip line resonator at 8 GHz to isolate the local oscillator (LO). Effective RF signal isolation is provided by the BPF, LPF and the resonator at the LO port, eliminating the losses through RF leakage and by wire-bonded connections amongst components.

HTS step-edge grain boundary technology is applied for the fabrication of the MMIC chip (Du *et al* 2014). The step-edge pattern is delineated on a 0.5 mm thick, 10 mm wide and 20 mm long MgO substrate. The techniques used are

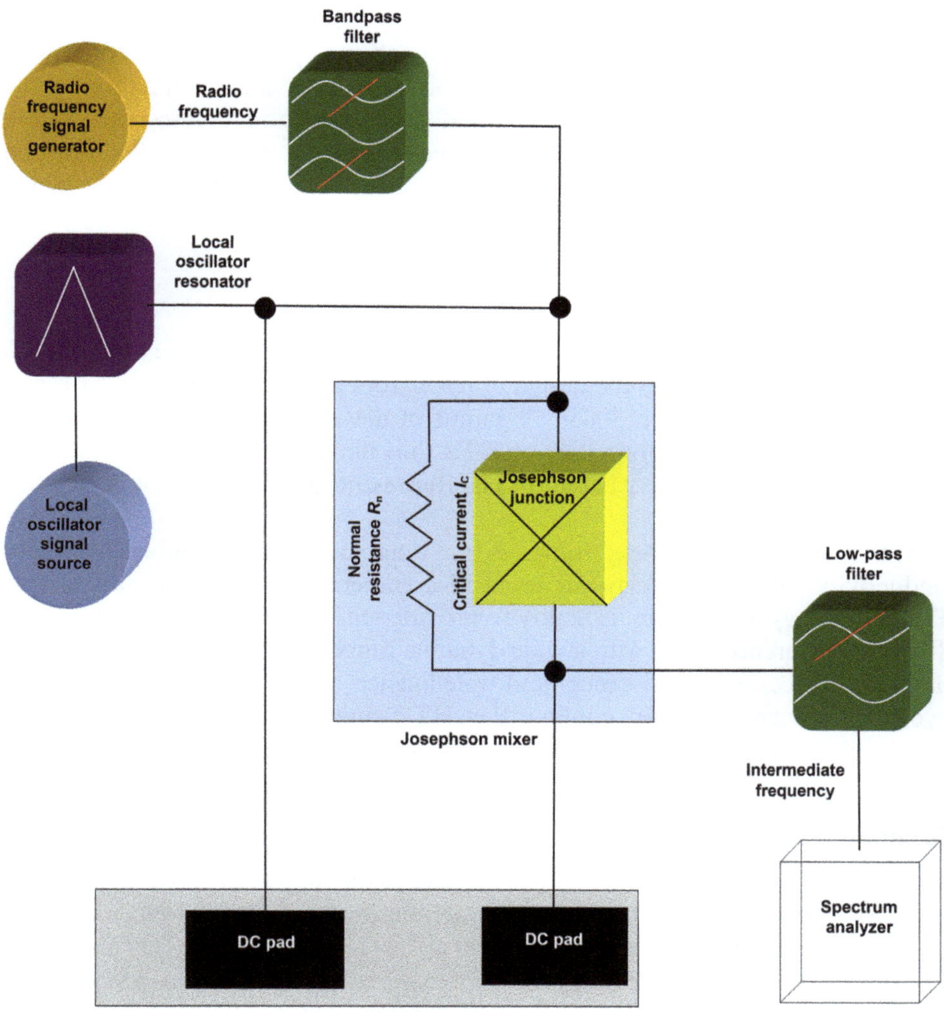

Figure 13.7. Frequency down-converter circuit.

photolithography, e-beam evaporation and argon ion beam etching, giving a step height of 0.4 μm and a step angle of 35°. A *c*-axis oriented YBCO film of thickness 0.22 μm is epitaxially deposited. Gold film of thickness 0.05 μm is deposited by e-beam evaporation. Patterning and etching the Au/YBCO film helps to form the junction. The gold film is removed from most regions, except for the contact pads and the lines feeding the filter. For packaging the MMIC chip, a gold-coated copper enclosure is used. Incorporating an RF blocking network and resistors in the DC biasing lines of the JJ prevents RF interference. The filters are linked to the connectors by silver epoxy. The chip is also grounded to the base plate by Ag-epoxy. The module has a total volume of 25 × 27 × 15 mm^3.

In comparison to a single Josephson mixer, the DC and RF characteristics showed a 7 dB improvement in conversion efficiency because of efficient RF coupling, isolation between the RF, IF and LO ports, and elimination of wire bonding. The MMIC circuit is integrated in a miniature cryo-cooler, and the complete assembly is packed inside a portable (350 mm × 350 mm × 250 mm) box. The circuit performed satisfactorily in an unshielded condition, confirming its usefulness in wireless communications.

13.9 Discussion and conclusions

Lowering of surface resistance of HTS materials by $\geqslant 10^3$-fold relative to cryo-cooled metallic conductors enables the design of resonators with quality factors $\geqslant 10^5$ in the unloaded condition (Simon 2003). A gamut of devices has been demonstrated by utilizing the exceptional properties of HTS thin films, notably high-Q resonators, filters, oscillators and mixers based on JJs. Representative examples of devices have been described.

HTS microwave microelectronics is an outcome of high-temperature superconductivity, offering widespread use and commercialization. The popularity of HTS technology arises from its ability to provide sensitivity and the rejection of out-of-band interference along with a wide dynamic range. All these virtues are essential for surveillance, reconnaissance and intelligence pursuits. The availability of cryogenic refrigerators for cooling the HTS materials at affordable costs is responsible for easy adoption of this technology in RF applications.

Review exercises

13.1. The resistance of an HTS increases with frequency. Give arguments in favor of using these materials in microwave circuits in place of normal conductors.

13.2. List the desirable properties of a material that qualify it as a substrate for the fabrication of microwave circuits. How far do the following materials meet these qualifications: MgO and alumina?

13.3. How is YBCO prepared by a solid-state reaction? What techniques are used for depositing thin films of this material?

13.4. Which material has a higher transition temperature: YBCO or TBCCO? Write the full forms of these acronyms.

13.5. What is the primary technology on which the design of HTS filters is based? Name four important parameters involved in design.

13.6. How is the deviation in the frequency response of a filter corrected? Why is the conventional method of mechanical tuning of filters not applicable to HTS filters?

13.7. How is cryogenic packaging of an HTS filter done? What precautions are taken to build a long-life, reliable Dewar?

13.8. What are the main steps in the design of an HTS filter for mobile telecommunications? How is the filter fabricated and packaged?

13.9. What are the four components of an all-HTS JJ-based frequency down-converter? Describe its fabrication and packaging. Discuss the characteristics of the circuit.

13.10. Name four types of microwave circuits realized using HTS thin films.

References

Anlage S M 2021 Microwave superconductivity *IEEE J. Microw.* **1** 389–402

Bai D D, Du J, Zhang T and He Y S 2013 A compact high temperature superconducting bandpass filter for integration with a Josephson mixer *J. Appl. Phys.* **114** 133906

Du J, Bai D D, Zhang T, Guo Y J, He Y S and Pegrum C 2014 Optimized conversion efficiency of a HTS MMIC Josephson down-converter *Supercond. Sci. Technol.* **2** 7105002

Du J, Wang J, Zhang T, Bai D, Guo Y J and He Y 2015 Demonstration of a portable HTS MMIC microwave receiver front-end *IEEE Trans. Appl. Supercond.* **25** 1500404

Du J, Zhang T, Guo Y J and Sun X W 2013 A high temperature superconducting monolithic microwave integrated Josephson down-converter with high conversion efficiency *Appl. Phys. Lett.* **102** 212602

Du J, Zhang T, Macfarlane J C, Guo Y J and Sun X W 2012 Monolithic HTS heterodyne Josephson frequency downconverter *Appl. Phys. Lett.* **100** 262604

Faisal W M 2012 High T_c superconducting fabrication of loop antenna *Alex. Eng. J.* **51** 171–83

Hong J-S, McErlean E P and Karyamapudi B M 2005 A high-temperature superconducting filter for future mobile telecommunication systems *IEEE Trans. Microw. Theory Tech.* **53** 1976–81

Ramesh R, Hwang D M, Barner J B, Nazar L, Ravi T S, Inam A, Dutta B, Wu X D and Venkatesan T 1990 Defect structure of laser deposited Y–Ba–Cu–O thin films on single crystal MgO substrate *J. Mater. Res.* **5** 704–16

Shen Z-Y, Wilker C, Pang P, Face D W, Carter C F and Harrington C M 1997 Power handling capability improvement of high-temperature superconducting microwave circuits *IEEE Trans. Appl. Supercond.* **7** 2446–53

Simon R 2003 High-temperature superconductor filter technology breaks new ground *Mobile Devices Des.* 28–37 http://mobiledevdesign.com/site-files/mobiledevdesign.com/files/archive/mobiledevdesign.com/images/archive/308RF_Simon28.pdf

Simon R W, Hammond R B, Berkowitz S J and Willemsen B A 2004 Superconducting microwave filter systems for cellular telephone base stations *Proc. IEEE* **92** 1585–96

Vendik O G, Vendik I B and Kholodniak D V 2000 Applications of high-temperature superconductors in microwave integrated circuits *Mater. Phys. Mech.* **2** 1515–24

Willemsen B A 2001 HTS filter subsystems for wireless telecommunications *IEEE Trans. Appl. Supercond.* **11** 60–7

Wooldridge I, Turner C W, Warburton P A and Romans E J 1999 Electrical tuning of passive HTS microwave devices using single crystal strontium titanate *IEEE Trans. Appl. Supercond.* **9** 3220–3

IOP Publishing

Extreme-Temperature and Harsh-Environment Electronics (Second Edition)
Physics, technology and applications
Vinod Kumar Khanna

Chapter 14

High-temperature superconductor-based power delivery

As the aging electricity infrastructure of metropolitan cities fails to cope with the burden of increasing power demands from customers, there is a dire need to face the challenge by introducing novel scientific ways of reducing the transmission losses incurred in sending electricity from generating plants to users. The widespread adoption of a high-temperature superconductor (HTS)-based power network is envisaged as a potent answer to the power crisis. In this chapter, the methods developed for producing current-carrying wires from brittle superconducting ceramics are described. Salient features of the first- and second-generation superconducting wires are mentioned. As the HTSs need to be incessantly cooled with liquid nitrogen, different architectures of the cables have evolved. The warm dielectric and cool dielectric designs are discussed. As the superconductors can readily undergo transition from low impedance to a high impedance state, they possess inherent fault limiting capabilities, which are utilized in resistive, shielded core and saturable-core devices for protecting against surges. Superconducting wire transformers offer several benefits, such as fault limitation and the absence of fire hazards, in addition to the incredibly low power losses.

14.1 Introduction

To cater to escalating electric power loads, attention must be paid to both:

(i) the generation of power, as well as
(ii) the transmission and distribution of power.

doi:10.1088/978-0-7503-5072-3ch14

As far as the transmission and distribution of power are concerned, the deployment of HTSs provides a convenient way of meeting this requirement. A systematic investigation and analysis of energy saving for power transmission by superconducting cables from a renewable energy source to a 100-MW-class data center showed that the energy lost in superconducting transmission is only 1.7% of that incurred in conventional AC transmission (Chen *et al* 2022). This calls for modernizing and upgrading the present power infrastructure. Let us therefore briefly review the existing power transmission structure to understand its shortcomings and limitations.

14.2 Conventional electrical power transmission

14.2.1 Transmission materials

Two elements constitute the backbone of power delivery. These are copper (electrical resistivity = $1.68 \times 10^{-8}\ \Omega$ m) and aluminum (resistivity = $2.65 \times 10^{-8}\ \Omega$ m). Copper is superior to aluminum due to its lower resistivity. Even though these resistivity values are small, they are finite and sufficient to introduce appreciable losses over long distances.

14.2.2 High-voltage transmission

Power transmission loss = I^2R (I = current and R = resistance of wire). To reduce the transmission loss, power is conveyed over long distances in the form of AC by stepping up its voltage and reducing the current. At the utility point, the power is stepped down to the needed lower voltage. Thus power transmission is essentially based on the principle of circumventing the resistive losses using step-up and step-down transformers. Large, high-voltage transformers are oil filled. The transformer coil is a hydrocarbon mineral coil composed of aromatics, paraffins, napthenes and olefins. It serves as an insulating medium between the internal parts and a coolant. In the case of short circuits, transformer fires causing oil vaporization and spreading are fearsome events, often accompanied by an explosion.

In addition to transformers, capacitive and inductive elements are used in power transmission to keep the waveform in synchronization. Further, it must be noted that the higher the voltage to which the current is boosted, the larger is the electromagnetic field surrounding the conducting wire, and therefore the larger are the losses incurred due to this field.

14.2.3 Overhead versus underground power delivery

Both modes have their advantages and drawbacks. The initial installation costs of overhead lines are lower than for underground lines. But overhead lines have to withstand the impact of weather attacks such as rain, storms, ice, etc. On the bright side, overhead lines are easily repaired and tapped. In underground lines, fault-finding becomes difficult because it is necessary to excavate the Earth and trace the fault location by trenching. Due to this reason, the duration of line outage may sometimes be longer for underground lines than for overhead lines.

14.3 HTS wires

The high-temperature ceramic materials are not only brittle, they are also fragile. They can provide flexibility at very small length scales only. Therefore, it is impossible to draw them out into long wires. The obvious solution is to take the help of either a metallic matrix or a thin bendable metallic substrate. These two approaches form the bases, respectively, for the first and second generation of HTS wires. The former employs a metallic matrix while the latter utilizes a thin metallic substrate as the supporting medium for superconducting particles. It must also be emphasized that the current conduction through the superconductor is influenced by the orientation of the crystalline structure. The rated currents are restricted by the grain boundaries of the crystalline microstructure. This microstructure also plays a decisive role in determining the magnetic field strength up to which the HTS wires remain superconducting. The value of this field strength is critical in energy technology applications. In view of the above constraints, the process of making long HTS wires is riddled with technological challenges, which have been faced by different companies and solved through various methods to achieve the goals.

14.3.1 First generation (1G) HTS wire

This wire has a bismuth-based multi-filamentary composition embedded in a silver matrix (figure 14.1). It is represented by the acronym BSCCO (bismuth–strontium–calcium–copper oxide $(Bi, Pb)_2Sr_2Ca_2Cu_3O_{10}$) or Bi-2223 (BINE 2010). It has a transition temperature $T_C = 110$ K. This kind of wire is commercially available in kilometer lengths. The oxide-powder-in-tube process can be used to produce reliable 1G HTS wires. In this process, the superconductive material BSCCO is crushed into a fine powder and filled into a hollow, malleable tube of a silver alloy. The tube is extruded to a thin silver wire from 35 mm to 2 mm, whereby the particles in the powder adjust to the smaller cross-section wire. In this way, several silver wires are formed. A large number of these silver wires containing the fine ceramic powder are combined together in a silver tube. The resulting silver tube is shaped into a multi-filament wire. This multi-filament wire is annealed in several stages at temperatures up to 950 °C. After undergoing annealing, the superconductive phase is arranged as a fine crystalline in parallel filaments. The metallic sleeves separate the individual filaments. By multiple rolling treatments carried out between successive annealing processes, the surfaces of the crystallites and the conducting CuO are aligned parallel to each other. The conductivity decreases if all the crystallite axes are not perfectly aligned.

The 1G HTS wire can carry 100 times more current than a copper wire of identical size but also costs 100 times the price of the copper wire because of its high silver content, ⩾60%. Hence, the extraordinary current-carrying advantage is offset by the high-cost disadvantage. The research focus was therefore diverted towards the second-generation HTS wire. Nonetheless, a superior version of BSCCO wire exhibiting high critical current and displaying more strength has been demonstrated with larger yield on a commercial scale by Sumitomo Electric Industries (SEI), than conventional BSCCO (Hirose *et al* 2006). This version is called dynamically innovative BSCCO (DI-BSCCO). It is produced by a specially developed controlled over pressure (CO-OP) process.

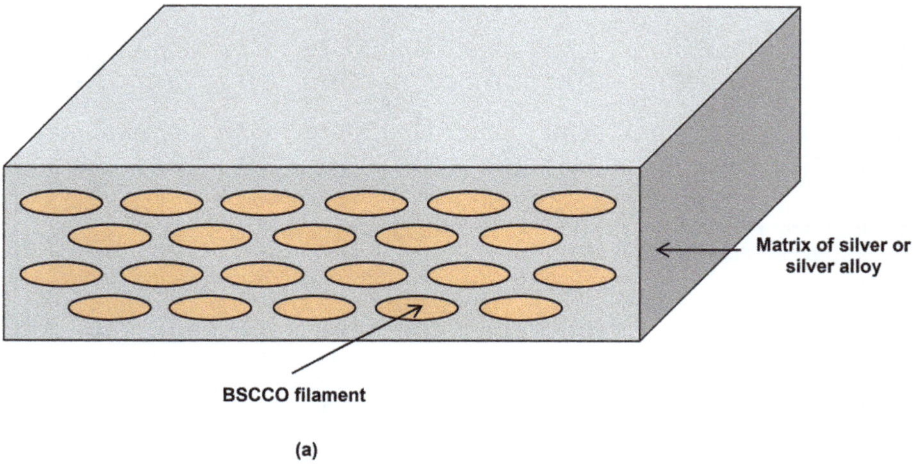

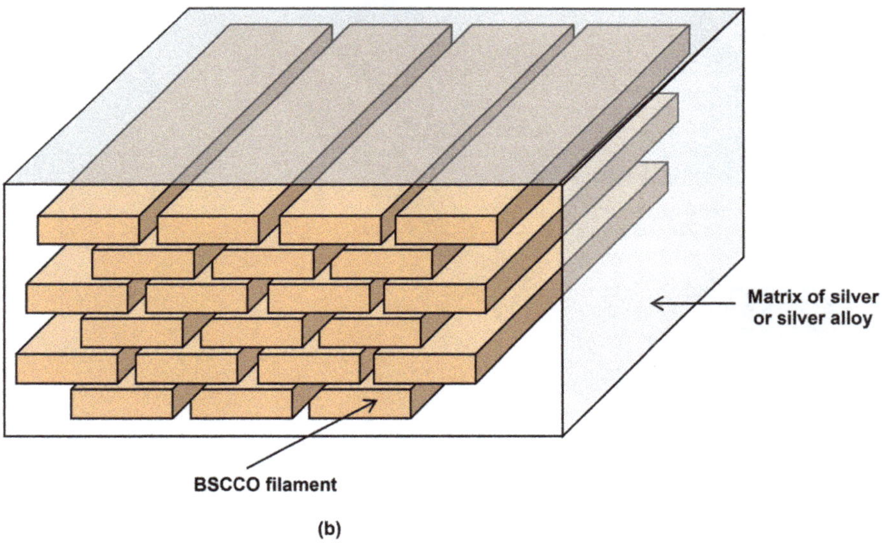

Figure 14.1. First generation superconducting wire: (a) front view and (b) view along the wire length.

14.3.2 Second-generation (2G) HTS wire

The 2G HTS wire is a thin-film rare earth barium copper oxide (ReBCO) tape where Re = yttrium, samarium, neodymium, gadolinium, etc. The 2G wire has the shape of a band while the 1G HTS wire was shaped like a wire. The 2G wire is less costly and can achieve higher current densities. It is more suitable for magnetic field applications.

There are several layers in this tape (figure 14.2): metal substrate–buffer layer–ReBCO HTS layer–stabilizing Ag layer–stabilizing Cu layer. In the SuperPower process, a thin Ni alloy substrate (50–100 μm) is electrochemically polished to a surface roughness <2 nm (Hazelton *et al* 2009). It is sufficiently smooth for the

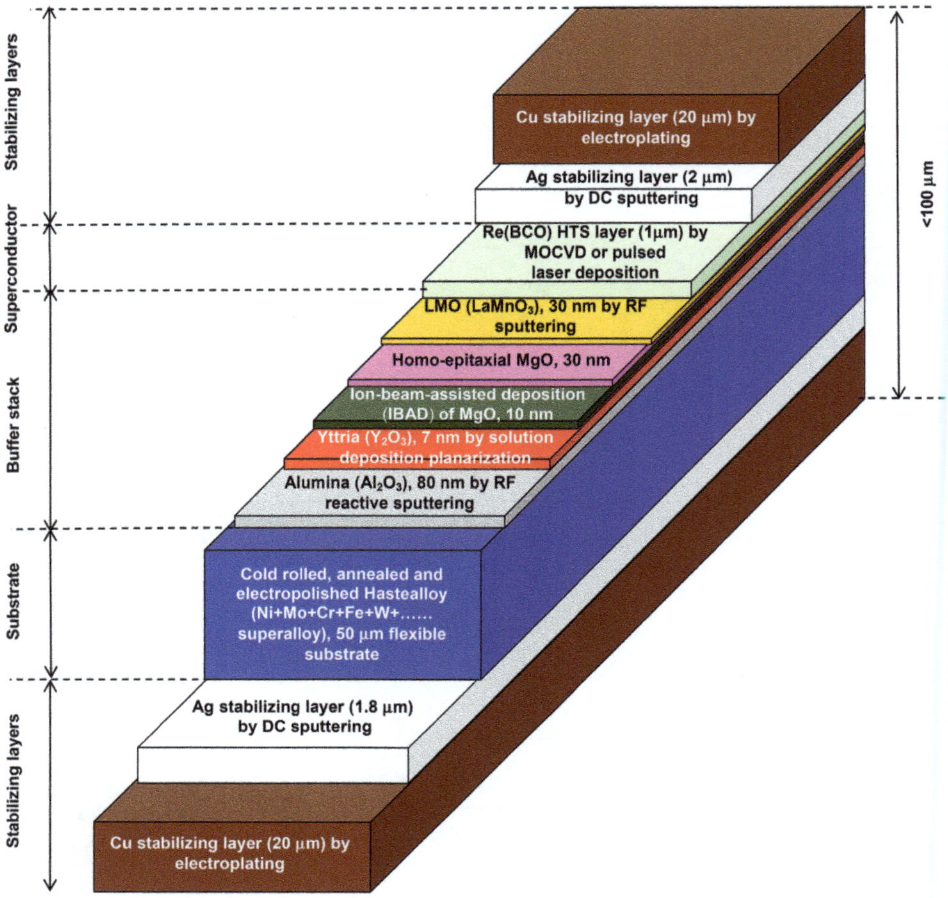

Figure 14.2. Second-generation superconducting wire.

formation of a 150 nm thick textured MgO-based buffer stack by ion beam-assisted deposition (IBAD)/sputtering. Then 1–5 µm thick ReBCO HTS film is formed by the metal–organic chemical vapor deposition (MOCVD) process. The deposition rate is 0.7 µm min^{-1}. A capping silver layer (2 µm) is sputtered for good electrical contact. The full structure is electroplated with a 2 × 20 µm surround copper layer for conductor stabilization. The thickness of the copper film is variable and is determined by the application.

The peak critical current at an angle ~5° between the magnetic field (1 T) and the surface of the film is 100 A cm^{-1} in width for 0.7 µm ZrGdYBCO and 158 A cm^{-1} in width for 0.7 µm SmYBCO (Hazelton *et al* 2009). The wires can tolerate 700 MPa of stress before the critical current is irreversibly degraded.

In the process of Sumitomo Electric Industries, the substrate is a long metal tape of high mechanical strength and low magnetism to keep AC loss small (Yamaguchi *et al* 2014). To prepare the textured metal substrate, a high-strength non-magnetic metal (100 µm thick) is laminated with Cu (20–50 µm thick) and the Cu film is plated with nickel (3 μm thick). The Cu film can be textured by thermal treatment while nickel does not allow copper to oxidize. RF sputtering of three films produces the intermediate layer on the textured metal covered substrate: CeO_2, cerium (IV) oxide, 0.1–0.2 μm thick as a seed layer; yttria stabilized zirconia (YSZ), 0.2–0.4 μm thick for preventing dispersion; and CeO_2, 0.1–0.2 μm thick for matching the lattice structure with the ensuing superconducting layer. The superconducting film $GdBa_2Cu_3O_y$ (1–4 μm thick) for power transmission is coated by pulse laser deposition. The stabilizing layer consists of a DC sputtered Ag layer (2–8 μm thick) followed by electroplated Cu film (20 μm thick). It also prevents the superconducting layer from suffering any mutilation. The critical current I_C is 200 A cm^{-1} width^{-1} over the 30 mm conductor with a maximum value of 500 A cm^{-1} width^{-1} in its central region (Yamaguchi *et al* 2014).

In the process of American Superconductor (Li *et al* 2009), 2G YBCO HTS wire is manufactured in continuous lengths by a full-scale reel-to-reel process. The process is based on an economical wide-strip technology. The wide-strip process using 4 cm wide strips commences with the deformation-texturing of a metal alloy substrate. The alloy used is Ni 5 at%W. The substrate is recrystallized in a reel-to-reel system. A cube-textured alloy substrate is thereby formed. A fast reactive sputtering technique is employed to deposit the buffer layers. These layers include: an Y_2O_3 seed layer (75 nm thick), an YSZ barrier layer (75 nm thick) and a CeO_2 capping layer (75 nm thick). After deposition of the buffer layers, an YBCO precursor film doped by rare earth is slot-die coated. The loading is 4800 mg $Y(Dy_{0.5})Ba_2Cu_3O_{7-\delta}$ m^{-2} of template corresponding to a calculated thickness of 0.8 μm $YBa_2Cu_3O_{7-\delta}$. It is pyrolyzed at a temperature below 600 °C. In a reel-to-reel system, it is transformed into an epitaxial $YBa_2Cu_3O_7$–δ film at 750 °C to 800 °C. Capping with a silver layer is followed by oxygenation at 500 °C. The produced strip of width 4 cm is divided into several thin wires and laminated to metallic stabilizers. As a result, a three-ply wire known as 344 superconductor is produced. If the metallic stabilizer is made of steel, the HTS wire is represented by 344S; if brass is used, it is 344B. In a pre-pilot device, the critical current achieved is $I_C > 100$ A (250 A cm^{-1} width^{-1}).

14.4 HTS cable designs

14.4.1 Single-phase warm dielectric HTS cable

At the center is a hollow flexible core through which liquid nitrogen flows (figure 14.3). The HTS wires are wound around this liquid nitrogen-cooled core and therefore are cooled by the liquid nitrogen. A vacuum jacket surrounds the HTS wires to thermally insulate them from the environment. Still outward is the dielectric layer for electrical insulation. Two HTS wires are necessary to complete the circuit: one for current inflow into the circuit and the other for current return from the circuit. In the region surrounding the warm dielectric HTS wire, a magnetic field exists. This magnetic field issue is absent in the cool dielectric HTS wire. Furthermore, the warm dielectric design has a shorter life span.

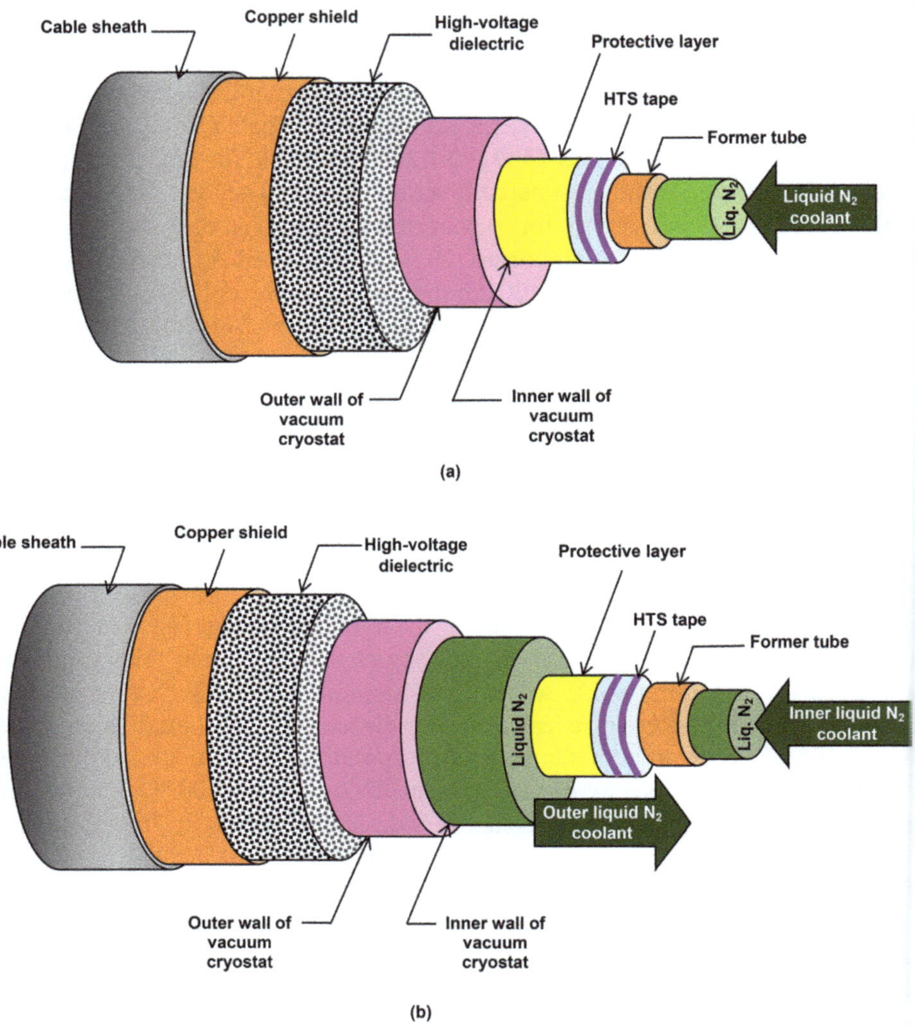

Figure 14.3. Two designs of a warm dielectric superconductor cable with: (a) inner liquid nitrogen coolant flow and (ii) inner and outer liquid nitrogen coolant flows.

14.4.2 Single-phase cool dielectric HTS cable

Recall the constructional features of a coaxial cable or coax used for electrical connections. It consists of a central core conductor surrounded by an insulating sheath. Over this sheath is the metallic conducting coating. A final dielectric sheath may enclose the above layers.

Like the coaxial cable, the cool dielectric HTS cable consists of a complex assembly comprising several tubular parts and sheaths (Demko *et al* 2000). It consists of an innermost flexible metal former through which liquid nitrogen is supplied (figure 14.4). The main HTS surrounds the metal former. It is the conductor carrying the supercurrent. The layer covering the main HTS contains the dielectric

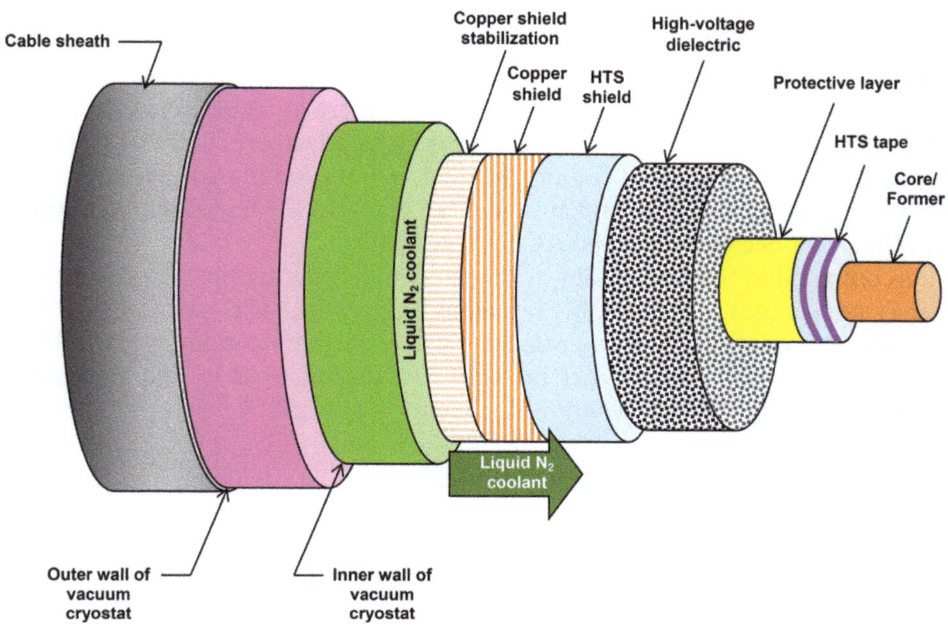

Figure 14.4. Cold dielectric superconductor cable.

and shielding layers for the low-temperature operation. The shielding HTS envelops these layers. This shielding HTS, which is separated from the main HTS by the dielectric material, carries the return current sent by the main HTS. Therefore, a single cool dielectric HTS cable is adequate for both forward and return supercurrents. The outermost layer is the flexible cryostat, the apparatus used to maintain low temperature for the enclosed components. Liquid nitrogen flows in the annular space between the shielding HTS and the flexible cryostat. The innermost flexible metal former and the outermost flexible cryostat are made of corrugated stainless-steel tubing. At one extremity of the cooling arrangement is a refrigeration unit. This unit provides a constant supply of liquid nitrogen.

The cool dielectric architecture is an enhancement over the warm dielectric design because only one cool dielectric cable is necessary instead of two warm dielectric cables. Moreover, it can carry higher current. Additionally, the external magnetic field produced by it is nearly zero. The life expectancy of the cool dielectric design is longer.

14.4.2.1 The parallel-flow cooling arrangement

Liquid nitrogen flows in the same direction in the innermost flexible metal former and annular region between the shielding HTS and the flexible cryostat. The path for returning of liquid nitrogen is a separate duct jacketed inside a vacuum enclosure.

14.4.2.2 The counter-flow cooling arrangement

The directions of flow of liquid nitrogen in the metal former and the annulus are opposite. Liquid nitrogen enters through the metal former. It cools the terminations as well as the HTS. Then, it returns through the annulus.

14.4.3 Flow rate, pressure drop and HTS cable temperatures

In both cooling arrangements, as the temperature of the HTS decreases, AC losses decrease due to the superconductivity effect while thermal losses increase due to the larger temperature difference between the superconductor and the surroundings according to Newton's law of cooling. Further, in both cooling arrangements, lower flow rates result in a higher temperature of the HTS. At the higher temperature, the critical current decreases and the AC loss increases.

As compared to the parallel flow arrangement, the counter-flow arrangement has a lower heat load. In addition, for the same length of the HT superconductor and flow rate, the counter-flow arrangement yields a higher temperature of the HTS and larger pressure drops. The larger pressure drop experienced in this arrangement imposes a limitation on the length of the counter-flow arrangement.

14.4.4 Three-phase cold dielectric HTS cable

For three-phase current supply, three phases are enclosed in a single cryogenic envelope, as shown in figure 14.5. They can also be put inside separate envelopes.

14.5 HTS fault current limiters

Sometimes under heavy load condition or during a lightning strike, a high-value fault current suddenly appears, damaging the installation. In these situations, fault current limiters (FCLs) come to the rescue. They are devices which limit the current to safe value fast enough to prevent damage. Superconducting FCLs (SFCLs) utilize the non-linear properties of superconducting materials, such as non-linearity in response towards temperature, current and magnetic field changes for restraining the current magnitudes. At a constant temperature and under a constant magnetic field, when the fault current becomes high, the increase in fault current makes a section of the superconductor so resistive that the heat produced cannot be removed locally. Through transference of excess heat to the neighboring regions of the superconductor, the temperature of these regions is raised. The combined effect of current and temperature effects is that the superconductor starts behaving as a normal conductor. This propagation of the normal conductor zone through a superconductor is known as quenching. Effectively, the material undergoes a transition from a low impedance superconducting state to a normal high-impedance conducting state in a fault episode.

14.5.1 Resistive SFCL

In a resistive SFCL (figure 14.6), the main current-carrying conductor under normal operating conditions of the transmission system is the superconducting material (Electric Power Research Institute 2009). When a fault occurs, the current increases, and the superconductor undergoes quenching, accompanied by an exponential rise in resistance. Due to the speedy increase in resistance of the superconductor, a voltage is produced across the superconductor. As a result, the current is transferred to a shunt consisting of a resistor and an inductor. Further increase in voltage across

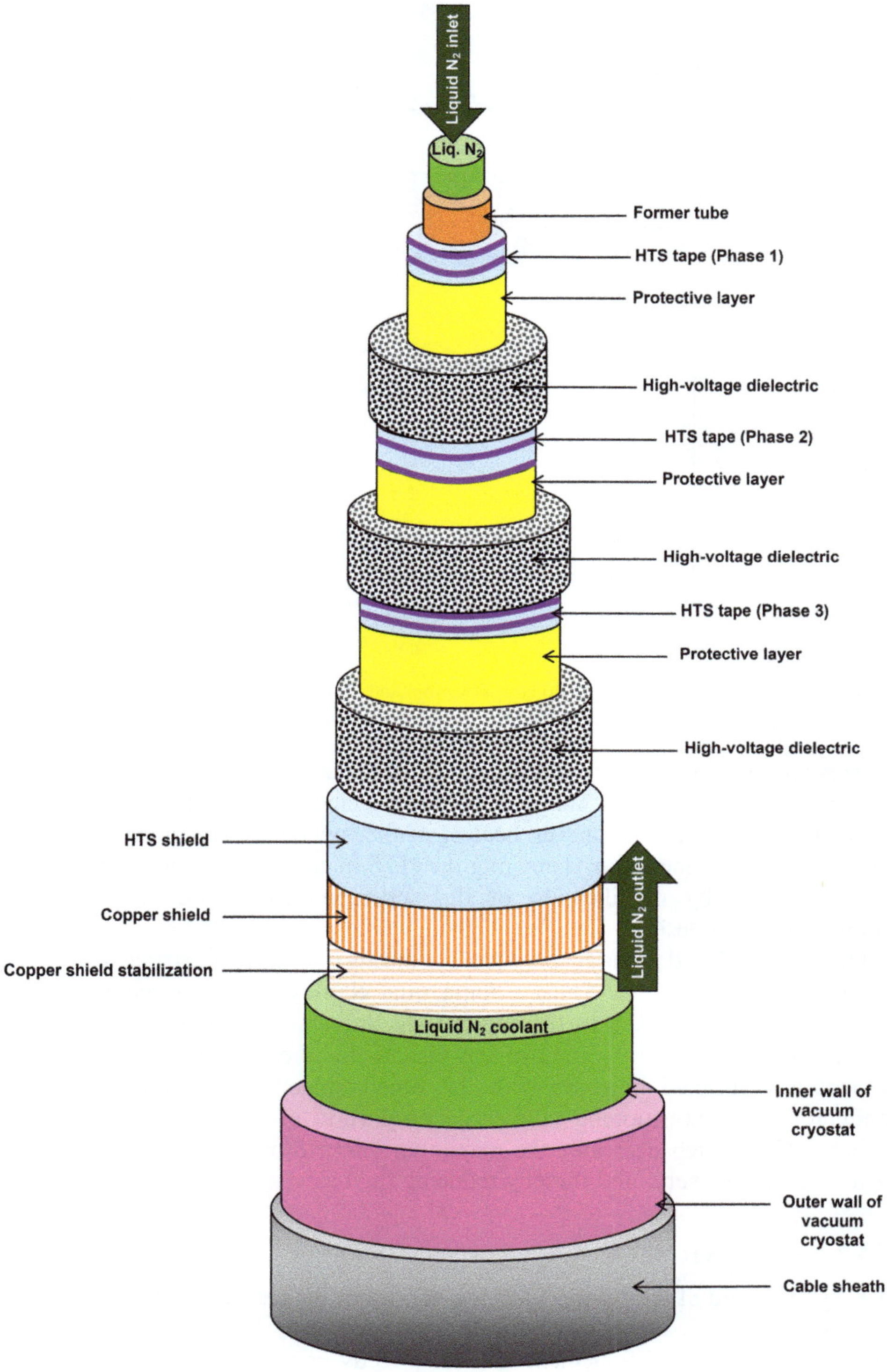

Figure 14.5. Triaxial cable with a cold dielectric arrangement.

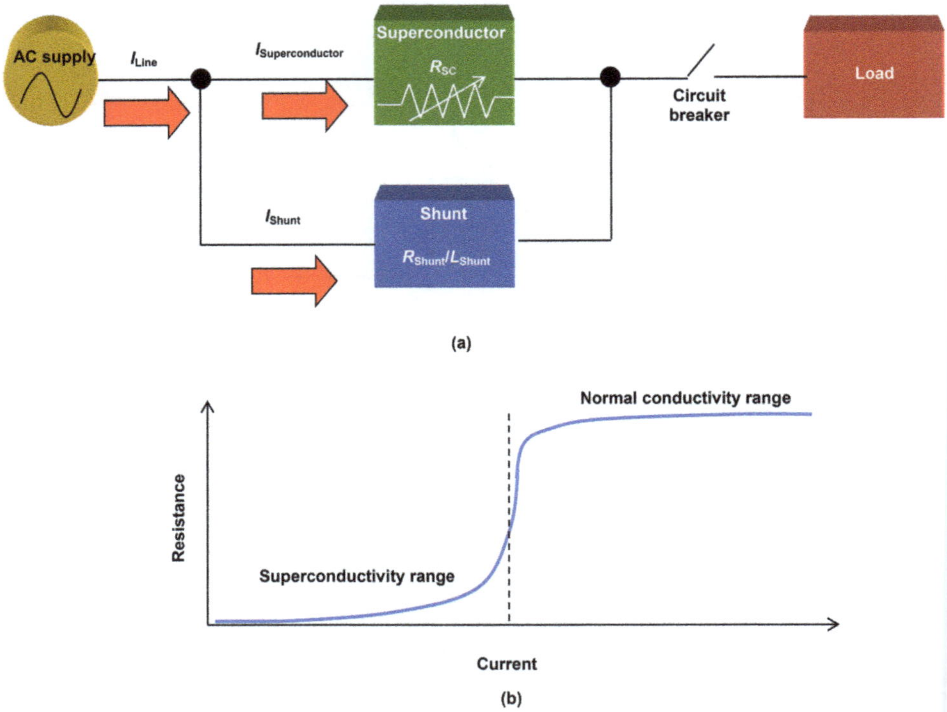

Figure 14.6. Resistive type SFCL: (a) circuit diagram and (b) its resistance–current characteristics.

the superconductor is thereby restricted. Thus, the superconductor acts as a switch responding in milliseconds. This switch causes a changeover of the load current to the shunt impedance. The uneven heating of the superconductor during quenching leads to hot spot generation, damaging the HTS material. To some extent, this has been mitigated by modification of the superconductor material process and innovative device designs.

The heat produced in the superconductor during quenching is carried away by the cryogenic cooling system. Thus, a finite time elapses before the cooling system restores the material back to the superconducting state and this time duration is referred to as the recovery time of the system. In some SFCLs, a fast-switching element is connected in series with the superconductor to separate the superconductor after transitioning of a large proportion of the current to the shunt element. This switch decreases the maximum temperature in the superconductor, hastening the recovery, and thereby reducing the recovery time.

14.5.2 Shielded-core SFCL

The shielded-core SFCL (figure 14.7) is a variant of the resistive SFCL in which the cryogenic cooling system is mechanically separated from the remaining circuit. The power line and the HTS element are not directly connected. They are indirectly connected through mutually coupled AC coils via a magnetic field. The SFCL device

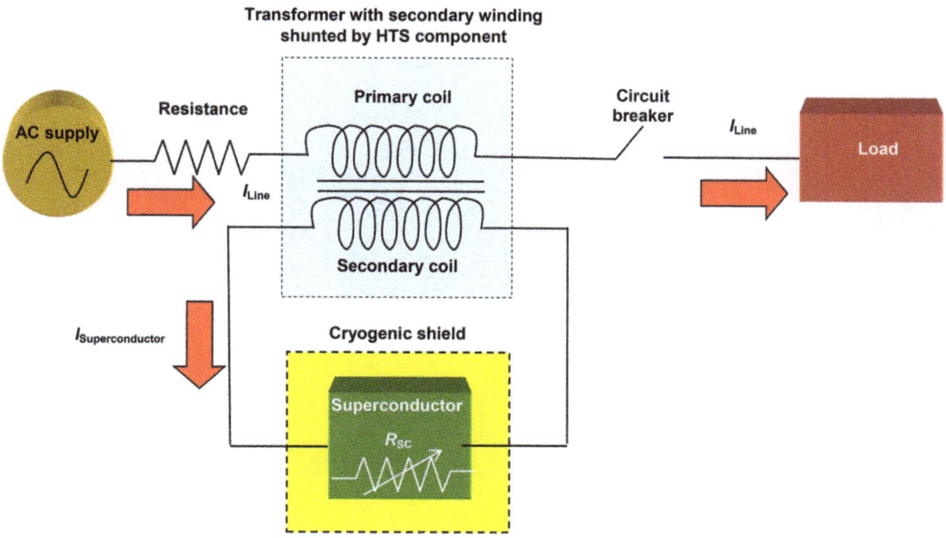

Figure 14.7. Shielded-core SFCL.

is similar to a transformer in construction. In this transformer, the secondary side is shunted by an HTS element. Whenever a fault current is produced, the current on the secondary side increases. Consequently, the HTS element quenches. The associated rising voltage across the primary coil is in such a direction that it opposes the fault current. By fiddling with the ratio of the number of turns in the transformer coils, the hot spot problem caused by non-uniform heating of the superconductor is avoided. However, like a resistive SFCL, the shielded-core SFCL must cool down after a current limiting action has taken place. A major disadvantage of the shielded-core SFCL is that it occupies four times the volume of a resistive SFCL. Also, it has four times the weight of the resistive SFCL.

14.5.3 Saturable-core SFCL

The saturable-core SFCL (figure 14.8) uses two iron cores and two AC windings for each phase. The AC windings comprise conventional conductors. These conductors are wound around the iron core forming an inductance in series with the AC line. An HTS winding carrying a constant current is also wound around the iron core. This winding provides a magnetic field. When the grid current is within normal limits, the HTS winding completely saturates the iron. Under this condition, its relative permeability is unity. For the coils carrying AC, the iron core acts like air. The AC impedance or inductive reactance resembles an air-core reactance component. But when the fault current is produced, the core is pulled out of saturation by the positive and negative peaks of current. Consequently, the line impedance increases during part of each half cycle. Hence, the peak fault current is appreciably reduced. Thus, the saturable-core SFCL is an iron-core reactive component having variable inductance. Normally, it is an air-core reactive component. Under a fault current

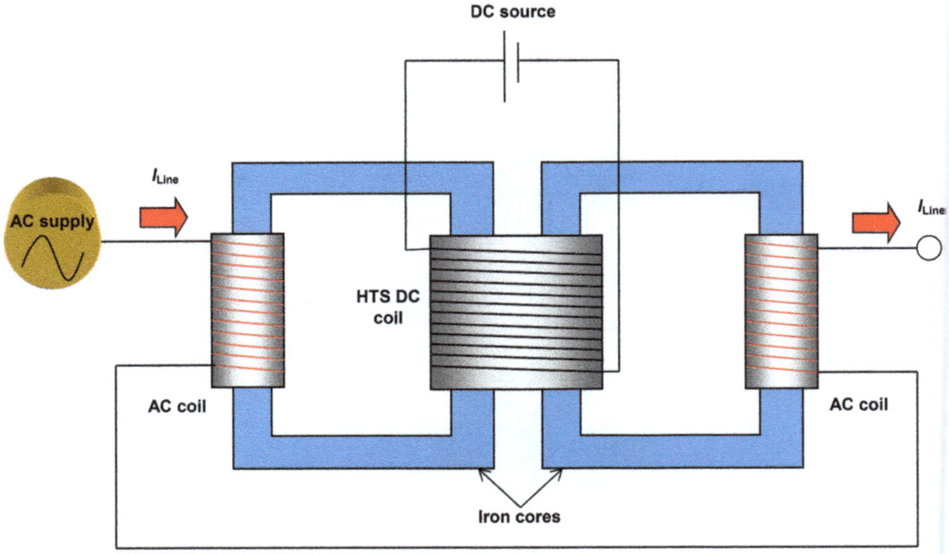

Figure 14.8. Saturable iron-core SFCL.

condition, its impedance becomes very high. The saturable-core SFCL differs from the resistive SFCL. It can control a series of faulty events successively without needing to cool down like a resistive SFCL. This happens because it does not quench. However, the volume and weight of the heavy iron core required to build a saturable-core SFCL is a serious impediment to its pervasive utilization.

14.6 HTS transformers

An HTS transformer is similar to an ordinary transformer except that the primary and secondary windings of the HTS transformer are made from zero-resistance HTS wires in place of the copper or aluminum windings used in ordinary transformers (US Department of Energy 2000). All the advantages of superconductivity are therefore derived in an HTS transformer as compared to an ordinary transformer. The HTS transformer is cooled down to 77 K in order that it operates in the superconductive mode, providing lower losses by approximately 30% at a 10–30 times higher current density than a regular transformer and with a 45% smaller weight than the same. An HTS transformer offers unmatched fault current limiting ability. But the biggest advantage of an HTS transformer is that it does not require any insulating oil coolant. The fire and environmental hazards of the oil coolant related to a conventional transformer are therefore automatically eliminated.

14.7 Discussion and conclusions

HTSs have changed the fundamental component in electrical and electronic circuits, namely, the 'conducting electrical wire'. Since the wire is a key component without which no circuit implementation is possible, the impact of superconductivity on electricity and electronics is expected to be profound and revolutionary. This chapter provided a glance at some of the ways that superconductors can reshape the electric power infrastructure and change the ways in which power is distributed and controlled. All these efforts pave the way towards maximizing the capabilities and thereby the available facilities while minimizing electricity bills and hence the financial liability of consumers.

Review exercises

14.1. Name the two elements used in making wires for transmitting electricity? Which one has a higher conductivity?

14.2. Why is power transmitted at high voltages and low currents? What is the role of transformers in power transmission? What types of hazards are possible due to transformer faults?

14.3. In what ways is overhead power transmission better than underground delivery? In what ways is it worse?

14.4. Why is it difficult to draw high-temperature superconducting materials into long wires? What are the two approaches followed in first and second-generation wire making from superconductors?

14.5. What is the acronym for the first generation HTS wire? What do the letters of the acronym represent?

14.6. What are the main steps in the oxide-powder-in-tube process of making 1G HTS wire? What is the main drawback of 1G HTS wire? What is a DI-BSCCO wire?

14.7. What is the acronym representing a 2G high-temperature superconducting wire? Mention three advantages of the 2G wire over the 1G wire.

14.8. How many layers of different materials are present in a 2G HTS wire? What are the names of these materials? What is the role of each material?

14.9. What are the main steps involved in the full-scale reel-to-reel process of making a 2G HTS wire? Describe the function of each step.

14.10. What is the substrate material used in a 2G HTS wire? What is the buffer layer and what is its function? What are the HTS materials used? What elements are used in the top stabilizing layers?

14.11. Explain with a diagram the construction of a warm dielectric HTS cable? What are its disadvantages?

14.12. Draw a labelled diagram and explain the constructional features of a cool dielectric HTS cable. Mention its advantages.

14.13. Explain the parallel-flow and counter-flow cooling arrangements used in a cool dielectric HTS cable. What is the length limitation of the counter-flow arrangement?

14.14. What is the function of the FCL in a power transmission circuit? What property of a superconductor helps in incorporating fault limitation in the circuit? What is quenching?

14.15. Draw the circuit diagram of a resistive SFCL. Explain its operating mechanism with reference to the circuit diagram. How is the recovery of the SFCL speeded up?

14.16. How does a shielded-core SFCL operate? Explain with a diagram.
14.17. What is the operating principle of a saturable-core SFCL? How does it differ from a resistive SFCL?
14.18. What is an HTS transformer? Point out the advantages offered by it over an ordinary transformer.

References

BINE 2010 *High temperature superconductors Projektinfo 06/10* www.bine.info/fileadmin/content/Publikationen/Englische_Infos/projekt_0610_engl_internetx.pdf

Chen X, Jiang S, Chen Y, Zou Z, Shen B, Lei Y, Zhang D, Zhang M and Gou H 2022 Energy-saving superconducting power delivery from renewable energy source to a 100-MW-class data center *Appl. Energy* **310** 118602

Demko J A, Lue J W, Gouge M J, Stovall J P, Butterworth Z, Sinha U and Hughey R L 2000 Practical AC loss and thermal considerations for HTS power transmission cable systems *Appl. Supercond. Conf. (Virginia Beach, VA, 17–22 September)* 141–4

Electric Power Research Institute 2009 Superconducting fault current limiters *Technology Watch* **1017793** www.suptech.com/pdf_products/faultcurrentlimiters.pdf

Hazelton D W, Selvamanickam V, Duval J M, Larbalestier D C, Markiewicz W D, Weijers H W and Holtz R L 2009 Recent developments in 2G HTS coil technology *IEEE Trans. Appl. Supercond.* **19** 2218–22

Hirose M, Yamada Y, Masuda T, Sato K-i and Hata R 2006 Study on commercialization of high-temperature superconductor *SEI Technical Review* No. 62, June 15–23

Li X, Rupich M W, Thieme C L H, Teplitsky M and Sathyamurthy S 2009 The development of second generation HTS wire at American Superconductor *IEEE Trans. Appl. Supercond.* **19** 3231–5

US Department of Energy 2000 Selected high-temperature superconducting electric power products *January* www.nrel.gov/docs/gen/fy00/supercon_products.pdf

Yamaguchi T, Shingai Y, Konishi M, Ohya M, Ashibe Y and Yumura H 2014 Large current and low AC loss high temperature superconducting power cable using REBCO wires *SEI Technical Review* No. 78, April 79–85

Part II

Harsh-environment electronics

Sub-part IIA

General considerations

IOP Publishing

Extreme-Temperature and Harsh-Environment Electronics
(Second Edition)
Physics, technology and applications
Vinod Kumar Khanna

Chapter 15

Humidity and contamination effects on electronics

High humidity is a primary initiator and booster of devastating effects on electronics. Great damage to electronics can be anticipated if moisture is allowed to seep into packages and reach the circuits. Contaminants introduced either during manufacturing or later by contact with handling workers or from the environment, partner with moisture to intensify the destruction. Mechanisms for destruction by corrosion are many and originate from different causes, e.g. electrochemical, anodic, cathodic, galvanic, etc. All these mechanisms, whereby high humidity along with dust, dirt and organic matter pose a hazard to the reliability of electronics, are dealt with in this chapter. The conditions favoring these mechanisms are mentioned. An understanding of these mechanisms is helpful to create conditions and devise suitable counteracting measures to obliterate their adversative consequences.

15.1 Introduction

Electronic devices and circuits have proliferated worldwide into newer applications, both indoors and outdoors. They have permeated into everyday activities. Reliable functioning of electronics is hampered by water in its different forms (Baylakoğlu *et al* 2021). Contamination of surfaces of electronic printed circuit boards degrades their insulation resistance instigating sporadic or permanent failure. Presence of ionic species in the dust particles further aggravates the situation causing short-circuiting in a humid environment (Song *et al* 2013).

doi:10.1088/978-0-7503-5072-3ch15

All this has led to mounting concerns on the effects of humidity on electronic equipment. The primary cause of this worry is the invariable association of a high-humidity environment with corrosion of electronic components or parts thereof. Ideally, the optimum humidity level prescribed for electronics is the 40%–60% relative humidity (RH) range.

Additionally, progressive miniaturization of components in computers, telecommunications and control systems has made them more susceptible to humidity-induced failures, even in gentler environments. In nanoelectronics, a flawed patch measuring a few nanometers is more detrimental than a several millimeter wide blotch in large steel structures. An automobile can function with the loss of several grams of material, but for an IC even a few picograms of material make the difference. To deal with this serious problem, the root mechanisms underpinning such failures must be elucidated.

15.2 Absolute and relative humidity

Moisture-induced failure modes exhibit a strong correlation with RH instead of absolute humidity (AH; Tencer and Moss 2002). There exists a threshold RH level at which a certain thickness of adsorbed water is formed on the surface of the component, leading to its non-performance. AH is the mass of water vapor (m) present in a given volume (V) of air, expressed in units of mg L^{-1}:

$$\mathrm{AH} = m/V. \tag{15.1}$$

The temperature of air is not taken into consideration in this definition. The RH at a given ambient temperature is a percentage representing the ratio of AH at that temperature to the maximum possible AH at the same temperature under saturation conditions. It is defined as a ratio of partial pressures: the partial pressure of water vapor $p_{\mathrm{H_2O}}$ present in the air at a given temperature to the equilibrium vapor pressure $p^*_{\mathrm{H_2O}}$ at that temperature:

$$\mathrm{RH} = p_{\mathrm{H_2O}}/p^*_{\mathrm{H_2O}}. \tag{15.2}$$

15.3 Relation between humidity, contamination and corrosion

Corrosion is the deterioration of the properties of a material, e.g. a metal, inorganic or organic compound, through chemical reactions of the material with environmental gases such as oxygen and water vapor. Adsorbed moisture is a key enabler and driver of corrosion. The rate of corrosion increases rapidly at RH > 60% and in the presence of contaminants such as dust and microbes. The high RH causes condensation of large quantities of water vapor on the surface of the material, e.g. in a humid environment, water droplets condense on circuit boards (figure 15.1). Ions can flow through the condensed water film, accelerating corrosion. Together with dust and dirt, they can trigger corrosion, which would not certainly take place if the circuit boards were clean. Very small quantities of surface contaminants suffice to initiate corrosion. The sources of these contaminants are:

(i) the manufacturing process steps in which the boards are exposed to gases, etching and plating solutions, soldering fluxes, etc.

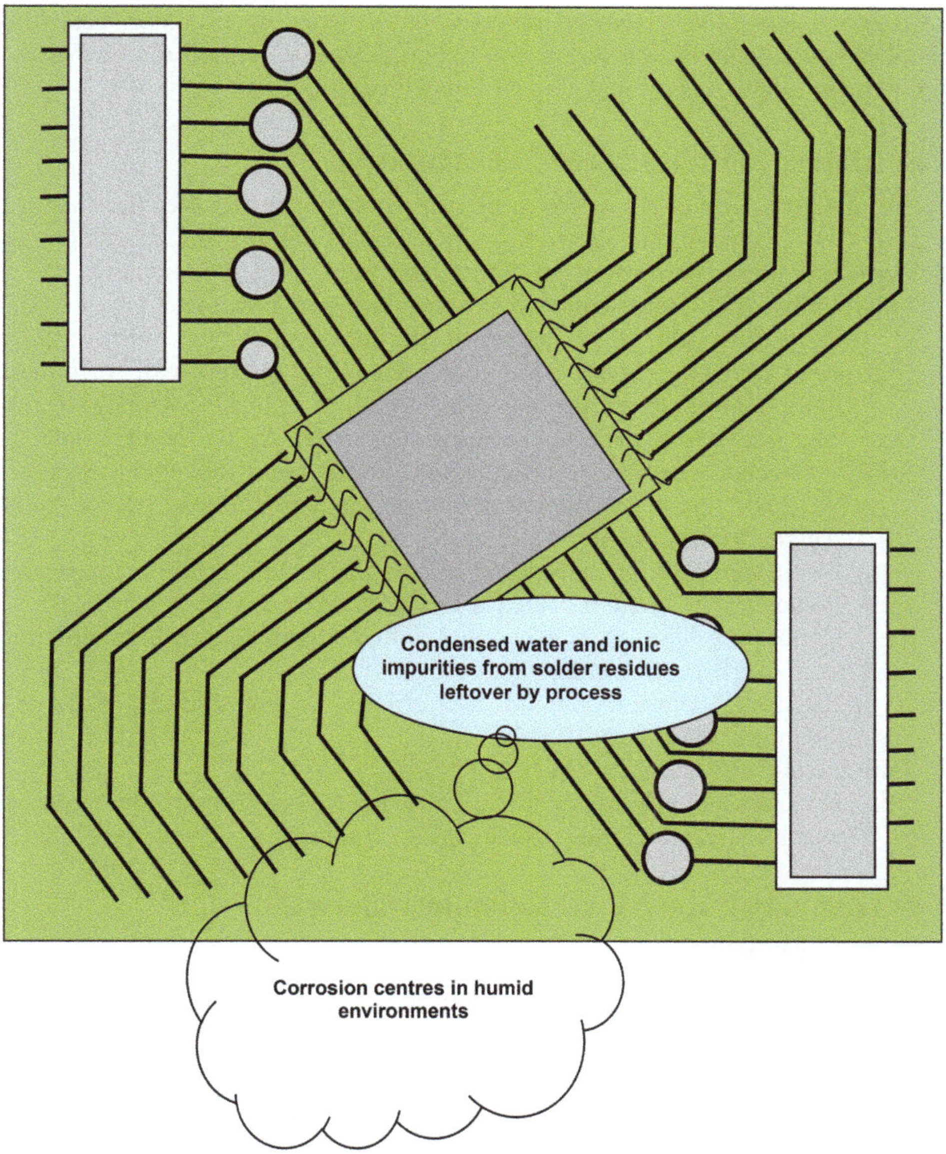

Figure 15.1. Humidity-promoted corrosion in electronic printed circuit boards.

(ii) contact with workers handling the boards and customers mounting them. Handling contamination comes from oils and grease on fingers, salts in sweat, skin flakes, cosmetics and creams on the skin, etc.

During processing, if any ion-releasing compounds are left over, they can promote corrosion. If not properly cleaned, chloride ions, salts and sulfur compounds can play havoc in the presence of water. Careful cleaning of the boards ensures avoidance of corrosion via ionic contents.

Corrosion looks like a natural outcome of the presence of water, heat, dust, microbes and oxygen on the surfaces of materials. The corrosion rate increases at high temperatures and becomes slow at low temperatures.

15.4 Metals and alloys used in electronics

It is the metallic parts of electronic equipment, which have to suffer the maximum onslaught of corrosion. In electronic chips, printed circuit boards (PCBs) and packages, a multiplicity of metals and alloys are used, some of which are:

(i) Chips of discrete semiconductor devices and integrated circuits (ICs): aluminum, gold, silver, copper in addition to silicon (a semi-metal).
(ii) Lead frames connecting the gold bumps on the chip with surrounding PCB: gold or silver-treated Cu/Zn37, CuFe2, FeNi42 or CuNiZn (on the chip side) and a solderable coating such as Ni–Au on the PCB side.
(iii) Wire bonding between gold bumps and the lead frame: gold or aluminum wire.
(iv) Wire bonding metal/aluminum pad on the chip and lead frame: Au/Au, Au/Ni, Au/Ag, Al/Al.
(v) Electronic chip packages: galvanized/chromated steel or magnesium. Polymer packages are coated with Cu/Ni or Al for EMI shielding.
(vi) Connecting lines on PCB: copper with solderable electroless Ni-immersion Au (IM Au) coating.
(vii) Electronic connectors: gold electroplated copper and its alloys, e g., CuSn6, CuBe2, CuNi10Sn2.
(viii) Lead-free solders: Sn–Ag–Cu.
(ix) Hard discs of computers: a platter made of Al electroplated with Ni over which there is a Co alloy with carbon coating.

15.5 Humidity-triggered corrosion mechanisms

15.5.1 Electrochemical corrosion

This corrosion, exemplified by the rusting of iron, is a two-step process:

(i) An atom at the surface of a metal dissolves in the condensed water film. The metal is left negatively charged. For a bivalent metal M,

$$\text{anodic reaction: } M(s) \rightarrow M^{2+}(aq) + 2e^{-}. \tag{15.3}$$

(ii) The electrons travel to the outside of the water droplet where they interact with an electron acceptor called the depolarizer, such as oxygen from the atmosphere,

$$\text{cathodic reaction: } O_2(g) + 2H_2O(l) + 4e^{-} \rightarrow 4OH^{-}(aq). \tag{15.4}$$

Inside the water droplet, the hydroxyl ions move inwards to react with metal ions:

$$M^{2+}(aq) + 2OH^{-}(aq) \rightarrow M(OH)_2(s). \tag{15.5}$$

The metal hydroxide quickly oxidizes:

$$4M(OH)_2(s) + O_2(g) \rightarrow 2M_2O_3{\cdot}H_2O(s) + 2H_2O(l). \quad (15.6)$$

In such atmospheric corrosion, changing weather conditions play a major role.

15.5.2 Anodic corrosion

This corrosion (figure 15.2) occurs when there is a potential difference between two conductors, such as between two soldering points on a PCB, connected by a thin film of liquid water (Lighting Global 2013). The metal ions dissolve from the anode. Depending on their stability in the aqueous solution, they either deposit on the cathode (e.g. Sn, Pb, Cu, Ag, etc) or may dissolve in water forming hydroxide, such as for aluminum, which is a non-migrating metal. The Sn, Pb, Cu or Ag ionic migration causes dendrite growth from the cathode metal towards the anode metal, bridging the gap between them and thus leading to a short-circuiting effect. Failures due to dendrite growth can take place between copper lines on a PCB.

Looking at the scenario of relentlessly shrinking dimensions, the spacing between the conductor lines is becoming very small, increasing the electric field between them. Higher electric fields hasten ionic migration. Micro-soldering processes carried out using low-residue, no-clean fluxes in inert gas ambience are gaining popularity. The long-term stability of circuits with narrow spacing between lines assembled using micro-soldering processes is not yet proven.

15.5.3 Galvanic corrosion

This corrosion, also called 'dissimilar metal corrosion', is a type of preferential corrosion of one metal relative to another when two dissimilar metals with distinctly different electrochemical potentials are brought into contact and are exposed to an electrolytic solution. The two different metals produce a potential difference driving the corrosion process without any help from an externally applied voltage. Water condensation from the environment or immersion of the object in water provides the electrolyte. The relative nobility of a metal is known from its corrosion potential, and can be obtained from the galvanic series (figure 15.3). The corrosive effect is

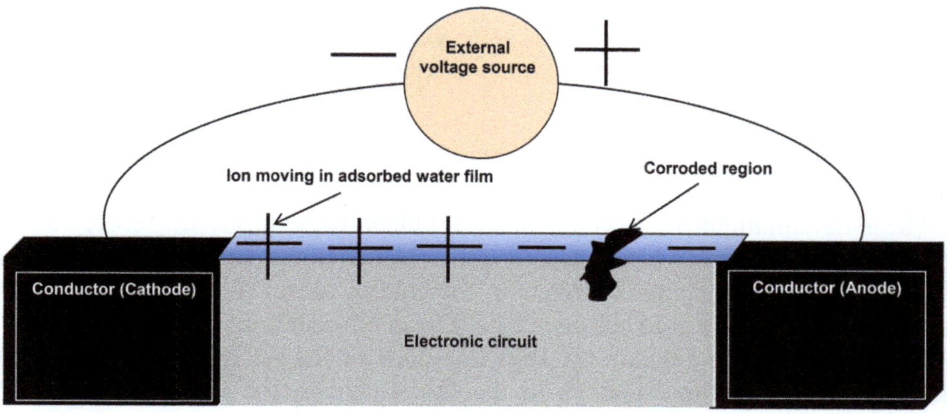

Figure 15.2. Anodic corrosion.

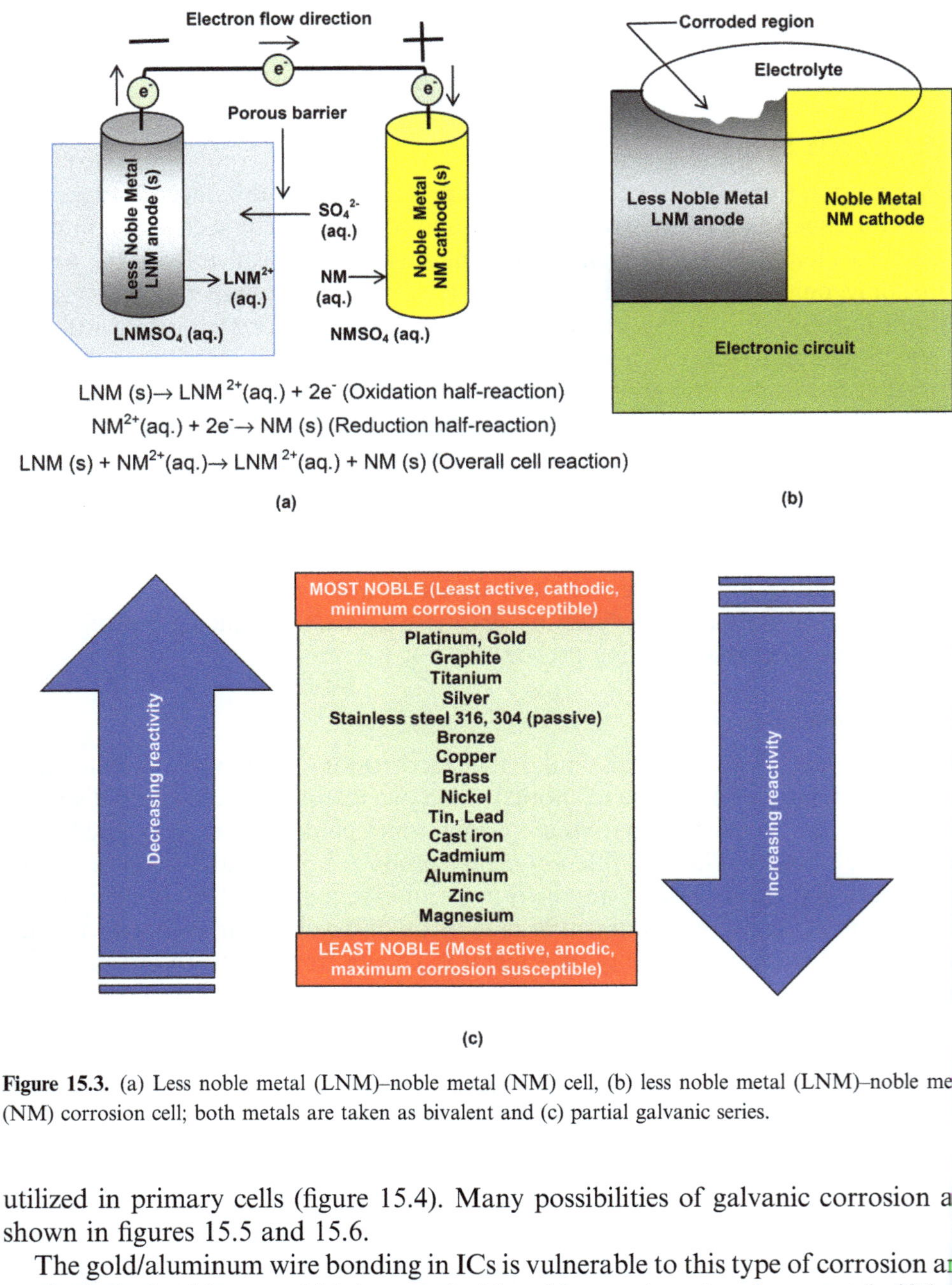

Figure 15.3. (a) Less noble metal (LNM)–noble metal (NM) cell, (b) less noble metal (LNM)–noble metal (NM) corrosion cell; both metals are taken as bivalent and (c) partial galvanic series.

utilized in primary cells (figure 15.4). Many possibilities of galvanic corrosion are shown in figures 15.5 and 15.6.

The gold/aluminum wire bonding in ICs is vulnerable to this type of corrosion and so also is the lead frame, which is coated with nobler coatings than base metals, if there are any cracks or mechanical defects in the coating metal. For touch connectors, graphite can substitute gold because of its slow anodic/cathodic reaction kinetics.

In a mobile phone keypad, the keys are made of IM Au and electroless nickel (EL Ni). If there is any damage to the IM Au layer, the EL Ni layer will be

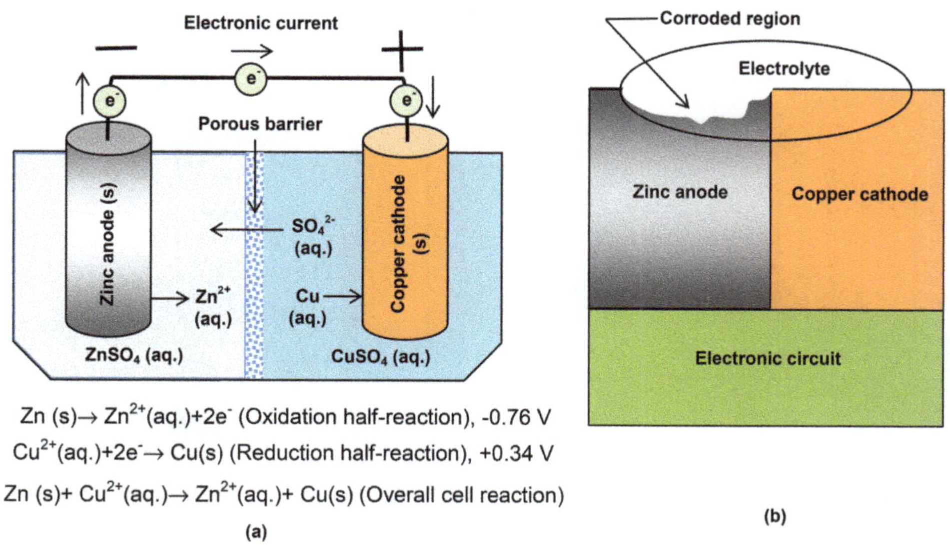

Figure 15.4. (a) Zn–Cu galvanic cell and (b) Zn–Cu corrosion cell.

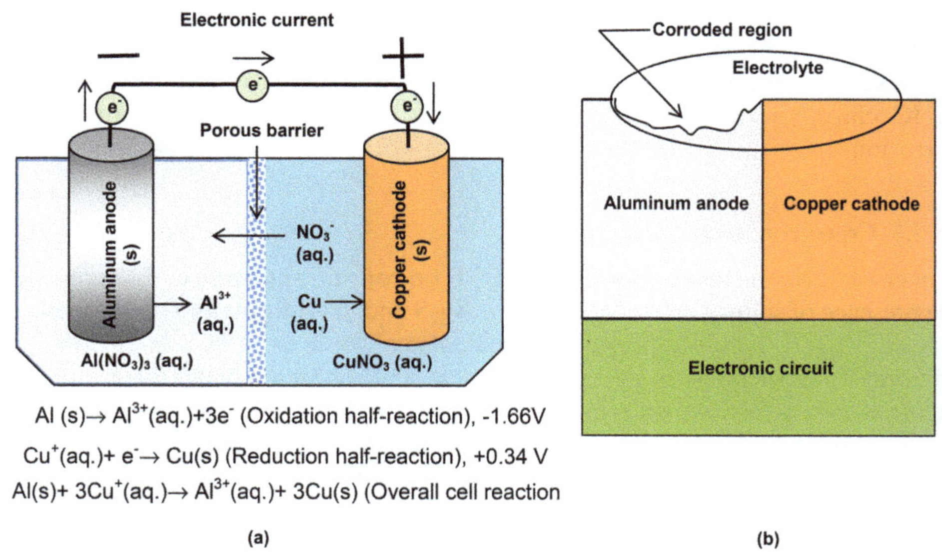

Figure 15.5. (a) Al–Cu galvanic cell and (b) Al–Cu corrosion cell.

corroded with IM Au, acting as the cathode. Pitting of the EL Ni will expose Cu and cause further damage.

15.5.4 Cathodic corrosion

A few metals used in electronics dissolve in acidic and alkaline solutions over a broad range of pH and potentials, e.g. aluminum and zinc are soluble in these

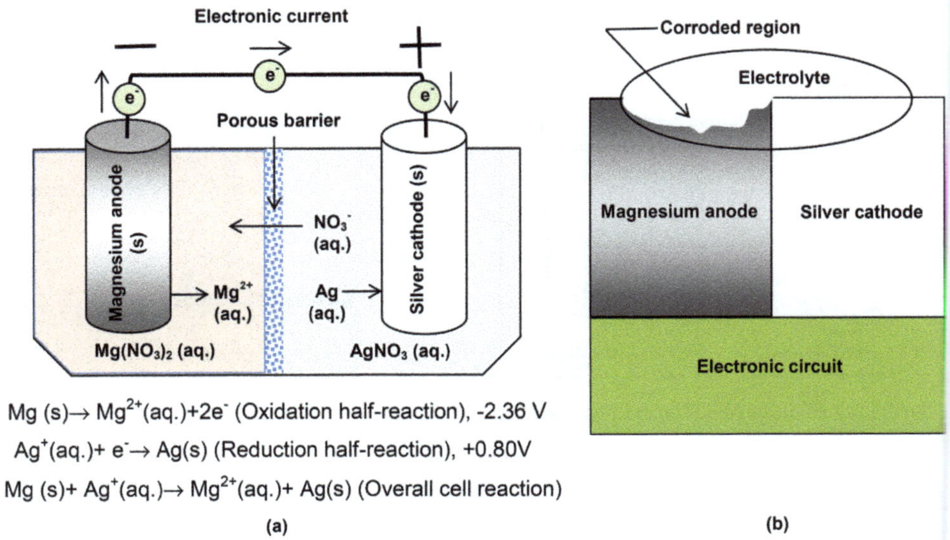

Figure 15.6. (a) Mg–Ag galvanic cell and (b) Mg–Ag corrosion cell.

solutions above the equilibrium dissolution potential. In an electrochemical cell deriving energy from redox reactions, called the galvanic cell (see above), OH^- ions are liberated by oxygen reduction at the cathode in accordance with equation (15.4). The resultant shift of pH towards the alkaline side enables dissolution of aluminum. In IC chips, the aluminum conducting lines fall prey to this type of cathodic corrosion.

15.5.5 Creep corrosion

Immersion silver as the surface finish of PCB on copper is prone to creep corrosion in the presence of sulfur and moisture due to Ag_2S and Cu_2S formation (Savolainen and Schueller 2012). Sulfur is present in the elemental form or as H_2S in environments adjacent to paper mills, as well as the cement and rubber manufacturing industries. Sulfur is also available near mining industries and waste-water treatment plants. Creep corrosion begins by dendritic growth and takes place equally in all directions.

Creep corrosion does not require any external potential difference to be applied to the PCB. The mechanism seems to be driven by a galvanic process because of the involvement of two metals, silver and copper, with copper behaving anodically with respect to silver and subject to more vigorous attack. The area of the anodic copper layer is much smaller than that of the cathodic silver covering layer. As a result, the attack on the anode metal is profoundly aggressive, compelling stronger corrosion reaction of copper with sulfur and moisture. The creep corrosion is described as a three-step process (Chen *et al* 2012):

(i) dissolution of the corrosion products in adsorbed water layers on the surface;
(ii) their movement by diffusion under the concentration gradient; and
(iii) their re-deposition at another site.

High-temperature organic solderability preservative (OSP) or anti-tarnish and lead-free hot air solder leveling (HASL) have been found to be good protection for mitigating creep corrosion. Proper surface cleanliness is of paramount importance. If silver finish is to be retained, several design modifications are necessary such as exclusion of metal features demarcated by solder mask (polymer coating on copper traces in PCB), thorough coverage of non-test vias with a solder mask, keeping smoothed rounded corners for component pads, etc.

15.5.6 Stray current corrosion

Exposing conductors to high electric and magnetic fields can cause high-magnitude stray current flow. Microwaves can induce turbulent liquid flow on a wet aluminum surface. The eddy currents flowing in the liquid film can aggravate corrosion problems.

15.5.7 The pop-corning effect

Sometimes, if a plastic encapsulated IC is not moisture baked before reflow soldering, the absorbed moisture pockets trapped inside may vaporize and expand by heating during the soldering process, thereby spoiling the substrate, die and the wire bonds (Chen and Li 2011). These are caused by built-up internal stresses, which may rupture the package and produce delamination between the molding compound and the lead frame. This phenomenon is known as the pop-corning effect from its resemblance to pop-corn, which dilates as the pressure builds up within the kernel accompanied by a small explosion. The pop-corning effect is easily avoided by moisture baking the plastic packaged microelectronic device or circuit before assembly and performing reflow within the stipulated time after moisture baking so that moisture has not gained re-entry into the IC.

15.6 Discussion and conclusions

High humidity leads to the formation of a liquid water film on the surfaces of electronic circuits. This water film has the same properties as bulk water in which electrolytic conduction takes place with ions as the charge carriers. Corrosion-causing ionic conduction on chip surface does not always need an externally applied potential difference. A class of corrosion reactions works on the electrochemical principle wherein the metal dissolves as positive ions and a depolarizer like oxygen from the atmosphere neutralizes the negative electronic charge on the surface. This class does not need that the circuit be powered. Another class of corrosion reactions is based on the galvanic cell principle. Such potential differences may be produced by galvanic cell formation between different metals on the chip. No external potentials are necessary. A third category is electrolytic ion migration, which happens under an applied voltage supply. Thus corrosion phenomena in electronics are driven by multiple factors.

Review exercises

15.1. Comment on the statement, 'the miniaturization of electronic components has increased the chance of corrosion-assisted failure'.
15.2. How does AH differ from RH?
15.3. Elucidate the interrelationship between ambient humidity, contamination and corrosion of electronics.
15.4. What metals are used in the following: (i) electronic chips, (ii) lead frames, (iii) bonding wires, (iv) electronic packages and (v) connecting lines on a PCB?
15.5. What metals are used in a computer hard disc?
15.6. Write down the chemical equations for the reactions taking place in an electrochemical corrosion process taking a bivalent metal as an example. Do these reactions require the existence of an externally applied voltage?
15.7. Explain the mechanism of PCB failure due to short-circuiting by anodic corrosion taking place by migration of Cu, Ag, Pb or Sn ions.
15.8. What is galvanic corrosion? Which parts of an IC are susceptible to galvanic corrosion and why?
15.9. How do the aluminum lines in an IC succumb to cathodic corrosion?
15.10. What is creep corrosion? Does it require the PCB to be connected to a power supply?
15.11. Describe the mechanism of creep corrosion. Why is it considered to be a three-step process? What are the preventive measures?
15.12. What is the pop-corning effect? With what type of IC packages does this effect take place? Why? How can it be avoided?

References

Baylakoğlu İ, Fortier A, Kyeong S, Ambat R, Conseil-Gudla H, Azarian M H and Pecht M G 2021 The detrimental effects of water on electronic devices *e-Prime – Adv. Electr. Eng. Electron. Energy* **1** 1–20

Chen C, Lee J C B, Chang G, Lin J, Hsieh C and Liao Huang J 2012 *The surface finish effect on the creep corrosion in PCB* www.smtnet.com/library/files/upload/creep-corrosion.pdf

Chen Y and Li P 2011 The 'pop-corn effect' of plastic encapsulated microelectronic devices and the typical case study *Int. Conf. Quality, Reliability, Risk, Maintenance, and Safety Engineering (Xi'an, June)* 482–5

Lighting Global 2013 Protection from the elements. Part III: corrosion of electronics *Technical Notes* Issue 14 September https://dev-lgla-merge.pantheonsite.io/wp-content/uploads/2013/09/63_issue14_part-iii_corrosion_technote_final.pdf

Savolainen P and Schueller R 2012 Creep corrosion of electronic assemblies in harsh environments *DfR Solutions IMAPS Nordic Annual Conf. Proc.* 54–8 www.dfrsolutions.com/publications/creep-corrosion-of-electronic-assemblies-in-harsh-environments/

Song B, Azarian M H and Pecht M G 2013 Effect of temperature and relative humidity on the impedance degradation of dust-contaminated electronics *J. Electrochem. Soc.* **160** C97

Tencer M and Moss J S 2002 Humidity management of outdoor electronic equipment: methods, pitfalls, and recommendations *IEEE Trans. Compon. Packag. Technol.* **25** 66–72

Chapter 16

Moisture and waterproof electronics

The exposure of electronic products to humidity and water from rain, sweat, washing, accidental dipping, etc, is a prominent cause of failure. Amongst the design approaches to protect from moisture are: evolving a fault-tolerant chip layout, enclosing heat-generating parts of the circuit in an air circulating chamber with the cooler parts confined in a watertight compartment, using the heat produced by the circuit to drive away moisture and making a careful choice of materials for boundary surfaces to avoid galvanic corrosion. From the practical viewpoint, various coating materials such as parylene, superhydrophobic coatings, volatile corrosion inhibitors and silicones are used to thwart moisture attacks. This chapter presents both design and technological methods to achieve water imperviousness of electronics.

16.1 Introduction

Many terms are interchangeably used with the word 'waterproof', e.g. watertight, water-repellant, submersible, sealed, etc. Severe pollution and high humidity levels have triggered shutdown accidents because thyristor voltage monitoring boards often fail in such conditions (Liu *et al* 2022).

Corrosion prevention is achievable at the design stage of a product by adroit and innovative design artifices. Depending on the intended application, the obvious solutions to waterproof electronics after chip fabrication include either encapsulation of the chip inside hermetic packages, using moisture-absorbent dessicants, covering the relevant surfaces with noble metals or protective films, or using volatile corrosion inhibitor coatings. This chapter will present a step-by-step description of the different strategies.

doi:10.1088/978-0-7503-5072-3ch16

16.2 Corrosion prevention by design

16.2.1 The fault-tolerant design approach

If the magnitudes of the current–voltage signals, leakage currents and impedances are fixed by design at values just sufficient for correct operation of the circuit, the design will be intolerant to any deviations. It will be easily offset by the smallest changes caused by moisture and therefore will be highly corrosion-sensitive. An expert designer will include the likely changes in series resistances of joints, whether they be present in switch contacts, relays, or other electromechanical connectors or at soldering points, in circuit boards and displays, that are expected to take place in the circuit in a humid environment. He/she shall do so by properly calculating the numerical tolerances using information from corrosion specialists. Corrosion risk is effectively reduced by minimization of the number of electromechanical connections. Notably, the connections consisting of male-ended plugs/pins and female-ended receptacles should be minimally used.

16.2.2 Air–gas contact minimization

Contact of polluted air containing dust and other undesirable particles with the circuit should be avoided. A simple mechanical cover helps in many instances. Wiring can be configured in a two-layer structure wherein the dense wiring containing small components is isolated from external airflow, whereas the wiring containing the power devices and heat-generating components receives strong airflow for efficient cooling. Thus a major part of the circuit is confined in an inert mass of air in which moisture cannot enter.

16.2.3 The tight dry encasing design

Electronics should always be maintained in a dry condition. Sometimes, the heat produced by the circuit itself may be used for this purpose. It aids in keeping the casing dry. Obviously, a circuit warmer than its surroundings is easy to keep dry. Moisture is swept away by the cooling air circulating over the circuit. If water condenses on the outer cover and drips into circuit boards, the circuit is at great risk. The designer should ensure that such dripping never happens.

16.2.4 A judicious choice of materials for boundary surfaces

In electronics, various boundary surfaces or interfaces exist in the microcircuits, switches, connectors, wiring and PCBs. In all probability, galvanic corrosion takes place at a spot where two metals are in contact if the atmospheric humidity is high. A high temperature further worsens the situation. Corrosion is minimum when two surfaces of the same metal come in contact. It is maximum if the contact metal surfaces are made from metals with widely different electric surface potentials, as seen from a glance at the electrochemical and galvanic series of metals. The use of different metals cannot be avoided in electronics, but the designer can exercise care in metal pair selection to choose metals which are the least likely to create trouble.

16.3 Parylene coatings

16.3.1 Parylene and its advantages

Parylene is a generic name. It stands for poly-para-xylylenes (Parylene Coatings and Applications: PCI 2017). Parylene coatings offer several advantages. The first is hydrophobicity. Next comes its chemical resistance to acids, caustic solutions and gases. The coating is pinhole free at a thickness >0.2 μm, serving as a thin, smooth, and chemically inert layer. Besides providing biostability and biocompatibility, it also gives thermal stability from −200 °C to +125 °C. Resistant to fungi and bacteria, it is sterilized by ethylene oxide (ETO) and gamma rays. It can be used for coating implantable medical electronics devices. It has high electrical impedance, providing excellent electrical isolation. It has a low dielectric constant. Last but not least, it can be deposited at room temperature in a vacuum system. It conforms to any surface clinging to it molecule by molecule with good adhesion. Bridging or pooling issues with liquid conformal coatings are avoided. Avoidance of a curing step helps to prevent curing stresses. No catalysts, solvents or plasticizers are required.

16.3.2 Types of parylene

There are several kinds of parylene, which differ in their moisture permeability, dielectric strength, biocompatibility and temperature-withstanding capability. The choice of parylene is decided by the requirements of the application at hand. Parylene N or parylene (natural) exhibits a high dielectric strength with a low loss tangent and a frequency-invariant dielectric constant. It has a linear structure. Each molecule is formed by the combination of carbon and hydrogen. Its penetrating power is very high. Substituting one aromatic hydrogen atom with a chlorine atom in parylene N produces parylene C. It is characterized by low permeability to water vapor and corrosive gases. It provides a pinhole free conformal insulating coating widely used for critical electronic devices. Parylene D is obtained by substituting two aromatic carbon atoms by two chlorine atoms in parylene N. It can withstand higher temperature than parylene C up to 125 °C. But it is not biocompatible. In parylene HT, a fluorine atom replaces the primary hydrogen atom of parylene N, extending its temperature limit up to 350 °C and making it UV-stable. Due to the small size of its molecule, it can reach every nook of the object to be coated.

16.3.3 The vapor deposition polymerization process for parylene coatings

A process resembling vacuum metallization is used to form parylene coatings (figure 16.1). But in contrast with vacuum metallization which is performed at a vacuum $\leqslant 10^{-5}$ Torr, parylene deposition is carried out at a vacuum level ~0.1 Torr, giving a mean free path ~0.1 cm at which all sides of the object are uniformly covered to produce a conformal coating (Parylene Engineering 2016). The process consists of three main steps:

(i) Vaporization of the solid dimer di-paraxylylene at 150 °C, 1 Torr.
(ii) Pyrolysis of the dimer at 680 °C, 0.5 Torr, giving a stable paraxylylene monomer by cleavage at the two methylene–methylene bonds.

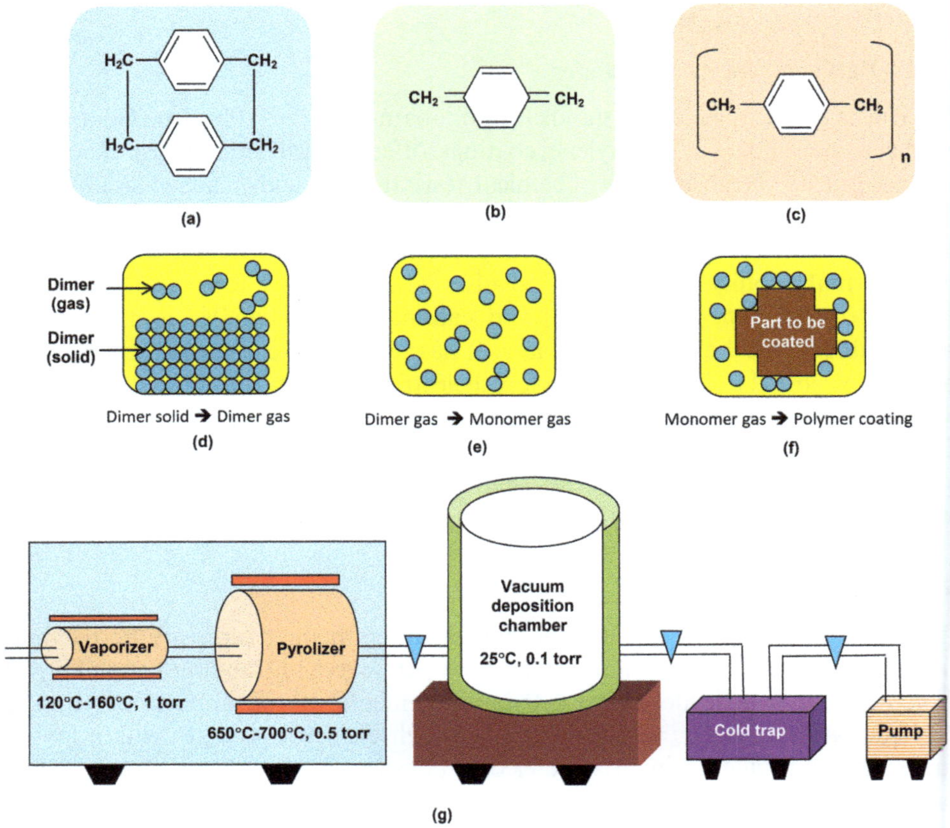

Figure 16.1. Vapor deposition polymerization: (a) dimer, [2, 2]paracyclophane, (b) monomer, paraxylylene, (c) polymer, poly[paraxylylene], (d) solid dimer to gaseous dimer conversion, (e) monomer gas, (f) parylene deposition on the workpiece and (g) the system for parylene deposition on the device.

(iii) Parylene deposition at 25 °C, 0.1 Torr, by adsorption and polymerization on the object. The deposition chamber is followed by a cold trap at −70 °C and a mechanical vacuum pump.

Two parameters determine the parylene thickness:
(i) the quantity of dimer vaporized; and
(ii) the dwelling time in the deposition chamber.

The achievable thickness accuracy is within ±5% of the targeted value. The parylene deposition rate is 0.2 mil h^{-1}. The deposition rate is directly proportional to the square of the monomer concentration. It varies inversely with the absolute temperature.

16.3.4 Typical electrical properties

For parylene N, the dielectric strength is 7000 V mil^{-1}, volume resistivity is 1.4×10^{17} Ω-cm, the dielectric constant at 60 Hz is 2.65 and the dissipation factor at the same frequency is 0.0002. Corresponding values for parylene C are 5600 V mil^{-1}, 8.8×10^{16} Ω-cm, 3.15 and 0.020.

16.3.5 Applications for corrosion prevention

Applications are as follows:

(i) As a protective coating for PCBs: 0.001″ thick parylene C intrudes and covers a 0.002″ space between the wire and the board.

(ii) Parylene thicknesses up to 0.4 μm are free from pinholes and can coat bonded wires and the conductor metal lines at the same thickness as the semiconductor surface. On the performance level, parylene coated ICs with transfer molded epoxide encapsulation are equivalent to those with hermetically sealed packaging.

(iii) As an electrically insulating coating for biomedical devices, protecting against the biological environment.

16.4 Superhydrophobic coatings

16.4.1 Concept of superhydrophobicity

A superhydrophobic or ultrahydrophobic coating is a nanoscopic coating, which forms a surface on the electronic component, that can be wetted with great difficulty. Examples include zinc oxide polystyrene nanocomposite or manganese dioxide polystyrene nanocomposite. Silica-based gels are also used.

An idea about the wettability of a solid surface by a liquid is obtained from the contact angle created by the liquid with the solid surface (Yuan and Lee 2013). Contact angle θ is defined as the angle subtended by the liquid–vapor interface with the liquid–solid interface, obtained by drawing a tangent from the contact point along the liquid–vapor interface in the profile of the liquid droplet (figure 16.2). The contact angle is an inverse measure of wettability. The smaller the contact angle, the larger is the tendency of the liquid to wet the solid. If $\theta = 0°$, the liquid completely spreads on the surface of the solid like a flat puddle at a rate limited by the viscosity of the liquid and the roughness of the solid. If $\theta < 90°$, the liquid is said to be a wetting liquid, the wetting is favored and the liquid will spread over a large area. The surface is called a hydrophilic surface. But when $\theta > 90°$, the liquid is non-wetting, the wetting is disfavored and the liquid tries to minimize its contact with the solid. It forms a compact liquid drop. The surface is termed hydrophobic. For contact angles $150° < \theta < 180°$, there is almost no contact between the liquid and the solid. The liquid forms a sphere or bead with minimal surface area and rolls off the surface without any adsorption. Such a surface is known as a superhydrophobic surface. The leaves of a lotus plant exhibit such an extreme degree of repellence of water. Hence, this effect is called the lotus effect. A superhydrophobic coating can create a moisture barrier far more effective than normal coatings.

16.4.2 Standard deposition techniques versus plasma processes

The standard coating techniques are spray coating, screen or spread coating, and dip or immersion coating in liquid. These techniques provide poor adhesion to

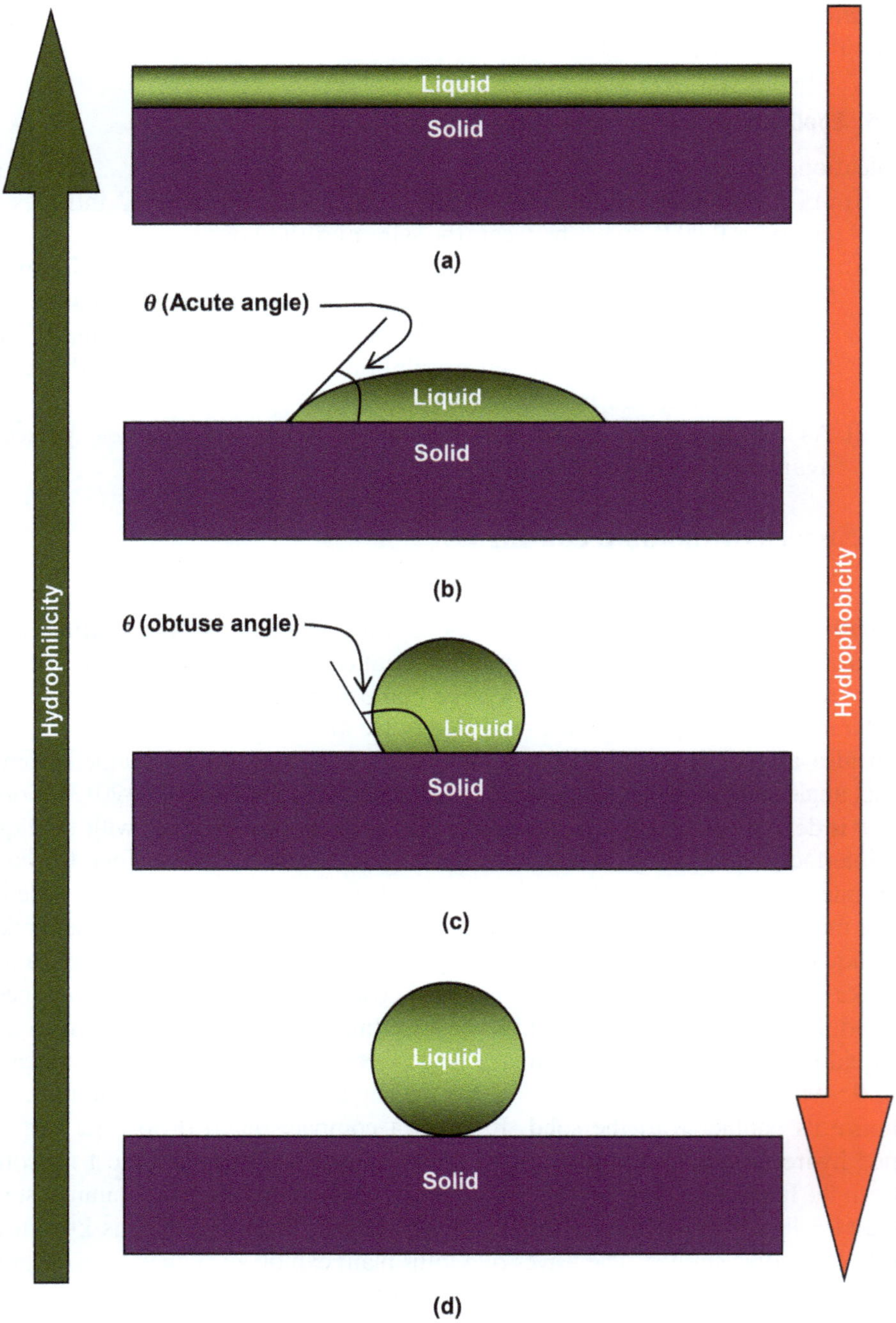

Figure 16.2. Contact angles θ for wetting of a solid by a liquid: (a) $\theta = 0°$, complete wetting; (b) $\theta < 90°$, high wetting and spreading; (c) $\theta > 90°$, low wetting and repelling; and (d) $\theta = 180°$, no wetting.

substrates, involve the use of a large quantity of coating material, and often need a high-temperature curing step to improve the quality of layers. The thickness of liquid coatings depends on the viscosity and working temperature/humidity, and is controllable with an accuracy of ±50% of the desired value. High-quality conformable coatings can be formed on objects of different shapes by plasma processes. Further, these coatings can be of nanoscale thickness so that they are virtually transparent and invisible. The plasma process is performed at room temperature. It is an environmentally benign process. In this process, ions, excited molecules and radicals produced by a RF electric field bombard the object to be coated, modifying its properties.

16.4.3 The main technologies

Three companies are intensively engrossed in research on nanodeposition processes, namely P2i, GVD and Semblant (Tulkoff and Hillman 2013). The key technologies are:

- P2i: pulsed plasma and halogenated polymer, specifically fluorocarbon.
- GVD: initiated CVD using PTFE (polytetrafluoroethylene) and silicone.
- Semblant: plasma deposition and halogenated hydrocarbon.

Hydrophobicity is related to the number and length of fluorocarbon groups, and their concentration on the surface. Whereas traditional plasma fragments the monomer, in the pulsed plasma approach the monomer structure is retained. Retention of this structure provides a high liquid repellency. The process is applied industrially using roll-to-roll plasma processing equipment (Rimmer 2014).

16.4.4 Applications

The main application is the reliability enhancement of electronic gadgetry, such as tablets and smartphones as well as headphones, headsets and hearing aids (P2i 2015).

16.5 Volatile corrosion inhibitor coatings

Volatile corrosion inhibitor coatings (VCIs) are organic compounds (figure 16.3) such as dicyclohexylammonium nitrite (DICHAN), $C_{12}H_{24}N_2O_2$; tolytriazole ($C_7H_7N_3$), etc. They come in solid form for easy handling (VCI; VCI2000 2016). They have sufficient vapor pressure ~10^{-3} –10^{-5} mm Hg at 21 °C, favoring easy vaporization (Vimala *et al* 2009). In an enclosed space, their vapors disseminate until equilibrium is attained, as determined by their partial pressure. Vaporization is followed by condensation and adsorption of vapors on a metal surface. Adsorbed VCI is in the form of microcrystals. Even very small quantities of water vapor cause dissolution of these crystals. As a result, protective ions are formed. These ions are also adsorbed on the surface of the metal. Consequently, an ultra-thin monomolecular passivating film is produced, covering the metal surface and breaking down the contact between the metal and electrolyte. This film prevents electron transport between anodic and cathodic areas on the metal, retarding the corrosion process. Thus the VCIs chemically condition the environment near the metal surface to make

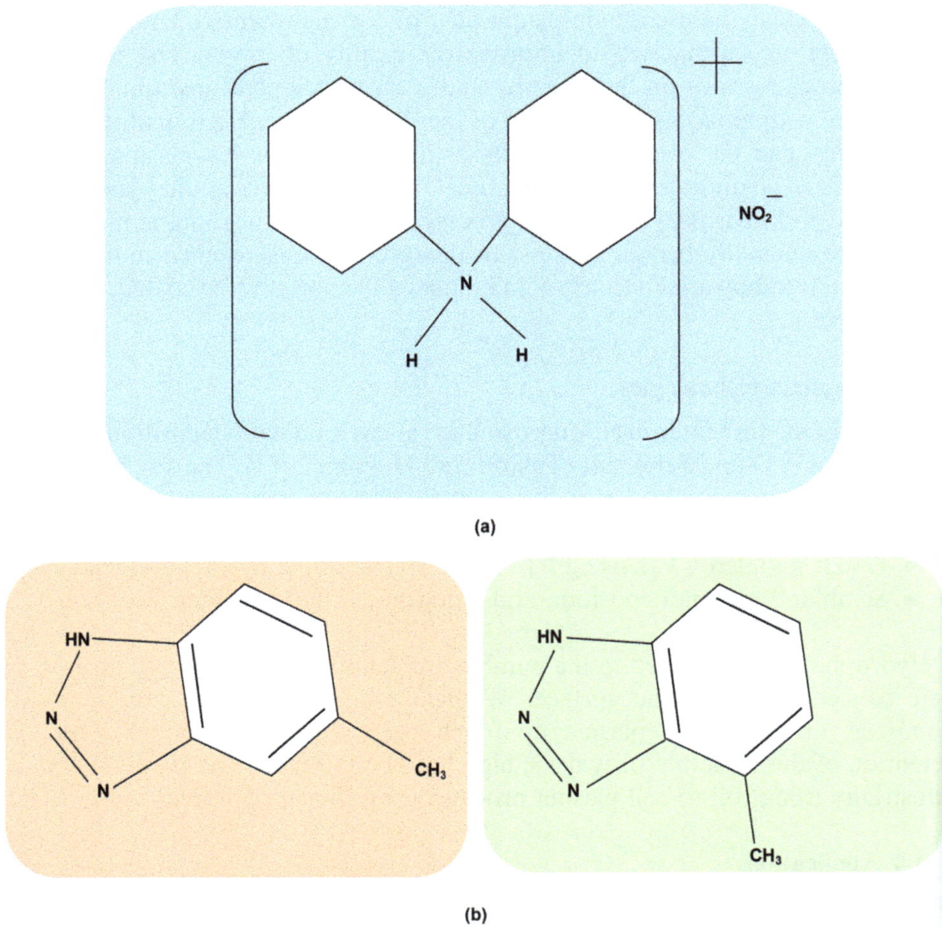

Figure 16.3. VCIs: (a) DICHAN and (b) tolyltriazole.

it less corrosive to the metal. It must be emphasized that the shielding layer formed by VCIs produces no interference with the resistivity of the underlying metal layer, which is crucial for an electronics application, and is in many respects superior to the indirect method of passivation by a coating. The VCI films are self-healing. In the case that the film is scratched from a portion of the metal, the VCI film from the adjoining region evaporates and deposits on the scratched portion (Prenosil 2001).

16.6 Silicones

The silicones, e.g. silicone rubber (figure 16.4), silicone resin, silicone oil, silicone grease, etc, are synthetic polymers containing long chains of silicon and oxygen atoms alternately, such as (–Si–O–Si–O–)$_n$. Organic groups, e.g. the methyl group –CH_3, attached to silicon atoms determine their properties. Their structural unit is R_2SiO where R is an organic group. Silicones contain carbon, hydrogen, oxygen and occasionally other elements as well. Silicones are known for their water repellence,

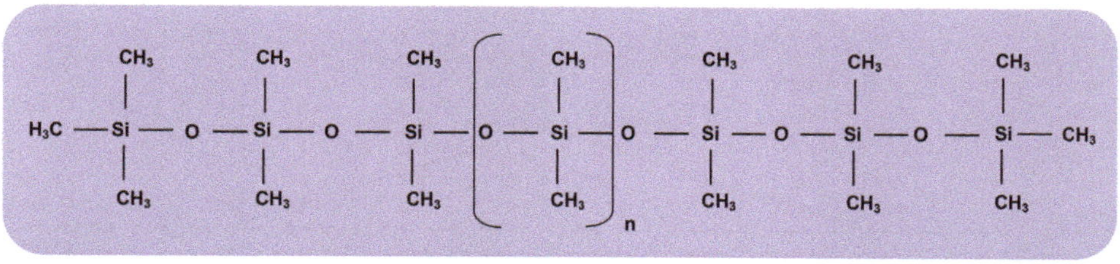

Figure 16.4. Silicone rubber.

chemical inertness, thermal stability (−100 °C to 250 °C), resistance to degradation by ozone and UV, and fluidic, resinous, lubricating, rubbery, electrical insulating, adhesive, stress-relieving, and shock and vibration absorbing properties. Silicones are prepared by decomposing halides of organic silicon compounds.

Silicone adhesives are of three types (DOW CORNING© (2000–6) Dow Corning Corporation):

(i) moisture cure for room temperature processing;
(ii) condensation cure offering room temperature and deep-section curing; and
(iii) heat cure for fast processing.

All the three varieties provide primerless adhesion with ceramics, reactive metals and plastics.

The one-part moisture-cure room temperature vulcanization (RTV) silicone is cured at room temperature in 30%–80% RH. With 24–72 h curing, >90% of physical properties are obtained. For condensation cure silicone in the two-component format, curing occurs at room temperature, resulting in good strength within an hour, but full properties are only attained after several days. The heat cure silicon is cured at ⩾100 °C. Thorough cleaning and/or degreasing of the surface is mandatory before silicone application.

RTV silicone is used for sealing electronic equipment and modules. It is used for fixing parts on PCBs of power supplies and cathode ray tubes (CRTs). Liquid crystal display (LCD)/LED modules are also assembled without affecting their optical properties. Condensation cure silicone is mainly used for gasketing and housing seals. Heat cure silicone is used for gasketing electronic modules. It is also used for sealing capacitors and electronic components; flyback transformers too are fixed with this silicone.

16.7 Discussion and conclusions

The necessary effort to make an electronic product water-impermeable commences right at the outset from its inception, and spans from product designing up to the complete fabrication and assembly level. Security against the harmful effects of dampness, humidity and water spillage can be provided by perfectly covering the product with water-repellant nanotechnological films, such as those possessing superhydrophobic properties. In less critical conditions, various other alternative coatings are applicable. Figure 16.5 summarizes the strategies to be adopted for making electronics safe in damp and wet conditions.

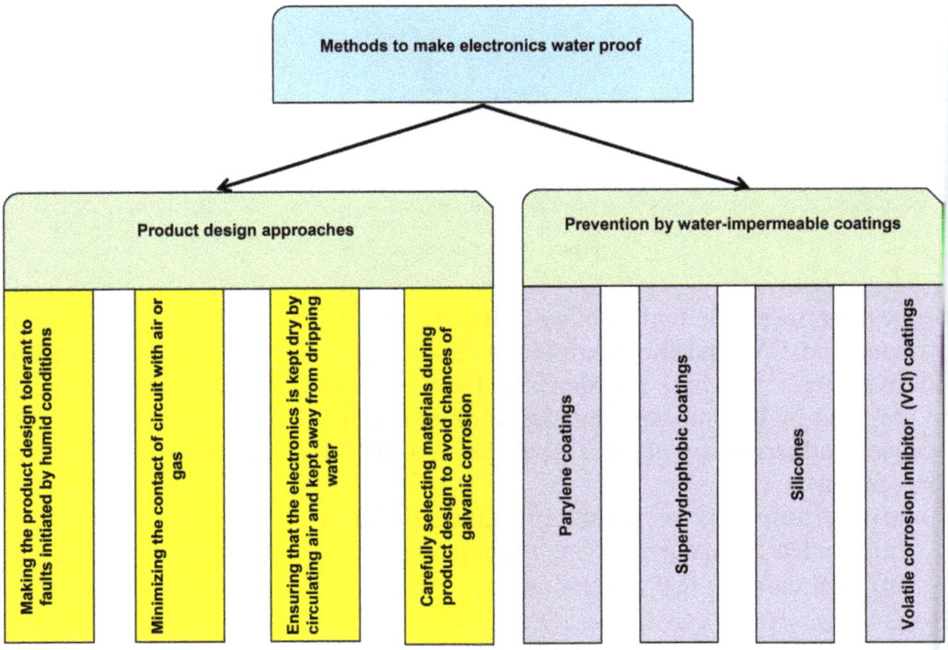

Figure 16.5. Schemes for moisture-free electronics.

Review exercises

16.1. What is meant by a fault-tolerant electronic design? Explain with examples.

16.2. How do the precautions taken regarding the choice of materials for boundary surfaces in an electronic product influence the chances of galvanic corrosion and impact its reliability?

16.3. What does parylene stand for? State five advantages of parylene as a coating material for protecting electronics from the harmful effects of moisture.

16.4. What are the salient features of the four types of parylene: parylene N, parylene C, parylene D and parylene HT?

16.5. How does the vapor deposition polymerization process for coating parylene differ from vacuum metallization? What are the three main steps in this process? How is parylene thickness controlled?

16.6. Which of the two parylenes, parylene N or parylene C, has a higher dielectric constant? Which one has the lower loss tangent?

16.7. Describe two corrosion prevention applications of parylene.

16.8. Define the contact angle of a liquid with a solid surface. How is it related to the wettability of the solid surface by the liquid?

16.9. Explain the following statements: (a) the contact angle of a liquid with a solid is <90°, (b) the contact angle is >90° and (c) the contact angle is >150° but <180°.

16.10. How does a hydrophobic surface differ from a hydrophilic surface in terms of: (a) contact angle and (b) wetting properties?

16.11. When is a surface said to be superhydrophobic? What is the lotus effect?
16.12. In what respects are the plasma processes for deposition of conformable coatings superior to conventional coating processes such as spraying and dip coating?
16.13. Describe three industrial approaches for the nanodeposition of superhydrophobic coatings, indicating the materials and process technologies used.
16.14. Give two examples of VCI coatings. What is the mechanism of corrosion prevention by these coatings?
16.15. Do VCI coatings alter the resistivity of the underlying metal? How does the self-healing property of a VCI help in repairing any abrasion of the coating from any portion?
16.16. What is a silicone? What is the basic structural unit in a silicone? What are the special properties of silicones which make them attractive as a coating material for electronic use?
16.17. How do the three types of silicones differ? What are the typical applications of the three types?

References

Dow Corning© 2000–6 Dow Corning Corporation information about Dow Corning® brand adhesives/sealants—silicones and electronics *Product Information* https://datasheet.octopart.com/227109-Dow-Corning-datasheet-572677.pdf

Liu Z, Wang L, Mei H, Cao B, Yin F, Xu Y, Zhang F, Jiang Y and Farzaneh M 2022 Effect of contamination and relative humidity on the function of thyristor voltage monitoring board *High Volt.* **7** 1165–74

P2i 2015 *Electronics specialists: P2i's focus is on the consumer electronics industry* www.p2i.com/electronics/

Parylene Coatings and Applications: PCI 2017 http://www.pcimag.com/articles/87074-parylene-coatings-and-applications

Parylene Engineering 2016 Deposition process: parylene *Parylene Engineering* www.paryleneengineering.com/parylene_deposition_process.htm

Prenosil M 2001 Volatile corrosion inhibitor coatings *CORTEC Corp. ORP Supplement to Materials Performance* 14–17 January

Rimmer N 2014 Vacuum plasma deposition of water and oil repellent nano-coatings *AIMCAL Web Coating and Handling Conf. 2014 Europe P2i* www.aimcal.org/uploads/4/6/6/9/46695933/rimmer_presentation.pdf

Tulkoff C and Hillman C 2013 Understanding nanocoating technology *Electronic System Technologies Conf. and Exhibition (May)* www.dfrsolutions.com/wp-content/uploads/2013/07/ESTC-Hillman-Understanding-New-Coating-Tech.pdf

VCI2000 2016 Anticorrosive VCI packaging *VCI* http://vci2000.com/technology/

Vimala J S, Natesan M and Rajendran S 2009 Corrosion and protection of electronic components in different environmental conditions—an overview *Open Corros. J.* **2** 105–13

Yuan Y and Lee T R 2013 Contact angle and wetting properties *Surface Science Techniques* ed G Bracco and B Holst (Berlin: Springer) pp 1–34

Chapter 17

Preventing chemical corrosion in electronics

The corrosion of electronic circuits has two principal sources: the leftover corrosive materials in the fabrication of microelectronic circuits and printed circuit boards (PCBs), and the interaction of corrosive gases in the atmosphere with the surfaces of metal layers of the circuits. Therefore, corrosion prevention must follow a two-pronged strategy, by identifying the possible reasons from both aspects of the problem and planning corrosion inhibition measures methodically. In this chapter, both the processing and environment-based likely causes of corrosion are indicated and remedial measures are described. Common corrosion prevention techniques, such as the application of conformal coatings, potting, plastic molding, hermetic packaging by ceramic and metal packages as well as by glass passivation, are elaborated.

17.1 Introduction

Microelectronic chip fabrication and packaging processes employ several metals and alloys, such as aluminum, copper, silver, tin, nickel and steel, to name a few (Salas *et al* 2012). They are used in conductive metal patterns, die attach and solder pastes, support frames, heat sinks, etc. All these metals/alloys have to bear the vagaries of atmospheric pollutants and climatic conditions. Corrosion is a phenomenon causing material deterioration that harms all metallic structures, and so necessitates intensive prevention and control measures (Rubin and Garcia 2020, Li *et al* 2021, Zerust 2022). In a corrosive environment, the metal surface is irreversibly damaged by chemical reactions via conversion of the metal into its more stable form such as metal oxide, hydroxide or sulfide (Harsimran *et al* 2021).

The escalation of cost-effective plastic packages replacing hermetically sealed types has provided easier access for corrosive gases to inner components because they can penetrate through most polymers. The main gases implicated in the corrosion of electronics are hydrogen sulfide, carbonyl sulfide, sulfur dioxide, nitrogen dioxide and chlorides.

17.2 Sulfidic and oxidation corrosion from environmental gases

The corrosion products formed on copper and silver surfaces are copper sulfides (Cu_2S and CuS) and silver sulfide (Ag_2S). The common reactions are:

$$2Cu + H_2S + \tfrac{1}{2}O_2 \rightarrow Cu_2S + H_2O \quad (17.1)$$

$$Cu + H_2S + \tfrac{1}{2}O_2 \rightarrow CuS + H_2O \quad (17.2)$$

$$2Ag + H_2S + \tfrac{1}{2}O_2 \rightarrow Ag_2S + H_2O. \quad (17.3)$$

In the presence of oxygen of the atmosphere, H_2S forms sulfuric acid, a strong corrosive agent:

$$H_2S + 2O_2 \rightarrow H_2SO_4. \quad (17.4)$$

The tarnish film on silver becomes darker with increasing thickness. The morphology of the grown film is non-uniform due to the formation of fern-shaped dendrites or filamentary whiskers.

Silver spontaneously oxidizes in air through the reaction:

$$4Ag + O_2 \rightarrow 2Ag_2O. \quad (17.5)$$

In an oxygen environment, copper oxidizes as follows:

$$4Cu + O_2 \rightarrow 2Cu_2O; 2Cu_2O + O_2 \rightarrow 4CuO. \quad (17.6)$$

The gold film covering a copper or silver layer can wear out. If the environment contains sulfur or chlorine, the corrosion material formed at the tattered regions will increase the contact resistance.

Due to the reactivity of lead, the lead component of lead–tin solder is readily oxidized in the atmosphere. When the weight percentage of tin is below 2%, the unstable lead oxides undergo further reaction with chloride, sulfate and phosphate ions leading to more wear.

17.3 Electrolytic ion migration and galvanic coupling

Aluminum metallization can corrode by ionic migration in an electric field using atmospheric moisture and halide contamination. In an applied electric potential gradient in a humid environment, silver ions migrate to the negatively charged surface forming silver dendrites, which can grow to sufficient lengths to close the gap between contacts, creating a short-circuiting condition.

In magnetic and magneto-optical memory devices, corrosion occurs at the sites where the carbon coating is damaged, exposing the Co-based layer and Ni–P substrate. Under an applied voltage between carbon and the metallic substrate, a galvanic couple is formed, dissolving the magnetic material.

17.4 Internal corrosion of integrated and printed circuit board circuits

The metals used in device fabrication may interact mutually. One example is the Al–Cu metallization in which intermetallic compounds like Al_2Cu are formed along the grain boundaries of aluminum. The intermetallic compound Al_2Cu acts as a cathodic site with respect to aluminum. Consequently, aluminum is dissolved and micropits are formed during rinsing after chemical etching.

The liquid and vapor-phase halogenated solvents used in integrated circuit (IC) and PCB manufacturing react with aluminum-containing devices meddling with their performance. If diluted with an aromatic alcohol, the halogenated solvent breaks down forming chlorine ions, which react with Al.

17.5 Fretting corrosion

Fretting is a kind of wearing away which takes place by rubbing (figure 17.1). Fretting corrosion impairs the functioning of switches using non-noble metals as contact materials. Gold does not form an oxide layer under exposure to atmosphere. But it is very expensive. So, one cannot afford to make switches from gold. Instead, tin-plated copper is used for making switches. When these switches experience vibration such as due to floor vibration or thermal stresses, mechanical rupturing of oxide layers occurs. It breaks down the oxide layers at the contact points. Fresh oxide layers are therefore formed. These are also destroyed in subsequent rupturing. Thus oxide debris gathers at the contact points. This debris increases the contact resistance. Demolition of the oxide film is caused by the wiping action of plugging/unplugging of the contact. Internal wire-to-wire contacts are not prone to this kind of threat. Notwithstanding, the internal wire-to-board contacts are confronted with a similar problem. They face the risk of fretting corrosion.

17.6 Tin whisker growth

Adoption of lead-free processes has led to tin whisker issues, which are not forms of corrosion but behave similarly. Thin, hard whiskers grow on tin-plated leads of the electronic components in reaction to surface compression stress in the tin coating. These whiskers short-circuit adjoining leads.

17.7 Minimizing corrosion risks

17.7.1 Using non-corrosive chemicals in device application and assembly

Flux residues have been pinpointed as a reliability threat to printed circuit board assemblies (PCBAs; Woolley 2013). In the early days, rosin-activated fluxes were used. These fluxes contained chlorine or bromine. These reactive halides were used primarily as a cleaning agent to remove the copper oxide layer, which grows on copper conductors in air, and inhibits the formation of an intimate contact between the solder and the copper layer. The removal of the oxide layer was essential for establishing a secure solder connection.

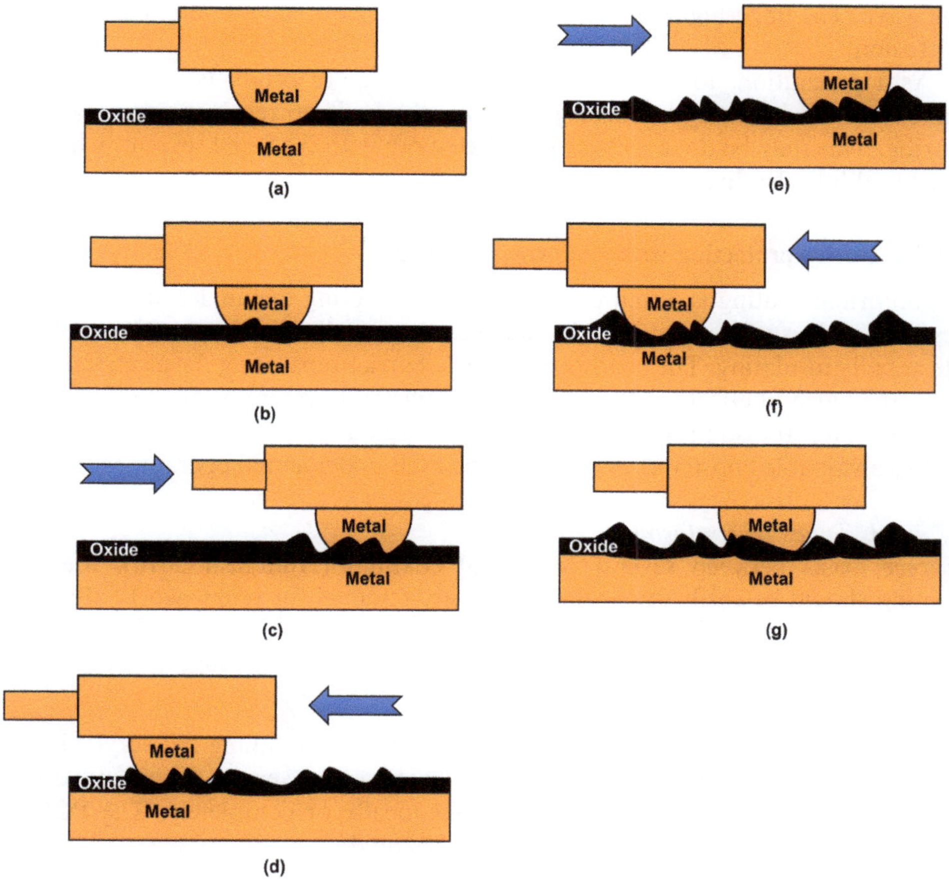

Figure 17.1. Fretting corrosion: (a) initial metal-to-metal contact, (b) interfacial native oxidation, (c) rightward micromotion, (d) leftward micromotion, (e) oxide debris pile-up on the right, (f) oxide debris pile-up on the left and (g) oxide film to oxide film contact after *N* cycles.

Several post-soldering cleaning methods evolved. But if the PCBA surface is not properly cleaned after soldering, remnant chlorine or bromine traces left on the surface lead to ionic contamination, which is a dominant cause of corrosion. Needless to say, the compounds used for cleaning the flux residues should be free from halides and phosphates.

To overcome the problem of leftover halide traces, restriction of hazardous substances (RoHS)-compliant low-solid 'no-clean fluxes' were introduced. These fluxes were not chlorinated. In place of the halogen element, they contained organic acids such as adipic acid, $C_6H_{10}O_4$, and citric acid, $C_6H_8O_7$, for cleaning purposes. They did not require any cleaning after the soldering step because the organic acids underwent decomposition at the soldering temperature. Cleaning was a non-issue during reflow soldering. However, during wave soldering, the flux wedged between the wave soldering pallet and the PCB is not bared to the molten solder. This happens because the pallet serves as a thermal impediment between the solder and the wave. The lingering flux, being acidic and hygroscopic in nature, participates in corrosion.

As a precaution, to avoid contamination of future products, the wave solder pallets should be always cleaned to remove the built-up residues acquired during fluxing. This may be done using a fiber-free cloth with the same solvent, which was used to dilute the flux.

17.7.2 Device protection with conformal coatings

A conformal coating is a thin polymeric film preserving the angles and topological features of the surface of a PCB. Its thickness is typically 25–100 μm. Generally, it is electrically insulating. The coating serves as a barrier separating the electrical circuit from the harmful effects of the hostile atmospheric gases, inimical chemical vapors and moisture. It also protects the circuit from organic attack by microbial action. It is thus a valuable booster of longevity of the PCB under actual operating conditions.

17.7.2.1 Types of conformal coatings

Choice exists between several conformal coatings (HumiSeal 2016). Ideally, a coating should be easily applicable and easily removal for repair work, allowing a simple process both for initial coating and for later repair of the PCB for replacement of any damaged components. But in practice, the coatings differ widely in terms of their removability.

(i) *Acrylic coatings.* These are plastic materials containing the acryloyl group obtained from acrylic acid. These coatings are very easily removed using specific stripping chemistries, either locally or completely. The PCB can then be repaired and its operation restored.

(ii) *Polyurethane coatings.* These are thermosetting/thermoplastic polymers containing organic units interlinked by carbamate or urethane –NH–(C=O)–O– groups. The higher chemical resistance of the polyurethane coatings than acrylic coatings makes their removal a little more difficult than the acrylics. Nonetheless, they are removable, permitting reworking of the PCB.

(iii) *Silicone coatings.* These are polymers constructed from recurring units of siloxane containing alternate S and O atoms, in combination with C, O and H atoms. These coatings are the most difficult to remove. By mechanical abrasion, they can be removed from small areas. So, at the most local repairing is possible.

(iv) *UV cured coatings.* These are the most chemically resistant and mechanically rugged of all coatings. Hence, repairing is extremely difficult. Removal by powder abrasion or by local melting in a small area can be attempted. Following repair, the board is re-coated.

(v) *Nanocomposite coatings.* Poysiloxane and carbon nanotube composite coatings produced by electrophoretic deposition are found to protect metals from corrosion (Bezkosty *et al* 2022).

17.7.2.2 Glob top

This is a variant of conformal coating (EPOXY Technology 2001). In appearance, it looks like a black spot on the circuit board. It is widely used in chip-on-board (COB) assembly as a protective dome over a fragile semiconductor die and associated wire bonding, thereby reinforcing it mechanically and making it tamper-proof, in addition to isolating it from the environmental effects and eliminating chances of static discharge.

Glob tops are of two types (figure 17.2): single part hemispherical and two-part dam-and-fill. In the former type of glob tops, a thixotropic material is used to erect a protective dome over the electronic component and wiring. The thixotropic property is a property exhibited by certain materials of liquid-like flowing (thin and less viscous) upon stirring, agitation or shaking, and thick and viscous behavior under static conditions.

In the latter type of glob tops, a dam is formed surrounding the electronic component using a thixotropic epoxy. Subsequently, the cavity is filled up using a low-viscosity solution. Thus it is a two-step process involving dam building and cavity filling. Both types of glob tops were found to sustain operation at 85 °C and 85% RH for 2000 h without any degradation (EPOXY Technology 2001), promising acceptable reliability assurance.

17.7.2.3 Application methods and inspection

The coatings are applied by dip coating, spraying and by robotic methods at defined locations. Voids or bubbles left in the film thus formed serve as a path for the ingress

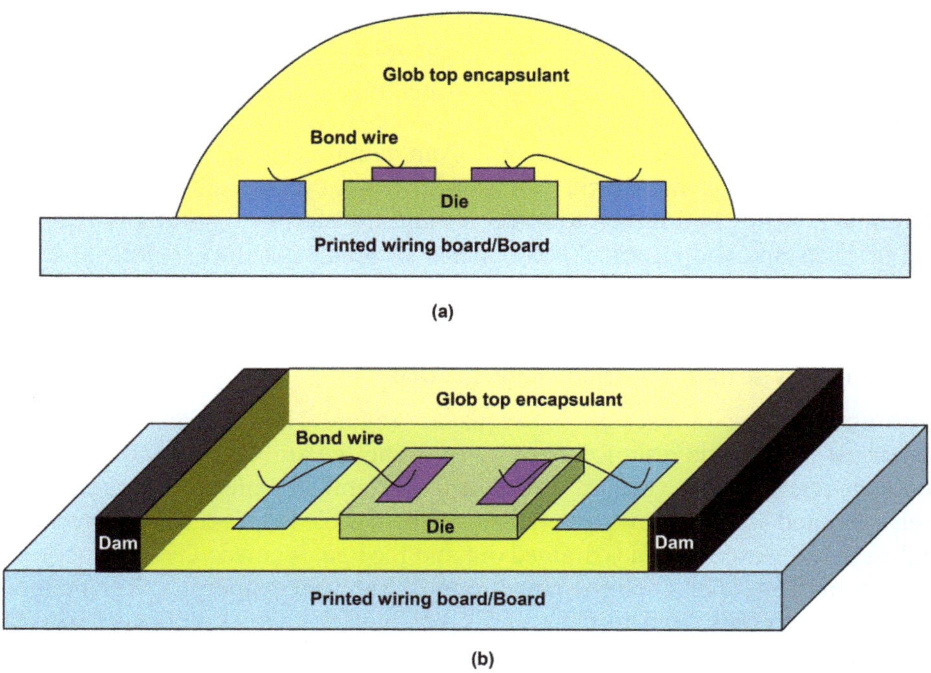

Figure 17.2. Types of glob tops: (a) hemispherical and (b) dam-and-fill.

of contaminants. By careful inspection, any defects in coating should be identified and immediately repaired. Neither any solder joints nor any sharp edges of components or their leads should be left bare. Otherwise the aggressive chemicals will intrude to the electronic devices and the whole purpose will be defeated.

17.8 Further protection methods

17.8.1 Potting or overmolding with a plastic

In applications demanding exceptional protection from environmental corrosion, the conformal coating is covered with a potting compound, which is a gelatinous substance for sealing and encapsulating the whole assembly (figure 17.3). The available potting compounds are one- or two-part formulations. The formulations are epoxies, silicones, polyurethanes and UV-curable substances. The liquid is poured over the circuit board and allowed to cure. It solidifies providing good mechanical strength to the electronic components. Immersing the assembly in a liquid resin is another way of potting. Taking it out and curing it completes the process. The electronic item is sometimes kept in a pre-fabricated potting shell. The shell is filled with the potting compound and the compound is cured. Then the item is taken out of the shell.

Void-free structures can be formed by a method called vacuum potting. To accomplish vacuum potting, the electronic item is placed in a vacuum chamber when the resin is in the liquid state, i.e. prior to its solidification. In this way, air is pumped out of the internal cavities. Then the vacuum is withdrawn and the item is exposed to atmospheric pressure. Under the influence of atmospheric pressure, the internal cavities collapse, forcing the liquid resin to fill any vacant spaces where it has not so far penetrated and thus leaving no voids. Vacuum potting is suitable for resins that polymerize, rather than those which work by the solvent evaporation mechanism.

The potting of the PCB can prevent it from damage even after immersion in salt water. Resistance against shock, vibration, thermal cycling and fungal attack are additional benefits. In addition to solenoids and transformers, potting is extensively used in electronic and electrical components such as capacitors, rectifiers, sensors, switches, ignition coils, motors, etc.

17.8.2 Porosity sealing or vacuum impregnation

In porosity sealing or vacuum impregnation (figure 17.4), the electronic item is submerged in a plastic resin followed by lowering the ambient pressure to vacuum level. On releasing the vacuum, i.e. raising the ambient pressure to atmospheric level, the liquid solution flows into the most inaccessible cavities. The electronic item is taken out, the excess resin is drained out and the item is cured. The polymeric resin hardens. In the case of a solvent-based resin, the solvent evaporates. The method is applied to solenoids, transformers and other high-voltage components to ensure good dielectric strength. It prevents the components from ionization-induced failures.

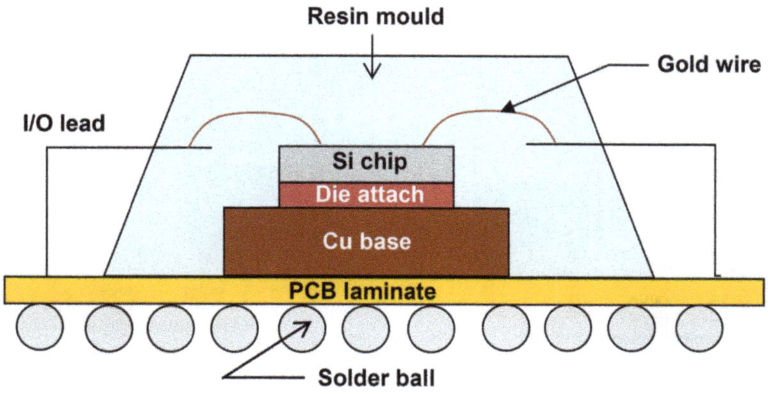

(a)

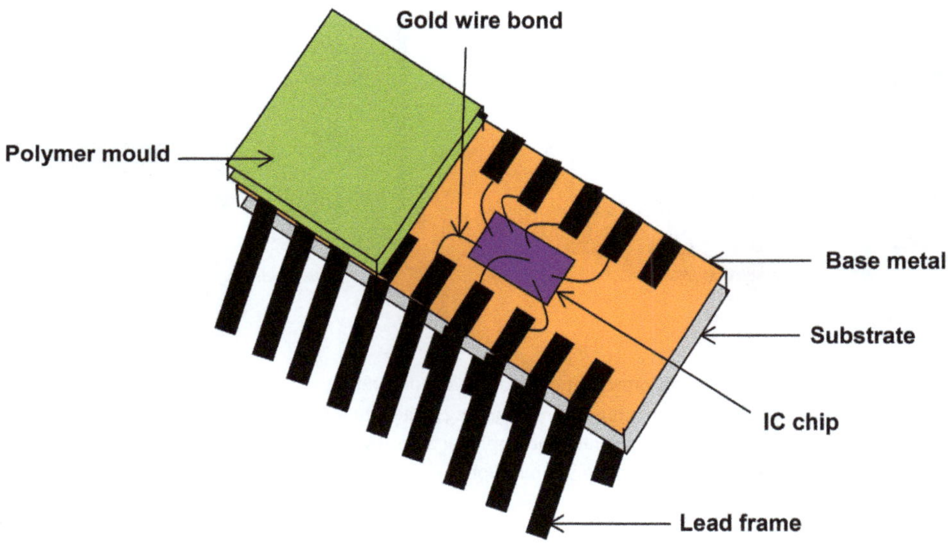

(b)

Figure 17.3. Plastic packages: (a) two leads and (b) multiple leads.

17.9 Hermetic packaging

'Hermetic sealing' refers to the quality of a container being airtight, completely prohibiting any entry or exit of air, oxygen, moisture and other gases. Perfect hermetic sealing is not possible because every material is permeable to gases, even though the permeability may be negligibly small for practical applications. A hermetic package can withstand higher operating temperatures than a plastic package, primarily due to the temperature-withstanding properties of the materials used in hermetic packaging. Different types of hermetic packages are described below.

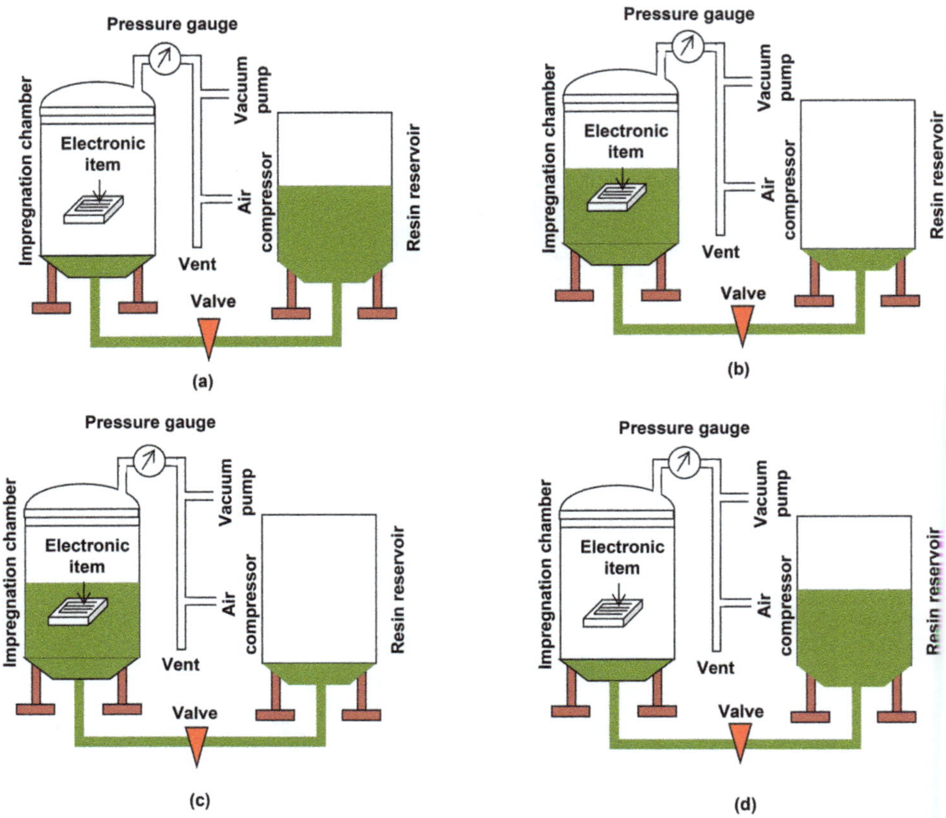

Figure 17.4. Steps of the vacuum impregnation process: (a) preheating the equipment, loading the electronic item into impregnation chamber, creating and maintaining a vacuum for 30 min; (b) flooding the impregnation chamber with resin; (c) releasing the vacuum and applying a positive pressure for 30 min and (d) releasing the pressure and discharging the resin back to the reservoir.

17.9.1 Multilayer ceramic packages

Metallization, lamination and firing of ceramic tape layers constitute the three main steps to form the body of the package (Texas Instruments 1999), see figure 17.5.

The multilayer ceramic tape construction helps the package designer to introduce useful electrical features into the package. Notable features include the power and ground planes, shield planes and controlled characteristic impedance of signal lines. The power and ground planes decrease inductance while the shield planes minimize crosstalk.

Nickel–gold electroplating is done on the metallized regions of the package. By soldering a metal ring into the metallized and plated seal ring, hermetic sealing is accomplished. These packages are therefore designated as solder-seal packages (figure 17.6).

Brazing is the process used for insertion of leads into the package body (figure 17.7).

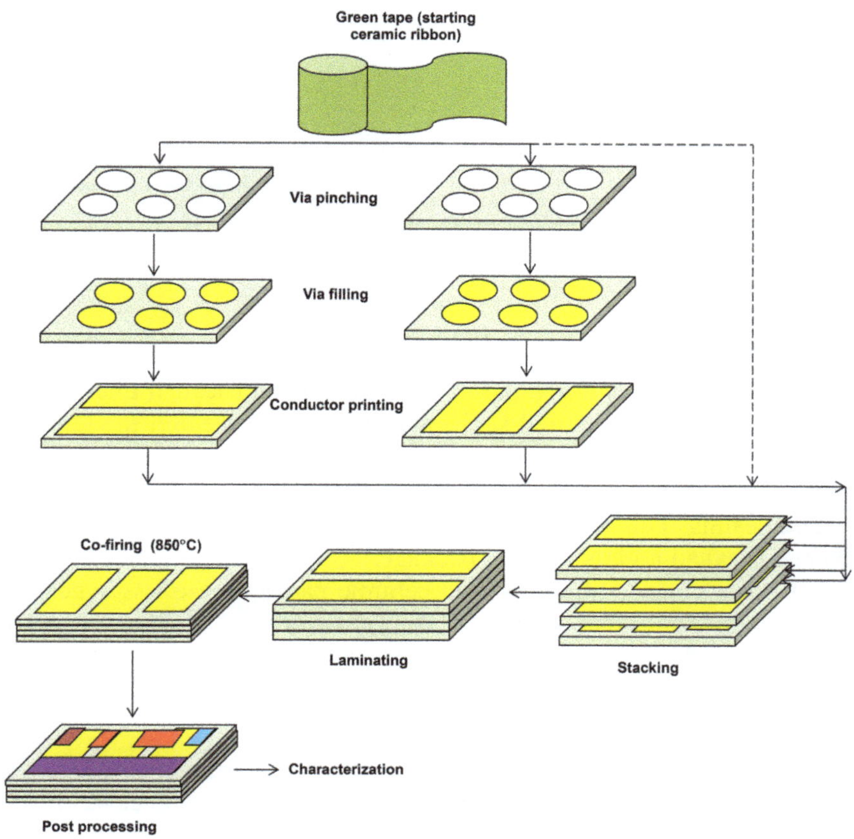

Figure 17.5. The low-temperature co-fired ceramic (LTCC) process.

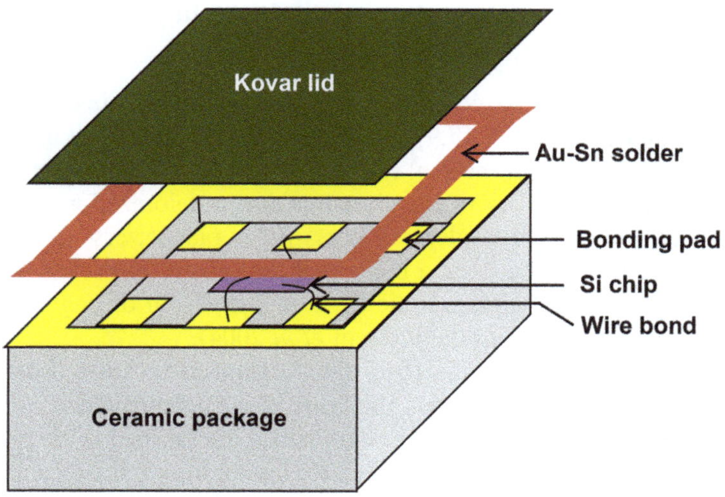

Figure 17.6. A solder-seal ceramic package.

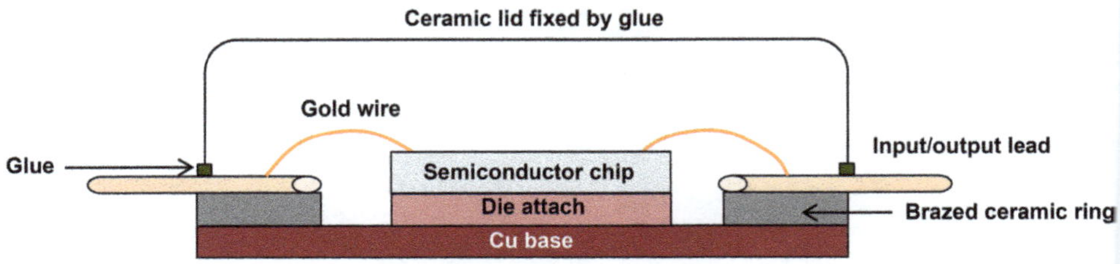

Figure 17.7. Insertion of leads into the package by brazing a ceramic ring with a copper base plate.

17.9.2 Pressed ceramic packages

These packages consist of three parts: the base, lid and lead frame. By pressing ceramic powder into the required shape and firing, the base and the lid are made. Upon the fired base and lid, glass paste is screened and fired. A separate lead frame is entrenched inside the base glass. The glass is melted over the base and the lead frame for hermetic sealing. This sealing method is known as glass frit sealing and therefore, the package is called a glass frit seal package.

From the cost consideration, pressed ceramic packages are less expensive than multilayer ceramic packages. However, their structural simplicity does not permit incorporation of additional electrical features.

17.9.3 Metal can packages

These packages have a metallic base (figure 17.8). The leads come out from a glass seal, which is either a compression seal or a matched seal. Resistance welding is used to attach a metal can or lid to the metallic base. Hermetic sealing is thereby achieved. The lead count in metal can packages is usually low, <24. Cost-wise, they are preferable to other packages.

17.10 Hermetic glass passivation of discrete high-voltage diodes, transistors and thyristors

Low alkali content, particularly Na^+ ion glasses provide high insulation resistance on p–n junction surfaces (figure 17.9). Thermal stress-induced cracking is minimized by matching the coefficient of thermal expansion of the semiconductor material with that of the glass used. A slurry of the glass powder in deionized water—suspended ceramic particles along with binders, dispersants or plasticizers—is applied to the surface. A relative motion between a blade and the substrate spreads the slurry in a thin uniform layer. A gel-type hermetic layer is formed on drying. The method is called doctor blading or tape casting (Berni *et al* 2004).

In another laser-based technique of forming a glass passivation layer, the part of the semiconductor substrate on which the layer is to be deposited is simultaneously irradiated with two laser beams: a laser beam which is absorbed in glass and another laser beam which is absorbed in the substrate (Iesaka and Yakushizi 1984).

A complementary metal–oxide–semiconductor (CMOS)-compatible process for hermetic packaging using low-temperature BSG involves plasma-enhanced e-beam

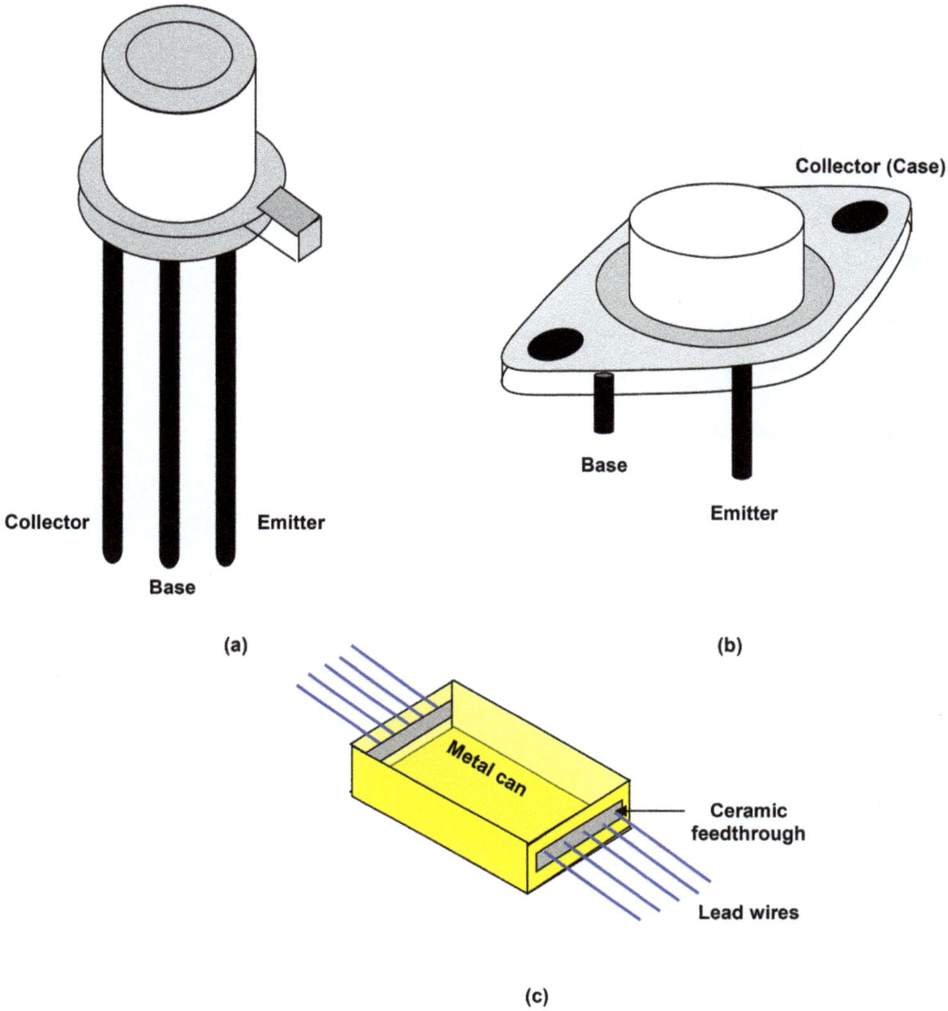

Figure 17.8. Metal packages: (a) TO-3, (b) TO-5 and (c) flat pack.

deposition of the glass layer at a temperature <100 °C followed by patterning using a photoresist lift-off technique (Leib *et al* 2009).

17.11 Discussion and conclusions

The package represents a barrier between the semiconductor chip/circuit/PCB and the environment, protecting the electronics from the detrimental effects of corrosive chemicals. Various options are available from the packaging technology side. One can select from a diversity of packages according to convenience. A wide range of coating materials is also available. These materials give more strength to packaging by providing additional protection. Taking care to reduce the remaining chemicals left on the chip and PCB surfaces is equally important to fabricating a reliable

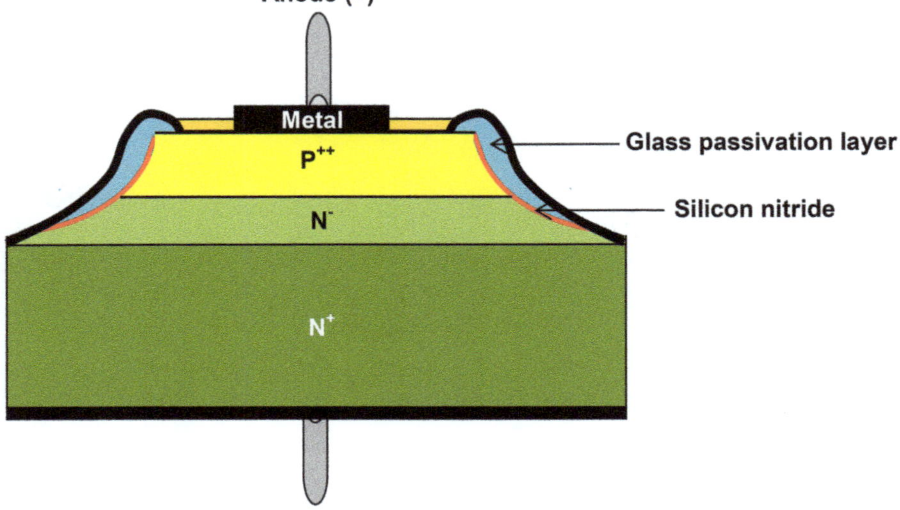

Figure 17.9. Glass-passivated high-voltage diode.

electronic product, because these residues act as corrosion-triggering centers during the operational life of the circuit.

Review exercises

17.1. Name a few metals used in the fabrication of microelectronic circuits. Write the names of three gases which are responsible for the corrosion of electronic circuits.

17.2. How does hydrogen sulfide react with copper and silver? Is the morphology of film growth on silver uniform?

17.3. How does corrosion of lead–tin solder take place in the atmosphere? What happens if the percentage of tin in lead–tin solder is less than 2% by weight?

17.4. How do silver ions cause short-circuiting problems in a humid climate? How are magnetic storage devices corroded when the carbon coating is abraded?

17.5. How do aluminum and copper interact mutually in Al–Cu metallization? How are micropits formed?

17.6. What is the role of chlorine in rosin fluxes? What happens if the residual flux left after soldering is not properly cleaned from the PCB surface?

17.7. What organic acids are used as cleaning agents in no-clean fluxes? Write their molecular formulae. Do they contain any halogens?

17.8. Why does the flux trapped between the wave soldering pallet and PCB trigger corrosion?

17.9. What is a conformal coating? How does it increase the reliability and lifespan of a PCB?
17.10. What are acrylics, polyurethane and silicone compounds? Discuss their relative merits/demerits in terms of the removability for PCB repair.
17.11. What is a glob top? Name the two types of glob top. How are they formed?
17.12. How do the voids or defects in conformal coatings lead to corrosion?
17.13. What is meant by potting an electronic component? How is this done?
17.14. How is vacuum potting done? How does it help in producing void-free structures?
17.15. What is vacuum impregnation? Name a few applications of vacuum impregnation.
17.16. What is meant by hermetic sealing? Can hermetic packages be used at higher temperatures than plastic packages?
17.17. How is the body of a multilayer ceramic package formed? What process is used for insertion of leads in the package? Why is this package called a solder-seal package?
17.18. What utility-enhancement electrical features can be introduced by multilayer ceramic tape construction of packages?
17.19. Why is the pressed ceramic package called a glass frit seal package? How does it compare with the multilayer ceramic package in terms of cost and provision of enhancement features?
17.20. How is hermetic sealing performed in a metal can package? How are the leads taken out?
17.21. Describe the doctor blading method of glass passivation?
17.22. Can glass passivation be performed in a CMOS-compatible process? How is this done?

References

Berni A, Mennig M and Schmidt H 2004 Doctor blade *Sol–Gel Technologies for Glass Producers and Users* ed M A Aegerter and M Mennig (Berlin: Springer) pp 89–92

Bezkosty P, Długoń E, Sowa M, Nizioł J, Jeleń P, Marchewka J, Błażewicz M and Sitarz M 2022 Corrosion resistance and electrical conductivity of hybrid coatings obtained from polysiloxane and carbon nanotubes by electrophoretic co-deposition *Int. J. Mol. Sci.* **23** 2897

EPOXY Technology 2001 Glob-top *Selected Application* www.epotek.com/site/files/brochures/pdfs/Glob_Top_SAS(1).pdf

Harsimran S, Santosh K and Rakesh K 2021 Overview of corrosion and its control: A critical review *Proceedings on Engineering Sciences* **3** 13–24

HumiSeal 2016 What is conformal coating *Chase Electronic Coatings* www.humiseal.com/conformal-coating/

Iesaka S and Yakushizi S 1984 Method of manufacturing a glass passivation semiconductor device *US Patent* US4476154 A

Leib J, Gyenge O, Hansen U, Maus S, Fischer T, Zoschke K and Toepper M 2009 Thin hermetic passivation of semiconductors using low temperature borosilicate glass—benchmark of a new wafer-level packaging technology *59th Electronic Components and Technology Conf. (San Diego, CA, 26–29 May)* 886–91

Li L, Chakik M and Prakash R 2021 A review of corrosion in aircraft structures and graphene-based sensors for advanced corrosion monitoring *Sensors (Basel)* **21** 2908

Rubin A and Garcia C 2020 Corrosion: effects on metals and electronic equipment, processes and prevention 1–13 AREPA www.arepa.com https://arepa.com/media/pkfbmjhx/arepa-white-paper-corrosion.pdf

Salas B V, Wiener M S, Badilla G L, Beltran M C, Zlatev R, Stoycheva M, de Dios Ocampo Diaz J, Osuna L V and Gaynor J T 2012 *H_2S Pollution and Its Effect on Corrosion of Electronic Components* (Rijeka: InTechOpen) ch 13 pp 263–86

Texas Instruments 1999 Hermetic packages *Texas Instruments Literature Number SNOA280* www.ti.com/lit/an/snoa280/snoa280.pdf

Woolley M 2013 Flux residues can cause corrosion on PCB assemblies *EDN Network* October, www.edn.com/electronics-blogs/all-aboard-/4423581/Flux-residues-can-cause-corrosion-on-PCB-assemblies

Zerust 2022 *Preventing electronics corrosion on computers, integrated circuits and microchips* https://zerustproducts.com/rust-prevention-tips/preventing-electronics-corrosion-on-computers-integrated-circuits-and-microchips/

Chapter 18

Radiation effects on electronics

Radiation is part and parcel of our environment. It is present in the skies, on the Earth and within our bodies. It is found naturally, but it is also a man-made hazard. One important effect of radiation on semiconductors is the total ionizing dose (TID) effect, which either causes ionization of lattice atoms or creates lattice defects by dislodgement of atoms from their regular lattice positions. Electron–hole pair (EHP) production by ionization impacts the performance of metal–oxide–semiconductor (MOS) devices through a shift in threshold voltage, transconductance degradation and leakage current enhancement. Lattice defect creation influences bipolar device operation via minority-carrier lifetime killing. Another significant class of radiation effects on semiconductor circuits is represented by single-event effects (SEEs), which are either temporary or permanent in nature. Temporary effects include single-event upset (SEU) and single-event transient (SET). The permanent effects are single-event latchup (SEL), single-event burnout (SEB) and single-event snapback (SES).

18.1 Introduction

Radiation is energy that is emitted from a source and travels through space or a material medium, either in the form of particles (the particle form of radiation) or waves (the wave form of radiation). The particle form of radiation consists of charged or uncharged particles such as electrons, protons, alpha particles, neutrons, etc. The wave form of radiation consists of electromagnetic waves such as visible light, x-rays, gamma rays, and so on. The former form of radiation may be either uniform or non-uniform depending on the distribution of particles in the beam, so

doi:10.1088/978-0-7503-5072-3ch18

that while part of the target electronic item is affected, the rest may not be influenced at all. The latter form is uniform in its impact on the target.

Particle and wave radiations impact the performance of state-of-the art semiconductor devices, notably microprocessors, SRAMs (static random access memories), DRAMs (dynamic random access memories) and flash memories leading to total dose and displacement damage, and introducing soft errors (Bagatin and Gerardin 2016). Upon bombardment by particulate radiation or on exposure to wave radiation, the conductivity of materials changes causing variations in the parameters of electronic devices (Goetz *et al* 2022).

Whether on Earth or in space, it is impossible to find a place without radiation. Radiation is present in every nook and corner of the Universe. Only its degree and energy vary. On Earth, we are protected to a large extent by the blanket of atmosphere enveloping our planet.

18.2 Sources of radiation

18.2.1 Natural radiation sources

Human beings living on Earth and so also the electronic equipment they use can never escape from natural radiation exposure. A primary source of radiation is the cosmic radiation or cosmic rays coming from the Sun and stars (USNRC 2014a). Also important are the van Allen radiation belts around the Earth containing trapped charged particles. Another source is terrestrial radiation, from radioactive materials such as uranium, thorium and radium inside the Earth's crust; radioactive carbon and potassium from plants and animals; and radon and thoron gases inhaled from the atmosphere. The third source is internal radiation due to the carbon-14, potassium-40 and lead-210 isotopes present within our bodies.

18.2.1.1 Cosmic rays

Noticeably impacting the functionality of electronic devices, these rays (figure 18.1) comprise high-energy radiation propagating at approximately the velocity of light with some fraction having energies up to 10^{14}–10^{15} eV. Their intensity increases with altitude above the Earth indicating their origin from outer space, beyond the Solar System. It also varies with latitude, suggesting presence of charged particles which are affected by the Earth's magnetic field. Compositionally, the primary cosmic rays, accelerated by the energy of detonation waves of supernova vestiges and striking the Earth's atmosphere, are composed of 99% nuclei of elements ranging from the lightest to the heaviest, and 1% electrons similar to β-particles. Of the nuclei, 90% are hydrogen nuclei or protons, 9% are helium nuclei or α-particles and the remaining 1% are nuclei of heavy elements. Antimatter particles such as anti-protons constitute a very small percentage (NASA Goddard Space Flight Center 2014). Particles with energies in the range 10^8–2×10^{10} eV cause the most damage. As the atmosphere acts as an armor against cosmic rays, the electronic gadgetry of aircraft flying at high altitudes and space vehicles is most susceptible to the detrimental effects of cosmic ray exposure.

18.2.1.2 Van Allen radiation belts

These are two belts or layers (figure 18.2), discovered by James Van Allen, extending at an altitude of 1000–6000 km above the Earth's surface and containing collections

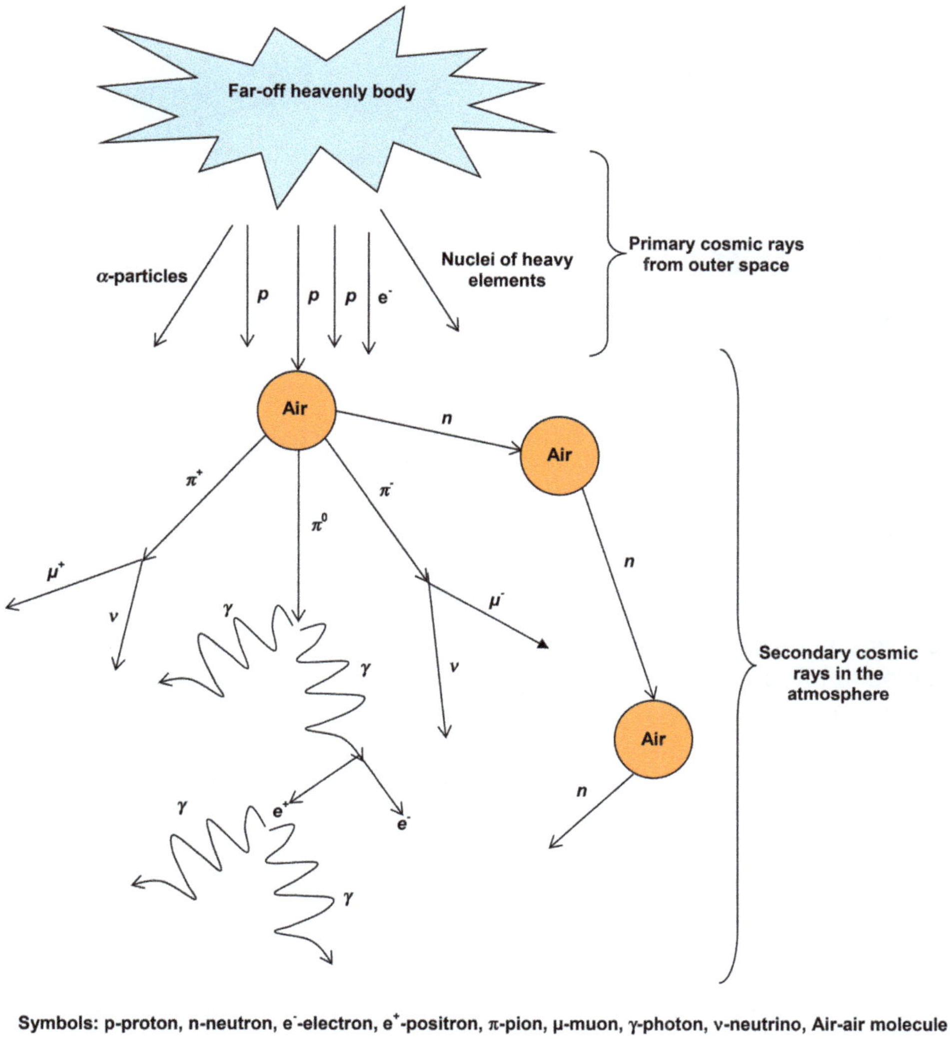

Figure 18.1. Primary and secondary cosmic rays.

of energetic particles, mainly electrons with energies up to 10^7 eV and protons having energies of hundreds of mega-electronvolts, from cosmic rays and solar wind, confined by the magnetic field of the Earth. Satellites spending a sizeable time in this region must take precautions against radiation (Fox 2014).

18.2.2 Man-made or artificial radiation sources

Major sources of artificial radiation are medical diagnostic x-rays such as radiography, mammography, computerized tomography and fluoroscopy; airport x-ray machines; nuclear medicine and radiation therapy using isotopes such as I-131, Tc-99m, Co-60, Ir-192, Cs-137, etc; nuclear reactors in power plants; nuclear

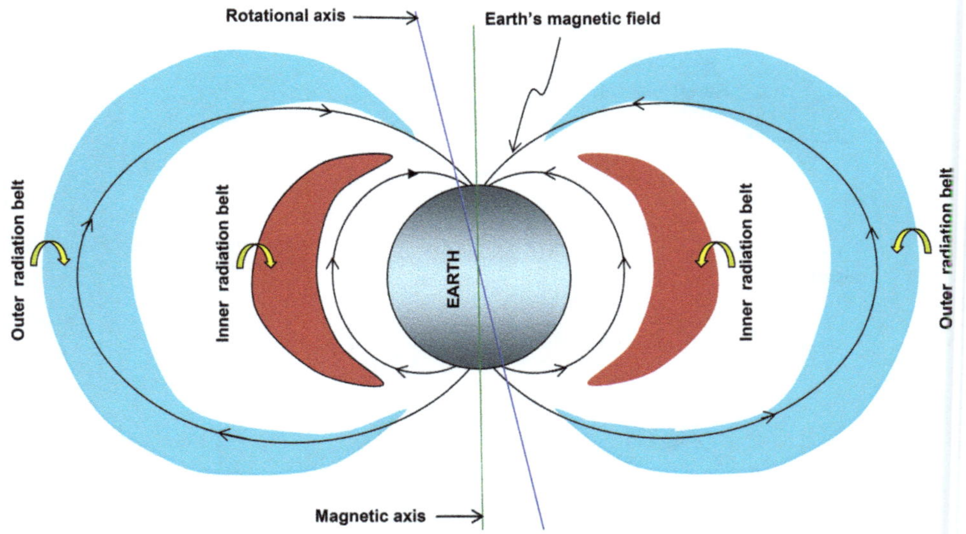

Figure 18.2. The Van Allen radiation belts.

accidents; nuclear bombs or weapons used in warfare; particle accelerators; in addition to tobacco (thorium), lantern mantles (thorium), luminous dials in watches (tritium), smoke detectors (americium), and so forth (USNRC 2014b).

The initial radiation released from a nuclear explosion during the first minute consists of mostly gamma and neutron radiation. Residual radiation over longer periods of time comes from the radioactive fallout, including the debris, fission products and soil.

18.3 Types of radiation effects

Based on extensive experimental studies, the effects of radiation on electronic devices and circuits are broadly classified into the three categories described below.

18.3.1 Total ionizing dose (TID) effect

The unit for measurement of TID is rad (radiation absorbed dose). The SI unit of TID is the Gray (Gy), 1 Gy = 100 rads. TID is a long-term failure mechanism. It accumulates over an extended period of time. To find the TID over this period, the total or cumulative radiation dose administered during the period is measured. The TID effect is observed as trapping of charges in oxides and the creation of defects by displacement damage DD, which occurs when the atoms are pushed out from their regular lattice positions. A change of the threshold voltage of a metal-oxide semiconductor field-effect transistor MOSFET, a decrease in its transconductance and increase of leakage current take place. In the case of a bipolar transistor, current gain is degraded.

18.3.2 Single-event effect

An SEE is an instantaneous temporary or permanent damage inflicted by particle strike. Soft errors, latchup or burnout follow. If SSE is like puncturing a football bladder, the TID effect bears resemblance to its aging over several years.

18.3.3 Dose-rate effect

The dose-rate effect is caused by the generation of photocurrent over the full circuit upon exposure to an extremely high dose rate over a very short time span. Latchup or burnout occurs accompanied by data loss.

18.4 Total dose effects

Particles and photons in the radiation interact with the crystal in two fundamental ways (figure 18.3):

(i) The ionization effect, by removal of the outer shell electrons, and creation of EHPs through the impact of charged particles or photons on the crystal.
(ii) DD, e.g., neutrons bumping against atoms in the crystal lattice dislodge them from their positions and move them into vacancy or interstitial sites. As a result, Schottky or Frenkel defects are produced in the crystal.

Effects (i) and (ii) can co-exist, e.g. a heavy particle can bombard the nucleus producing charged particles having sufficient energy to cause ionization by collision.

18.4.1 Gamma-ray effects

Gamma rays predominantly cause the ionization effect (figure 18.4). EHPs are produced in semiconductors such as silicon, as well as insulators such as silicon dioxide. In silicon dioxide, the mobility of electrons is much higher than that of holes by four orders of magnitude. So, electrons move fast under an applied electric field leaving holes behind. These remnant holes create a charge build-up in SiO_2 and at the Si–SiO_2 interface, which has two effects:

(i) *Production of hole traps*. The subgroup of holes accumulated in SiO_2 serve as traps. Some of these positive traps recombine while others anneal out with time, but the remaining traps persist as electrically active defect centers for years.
(ii) *Interface state generation at the Si/SiO_2 interface*. The subgroup of holes lying near the Si–SiO_2 interface causes re-arrangement of atomic bonds at the Si/SiO_2 interface. Consequently, new interface states are produced.

18.4.1.1 Threshold voltage shifts in NMOS and PMOS transistors

The threshold voltage of an n-channel MOS (NMOS) transistor is positive. The build-up positive hole charges discussed above resemble a type of persistent positive biasing of the gate. They produce the same effect as the positive gate voltage.

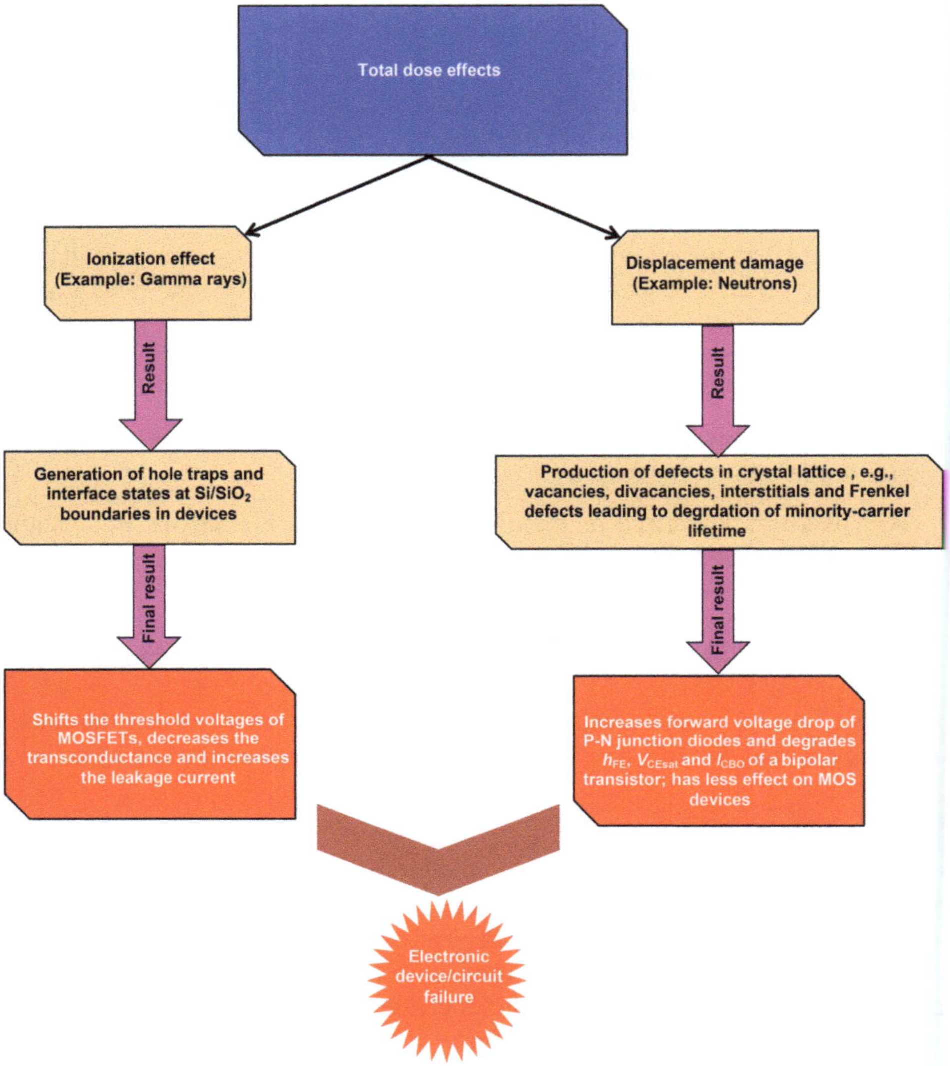

Figure 18.3. Total dose effects at a glance.

Therefore, less gate voltage has to be applied to produce the same drain–source current. Hence, the threshold voltage of an NMOS transistor decreases, i.e. the change in threshold voltage ΔV_{Th} is negative.

A PMOS transistor has a negative threshold voltage. To counteract the positive hole charge, more negative voltage must be applied. Only then is it possible to obtain the same drain–source current as without positive charge. Therefore, threshold voltage increases. This means that the change in threshold voltage ΔV_{Th} is positive. Thus the operations of NMOS and p-channel (PMOS) transistors are affected in the opposite directions.

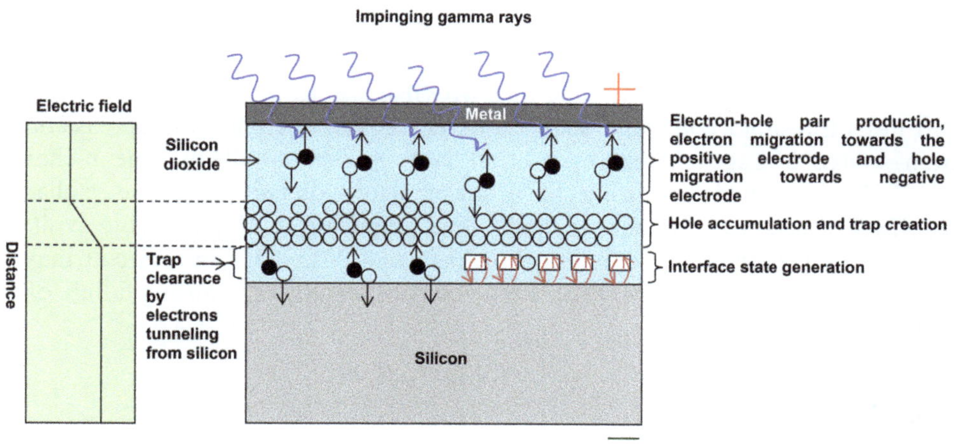

Figure 18.4. The effect of gamma rays on a MOS structure. The electric field disturbance caused by the sheet of hole charge is shown on the left.

18.4.1.2 Superior radiation hardness of the PMOS over the NMOS transistor

The position of the built-up sheet of charge depends on the sign of the gate voltage. The smaller the distance between the gate terminal and charge sheet, the smaller is the influence of the charge sheet on silicon for inversion layer formation, because the charge sheet is located far away from the silicon. Thus the additional electric field due to the charge sheet is smaller, as is the shift in threshold voltage. In an NMOS transistor, the charge sheet is repelled a larger distance away from the positive gate terminal towards silicon. Hence, the additional electric field is also larger and a substantial threshold voltage shift results. For a similar reason, a comparatively insignificant threshold voltage shift is observed in a PMOS transistor. The obvious implication is that a PMOS transistor is more radiation-hard than an NMOS transistor. PMOS transistors are used as dosimeters in radiotherapy for accurate measurement of the gamma radiation doses delivered to patients.

18.4.1.3 Decrease of transconductance g_m and increase of leakage current

The charges trapped in the silicon dioxide and at the silicon–silicon dioxide interface decrease the carrier mobility in accordance with equation

$$\mu_f = \mu_i / (1 + \alpha_{it}\Delta N_{it} + \alpha_{ot}\Delta N_{ot}), \quad (18.1)$$

where μ_f is the final mobility, μ_i is the initial mobility, ΔN_{it} is the interface trap density, ΔN_{ot} is the oxide-trapped charge density, α_{it} is the coefficient for the interface states and α_{ot} is the coefficient for oxide-trapped charges. The mobility reduction brings down the transconductance of the device.

The aforementioned positive space charges accumulate n-type surface layers. They also invert p-type surface layers. The creation of such parasitic conduction paths or channels between the source and drain, which are not under gate control, increases the MOSFET leakage currents. The leakage current flowing between a

MOS transistor and other circuit components is also increased. The resultant breakdown voltage degradation calls for incorporation of guard ring structures in the design. Overall, the functioning of a MOSFET is adversely affected through a shift in its threshold voltage, a decrease in its transconductance, an increase of the drain–source leakage current and a fall in breakdown voltage. The n-channel MOSFET is turned on easily at a lower gate voltage while the p-channel MOSFET becomes difficult to turn on, requiring a higher gate voltage. Eventually, this leads to a situation in which an NMOSFET in a circuit may be permanently on while the PMOSFET may be permanently off, causing failure of the circuit.

18.4.2 Neutron effects

Neutrons are heavy particles. They are also charge-less. When a high-energy neutron impinges on a semiconductor device, it imparts energy to the atom struck through an elastic scattering process. An elastic scattering is like a collision between two hard elastic spheres in which the kinetic energy of the neutron is conserved with modification of its direction of motion. Consequent, upon this impact, the target atom may be kicked off from its lattice position and may occupy another position, either at a lattice site or between atoms. The recoiling atom of the semiconductor lattice may collide with other lattice atoms and trigger further collisions, initiating a cascade of collisions, depending on its energy, or it may participate in an EHP production process. On the whole, the neutron generates an assortment of lattice defects, including vacancies, divacancies, interstitials, Frenkel defects, etc (figure 18.5). The creation of defects is tantamount to the introduction of additional energy levels deep within the bandgap of the semiconductor; hence called 'deep-level defects'. The defects facilitate recombination of charge carriers by acting as stepping-stones for carrier recombination. In this way, they decrease the minority-carrier lifetime of the semiconductor. Operation of bipolar junction transistors (BJTs) is impaired because the carrier lifetime reduction leads to degradation of common-emitter current gain h_{FE} via enhanced recombination in the base. For this reason, narrow base width BJTs, particularly those for high-frequency operation, tend to be more radiation resistant. Furthermore, due to the loss of free charge carriers through increased recombination favored by a low carrier lifetime, the silicon resistivity increases, raising the saturation voltage V_{CEsat} of the transistor. Additionally, the collector-base leakage current I_{CBO} increases due to charge build-up in the SiO_2 layer adjoining the junction. Thus the h_{FE}, V_{CEsat} and I_{CBO} parameters of a BJT deteriorate upon neutron irradiation.

The decrease in carrier lifetime from its initial value τ_i prior to neutron bombardment, to the final value τ_f after bombardment with neutrons of particle fluence ϕ, is expressed as (Makowski 2006)

$$\Delta(1/\tau) = 1/\tau_f - 1/\tau_i = K_\tau \phi, \tag{18.2}$$

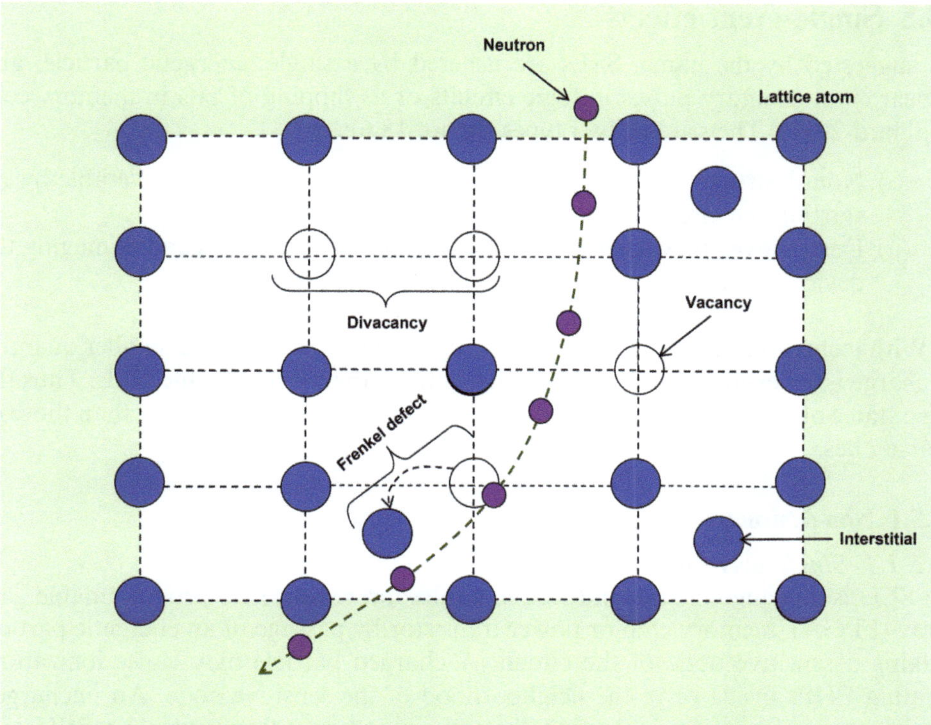

Figure 18.5. Atomic displacement defects such as vacancies, divacancies, interstitials, and Frenkel defects produced in a semiconductor crystal by neutron bombardment.

where K_τ is the damage coefficient associated with carrier lifetime. The consequential decrease in current gain h_{FE} of a p–n–p transistor is written as (Makowski 2006)

$$\Delta(1/h_{FE}) = (1/2)(W_B^2/D_{pB}) \times K_\tau\phi = K_{FE}\phi, \tag{18.3}$$

where W_B is the base width, D_{pB} is the diffusion coefficient of holes in the base and K_{FE} is the damage coefficient related to current gain. The larger the neutron fluence ϕ, the greater is the degradation in current gain.

Contrary to the high-energy neutron, a low-energy neutron may be captured by the nucleus of a lattice atom resulting in an excited atomic nucleus. This is the neutron capture mechanism. During de-activation of the excited nucleus, gamma rays are produced which produce EHPs by ionizing the atoms on their way.

Fortunately, MOSFETs are majority-carrier devices. So, their operation is little influenced by carrier lifetime killing. Whereas bipolar devices start showing pronounced changes in electrical parameters at neutron fluxes of 10^{10}–10^{11} neutrons/cm^2, CMOS transistors are imperceptibly affected up to 10^{15} neutrons/cm^2. Like the BJTs, other minority-carrier devices such as solar cells have to cope with the problems of generation–recombination centers, which are trivial issues for MOSFETs.

18.5 Single-event effects

As suggested by the name, SSEs are ushered by a single, energetic particle, and appear as momentary pulses in logic circuits or as flipping of bits in memory cells (Gaillard 2011). They are of two types (figure 18.6):

(i) Non-destructive, appearing as soft or temporary errors, recoverable by re-starting the circuit or re-processing the data.
(ii) Destructive, appearing as firm, hard or permanent errors and damaging the device or circuit.

With technological advancement towards smaller feature sizes, a smaller quantity of charge is necessary to store a logic high state than in primitive methods. Thus the logic states of ultra-small dimension technologies are more easily upset than those of coarse ones.

18.5.1 Non-destructive SEEs

18.5.1.1 Single-event upset
An SEU is a change of state induced in a microprocessor, field programmable gate array (FPGA), memory chip or power transistor by passage of an energetic particle striking a sensitive node of the circuit. A charged particle may cause ionization, creating EHPs inside or in the neighborhood of the sensitive node. An uncharged particle may also cause ionization through its ionizing fragments. An SEU is a

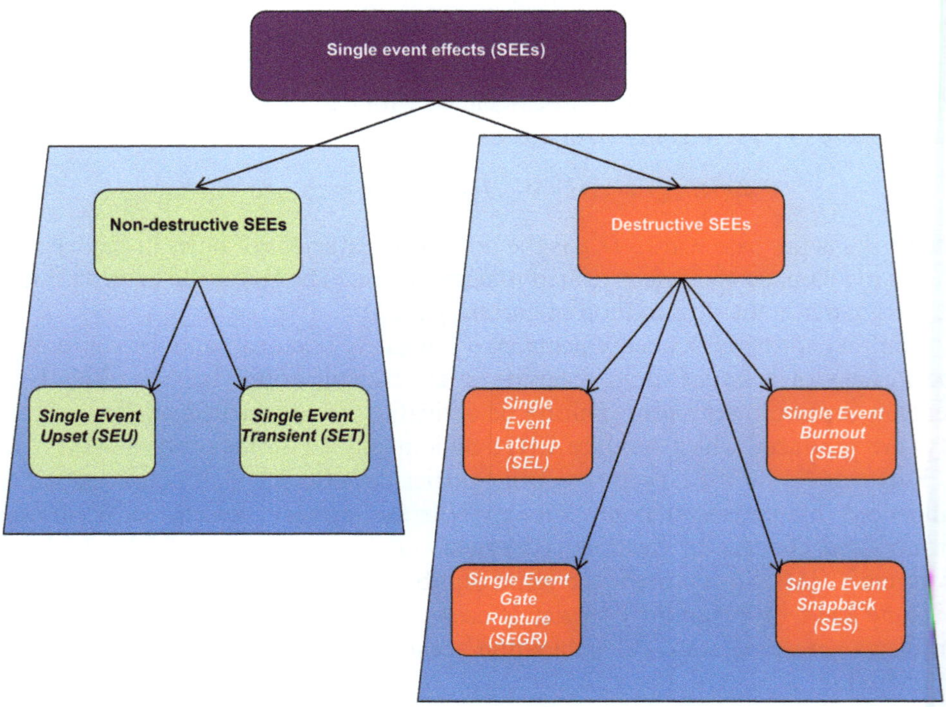

Figure 18.6. Classification of SEEs on semiconductor electronics.

transient error, which is corrected by resetting the device. In a multiple-bit SEU, two or more bits are affected concomitantly. An SEU of more serious concern is the single-event functional interrupt (SEFI). The SEFI stops device operation. A power reset is required to restore its functionality (Holbert 2006).

18.5.1.2 Single-event transient

A SET is a spurious voltage or current pulse in a voltage comparator, an operational amplifier (OP-AMP) or a digital circuit, which is triggered by the charge collected during the ionization event. This pulse travels through the analog or digital circuit, causing a perturbation in the output signal.

18.5.2 Destructive SSEs

18.5.2.1 Single-event latchup

SEL is a condition damaging the device, which is provoked by a single-event current state. It is harmful for CMOS circuitry, which contains NMOS and PMOS transistors. Inside the CMOS structure are parasitic n–p–n and p–n–p transistors (figure 18.7). These transistors constitute a parasitic four-layer n–p–n–p or p–n–p–n thyristor. Normally, the reverse biasing of the n-well–substrate and p-well–substrate junctions keeps the thyristor in a disabled condition. But the thyristor can be triggered into action by irradiation if the product of current gains of the two transistors exceeds unity, i.e.

$$\beta_{\text{npn}} \times \beta_{\text{pnp}} > 1, \tag{18.4}$$

together with injection of adequate currents into the base–emitter junctions of the transistors, and provision of holding current by the power supply. Whenever the above conditions are complied with, regenerative thyristor action takes place, causing latchup. The latchup can be cleared by resetting through power on–off to

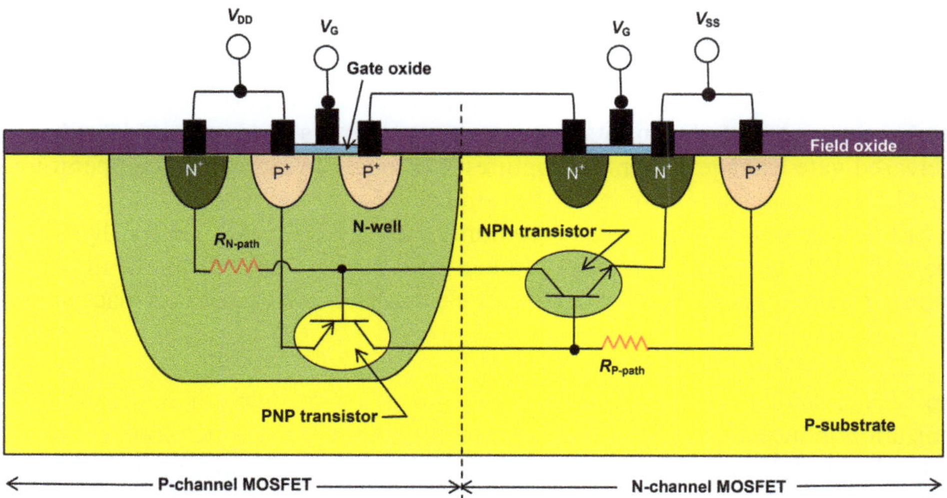

Figure 18.7. Parasitic p–n–p and n–p–n bipolar transistors constituting the thyristor, which is responsible for regenerative action causing latchup in an n-well CMOS device.

disconnect the supply. If not done immediately, excessive overheating due to high current flow may lead to catastrophic failure and irreversible damage to metallization or bond wires. Latchup is a strong function of temperature. Its susceptibility increases as temperature increases.

18.5.2.2 Single-event burnout

SEB is a damaging condition encountered due to high current flow in high-power MOSFETs, such as vertical double-diffused MOS (DMOS) transistors, caused by a heavy-ion component of cosmic rays and solar flares. Suppose a vertical DMOS is in the off-state. It is blocking a high drain–source voltage. Let a heavy ion pass through the transistor structure across the source region. As it leaves behind a tail of EHPs along its path, and because a large drain–source voltage is applied, a high current density ~10^4 A cm^{-2} is generated in the parasitic bipolar transistor below the source. As soon as the voltage drop across the base–emitter junction of the parasitic BJT is >0.7 V, it is turned on by avalanche multiplication of collector current. Forward-bias second breakdown is initiated. The localized junction heating leads to burning of the device, provided sufficient short-circuit energy is available.

Power MOSFETs are highly vulnerable to SEB at $V_{DSS} \geqslant 100$ V. Nonetheless, SEBs have also occurred at $V_{DSS} < 30$ V. In the beginning, SEBs were considered to be problems afflicting MOS devices only. But SEBs are not confined to MOSFETs alone and have also taken place in BJTs at voltages >400 V and thyristors at voltages >1000 V (Foutz 2005).

18.5.2.3 Single-event gate rupture

A single-event gate rupture (SEGR) is a destructive burnout of the gate dielectric consequent upon the formation of a conducting path through it by the passage of a heavy ion traversing across the gate insulator through the neck region of a MOSFET cell, evading the p-regions and moving through the n-epilayer, applying the drain potential to the gate. A plasma filament of EHPs is produced in the n-epilayer. The electrons in the plasma quickly diffuse away. But the slower holes pile up and this hole accumulation generates a transient electric field of magnitude exceeding the breakdown voltage of the gate insulator. If this lasts for a short duration, the device may recover, but if it continues too long, it destroys the gate oxide either via increased gate leakage current or ruptures it through an increase of temperature in the vicinity, completely damaging the same.

SEGR is witnessed in non-volatile memories such as EEPROMs (electrically-erasable programmable read-only memories) during write or erase operations. In these operations, high voltages are applied to the gate electrodes. SEGR also takes place in BJTs.

18.5.2.4 Single-event snapback

An SES is similar to an SEL, except that no p–n–p–n thyristor is involved. The ionization induced by the energetic particle turns on the parasitic bipolar transistor of the MOSFET structure. Recovery is possible by properly sequencing the signals without reduction of the power supply. SES may lead to circuit destruction if the local current density is very high.

18.6 Discussion and conclusions

Both MOS and bipolar devices, and associated circuits are prone to the radiation effects of either short-term or long-term nature. Changes in MOSFET parameters upon radiation exposure are manifested in CMOS circuits as variations in low and high digital logic levels, a decrease in output current, lengthening of propagation delay and upswing of quiescent current values. Digital circuits are vulnerable to SEEs of short-lived or long-lasting nature, mainly SEU and SEL.

Review exercises

18.1. What are the two principal forms of radiation? Give examples.
18.2. Name and describe the three natural sources of radiation.
18.3. What leads us to believe that cosmic rays come from outer space?
18.4. What argument leads to the inference that cosmic rays contain charged particles?
18.5. What protects us from the harmful effects of cosmic radiation on Earth's surface?
18.6. What is the typical composition of primary cosmic rays?
18.7. What are the Van Allen radiation belts? What particles are they composed of?
18.8. Give five examples of man-made radiation sources?
18.9. What is the composition of initial radiation from a nuclear explosion?
18.10. What is the TID effect of radiation on semiconductor devices and circuits? What are the different forms in which it is generally observed? What electrical parameters of MOS and bipolar devices undergo degradation by TID?
18.11. What are SEEs and dose-rate effects of radiation on semiconductors?
18.12. What happens when a charged particle or photon strikes a semiconductor crystal?
18.13. What happens when a semiconductor crystal is subjected to neutron bombardment?
18.14. Gamma rays are striking a layer of silicon dioxide. What effects are observed within the silicon dioxide layer and at the silicon–silicon dioxide interface?
18.15. What are the effects of positive charge build-up by gamma-ray irradiation of a silicon dioxide gate insulator on the threshold voltage of: (a) an n-channel MOSFET and (b) a p-channel MOSFET? Why is the PMOS transistor more radiation-hard than the NMOS transistor?
18.16. What is the effect of gamma-ray irradiation on the transconductance and leakage current of a MOSFET device?
18.17. How are deep-level effects created in a semiconductor crystal by bombardment with high-energy neutrons? How does the neutron-created damage affect the minority-carrier lifetime in the semiconductor?
18.18. How does a low-energy neutron interact with a semiconductor crystal? Does this interaction affect the carrier lifetime or produce a different result?

18.19. What electrical parameters of a bipolar junction transistor are degraded by a neutron strike and how?
18.20. Why are MOSFETs more immune to the damaging effects of neutrons than bipolar devices?
18.21. What are SEEs of radiation on semiconductor circuits? What are the two types of SEEs? Why do small geometry devices fall prey to SEEs more easily than the large geometry ones?
18.22. Differentiate between an SEU and an SET.
18.23. How does SEL take place in a CMOS structure? Explain with a diagram.
18.24. What types of MOSFETs are susceptible to SEB? How does this burnout take place?
18.25. How is the gate dielectric of a MOSFET ruptured in an SEE? Give examples of MOSFET-based structures prone to this type of damage.
18.26. In what way is SES different from SEL? Is it recoverable? How?

References

Bagatin M and Gerardin S (ed) 2016 Introduction to the effects of radiation on electronic devices *Ionizing Radiation Effects in Electronics: From Memories to Imagers* (Boca Raton, FL: CRC Press) pp 1–22

Foutz J 2005 *Power transistor single event burnout* www.smpstech.com/power-mosfet-single-event-burnout.htm

Fox K C 2014 NASA's Van Allen probes spot an impenetrable barrier in space *NASA* www.nasa.gov/content/goddard/van-allen-probes-spot-impenetrable-barrier-in-space

Gaillard R 2011 Single event effects: mechanisms and classification *Soft Errors in Modern Electronic Systems* ed M Nicolaidis (Berlin: Springer) pp 27–54

Goetz K C, Reed F K, Taylor N and Ezell N D B 2022 Considerations of radiation-hardened electronics for alpha detectors in molten salt reactors *Oak Ridge National Laboratory* ORNL/TM-2022/2745 p 5

Holbert K E 2006 *Single event effects* http://holbert.faculty.asu.edu/eee560/see.html

Makowski D 2006 The impact of radiation on electronic devices with the special consideration of neutron and gamma radiation monitoring *PhD Thesis* Technical University of Lodz, Łódź, Poland pp 13, 14, 20, 21

NASA Goddard Space Flight Center 2014 *What are cosmic rays?* http://imagine.gsfc.nasa.gov/science/objects/cosmic_rays1.html

USNRC 2014a Natural background sources *USNCR* www.nrc.gov/about-nrc/radiation/around-us/sources/nat-bg-sources.html

USNRC 2014b Natural and man-made radiation sources: reactor concepts manual *USNRC Technical Training Center* **0703** 6-1–2 www.nrc.gov/reading-rm/basic-ref/students/for-educators/06.pdf

Chapter 19

Radiation-hardened electronics

Various techniques used to immunize electronic devices and circuits against detrimental radiation effects are described. These techniques are subdivided into two categories: process-based and design-based. The principal process-based techniques include the implementation of triple-well and SOI CMOS technologies, in addition to impurity diffusion profile tailoring and carrier lifetime killing. Amongst the design-based schemes, mention may be made of using annular MOSFET geometry, channel stoppers and guard ring structures. The channel aspect ratio may be increased to control the dissipation of charge. Spatial and temporal redundancy concepts may be applied to make devices suitable for operation in radiation-contaminated environments. Single-event upset (SEU)-resistant flip-flops and RAM structures are designed using dual-interlocked-storage-cell (DICE) as memory cells. As the radiation-hardened electronics constitutes a small segment of the total chip market, the fabrication of radiation-hardened designed structures by commercial complementary metal–oxide–semiconductor (CMOS) processes has received more acceptance than the adoption of high-cost technologies.

Scaling in feature size has made combinational logic circuits in CMOS technology more vulnerable to erratic behavior in a radiation environment (Sharma and Rajawat 2018). Insight into radiation effects on electronics and methods for their mitigation is necessary to build high-performance, fault-tolerant electronic components for spacecrafts and satellite systems (Velazco *et al* 2019). Electronic circuits used in nuclear power plants, spacecrafts and satellites must be designed in such a way that they continue to work accurately in a radiation-filled environment (Adupa *et al* 2020).

doi:10.1088/978-0-7503-5072-3ch19

19.1 The meaning of 'radiation hardening'

To radiation harden a device or circuit, or to make it 'rad-hard', means that it was designed or fabricated to withstand a certain amount of radiation without malfunctioning. The necessary procedures are executed on the device/circuit during the design or fabrication stage to ensure proper functioning. Sometimes, the term 'radiation hardening' is used in a different context. It implies that the device/circuit has undergone a series of post-fabrication tests to check its survivability and correct operation in radiation-rich environments. Thus the term may have different connotations depending on the contextual setting.

In this chapter, we shall consider radiation hardening as a set of techniques used to mitigate radiation effects in semiconductor devices by developing immunity against radiation onslaught from the very beginning of device conception to the last phase of their delivery to the user. Radiation-hardened electronics is thus the circuitry or equipment in which these immunization steps are fully taken care of.

19.2 Radiation hardening by process (RHBP)

19.2.1 Reduction of space charge formation in silicon dioxide layers

To minimize the radiation-induced positive space charge buildup in oxide layers, one strategy is to grow thin oxide layers so that a smaller hole population is created and trapped. In reference to the contemporary downscaling of metal–organic semiconductor field-effect transistor (MOSFET) dimensions, this oxide thinning is a boon because further miniaturization is aided by reduced gate oxide thicknesses. So, with unabated shrinkage of transistor sizes, radiation hardening of gate oxides is in harmony with technological trends, and requires no major effort. The major bottleneck is the prevention of charging of the thicker field oxides of MOSFETs. This is achieved by decreasing the mechanical stress in the oxides as well as ion-implantation induced damage, by using a smaller quantity of hydrogen in the process and by using lower process temperatures after gate formation (Myers 1998).

19.2.2 Impurity profile tailoring and carrier lifetime control

Increasing the substrate doping and adjusting the impurity doping in MOS-well regions helps to decrease the single-event upset (SEU) and single-event transient (SET) sensitivity. Single-event latchup (SEL) immunity is enhanced by preventing positive feedback action of parasitic transistors in a CMOS structure by choosing an appropriate doping profile and decreasing the minority carrier lifetime in the base region.

19.2.3 Triple-well CMOS technology

There are four CMOS processes: n-well, p-well, twin-well and triple-well (or deep n-well), as depicted in figure 19.1. Some of the electrons produced by radiation exposure are swept away into the deep n-well. Many of these electrons fail to reach the drain of the n-channel MOSFET. As the drain of the NMOS device either does not collect them at all or collects a very small number, these electrons are unable to disturb its operation. The likelihood of an SEU is thereby reduced. In other words,

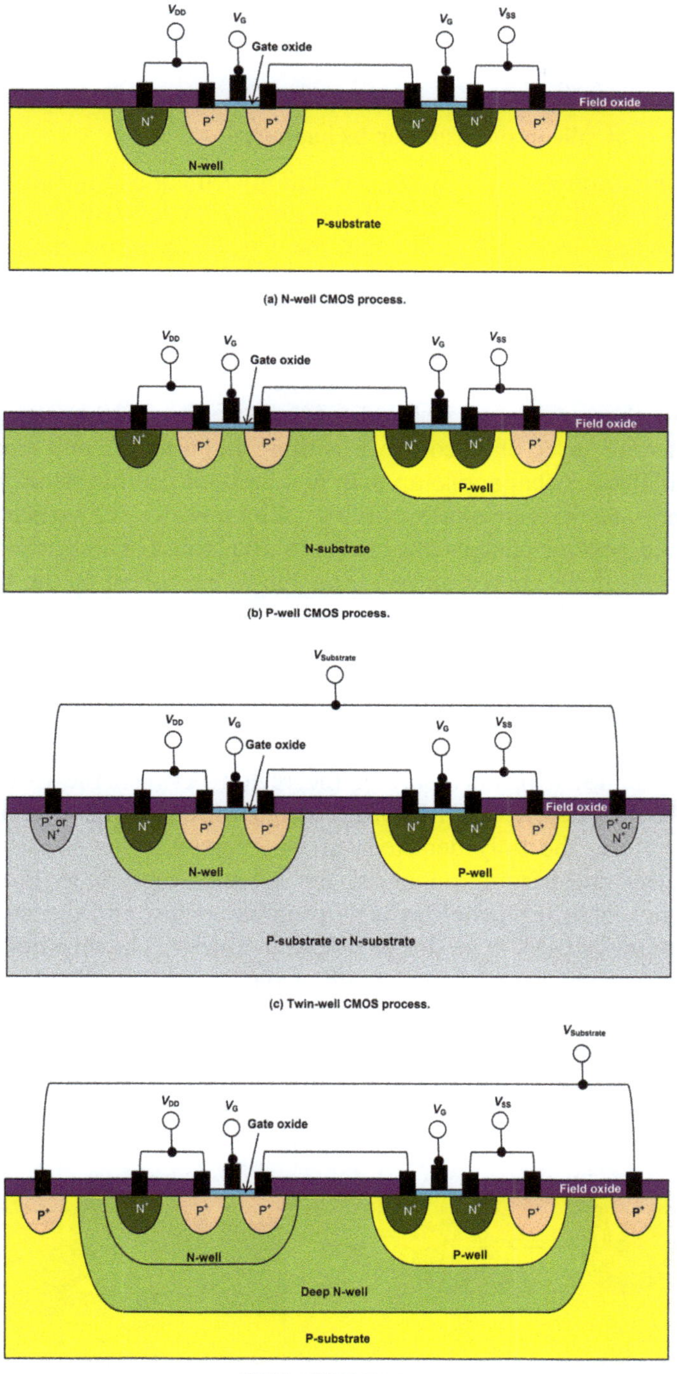

Figure 19.1. Four kinds of CMOS processes: (a) n-well, (b) p-well, (c) twin-well and (d) triple well. V_{DD} = drain supply, V_{SS} = source supply, V_G = gate supply and $V_{Substrate}$ = substrate supply.

the upset is prevented in most cases by strengthening the isolation of the n-channel MOSFET in the p-substrate.

19.2.4 Adoption of silicon-on-insulator technology

Two major shortcomings of the junction isolation technique are readily overcome by the adoption of silicon-on-insulator (SOI) technology. These shortcomings are leakage current and latchup. In the SOI structure, there are neither any leakage paths along device edges nor any p–n–p and n–p–n transistors. SOI technology guarantees freedom from both these problems. Susceptibility to single-event effects (SEEs) is greatly reduced in devices fabricated using SOI wafers, because these devices have a much smaller volume for collection of charges and energy due to irradiation, in comparison to those made on bulk silicon wafers (figure 19.2). By confining the transistor in a thin layer of silicon over a thin insulator, a much smaller sensitive volume is available for charge pile-up than in a bulk silicon transistor having a larger sensitive volume in a thick substrate, offering a long trajectory of particle movement in which undesired positive charges are produced and stored. The smaller this volume, the smaller is the hole charge storage and the more robust is the device against radiation effects. In addition to improving radiation hardness, SOI technology also reduces capacitive loading. The capacitance between the source and substrate, and also that between the drain and substrate, is decreased, offering high-speed operation. Power consumption too is less. The noise level is low and so is the cross-talk.

There are two offshoots of SOI MOSFETs. In the partially depleted SOI-MOSFET (figure 19.3(a)), only part of the region between the gate and BOX is depleted. In the fully depleted SOI-MOSFET (figure 19.3(b)), this region is completely depleted. Therefore, the fully depleted version represents the smallest volume for charge generation and has a minimal risk for radiation-induced malfunction.

The drawback of SOI technology is its prohibitive cost and the accumulation of positive charge in the BOX layer due to radiation impact. The threshold voltage shift caused by this positive charge in a partially depleted transistor leads to inversion of the bottom part of the channel (back gate side), increasing the drain–source leakage current. The unwarranted leakage current causes failure of the circuit.

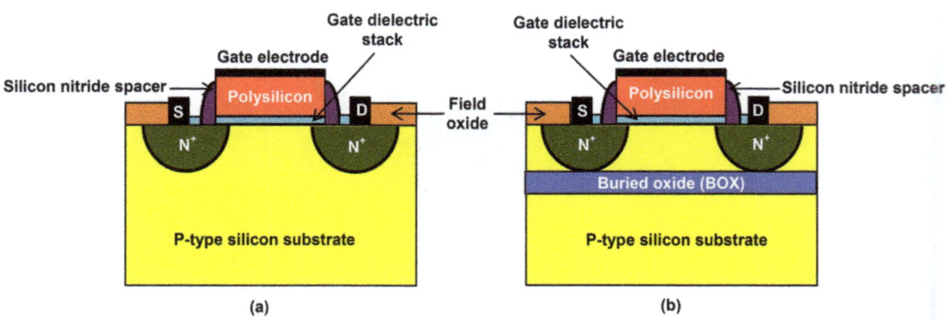

Figure 19.2. Planar MOSFETs on (a) a bulk silicon wafer and (b) an SOI wafer.

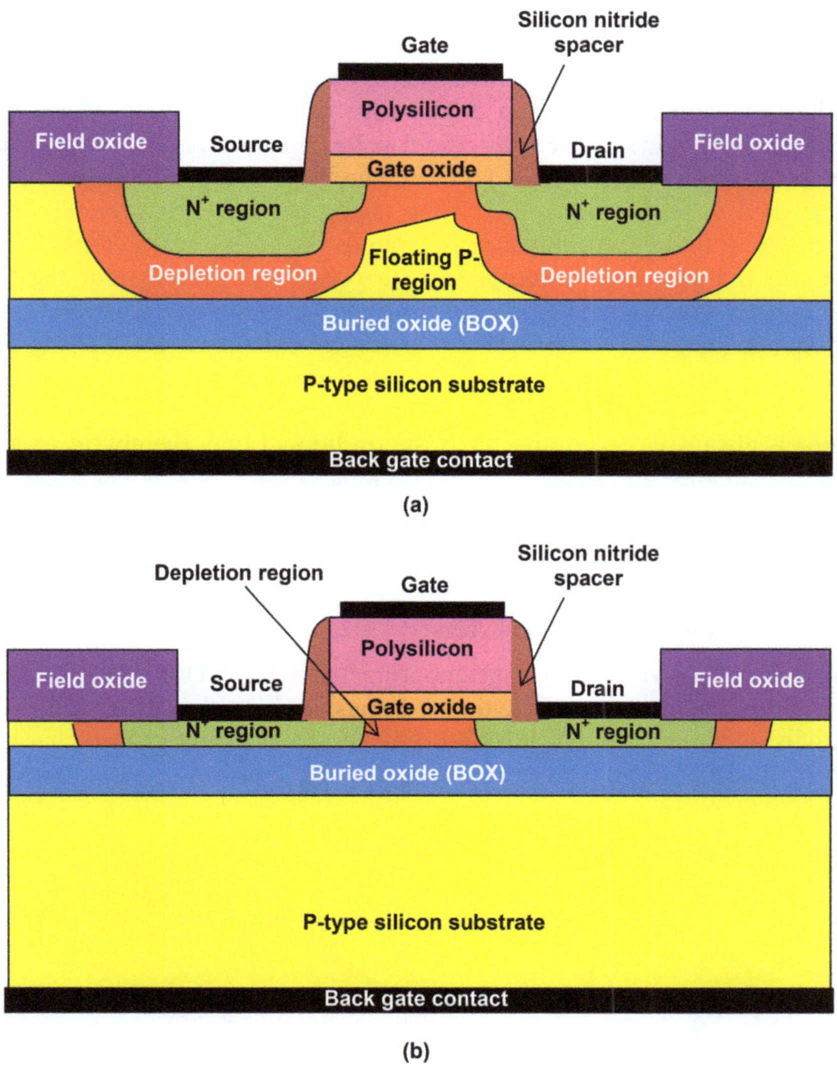

Figure 19.3. SOI-MOSFET structures: (a) a partially depleted MOSFET (PD-MOSFET) and (b) a fully depleted MOSFET (FD-MOSFET).

19.3 Radiation hardening by design

The radiation hardening by design (RHBD) methodology combines the advantages offered by the introduction of manufacturing processes mitigating radiation effects in commercial state-of-the-art foundries, together with novel device and circuit layouts, geometries and topologies to enhance the radiation tolerance. Very often, a price has to be paid for such designs in terms of increased power consumption, larger chip area and lower operating frequency.

19.3.1 Edgeless or annular MOSFETs

Traditional MOSFET design geometry (figure 19.4) has the drawback that holes trapped in the bird's beak region at the edge of the MOSFET, where the gate overlaps the field oxide, cause a threshold voltage shift and contribute to leakage current flow along the gate edges. To overcome these issues, an edgeless MOSFET design is used in which either the source region completely surrounds the drain region or the drain region completely surrounds the source region (figure 19.5). In either case, the gate region completely surrounds the source/drain region. For circular geometry, the gate region circumscribes the source/drain region.

The enclosed geometry transistor eliminates the leakage current flowing between the drain and source along the edges of the gate in conventional geometry, due to the ionization-induced charge generation by irradiation. On the downside, the annular transistor is always larger in area than the traditional one, thereby increasing the intrinsic capacitance and decreasing the packing density. The larger input capacitance prolongs the switching time, slowing down the device. Therefore, annular transistors mainly find use in radiation environments.

19.3.2 Channel stoppers and guard rings

Leakage currents flowing between neighboring n-channel MOSFETs are stopped by forming p-type channel stoppers and guard rings. A channel stopper (figure 19.6(a)) is a heavily doped region, which cannot be easily inverted by surface charges. The guard ring (figure 19.6(b)) is a deeper diffusion of the same polarity and larger radius

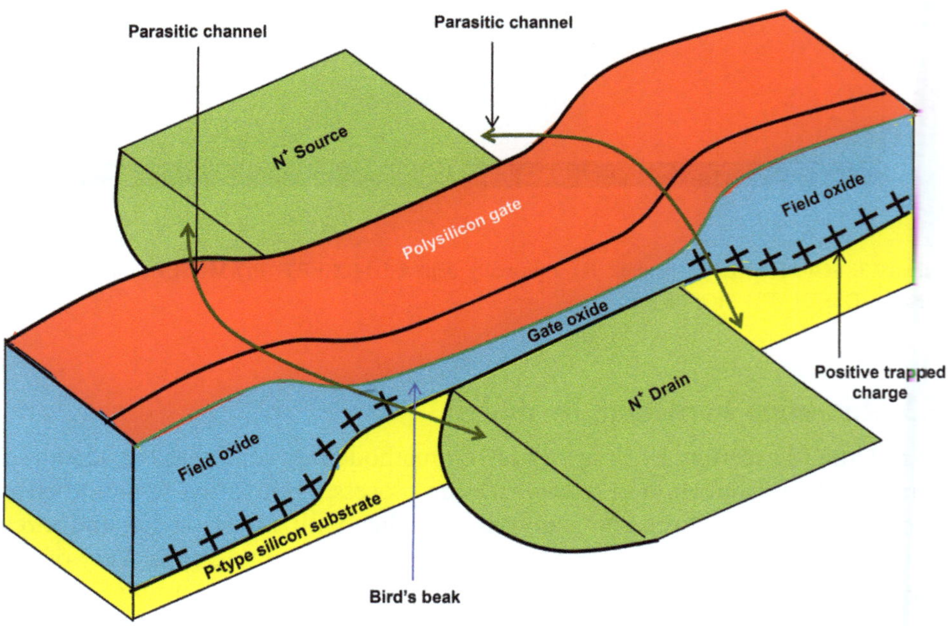

Figure 19.4. Bird's beak formation in the MOSFET isolation region and the holes trapped inside.

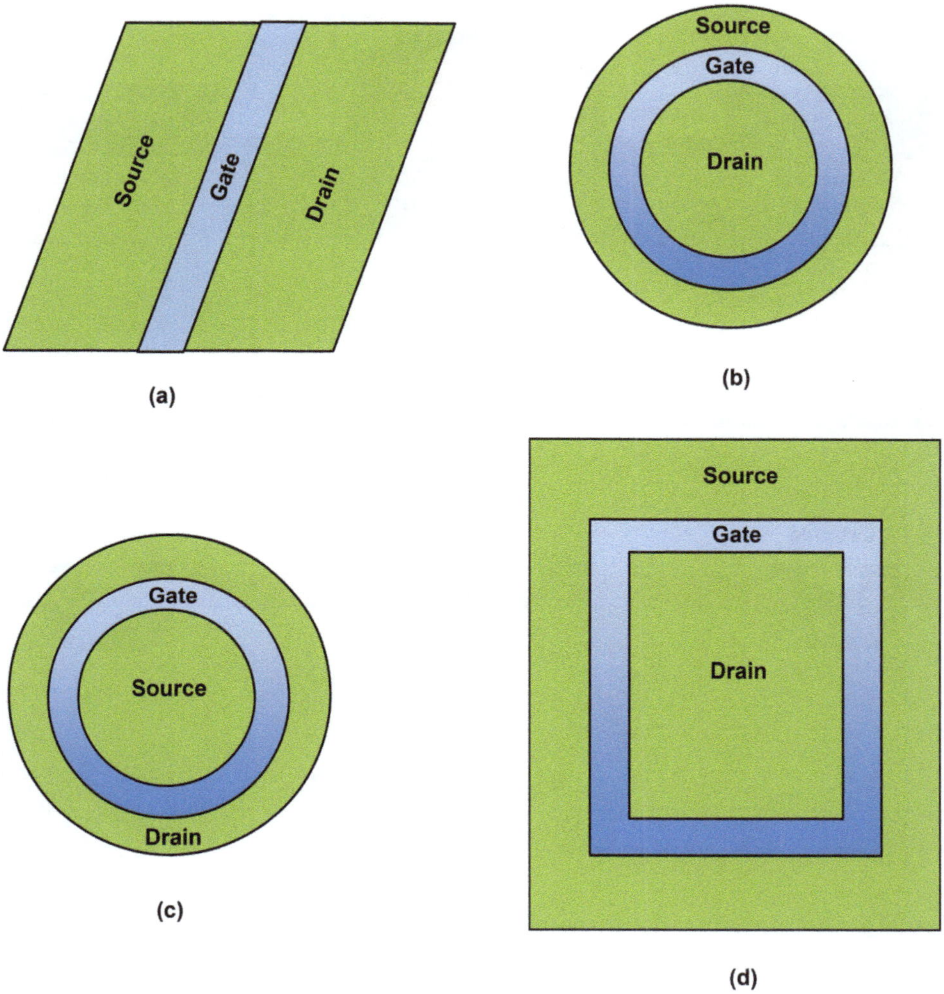

Figure 19.5. MOSFET geometries: (a) conventional and (b), (c), (d) annular edgeless; (b) circular geometry with an external source and internal drain; (c) circular geometry with an external drain and internal source; and (d) rectangular geometry.

of curvature, and therefore higher breakdown voltage, surrounding a shallow diffused junction of smaller radius of curvature and hence lower breakdown voltage. The deep diffused guard ring improves the breakdown voltage of the shallow junction.

19.3.3 Controlling the charge dissipation by increasing the channel width to the channel length ratio

The charge Δq collected at a circuit node due to a particle strike is related to the effective capacitance C of the node and the change in node voltage ΔV caused by the strike, through the familiar equation

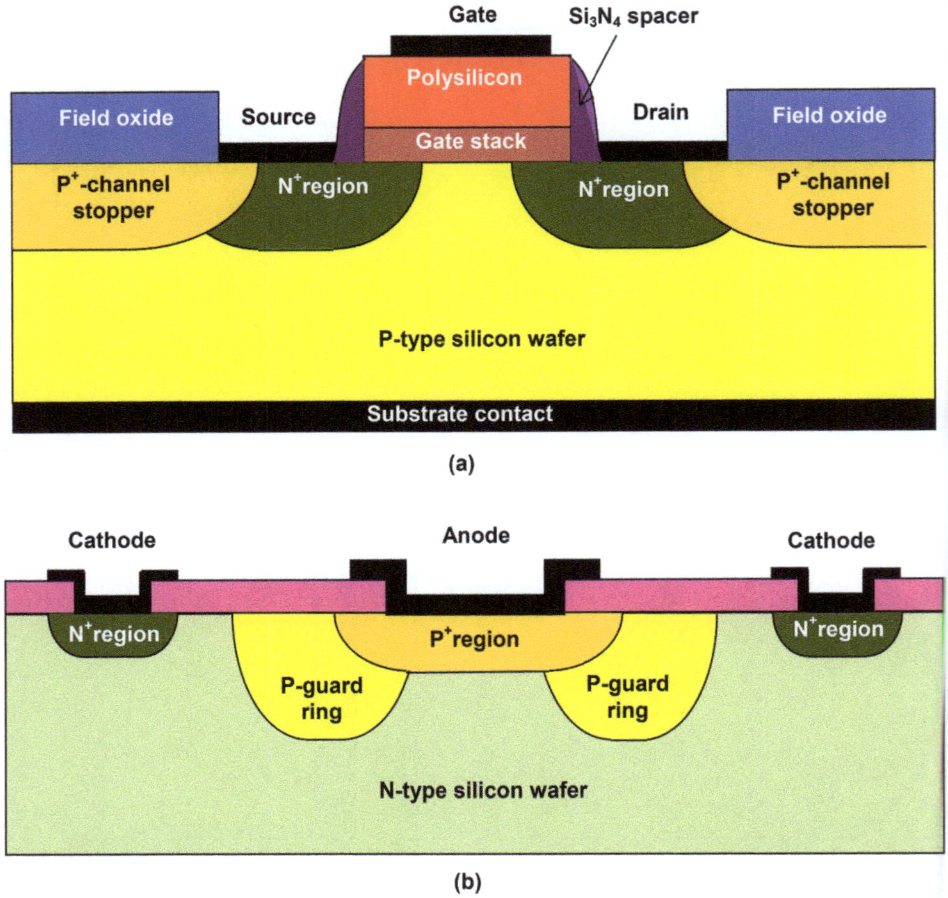

Figure 19.6. Channel stopper and guard ring structures: (a) a channel stopper in an n-channel MOSFET diode and (b) a guard ring in a p–n junction.

$$\Delta q = C\Delta V. \tag{19.1}$$

The impact of MOSFET downscaling is that the minimum value of C has decreased with the square of feature size. However, Δq has remained the same. The implication is that the disturbance ΔV in node voltage has drastically increased. Additionally, integrated circuit (IC) power supply voltages have continuously decreased. The decrease has made the proportionate consequence of the ΔV upsurge even more pronounced by comparison.

Insofar as the influence of the operating frequency value on SSEs is concerned, we note that

$$\Delta q = I\Delta t, \tag{19.2}$$

where I is the maximum charging/discharging current at the node and Δt is the time taken by the charge to dissipate from the node. The present trend is that the

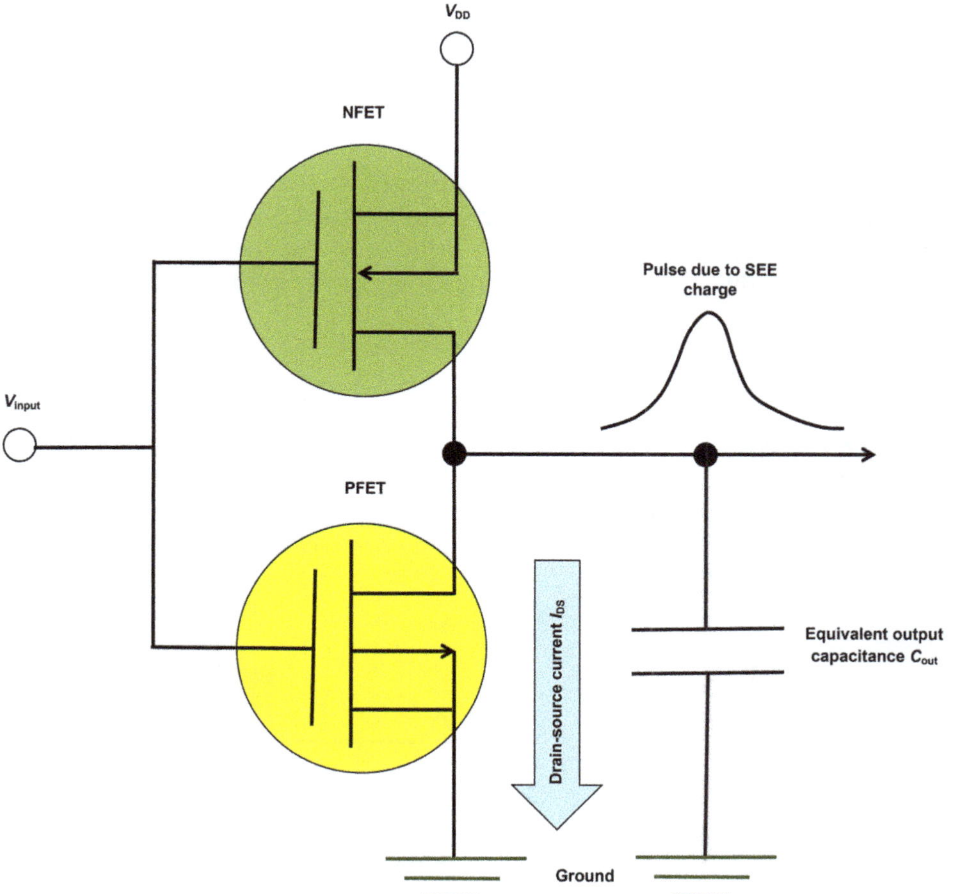

Figure 19.7. Application of the principle of charge dissipation to a digital inverter circuit.

frequency f of the clock signal of ICs is continuously increasing. So, the periodic time T of the clock signal, being the reciprocal of frequency, is decreasing. As $T \rightarrow \Delta t$ and T becomes $<\Delta t$, the chances of an SEU moving through the remaining circuit increase. Thus, there is a greater danger of SEU in higher operating frequency circuits, causing more uncertainty.

Reduction of Δt appears to be the obvious remedy to obviate SEU. This is accomplished by applying the principle of charge dissipation (Holman 2004). For this purpose, the maximum charging/discharging current I at a critical node is increased. In the inverter circuit shown in figure 19.7, the logic high signal turns on the n-channel MOSFET while the p-channel MOSFET is turned off. Any charge deposited on the output capacitance C_{out} is discharged through the n-channel MOSFET. The strategy followed is to increase the ratio of channel width/channel length to raise the maximum drain–source current I_{DS}. This enables the dissipation of total change in capacitor voltage due to a particle strike before the succeeding stage can disseminate the resulting transitory voltage as a bit error. It should be noted that the above strategy for increasing *I* has the disadvantage of a larger area and more power consumption of the circuit.

19.3.4 Temporal filtering

This involves insertion of resistive and/or capacitive components at critical nodes of the circuit (figure 19.8). The circuit modification provides low-pass filtering of the fast voltage pulses generated by SSEs. Principally due to the area and speed disadvantages of this method, it is rarely used in high-performance circuits.

19.3.5 Spatial redundancy

A combinational logic circuit with triple modular redundancy (TMR) is implemented using three physically separated data paths to feed logic circuits *X*, *Y* and *Z* (figure 19.9). The aim is to ensure that any single particle strike cannot disturb more than one logic circuit (Holman 2004). A voting circuit hardened against errors produces a valid result only when a minimum of two calculations is in agreement.

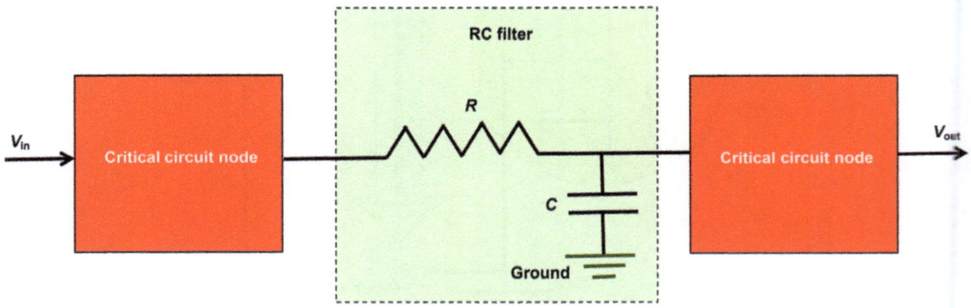

Figure 19.8. Temporal (low-pass) filtering.

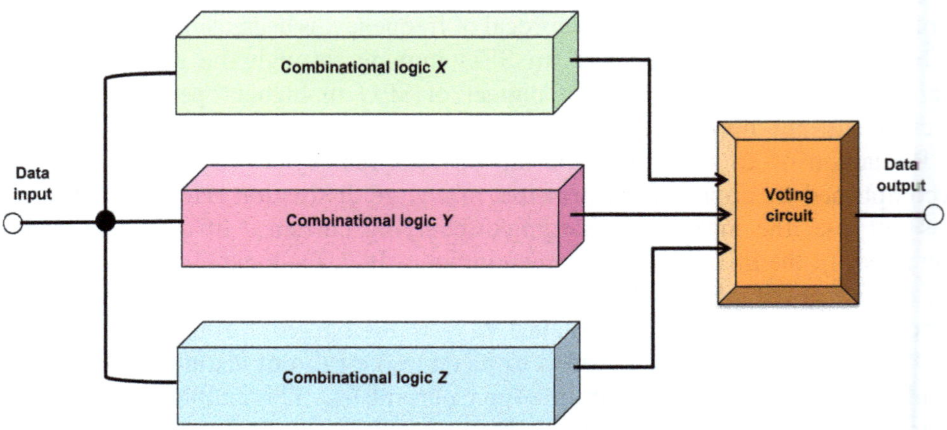

Figure 19.9. Spatial TMR.

The circuit provides protection against SEEs with penalty of larger area consumption.

TMR techniques involving storage cell replication and majority voting are also applicable to sequential logic circuits, such as flip-flops and registers, at the cost of system level overhead and power dissipation.

In TMR-based circuits, SEU tolerance may be lost due to error latency, i.e. a delay between the stimulus receipt and the response to it. A first upset is tolerated. But thereafter a second upset may take place before recovery from the effect of the first upset. This leaves the system prone to interconnected dual errors.

19.3.6 Temporal redundancy

Similar to the protection provided by TMR with respect to SEU by separation of data signals in space, temporal redundancy does so by separating the data signals on a time scale (Holman 2004). Figure 19.10 illustrates triple-path temporal redundancy implementation in an SEU-resistant data latch. The circuit consists of a dual input multiplexor connected in a feedback loop, delay circuits and a voting circuit. There are three separate data paths to the voting circuit. Two of these paths insert time delays, one time delay is Δt and the other time delay is $2\Delta t$, where Δt is the time during which the maximum charge of a particle strike can be dissipated. The penalization of the circuit is that a waiting period = $2\Delta t$ seconds must be allowed to

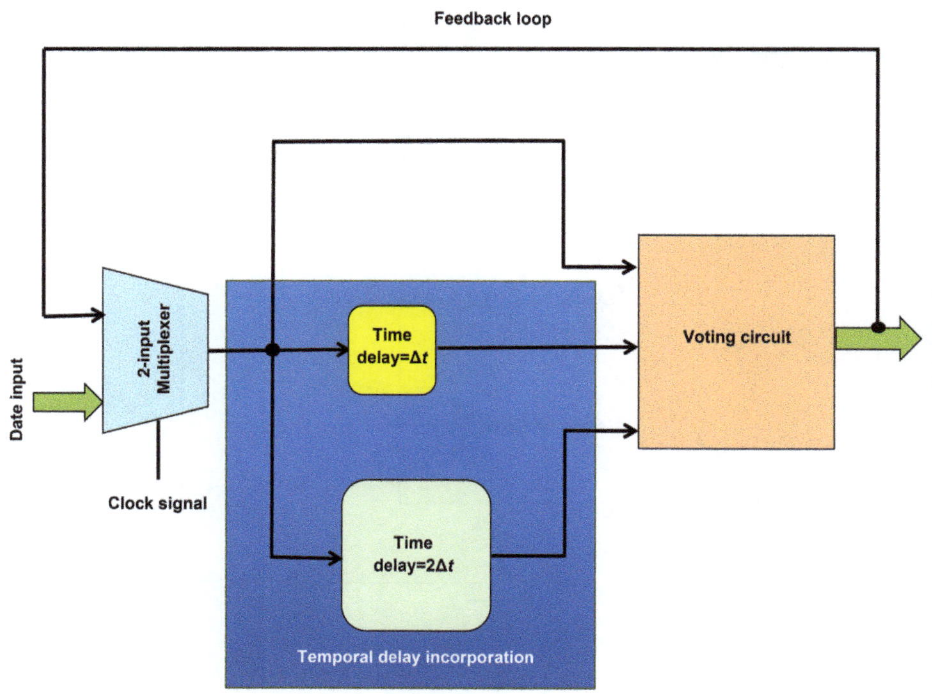

Figure 19.10. Three-path temporal redundancy; Δt = time taken to dissipate the maximum anticipated charge due to ion strike.

execute a calculation. At a frequency of 50 MHz this latency is small, but at higher frequencies it becomes unacceptable.

19.3.7 Dual interlocked storage cell

The DICE memory cell is used to make CMOS flip-flops and static RAMs resistant to SEU (Calin *et al* 1996, Mukherjee 2008). It is a redundant structure containing four internal nodes V_0, V_1, V_2, V_3. Data are stored on these nodes. Alternate logic levels are always present on the nodes. Logic 0 is represented by 0101, i.e. $V_0V_1V_2V_3 = 0101$. Logic 1 is represented by 1010, i.e. $V_0V_1V_2V_3 = 1010$. To perform read and write operations, the enable input C is asserted and data accession is performed through bidirectional lines D, $\overline{D}$.

The DICE structure (figure 19.11) consists of two horizontal cross-coupled inverter latch structures (N_0, P_1) and (N_2, P_3). These horizontal latch structures are connected by vertical bidirectional feedback inverters (N_1, P_2) and (N_3, P_0). The inverters are p-channel or n-channel MOS transistors, as indicated by the corresponding p and n symbols. These transistors are arranged in feedback loops. The PMOS transistors (P_0, P_1, P_2, P_3) constitute a clockwise loop. Likewise, the NMOS

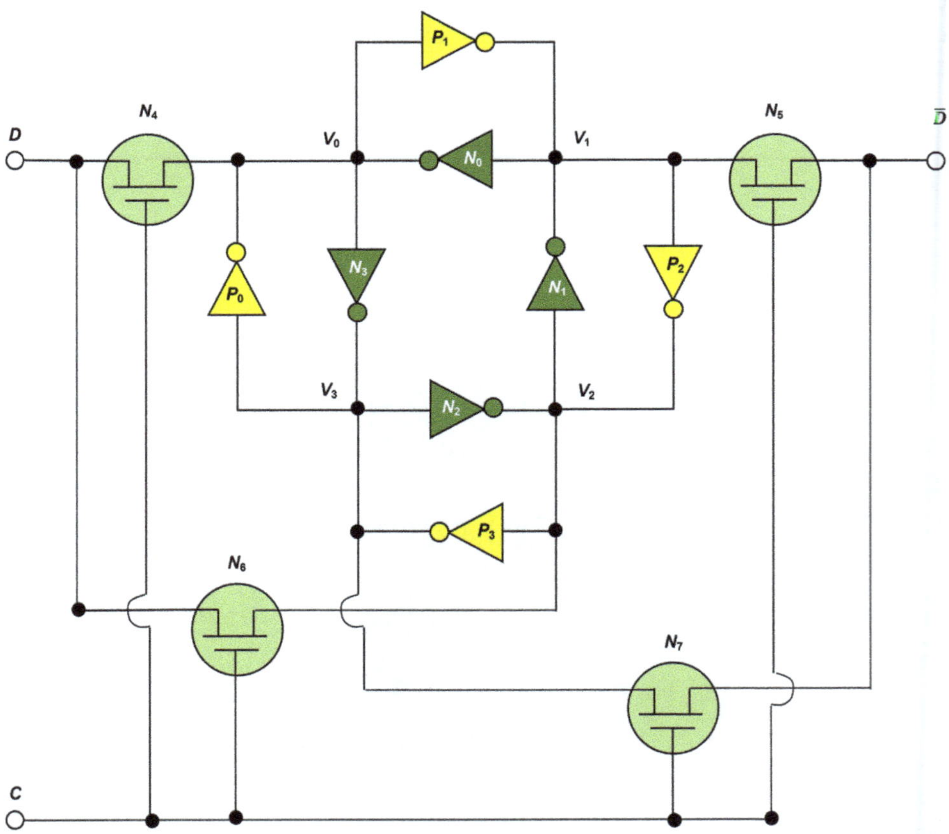

Figure 19.11. The concept of the DICE using a four-node redundant layout.

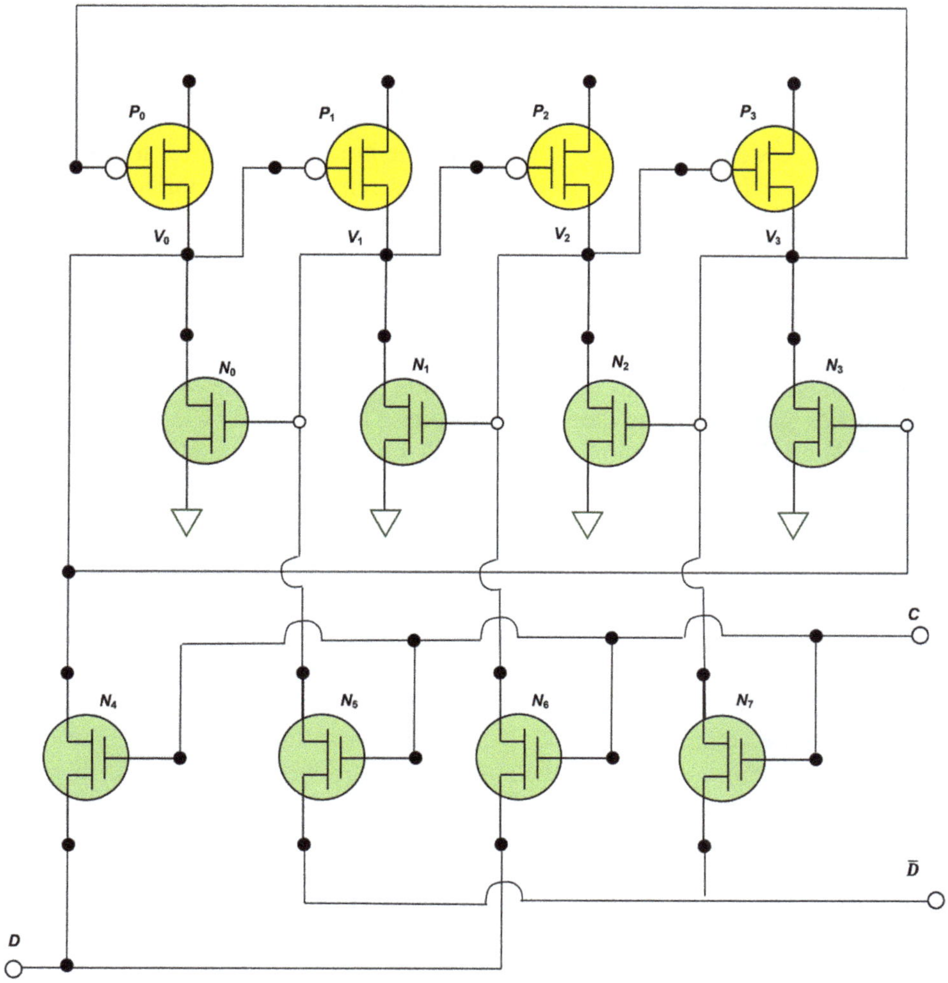

Figure 19.12. Transistor-level diagram of the DICE.

transistors (N_0, N_1, N_2, N_3) represent a counterclockwise loop. The DICE circuit showing the PMOS and NMOS transistors is given in figure 19.12.

The DICE is based on the redundancy approach. By using two latch sections that store the same data, redundancy in the circuit maintains a source of uncorrupted data after an SEU. The data in the uncorrupted section furnish feedback for the restoring state for the recovery of corrupted data. Immunity to SEU is obtained through dual node feedback control. This implies that two adjoining nodes situated on the opposite diagonal govern the logic states of each of the four nodes of the cell. There is no direct interdependence of two nodes on each diagonal. Two nodes of the other diagonal control the states of these nodes.

Let us consider a node V_i ($i = 0,1,2,3$). Here the subscript i is a modulo 4 number, e.g. $0 \equiv 0$ modulo 4, $1 \equiv 1$ modulo 4, $2 \equiv 2$ modulo 4, $3 \equiv 3$ modulo 4, $4 \equiv 4$ modulo 4, but $-1 \equiv -1 \times 4 + 3 = 3$ modulo 4. With the help of single-transistor complementary feedback connections through transistors N_{i-1} and P_{i+1}, any node V_i controls two complementary nodes on the opposite diagonal V_{i-1} and V_{i+1}. For $i = 0$, using transistors N_3 and P_1, the node V_0 controls the nodes V_3 and V_1.

In the logic state 0, i.e., $V_0V_1V_2V_3 = 0101$, the horizontal inverter loops (N_0, P_1) and (N_2, P_3) are in the conduction state. They represent two horizontal latches. These latches store the same data at their nodes (V_0, V_1) and (V_2, V_3). Under this condition, the pairs of transistors (N_1, P_2) and (N_3, P_0) of the vertical inverters are non-conducting. They perform the role of an interlocking function by isolating the two horizontal latches from each other.

Similarly, in the logic state 1, i.e., $V_0V_1V_2V_3 = 1010$, the vertical inverter loops (N_1, P_2) and (N_3, P_0) are in the conduction state. They represent two vertical latches. These latches store the same data at their nodes (V_0, V_1) and (V_2, V_3). Under this condition, the pairs of transistors (N_0, P_1) and (N_2, P_3) of the horizontal inverters are non-conducting. They perform the role of interlocking function by isolating the two vertical latches from each other.

As the four nodes V_0, V_1, V_2, V_3 are sensitive to particle strikes, let us examine their immunity against radiation by taking the node V_i as an example.

Case I. Suppose a negative upset pulse is produced at the node V_i. This negative pulse will induce a positive pulse disturbance at node V_{i+1} by feedback via PMOS transistor P_{i+1}. But it will be unable to exert any influence on the same logic state stored at the node V_{i-1}. This happens due to blockage of the feedback transistor N_{i-1} by the negative upset pulse at node V_i. Furthermore, the positive pulse disturbance induced at node V_{i+1} will not be transmitted onward through PMOS transistor P_{i+2}. Isolation of the nodes V_{i-1} and V_{i+2} helps in preservation of their logic states. Thus logic state changes take place only at the two nodes V_{i-1} and V_{i+1}. These changes are temporary. After the upset transient has departed, state-reinforcing feedback occurs by the nodes V_{i-1} and V_{i+2} via transistors P_i and N_{i+1}.

Case II. Suppose a positive upset pulse is produced at the node V_i. The perturbation can be analyzed in a similar manner to case I. The positive pulse will disturb the node V_{i-1} via NMOS transistor N_{i-1}. After the upset transient has disappeared, the nodes V_{i-1} and V_{i+2} will capacitively maintain their states. They will restore the correct logic states of the two perturbed nodes via transistors N_i and P_{i+1}.

Case III. If a single particle flips two sensitive nodes of the cell together at the same time, which store the same logic state, e.g. either nodes (V_0, V_2) or (V_1, V_3), recovery of the cell is not possible. Spacing apart the drain regions of the node pairs that are simultaneously sensitive on the cell layout can diminish the probability of occurrence of such a failure. This will minimize the chances of collection of the threshold charge that is adequate for upsetting the working of the cell.

It may be noted that ensuing latches may capture any glitch that occurs at the output of the cell. Although the cell itself can recover, such a glitch cannot be prevented from proceeding further.

A DICE may be 1.7–2.0 times larger than original cell, but this factor can be reduced to 1.5 by careful design.

19.4 Discussion and conclusions

The process and design techniques for realizing radiation-tolerant ICs have been discussed. The radiation-hardened IC market constitutes only a small segment of the total IC outlay, which is dominated by the commercial IC market. Therefore, the trend of the semiconductor industry is towards reinforcing RHBD techniques using commercial IC processes instead of spending more capital on radiation hardening IC processes.

Review exercises

19.1. Elucidate the meaning of the term 'radiation-hardened electronics'. Distinguish between RHBP and RHBD.

19.2. How is the technological trend of downscaling MOS devices to smaller dimensions in synchrony with the requirement of a small volume of gate oxide for radiation hardening? Explain. How will you protect the thicker field oxides from radiation effects?

19.3. How do the adjustments of the impurity doping profiles and minority carrier lifetime help in offsetting radiation effects?

19.4. How do the n-well, p-well and triple-well CMOS processes differ? In what ways is the triple-well process helpful in avoiding SEU?

19.5. How does SOI technology overcome leakage current and latchup problems in semiconductor devices caused by radiation exposure?

19.6. What is the basic reason which makes SOI-based MOSFETs more immune to SEEs in comparison to planar bulk silicon MOSFETs? Besides radiation hardness, what are the additional advantages of adopting SOI technology?

19.7. What are the two types of SOI MOSFETs? Which type is more radiation hard? Why?

19.8. How does the positive charge in the BOX layer of a partially depleted SOI-MOSFET increase its leakage current?

19.9. What is the main idea behind RHBD? What sacrifices have to be made if RHBD is done?

19.10. How do the holes trapped in the bird's beak region of a conventional MOSFET degrade its performance? How does the edgeless MOSFET design avoid this degradation? What is the penalty paid?

19.11. What is a channel stopper? How does it decrease the leakage current flow?

19.12. What is a guard ring? How does it increase the breakdown voltage of a shallow diffused p–n junction?

19.13. How does increasing the channel width-to-channel length ratio of a MOSFET aid in obviating SEU?
19.14. What is the temporal filtering technique for SEE mitigation? Why is it unpopular?
19.15. Explain the concept of triple modular redundancy? How does it decrease proneness to SEEs?
19.16. How is the likelihood of SEUs decreased by triple-path temporal redundancy implementation? Explain with reference to a data latch circuit.
19.17. Draw the circuit diagram of a DICE memory cell. Explain its operation when a negative/positive upset pulse is produced at an internal node and if a single particle flips two sensitive nodes of the cell simultaneously.

References

Adupa C, Prakash T C, Rao P R, Kumar J T and Prasad C R 2020 Radiation hardened circuits in multiple harsh environments *IOP Conf. Ser.: Mater. Sci. Eng.* **981** 1 032044

Calin T, Nicolaidis M and Velazco R 1996 Upset hardened memory design for submicron CMOS technology *IEEE Trans. Nucl. Sci.* **43** 2874–8

Holman W T 2004 Radiation-tolerant design for high performance mixed-signal circuits *Radiation Effects and Soft Errors in Integrated Circuits and Electronic Devices* ed R D Schrimpf and D M Fleetwood (Singapore: World Scientific) pp 767–7

Mukherjee S 2008 *Architecture Design for Soft Errors* (Amsterdam: Elsevier) pp 72–3

Myers D R 1998 A brief discussion of radiation hardening of CMOS microelectronics *SciTech* www.osti.gov/scitech/servlets/purl/2453

Sharma V and Rajawat A 2018 Review of approaches for radiation hardened combinational logic in CMOS silicon technology *IETE Tech. Rev.* **35** 562–73

Velazco R, McMorrow D and Estela J 2019 *Radiation Effects on Integrated Circuits and Systems for Space Applications* (Cham: Springer Nature Switzerland AG)

IOP Publishing

Extreme-Temperature and Harsh-Environment Electronics
(Second Edition)
Physics, technology and applications
Vinod Kumar Khanna

Chapter 20

Vibration-tolerant electronics

Since there is a limit to the ultimate strength of materials used in the fabrication of electronic devices and circuits, the logical way to protect an electronic assembly from the deleterious effects of shock and vibration is to isolate the device/assembly from the vibrating platform supporting it. This has led to the development of passive and active techniques of vibration isolation. This chapter attempts to describe some of the opportunities at hand under both categories of vibration isolators, and provides glimpses of the applications where they are deployable. The mathematical theory of passive vibration isolation is developed, starting from free undamped vibrations and moving to forced damped vibrations. It is shown how at frequencies above the natural frequency of the system, the vibration isolator acts as a low-pass mechanical filter. The clear-cut superiority of active vibration isolators at very low frequencies is emphasized, where they are used in a plethora of sophisticated high-precision instrumentation and machines, notably nanoscale microscopies, lithography equipment for semiconductor device fabrication, and so forth.

20.1 Vibration is omnipresent

The moment an electronic device leaves the factory on its way to the showroom or the customer, its survivability is time and again put to test against two adversaries: shock and vibration. Whether it is transported by road or through aerial or water routes, vibration is always a serious threat to it, although to different degrees. The applications of electronic equipment are diverse, covering domestic, office, industrial, military and space deployment. In a building, everything vibrates in the presence of heavy machinery, high-voltage AC (HVAC) transmission, nearby traffic

doi:10.1088/978-0-7503-5072-3ch20

and weather effects. In the industrial sector, shocks of 5–10 g may be encountered, and in military and space applications, shocks >100 g are expected. Even if handled with the utmost care, the chances of an accidental fall anywhere are never ruled out.

Vibration fatigue is a major failure mechanism of electronic devices. Frequently vibrations of electronic components and packages couple with each other amplifying the stress and causing early failure (Yang *et al* 2015). The DC resistance of lithium-ion cells is observed to increase due to vibration while the 1C capacity decreases (Zhang *et al* 2017). The C-rate is the unit of the speed of charging or discharging of a battery, e.g., a C-rate of 1C means the battery is charged from 0% to 100% in 1 h, it also implies 1 h discharge from 100% to 0%.

Vibration resistance is the limiting acceleration or amplitude in a given frequency range up to which an electronic device will operate as per designed specifications during its working life without showing any malfunctioning or suffering any damage (ETS Solutions 2021). Therefore, vibration-withstanding capability has to be built in the equipment to enable it to face real world situations. Structural integrity of electronic devices/systems must be ensured by collaborative efforts of test and simulation engineers (Tormala 2022). Figure 20.1 illustrates the progress of vibration-induced deterioration in microelectronic circuits.

20.2 Random and sinusoidal vibrations

Practically, vibrations are of random nature consisting of several frequencies. Electronic equipment loaded in a truck moving over a highway route experiences a random vibration according to the roughness and undulations of the surface of the road, whereas equipment placed near a rotating electric pump motor feels sinusoidal vibrations. A robust vibration-enduring chassis is required for protection against both types of vibration.

20.3 Countering vibration effects

There are two ways to counter vibration effects. One way is to make the devices and connections sufficiently sturdy and durable, so that they do not crumble under vibration attacks. But there is a limit to the ultimate strength achievable. The other solution is to prevent the vibration from harming the fragile parts of the equipment. Isolating the delicate parts from vibration effects can help to protect them, and this line of thought leads us to the interesting field of vibration isolators. A plethora of vibration isolators exist and these can be applied, as per convenience.

20.4 Passive and active vibration isolators

Vibration isolators can be subdivided broadly into two classes. Passive isolators are conventional systems using springs and dampers (figure 20.2), which do not need any electrical power. Active vibration isolators use an electronic circuit and work on the feedback/feedforward principle to apply the calculated cancellation force of correct magnitude to impede the vibration. Let us begin with the passive vibration isolators. Figure 20.3 provides a glimpse of different vibration isolation systems.

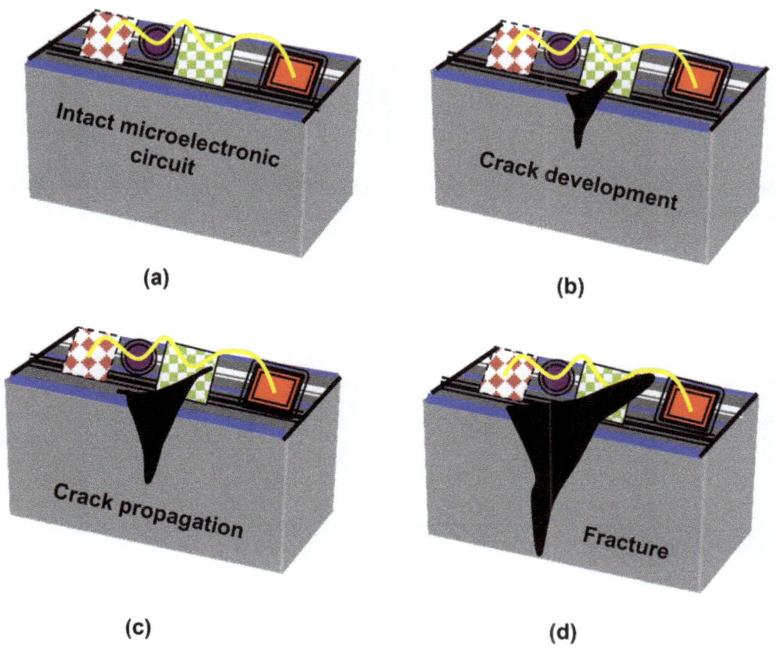

Figure 20.1. Stages in the vibration-caused breakage of a microelectronic circuit: (a) original circuit, (b) crack initiation, (c) crack spreading and (d) complete damage.

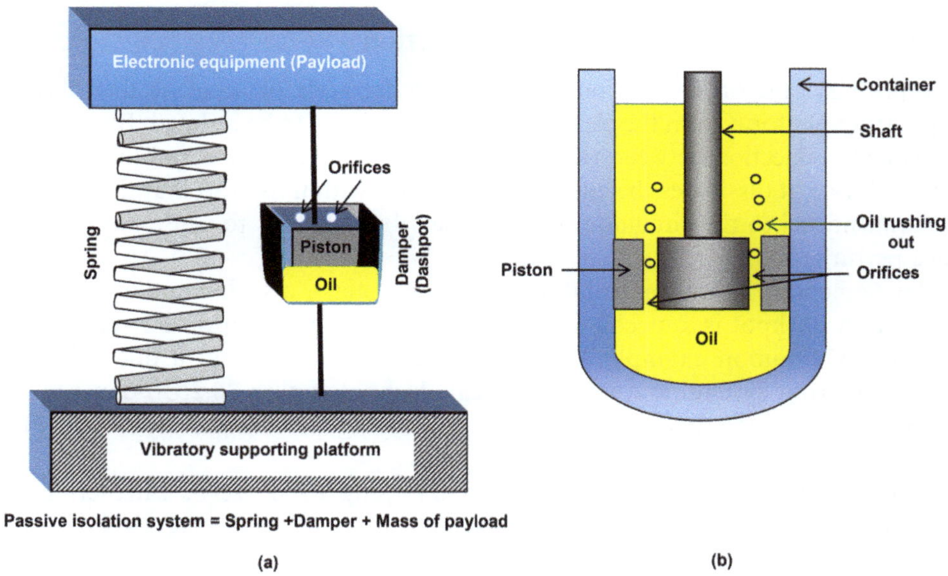

Figure 20.2. (a) Passive isolator and (b) dashpot.

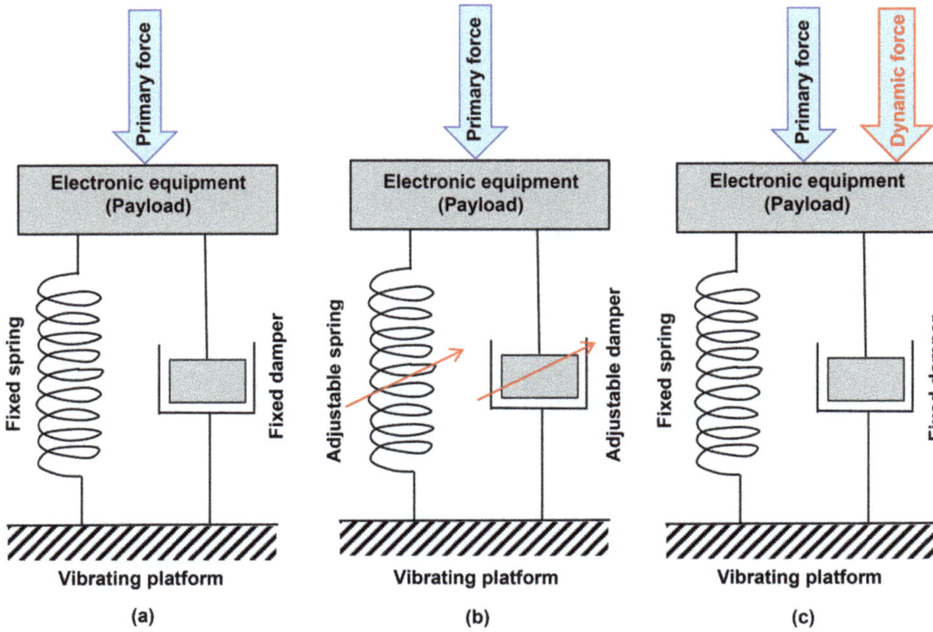

Figure 20.3. Vibration isolation systems: (a) fully passive with fixed values of spring and damper; (b) adaptive passive or semi-active with a variable spring stiffness and damping property to minimize the vibrations; and (c) fully active with dynamic forces applied to mitigate vibrations.

20.5 Theory of passive vibration isolation

The primary role of a vibration isolator is to decouple or separate the vibrating supporting platform from the electronic circuit or device to be isolated. By providing a resilient connection between the vibrating supporting platform and the electronic circuit/device, it dissipates the energy of vibrations within itself so that this energy is not transmitted to the circuit/device, and the vibration fails to cause any performance perturbation.

An elementary vibration isolator consists of a mass, a spring and a dashpot or damper. A dashpot is a mechanical device to constrain vibration through viscous friction. A common example of a dashpot is the one used in a door closer to prevent it from slamming loudly. A linear dashpot (figure 20.2(b)) consists of a piston moving in a hydraulic cylinder containing pressurized hydraulic fluid. Typically, oil is used as the fluid.

For simplification, let us assume the damping to be zero. Further, let us assume that the vibrations take place in the vertical direction x only. Displacement in the downward direction is taken as positive. In the upward direction it is taken as negative. Let us analyze the free vibrations of this system without damping.

20.5.1 Case I: free undamped vibrations

For an undamped system (figure 20.4), if the payload (electronic equipment) has a mass m kg and the spring has stiffness k N m^{-1}, and if the deformation of the spring in static equilibrium is d, the force F acting on the spring is

$$F = kd \tag{20.1}$$

according to Hooke's law.

If g is the acceleration due to gravity, applying Newton's second law of motion, the gravitational force acting on mass m is

$$F = mg. \tag{20.2}$$

This force is also represented by F because the spring force = gravitational force.

From equations (20.1) and (20.2), we have

$$F = kd = mg. \tag{20.3}$$

Applying Newton's second law of motion to the mass m, we have

$$m\frac{\mathrm{d}^2x}{\mathrm{d}t^2} = \sum F = mg - k(d + x) = mg - kd - kx = mg - mg - kx = -kx$$

or

$$m\frac{\mathrm{d}^2x}{\mathrm{d}t^2} + kx = 0. \tag{20.4}$$

This is a homogeneous second order linear differential equation with the solution

$$x(t) = \exp(\ pt). \tag{20.5}$$

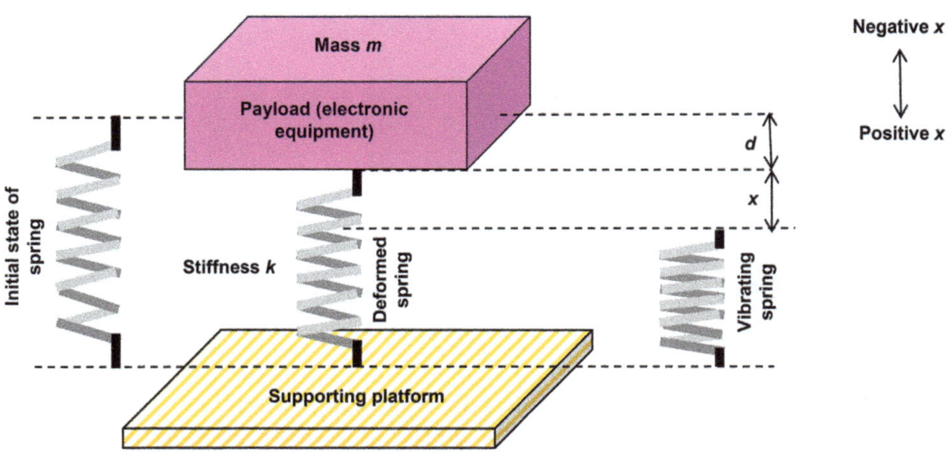

Figure 20.4. Free undamped vibrations of an electronic device.

Since

$$\begin{aligned}\frac{\mathrm{d}^2x}{\mathrm{d}t^2}&=\frac{\mathrm{d}^2\{\exp(pt)\}}{\mathrm{d}t^2}=\frac{\mathrm{d}}{\mathrm{d}t}\left[\frac{\mathrm{d}\{\exp(pt)\}}{\mathrm{d}t}\right]=\frac{\mathrm{d}}{\mathrm{d}t}\{p\exp(pt)\}\\&=p\frac{\mathrm{d}}{\mathrm{d}t}\{\exp(pt)\}=p\times p\ \exp(pt)=p^2\exp(pt)\\&\therefore mp^2\exp(pt)+k\ \exp(pt)=0\end{aligned}\tag{20.6}$$

or

$$mp^2+k=0$$

or

$$p^2=-k/m$$

or

$$p=\pm\mathrm{i}\sqrt{k/m}=\pm\mathrm{i}\omega_\mathrm{n}\tag{20.7}$$

by putting

$$\omega_\mathrm{n}=\sqrt{k/m}.\tag{20.8}$$

Hence the solution is

$$\begin{aligned}x(t)&=C\ \exp(+\mathrm{i}\omega_\mathrm{n}t)+D\ \exp(-\mathrm{i}\omega_\mathrm{n}t)\\&=C\{\cos(\omega_\mathrm{n}t)+\mathrm{i}\sin(\omega_\mathrm{n}t)\}+D\{\cos(\omega_\mathrm{n}t)-\mathrm{i}\sin(\omega_\mathrm{n}t)\}\\&=\mathrm{i}(C-D)\sin(\omega_\mathrm{n}t)+(C+D)\cos(\omega_\mathrm{n}t)=A\ \sin(\omega_\mathrm{n}t)\\&\qquad+B\ \cos(\omega_\mathrm{n}t),\end{aligned}\tag{20.9}$$

where the arbitrary constants A and B are obtained from the initial conditions.

$$\begin{aligned}&\text{At } t=0,\ x(t)=x(0);\ \text{hence } x(0)=A\ \sin(0)+B\ \cos(0)=0+B=B\\&\quad\text{or, } B=x(0).\end{aligned}\tag{20.10}$$

$$\text{At } t=0,\quad \dot{x}(t)=\dot{x}(0).\tag{20.11}$$

Since,

$$\dot{x}(t)=A\omega_\mathrm{n}\ \cos(\omega_\mathrm{n}t)-B\omega_\mathrm{n}\ \sin(\omega_\mathrm{n}t)\tag{20.12}$$

$$\dot{x}(0)=A\omega_\mathrm{n}\ \cos(0)-B\omega_\mathrm{n}\ \sin(0)=A\omega_\mathrm{n}-0=A\omega_\mathrm{n}\tag{20.13}$$

$$\therefore A=\dot{x}(0)/\omega_\mathrm{n}.\tag{20.14}$$

Substituting the values of A and B into equation (20.9), we obtain

$$x(t)=\{\dot{x}(0)/\omega_\mathrm{n}\}\sin(\omega_\mathrm{n}t)+x(0)\cos(\omega_\mathrm{n}t)\tag{20.15}$$

where

$$\begin{aligned}\omega_n &= \sqrt{k/m} = \sqrt{\text{stiffness/mass}} \\ &= \text{natural angular frequency of the system.}\end{aligned} \tag{20.16}$$

Hence, the natural frequency f_n is given by

$$f_n = \frac{\omega_n}{2\pi} = \frac{1}{2\pi}\sqrt{\frac{k}{m}}. \tag{20.17}$$

This frequency is the frequency at which the system will vibrate if it is pulled from its equilibrium static position and released.

20.5.2 Case II: forced undamped vibrations

Suppose the supporting platform undergoes sinusoidal vibrations under an excitation frequency Ω_s, and a sinusoidal applied force

$$F = F_0 \sin(\Omega_s t), \tag{20.18}$$

where F_0 is the peak value of the force waveform following a trigonometric sine curve. The vibrating supporting platform is shown in figure 20.5.

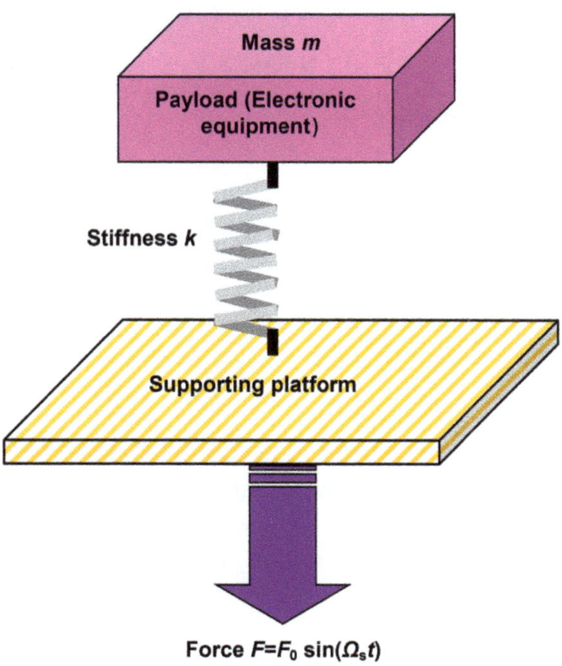

Figure 20.5. Forced vibrations of an electronic equipment without any damping.

Then equation (20.4) is modified to

$$\begin{aligned} m\frac{d^2x}{dt^2} &= \sum F = mg + F_0 \sin(\Omega_s t) - k(d + x) \\ &= mg + F_0 \sin(\Omega_s t) - kd - kx \\ &= mg + F_0 \sin(\Omega_s t) - mg - kx = F_0 \sin(\Omega_s t) - kx \end{aligned} \tag{20.19}$$

$$\begin{aligned} \text{or } m\frac{d^2x}{dt^2} + kx &= F_0 \sin(\Omega_s t) \\ \text{or } \frac{d^2x}{dt^2} + \frac{k}{m}x &= \frac{F_0}{m} \sin(\Omega_s t) \\ \text{or } \frac{d^2x}{dt^2} + \omega_n^2 x &= \frac{F_0}{m} \sin(\Omega_s t), \end{aligned} \tag{20.20}$$

from equation (20.16).

The solution of this second order differential equation comprises two parts: the complementary solution and the particular solution. To find the complementary solution, the right-hand side of the equation is set to zero obtaining

$$\frac{d^2x}{dt^2} + \omega_n^2 x = 0, \tag{20.21}$$

which turns out to be the same as equation (20.4) with $\omega_n = \sqrt{(k/m)}$. Therefore, it has the same solution as for the case of free vibration.

To find the particular solution, let us try the following value as a solution

$$x = X_0 \sin(\Omega_s t). \tag{20.22}$$

Differentiating twice with respect to t, we obtain

$$\frac{dx}{dt} = \Omega_s X_0 \cos(\Omega_s t) \tag{20.23}$$

$$\frac{d^2x}{dt^2} = -\Omega_s^2 X_0 \sin(\Omega_s t). \tag{20.24}$$

Therefore, from equation (20.20),

$$-\Omega_s^2 X_0 \sin(\Omega_s t) + \omega_n^2 X_0 \sin(\Omega_s t) = \frac{F_0}{m} \sin(\Omega_s t). \tag{20.25}$$

Dividing both sides by $\sin(\Omega_s t) \neq 0$, we obtain

$$\begin{aligned} \text{or } -\Omega_s^2 X_0 + \omega_n^2 X_0 &= \frac{F_0}{m} \\ \text{or } (\omega_n^2 - \Omega_s^2) X_0 &= \frac{F_0}{m}. \end{aligned} \tag{20.26}$$

Dividing throughout by ω_n^2,

$$\frac{(\omega_n^2 - \Omega_s^2)X_0}{\omega_n^2} = \frac{1}{\omega_n^2} \times \frac{F_0}{m} = \frac{1}{k/m} \times \frac{F_0}{m} = \frac{m}{k} \times \frac{F_0}{m} = \frac{F_0}{k} \tag{20.27}$$

$$\therefore X_0 = \frac{\omega_n^2}{\omega_n^2 - \Omega_s^2} \times \frac{F_0}{k} = \frac{F_0/k}{1 - \Omega_s^2/\omega_n^2} = \frac{F_0/k}{1 - r^2}, \tag{20.28}$$

where we have put

$$r = \Omega_s/\omega_n. \tag{20.29}$$

The force transmitted through the spring is written as

$$P = kx = kX_0 \sin(\Omega_s t) = P_0 \sin(\Omega_s t), \tag{20.30}$$

where

$$P_0 = kX_0 \tag{20.31}$$

is the amplitude of the transmitted force.

Transmissibility T is defined as the ratio of two amplitudes, namely, the amplitude of vibration of the electronic circuit/device to the amplitude of vibration of the supporting platform. It is expressed as

$$\begin{aligned} T &= \frac{\text{amplitude of the transmitted force}}{\text{amplitude of the applied force}} = \left|\frac{P_0}{F_0}\right| = \left|\frac{kX_0}{F_0}\right| \\ &= \left|\frac{k}{F_0} \times \frac{F_0/k}{1 - r^2}\right| = \left|\frac{1}{1 - r^2}\right|. \end{aligned} \tag{20.32}$$

The effectiveness of the vibration isolator is given by the equation

$$\%\ \text{isolation} = (1 - T) \times 100. \tag{20.33}$$

A high value of T signifies poor isolation. On the opposite side, a low value of T indicates higher percentage isolation. The transmissibility curve showing the plot of transmissibility against frequency is sketched in figure 20.6. From the diagram, it may be noted that at frequencies Ω_s of the supporting platform which are lower than the natural frequency ω_n of the system, transmissibility is approximately 1. This means that the amplitude of vibration of the electronic circuit/device is the same as that of the supporting platform. At the above frequency values, the incoming vibration energy of the supporting platform is neither amplified nor decreased. The vibrating supporting platform and the electronic circuit/device vibrate in phase.

When the frequency Ω_s of vibration of the supporting platform is equal to the natural frequency ω_n of the system, the transmissibility reaches a peak value. This is the resonance condition under which the electronic circuit/device vibrates with maximum amplitude. The vibration energy of the supporting platform has been amplified. A transmissibility value >1 indicates amplification. The phase of the electronic circuit/device is shifted by 90° with respect to that of the supporting platform.

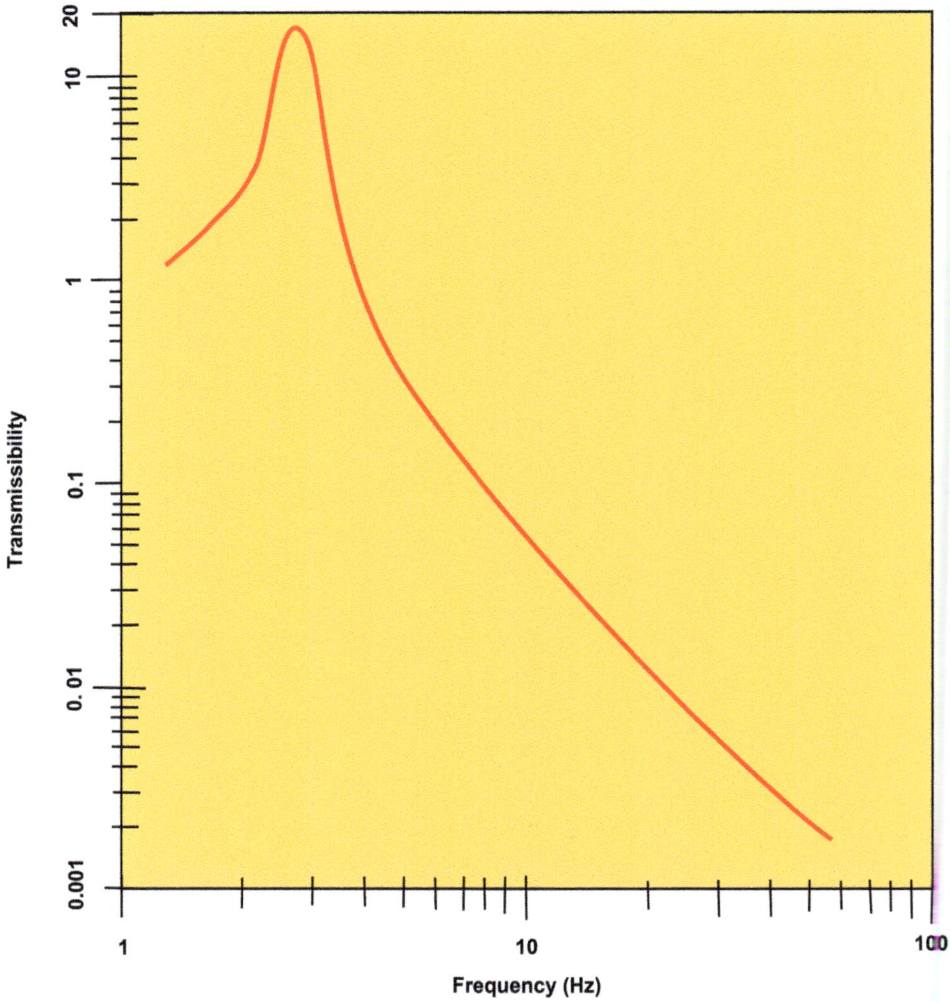

Figure 20.6. Transmissibility curve.

Now if the frequency Ω_s of vibration of the supporting platform exceeds the natural frequency ω_n of the system, the transmissibility starts to decrease and falls considerably at high frequencies. A transmissibility value <1 suggests attenuation. The electronic circuit/device vibrates in anti-phase with the vibration source.

Thus lower frequencies Ω_s of vibration of the supporting platform are conveyed to the electronic circuit/device as such, producing vibrations of the same displacement in it as in the supporting platform. But frequencies Ω_s higher than the natural frequency ω_n of the system are weakened, so that the electronic circuit/device vibrates with progressively smaller amplitudes. This essentially means that higher frequencies of vibration of the supporting platform are prevented from affecting the electronic circuit/device; hence the electronic circuit/device is successfully separated from the fast vibrating supporting platform. Another noteworthy inference is that the higher frequencies have been filtered. So, the behavior of the vibration isolator is analogous to that of a low-pass mechanical filter. Obviously, the best vibration isolation system will be one with a very low natural frequency because all higher frequencies will be sieved off.

For effective vibration isolation, the ratio (Ω_s/ω_n) should be >1.4 and ideally, it should be >2–3.

20.5.3 Case III: forced vibrations with viscous damping

For this situation, the equation of motion for undamped forced vibrations is modified as follows by including the damping effect through the term $c(\mathrm{d}x/\mathrm{d}t)$

$$m\frac{\mathrm{d}^2x}{\mathrm{d}t^2} + c\frac{\mathrm{d}x}{\mathrm{d}t} + kx = F_0\ \sin(\Omega_s t), \tag{20.34}$$

where c is the damping coefficient (figure 20.7).

By dividing both sides by m, this equation is rewritten in the form

$$\frac{\mathrm{d}^2x}{\mathrm{d}t^2} + \frac{c}{m}\frac{\mathrm{d}x}{\mathrm{d}t} + \frac{k}{m}x = \frac{F_0}{m}\ \sin(\Omega_s t) \tag{20.35}$$

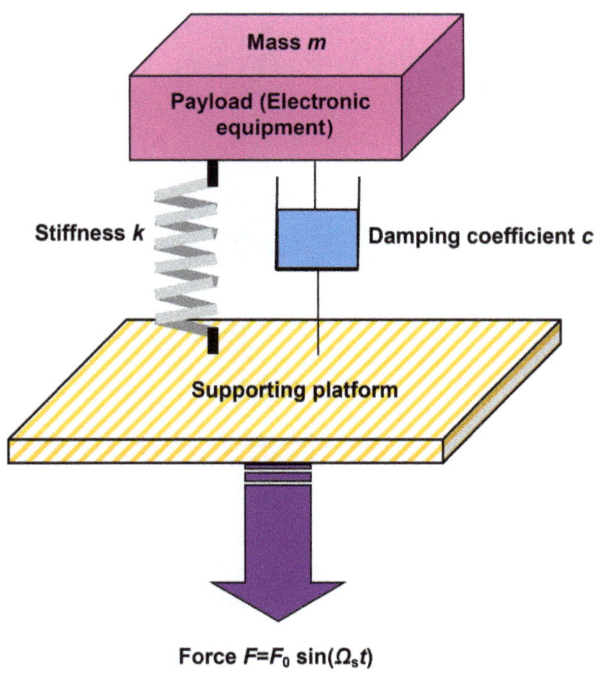

Figure 20.7. Forced vibrations of an electronic device with viscous damping.

$$\text{or } \frac{\mathrm{d}^2x}{\mathrm{d}t^2} + \frac{2c}{2\sqrt{km}} \times \sqrt{\frac{k}{m}}\frac{\mathrm{d}x}{\mathrm{d}t} + \frac{k}{m}x = \frac{F_0}{m}\sin(\Omega_s t)$$
$$\text{or } \frac{\mathrm{d}^2x}{\mathrm{d}t^2} + 2\zeta\omega_n\frac{\mathrm{d}x}{\mathrm{d}t} + \omega_n^2 x = \frac{F_0}{m}\sin(\Omega_s t), \tag{20.36}$$

where we have put

$$\frac{c}{2\sqrt{km}} = \zeta \tag{20.37}$$

and

$$\sqrt{\frac{k}{m}} = \omega_n. \tag{20.38}$$

Let us try a particular solution to this equation of the form

$$x = X_0 \sin(\Omega_s t - \phi). \tag{20.39}$$

Then,

$$\frac{\mathrm{d}x}{\mathrm{d}t} = \Omega_s X_0 \cos(\Omega_s t - \phi) \tag{20.40}$$

and

$$\frac{\mathrm{d}^2x}{\mathrm{d}t^2} = -\Omega_s^2 X_0 \sin(\Omega_s t - \phi). \tag{20.41}$$

Hence, equation (20.36) becomes

$$-\Omega_s^2 X_0 \sin(\Omega_s t - \phi) + 2\zeta\omega_n\Omega_s X_0 \cos(\Omega_s t - \phi)$$
$$+ \omega_n^2 X_0 \sin(\Omega_s t - \phi) = \frac{F_0}{m}\sin(\Omega_s t) \tag{20.42}$$

$$\text{or, } (\omega_n^2 - \Omega_s^2)X_0 \sin(\Omega_s t - \phi) + 2\zeta\omega_n\Omega_s X_0 \cos(\Omega_s t - \phi) = \frac{F_0}{m}\sin(\Omega_s t)$$
$$\text{or, } (\omega_n^2 - \Omega_s^2)X_0(\frac{m}{F_0}) \sin(\Omega_s t - \phi) + 2\zeta\omega_n\Omega_s X_0(\frac{m}{F_0})\cos(\Omega_s t - \phi) = \sin(\Omega_s t). \tag{20.43}$$

Let us put

$$(\omega_n^2 - \Omega_s^2)X_0\left(\frac{m}{F_0}\right) = \cos\,\phi \tag{20.44}$$

and

$$2\zeta\omega_n\Omega_s X_0\left(\frac{m}{F_0}\right) = \sin\phi. \tag{20.45}$$

Then

$$\cos\phi\,\sin(\Omega_s t - \phi) + \sin\phi\,\cos(\Omega_s t - \phi) = \sin(\Omega_s t) \tag{20.46}$$

$$\text{or}\ \sin(\phi + \Omega_s t - \phi) = \sin(\Omega_s t), \tag{20.47}$$

which means that the equation is satisfied. Now,

$$\begin{aligned}\cos^2\phi + \sin^2\phi = 1 &= \left\{\left(\omega_n^2 - \Omega_s^2\right)X_0\left(\frac{m}{F_0}\right)\right\}^2 + \left\{2\zeta\omega_n\Omega_s X_0\left(\frac{m}{F_0}\right)\right\}^2 \\ &= X_0^2\left(\frac{m}{F_0}\right)^2\{(\omega_n^2 - \Omega_s^2)^2 + (2\zeta\omega_n\Omega_s)^2\}\end{aligned} \tag{20.48}$$

$$\begin{aligned}\therefore X_0^2 &= \frac{1}{\left(\frac{m}{F_0}\right)^2\{(\omega_n^2 - \Omega_s^2)^2 + (2\zeta\omega_n\Omega_s)^2\}} \\ &= \frac{\left(\frac{F_0}{m}\right)^2}{\{(\omega_n^2 - \Omega_s^2)^2 + (2\zeta\omega_n\Omega_s)^2\}} \\ &= \frac{F_0^2(k/m)^2 \times (1/k)^2}{\{(\omega_n^2 - \Omega_s^2)^2 + (2\zeta\omega_n\Omega_s)^2\}} = \frac{F_0^2\omega_n{}^4 \times (1/k)^2}{\{(\omega_n^2 - \Omega_s^2)^2 + (2\zeta\omega_n\Omega_s)^2\}}\end{aligned} \tag{20.49}$$

giving

$$\begin{aligned}X_0 &= \frac{F_0\omega_n^2/k}{\sqrt{(\omega_n^2 - \Omega_s^2)^2 + (2\zeta\omega_n\Omega_s)^2}} \\ &= \frac{F_0/k}{\sqrt{\{(\omega_n^2 - \Omega_s^2)^2 + (2\zeta\omega_n\Omega_s)^2\}/\omega_n{}^4}} \\ &= \frac{F_0/k}{\sqrt{(1 - \Omega_s^2/\omega_n^2)^2 + (2\zeta\Omega_s/\omega_n)^2}} = \frac{F_0/k}{\sqrt{(1 - r^2)^2 + (2\zeta r)^2}}.\end{aligned} \tag{20.50}$$

Also,

$$\begin{aligned}\frac{\sin\ \phi}{\cos\ \phi} = \tan\phi &= \frac{2\zeta\omega_n\Omega_s X_0\left(\frac{m}{F_0}\right)}{(\omega_n^2 - \Omega_s^2)X_0\left(\frac{m}{F_0}\right)} = \frac{2\zeta\omega_n\Omega_s}{\omega_n^2 - \Omega_s^2} = \frac{(2\zeta\omega_n\Omega_s)/\omega_n^2}{(\omega_n^2 - \Omega_s^2)/\omega_n^2} \\ &= \frac{2\zeta(\Omega_s/\omega_n)}{1 - \Omega_s^2/\omega_n^2} = \frac{2\zeta r}{1 - r^2}.\end{aligned} \tag{20.51}$$

Force transmitted to the electronic circuit/device is

$$
\begin{aligned}
P &= c\frac{\mathrm{d}x}{\mathrm{d}t} + kx = c\Omega_s X_0 \cos(\Omega_s t - \phi) + kX_0 \sin(\Omega_s t - \phi) \\
&= \frac{c\Omega_s(F_0/k)}{\sqrt{(1-r^2)^2 + (2\zeta r)^2}} \cos(\Omega_s t - \phi) + \frac{k(F_0/k)}{\sqrt{(1-r^2)^2 + (2\zeta r)^2}} \sin(\Omega_s t - \phi) \\
&= \frac{F_0}{\sqrt{(1-r^2)^2 + (2\zeta r)^2}} \sin(\Omega_s t - \phi) + \frac{c\Omega_s(F_0/k)}{\sqrt{(1-r^2)^2 + (2\zeta r)^2}} \cos(\Omega_s t - \phi).
\end{aligned} \tag{20.52}
$$

This is of the form

$$
\begin{aligned}
P &= P_0 \sin(\Omega_s t - \phi + \delta) = P_0 \sin(\Omega_s t - \phi)\cos\delta + P_0 \cos(\Omega_s t - \phi)\sin\delta \\
&= P_0 \cos\delta \sin(\Omega_s t - \phi) + P_0 \sin\delta \cos(\Omega_s t - \phi).
\end{aligned} \tag{20.53}
$$

Comparing equations (20.52) and (20.53), we can write

$$
\frac{F_0}{\sqrt{(1-r^2)^2 + (2\zeta r)^2}} = P_0 \cos\delta \tag{20.54}
$$

and

$$
\frac{c\Omega_s(F_0/k)}{\sqrt{(1-r^2)^2 + (2\zeta r)^2}} = P_0 \sin\delta \tag{20.55}
$$

$$
\therefore \frac{F_0^2 + \{c\Omega_s(F_0/k)\}^2}{(1-r^2)^2 + (2\zeta r)^2} = P_0^2(\cos^2\delta + \sin^2\delta) = P_0^2(1) \tag{20.56}
$$

or

$$
\begin{aligned}
P_0 &= \sqrt{\frac{F_0^2 + \{c\Omega_s(F_0/k)\}^2}{(1-r^2)^2 + (2\zeta r)^2}} = F_0\sqrt{\frac{1 + \{c\Omega_s(1/k)\}^2}{(1-r^2)^2 + (2\zeta r)^2}} \\
&= F_0\sqrt{\frac{1 + \{\frac{2c\Omega_s}{2k}\}^2}{(1-r^2)^2 + (2\zeta r)^2}} \\
&= F_0\sqrt{\frac{1 + \{\frac{2c\Omega_s}{2\sqrt{km}\sqrt{\frac{k}{m}}}\}^2}{(1-r^2)^2 + (2\zeta r)^2}} = F_0\sqrt{\frac{1 + \{\frac{2\zeta\Omega_s}{\omega_n}\}^2}{(1-r^2)^2 + (2\zeta r)^2}} \\
&= F_0\sqrt{\frac{1 + (2\zeta r)^2}{(1-r^2)^2 + (2\zeta r)^2}},
\end{aligned} \tag{20.57}
$$

where equations (20.16), (20.29) and (20.37) have been applied.

Further,

$$\frac{P_0 \sin\delta}{P_0 \cos\delta} = \tan\delta = \frac{\dfrac{c\Omega_s(F_0/k)}{\sqrt{(1-r^2)^2+(2\zeta r)^2}}}{\dfrac{F_0}{\sqrt{(1-r^2)^2+(2\zeta r)^2}}} = \frac{c\Omega_s(F_0/k)}{F_0} = c\Omega_s/k = \frac{2c\Omega_s}{2\sqrt{km}\sqrt{\frac{k}{m}}} \tag{20.58}$$
$$= 2\zeta\frac{\Omega_s}{\omega_n} = 2\zeta r$$

by applying equations (20.16), (20.29) and (20.37).

From equation (20.57), the transmissibility T is

$$\boldsymbol{T} = \frac{P_0}{F_0} = F_0\frac{\sqrt{\dfrac{1+(2\zeta r)^2}{(1-r^2)^2+(2\zeta r)^2}}}{F_0} = \sqrt{\frac{1+(2\zeta r)^2}{(1-r^2)^2+(2\zeta r)^2}}. \tag{20.59}$$

The phenomenon of damping is correlated with the resonance condition. The reason for this statement is that damping has a pronounced effect on the transmissibility value at resonance or in its neighborhood. Hence, damping may be deemed essentially to be an effect associated with resonance. At resonance, the value of $r = 1$. Then the transmissibility attains its maximum magnitude. So, under the resonance condition, equation (20.59) reduces to the simplified form

$$T_{\text{maximum}} = \sqrt{\frac{1+(2\zeta)^2}{(1-1)^2+(2\zeta)^2}} = \sqrt{\frac{1+(2\zeta)^2}{0+(2\zeta)^2}} = \sqrt{1+\frac{1}{(2\zeta)^2}}$$
$$\approx \sqrt{\frac{1}{(2\zeta)^2}} = \frac{1}{2\zeta}. \tag{20.60}$$

At resonance, the height of the transmissibility peak is determined by damping. In the absence of damping, the peak value of transmissibility at resonance is infinity. Moreover, without damping the vibrations will perpetuate forever and will not cease after withdrawal of the exciting force. In practice, systems cannot be made free from damping. Damping is always present in all systems. Only the degree of damping varies. The value of damping may range from very small to very large.

20.6 Mechanical spring vibration isolators

This is the kind of vibration isolators which immediately come to mind, because we are accustomed to their widespread deployment in motor vehicles such as cars, trucks and buses. They absorb vibration and provide us a comfortable journey on rough and bumpy roads.

20.7 Air-spring vibration isolators

A tube filled with air serves as a cushion. This cushioning effect is utilized to protect the overlying equipment from damage. Air springs are manually filled air bladders that can be filled with a hand-operated pump. A compressed air tank can also be used to replenish the air.

Air-spring vibration isolators are of two types:

(i) *Bellows air bags.* These are configured in single, double, or triple convoluted chambers. The chambers are made of durable reinforced rubber with the capability to load and carry large weights. Bellow air bags find heavyweight applications where sufficient space is obtainable to accommodate them.

(ii) *Rolling sleeve air bags.* They contain an internally mounted sleeve with an internally molded bead and covered with an air bag, which can be inflated. The assembly provides a smaller diameter than the bellows design. They are useful to support lighter weights. They are suitable when vacant space is scarce.

20.8 Wire-rope isolators

These isolators are based on the spring action of a rope coil over which the crucial electronic part is suspended.

20.9 Elastomeric isolators

Elastomer is a generic term applied to all types of rubber. These rubbers may be natural or synthetic. The elastomers are elastic polymers, e.g. India rubber, silicone rubber, fluorosilicone rubber, butyl rubber, etc. Elastomeric isolators utilize the suppleness and bounciness provided by a rubbery compound to recover its size and shape when deformed. The homogeneous nature of elastomers allows them to be shaped into compact forms such as planar, laminate, cylindrical, etc.

Natural rubber serves as a baseline for comparing the relative performance of different elastomers. It has high strength, excellent fatigue properties and provides low-to-medium damping. But it suffers from the drawback of a narrow operating temperature range from −18 °C to +82 °C. It tends to stiffen at lower temperatures. Several silicone elastomers have been developed with broader temperature ranges.

20.10 Negative stiffness isolators

Hold a flexible plastic scale upright on the table (figure 20.8). Press it downwards by hand by applying a force at its top end. It bends (May 2001). As soon as the force is released, the plastic ruler regains its original straight shape. The recovery force acts opposite to the deformation force. This is positive stiffness. Now suppose when the ruler is deformed, a finger is pressed in the middle of the ruler against its bulging side. Then the ruler acquires the *S*-shape shown in the diagram. If the finger pressure is maintained the ruler bulges out on the opposite side. This is a case of negative stiffness. A force acts on the ruler in the same direction as the deforming force and aids the buckling.

To utilize the negative stiffness effect, generally three isolators are organized in series (Platus 1991). A tilt-motion isolator is perched above a horizontal motion isolator placed over a vertical motion isolator.

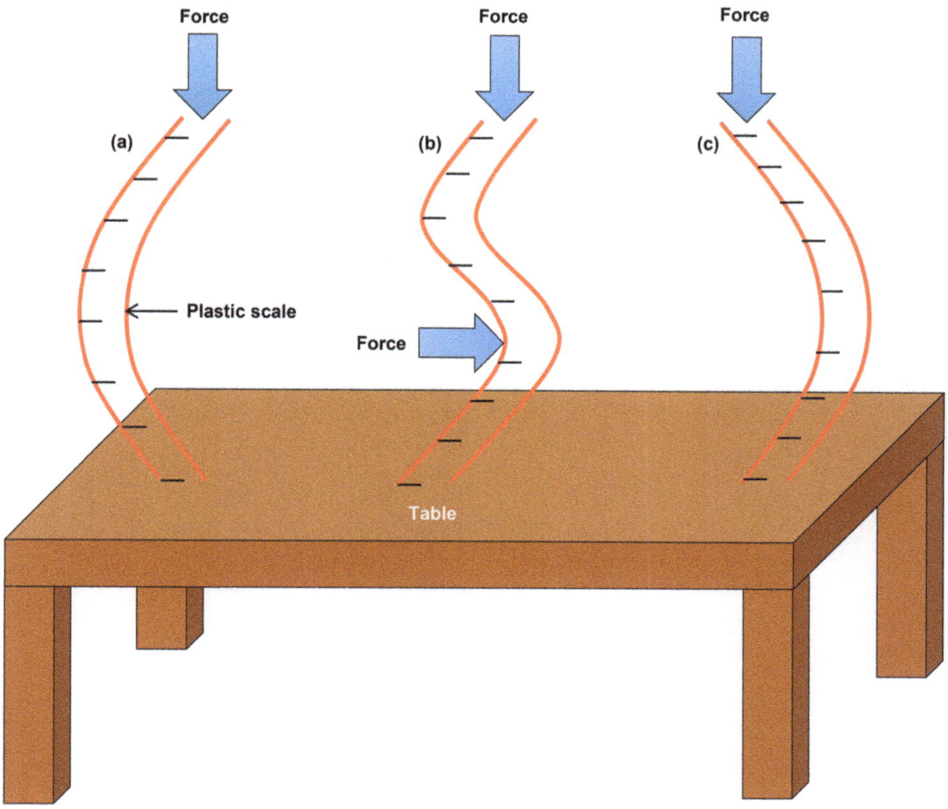

Figure 20.8. Negative thickness effect in the deformation of a plastic scale: (a) a ruler pressed from the top, (b) a ruler pressed from the top and with a finger applied at the middle and (c) subsequent bending of the ruler towards the opposite side.

The vertical motion isolator (figure 20.9) comprises a customary spring squeezed by the weight W to the working point. This spring is connected to a structure made of two bars. It is these bars which provide the negative thickness effect. The bars are hinged at the center. At the extremities, they rest on pivots. Compression forces F, F are applied to the bars. The stiffness k_{Vertical} of the isolator is expressed as

$$k_{\text{Vertical}} = k_{\text{Spring}} - k_{\text{Bar effect}}. \tag{20.61}$$

$k_{\text{Bar effect}}$ depends on the length of the bars and the force F. The resulting stiffness of the isolator can be reduced to zero while the weight W is supported by the spring.

The horizontal motion isolator (figure 20.10) consists of two beam-column isolators. Considering single isolators, each isolator works in the same manner as two fixed-free beam columns. These beam columns are loaded in the axial direction by the weight W. The horizontal stiffness of the beam columns in the absence of weight load is k_{Spring}. Application of the weight load decreases the lateral bending stiffness by the beam–column effect. The arrangement behaves as a horizontal

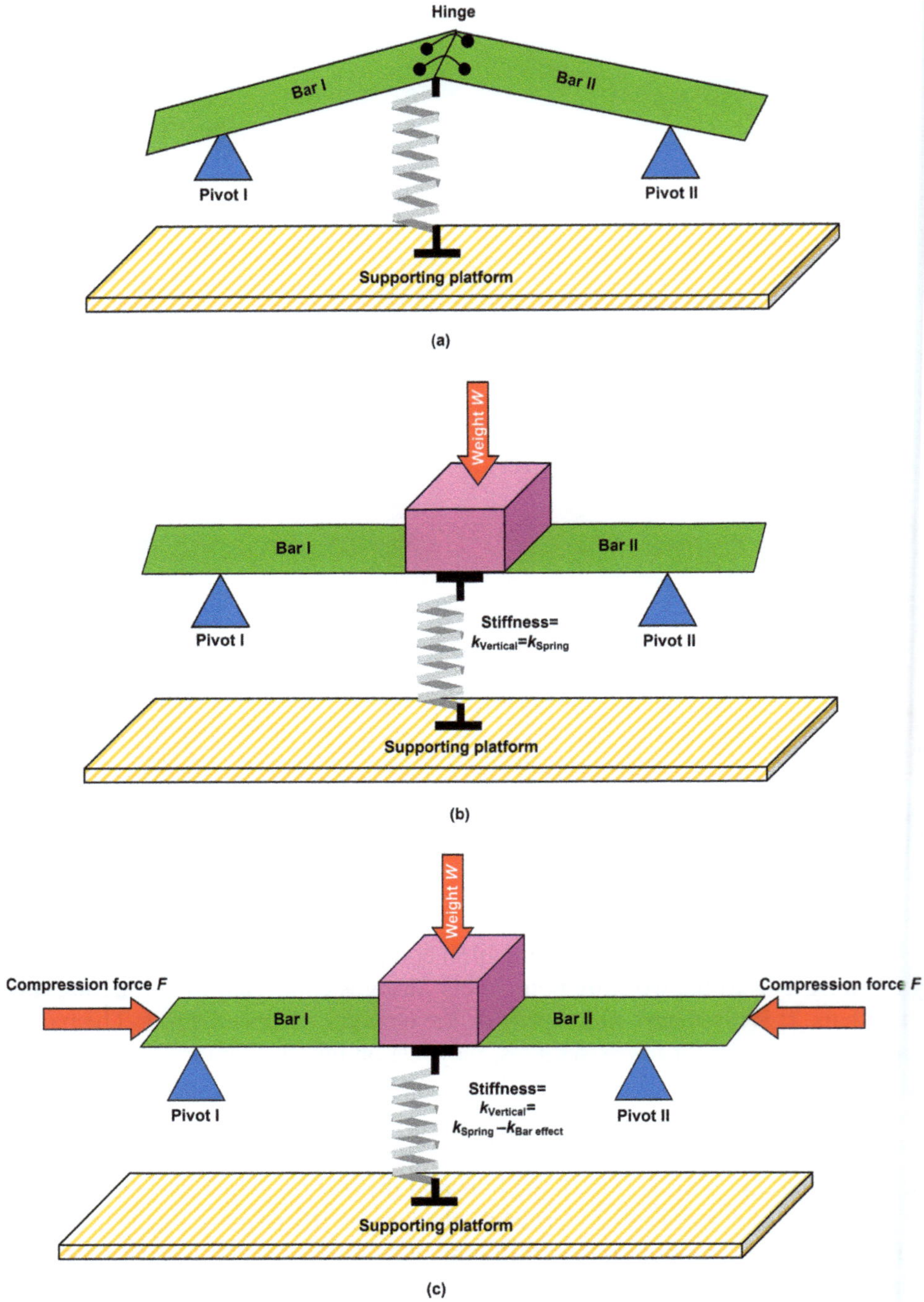

Figure 20.9. Vertical motion isolation concept: (a) unloaded condition, (b) on placing weight W and (c) on applying compressive forces F, F.

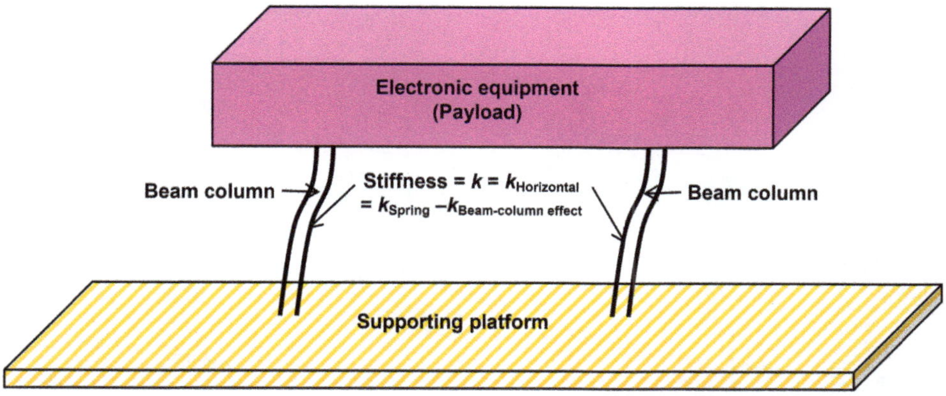

Figure 20.10. Principle of horizontal motion isolation.

spring combined with a negative stiffness mechanism. The horizontal stiffness $k_{\text{Horizontal}}$ is given by

$$k_{\text{Horizontal}} = k_{\text{Spring}} - k_{\text{Beam-column effect}}. \tag{20.62}$$

By loading the beam columns in such a way that they attain their critical buckling load, the horizontal stiffness can be made nearly zero.

20.11 Active vibration isolators

All the above isolators belong to the category of passive isolators, which do not use any power supply. We now look at active vibration isolators (Yoshioka and Murai 2002).

20.11.1 Working

Feedback active vibration isolators are reactive in nature. They work by modifying the condition of the vibrating platform on the basis of data retrieved (Shen *et al* 2013). They use a piezoelectric sensor such as an accelerometer to measure the vibration. Along with this sensor, they employ a force actuator such as a loud-speaker voice coil to deliver a counterforce or an anti-phase signal on the vibrating object to oppose the vibration (figure 20.11). In this kind of isolator, a continuous vigil is kept on the platform to be controlled and the required counterforce is applied in accordance with the instantaneous vibration level. In contrast, feedforward active vibration isolators are anticipative in nature. They are deployed in applications after detailed information on the vibrating characteristics of the platform has been acquired. Only then is it possible to send appropriate pre-decided signals to control the vibrations.

An active air isolator has a compressor with servo valves (Shaidani 2008). These valves either feed air into the bladder or bleed air from it to maintain the payload at zero vibration level.

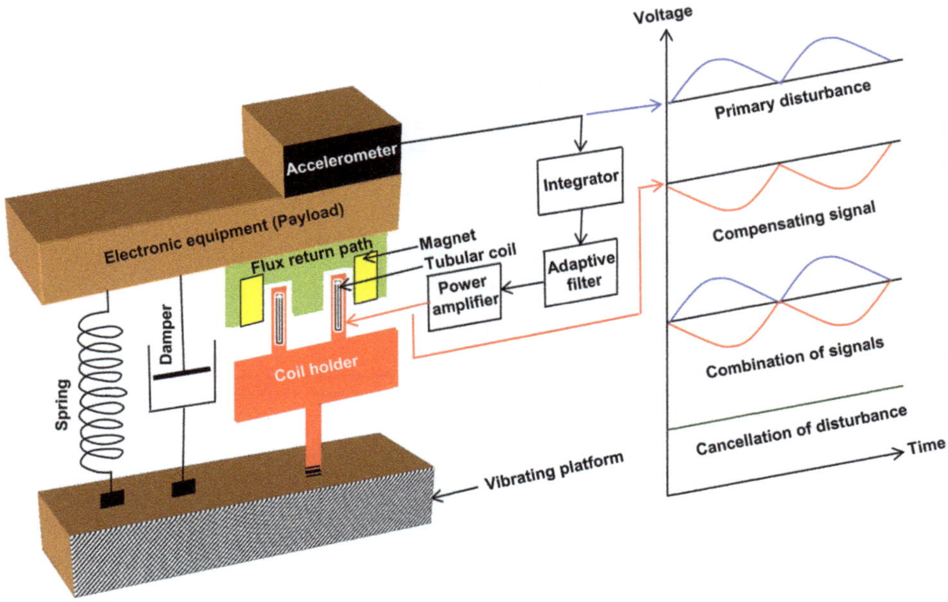

Figure 20.11. Main parts of an active vibration isolator.

20.11.2 Advantages

A major advantage in favor of active isolation systems over passive systems concerns their ability to isolate very low frequencies <10 Hz and even sub-hertz. Such vibrations are common in buildings near subways, and can lead to damage through high-amplitude vibrations via resonances. These low-frequency resonances are easily terminated through closed-loop design by feeding the necessary compensation. Active isolation systems perform better at frequencies <100 Hz; the isolation range is >0.7 Hz. They have neither any resonance nor amplify any vibrations at any frequency. Above 100 Hz, passive systems are useful; the isolation range is >5 Hz. Passive pneumatic systems are known to amplify vibrations instead of suppressing them in the frequency range 1–8 Hz. At frequencies >30 Hz, 98%–99% isolation is achievable with both active and passive systems. Another favorable feature is the increased stiffness, >100 times, of these systems, providing superior directional and positional stability for a precision instrument. Softer systems take a longer time to settle down once disturbed and are also influenced by air currents. Active systems settle down in 10–20 ms, whereas passive ones have a longer settling time ~2–10 s. Further, it is possible to adjust the active system robotically, without any human intervention. This adjustment may be performed for a varying load. It can also be performed in the case in which the load is spread out on the platform. However, active systems are more expensive than passive systems.

20.11.3 Applications

Active vibration isolation systems come in a variety of form factors and sizes. They can debilitate vibrations in various translational and rotational modes. They can do so in real time in all six degrees of freedom (x, y, z: roll, pitch and yaw). They are widely used in atomic force and scanning tunneling microscopy (Lan *et al* 2004) where atomic scale resolution is at stake; space telescopes, laser communication systems, as well as in interferometry and metrology. An example of the use of an active system is in lithography process in the semiconductor industry. The silicon wafer is placed on a heavy stage and positioned with great accuracy. The resolution levels are reaching 1 nm and even the slightest vibration can undermine the process. An active system is immensely useful for such critical processes. Similar remarks apply to all manufacturing processes and scientific experiments where such low-dimensional structures are involved.

20.12 Discussion and conclusions

It is essential to design, evaluate and test electronic equipment in various vibration environments that can cause failure during operation (Steinberg 2000). To prevent damage to the delicate electronic equipment from all-prevailing and unavoidable vibration conditions, it is necessary to design ingenious vibration isolation mechanisms. The techniques vary with the intended field of operation of the device, cost considerations and the accuracy expectations of the customer. Obviously, instruments aiming at nanometer accuracy cannot tolerate the least disturbing vibrations, whereas those for crude measurements have a much higher level of endurance. Therefore, each application has a unique solution and no versatile design philosophy of universal applicability can be laid out.

Review exercises

20.1. Why is it necessary to build vibration tolerance capability into electronic equipment? How does sinusoidal vibration differ from random vibration?

20.2. What are the two approaches followed to counter the effects of vibration on electronics? However strong an electronic device may be built to withstand vibration, it is necessary to isolate it from vibration. Why? What are the two categories of vibration isolators? How do they differ?

20.3. What is the main function of a vibration isolator? What are the chief components of a vibration isolator? What is a dashpot?

20.4. Frame and solve the differential equation for free undamped vibrations of a payload of mass m supported by a spring of stiffness k. What are the vital parameters which decide the natural frequency of the system?

20.5. Write and solve the differential equation for forced vibrations of a vibration isolation system consisting of a spring and a damper. How does it differ from that of free vibration system?

20.6. What is meant by transmissibility of the vibration isolator? A vibration isolator has a transmissibility of 0.2. What does this low value of transmissibility indicate? What is the meaning of a transmissibility value = 1?

20.7. Derive an equation for the transmissibility of a vibration isolator in terms of the natural frequency of the payload and the exciting frequency of the platform. How is transmissibility affected by the relative magnitudes of these frequencies?
20.8. The ratio of magnitudes of the natural frequency of the payload to the exciting frequency of the platform is 0.2. Is the vibration isolation effective? Why?
20.9. A vibration isolation system has a damping value of zero. What is the transmissibility of vibrations at resonance? Hence, explain the role of damping in determining transmissibility.
20.10. What is an air-spring vibration isolator? Name the two types of air-spring isolators and describe their operation.
20.11. What is an elastomer? What is the advantage of synthetic silicone elastomers over natural rubber?
20.12. What is meant by positive stiffness and negative stiffness? Explain with an example.
20.13. Explain with a diagram the working of a vertical motion isolator. What are the parameters on which vertical stiffness depends?
20.14. Draw the diagram of a horizontal motion isolator and describe its operation. Indicate the parameters deciding the horizontal stiffness.
20.15. How does a feedback active vibration isolator differ from a feedforward vibration isolator? In what situations are the feedback and feedforward isolation concepts applied?
20.16. What ranges of the vibration frequencies are accommodated by active and passive vibration isolator systems? What will happen if a passive system is used for controlling very low frequency vibrations?
20.17. How do passive and active vibration isolators differ regarding the stiffness parameter? How does the difference in stiffness translate into a difference in settling time after a disturbance from mean position?
20.18. Give examples illustrating the desperate necessity of an active vibration isolator? Justify the criticality of the situations.

References

ETS Solutions 2021 *Vibration resistance of electrical equipment* https://etssolution-asia.com/blog/vibration-resistance-of-electrical-equipment

Lan K J, Yen J-Y and Kramar J A 2004 Active vibration isolation for a long range scanning tunneling microscope *Asian J. Control* **6** 179–86

May M 2001 Getting more stiffness with less *Am. Sci.* November–December www.americanscientist.org/issues/pub/getting-more-stiffness-with-less

Platus D L 1991 Negative-stiffness-mechanism isolation systems *Proc. SPIE* **1619** 44–54

Shaidani H 2008 Vibration isolation in cleanrooms: a system for virtually every application *Control. Environ. Mag.* January, 7 pages

Shen H, Wang C, Li L and Chen L 2013 Prototyping a compact system for active vibration isolation using piezoelectric sensors and actuators *Rev. Sci. Instrum.* **84** 055002

Steinberg D S 2000 *Vibration Analysis for Electronic Equipment* 3rd edn (New York: Wiley-Interscience)

Tormala K 2022 *Shake, drop, and test! Simulation and test for electronics reliability* https://blogs.sw.siemens.com/electronics-semiconductors/2022/08/03/vibration-test-and-simulation-in-electronics-design/

Yang L, Chen Y, Yuan Z and Chen L 2015 Effect of vibration transmissibility on fatigue lifetime of electronic devices *Industrial Engineering, Management Science and Applications 2015Lecture Notes in Electrical Engineering* vol 349 ed M Gen, K Kim, X Huang and Y Hiroshi (Berlin: Springer)

Yoshioka H and Murai N 2002 An active microvibration isolation system *Proc. 7th Int. Workshop Accelerator Alignment* pp 388–401

Zhang L, Ning Z, Peng H, Mu Z and Sun C 2017 Effects of vibration on the electrical performance of lithium-ion cells based on mathematical statistics *Appl. Sci.* **7** 1–21

Part II

Harsh-environment electronics

Sub-part IIB

Application-specific robust electronics techniques

Chapter 21

Making electronics compatible with electromagnetic interference environments

Connotations of the terms 'electromagnetic interference (EMI)' and 'electromagnetic compatibility (EMC)' are elucidated. Electromagnetic interference is classified according to the sources, production mechanisms, duration and bandwidth of the EMI signal. The noise generated by EMI is explained with examples of its impact on the performance of electrical equipment. The inability of single-ended transmission of signals to cope with EMI noise is highlighted and the necessity of differential-mode transmission is emphasized. Common-mode and differential-mode signals, and the associated common-mode and differential-mode EMI are distinguished. Twisted pair cables are effective in mitigation of both self and external EMI. The origin of common-mode interference by common impedance coupling of two circuits sharing a common return path is discussed. Removal of differential-mode noise by using resistance–capacitance, capacitance–inductance and capacitance–inductance–capacitance filters is described. Common-mode choke coils are used for suppression of noise common to two or more power or data lines. Improper grounding of circuits often leads to problems such as the formation of ground loops. Single-point, multi-point and hybrid grounding schemes are outlined, and their relative merits and demerits are elaborated. Often the sensitive electronic circuits need to be shielded from EMI. Electrical shielding can be provided by enclosing the electronic circuits inside a metal box. The enclosure called a Faraday cage works on the cancellation of the external electric field with an oppositely-directed polarization field created on the cage. Shielded cables with foil, braid and combined foil and braid shields also provide effective protection against EMI. Grounding of shielded cables, whether to be done at a single end or both ends is determined by the objective of shielding, and is decided by knowing whether electrical or magnetic shielding is required.

doi:10.1088/978-0-7503-5072-3ch21

The evolution of MOSFETs in 1950s, the integrated circuits in 1960s and the microprocessor chips in 1970s together with the increasing demands on voice, picture and data communications led to the onslaught of an electromagnetic pollution of the environment (Zhao *et al* 2021).

Cyber physical systems are new-generation, integrated digital systems with decision-making capability. They combine sensor networks with embedded computing to control the physical environment. Electric vehicles will serve as viable options to steer transport away from fossil vehicles. Smart grids are electricity distribution networks utilizing digital technologies to monitor the varying demands of consumers.

The progress in these areas is riddled with complexities because the low-voltage internet nodes are often located in close proximity to power conditioning equipment, raising issues of electromagnetic interference at device and system levels (figure 21.1). This chapter deals with strategies to cope up with challenges from electromagnetic interference for electronics.

21.1 Electromagnetic interference

Electromagnetic interference (EMI) is the disturbance caused on the operation of an electronic circuit by the electromagnetic field created by it during its normal functioning or by the electromagnetic fields of neighbouring electronic circuits. This interference is the sum-total disturbance arising from the circuit itself and from adjacent circuits. EMI covers the full electromagnetic spectrum from as low as DC to 300 GHz including radio and microwaves.

$$\begin{aligned}\text{EMI} = \text{Disturbance from self} + \text{Disturbance from neighbors}\\ + \text{Disturbance to neighbors}\end{aligned} \tag{21.1}$$

The disturbance from EMI is detrimental to the overall performance of the concerned circuit. It sometimes leads to small excursions from the normal working consistent with the designed objective but may also cause serious breakdown of circuit function.

21.2 Electromagnetic compatibility

Electromagnetic compatibility (EMC) expresses the ability of an electronic circuit to work without causing any disturbing electromagnetic influence to its own activity, without suffering from any degradation due to electromagnetic fields of the surrounding circuits in the environment for which it is designed, and without producing any harmful electromagnetic effects on those surrounding circuits. This ability is the combination of abilities of the circuit for working satisfactorily in the circuit's own field, and in the field of adjacent circuits without exerting any adverse effects on adjacent circuits.

$$\begin{aligned}\text{EMC} = \text{No self} - \text{disturbance} + \text{No disturbance from neighbors}\\ + \text{No disturbance to neighbors}\end{aligned} \tag{21.2}$$

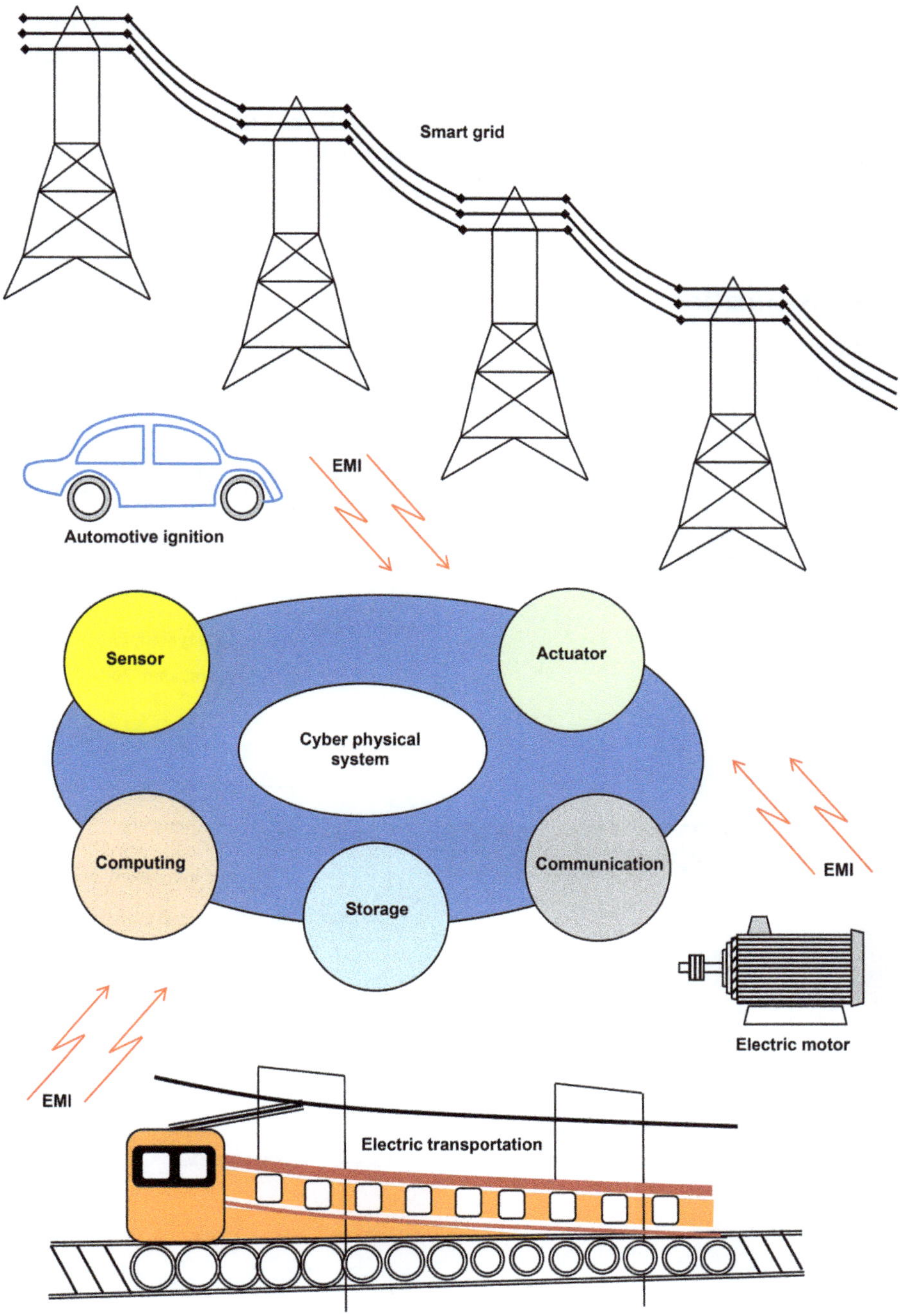

Figure 21.1. EMI flare-up from close proximity of logic circuits and power systems in the congested locations of cyber physical systems, smart grid, electric transportation, automotive ignition, electric motor and other electrical appliances.

In easy terms, an EMC-compliant circuit is a circuit which is not disturbed by its own electromagnetic fields; it is neither upset from electromagnetic fields of neighboring circuits, nor produces any disturbing electromagnetic fields for neighboring circuits.

Electromagnetic compatibility is the branch of science and engineering concerned with designing and fabricating equipment in such a way that they are safe from any electromagnetic interference from other equipment and also the interferences generated by the equipment are within prescribed limits (Christopoulos 2022).

21.3 Classification of EMI

A systematic classification of EMI can be made from the viewpoint of sources of EMI, its production mechanisms, and duration and bandwidth of the interference signal (figure 21.2).

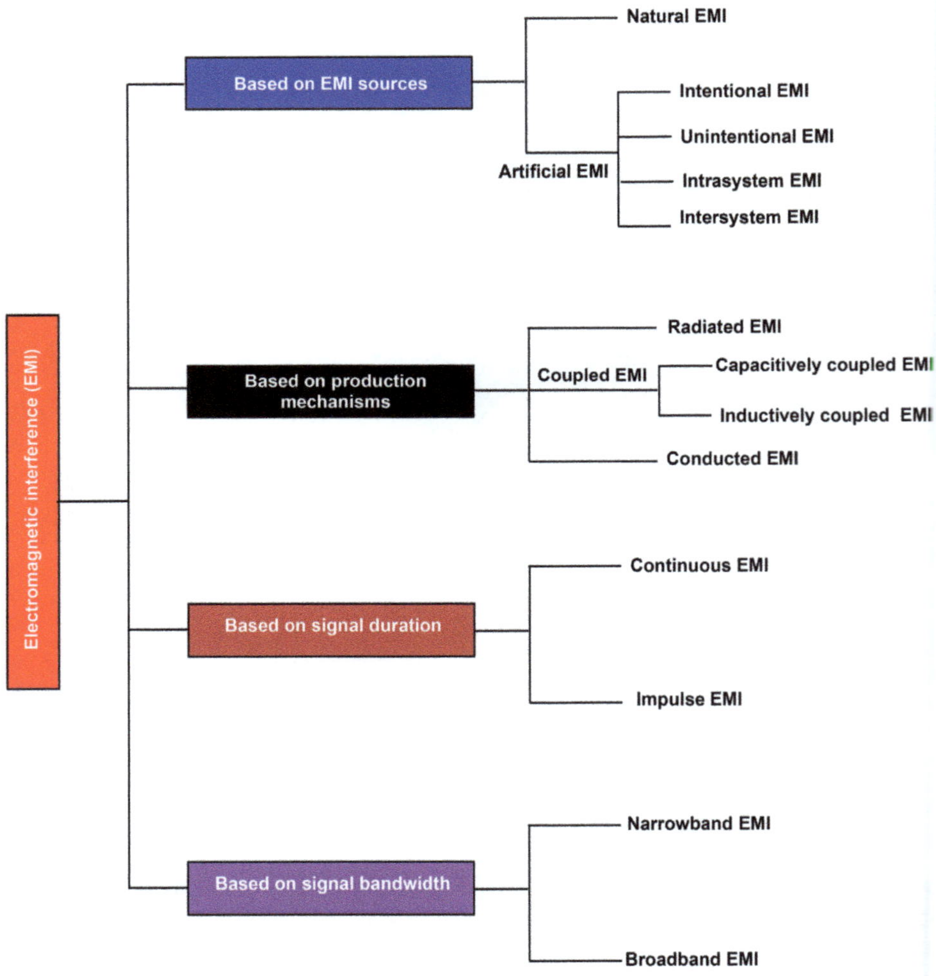

Figure 21.2. Classification of electromagnetic interference on the basis of EMI sources, production mechanisms, signal duration and bandwidth.

21.3.1 Sources of EMI

EMI sources can be broadly classified into two groups: natural and artificial or man-made:

$$\text{EMI sources} = \text{Natural sources} + \text{Artificial sources} \tag{21.3}$$

21.3.1.1 Natural EMI sources

These sources range from terrestrial sources to celestial radiation from unidentified sources, inside or beyond the Milky Way Galaxy, e.g., lightning discharge in Earth atmosphere during a thunderstorm, electrostatic discharge by contact/dielectric breakdown, solar magnetic storms, cosmic or galactic radio noise.

21.3.1.2 Artificial EMI sources

A few human-made sources are: utility grid transmission lines, power conditioning circuits, power supplies, radio/TV, cell phones, computers, airport RADAR, switch gear, train signaling and control systems, automotive ignition, electric motors, fluorescent lamps, arcing equipment, life support ventilators, defibrillators, motorized wheelchairs, medical imaging equipment like x-ray machines, CT and MRI scanners; telemetry, etc. Many electrical and electronic machines fall in this list.

Depending on whether EMI is deliberately introduced or inadvertently happening, artificial EMI sources are of following two types:

(i) Intentional EMI: It is deliberately generated EMI such as that produced by radars, jammers, etc during electronic warfare (Bäckström 2006, Radasky and Bäckström 2014).

(ii) Unintentional EMI: It is EMI produced during the operation of a machine which is not designed to radiate electromagnetic energy. Power cables, DC motors, welding machines, computers, and automotive engine ignition systems produce unwanted EMI although they are not designed and constructed to do so.

From a systemic viewpoint, electromagnetic interference may be looked upon from the perspective of happening either within a system or between different systems. EMI raises grave concerns when a compact installation in an aircraft or satellite carries several closely packed electrical and electronic circuits. EMI is also an issue of serious concern in a situation such as operation of a personal computer at home distant from a high-power radio transmitter. But the separation distance between the EMI source and the circuit in use differs in the two cases. Hence an intrasystem or an intersystem approach may be followed in deciding the boundary of electromagnetic interference range for electronic equipment in relation to the class of environment, whether it is industrial, commercial or domestic, and the criticality of operation (Williams 2017).

Two types of artificial EMI from this viewpoint are:

(i) Intrasystem EMI: Voltage or current spikes in power lines produce EMI which hampers the system's own working. This is intrasystem EMI.

(ii) Intersystem EMI: Electronic equipment working in a broad frequency ranging from 50 Hz to several gigahertz produce EMI which affects other systems. It is called intersystem EMI.

21.3.2 EMI production mechanisms

According to EMI production mechanisms, there are three kinds of EMI (figure 21.3) as elucidated in ensuing subsections.

21.3.2.1 Radiated EMI

It is caused by radiation of electromagnetic waves from a source circuit to a victim circuit without any physical contact between them. The two circuits may be positioned near each other or placed far apart.

21.3.2.2 Coupled EMI

It is capacitively or magnetically-induced EMI:

(i) capacitive coupling through charge transference to the victim circuit by a voltage variation in the source circuit,

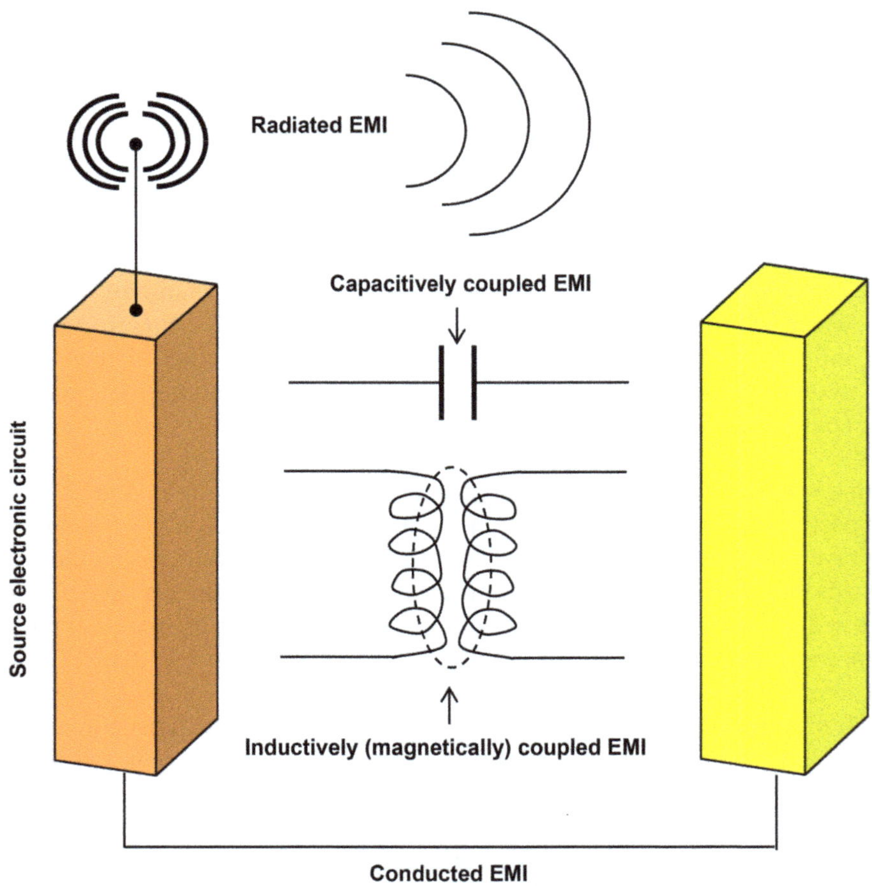

Figure 21.3. Different components contributing to the electromagnetic interference: radiated EMI, capacitively coupled EMI, inductively coupled EMI and conducted EMI, caused in a victim electronic circuit by a source electronic circuit.

(ii) magnetic coupling through induction of current in the victim circuit by a varying magnetic field in the source circuit.

21.3.2.3 Conducted EMI
It is produced by electrical connection between source circuit and victim circuit through wires. When interference signals in the two conductors, e.g., the positive and negative cables, are in phase, the EMI is said to be common-mode EMI; otherwise, it is differential-mode EMI.

$$\begin{aligned}\text{Net EMI} &= \text{Radiated EMI} + \text{Coupled EMI} + \text{Conducted EMI} \\ &= \text{Radiated EMI} + \text{Capacitively Coupled EMI} \\ &+ \text{Inductively Coupled EMI} + \text{Conducted EMI}\end{aligned} \tag{21.4}$$

21.3.3 Duration of EMI

Depending on the duration of interference signal, EMI is of two types: continuous or impulse EMI.

21.3.3.1 Continuous EMI
It is constantly experienced over a long-time interval.

21.3.3.2 Impulse EMI
It extends over a short period of time or intermittently.

21.3.4 Bandwidth of EMI

Looking at the spectral content in the noise signal, it has different bandwidths. It is subdivided into narrowband and broadband noises.

21.3.4.1 Narrowband EMI
It happens at a single frequency or over a small range of frequencies.

21.3.4.2 Broadband EMI
It occurs over a large range of frequencies.

21.4 Effects of EMI

21.4.1 EMI noise signal

In all cases, EMI whether coming from the natural environment, internally generated in a circuit, or emitted by an exterior circuit source, is observed as an undesirable electrical signal in the user's circuit. This undesired signal is known as the EMI noise signal because it tries to overshadow the intended electrical signal and reduce its effect. It is a form of electromagnetic noise pollution.

EMI is detrimental to the overall performance of the affected circuit. It sometimes leads to minor deviations from the regular working in line with the planned scheme but may also cause appalling malfunctioning. It is not only a cause of inconvenience and annoyance, but may act as a reason for disruption or catastrophic failure of equipment. It may play havoc in the victim circuit (Weston 2017).

21.4.2 Examples of disablement of equipment functions by EMI

A few examples are given below:

(i) Bit errors in digital data and distortion of analog signals.
(ii) Sun outage, the interruption of geostationary satellite signals in TV reception.
(iii) Humming and buzzing sound causing difficulty in audibility of talk during a telephonic conversation.
(iv) Momentary flicker in TV when a mixer is switched on in the kitchen or the floor is vacuum cleaned.
(v) A car radio picking up two broadcast stations simultaneously.
(vi) Hinderance of WI-FI connectivity upon turning on the microwave oven.
(vii) Range reduction and fall in detection probability in radar search.
(viii) Range and angular errors in navigation systems.

To avoid such mis-happenings, EMC regulations are imposed on electronic equipment by government agencies (Paul 2006). Compliance with these requirements is compulsory for marketing of the product. The compliance is tested by measuring the radiated and conducted emissions from the product. Additionally, requirements are voluntarily levied internally by manufacturers for sake of customer satisfaction. Specifications for military products are more critical than those for commercial items.

21.5 Single-ended and differential transmission of signals

21.5.1 Single-ended transmission of signals

A voltage V_{Signal} with respect to transmitter ground is applied to the line-driver, which is an amplifier at the transmitter (figure 21.4(a)). The voltage V_{Signal} with respect to receiver ground is perceived by the amplifier at the receiver and delivered to the circuits in the receiver system. The transmitter and receiver grounds are connected through a cable to ensure their equality.

The ideal EMI-free environment was assumed while drawing figure 21.4(a). The real-world scenario differs from this ideal situation because of the presence of EMI sources. Then the situation is displayed in figure 21.4(b) where the input signal at the transmitter was not noisy but is contaminated with EMI noise $N(t)$ on the path to receiver; here the noise is taken to be varying with time t. The arrangement does not have any means of removing the EMI in the signal which therefore propagates to the receiver. The noise in the received signal persists and spoils the performance of the receiver.

21.5.2 Differential transmission of signals

The inherent problem in single-ended transmission is solved by adopting a differential transmission scheme (figure 21.5). Differential-mode signal transmission is the

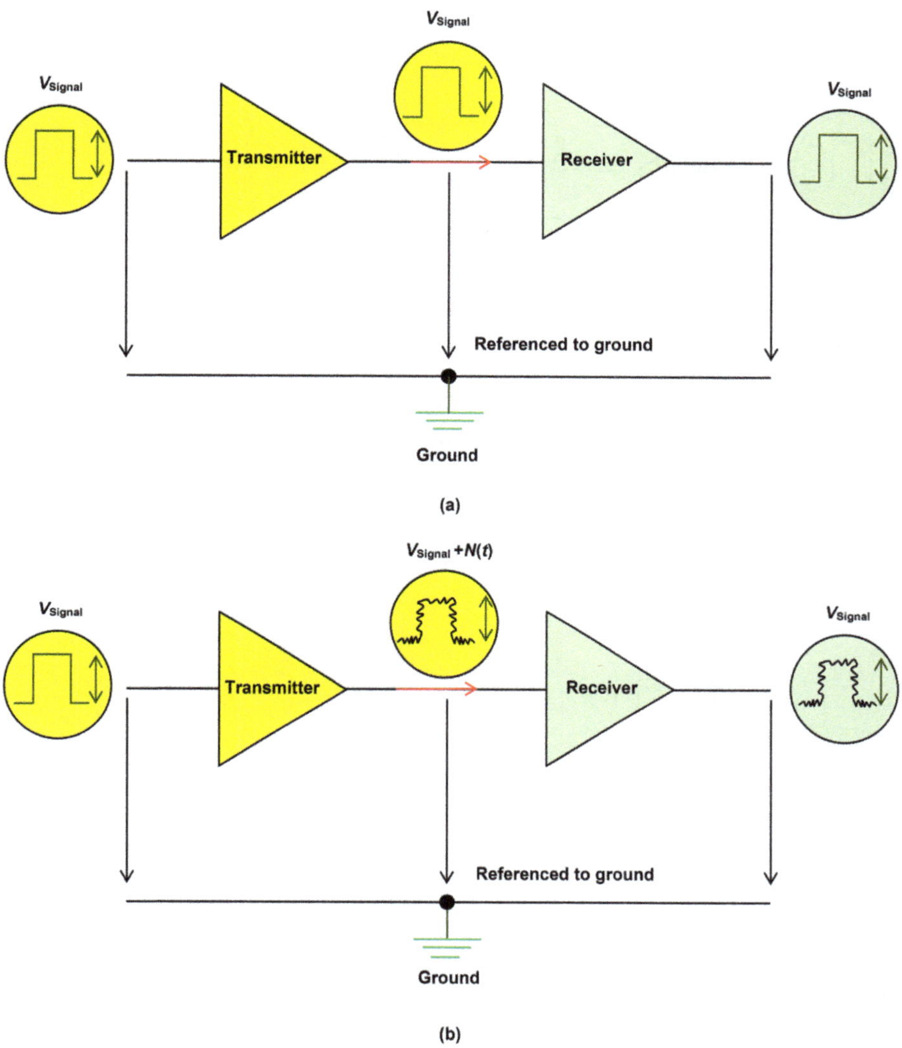

Figure 21.4. Layout of single-ended signal transmission: (a) pure noise-free signal and (b) noisy signal.

mainstream data transmission technique used in high-speed telecommunications. In this technique, two voltage signals of equal magnitude V_+, V_- carrying currents I_+, I_- but opposite in phase (ϕ_+, ϕ_-) are sent on two separate wires using the potential difference between the wires.

The line-driver of the transmitter has two outputs: a non-inverting output and an inverting output. The non-inverting output terminal is identified with a plus sign. The inverting output is identified with a minus sign and a small circle on the schematic symbol.

The signal V_+ suffixed with a positive sign is the same as the signal used in the single-ended input transmission. It is the non-inverting output of the line-driver at the transmitter. The signal V_- suffixed with a negative sign is the negative replica of

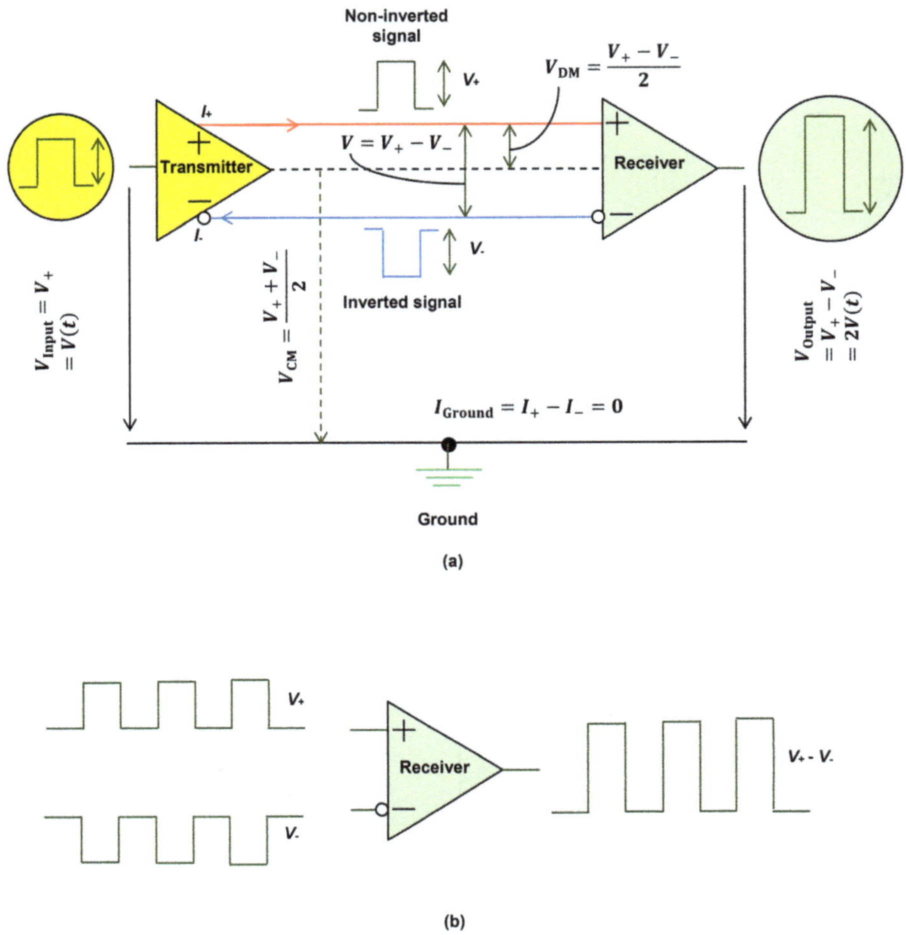

Figure 21.5. Differential signal transmission of a pure noise-free signal $V(t)$: (a) the circuit arrangement and (b) the waveforms at the input and output of the receiver amplifier.

the signal used in the single-ended input transmission. It is the inverting output of the line-driver at the transmitter.

The signal at the output of the transmitter is delivered through two separate wires to the receiver, which has two inputs, labeled as the non-inverting and inverting inputs corresponding to the two outputs of the transmitter. The signal V_+ is applied at the non-inverting input of the receiver while the signal V_- is given to its inverting input. The receiver performs the task of subtraction of the signals so that the signal at the output of the receiver is

$$V_{\text{Output}} = V_+ - V_- = V_+ - (-V_+) = V_+ + V_+ = 2V_+ \tag{21.5}$$

If the signal V_+ is a time-dependent signal it is written as

$$V_+ = V(t) \tag{21.6}$$

and

$$V_{-} = -V(t) \tag{21.7}$$

Equation (21.5) reduces to

$$V_{\text{Output}} = V_{+} - V_{-} = V(t) - \{-V(t)\} = V(t) + V(t) = 2V(t) \tag{21.8}$$

The above equation is valid for the ideal circumstance in which EMI is absent. How does the differential transmission arrangement respond to the introduction of EMI? Considering the EMI noise as a time-dependent variable and denoting it by $N(t)$, it is obvious that noise will affect both V_{+} and V_{-} signals equally (figure 21.6). Addition of the noise signal $N(t)$ to the signals V_{+} and V_{-} enables us to write equations (21.6) and (21.7) for EMI-affected situation as

$$V_{+} = V(t) + N(t) \tag{21.9}$$

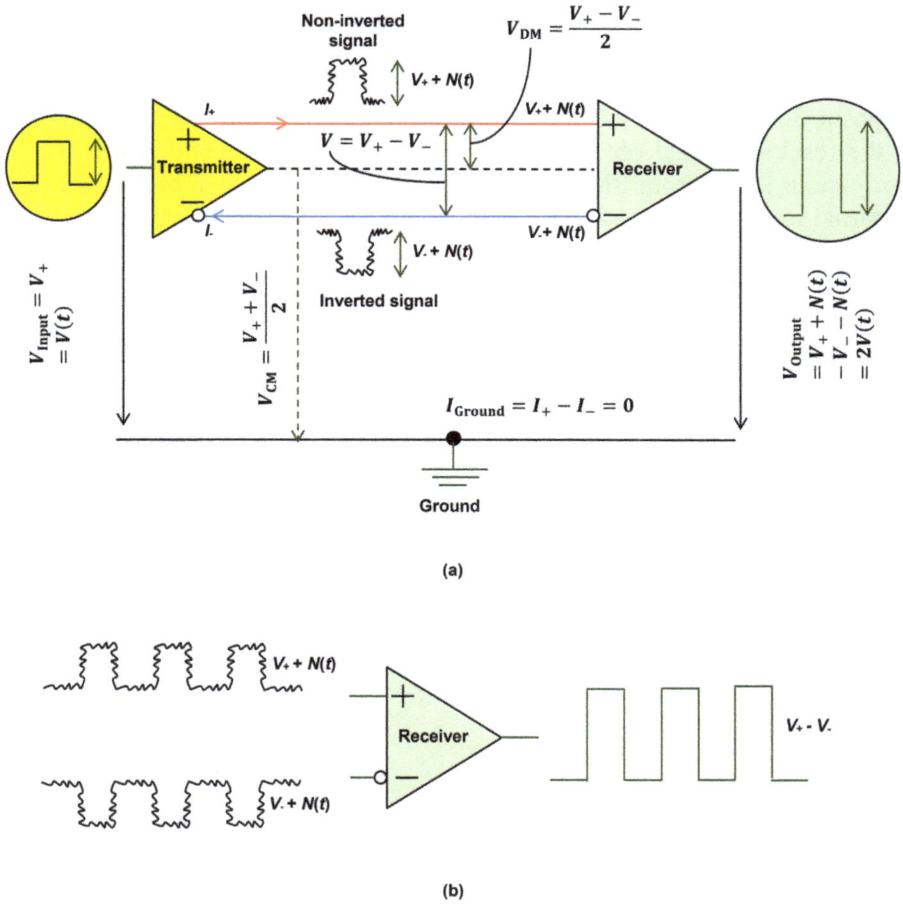

Figure 21.6. Differential transmission of a signal $V(t)$ contaminated with noise $N(t)$ on its route: (a) the circuit arrangement and (b) the input and output waveforms of the receiver amplifier showing that the input waveforms are affected by noise but the output waveform does not contain any noise.

$$V_{-} = -V(t) + N(t) \tag{21.10}$$

From equations (21.5), (21.9) and (21.10), the output signal for the differential signal transmission case is

$$\begin{aligned} V_{\text{Output}} &= V_{+} - V_{-} = V(t) + N(t) - \{-V(t) + N(t)\} \\ &= V(t) + N(t) + V(t) - N(t) = 2V(t) \end{aligned} \tag{21.11}$$

which is the same result as in equation (21.8) obtained for EMI-free environment.

21.5.3 Effects of EMI currents induced in the wires by magnetic fields generated around them during high-frequency differential current flow

When high-frequency differential currents I_+, I_- flow in the wires, time-varying magnetic fields are produced around them (figure 21.7). The changing magnetic field around one wire induces a current flow in the other wire and vice versa. These are the EMI currents in the wires. But we must note that at a particular instant, the high-frequency currents are flowing in opposite directions in the two wires. So, the EMI currents induced by magnetic fluxes of the high-frequency currents are also directed in opposite directions. These induced EMI currents cancel the effect of each other so that the resultant signal obtained as a difference in potentials between the transmitted signals is free from EMI. Further, any EMI due to signal harmonics or from an external source affects both the signals equally, and its effects will be dampened in the same way. Owing to this EMI immunity, differential mode is the preferred data transmission technique in wide domestic and industrial practice.

21.6 Differential- and common-mode voltages

Suppose two signals of different voltage magnitudes V_1, V_2 and dissimilar phases ϕ_1, ϕ_2 are flowing in the two wires of a two-wire cable. The signals are referenced with respect to a local common or ground terminal. Then the differential-mode

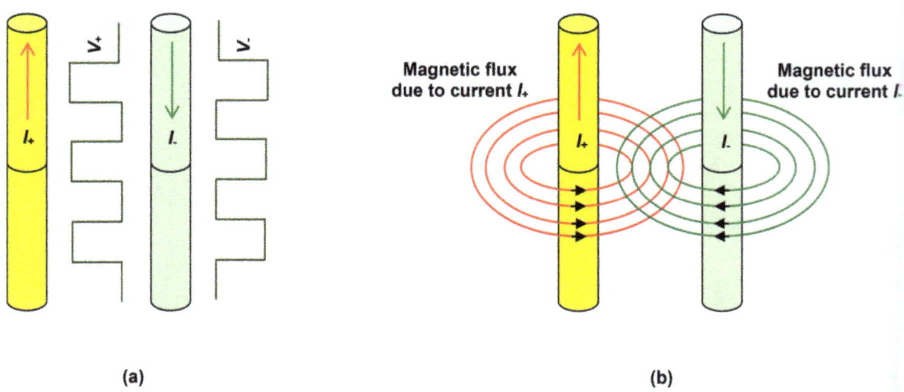

Figure 21.7. Cancellation of magnetic fluxes produced by oppositely-directed currents I_+, I_- flowing during differential transmission: (a) current directions and associated voltage waveforms V_+, V_- in the two current-carrying conductors, and (b) current directions and accompanying magnetic fluxes in the two conductors.

voltage V_{DM} for the two signals considered is defined as an average voltage equal to half the voltage difference between the two signals as

$$V_{\mathrm{DM}} = \frac{V_1 - V_2}{2} \tag{21.12}$$

The common-mode voltage V_{CM} for the signals under examination is defined as the average voltage equal to half the sum of voltages of the two signals as

$$V_{\mathrm{CM}} = \frac{V_1 + V_2}{2} \tag{21.13}$$

For the special case in which

$$V_1 = V_+ \text{ and } V_2 = V_- \tag{21.14}$$

equation (21.12) becomes

$$V_{\mathrm{DM}} = \frac{V_+ - V_-}{2} \tag{21.15}$$

and equation (21.13) reduces to

$$V_{\mathrm{CM}} = \frac{V_+ + V_-}{2} \tag{21.16}$$

V_{DM} and V_{CM} are marked in figures 21.5(a) and 21.6(a).

21.7 Differential-mode interference

21.7.1 Cause of differential-mode interference

Differential-mode interference arises when an interference source affects only one of the two wires in the 2-wire circuit. Such an incident happens in absence of close coupling between the wires. The interference thus produced circulates in a loop. Differential interference is also produced in power supply or load components. In switched mode power supplies, ground currents are usually responsible for interference which is generated through imbalances in the circuit. With the introduction of interference by the power supply, within the load or in the wires, the current fed to the load is not equal to the current in the return path.

21.7.2 Differential-mode noise voltage

An interference noise voltage is developed across the power supply lines (figure 21.8(a)).

21.7.3 Differential-mode noise current

The interference noise current flows in the signal wire and neutral wire in same direction along the same path as the power supply current, i.e., the interference noise current is in series with power supply current. As a result, the current entering the load and the current flowing in the return path are different.

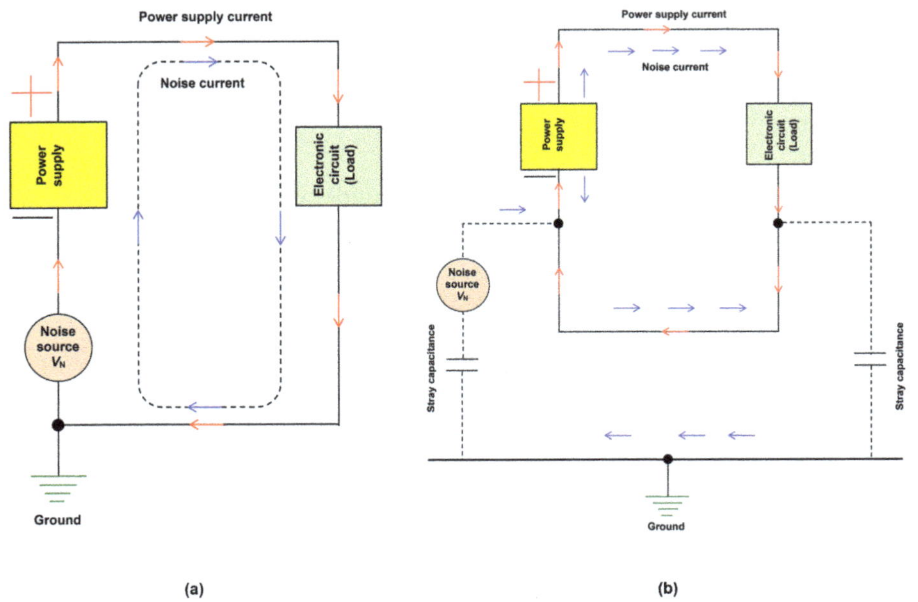

Figure 21.8. Two types of noise: (a) differential-mode noise in which the noise voltage V_N appears across the power supply lines, and the direction of flow of noise current is the same as that of power supply current, (b) common-mode noise in which the noise voltage V_N appears between a power supply terminal and ground, and the directions of flow of noise current on the positive and negative sides of power supply are same.

21.8 Common-mode interference

21.8.1 Cause of common-mode interference

Common-mode interference is induced when the same signal is picked up by two wires in a two-wire system. The interference signal flows in the signal wire and neutral wire in the same direction, i.e., the interference current is in series with the power supply current. The principal source of common-mode noise is the potential difference between two ground points situated at a distance. This potential difference initiates a parallel path for interference current flow facilitated by a parasitic capacitance back to the chassis ground. Ungrounded sources too are interference creators. An ungrounded analog signal wire acts as an antenna radiating the interference signal. This signal is equally coupled to both the wires in a two-wire cable.

21.8.2 Common-mode interference noise voltage

Unlike differential-mode interference noise voltage, common-mode interference noise voltage does not build up across the power supply lines. Common-mode interference noise voltage is a voltage between power supply line and ground (figure 21.8(b)).

21.8.3 Common-mode interference noise current

Unlike differential-mode interference noise currents, common-mode interference noise currents flow in the same direction on the positive and negative sides of the power supply.

21.9 Twisted pair cable for common-mode EMI noise rejection

21.9.1 The twisted wires

A popular cable designed for common-mode EMI noise rejection is the twisted pair (TP) cable (figure 21.9). The TP cable is a type of guided transmission medium consisting of two insulated copper wires twisted together to form a cable. One copper wire carries the current while the other acts as the return path and ground reference.

21.9.2 Magnetic fields and induced currents

Suppose that the current flowing in the wires is an alternating current. This current is associated with a changing magnetic field. Each wire is surrounded by a changing magnetic field encircling the wire. The wires residing in this changing magnetic field act like the secondary coils of a transformer and a current is induced in the wires by the magnetic field. Even if there is a DC current in the wires, and there is a small physical motion of wires, e.g., by vibrations, current induction is unavoidable.

21.9.3 Induced current cancellation

In the case of differential-mode transmission, the signal currents flowing in the two wires are in opposite directions. So, the magnetic fields around the wires are also oppositely directed and so also the currents induced by these magnetic fields. As the directions of currents in the two wires are opposite, they cancel out. This cancellation effect reduces the EMI.

21.9.4 Untwisted wires

What happens with untwisted wires? One wire is nearer to the EMI field source than the other (figure 21.10(a)). The two wires receive nearly equal induction but not exactly the same induction. The extent of current counter-balancing is less. So, the weakening of EMI effect is also diminished. Hence untwisted wires lose efficiency in EMI reduction. There is always some residual unbalanced EMI current.

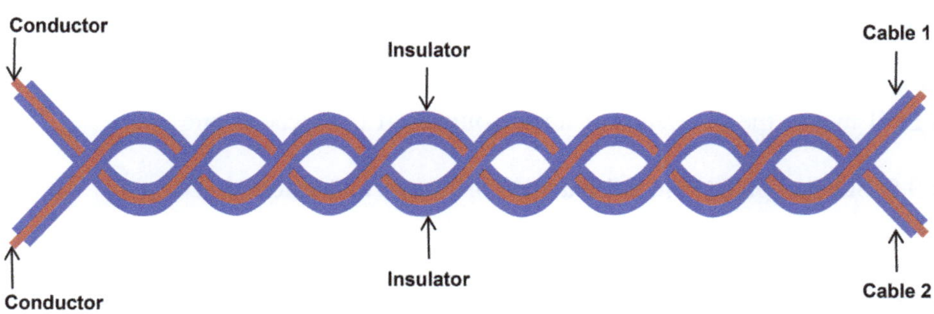

Figure 21.9. A twisted pair cable consisting of cable 1 in which there is a conducting wire covered by insulating coating and another similar cable 2 made of insulated conducting wire, and the two cables 1, and 2 are twisted together to form the pair.

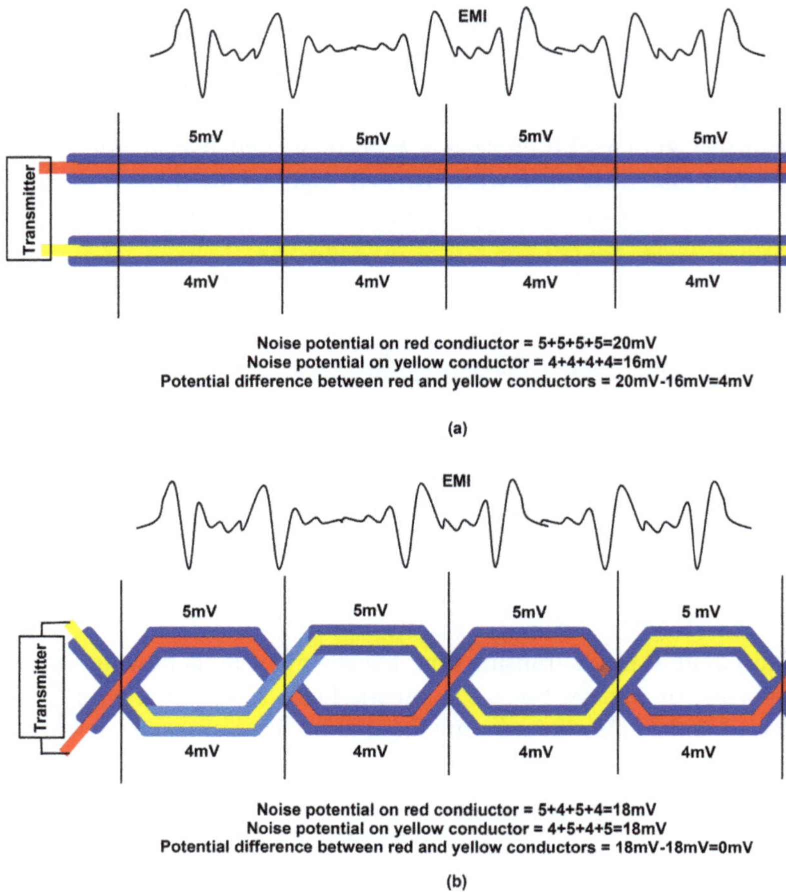

Figure 21.10. Examination of the effect of distance of EMI source on the noise effects in: (a) a cable made of parallel wires and (b) a twisted wire cable. In the parallel wire pair, the EMI induces more noise voltage per unit length in the red wire lying in vicinity of the EMI source while the noise induced pick-up voltage per unit length is considerably less for the blue wire lying far apart from the EMI source. Total difference of noise voltages between the two wires at the receiver end = 4 mV.

In the twisted wire pair, the twists of red and blue wires alternately change positions coming closer to the EMI source and moving away from it. So, both the red and blue wires experience the same exposure to EMI. This equality of exposure to EMI makes the difference of noise voltages at the receiver end = 0.

21.9.5 Subdual of EMI in twisted wires from self and external EMI

Not only is the mutual interference between signals in the two wires avoided, the twisting of wires also helps in mitigating external EMI because each wire is equally susceptible to unwanted external EMI signal as its twisted partner. Over each twist cycle, the wires alternate in proximity to the EMI field. Hence, by twisting the wires together, we have ensured that annulment effect is enhanced to reach maximum (figure 21.10(b)). So, a common-mode EMI signal will couple to each partner equally and the equal EMI signals will get cancelled. Thus, the twisted pair automatically discards common-mode EMI noise.

21.9.6 Explanation of distance effect on noise creation in untwisted and twisted wires with assumed noise potentials per unit length

The effect of EMI on the wire located close to it is greater than that on the distant wire. For comparing figure 21.10(a) and figure 21.10(b), the EMI noise voltage is arbitrarily assumed to be distributed at 5 mV per unit length for the nearby wire and 4 mV per unit length for the distant wire. In figure 21.10(a), the potential at the destination side = $5 \times 4 = 20$ mV for the nearby red wire and $4 \times 4 = 16$ mV for the distant yellow wire, Hence the difference of potential between the two wires at the destination end = $20 - 16 = 4$ mV. In figure 21.10(b), the potential for the red wire at the destination side = $5 + 4 + 5 + 4 = 18$ V while the same for the yellow wire = $4 + 5 + 4 + 5 = 18$ V. Hence the difference of potential between the red and yellow wires at the destination side = 18 V − 18 V = 0 V.

It is seen that the total noise voltage is 4 mV in the untwisted parallel wire cable but it is zero for the twisted pair cable. The reason for the larger effect of noise in figure 21.10(a) than figure 21.10(b) is that the red wire is always the nearby wire and the yellow wire is always the distant wire in this case but in figure 21.10(b), the red and yellow wires alternately become nearby and distant wires. This alternation of roles accounts for the difference in the effect of EMI on the two cables.

21.9.7 EMI not stopped, only weakened

EMI production is not stopped, only its effect is attenuated by counter-balancing of opposing magnetic fields, irrespective of whether the EMI is caused by currents flowing in the two wires themselves or from current flow in a nearby wire (figure 21.11). EMI currents are not destroyed. They are flowing as before. The twisting is meant to guarantee that equal EMI currents are induced in both the wires to achieve ideal balancing of currents. In other words, by twisting the wires, the induced noise current flows in one direction in some segments of the wires while it flows in opposite direction in an equal number of other segments. Hence the noise current becomes zero or much lower than that observed in the untwisted straight wires.

21.9.8 Applications of twisted wire cables

Twisted pair cables are used in telephone lines, local area networks, and for analog and digital signal communication.

21.10 Common-mode interference from common impedance coupling

Common-mode interference is produced when two circuits share the same ground impedance (Learn EMC 2022). The EMI generation mechanism is called common impedance coupling. Figure 21.12 shows two circuits A and B sharing a common return path. Circuit A having a voltage source V_{SA} with resistance R_{SA} supplies a current I_A. This current flows through a load resistance R_{LA} and returns through a

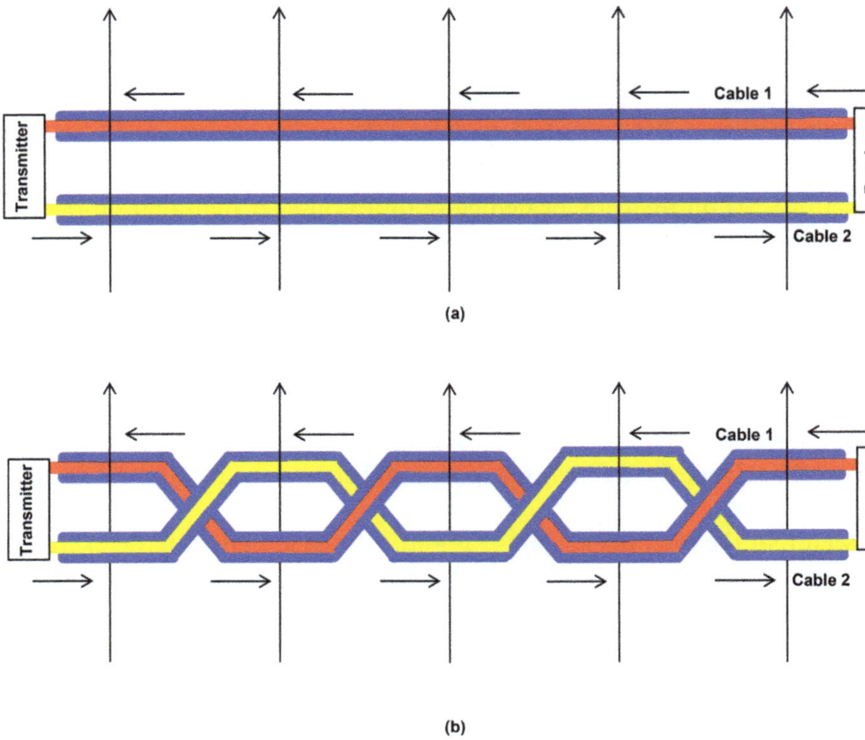

Figure 21.11. Analysis of the effect of an EMI source in the form of a changing magnetic field on: (a) a cable made of untwisted parallel wires and (b) a twisted wire cable. In part (a), the electric current induced by the magnetic field on the red wire is always in one direction throughout the wire while that on the yellow wire is always in the opposite direction all through the wire leading to a net noise current. In part (b), the electric current induced in a given wire e.g., the red or yellow wire flows in one direction in one segment of the wire but in reverse direction in its next closely-spaced segment. This reversal of current between any two adjacent segments makes the currents cancel each other so that the total noise current becomes zero. Similar cancellation effect is not possible in distant parallel wires of part (a).

path with resistance R_{Return} to the voltage source V_{SA}. In the circuit B, the voltage source is V_{SB} with resistance R_{SB}. It sends a current I_{B} through a load resistance R_{LB}, which returns through the same path of resistance R_{Return} as followed by current I_{A} to the voltage source V_{SB}. The notable feature of the two circuits is that the return path and the return path resistance R_{Return} are same for the circuits A and B. It is this feature of return path sharing which is the cause of all problems because it makes the two circuits mutually interdependent. It does not allow any one of them, neither A nor B, to work serenely unruffled by the other. Whenever a current flows in the first circuit A, a voltage appears across the load resistance R_{LB} in the second circuit B. Conversely, a current flow in the second circuit B leads to the appearance of a voltage across the load resistance R_{LA} in the first circuit A. Let us see how the two circuits A and B affect each other.

To express the degree of electromagnetic coupling between two circuits, a crosstalk parameter $Xtalk_{\text{BA}}$ is defined as the ratio of the coupled EMI voltage in

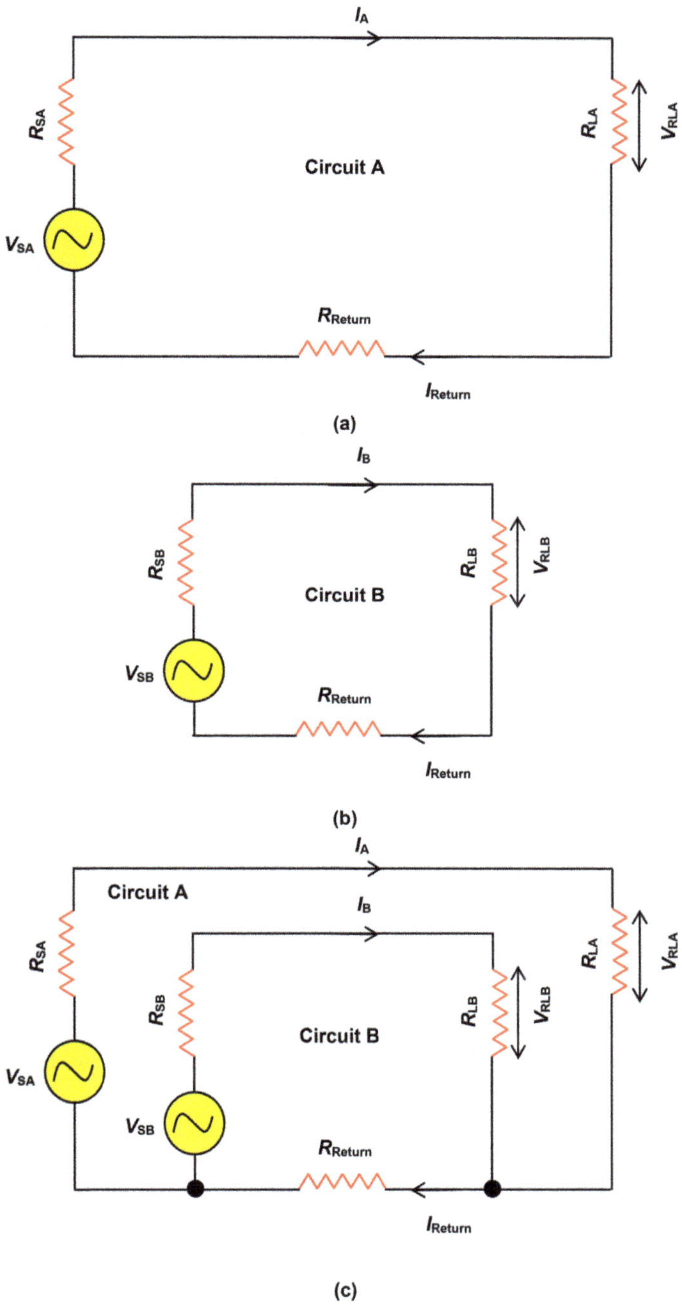

Figure 21.12. Analysis of common-mode interference originating from common-impedance coupling between two circuits A and B: (a) circuit A alone, (b) circuit B alone and (c) circuits A and B joined together when they have the same return path for signals.

the receptor circuit B for a given signal voltage in source circuit A. In dB scale, the crosstalk parameter is written as

$$XTalk_{BA} = 20\log_{10}\left|\frac{V_{RLB}}{V_{RLA}}\right|_{VSB=0} \tag{21.17}$$

where V_{RLB} is the voltage across the load resistance R_{LB} of the receptor circuit B, V_{RLA} is the voltage across the load resistance R_{LA} of the source circuit A. The crosstalk is obtained under the condition that the source voltage V_{SB} in circuit B is zero.

Applying Kirchoff's voltage law to the current loop of circuit B we get

$$V_{SB} + I_B R_{SB} + I_B R_{LB} + (I_A + I_B)R_{Return} = 0 \tag{21.18}$$

In equation (21.18) we put $V_{SB} = 0$ following the definition of $Xtalk_{BA}$ given in equation (21.17) to obtain

$$I_B R_{SB} + I_B R_{LB} + (I_A + I_B)R_{Return} = 0 \tag{21.19}$$

or,

$$I_B(R_{SB} + R_{LB} + R_{Return}) = -I_A R_{Return} \tag{21.20}$$

or,

$$I_B = -\frac{R_{Return}}{R_{SB} + R_{LB} + R_{Return}} I_A \tag{21.21}$$

But

$$I_A = \frac{V_{RLA}}{R_{LA}} \tag{21.22}$$

and

$$I_B = \frac{V_{RLB}}{R_{LB}} \tag{21.23}$$

Putting the values of I_A and I_B from equations (21.22) and (21.23) in equation (21.21), we have

$$\frac{V_{RLB}}{R_{LB}} = -\frac{R_{Return}}{R_{SB} + R_{LB} + R_{Return}}\left(\frac{V_{RLA}}{R_{LA}}\right) \tag{21.24}$$

or,

$$\frac{V_{RLB}}{V_{RLA}} = -\frac{R_{Return}}{R_{SB} + R_{LB} + R_{Return}}\left(\frac{R_{LB}}{R_{LA}}\right) \tag{21.25}$$

Since this equation is derived under the condition $V_{SB} = 0$, we can write

$$\left|\frac{V_{RLB}}{V_{RLA}}\right|_{V_{SB}=0} = \left|-\frac{R_{Return}}{R_{SB} + R_{LB} + R_{Return}}\left(\frac{R_{LB}}{R_{LA}}\right)\right| = \frac{R_{Return}}{R_{SB} + R_{LB} + R_{Return}}\left(\frac{R_{LB}}{R_{LA}}\right) \tag{21.26}$$

From equation (21.26) and the definition of $Xtalk_{\mathrm{BA}}$ in equation (21.17) it is easy to see that

$$\begin{aligned} XTalk_{\mathrm{BA}} &= 20\log_{10}\left|\frac{V_{\mathrm{RLB}}}{V_{\mathrm{RLA}}}\right|_{V\mathrm{SB}=0} \\ &= 20\log_{10}\left\{\frac{R_{Return}}{R_{\mathrm{SB}}+R_{\mathrm{LB}}+R_{\mathrm{Return}}}\left(\frac{R_{\mathrm{LB}}}{R_{\mathrm{LA}}}\right)\right\} \end{aligned} \quad (21.27)$$

Putting typical values, $R_{\mathrm{Return}} = 50$ milliohms, $R_{\mathrm{SA}} = R_{\mathrm{LA}} = R_{\mathrm{SA}} = R_{\mathrm{SB}} = 10\ \Omega$ in equation (21.27) we get

$$\begin{aligned} XTalk_{\mathrm{BA}} &= 20\log_{10}\left\{\frac{50\times10^{-3}}{10+10+50\times10^{-3}}\left(\frac{10}{10}\right)\right\} \\ &= 20\log_{10}\left\{\frac{50\times10^{-3}}{20.05}\right\} \\ &= 20\log_{10}(0.002494) = 20\times-2.6 = -52\ \mathrm{dB} \end{aligned} \quad (21.28)$$

$$\therefore \frac{\text{EMI voltage}}{\text{Signal voltage}} = 10^{-\frac{52}{20}} = 10^{-2.6} = 0.0025 \quad (21.29)$$

by applying the definition of crosstalk parameter.

For a signal voltage = 5 V,

$$\text{EMI voltage} = 0.0025\times5\ \mathrm{V} = 0.0125\ \mathrm{V} = 12.5\times10^{-3}\ \mathrm{V} = 12.5\ \mathrm{mV} \quad (21.30)$$

21.11 Combined EMI noise

EMI is an indicator of the amount of noise produced when a circuit is operated, and

$$\begin{aligned} \text{Combined EMI noise} = \ & \text{Differential} - \text{Mode EMI Noise} \\ & + \text{Common} \\ & - \text{Mode EMI Noise} + \text{Radiation} \\ & - \text{induced EMI noise} \end{aligned} \quad (21.31)$$

A circuit showing good compliance with EMI/EMC generates less EMI noise. Reduction or elimination of EMI is done by three techniques: filtering, grounding and shielding. For judiciously selecting the suitable noise suppression technique, it is necessary to know the source of noise and the type of noise. Then only noise combating can be planned properly to reduce it to a tolerable minimum level.

21.12 Filters for EMI noise suppression

21.12.1 Differential-mode EMI noise filter

Depending on the noise frequency f, a low-pass, high-pass or band-pass filter is used. The filter circuit contains passive components, capacitors and inductors (figure 21.13). Capacitors block DC and low-frequency currents while offering a low impedance path to high-frequency noise signals because the capacitive reactance at a frequency f is

$$X_C = 1/2\pi fC \tag{21.32}$$

for a capacitor of capacitance C. These high-frequency noise signals flow back to the power supply ground (figure 21.13(a)). The inductors allow DC

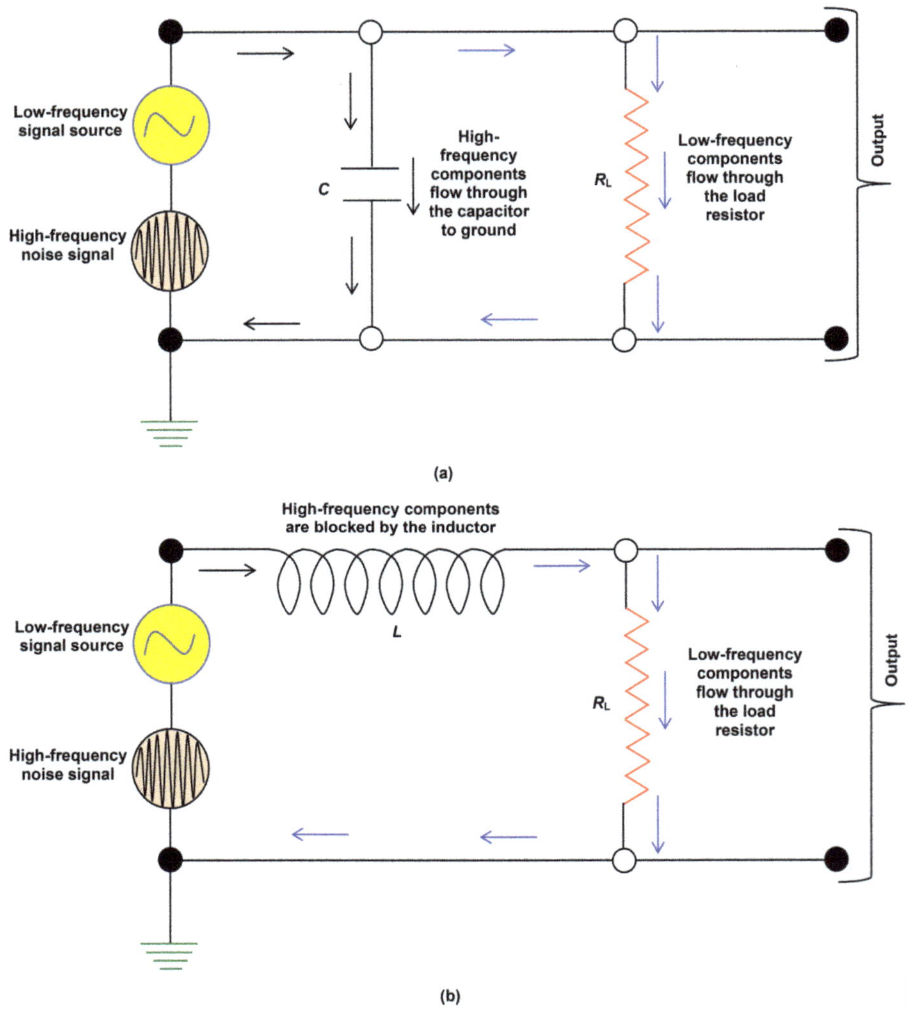

Figure 21.13. Capacitor and inductor filter circuits: (a) capacitor filter and (b) inductor or choke filter. In part (a), the capacitor offers a small resistance to high-frequency components of the signal and high resistance to low-frequency components; the former components only reach the load resistor while the latter components are blocked by the capacitor. In part (b), the inductor offers a large resistance to high-frequency components and a small resistance to low-frequency components; the latter components only flow to reach the load resistor while the former components are stopped by the inductor.

current flow. Also, the inductive reactance for an inductor of inductance L being

$$X_L = 2\pi fL \tag{21.33}$$

is small at low frequencies. Hence, low-frequency current encounters low impedance while high-frequency current experiences greater opposition (figure 21.13(b)).

Generally, for the 50/60 Hz mains signal, the EMI filter is a low-pass filter (figure 21.14). In a differential-mode C–L filter (figure 21.14(a)), the capacitor is placed between the signal source V_+ and V_- lines while the inductor is placed in series with signal V_+ or V_- line. The values of the inductors or capacitors used in the filter are determined by the impedance of the source, load and the cut-off frequency.

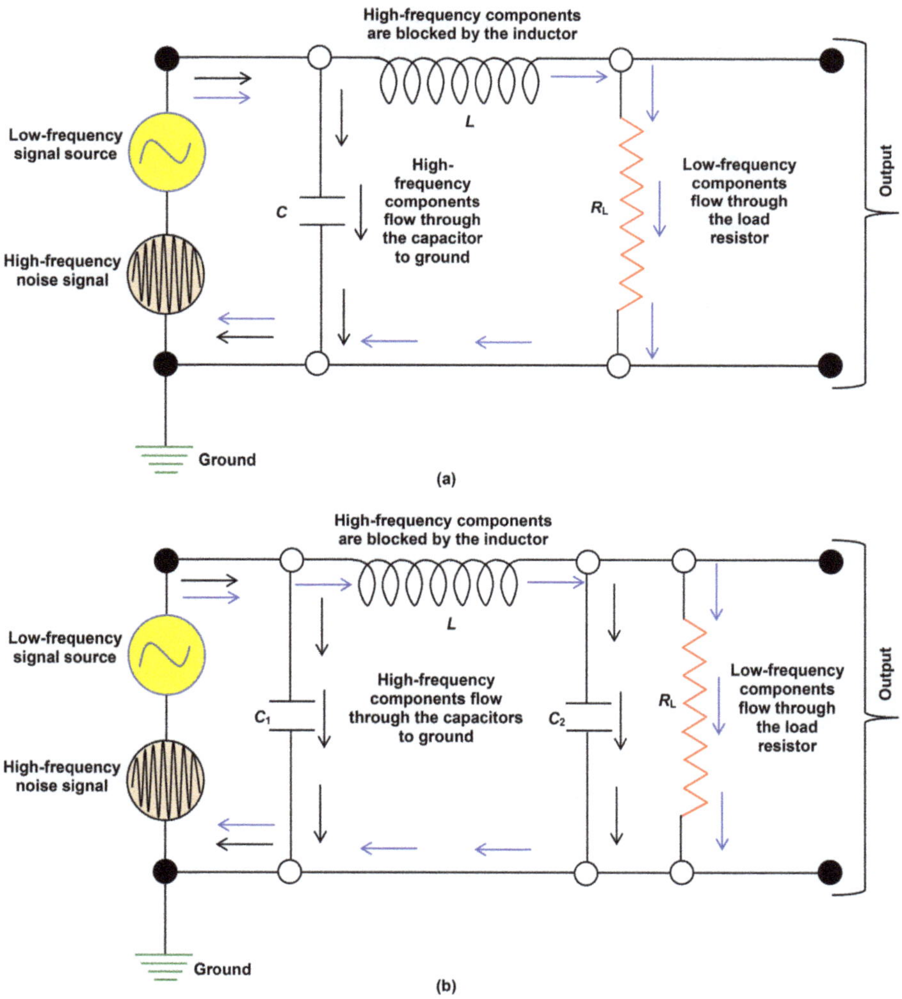

Figure 21.14. Configurations of differential-mode low-pass EMI filters: (a) a C–L filter consisting of a shunt capacitance C and series inductance L, and (b) a C–L–C or π-filter consisting of two shunt capacitances C_1 and C_2 and one series inductance L.

The cut-off frequency is the maximum frequency that is allowed to pass through the filter.

A π-filter is a passive low-pass filter consisting of two capacitors C_1, C_2 connected to ground with a series inductor L between them (figure 21.14(b)). The components are arranged in the shape of the Greek letter π, hence the name. A large fraction of the EMI noise currents is bypassed by the capacitor C_1 in the input stage. The remaining EMI noise currents are removed on passing through the inductor coil L and the capacitor C_2 connected in parallel across the load.

21.12.2 Common-mode EMI noise filter

A common-mode EMI noise filter consists of two magnetically coupled coils (figure 21.15(a)). One coil is connected to V_+ terminal 1 of the differential-mode transmission system while the other coil is connected to its V_- terminal 2 (Murata

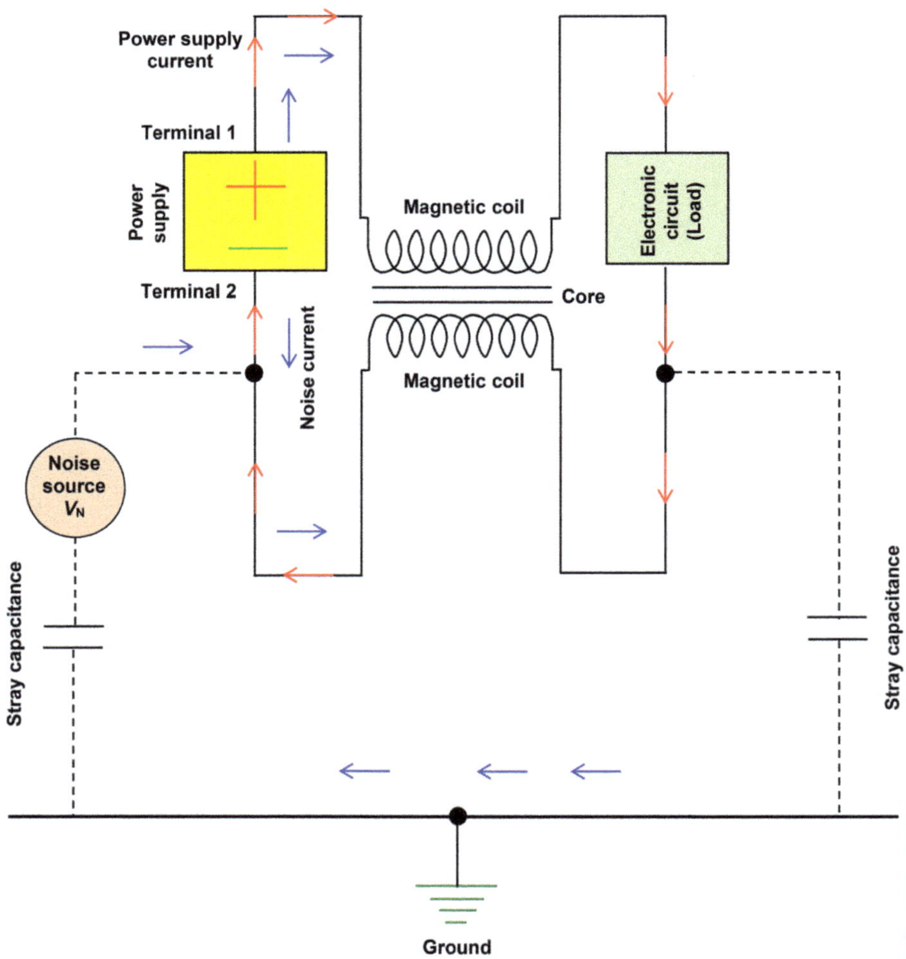

Figure 21.15. A common-mode filter for EMI noise.

Manufacturing Co. Ltd 1998). Its capability to remove common-mode EMI noise signal and allow differential-mode EMI noise signal to pass through it depends on the difference of its interaction with these two types of signals.

When a common-mode signal enters the filter (figures 21.16(a) and (b)), the magnetic fluxes associated with the currents flowing in the two wires are in phase. The in-phase fluxes add up to create an intense flux setting up a large current in opposite direction to the common-mode signal. This is equivalent to an increase in impedance to common-mode signal. The impedance of the filter made of inductive reactance is called common-mode impedance. The increased common-mode impedance blocks the common-mode signal.

But when a differential-mode signal enters the filter (figures 21.16(c) and (d)), there is a phase difference of 180° between the currents in the two wires. The magnetic fluxes produced by these anti-phase currents act in opposite directions, and the resultant flux is zero. Consequently, there is no flux to set up a current in opposite direction to the differential-mode signal. This is tantamount to a decrease in impedance to the differential-mode signal. Hence, the filter offers a low impedance path allowing a differential-mode signal to flow easily through it.

Thus, for a common-mode signal, the filter circuit becomes a high-impedance inductor while for a differential-mode signal, it acts as a low-impedance transmission line. The value of impedance at 100 MHz is taken as an indicator of the ability of the filter to reject common-mode noise.

21.13 Grounding

Ground is the zero-volt reference for a circuit. Ideally, ground wire is not a path of return of current because when appreciable current flows through this wire, a voltage drop is incurred across it, and it can no longer act as a zero-volt reference potential. But often the ground wire is used as a current return path. Then the current return wires are mistakenly labeled as ground wires. However, when using the ground wire as a current return path, its operation must be carefully analyzed to ascertain the extent of deviation from the standard grounding condition of acting as a zero-volt reference, and whether this non-conformity with the grounding condition impacts the satisfactory operation of the circuit to unacceptable levels.

21.13.1 Ground loops, and a simplified ground loop circuit

A ground loop is an unintentionally formed undesirable current path caused by sharing of a common ground point by two or more circuits. Continuing with the discussion on common-impedance coupling in section 21.10, let us look at the ground loop circuit in figure 21.17 where two circuits A and B with voltage sources V_A and V_B share a common path to ground. This path has a non-zero resistance R_G. Due to the fact that $R_G \neq 0$, the resistors R_G and R_A constitute a voltage divider. If I_A is the current flowing from circuit A through R_G, a voltage drop

$$V_G = I_A R_G = \frac{V_A}{R_A + R_G} R_G \tag{21.34}$$

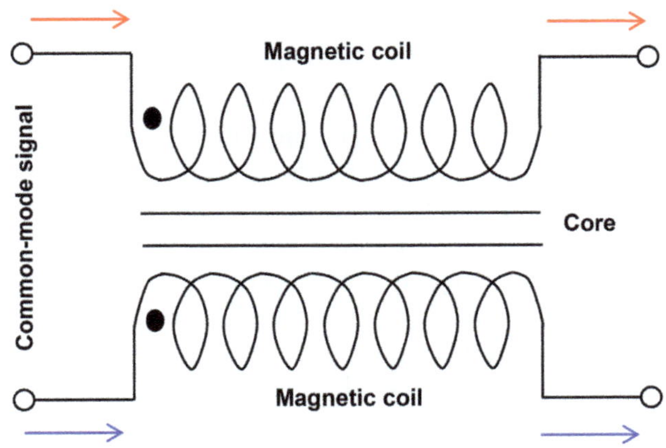

(a)

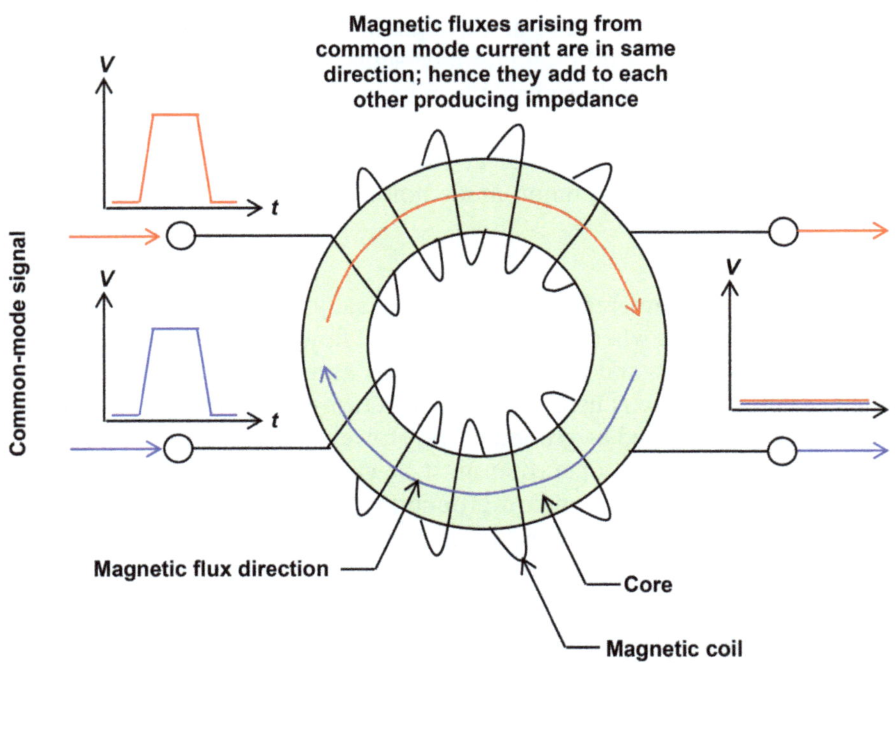

(b)

Figure 21.16. Changing roles of a common-mode filter: (a) and (b) behaving as a large impedance inductor offering high opposition to a common-mode signal, and (c) and (d) behaving as a conducting wire facilitating the transmission of a differential-mode signal.

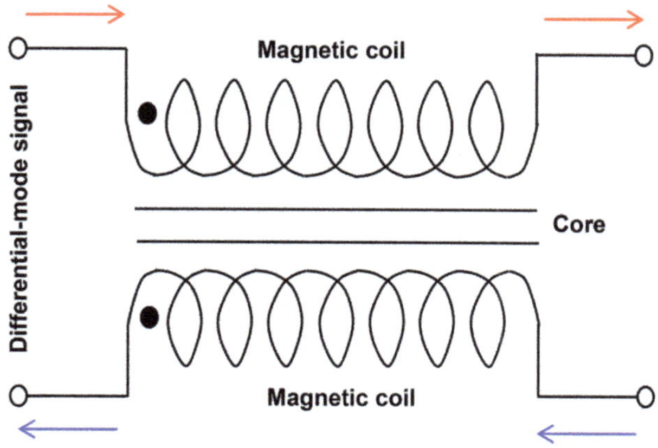

(c)

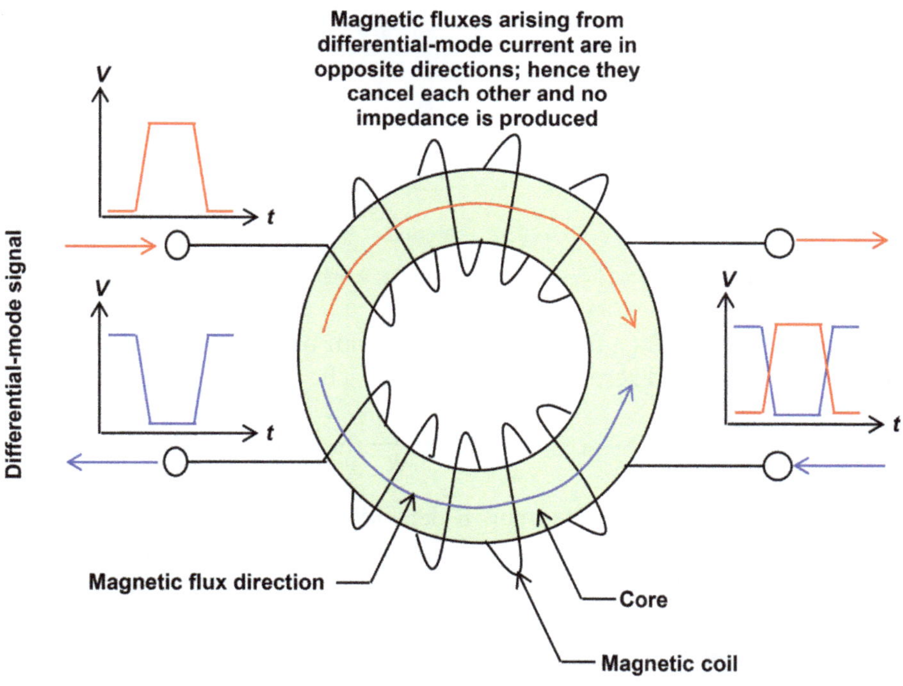

(d)

Figure 21.16. (Continued.)

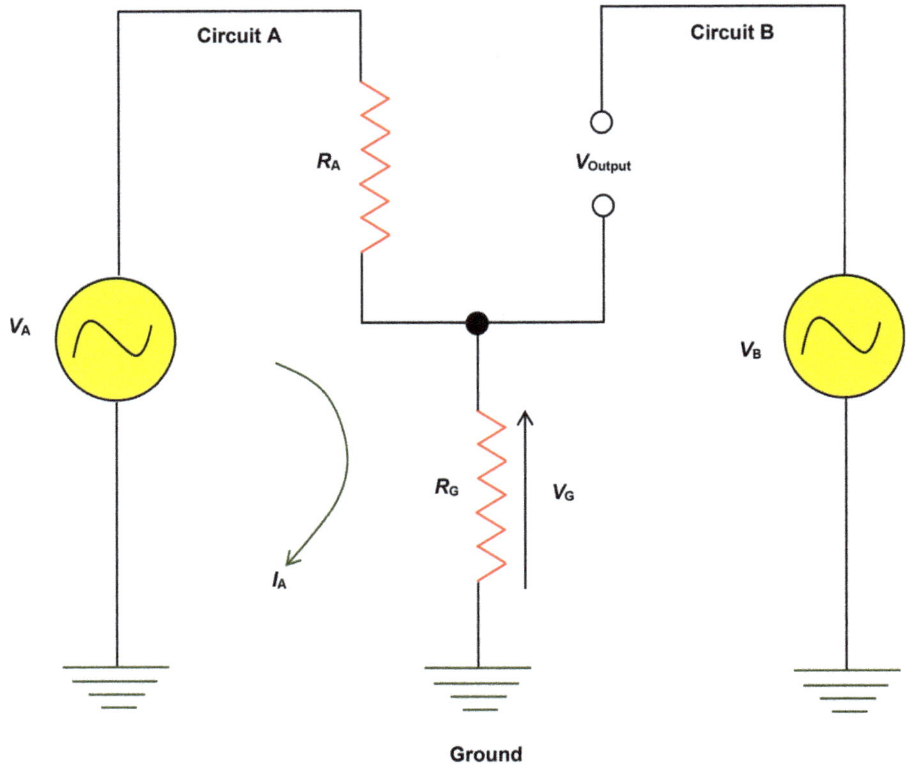

Figure 21.17. Formation of a ground loop between two circuits A and B which have the same path to ground.

is produced across R_G. As a result, the ground point of the two circuits A and B no longer acts as a standard zero-volt ground point. The output voltage of circuit B is

$$V_{\text{Output}} = V_B - V_G = V_B - \frac{V_A}{R_A + R_G} R_G \tag{21.35}$$

which means that the output of circuit B depends on circuit A. If circuit A carries 50/60 Hz AC and circuit B is an audio system with speakers, a disturbing 50/60 Hz hum is heard in the speakers.

21.13.2 Induction of interference currents by stray magnetic fields

Magnetic fields arising from AC flow in a nearby cable cause induction of current in a loop of conductive material. According to Faraday's law of electromagnetic induction, any time-varying magnetic field passing through a loop of wire induces an electromotive force in the loop resulting in current flow in the loop. The larger the area of the loop, the higher the magnetic field intensity and higher the frequency of current, the more is the induced EMF. Since the wire has a finite resistance, a voltage drop is produced when a current flows through it. So, the ground loop consisting of a wire connecting two circuits A and B adds a voltage = voltage difference between the two ground points to the proper signal in accordance with Ohm's law, which is indistinguishable from the proper signal. An EMI voltage source is superimposed on the proper voltage source.

21.14 Grounding approaches

There are three principal grounding approaches: single-point, multi-point and hybrid grounding (Mari 2021). When using any of these approaches, the impedances introduced by interconnections of grounding points must be carefully considered because EMI effects can arise from common impedance coupling. The limitations imposed by the particular approach used cause susceptibility to EMI.

21.14.1 Single-point grounding

This type of grounding is preferred in low-frequency analog circuits at frequencies < 1 MHz. It has two forms: series and parallel.

21.14.1.1 Single-point grounding: series connection

Also called common ground or daisy chain configuration, it consists of a series connection of ground points a, b, c, ... of different circuits A, B, C, ... (figure 21.18). Here, return current paths from several isolated circuits A, B, C, ... are connected through wires W_A, W_B, W_C, ...W_N of impedances Z_A, Z_B, Z_C, ...Z_N to different points a, b, c,....on a wire W, and this wire W is connected to the ground point G.

$$V_a = (I_A + I_B + I_C)Z_A = I_AZ_A + (I_B + I_C)Z_A \tag{21.36}$$

V_a, the ground potential of circuit A depends on return currents from circuits A, B and C, and impedance Z_A. This indicates that circuits A, B and C are interacting with each other. Clearly, the series connection is susceptible to common-mode noise.

$$\begin{aligned} V_b &= (I_A + I_B + I_C)Z_A + (I_B + I_C)Z_B \\ &= I_B(Z_A + Z_B) + I_C(Z_A + Z_B) + I_AZ_A \end{aligned} \tag{21.37}$$

V_b, the ground potential of circuit B depends on return currents from circuits A, B and C, and impedances Z_A and Z_B.

$$\begin{aligned} V_c &= (I_A + I_B + I_C)Z_A + (I_B + I_C)Z_B + I_CZ_C \\ &= I_C(Z_A + Z_B + Z_C) + I_B(Z_A + Z_B) + I_AZ_A \end{aligned} \tag{21.38}$$

V_c, the ground potential of circuit C depends on return currents from circuits A, B and C, and impedances Z_A, Z_B and Z_C. Also,

$$V_c > V_b > V_a \tag{21.39}$$

or,

$$\text{Ground potential at point C} > \text{Ground potential at point B} > \text{Ground potential at point A} \tag{21.40}$$

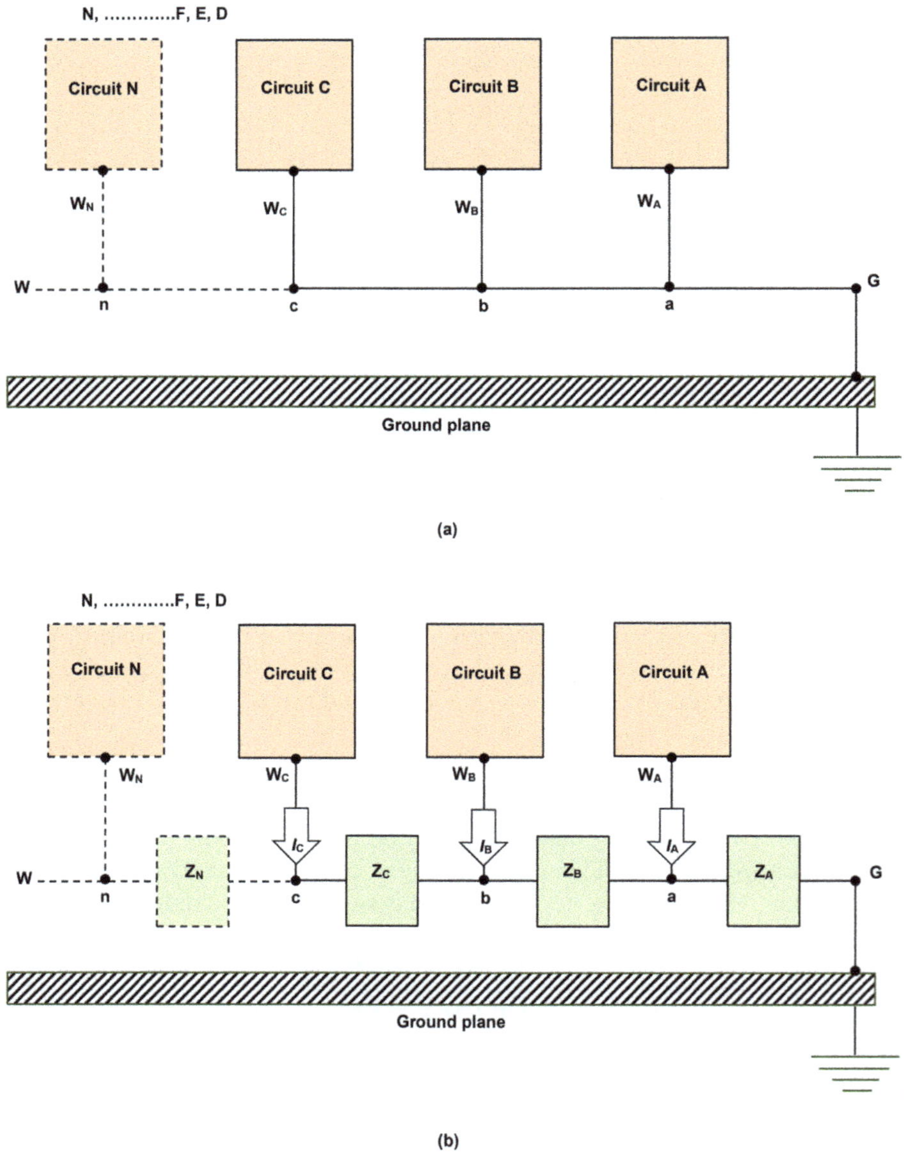

Figure 21.18. Single-point grounding with series connection of grounds of individual circuits: (a) the original circuit, and the (b) the circuit drawn again showing the impedances of the connecting wires and the currents flowing through them. The circuits A, B and C are connected through wires W_A, W_B and W_C (of impedances Z_A, Z_B and Z_C passing currents I_A, I_B and I_C) to a wire W at points a, b, c with the wire W connected to the point G. Point G is connected to the ground plane with a wire of negligible impedance.

Thus, ground potential at the point 'a' varies not only with the return current I_A, but also return currents I_B and I_C related to circuits B and C, and will be influenced by changes in circuits B and C. Similarly, ground potentials at points 'b' and 'c' are affected not only by variations in circuits B and C but also by changes in circuit A. None of the ground potentials V_a, V_b or V_c is independent.

The condition for grounding,

$$V_c = V_b = V_a = 0 \tag{21.41}$$

is approximately true at low currents only for small impedances Z_A, Z_B and Z_C. As current increases, voltage drops in different parts of the circuit affect the potentials V_a, V_b and V_c causing them to move away from zero. Ground potential at point 'a' is closest to zero and it is distanced from zero on moving towards points b, c, … Hence the circuit most critically needing accurate grounding should be placed at the location of circuit A.

In high-frequency circuits, fast switching actions produce large current pulses. So, such circuits should not use single-point series grounding, particularly circuits involving both power and control signals because high-energy impulses of the power circuit will adversely impact the working of the low-energy control circuit. Notwithstanding these disadvantages, this grounding circuit is widely used as it offers easy low-cost implementation.

21.14.1.2 Single-point grounding: parallel connection

For eliminating the common impedances occurring in the equations for potentials V_b and V_c, …. at ground points b and c, …., return current paths from isolated circuits A, B, C, …. are connected through wires W_A, W_B, W_C, …. to the same point, the ground point G (figure 21.19).

The equations for this circuit are:

$$V_a = I_A Z_A \tag{21.42}$$

$$V_b = I_B Z_B \tag{21.43}$$

and

$$V_c = I_C Z_C \tag{21.44}$$

Ground potential V_a at point 'a' depends only on return current I_A and impedance Z_A. It is not affected by I_B or Z_B. Similar remarks apply to ground potentials V_b and V_c at points 'b' and 'c'. The ground potentials V_a, V_b and V_c are independent of each other. Hence, this connection is not prone to common-mode noise, which is distinctly an advantage of this connection over the series circuit. However, the grounding condition of equation (21.41) is only met when the currents I_A, I_B and I_C as well as impedances Z_A, Z_B and Z_C have small values.

The single-point parallel grounding connection is most appropriate for low frequencies because the currents from circuits A, B, C, …. are not interlinked. Due to the inductance L of the wire connecting any circuit to ground, it is associated with an impedance ωL which increases with angular frequency ω. When the wire length equals odd multiples of 1/4 wavelength of the signal, resonance takes place and the ground impedance increases to a high value. Hence the wire connecting any circuit to ground should be taken several times shorter than the wavelength λ corresponding to maximum frequency of operation. Furthermore, this type of connection is expensive and bulky requiring extensive wiring in complex equipment.

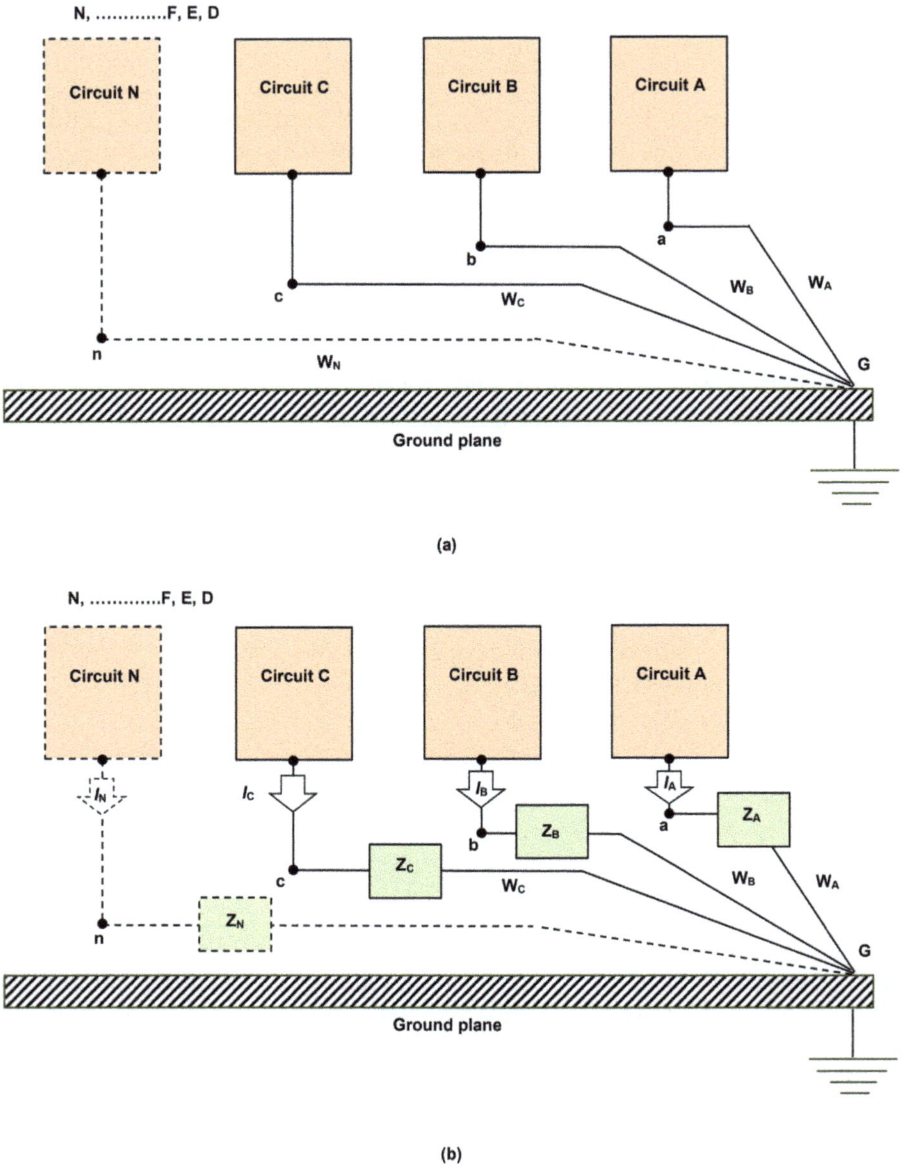

Figure 21.19. Single-point grounding in which the grounds of the different circuits are connected in parallel: (a) the original circuit and (b) the circuit redrawn to show the impedances of the connecting wires and the currents flowing in those wires. The same symbols are used as in figure 21.18.

21.14.2 Multi-point grounding

Return current paths from separate circuits A, B, C, ... are connected through short length low-resistance wires to the ground plane. The ground plane is the common low-impedance plane for termination of grounding wires (figure 21.20). This plane is

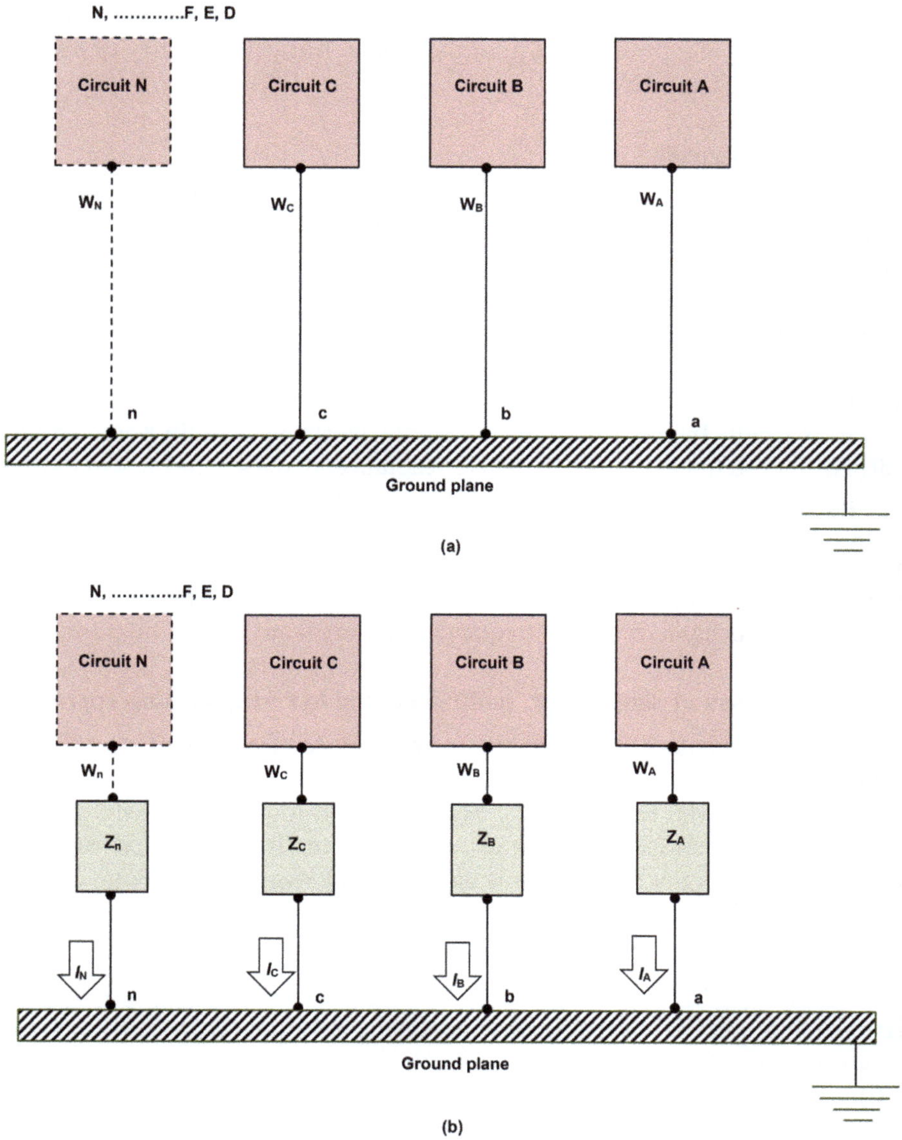

Figure 21.20. Multi-point grounding showing: (a) the circuit comprising circuits A, B, C, …, N connected to the ground plane at points a, b, c.… n through wires W_A, W_B, W_C, …, W_N; and (b) the circuit diagram showing impedances Z_A, Z_B, Z_C, …, Z_N of the wires W_A, W_B, W_C, …, W_N passing currents I_A, I_B, I_C, …, I_N.

connected to the chassis which in turn is bound to the metal frame enclosing the equipment. A solid piece of metal can act as a ground plane.

Multi-point grounding arrangement provides multiple parallel paths connected to ground plane. There is no single ground point G but a multiplicity of ground points G_1, G_2, G_3 distributed on the ground plane. The individual circuits are connected to separate ground points. Moreover, the wire lengths for connections are sufficiently short < (1/20) λ to avoid resonance effects.

Multi-point grounding is used in high-frequency circuits >10 MHz. High-speed digital circuits too are grounded at multiple points. Intricate circuitry of equipment is simplified by this approach.

21.14.3 Hybrid grounding

It is a mixture of single-point and multi-point grounding which seeks to provide low noise at reduced cost (figure 21.21). Consider the hybrid grounding circuit using capacitors (figure 21.21(a)). At low frequencies, the capacitive reactances $X_C = 1/\omega C$ are high. Then the circuit works as a series single-point grounding circuit. As frequency increases, the capacitive reactances fall, and the circuit becomes a multi-point grounding circuit. The reverse is the case with the hybrid grounding circuit using inductors (figure 21.21(b)). At low frequencies, the inductive reactance $X_L = \omega L$ is small. In this situation, the circuit works as a multi-point grounding circuit. At high frequencies, the inductive reactance increases. In this circumstance, the circuit function is transformed to that of a series single-point grounding circuit.

In essence, the function of a hybrid grounding system changes with the operating frequency between single-point and multi-point types. These changes are governed by the frequency-dependent reactances present in the circuit, whether capacitive or inductive. Their behaviour at low frequencies differs from that at high frequencies.

21.14.4 Comparison of single-point, multi-point and hybrid grounding approaches

Series or parallel single-point grounding works best for frequencies <1 MHz. They suffer from high ground loop currents. Multi-point grounding is good for frequencies >1 MHz and even more than 10 MHz. Ground loop currents are minimized. Maximum EMI suppression is obtained when wire lengths are kept extremely short. Hybrid grounding builds a compromising situation between single-point and multi-point grounding in the same system with their accompanying merits and demerits.

21.15 EMI shielding

EMI shielding implies partially or completely enveloping an EMI emitter or susceptor (Tong 2009). It has two main goals:

(i) to protect an electronics circuit from EMI caused by surrounding circuits, and
(ii) to prevent the surrounding circuits from the EMI influence of the protected electronics circuit, i.e., neither allowing any interference nor interfering with others.

21.15.1 Shielding efficiency

The parameter measuring the capability of a material to oppose electromagnetic energy is the shielding efficiency η defined in terms of incident power P_I and transmitted power P_T through the material as (Shukla 2019)

$$\eta = 10 \log \frac{P_I}{P_T} \tag{21.45}$$

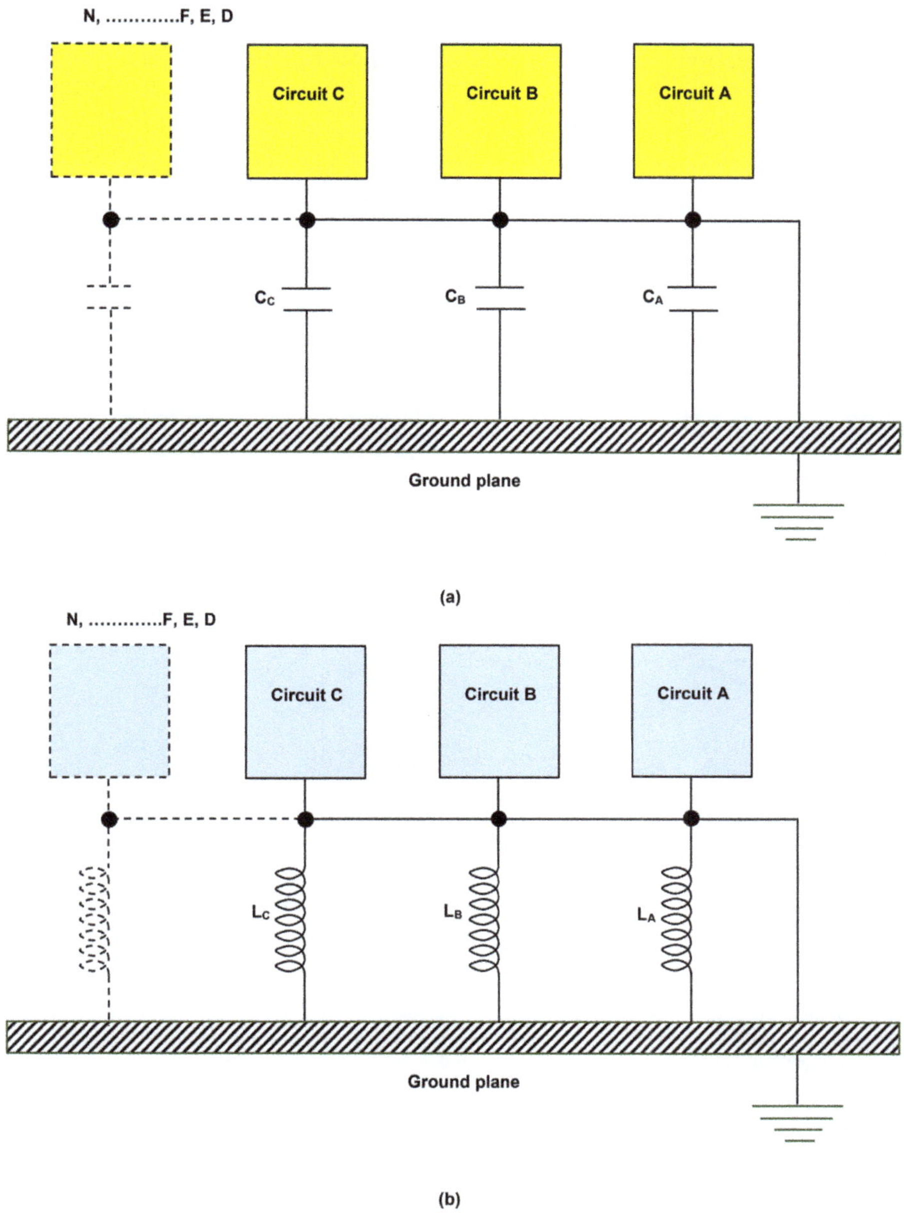

Figure 21.21. Hybrid grounding configurations: (a) configuration made by using capacitors and (b) configuration with inductors. For figure 21.21(a): At low frequencies, the capacitors offer high reactances. Then the circuit behaves as a single-point grounding circuit in series connection. At high frequencies the capacitors offer low reactances. Then they look like conducting wires and the circuit acts as a multi-point grounding circuit. For figure 21.21(b): At low frequencies, inductive reactances are small. So, the circuit becomes a multi-point grounding circuit. At high frequencies, inductive reactances are large. The circuit behavior resembles a series-connected circuit with single-point grounding. Thus, the inductive circuit shows opposite behavior to the capacitive circuit.

The shielding efficiency comprises three components (figure 21.22), namely, the efficiency of reflection η_{R}, efficiency of absorption η_{A}, and efficiency of multiple reflections η_{MR}:

$$\eta = \eta_{\mathrm{R}} + \eta_{\mathrm{A}} + \eta_{\mathrm{MR}} \tag{21.46}$$

The component η_{MR} arises because when the wave reaches the second surface of the material slab it is partly reflected and partly transmitted. The reflected wave bounces back from the first surface and is again reflected/transmitted. This process occurs repeatedly until the wave is attenuated. It can be neglected for thicker slabs of material so that

$$\eta \approx \eta_{\mathrm{R}} + \eta_{\mathrm{A}} \tag{21.47}$$

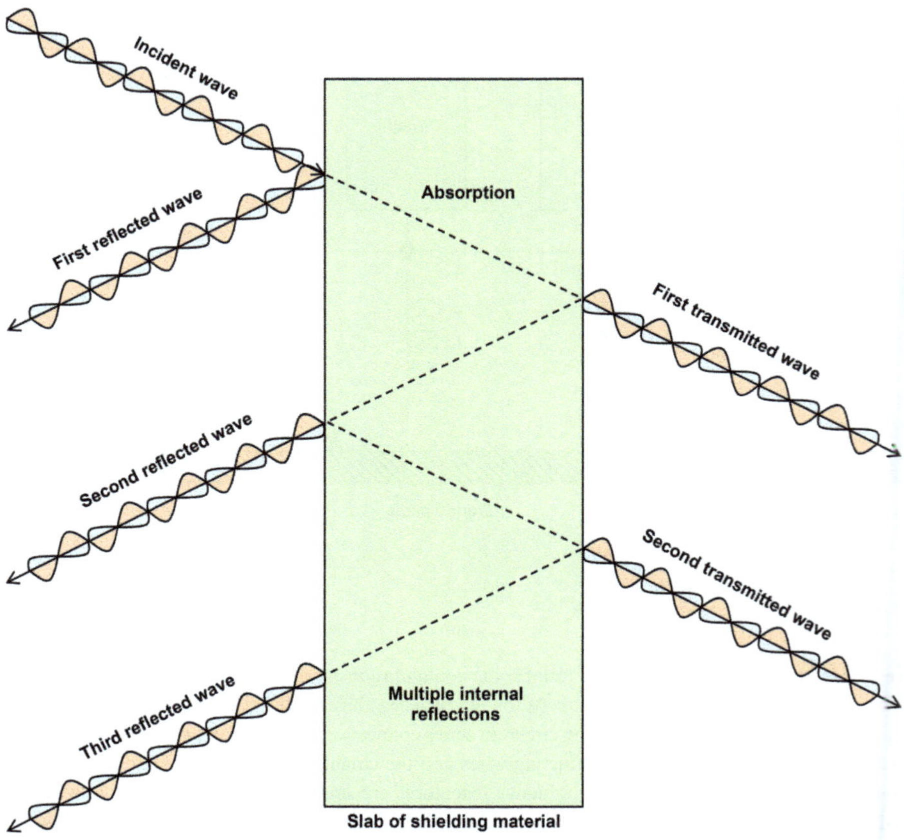

Figure 21.22. Mechanism of EMI shielding showing the components generated by reflection, absorption, transmission and multiple internal reflections when an incident electromagnetic wave strikes a medium of finite dimensions.

21.15.2 Shielding materials

The primary shielding mechanism is represented by the term η_R due to reflection. This term decreases with frequency f of the signal for given conductivity σ and permeability μ of the material. At a constant frequency, it is proportional to the ratio σ/μ. The σ and μ values of the material are crucial parameters influencing the shielding efficiency. Obviously, a high conductivity material containing a larger number of free charge carriers will be more effective for electromagnetic shielding. The absorption term η_A is dependent on thickness of the material taken besides its σ and μ.

Enclosures for EMI shielding are generally made of mumetal (nickel–iron–copper–chromium alloy), brass (copper–zinc alloy), stainless steel (iron–chromium–carbon alloy), nickel, aluminum, silver, etc (Geetha *et al* 2009). Apart from metals and alloys, plastics with modified electrical conductivity are employed, e.g., plastics coated with conducting films, plastics compounded with conductive fillers such as graphene and carbon nanotubes (González *et al* 2016), and also intrinsically conductive polymers.

21.15.3 The Faraday cage

The most obvious way of protection of a sensitive electronic circuit is to completely cover it with a box of conductive material such as a metal. Then EMI caused by adjoining circuits during their operation will induce a charge on the surface of the box. This charge will lie on the surface of the box. It will be confined to the surface of the box only and will not enter the box. So, the sensitive electronic circuit mounted below the surface of the box will not be affected by the EMI.

Gauss' law of electrostatics provides the explanation for the residence of charge on the surface of a charged conductor. Consider an arbitrary Gaussian surface inside the surface of the charged conductor. Imagine it to be at an infinitesimal distance from the surface (figure 21.23). The electric flux through the Gaussian surface is zero

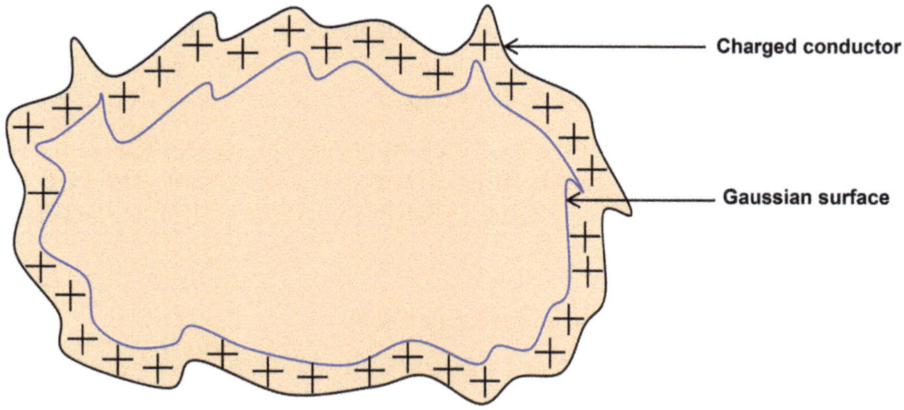

Figure 21.23. Drawing a Gaussian surface inside a charged conductor to explain that any charge given to a conductor will only reside on its outer surface.

because the electric field is zero everywhere inside the conductor. But the Gaussian surface was selected just below the surface of the conductor. So, there is no charge inside the surface of the conductor. Any charge placed anywhere inside the conductor always migrates to its surface.

The cycle of events described in the case of an EMI striking the metallic box from outside is repeated when the sensitive electronic circuit inside the box radiates electromagnetic waves which may impair the working of adjoining circuits outside the box. This time too, the induced charge is restricted to the surface of the box and is unable to upset the functioning of any circuit located beyond the surface of the box.

The operating principle of the metallic box for EMI protection is known as the principle of Faraday cage. The metallic box is impervious to both static or non-static fields. The Faraday cage (figure 21.24) can be made of a mesh of wire. It can also be made from coils of wires, thin sheets or foils of metals in a parallelopiped, spherical or cylindrical shape.

Because charge always remains on the surface of the Faraday cage, the cage does not need to be grounded for its operation. But beware! Grounding is essential for safety reasons because when anyone touches the surface, the accumulated charge will pass through the person's body to ground giving a painful shock.

The Faraday cage is used for testing the electromagnetic compatibility of an electronic circuit. The circuit is placed on a wooden platform inside the Faraday cage. A sensitive antenna is mounted in this cage. The circuit under test is rotated through an angle of 360°. The pattern and strength of electromagnetic radiation emitted by the circuit are measured from the signal received at the antenna. If these emissions are found to be below the set guidelines, the circuit is qualified to have passed the EMC test.

EMC standards have been set by organizations such as International Special Committee on Radio Interference or Comité International Spécial des Perturbations Radioélectriques (CISPR), Federal Communications Commission (FCC), and European Committee for Standardization or Comité Européen de Normalisation (CEN). Any circuit which is marketed must fulfil the limits laid down in these standards for radiated and conducted electromagnetic interference.

21.15.4 Board level shielding (BLS) for PCBs

The shielding consists of a PCB with a built-in ground plane. A metal can is fixed over the sensitive electronic circuit. It is known as the shield can. The PCB shield cans are available as stamped one-piece or two-piece (frame and cover) nickel–silver or tin cans.

21.15.5 Unshielded and shielded twisted pair cables

There are two kinds of twisted pair cables:

(i) unshielded twisted pair (UTP) cable which does not have a protective shield, and (ii) shielded twisted pair (STP) cable which has a copper braid mesh covering (figure 21.25).

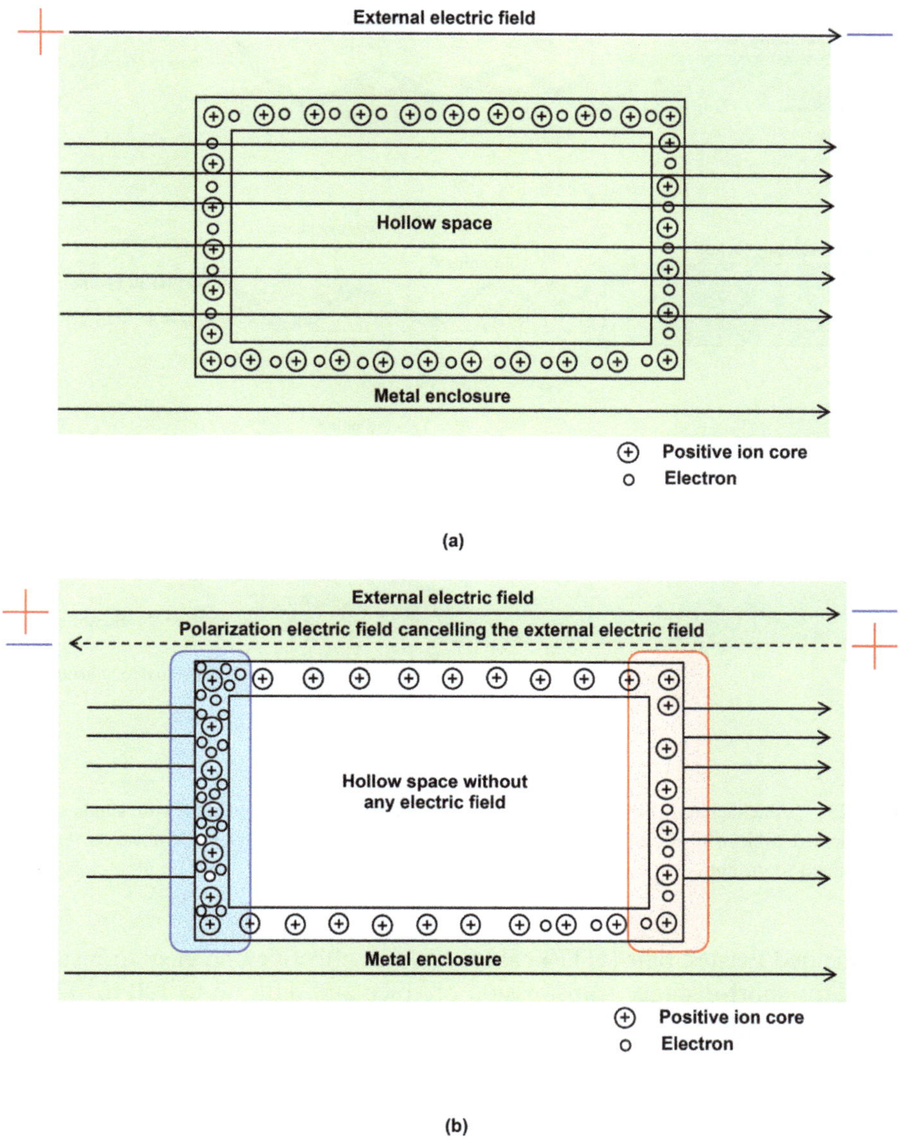

Figure 21.24. Cancellation of an external electric field applied on a Faraday cage by the induced electrical polarization in the cage: (a) the situation assuming that there is no effect of the cage and (b) the actual situation accounting for the electric polarization produced on the cage.

As said earlier, any current induced in the copper braid mesh covering of the cable is carried to ground and so cannot affect the main signal flowing in the twisted pair cable, which already has been explained in section 21.9 as an EMI inhibitor structure owing to oppositely-flowing currents and converse fluxes during differential-mode signal transmission. So, the shield is an additional armor against EMI.

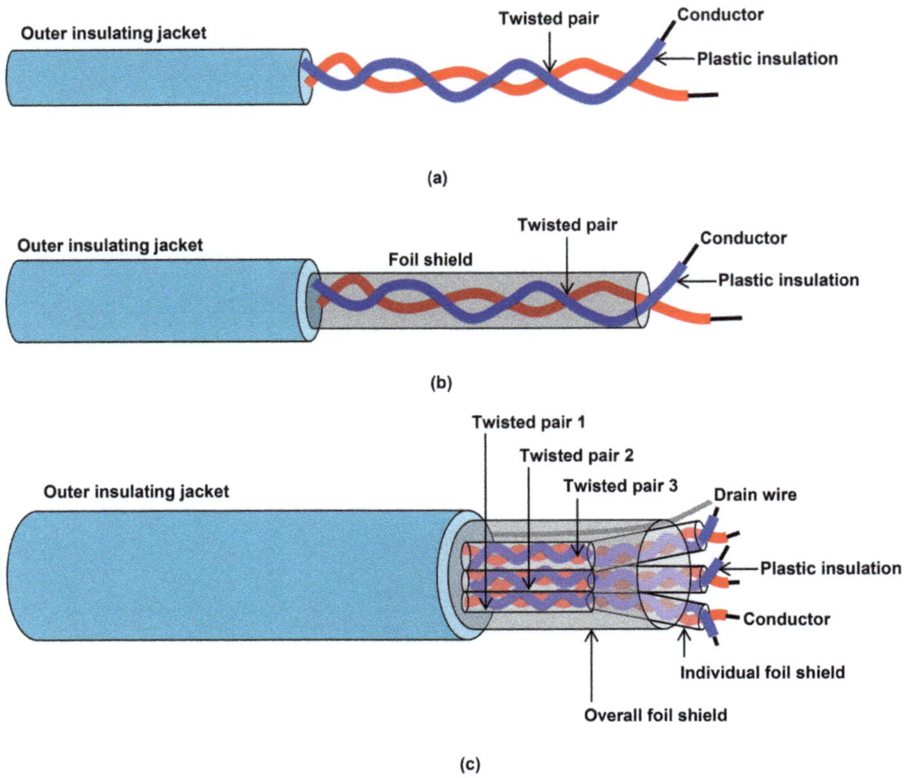

Figure 21.25. Unshielded and shielded cables: (a) unshielded cable, (b) shielded cable with single twisted pair, and (c) shielded cable with multiple twisted pairs. The drain wire carries away the noise current from the electrical shield to ground.

Unshielded twisted pair (UTP) cables are cost-effective and easy to install. They are used for short-distance transmission of voice and data up to 100 m. They have limited bandwidth.

Shielded twisted pair (STP) cables are expensive cables installed with metallic shielding grounded at both ends. They are used for long-distance voice and data communication. They require more frequent maintenance but provide larger bandwidth.

21.15.6 Types of shielded twisted pair cables

A shielded cable consisting of one or more insulated conducting wires enclosed in a conductive covering protecting it against EMI is made in three types shown in figure 21.26.

21.15.6.1 Foil shield

The shield is made of a foil of Al/polyester or Al/kapton foil (Multicable Corporation 2017–22). The conductor is tinned copper drain wire (individual or

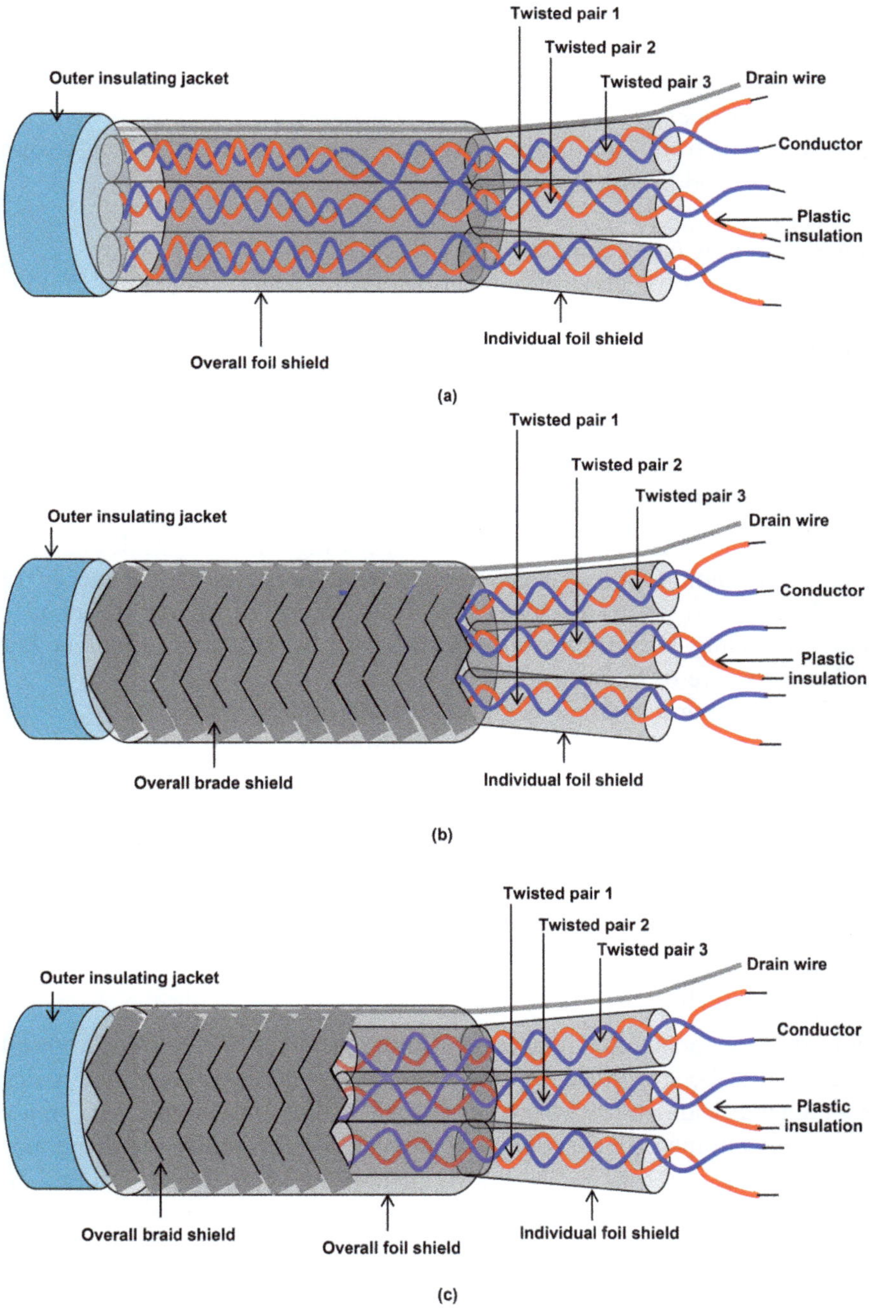

Figure 21.26. Different types of shielded cables: (a) foil-shielded cable, (b) braid-shielded cable and (c) mixed (foil + braid) shielded cable.

twisted pairs). The coverage of shield is 100%. The shield is useful for frequencies >15 kHz.

21.15.6.2 Braid shield

A mesh is woven from bare, tinned, Ag- or Ni-plated Cu wires. The coverage is 70%–95%. It is used at low frequencies <15 kHz.

21.15.6.3 Foil plus braid shield

It is the best shield protecting over much wider frequency range than braid or foil shields.

A shielded cable protects the transmitted signal in two ways:

(i) by reflecting EMI, and
(ii) by picking any penetrating electromagnetic radiation and sending it to ground.

21.16 Grounding of shielded cables

The grounding is done by examining and deciding the end result that is to be achieved by grounding. In near-field energy transfer or non-radiative energy transfer taking place across short distances, the transference of energy takes place by capacitive coupling between metal electrodes or by inductive coupling between coils. Two cases are of interest:

(i) low-speed or low-frequency applications, and
(ii) high-speed or high-frequency applications.

The difference between these two modes is that in the low-frequency mode, the energy surrounding the conductors is mainly electrical while in the high-frequency mode, it is largely magnetic. Naturally, therefore, the shield must work as an electrical shield in case (i) and as a magnetic shield in case (ii).

21.16.1 Electrical shielding

In low-frequency applications, electrical shielding is necessary. In these applications, the shield must be grounded at one end only. The expanding and collapsing electromagnetic fields emitted by sources such as the high-voltage AC mains line or the switching of mechanical relays will induce EMI currents in the shield material. The shield will act as a sink for these EMI currents and pass them to ground. So, the current flowing in the wire enclosed within the shield will not be affected by this EMI.

It may appear that grounding both ends of the shield will provide double protection and further improve protection. But this is not true. Grounding both ends of the shield will only make it more prone to formation of ground loop resulting in a new source of EMI. Because the shield material has a non-zero resistance, a difference of potential will be generated between the two grounded ends. A current will flow from the high potential to the low potential point. This EMI current will induce a current in the protected wire inside the shield carrying the signal current.

Consequently, the signal current will be affected by this EMI. So, grounding at one end suffices. Grounding at both ends causes an EMI issue.

21.16.2 Magnetic shielding

In high-frequency applications, magnetic shielding is required. The conducting shield is an electrical shield only when grounded at one end. But it can provide magnetic shielding when grounded at both ends. How? Grounding both ends allows circulation of high frequency currents in the shield. These high-frequency currents counteract the currents flowing in the protected signal conductors because they produce magnetic fields in opposite directions to the magnetic fields emitted by the signal conductors. The net result is that a magnetic shielding effect is obtained. Thus, the magnetic shield will function satisfactorily if current is permitted to enter or exit from both ends of the shield, which is indeed what double grounding has done.

21.16.3 Considerations for a shielded cable grounded at both ends

For proper operation of a shielded cable grounded at both ends, the shield should not be used as a return path for current. Radio-frequency communication signals and low-frequency, high-power currents should be routed on separate paths isolated from the chassis. If the EMI currents are produced at DC or low frequencies, they can be shunted by an RF capacitor connected to ground. Since the capacitor has a high impedance to DC or low-frequency power currents, no ground loop is created. Another way is to use a system comprising differential transmitter and receiver. Then the EMI current induced in the protected conductors will be a common-mode signal. It will be rejected by the common-mode rejection ratio of the receiver. The common-mode rejection ratio (CMRR) of a differential amplifier is the ratio of its differential-mode voltage gain to the common-mode gain, expressing the ability of the amplifier to suppress signals common to its two inputs, i.e., the common-mode signals.

21.17 Discussion and conclusions

Proliferation of electrical and electronic equipment and their permeation into our everyday lives has placed increasing load on the requirement for guaranteeing that any equipment does not disturb the operation of those in its neighbourhood. Issues of electromagnetic interference and compatibility therefore need utmost attention. The preventive steps although elementary in nature have a far-reaching significance. Grounding, filtering and shielding are the three keys to achieve EMC.

Review exercises

21.1. What are cyber physical systems and smart grids? Why is electric transportation being preferred over the existing transport system?

21.2. Explain how the expansion of cyber physical systems, smart grids and electrical transportation needs increased care in circuit layout to avoid electromagnetic interference problems.

21.3. Define the terms, 'electromagnetic interference' and 'electromagnetic compatibility'.
21.4. Give five examples of natural EMI sources.
21.5. Cite five examples of artificial EMI sources.
21.6. Distinguish between intentional EMI and unintentional EMI.
21.7. Differentiate between intrasystem EMI and intersystem EMI.
21.8. What are the three mechanisms of EMI production? Explain these mechanisms with a diagram.
21.9. What is meant by continuous EMI and impulse EMI?
21.10. How does narrowband EMI differ from broadband EMI?
21.11. What is EMI noise signal? How does it impact the performance of electronic equipment? Illustrate with a few examples.
21.12. What is meant by Sun outage of TV reception?
21.13. Is it necessary that the transmitter and receiver grounds be connected during: (a) single-ended transmission and (b) differential transmission of signals? Explain giving reasons.
21.14. An electrical signal is conveyed from a station A to a station B by differential-mode transmission. On the way the signal is affected by noise. Will the signal received at station B be noisy? If the same signal is sent by single-ended transmission, will the signal at B contain noise introduced on the way?
21.15. How does differential-mode transmission get rid of any noise creeping into the signal?
21.16. Suppose a signal $V = V(t) + N(t)$ is sent by single-ended transmission from a transmitter to a receiver. Which of the following statements is true:

(a) The signal at the receiver is

$$V = V(t) \tag{21.48}$$

(b) The signal at the receiver is

$$V = V(t) + N(t) \tag{21.49}$$

21.17. If a signal $V_+ = V(t) + N(t)$ containing the noise $N(t)$ is sent by differential transmission to a receiver, the signal at the output of the receiver is:

(a) $$V_{\text{Output}} = V(t) + N(t) \tag{21.50}$$

(b) $$V_{\text{Output}} = V(t) \tag{21.51}$$

$$(c)\ V_{\text{Output}} = 2V(t) \tag{21.52}$$

Which of (a), (b) and (c) is correct and why?

21.18. Suppose a high-frequency current signal is sent from circuit X to circuit Y through wires 1 and 2. A varying electric current is associated with a changing magnetic field and a changing magnetic field engenders an electric field. How will the EMI current induced in wire 1 by current flowing in wire 2 and conversely the EMI current induced in wire 2 by the current in wire 1 interact with each other in the case of:

(a) single-ended transmission, and

(b) differential-mode transmission?

On the basis of your answer, justify whether transmission method (a) or method (b) gives better performance?

21.19. Two signals V_1, V_2 having phases ϕ_1, ϕ_2 are flowing between two points P and Q. Write down the equations for differential-mode voltage V_{DM} and common-mode voltage V_{CM} for the signals.

21.20. Differential-mode interference is produced when:

(a) the EMI source affects only one of the two wires in the two-wire circuit,

(b) the EMI source affects both the wires.

Choose the correct answer.

21.21. Common-mode interference occurs when:

(a) the noise source affects only one wire,

(b) the noise source affects both the wires of the two-wire system.

Choose the correct option.

21.22. Considering a power supply feeding a circuit, show on a diagram the paths followed by EMI noise current for: (a) differential-mode interference and (b) common-mode interference.

21.23. Correct the following statements:

(a) Differential-mode noise current flows in parallel with the power supply current.

(b) Common-mode noise voltage does not act between power supply line and ground.

21.24. There are two cables A and B. Cable A is an untwisted pair cable and cable B is a twisted pair cable. Alternating current is sent in differential mode in cables A and B. The varying current produces a changing magnetic field. The changing magnetic field around one wire of a cable induces a current in the second wire of the same cable. Explain giving arguments which cable, A or B, will have more EMI noise current.

21.25. Assume that an external EMI source produces an EMI noise signal of 5 mV per unit length on the wire of a two-wire cable lying near the source and 4 mV per unit length on the distant wire of the cable. Consider two cables: an untwisted pair cable and a twisted pair cable. Divide both cables into four segments of unit length and take into account the effects of EMI noise on the wires of the two cables to determine the difference of potential that will be created between the two wires of each cable at the destination ends of the two cables.

21.26. Consider an external changing magnetic field inducing currents in the wires of an: (a) untwisted pair cable and (b) twisted pair cable. Draw diagrams of both type of cables, indicate the direction of the varying magnetic field and show the directions of induced currents in the wires of the cables. What significant information is revealed from this diagram regarding the difference in behaviors of the two types of cables in a magnetic field?

21.27. What is common impedance coupling? When does it happen?

21.28. Define the crosstalk parameter $Xtalk_{BA}$ between a receptor circuit B and source circuit A. Write the equation for the crosstalk parameter in terms of the load resistances R_{LB} and R_{LA} in circuits B and A, the return path resistance R_{Return} and voltage source resistance R_{SB} of circuit B.

21.29. Explain the operation of:
  (a) a capacitor filter, and
  (b) an inductor filter.

Draw the circuit diagrams of both filters.

21.30. Explain with circuit diagrams the suppression of differential-mode EMI noise by a:
  (a) C–L filter, and
  (b) π filter.

21.31. Draw the circuit diagram of a common-mode EMI filter and explain how does it block a common-mode signal while it allows easy passage to a differential-mode signal.

21.32. What is meant by the ground point in an electrical circuit? Why it is incorrect to refer to the ground wire as a return path for current in the circuit?

21.33. How is a ground loop formed in a circuit? Consider the formation of ground loop between two circuits A and B, and show how the output voltage of circuit B depends on the circuit A.

21.34. Explain how a ground loop between two circuits adds a voltage having a magnitude equal to the voltage difference between the two ground points.

21.35. Name the three principal approaches of circuit grounding, and describe the salient features of each approach.

21.36. Draw a series-connected single-point grounding arrangement of three circuits A, B and C in which the current return paths of the three circuits are connected at points a, b and c to a wire joined to the ground point, and write down the equations expressing the dependence of the ground potentials of circuits A, B and C on the return currents from circuits A,

B and C and the impedances Z_A, Z_B and Z_C of the wires connecting the circuits A, B and C to the points a, b and c.

In the above circuit analysis, which of the three ground potentials has the highest value and which is lowest in value: ground potential at point C (V_c), ground potential at point B (V_b) and ground potential point A (V_a). Write the three ground potentials in ascending order.

21.37. Why should single-point series grounding not be used in high-frequency circuits involving control and power signals?

21.38. Why is a single-point series grounding circuit frequently used despite its disadvantages?

21.39. Why is single-point series grounding prone to common-mode noise while single-point parallel grounding is not susceptible to such noise?

21.40. What advantage is derived by directly connecting the return paths from different circuits to the ground point, as done in single-point parallel grounding, instead of first connecting them to a wire and connecting this wire to the ground point, as done in a single-point series grounding circuit?

21.41. In a single-point parallel grounding circuit, what happens to the ground impedance when the length of the wire connecting any circuit to ground becomes an odd multiple of a quarter wavelength of the electrical signal. Is the ground impedance acceptable under this condition?

21.42. In what respect is single-point parallel grounding better than single-point series grounding for circuits working at low frequencies?

21.43. In a grounding arrangement, the return paths of currents from different circuits are connected by short wires of low resistances to the ground plane. What is this circuit arrangement called?

21.44. For what frequency range of electrical signals should multi-point grounding be used?

21.45. What type of grounding scheme is suitable for high-speed digital circuits?

21.46. Name the grounding arrangement which allows availing of the advantages of both single-point and multi-point grounding in the same circuit.

21.47. Draw a hybrid grounding circuit using capacitors and explain its operation.

21.48. How will you implement hybrid grounding using inductors? Draw the circuit diagram and describe the working of the circuit.

21.49. Mention the two prime objectives of EMI shielding.

21.50. Sketch an arbitrarily-shaped Gaussian surface inside a charged conductor. Explain why electric charge should reside only on the outer surface of a charged conductor. Remember that electric field is zero everywhere inside the conductor because if it exists, charges will move to neutralize the potential difference and make the field zero.

21.51. Will the operation of a Faraday cage be impaired if it is not grounded? Is it essential to ground a Faraday cage from human safety considerations?

21.52. Draw the diagram of a Faraday cage and explain how the electric polarization field built on it cancels the external electric field.

21.53. How is electrical shielding done at the board level for printed circuit boards?
21.54. What is the primary function of the metal shield covering a cable?
21.55. Compare the relative advantages/disadvantages of unshielded and shielded cables. For what type of applications are the two types of cables used?
21.56. Draw the diagrams of:
 (a) a foil-shielded twisted pair cable,
 (b) a braid-shielded twisted pair cable and
 (c) foil plus braid-shielded cable.
 What is the role of the drain wire in these cables. For what frequency ranges of electrical signal are these cables suitable?
21.57. In low-speed near-field energy transfer, the energy surrounding the conductors is mainly:
 (a) electrical, or
 (b) magnetic.
 Which is correct: (a) or (b)? Suggest the type of shielding necessary for this situation.
21.58. In high-speed near-field energy transfer, the energy surrounding the conductors is mainly:
 (a) electrical, or
 (b) magnetic.
 Which is correct: (a) or (b)? What type of shielding is necessary in this circumstance?
21.59. Why should an electrical shield be grounded at one end only? What will be the harm if it is grounded at both ends?
21.60. How does grounding of a metal shield at both ends make it an effective shield against magnetic fields?
21.61. Can a magnetic shield grounded at both ends be used as a current return path? Mention two precautions which should be taken with magnetic shielding for preventing EMI.

References

Bäckström M 2006 The threat from intentional EMI against the civil technical infrastructure, reprint from ESW2006 *3rd European Survivability Workshop (Toulouse, 16–19 May 2006)* 1–10

Christopoulos C 2022 *Principles and Techniques of Electromagnetic Compatibility* (Boca Raton, FL: CRC Press) p 1

Geetha S, Satheesh Kumar K K, Rao C R K, Vijayan M and Trivedi D C 2009 EMI shielding: methods and materials—a review *J. Appl. Polym. Sci.* **112** 2073–86

González M, Mokry G, de Nicolás M and Pozuelo J B J 2016 Carbon nanotube composites as electromagnetic shielding materials in GHz range *Carbon Nanotubes—Current Progress of their Polymer Composites* ed M R Berber and I H Hafez (London: IntechOpen) ch 11

Learn EMC 2022 *Common impedance coupling* https://learnemc.com/common-impedance-coupling#:~:text=If%20you've%20ever%20noticed,of%20their%20respective%20current%20paths

Mari L 2021 Single-point and multi-point signal grounding https://eepower.com/technical-articles/single-point-and-multi-point-signal-grounding/#

Multicable Corporation 2017–22 Signal interference and cable shielding https://multicable.com/resources/reference-data/signal-interference-and-cable-shielding/

Murata Manufacturing Co. Ltd 1998 Murata, No. TE04EA-1, Noise suppression by EMI filtering: Basics of EMI filters pp 26–27 http://datasheet.datasheetarchive.com/originals/library/Datasheets-UEA1/DSAFRAZ008703.pdf

Paul C R 2006 *Introduction to Electromagnetic Compatibility* (Hoboken, NJ: Wiley-Interscience) pp 49–90

Radasky W A and Bäckström M 2014 Brief historical review and bibliography for intentional electromagnetic interference (IEMI) *2014 XXXIth URSI General Assembly and Scientific Symp. (URSI GASS) (Beijing, 16–23 August 2014)* 1–4

Shukla V 2019 Review of electromagnetic interference shielding materials fabricated by iron ingredients *Nanoscale Adv.* **1** 1640–71

Tong X C 2009 *Advanced Materials and Design for Electromagnetic Interference Shielding* (Boca Raton, FL: CRC Press) p 11

Weston D A 2017 *Electromagnetic Compatibility: Methods, Analysis, Circuits and Measurement* (Boca Raton, FL: CRC Press) p 1

Williams T 2017 *EMC for Product Designers* (Oxford: Newnes) pp 10–1

Zhao Y, Yan W, Sun J, Zhou M and Meng Z 2021 *Electromagnetic Compatibility: Principles and Applications* (Beijing/Singapore: Science Press/Springer Nature Singapore Pte Ltd) p 1

Chapter 22

Developing sensor capabilities for aggressive environments

Each harsh environment involves a mix-up of parameters to be monitored. Sensors must be designed to work at elevated temperatures. The new energy systems place increased demands on the sensors. A wide variety of harsh environment conditions and suitable sensors for these conditions are invoked. Accelerometers described include SiC piezoresistive accelerometer, YCOB piezoelectric crystal-based compression-mode and shear-mode accelerometers, and optical accelerometer. 3C-SiC on-glass and fiber optic flow sensors working on hot-wire anemometry principle are discussed. Among pressure sensors, a SiC capacitive pressure sensor, a sensor using Pt thin-film strain gauges on a sapphire membrane, a ceramic nanofiber flexible sensor and an all-SiC fiber optic sensor are considered. Silicon-on-insulator (SOI) diode temperature sensor is a useful sensor for *in situ* monitoring of integrated circuit (IC) temperatures. Two types of wirelessly accessed temperature sensors (LTCC LC and langasite SAW resonator) are delved into. Humidity sensing is done by an approach based on difference in thermal conductivity of dry and moist air obtained from temperature-resistance characteristics of Pt meander patterns. Another moisture sensor working on optical principles rests on the expansion of a hydrogel film with adsorption of water vapor. Of immense value are the oxygen lambda sensor for measuring the air/fuel ratio of automobiles, the mixed potential CO sensor as well as SiC field-effect transistor (FET) sensor for air pollution control. Relative prospects and limitations of various harsh environment sensors are compared.

doi:10.1088/978-0-7503-5072-3ch22

A harsh environment is a mix-up situation for sensors with regard to the parameters to be watched. Frequently sensors cannot be installed at the points of measurement. Often sensors are required to work at high temperatures. Particularly, new clean energy systems need more stringent surveillance by sensors laying heavy responsibility on sensors.

Let us elaborate our thoughts on these topics. GaAs, SiC, GaN and diamond sensors capable of working at high temperatures were touched upon earlier in the book. Additional sensors for harsh environments will be explored in this chapter reinforcing the previous descriptions of GaAs photodetectors in section 7.15, SiC sensors in section 8.12, GaN sensors in section 9.10, and diamond sensors in section 10.1.

22.1 Disorganized scenario in a harsh environment, and denial of accessibility to the sensor

A harsh environment condition represents a bewildered situation intertwined with a variety of physical and chemical parameters, involving high temperature and high pressure, in the presence of corrosive materials and electromagnetic pollution (Fahrner *et al* 2001). Often a combination of parameters needs to be sensed, e.g., high temperature and high pressure in a boiler; high temperature, high pressure and explosive gases in the combustion chamber of an automobile; high temperature along with nuclear radiation and strong electromagnetic fields in a warfare zone, etc.

In many situations, it is difficult to mount the sensor at the site of measurement and connect its wiring to the instrument, e.g., a sensor cannot be fixed and wired inside the ignition chamber of an engine, or fitted on the shaft of a running drill machine. In these situations, recourse is taken to remote sensing.

22.2 High-temperature sensors

Sensors for detecting various environmental parameters and having capability of withstanding high temperatures >800 °C are known as high-temperature (HT) sensors. They are used in automotive, aerospace and energy sectors. The sensors are applied to monitor the health of the structures. The structures include turbine engines, internal combustion engines and electricity generation plants. The objective is intelligent design of propulsion system to improve safety. The timing and shapes of pulses for combustion parts, e.g., the injectors and valves of automotives require careful monitoring of repetitive temperature cycles. For these applications, the sensors must be able to work at 500 °C–1000 °C. The working duration is 100 000 h. Moreover, the sensor must be placed as close to the hot environment as possible to ensure precision in measurement (Kim *et al* 2012).

22.3 Need of tightly monitoring energy systems aggravates burden on sensors

To achieve the goal of clean energy, next-generation energy systems work under more severe conditions of conversion of fossil fuels into heat and power than conventional combustion/steam cycles. These systems critically depend on harsh environment processes for energy storage and decarbonization requirements. With increase in electrification of transportation, the processes must be more strictly monitored and more efficiently controlled demanding advanced sensing technologies operating reliably, in harmony, and in real time at temperatures >1600 °C, pressures up to 340 atmosphere and under cyclic loading. Materials used for sensing and the design of the sensor are the major considerations in development of sensors (Romanosky and Maley 2013).

We now move on to sensors for specific parameters, e.g., acceleration, flow, pressure, temperature, humidity and gases.

22.4 Accelerometers

22.4.1 All 4H-SiC MEMS piezoresistive accelerometer

Silicon experiences plastic deformation at temperatures ⩾500 °C. Plastic deformation or plasticity of a material is its ability to undergo permanent change in shape when stressed beyond a limit called the limit of elasticity without any rupture or damage. Elastic deformation or elasticity is the reversible change in shape of a material, which is stressed by a force smaller than elastic limit. Thermal limitation makes silicon accelerometers unsuitable for high-temperature environments around 2000 °C such as the temperature of turbine blades of an aeronautical engine.

Silicon carbide is a mechanically hard, highly-temperature resistant material with melting point 2730 °C. Integrated electronics for 400 °C–600 °C temperature range is realized with SiC. Owing to its superior thermomechanical properties, and inert chemistry, SiC is more suited for fabricating accelerometers for these applications.

A microelectromechanical systems (MEMS) technology-based SiC accelerometer working on piezoresistive principle helps to surmount the high temperature constraint of silicon accelerometer (Zhai *et al* 2022a, 2022b). It is a quad-beam accelerometer whose top view is shown in figure 22.1. Resistors R_1, R_3 are formed at the junctions between the beams and the support frame while resistors R_2, R_4 are made at the junctions between the beams and the proof mass. Resistors R_1, R_2, R_3, R_4 made at the junctions of beams 1 and 4 with support frame and proof mass constitute a Wheatstone bridge. Similarly, the resistors R_1, R_2, R_3, R_4 made at the junctions of beams 2 and 3 with support frame and proof mass constitute another Wheatstone bridge. V_{IN+}, V_{IN-} are the input voltage terminals of the Wheatstone bridge. V_{OUT+}, V_{OUT-} are the output voltage terminals. Thus, the accelerometer consists of a proof mass (5000 μm × 6000 μm × 80 μm), four support beams, each of dimensions 1250 μm × 200 μm × 20 μm, one fixed frame and eight resistors configured in two independent Wheatstone bridge circuits. One Wheatstone bridge is enclosed in dotted lines and clearly redrawn above the top view of accelerometer to show the electrical connections. Any one bridge can be used for measurements.

Figure 22.2 shows the schematic cross-section of the sensor. In the cross-sectional diagram, the accelerometer is seen as comprising a proof mass suspended on its two sides by beams from a fixed frame. Also seen are the piezoresistors formed at the maximum stress locations, namely the junction of the beam and frame and that of the beam with the proof mass.

The device is sensitive to acceleration in the Z-direction. When the accelerometer is subjected to an acceleration in the Z-direction, the proof mass is deflected from its

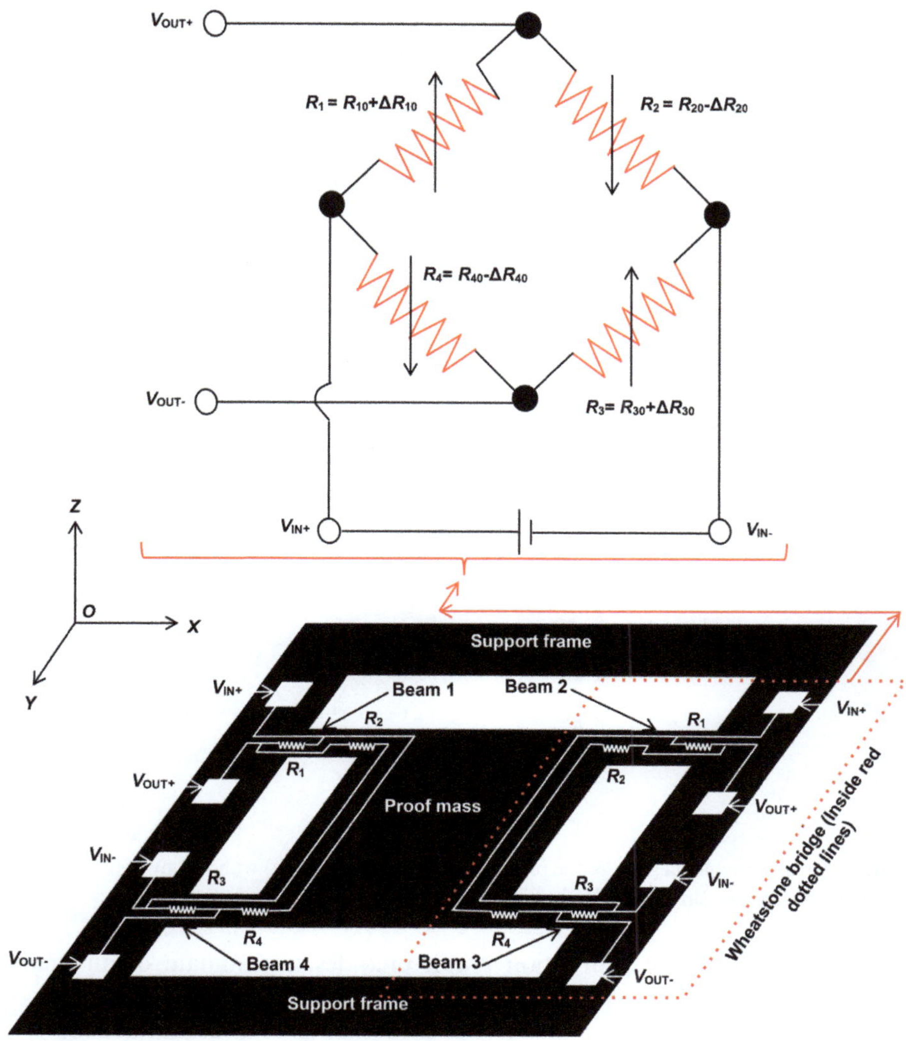

Figure 22.1. Top view of the quad-beam piezoresistive accelerometer showing the support frame and proof mass suspended from four beams numbered 1, 2, 3, 4. The R_0 symbols denote initial values of resistors and ΔR_0 symbols represent changes in the values of resistors under acceleration.

position. Due to deflection of the proof mass, bending stresses are produced in the support beams. When the beams undergo bending stresses, the piezoresistors made on the joints of the beam with the frame and the proof mass show changes in resistance values. The piezoresistor located on the joint of the beam with proof mass experiences tensile stress and increases in value while that on the joint of the beam with the frame feels compressive stress and decreases in value. These stresses alter the resistances R_1, R_2, R_3, R_4 in directions indicated by arrows in figure 22.1, which are shown pointing upwards for increase in resistance and downwards for decrease in resistance. Variations in resistances lead to changes in the output voltage V_{OUT} of the bridge. The accelerometer is calibrated by measuring the output voltage as a function of the

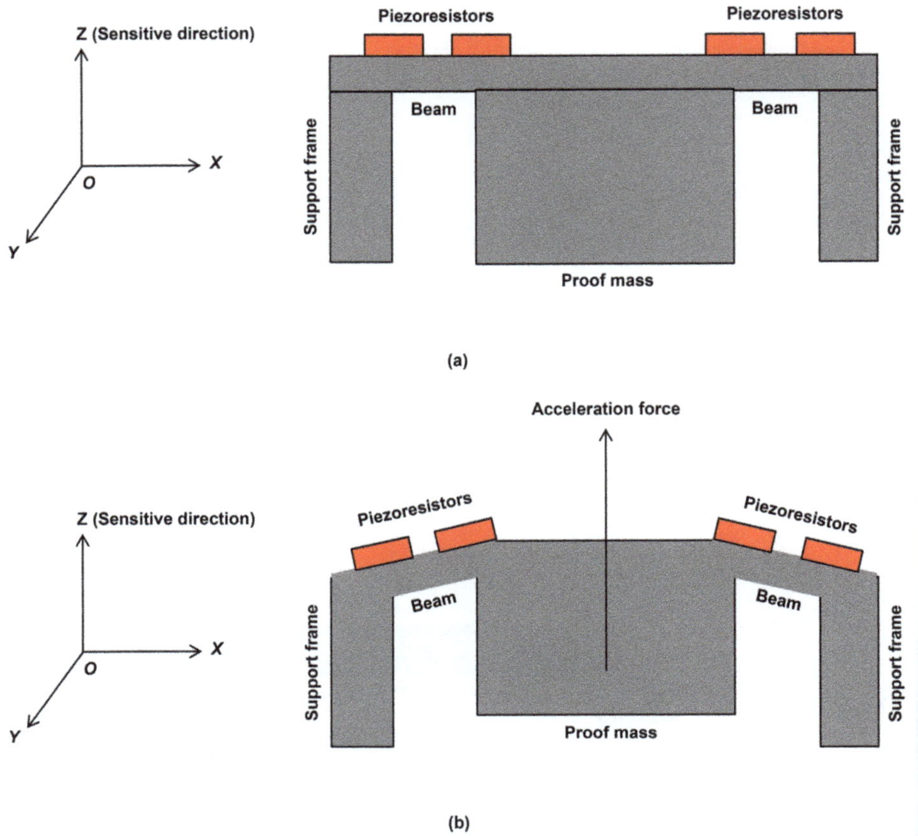

Figure 22.2. Cross-sectional diagram of the piezoresistive accelerometer: (a) without any applied acceleration and (b) with acceleration force in the *Z*-direction.

applied acceleration. The calibration graph provides an estimate of an unknown acceleration from the V_{OUT} measurement. The dynamic sensitivity of the accelerometer is 0.51 mV g^{-1}. It has a linearity of 99.8% (Zhai *et al* 2022a, 2022b).

22.4.2 Piezoelectric $YCa_4O(BO_3)_3$ (YCOB) single-crystal-based accelerometer

Piezoelectric accelerometers offer advantages of structural simplicity and fast response. They are also easily integrable with electronics (Jiang *et al* 2014). Several piezoelectric materials have been considered for fabricating accelerometers for high-temperature use. Quartz shows high losses at temperatures >350 °C. Transition of quartz takes place from its α to β phase at 573 °C (Kim *et al* 2011, 2012). $LiNbO_3$ starts decomposing at 300 °C losing oxygen and thereby resulting in a short lifetime of 10 days at 400 °C which decreases to 0.1 day at 450 °C. The plummeting of mechanical quality factor restricts usage of $GaPO_4$ to temperatures <700 °C. Langasite crystals exhibit no phase transition below their melting point (1470 °C). However, both the resistivity and quality factor decline at high temperatures because of diffusion of oxygen vacancies in the lattice. Piezoelectric properties of AlN thin film are maintained up to 1170 °C. Also, AlN has a high melting point (2200 °C). Yttrium calcium oxyborate with formula $YCa_4O(BO_3)_3$, written as YCOB, does not suffer phase transformation at temperatures less than its melting point (1500 °C) unlike quartz and $GaPO_4$. Moreover, it has a high resistivity (10 MΩ-cm) at 1000 °C. This resistivity value is 2–3 orders of magnitude higher than that of langasite crystal. Further, its piezoelectric properties exhibit stability at elevated temperatures (Zhang *et al* 2010).

22.4.2.1 Compression and shear-mode piezoelectric accelerometers

Two types of accelerometers must be distinguished: compression-mode and shear-mode types.

In a compression-mode accelerometer, the piezoelectric element, the electrodes and the proof mass are stacked one on top of another. A screw is passed through this assembly. The screw is tightened to a predesigned load value. During operation the piezoelectric element undergoes compression and expansion as the proof mass presses against it or the pressure due to proof mass is released; hence the name 'compression mode'.

In a shear-mode accelerometer, the piezoelectric element(s) are mounted radially around a fixed central post, and the proof mass is held around these piezoelectric elements with metal bands tightened with a screw. Under acceleration, shear forces come into play between the piezoelectric material and the proof mass; hence it is called 'shear mode'.

Although compression-mode accelerometers belong to the old technology, they are very robust and reliable showing very small drift over long periods of time.

22.4.2.2 Compression-mode piezoelectric accelerometer

This consists of a base plate separated from the piezoelectric element by an electrically insulating layer (figure 22.3). Over this insulating layer lies the piezo-electric crystal sandwiched between metal electrodes. The proof mass is placed above this sandwich and fixed by a preloading bolt in a housing by tightening the nut. An alumina rod connected to the base plate transfers the acceleration from a vibration source to the accelerometer. The electrical output of the accelerometer is measured by connecting wires across the metal electrodes.

The YCOB piezoelectric crystal has the dimensions: 15 mm × 7 mm × 2 mm. The proof mass is made of a nickel-chromium-based superalloy known as inconel known for its high temperature sustaining capability; and resistance to corrosion, oxidation, and creep. It has a coefficient of thermal expansion ~10 ppm K^{-1} matching well with that of YCOB and steel. This is necessary because the preloading bolt/nut and the housing are made of stainless steel. 50 μm thick platinum foils are used for the top and bottom electrodes.

Figure 22.4 shows how the output signal of the accelerometer changes when it is subjected to acceleration. A piezoelectric crystal with two electrodes and a current meter connected between the electrodes is shown. The piezoelectric current flows in opposite directions upon compression of the crystal and upon release of the compressive force.

Over a temperature range from room temperature to 1000 °C and frequency range 100–600 Hz, the sensor shows a stable sensitivity of 2.4 pC g^{-1}. Its behavior is also stable during storage for an extended period of time >3 h at 900 °C (Zhang *et al* 2010).

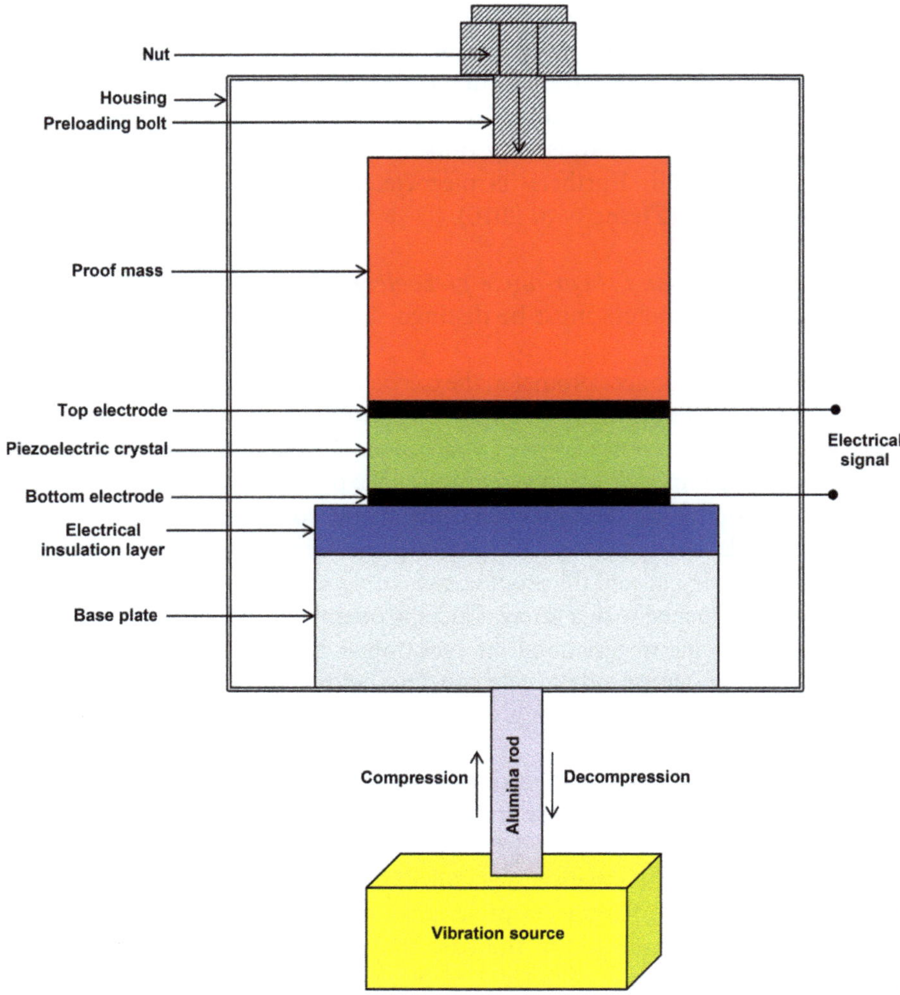

Figure 22.3. Compression-mode piezoelectric accelerometer.

22.4.2.3 Shear-mode piezoelectric accelerometer

It consists of a base plate with a center post (figure 22.5). The piezoelectric crystal is wrapped coaxially around this post (Kim *et al* 2011). Surrounding the crystal is the proof mass layer. A retaining ring is tightened around the assembly with a washer and a nut to hold the components in place. The components are enclosed inside a housing fixed to the base plate. An alumina rod is attached to the center post to transfer the acceleration from a vibration source to the accelerometer. The electrical signal is recorded between the center post and the proof mass.

The piezoelectric crystal is (YXt)-30° cut YCOB single crystal. Its dimensions are 10 mm × 12 mm × 1 mm. The proof mass, center post, housing and wiring are all made of inconel 601 alloy (Ni: 58%–63%, Cr: 21%–25%, Al: 1%–1.7%, balance Fe). Four pieces of YCOB crystal arranged on both sides of the center post are firmly

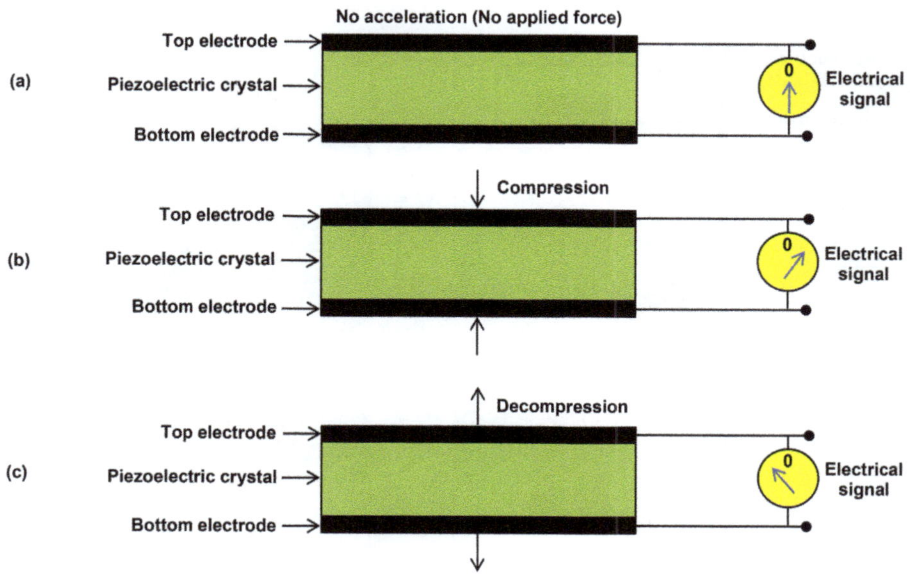

Figure 22.4. Effects of acceleration forces on the output signal of compression-mode accelerometer: (a) zero output signal in absence of acceleration, (b) electrical current flow when the acceleration and inertial forces compress the piezoelectric crystal and (c) electric current flow when the compression forces are released causing the crystal to expand.

secured with the help of proof masses by a bolt. The electrode-less design of the accelerometer has an inherent advantage. Absence of thin film electrodes and conductive adhesives makes the accelerometer capable of withstanding high temperature without suffering any oxidation and corrosion.

Figure 22.6 shows the influence of acceleration on the output signal of the accelerometer. Upon acceleration, shearing takes place between the proof mass and the piezoelectric crystal producing a charge and hence current flow. A current meter is connected between the proof mass and the piezoelectric crystal. The direction of current depends on the direction of shearing.

In the tested frequency range 80–1000 Hz and between room temperature and 1000 °C, the accelerometer shows a sensitivity of 5.4–5.7 pC g^{-1}, which is maintained during 4 h storage at 1000 °C (Kim *et al* 2011). In another experiment, the prototype accelerometer tested between 50 and 350 Hz and from room temperature to 1000 °C shows a sensitivity of 5.9 pC g^{-1}, and this value remains constant for 9 h at 1000 °C (Kim *et al* 2012).

22.4.3 Optical accelerometer

The optical accelerometer is a quad-beam accelerometer (Plaza *et al* 2004). The accelerometer (figure 22.7(a)) consists of four beams, viz., beam 1, beam 2, beam 3 and beam 4. Each beam is anchored to the support frame on one side and attached to the proof mass on the other side. The accelerometer is glued to a piece of aluminum. On the top surface of the accelerometer mounted on aluminum piece, a

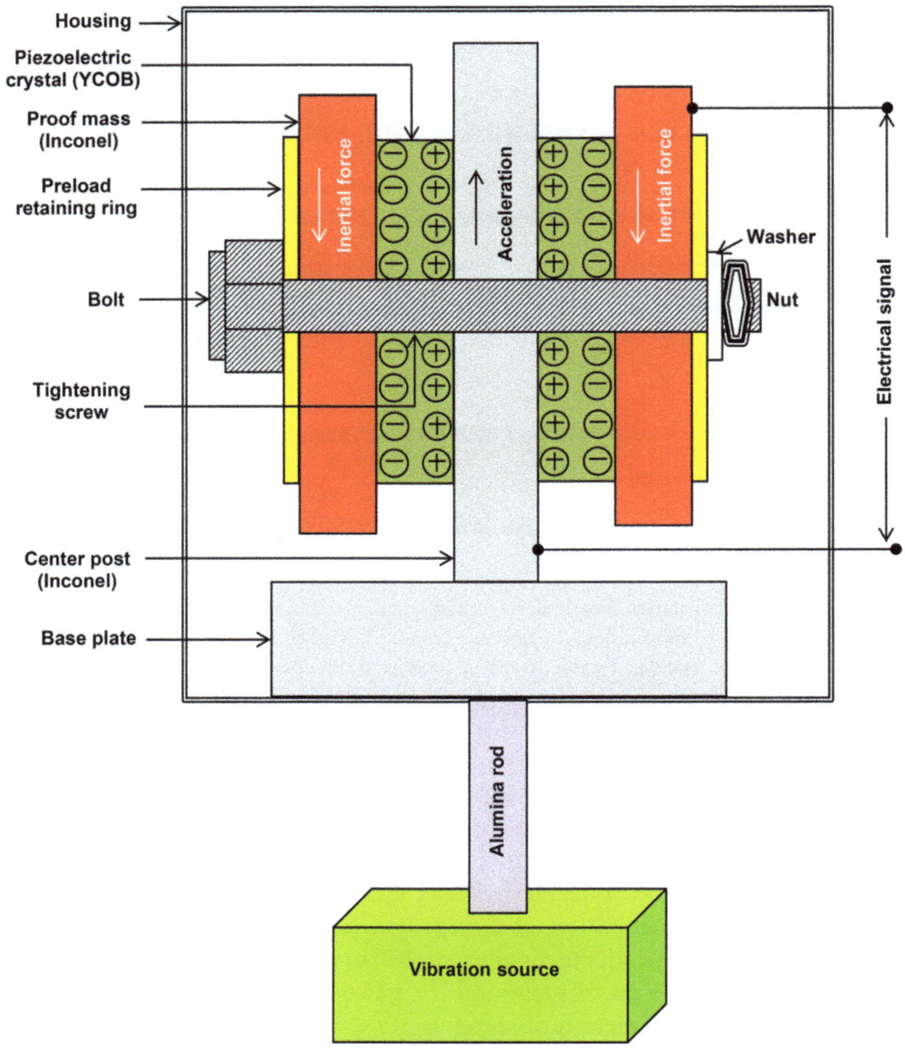

Figure 22.5. Shear-mode accelerometer.

three-segment waveguide is laid down with two segments resting on the support frame and one segment on the proof mass. The three waveguide segments are arranged along one straight line and are aligned with each other.

For measurement (figure 22.7(b)), two optical fibers are placed on V-groove chips on opposite sides of the accelerometer. They are glued to pieces of aluminum to facilitate alignment with the waveguide on the proof mass of the accelerometer. Light enters through the input optical fiber, strikes the input waveguide, passes through the waveguide on the proof mass, falls on the output waveguide, and finally reaches the output optical fiber.

Figure 22.8 shows how the acceleration affects the intensity of output light from the accelerometer due to misalignment of the waveguides. Sensitivity is given

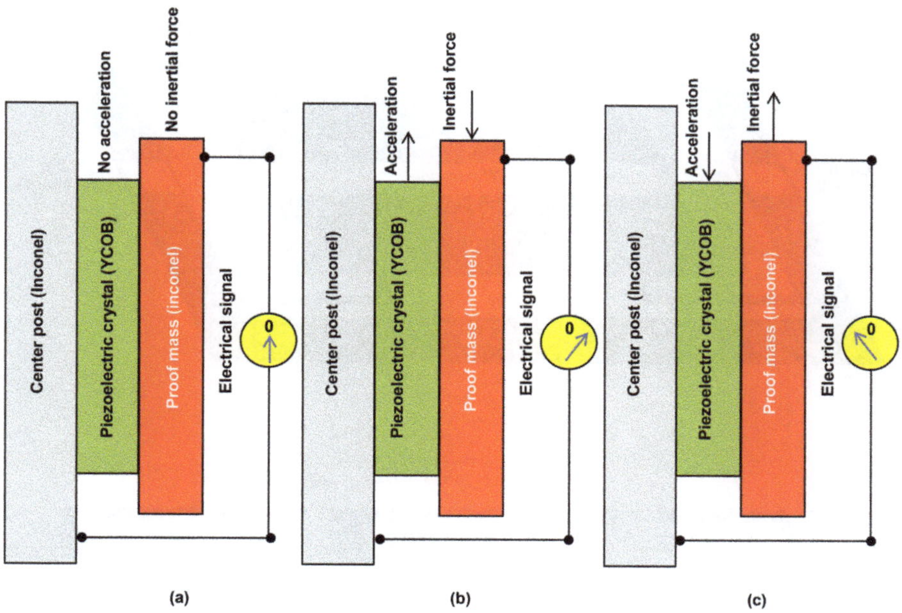

Figure 22.6. Effects of acceleration forces on the output signal of shear-mode accelerometer: (a) zero output signal in absence of acceleration, (b) electrical current flow for shearing caused by acceleration and inertial forces in one direction and (c) electric current flow when shearing takes place in reverse direction.

in dB g^{-1} because optical losses are measured as a function of misalignment. Measurements are performed at 633 nm. For positive acceleration, the optical sensitivity is 2.3 dB g^{-1}. For negative acceleration, it is 1.7 dB g^{-1}. Positive and negative accelerations refer to positive and negative misalignments. Positive acceleration causes the proof mass to move upwards, whereas negative acceleration makes it move downwards.

Advantages of the optical method of acceleration measurement are (Plaza *et al* 2004):

(i) Ability to sense in an EMI environment: This sensor combines micromachined silicon structure with integrated optics. The combination is immune to EMI. So, the sensor can be placed in an environment where EMI is expected, e.g., in electric power generation stations or near power transformers. Intense electromagnetic fields exist in/around these power systems. Ability to work in an EMI environment is a distinct advantage of optical sensor over the electrical sensor whose operation is impaired in these situations.

(ii) Remote sensing ability: The source of light and the photodetector can be located far away from the measurement zone. The ability of accessing and measuring from a distance is particularly useful when the measurement zone is an unsafe area such as a hazardous region with littered radioactive waste where human exposure must be avoided or a place where explosion may be triggered by electrical spark rendering electrical methods unusable (Plaza *et al* 2004).

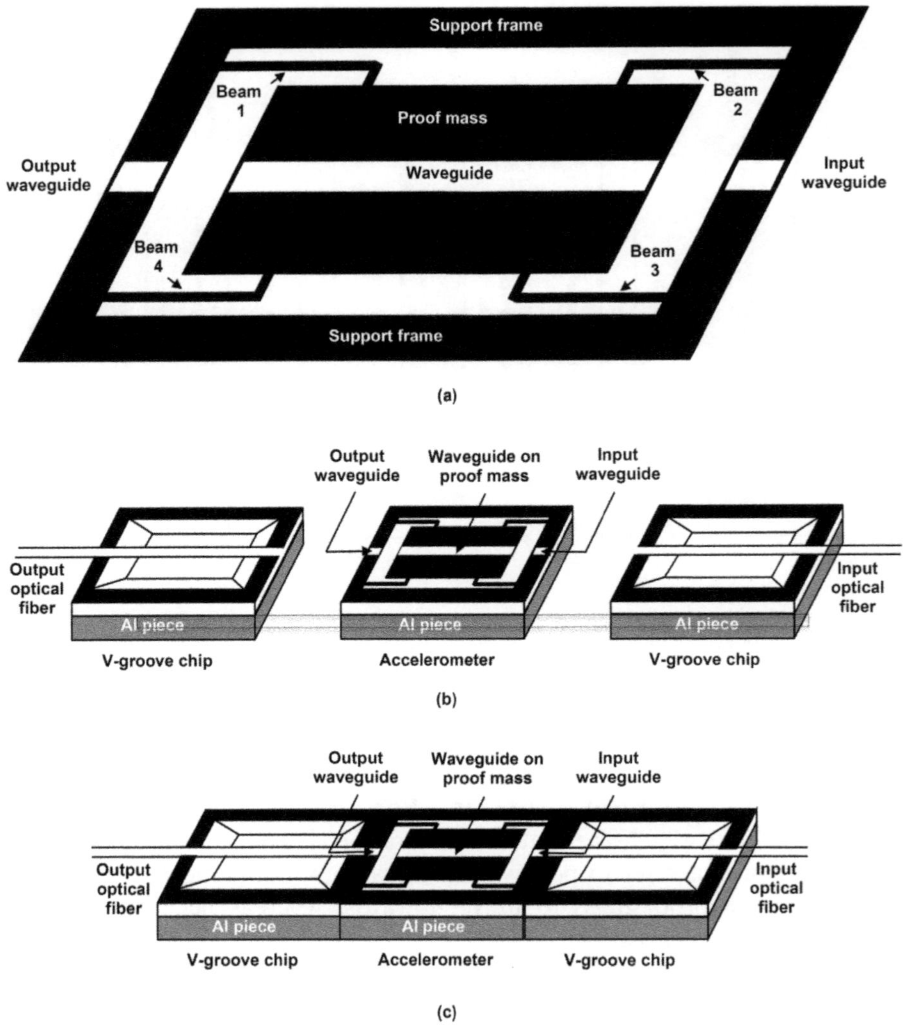

Figure 22.7. Optical accelerometer: (a) quad-beam accelerometer with optical waveguides, (b) setup of the separated three parts of the accelerometer, and (c) alignment and joining of three parts for measurements.

22.5 Flow sensors

Flow sensors are in great demand in automotive, aerospace, chemical and energy industries for applications such as combustion optimization and control of gas emissions, chemical process control, etc (Balakrishnan *et al* 2017).

22.5.1 3C-SiC on-glass-based thermal flow sensor

The 3C-SiC on-glass-based thermal flow sensor works on the principle of hot-wire anemometry in which the flow rate of a fluid is determined from the heat exchange between a heated element and a nearby-located temperature sensor in the fluid stream.

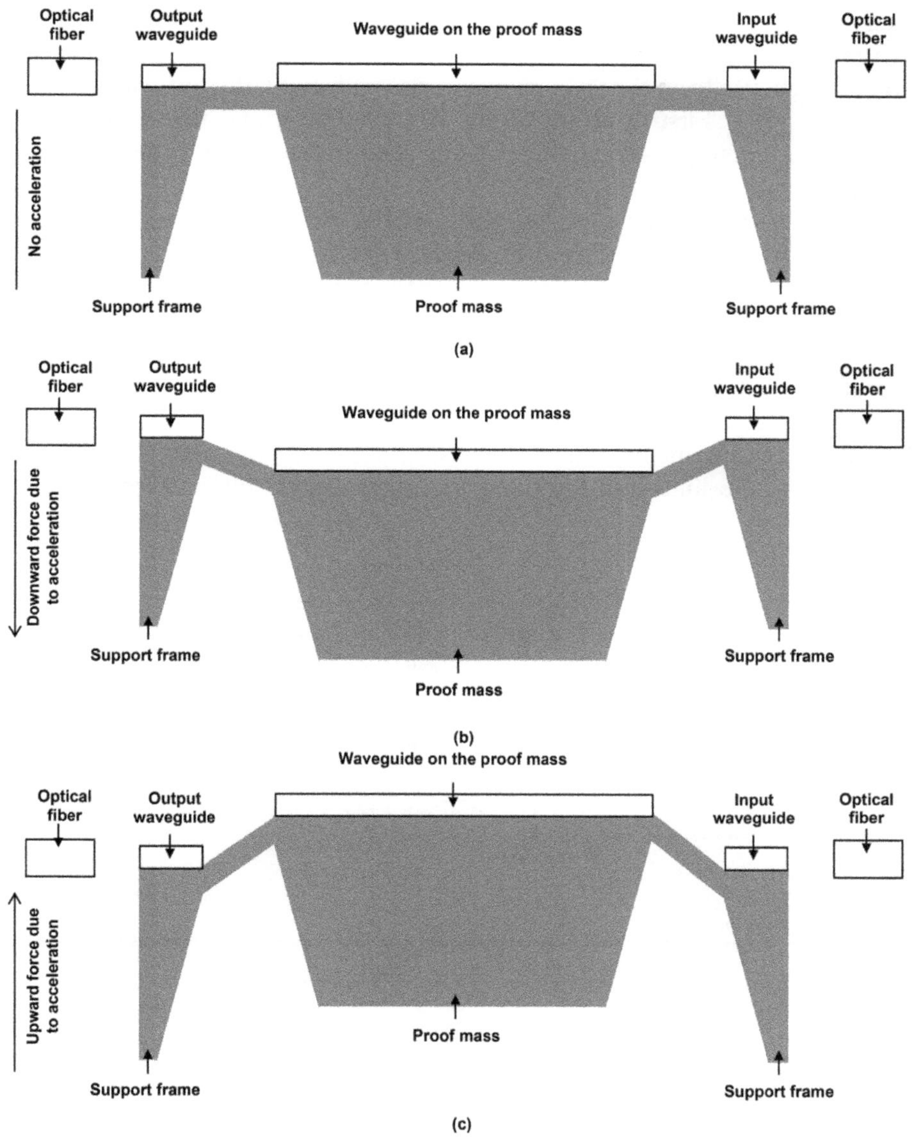

Figure 22.8. Variation of intensity of light passing through the accelerometer with acceleration, showing three cases: (a) without acceleration: the optical fibers and the waveguides are all aligned so that optical losses are minimum and intensity of light received in the optical fiber on the output side is maximum, (b) force due to acceleration acting in downwards direction: the force pushes the proof mass downwards causing misalignment of the waveguides with the optical fibers so that intensity of output light decreases, (c) force due to acceleration acting in upwards direction: the force deflects the proof mass upwards affecting the intensity of output light in a similar way to that in (b).

Three polytypes of SiC are commonly used in electronics: 4H-SiC, 6H-SiC and 3C-SiC. Of these three polytypes, only 3C-SiC can be directly grown on a silicon substrate by heteroepitaxy. But when SiC grown over Si is exposed to a high temperature, the conductivity of underlying silicon substrate increases around 200 °C.

Current leakage starts between Si and SiC. Further, the yield strength of silicon decreases above 450 °C impacting the device performance. So, for making an HT sensor, it is better to form the SiC film over a glass substrate. But SiC cannot be grown directly over glass. First growing 3C-SiC film on silicon and then transferring the grown 3C-SiC film to glass by anodic bonding helps to resolve this issue (Balakrishnan *et al* 2018).

Figure 22.9 shows the chip of the thermal flow sensor and its electrical connections. As shown, the chip is made on a glass substrate. Over the surface of the glass substrate, there is a 3C-SiC heater in the central region. On both sides of the heater, there are temperature sensors. The sensors are also made of 3C-SiC. From experimental characterization of thermo-resistive properties of 3C-SiC on-glass, the temperature coefficient of resistance of 3C-SiC has a negative value. This value is −20 716 ppm K^{-1}.

For flow measurements, the heater is turned on. One sensor measures the temperature of the inflowing fluid and the other sensor gives the temperature of

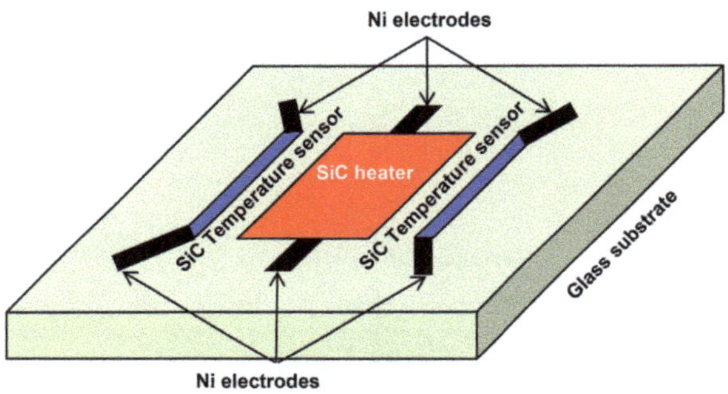

(a)

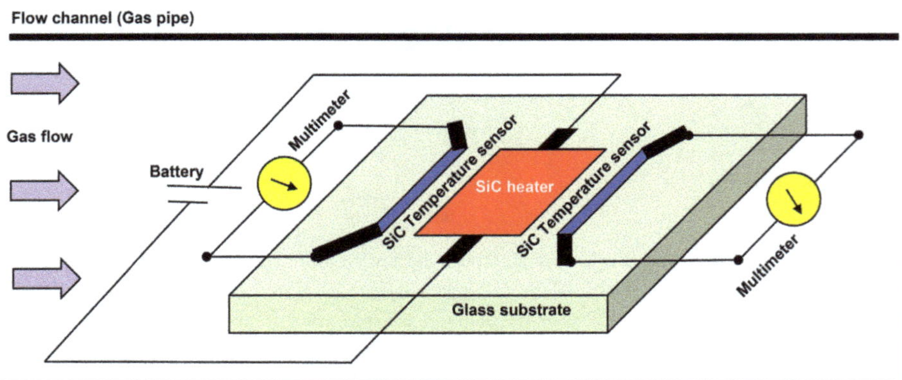

(b)

Figure 22.9. 3C-SiC-on-glass thermal flow sensor: (a) the sensor chip consisting of a SiC heater and two SiC temperature sensors on a glass substrate for measuring the downstream and upstream gas temperatures. Both the heater and the sensors have nickel electrodes; and (b) the electrical connections of the sensor for flow measurement in a gas pipe showing the heater supplying current from a power source and the electrical output of the temperature sensors read from multimeters.

the outflowing fluid. The outgoing fluid is at a higher temperature because of the heat exchange between the heater and the incoming fluid. The heat interchange itself is governed by the gas flow rate. The sensitivity is the output relative resistance change ($\Delta R/R$) per unit flow rate in m s^{-1}. For a 3C-SiC heater of size (1000 μm × 1000 μm) with power consumption 133.5 mW, and turbulent flow velocity of 7.4 m s^{-1}, the sensitivity is 0.091 s m^{-1} (Balakrishnan *et al* 2018).

22.5.2 Fiber optic flow sensor

Fiber optic sensors are well suited to environments at high temperatures, those containing corrosive materials and those in which strong electromagnetic fields are present. Like the 3C-SiC on-glass-based thermal flow sensor described in section 22.5.1, the fiber optic flow sensor too works on the principle of hot-wire anemometry (Chen *et al* 2014).

The optical flow sensor is shown in figure 22.10. It consists of an in-fiber heating element in the form of HAF (high-attenuation fiber), an RFBG (regenerated fiber Bragg grating) inscribed in HAF to measure the temperature of the heating element, and an RFBG in SMF-28 (single mode optical fiber-28) located upstream of the flowing gas to measure the temperature of gaseous ambience surrounding the heating element.

For applying the optical hot-wire anemometry principle, the influence of gas flow on the transference of heat from the optically heated HAF element to the RFBG inscribed in HAF is assessed to calculate the gas flow rate. The sensor uses in-fiber light for performing the dual function:

(i) as a source of heat in the HAF heating element for optical heating, and
(ii) as a source of light in the RFBG sensor for temperature measurement, i.e., as an interrogation source of light.

Reliable flow measurements are performed in a wide temperature range spanning from room temperature to 800 °C, and also in a broad flow rate of nitrogen 0.066 m s^{-1} to 0.66 m s^{-1} (Chen *et al* 2014).

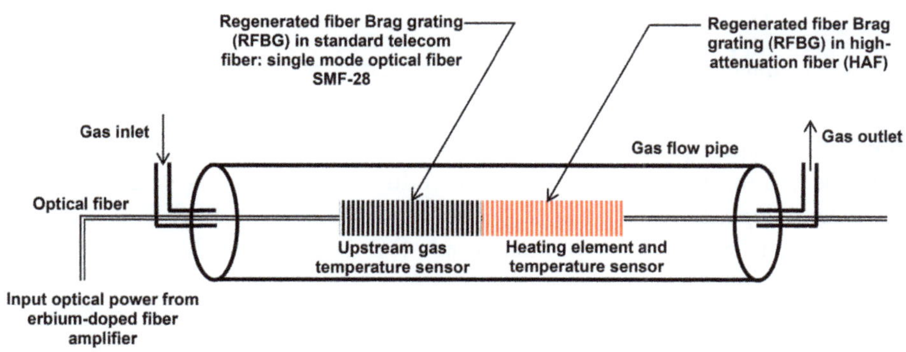

Figure 22.10. All-optical flow sensor.

22.6 Pressure sensors

Applications such as oil prospecting and combustion control of gas turbines of aircraft require pressure sensors capable of working at temperatures >220 °C.

A piezoresistive pressure sensor with integrated signal conditioning is fabricated by MEMS technology using SOI material (Yao *et al* 2016). It works from −50 °C to +220 °C.

Brittle-to-ductile transition takes place in silicon at a temperature between 520 °C and 600 °C. The stressed single-crystal silicon is susceptible to creep, and continuous deformation may lead to unacceptable changes in dimensions (Ren *et al* 2016).

After silicon falls, we look towards silicon carbide displaying stable electronic properties up to 600 °C. Between 350 °C and 600 °C, SiC piezoresistive and capacitive pressure sensors work satisfactorily (Wieczorek *et al* 2007, Marsi *et al* 2015, Beker *et al* 2018, Nguyen *et al* 2018).

22.6.1 Silicon carbide capacitive pressure sensor

Figure 22.11 shows the SiC pressure sensor. The top and the bottom electrodes of the sensor are made of N-type heavily-doped silicon carbide films formed by LPCVD (low pressure chemical vapor deposition) (Jin 2011, Jin *et al* 2011, 2012). LPCVD low-stress silicon nitride (LSN) isolates the substrate from the capacitor. LPCVD low-temperature oxide (LTO) is used as a sacrificial layer to release the diaphragm. The electrodes are made from Ta/$TiSi_2$/Pt metallization deposited by sputtering. $TiSi_2$ acts as a diffusion barrier between tantalum and platinum. An LTO ring around the diaphragm seals the cavity.

When subjected to a pressure, the 2.7 μm thick poly-SiC sensing diaphragm is deformed and the gap between the top and bottom electrodes of the sensor decreases resulting in an increase in capacitance. The capacitance variation with pressure is used to measure unknown pressures.

To study the effect of thermal mismatch between the substrate and SiC, sensors are fabricated on Si and poly-SiC substrates. The effect of substrate material on sensor performance is found to be of little significance (Jin *et al* 2011). In the sensor for the 6.9 MPa pressure range, the gap between the electrodes is 1.5 μm and the radius of the diaphragm is 70 μm. The sensor shows a sensitivity of 0.03 fF Pa^{-1}, and the nonlinearity is 4% at 500 °C in contact mode operation (Jin *et al* 2012).

22.6.2 Micromachined pressure sensor with sapphire membrane and platinum thin film strain gauges

Two main reliability problems associated with SiC pressure sensors are:

(i) the reliability of the metal–semiconductor contacts for wide temperature excursions, and
(ii) the reliability of materials used for packaging the sensor at high temperatures.

To solve these problems, sapphire is chosen as the material for the pressure sensor diaphragm (Fricke *et al* 2012).

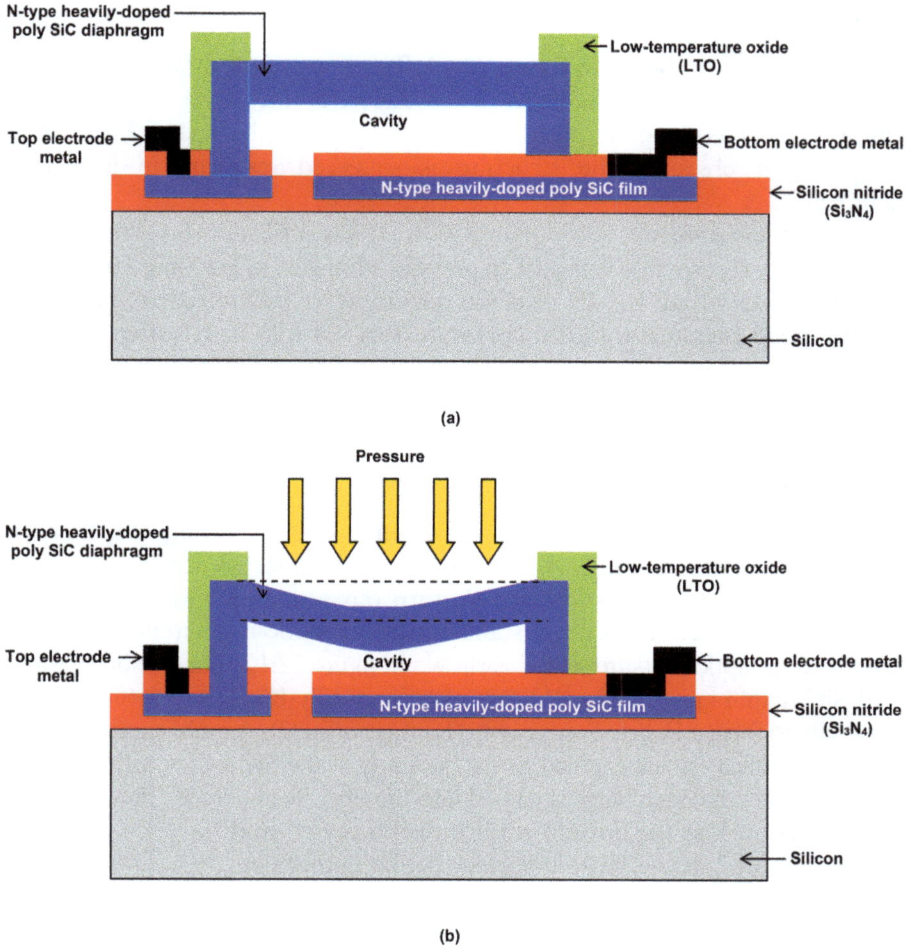

Figure 22.11. SiC capacitive pressure sensor made on a silicon substrate by surface micromachining: (a) in the absence of applied pressure and (b) with pressure applied on the diaphragm.

Solution to problem (i): A micromachined pressure sensor is formed with Pt strain gauges on a sapphire diaphragm without any metal–semiconductor contacts to ensure long-term stability at high temperatures. Two methods are tried for making the diaphragm with a targeted thickness of 200 μm, namely, abrasive drilling and ultrasonic-assisted drilling. The latter technique saves time and reduces cost. 1 μm thick platinum thin film is deposited in Ar atmosphere after chemically activating the sapphire surface by RF cleaning. No adhesion promoter is used. Platinum thin-film strain gauges are made in a meander-shaped design over the diaphragm.
Solution to problem (ii): during packaging, a Ni-based superalloy material is used for the package body. The contacts are made of platinum and MACOR-high temperature machinable glass ceramic is used for electrical insulation.

The sensor is characterized up to 440 °C temperature and 30 bar pressure. Its sensitivity is 10 $\mu V\ V^{-1}\ bar^{-1}$ (Fricke *et al* 2012).

22.6.3 Ceramic nanofiber-based flexible pressure sensor

The flexible sensor is shown in figure 22.12. The sensor is a sandwich structure in which a network of TiO_2 nanofibers is the dielectric and silver nanowires-coated polyimide is used as the material for the two conductive electrodes (Fu *et al* 2020). On applying a pressure, the structure is squeezed resulting in a change in capacitance.

The sensor works satisfactorily up to 370 °C. It has sufficient endurance to survive in butane flame. It responds very fast to pressure changes, in less than 16 ms. Its limit of detection is less than 0.8 Pa and the sensitivity = relative capacitance change per unit applied pressure = $(\Delta C/C)$/pressure P = 4.4 kPa^{-1}. It suffers low fatigue after 50 000 cycles of pressure loading and unloading. It is suitable as a wearable device for monitoring health in real time.

22.6.4 All SiC fiber optic pressure sensor

Electrical pressure sensors face many problems, notably strong dependence on temperature and susceptibility to electromagnetic interference besides lack of repeatability and long-term stability. We turn our attention to fiber optic sensors when tolerance to high temperatures and electromagnetic interference are essential.

The SiC fiber optic pressure sensor is shown in figure 22.13. The sensor consists of a thinned SiC diaphragm (thickness 200 μm) hermetically bonded to a SiC wafer with etched cavity to form a Fabry–Perot cavity (Liang *et al* 2022). A multi-mode optical fiber inserted into a silica ferrule is glued to the backside of the SiC wafer with an adhesive. White light from a halogen lamp is passed into the fiber. It undergoes two reflections:

(i) reflection 1 at the bottom surface of the cavity, and
(ii) reflection 2 at the bottom surface of the diaphragm.

The two reflected beams interfere between themselves forming an interference spectrum. The interference signal is fed to a spectrometer to extract the length L of the cavity. The cavity length varies with the applied pressure. The pressure is monitored from the changes in cavity length L.

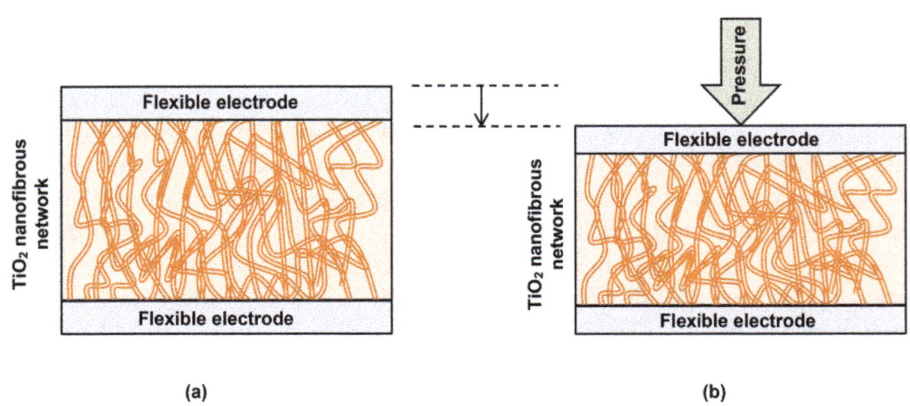

Figure 22.12. TiO_2 nanofiber capacitive pressure sensor.

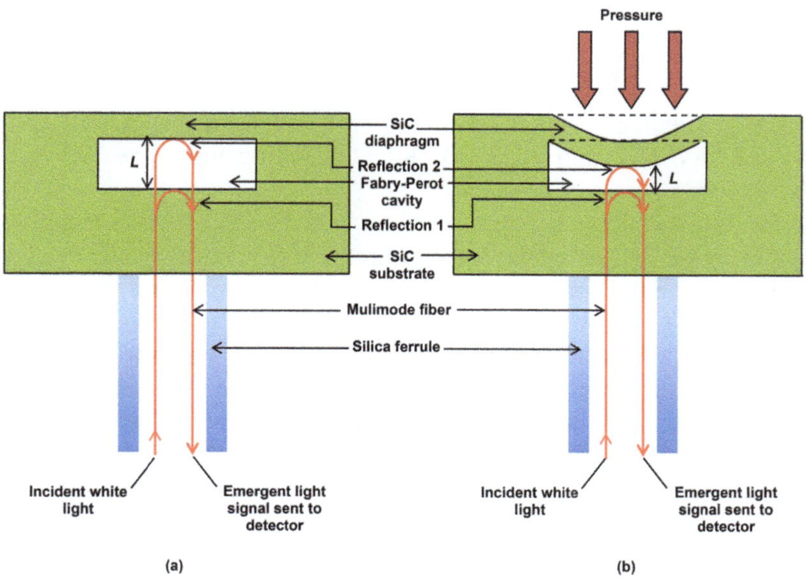

Figure 22.13. All-SiC fiber optic pressure sensor: (a) without applied pressure and (b) under applied pressure.

The sensor has a linear cavity length characteristic from 0 to 800 kPa and 23 °C–400 °C. At 23 °C, the cavity length decreases from 14.8 μm at 0 kPa to 13.3 μm at 800 kPa. At 400 °C, it decreases from 15.05 μm at 0 kPa to 13.58 μm at 800 kPa (Liang *et al* 2022).

22.7 Temperature sensors

22.7.1 SOI diode temperature sensor

Peak allowed junction temperature for CMOS ICs has climbed from 150 °C to 175 °C, and then to 200 °C using electromigration-resistant metals, copper or tungsten (Santra *et al* 2008, 2010). SOI technology has reduced leakage currents and parasitic capacitance due to isolation from bulk silicon. It has also minimized latch-up triggering due to isolation of N-well and P-well structures, thereby uplifting the allowed temperature limit to 225 °C–250 °C for applications in automotive industry; control of motors in power electronic systems; in power supplies, medical equipment, and nuclear plants. Discrete temperature sensors such as platinum resistance thermometers, thermocouples and thermistors guarantee high accuracy of measurements in a broad temperature range up to 2000 K, and are therefore immensely popular. But the need of effective management of temperature of power-hungry complementary metal–oxide–semiconductor (CMOS) circuits make on-chip temperature sensors necessary for *in situ* measurements. To cater to such requirements, an integrated SOI temperature sensor is most useful. Along with the temperature sensor, the signal processing circuit and intelligence hardware is also included in the IC for interpretation of the temperature reading and initiation of proper action in case of unwanted temperature excursions which may cause chip burnout.

The diode temperature sensor works on the principle: when a diode is driven by a constant forward current, a forward voltage drop is produced across the diode. This voltage drop shows a linear variation with temperature. The calculated value of temperature gradient of a silicon diode is -2.1 mV K^{-1}, but in practice it varies between -1.2 mV K^{-1} and -2.2 mV K^{-1} depending upon the forward current (Udrea *et al* 2008).

Figure 22.14(a) depicts the structure of an SOI diode in CMOS technology. The device has an $N^+/P^-/P^+$ gated diode structure. It is fabricated in the device layer of the SOI wafer above the buried oxide, and is separated from other parts of the circuit by shallow trench isolation (STI).

The practical application of the diode as a temperature sensor in a microheater-embedded gas sensor is illustrated in figure 22.14(b). The diode is a part of CMOS circuit. It is suspended on a membrane which is the buried oxide of the SOI wafer. The membrane is formed by deep reactive ion etching of silicon. The etching is terminated on reaching the buried oxide layer which acts as an etch-stop layer. The thermodiode is embedded inside an SiO_2 layer. The contacts to the thermodiode are

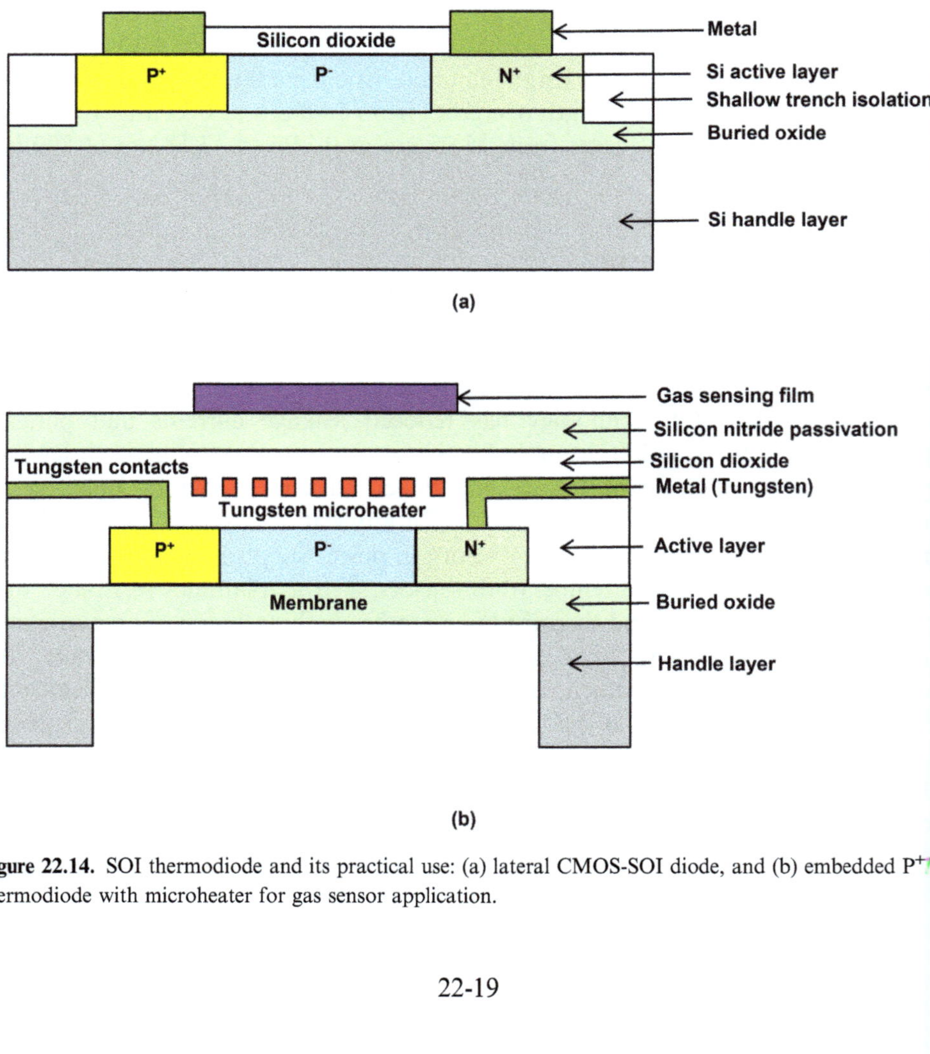

Figure 22.14. SOI thermodiode and its practical use: (a) lateral CMOS-SOI diode, and (b) embedded P^+/N^+ thermodiode with microheater for gas sensor application.

made of tungsten. Tungsten is preferred over aluminum as a metallization layer because of its high temperature withstanding capability. The melting point of tungsten (3422 °C) is much higher than that of aluminum (660.3 °C). Moreover, tungsten is comparatively less prone to degradation by electromigration than aluminum. The microheater embedded above the diode in the SiO_2 layer also uses tungsten as the resistive heating material. The thermodiode can operate in a broad temperature range from −200 °C to +700 °C maintaining a linear characteristic (De Luca *et al* 2013).

Experimental, analytical and numerical studies of nonlinearity of characteristic have revealed that the SOI diode temperature sensor performs well in a wide range of temperatures from 80 K to 1050 K (De Luca *et al* 2015).

22.7.2 LTCC wireless temperature sensor

Low-temperature cofired ceramic (LTCC) technology uses ceramic substrate tapes to produce multilayer circuits. In this technology, conductive, resistive and insulating pastes are applied on the tapes. The tapes are laminated together and fired at a low temperature ~850 °C forming a monolithic, hermetic structure. The LTCC module is very robust against mechanical and thermal stresses, has a high dielectric strength and displays long-lasting physical and chemical stability exceeding 20 years. The outstanding stability bestows on the LTCC device the ability to cope with high temperatures and corrosive atmospheres.

Figure 22.15 shows the LTCC wireless temperature sensor. The sensor consists of a planar spiral inductor connected to a temperature-dependent capacitor made from a $PbNb_2O_6$-based ferroelectric ceramic as dielectric (Tan *et al* 2014). The relative permittivity of the ferroelectric ceramic varies with temperature. The resultant shift in resonant frequency of the LC circuit is contactless measured from the impedance parameters of an external antenna. From room temperature to 430 °C, the sensitivity is −5.75 kHz $°C^{-1}$. In the 430 °C–700 °C temperature range, it is −16.67 kHz $°C^{-1}$.

22.7.3 Langasite SAW resonator-based high temperature sensor

The primary advantage of surface acoustic wave (SAW) sensors is that they allow wireless interrogation. No battery or embedded electronic circuit is necessary. This unique feature makes them popular in applications in which wiring is difficult. They are used in measuring physical parameters of moving parts at high temperatures in metallurgy, glass, oil extraction and other industries.

The SAW temperature sensor is shown in figure 22.16. The sensor consists of a substrate of a piezoelectric material, here langasite, with an interdigitated (IDT) electrode pattern and clamped/brazed antennas in the middle of the substrate, and reflector patterns on the two sides of this central IDT pattern (Wall *et al* 2015). Also shown is a transmitter sending RF pulse to the sensor. The received signal is analysed by a reader to display the temperature to which the sensor was exposed.

The principle of temperature measurement by a SAW sensor is: electromagnetic waves sent from a transmitter induce surface acoustic waves in the substrate. The

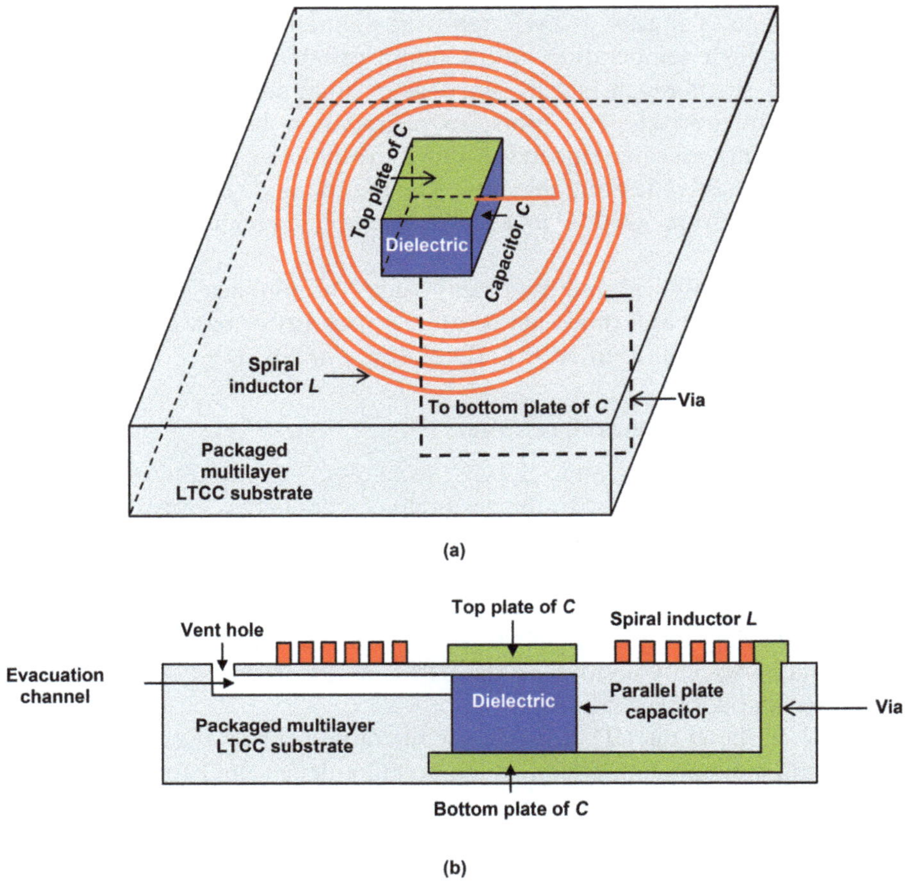

Figure 22.15. LTCC-technology-based LC resonant circuit temperature sensor; (a) 3D view and (b) cross-sectional diagram.

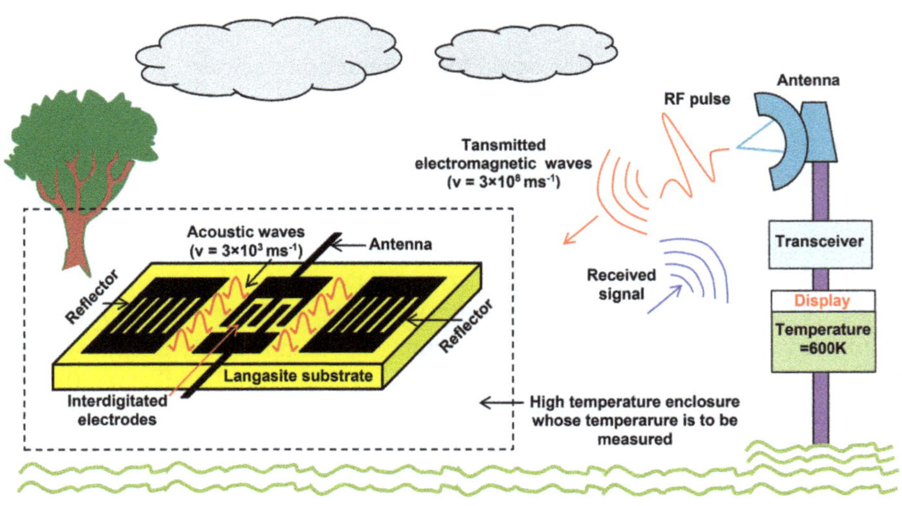

Figure 22.16. Wirelessly accessed SAW (surface acoustic wave) high-temperature sensor.

acoustic energy is stored in the resonator. After stoppage of transmitted electromagnetic waves, the acoustic energy stored in the resonator is backscattered. The frequency of this backscattered signal depends on the temperature to which the SAW sensor has been exposed because the resonant frequency of the device is a function of temperature. Thus, temperature sensing by a SAW device entails electrical induction of surface acoustic waves in a piezoelectric material substrate, and reconversion of the acoustic energy affected by the temperature of the SAW sensor into an electrical signal to extract the temperature of the environment in which the sensor was placed.

Explaining more elaborately, a short duration RF pulse called the interrogation pulse is sent to energize the resonator. This happens when the frequency spectrum of RF pulse overlaps with the resonance frequency of the resonator. After the interrogation pulse ends, a long ringing signal is broadcast by the resonator. The received signal is subjected to amplification, filtration, digitization and analysis to find the frequency of the resonator. The unknown temperature is determined from the stored data of sensor calibration and the obtained resonator frequency. For wireless measurements, the frequency chosen is 433 MHz and a suitable antenna is optimized. Measurements are performed up to 700 °C. Stability is tested by temperature cycling between 50 °C and 650 °C, and keeping at 600 °C for 1000 h dwell time.

Important provisions made in this sensor to achieve high temperature operation capability are:

(i) Langasite is chosen as the piezoelectric material because it shows no phase transitions up to its melting point (1470 °C). Langasite represents lanthanum gallium silicate, with the acronym LGS and chemical formula $La_3Ga_5SiO_{14}$.

(ii) Particular care is taken regarding metallization. Pure platinum shows dewetting at 600 °C. To prevent dewetting, platinum metallization is modified. By annealing at a high temperature, an Al/Al_xO_y surface layer is produced on platinum yielding a multilayer stack $Al/Al_xO_y/Pt$ structure which does not agglomerate at elevated temperatures.

(iii) A sturdy and reliable package known as eHTP (extremely high temperature package) is developed. It is a high temperature cofired ceramic (HTCC) package in which the tungsten metallization is replaced by the platinum-based metallization which is well suited to interconnection by soldering, wire bonding and brazing (Wall *et al* 2015).

22.7.4 Sapphire fiber Bragg grating as temperature sensor

Monitoring crystal growth, molten metal or ceramic processing as well as optimization of combustion engine processes involves measurements of high temperatures above 1000 °C. Although noble metal thermocouples are useful up to 1700 °C, they are prone to drift, and require frequent re-calibration and replacement. Moreover, a metallic sensor cannot be used in an electromagnetic environment. Pyrometers are applied at temperatures >2000 °C but their usage is limited by the fact that their

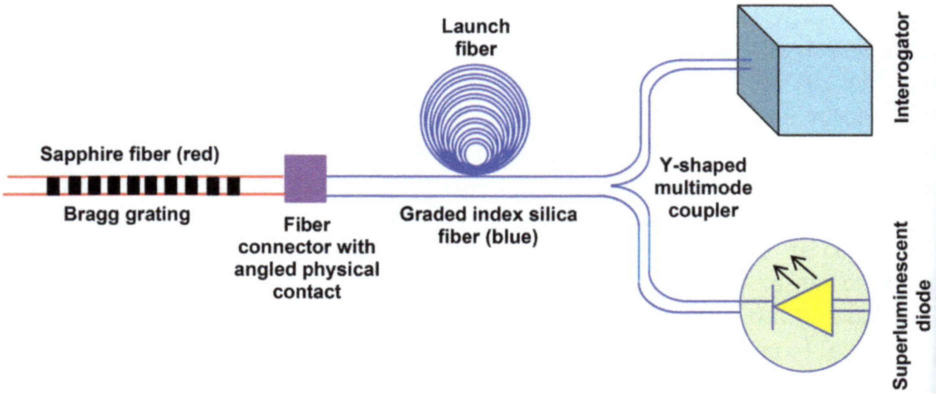

Figure 22.17. Setup of SFBG (sapphire fiber Bragg grating) high-temperature sensor.

direct optical access to the hot surface is essential. Fiber optic sensors are not affected by electromagnetic fields and can be used at high temperatures. But the fused silica used in conventional sensors has a glass transition temperature ~1260 °C. The next option is single-crystal sapphire in the form of a fiber. The fiber is grown from the melt. It can be used up to its melting point ~2050 °C. For measuring temperature with a fiber, a Bragg grating is inscribed in the fiber with the help of a femtosecond laser source.

The layout of a sapphire fiber Bragg grating temperature sensor is drawn in figure 22.17. A superluminescent diode (SLD) with maximum intensity at 1545 nm and a fiber Bragg grating (FBG) interrogator are connected to a Y-shaped fiber coupler (Habisreuther *et al* 2015). A 50 m launch cable works as a mode scrambler making possible the homogeneous excitation of the guided modes in a multi-mode single-crystalline fiber of sapphire. FC/APC (fiber connector with angled physical contact) plugs are used for connecting the silica fiber with sapphire fiber. In the sapphire fiber, a Bragg grating is inscribed. A stable, smooth reflection spectrum is produced. The spectrum contains the temperature information. The graph of reflected intensity from the FBG against wavelength shows several peaks between room temperature to 1900 °C. The positions of the peaks depend on temperature. The reflected intensity signal is processed. The variation of Bragg wavelength with temperature is described by a second order polynomial. The shift in Bragg wavelength with temperature enables fast temperature diagnostics.

The sensor is used for temperature measurements from room temperature to 1900 °C with a resolution better than ±2 K. Fast dynamic temperature monitoring can be done at a rate of 20 Hz (Habisreuther *et al* 2015).

22.8 Humidity sensors

22.8.1 Micromachined humidity sensor

Conductive type humidity sensors using porous oxide semiconductors, e.g., $MgCr_2O_4$–TiO_2 are highly sensitive to humidity, but their basic operating mechanism involves the change in their surface conductivity by absorption of moisture on the surface. Contamination of the active surface causes instability and drift of sensor characteristics leading to fast aging (Lee *et al* 2001).

Capacitive type humidity sensors work on the change in dielectric properties of a moisture-sensitive material such as a polyimide film with relative humidity. The hygroscopic polyimide film is not chemically stable over an extended period of time. These observations indicate the dire necessity of a rugged and reliable humidity sensor for food processing industries where the fumes of the cooked food condense on the sensor surface, and may cause sensor failure.

The sensor (figure 22.18) consists of two Ti/Pt meander resistors 1 and 2 on a thin membrane of silicon oxide–silicon nitride–silicon oxide (ONO). The resistance–temperature characteristics of platinum thin film are applied to determine the difference in thermal conductivity between moist air and dry air. The chip is encapsulated in a transistor outline (TO) package. It is mounted on a metal stem platform with a hole below the resistors 1. Resistors 1 act as the sensitive component while resistors 2 serve as the compensating component. The space above the chip is filled with nitrogen. The resistors are heated to a temperature of 250 °C by connecting a DC supply. Moist air is introduced through the hole. Resistors 1 are cooled due to the presence of water vapor resulting in fall in resistance depending on relative humidity. The resistors 1 and 2 can be connected in a bridge circuit to obtain

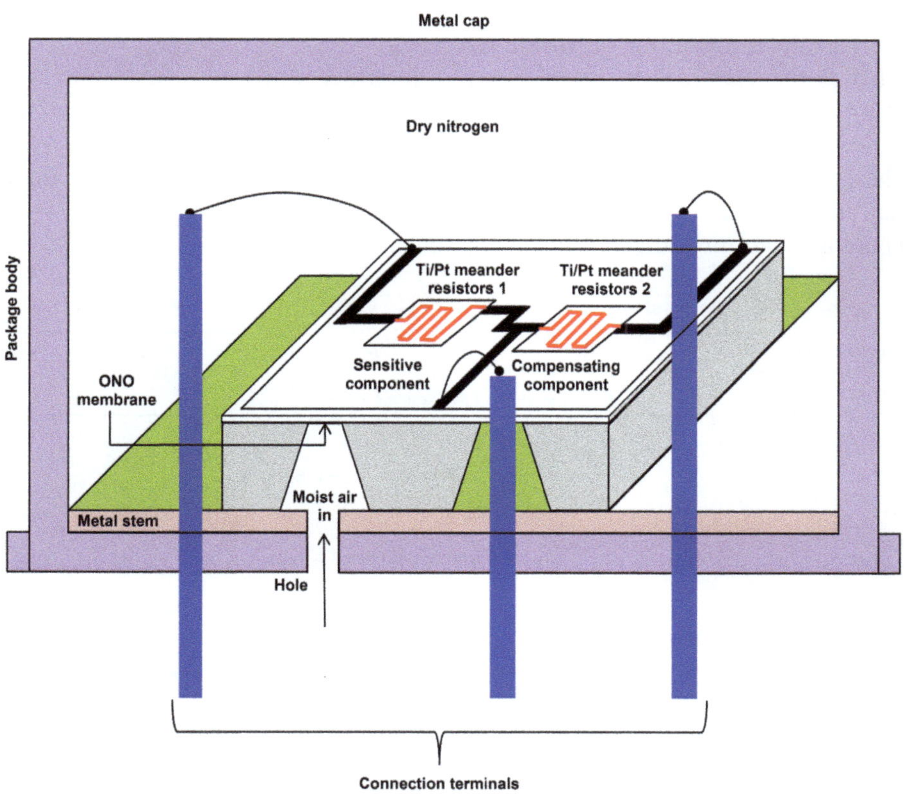

Figure 22.18. Micromachined humidity sensor.

a temperature-compensated signal because ambient temperature affects resistors 1 and 2 equally.

The sensitivity of the sensor is 0.054 mV RH^{-1}. Its response time is 25 s. The nonlinearity is 1% over full scale. Between 40% and 90% relative humidity, the hysteresis is less than 2%. Stability of the sensor is examined by exposure to a high relative humidity 95% at 150 °C/−70 °C (Lee *et al* 2001).

22.8.2 Optical humidity sensor based on hydrogel thin film expansion

When performing a humidity measurement in a corrosive or explosive environment, an electrical or electronic type of sensor must be avoided. For these applications, a sensor working on an optical principle is safer. The swelling of a hydrogel film is optically examined to ascertain the variation in thickness of the film caused by changes in relative humidity of the surrounding atmosphere (Buchberger *et al* 2018, 2019).

The optical humidity sensor (figure 22.19) consists of a 600 nm thick pHEMA hydrogel thin film formed on a sapphire substrate by initiated chemical vapor deposition (iCVD). By pHEMA is meant the hydrogel of poly (2-hydroxyethyl methacrylate). The hydrogen thin film swells in a humid environment. The relative humidity level is correlated with change in thickness of the hydrogel film. Light falling on the film is reflected three times:

(i) at the ambient air–substrate interface,
(ii) at the substrate–hydrogel thin film interface, and
(iii) at the hydrogel thin film–moist air interface.

The three reflected beams R_1, R_2, R_3 superimpose on each other and undergo optical interference.

There are two implementations of the sensor:

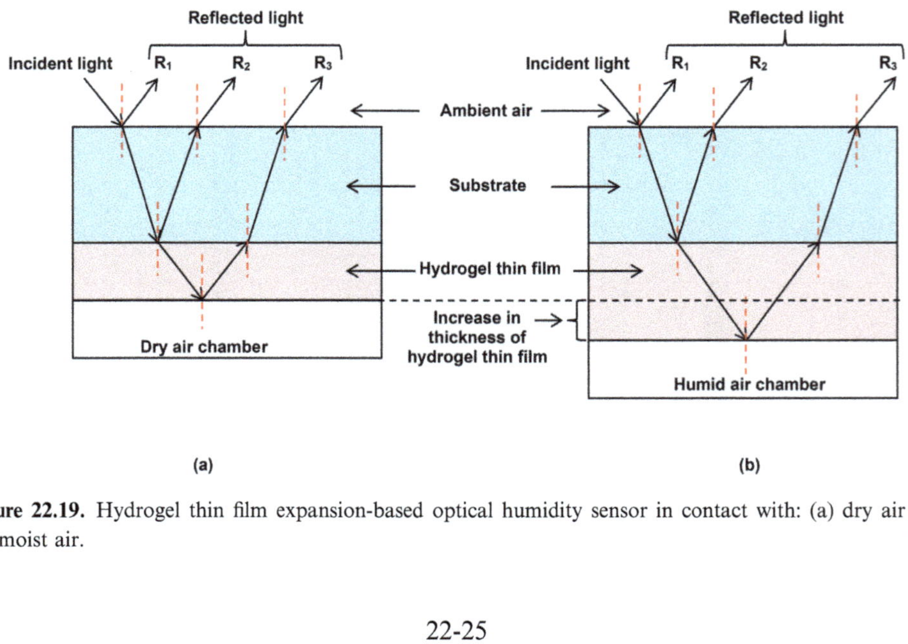

Figure 22.19. Hydrogel thin film expansion-based optical humidity sensor in contact with: (a) dry air and (b) moist air.

(i) In one implementation of the sensor, a laser source is used along with a photodetector. The relative change in thickness is measured. For a larger increase in thickness, there is ambiguity in result so that the approach is restricted up to a certain maximum relative humidity.
(ii) In another implementation, a broadband light source is used. For every value of thickness, a wavelength-dependent interference pattern is obtained. This pattern is unique for each thickness value. An absolute value of thickness is derived from the recorded reflectance spectrum in the 3%–97% relative humidity range. The response time of the sensor for an abrupt change of humidity is less than 2.5 s, which means that it is an incredibly fast humidity sensor (Buchberger *et al* 2018, 2019).

22.9 Gas sensors

22.9.1 TiO_2–ZrO_2 oxygen lambda sensors

Anthropogenic sources such as transportation systems contribute tremendously to atmospheric pollution through emission of gaseous products produced by the combustion of fossil fuels from their exhausts. Air/fuel ratio of automobiles is efficiently controlled to prevent deterioration of air quality. To achieve this aim, oxygen sensors are used to measure the air/fuel ratio of the exhaust gases ejected from the catalytic converter. The catalytic converter fitted in the exhaust pipe of the automobile engine uses a catalyst like platinum or palladium to convert the harmful nitrogen oxides, carbon monoxide and hydrocarbons into less toxic gases like nitrogen, carbon dioxide and water vapor (Lari *et al* 2008).

Lamba sensor also known as lambda probe is an oxygen sensor. Lambda is the ratio of amount of oxygen actually present in a combustion chamber to the amount of oxygen required for ideal combustion. A value of $\lambda = 0.97$ or 1.03 is normal and okay.

The lambda sensor is fitted in the exhaust pipe of the engine. It measures the oxygen concentration in the exhaust. The sensor output is fed to the engine control to maintain the optimum air/fuel ratio for clean combustion. Thus, the sensor determines whether the air/fuel ratio is lean, rich or optimal. The ideal fuel ratio for a gasoline engine is 14.7:1, which means 14.7 parts of air are required for 1 part of fuel. A mixture of air and fuel in the ratio 14.7:1 is referred to as a stoichiometric mixture. With this ratio, all the fuel is burnt and there is no excess air.

The sensor (figure 22.20) consists of a YSZ (yttria-stabilized zirconia) pellet with two parallel Pt electrodes on its opposite surfaces, one of which is coated with $Ti_{0.75}Zr_{0.25}O_2$ abbreviated as Ti–Zr to act as a solid-state reference. The electromotive force of the sensor changes around $\lambda = 1$ on exposure to a step change between two gas mixtures, namely, rich and lean gas mixtures representative of $\lambda = 0.8$ and 1.2. The amplitude of the transition remains constant in a temperature range of 350 °C–650 °C so that the lambda sensor is temperature independent (Lari *et al* 2008).

22.9.2 Mixed potential CO sensor

Accurate monitoring of fuel combustion is necessary to attain high efficiency of fossil fuel burning for generation of heat to operate boilers in power plants. This can

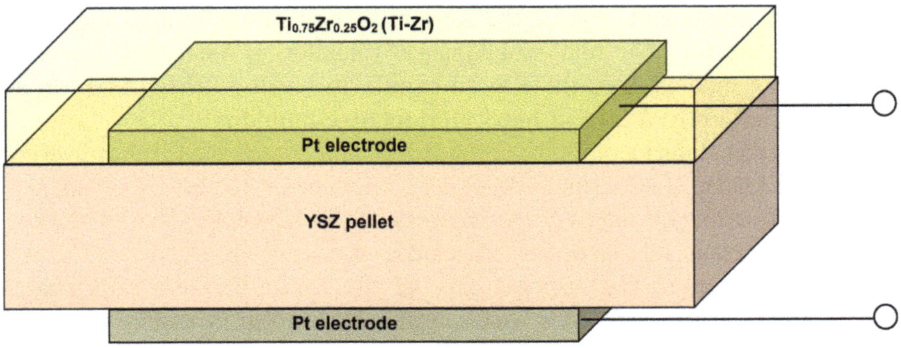

Figure 22.20. TiO_2–ZrO_2 oxygen lambda sensor.

be accomplished when the sensor is placed at the combustion site because then only the sensor can supply exact real-time data to the control system to regulate the fuel consumption. Obviously, such a sensor should be able to withstand temperatures ~1000 °C–1500 °C. Depending on the type of fuel, the optimum condition for boilers is to operate in 1%–2% oxygen and 100–200 ppm CO. So, the sensor should work in conditions of hundreds of ppm of CO under 0.3%–3% oxygen at 1000 °C (Wang *et al* 2020).

Figure 22.21(a) shows the (YSZ)-based mixed potential CO sensor and the testing setup. On the sensing side of the YSZ electrolyte, there is a NiO electrode and a Pt electrode. On the reference side, there is a Pt electrode. The sensing side is exposed to target gas while the reference side is kept in air. In the testing setup (figure 22.21(b)) the sensor pellet is pressed between two silica tubes using mica O-rings to prevent leakage. ($CO+N_2$+air) mixture is flowed through an alumina tube on the sensing side while air is flowed on the reference side through another alumina tube. The voltage signals are measured keeping the working electrodes connected to NiO electrode and Pt electrode on the sensing/target gas side while the reference electrode is connected to Pt electrode on the reference/air side.

Mixed potential is a potential resulting from two or more electrochemical redox reactions. The mixed potential of an electrode is set up through electrochemical redox reactions taking place over the electrode. It is the electrode potential resulting from the simultaneous action of more than one redox couple when the net electrode current is zero. In the case of the mixed potential CO sensor, the mixed potential is established due to oxidation of CO and reduction of oxygen.

Tests carried out under 0.5%–3% oxygen at 1000 °C show that NiO is highly sensitive to CO producing a mixed potential of 36 mV at 1000 ppm concentration. Exposure to 10% CO_2 does not affect the sensor response. Presence of CH_4 causes fluctuations in output signal but is unable to vitiate the average output signal. The sensor performs well in an 11 day stability test in 200 ppm CO at 2% oxygen and 1000 °C (Wang *et al* 2020).

22.9.3 SiC FET sensor for NO, NH_3, O_2, CO, and SO_2

Flue gas is the exhaust gas from a fireplace, furnace, boiler, etc released to the atmosphere through a pipe or channel (the flue). It contains oxygen (O_2), carbon monoxide (CO),

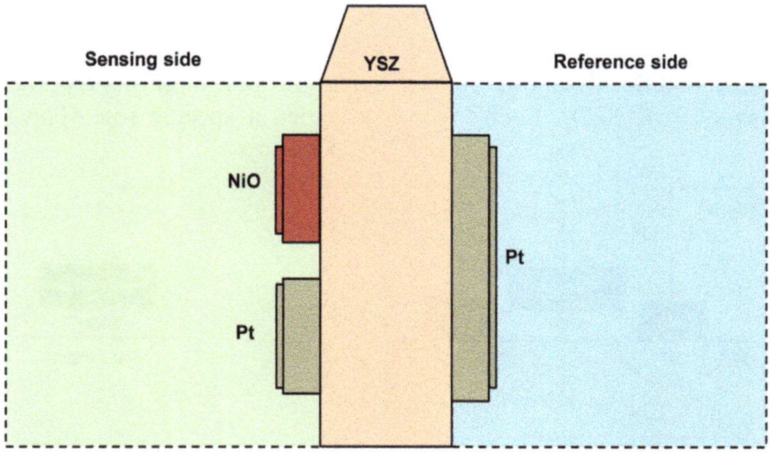

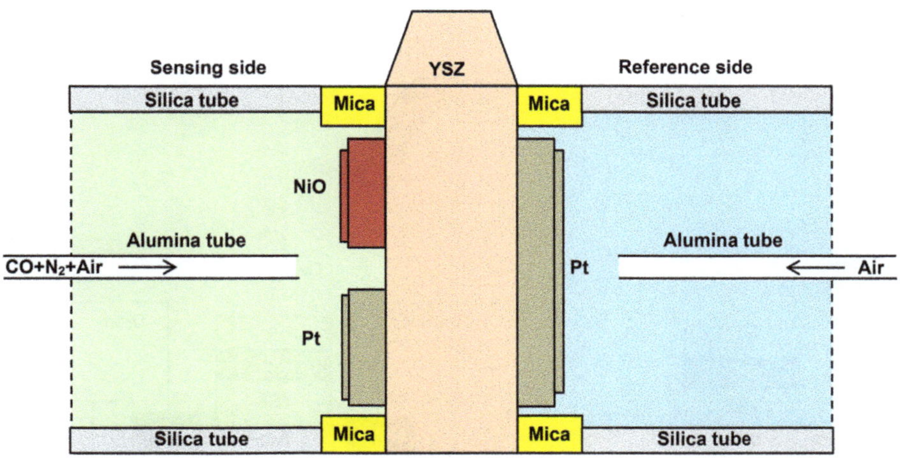

Figure 22.21. CO sensor and its testing setup: (a) yttrium-stabilized zirconia (YSZ)-based mixed potential CO sensor made with a YSZ electrolyte, and (b) its two-compartment laboratory testing setup.

oxides of nitrogen (NO_x), sulfur dioxide (SO_2), hydrocarbons (HCs), ammonia (NH_3), dust, soot, etc. Flue gas must comply with government regulations regarding the maximum limits of polluting gases and particulate matter. Before exiting to the atmosphere, the flue gas is passed through a cleaning system consisting of gas scrubbers and dust filters.

Sensors used for monitoring exhaust emissions include solid electrolyte gas sensors, e.g., the lambda sensor for oxygen and the resistive type metal–oxide–semiconductor (MOS) sensor, e.g., the TiO_2-based MOS sensor. A field-effect transistor made from wide bandgap semiconductor silicon carbide is a viable sensor option for automotive exhaust emission control.

The SiC FET sensor (figure 22.22(b)) is the modified version of the SiC FET (figure 22.22(a)). During modification, a gate stack consisting of three layers, namely, yttria-stabilized zirconia, nickel oxide and platinum has replaced the gate metal (Sasago *et al* 2020). Each layer is assigned a specific role. The NiO layer

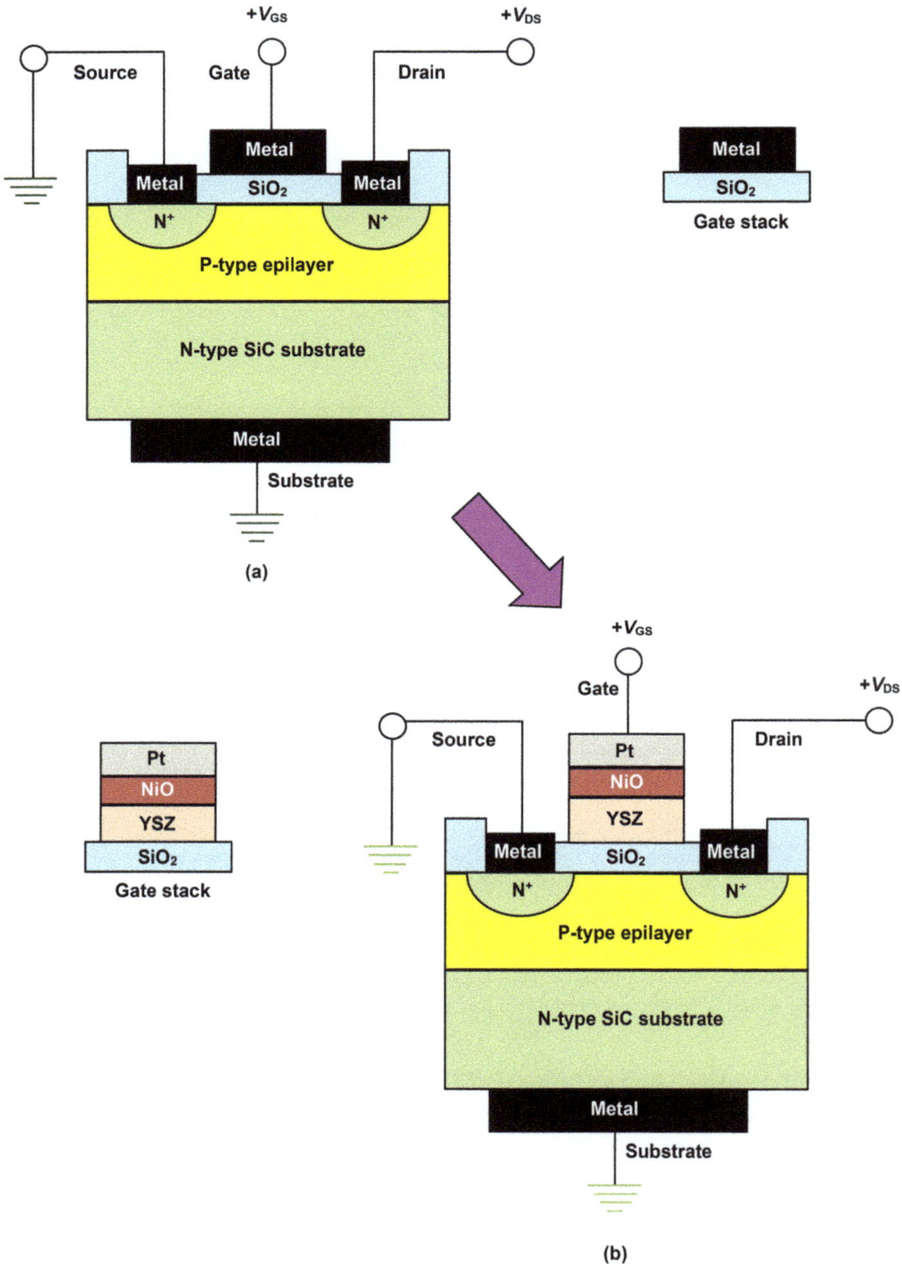

Figure 22.22. SiC FET gas sensor for NO, O_2, NH_3, CO, and SO_2: The SiC FET structure in (a) is revised for gas sensing by modifying the gate stack. The sensor in (b) consists of a triple-layer gate stack consisting of yttria-stabilized zirconia (YSZ), nickel oxide (NiO), and platinum (Pt).

improves the heat resistance of the sensor making it possible for the SiC substrate device to operate up to 800 °C. It also increases the sensitivity towards detection of NO. YSZ is an oxygen ion conductor. Platinum is widely used as an electrode in gas sensors.

The working mechanism of the sensor is: when the sensor is exposed to a gaseous environment, interaction takes place between the gas species and the material of gate insulator. Through this interaction, the gate-to-substrate electric field is modulated, changing the drain–source current. Taking the case of a hydrogen-containing gas and Pt gate, the gas is adsorbed by dissociation on the Pt gate surface producing atomic hydrogen which diffuses through the platinum film to reach the oxygen atoms on gate oxide producing hydroxide groups. As a consequence, the polarization of the metal/oxide interface occurs and hence the electric field at the gate-to-semiconductor interface is altered (Andersson *et al* 2018).

Threshold voltage of the field-effect transistor varies with concentration of target gas and temperature of measurement. The threshold voltage decreases when the target gas is NH_3 (negative direction shift) but increases when the target gas is NO (positive direction shift). For oxygen, the threshold voltage shifts in the positive direction at 115 °C. The amount of shift depends on the concentration of oxygen.

The gases O_2, NO, CO, SO_2, and NH_3 can be detected only at high temperature. So, a heater must be placed close to the sensor, or the sensor and the heater must be integrated in a module (Sasago *et al* 2020).

22.10 Discussions and conclusions

Harsh environment conditions for sensor deployment are many and varied. Each environment has its own individual requirements, e.g., one environment may need accelerometer or pressure sensor or the sensor for temperature itself capable of working up to very high temperatures; another environment may require a humidity sensor that can give correct values in an atmosphere contaminated with chemical fumes and vapors; a third environment may need to find these parameters in a situation where explosive materials are suspected. Off-the-shelf devices are unable to meet these challenges. A sensor may be selected from among the different types of sensors described in this chapter to perform the measurement depending on availability.

Design and development of sensors for harsh environments essentially involves the proper choice of materials that do not degrade on exposure to the host environment and use of the suitable technique, electrical, optical or wireless remote sensing, as deemed safe and reliable. If the decision goes haywire, the sensor may age fast and its characteristics may drift leading to erroneous output, or the sensor may be damaged resulting in no output signal.

Review exercises

22.1. Explain giving examples the statement, ‘A harsh environment is a mix-up situation for sensors’.

22.2. Under what condition is remote sensing resorted to?

22.3. What is a high-temperature sensor? Where is it used?
22.4. What do new generation energy systems expect from sensors?
22.5. What do you understand by plastic deformation of a material? How does it differ from elastic deformation? At what temperature does silicon show plastic deformation?
22.6. Compare silicon with silicon carbide as a material for sensors used in harsh environments.
22.7. What locations in a quad-beam accelerometer feel maximum stress when the device is subjected to acceleration? How is the knowledge of these locations utilized in deciding the placement of piezoresistors for pressure sensing? Draw a diagram marking the locations.
22.8. How does a piezoresistor placed at the joint of the beam with proof mass of the accelerometer change during acceleration? What happens to the resistor placed at the joint of the beam with the sensor frame?
22.9. What is the unit of sensitivity of a piezoresistive accelerometer using a Wheatstone bridge arrangement of resistors?
22.10. Why can quartz, lithium niobate and gallium phosphate not be used as piezoelectric materials in an accelerometer intended to be used at high temperatures?
22.11. What does YCOB stand for? Justify, giving reasons, the use of YCOB as a material for a piezoelectric accelerometer for high-temperature applications.
22.12. Explain the difference in construction of compression-mode and shear-mode accelerometers, and correlate their working mechanisms with their names.
22.13. What are the reasons for using inconel as a proof mass material in accelerometers? What is the composition of this alloy?
22.14. What is gained by making the shear-mode accelerometer in an electrode-less design?
22.15. In what units is the sensitivity of a piezoelectric accelerometer expressed?
22.16. What modifications are made in a quad-beam accelerometer to make it an optical electrometer? Explain with a diagram.
22.17. Give two advantages of the optical accelerometer over the micro-machined silicon accelerometer.
22.18. Why is sensitivity of an optical accelerometer measured in dB g^{-1}?
22.19. Why is it better to make a thermal flow sensor using SiC film-over-glass structure than SiC film over silicon growth substrate?
22.20. Why is 3C-SiC polytype preferred over 4H-SiC and 6H-SiC polytypes for making the SiC flow sensor?
22.21. Why is the sensitivity of a flow sensor expressed in units of reciprocal flow rate?
22.22. In a fiber optic flow sensor, what are the roles of high attenuation fiber and regenerated fiber Bragg grating? How is the hot-wire anemometry principle applied in this sensor?

22.23. What are the two functions performed by in-fiber light in a fiber optic flow sensor?
22.24. Which sensor is better in an environment in which strong electromagnetic fields exist, electrical or optical?
22.25. Which sensor is better in an environment in which explosive materials may be present, electrical or optical?
22.26. Why do we move away from silicon towards silicon carbide for fabricating a pressure sensor for harsh environments? Give reasons.
22.27. Explain with the help of a diagram the construction and working of a silicon carbide capacitive pressure sensor. Does it make any difference if SiC pressure sensors are fabricated on Si or poly-SiC substrates?
22.28. What is the function of $TiSi_2$ layer in a $Ta/TiSi_2/Pt$ metallization?
22.29. Point out two problems with a silicon carbide pressure sensor at high temperatures, and show how they are solved by making platinum thin film strain gauges on a sapphire membrane?
22.30. Why is sensitivity of a capacitive pressure sensor expressed in units of kPa^{-1}?
22.31. What are: (i) dielectric material, and (ii) electrode material used to make a flexible ceramic pressure sensor?
22.32. Mention two shortcomings of an electrical pressure sensor which constrain us to turn our attention towards fiber optic sensors?
22.33. What is the name of a cavity made in a fiber optic SiC pressure sensor and what optical principle is utilized for operation of this sensor? What parameter of the cavity is related to pressure? How is pressure value extracted from optical measurements?
22.34. Mention two advantages of SOI-CMOS technology over bulk CMOS technology?
22.35. Justify the need of an on-chip temperature sensor integrated with signal processing in CMOS circuits.
22.36. Draw and explain the diagram of a lateral CMOS-SOI diode. What property of the diode is exploited for sensing temperature? Mention one practical application of this temperature sensor.
22.37. Why is tungsten metallization preferred over aluminum metallization in a high temperature SOI diode temperature sensor?
22.38. What is the temperature range over which an SOI diode temperature sensor works satisfactorily?
22.39. What is LTCC technology? Write a few advantages of this technology.
22.40. Outline the working principle of an LTCC temperature sensor. What is the unit of sensitivity for this sensor?
22.41. Mention two advantages of SAW sensors. For what special applications are these sensors very useful?
22.42. Explain with a diagram of a SAW sensor and its associated wireless system how temperature measurements are made with this sensor?

22.43. Write the acronym and the chemical formula for langasite. Why is langasite chosen as the substrate material of a high-temperature SAW sensor?
22.44. What is the problem with tungsten metallization at high temperature? How is this problem solved by using a multilayer metallization scheme? Write this scheme.
22.45. What are the shortfalls of thermocouples and pyrometers for high-temperature measurements? In what respects do fiber optic sensors have an edge over them?
22.46. Up to what temperature can single-crystal sapphire in the form of a fiber be used as a temperature sensor? What optical component is inscribed in the fiber for measuring temperature?
22.47. Draw the layout of a sapphire fiber Bragg grating temperature sensor and describe how the temperature is obtained using this sensor?
22.48. What hinders the use of conductive and capacitive humidity sensors in a food processing environment? Explain with a diagram the working of a sensor suitable for humidity measurements in such an environment.
22.49. Name a sensor which is suitable for humidity measurements in a corrosive or explosive environment, and explain its working with a diagram.
22.50. What do you understand by pHEMA hydrogel and iCVD?
22.51. Why is the hydrogel humidity sensor said to be a fast humidity sensor?
22.52. What is a lambda sensor? Where is it fitted in an automobile? What is its function?
22.53. What is the optimum condition for boiler operation with regard to oxygen and CO presence? What is the condition under which a CO sensor may be required to work in this situation?
22.54. What electrochemical redox reactions taking place in a CO sensor give rise to the mixed potential of the CO electrode?
22.55. Draw the diagrams of a mixed potential CO sensor and its laboratory test setup. Explain with reference to the diagrams how the CO measurements are carried out?
22.56. What is meant by flue gas? Name the principal gases present in it. How must flue gas comply with government regulations? What measures are taken to achieve compliance?
22.57. Draw the cross-sectional diagram of a silicon carbide field-effect transistor and modify the gate stack to make a gas sensor. What is the gas-sensitive parameter of this device? Describe the operating mechanism of the gas sensor.

References

Andersson M, Khajavizadeh L and Spetz A L 2018 On the applicability of silicon carbide-based field effect sensors in the control of exhaust/flue gas after-treatment systems *Proc.* **2** 1–5

Balakrishnan V, Dinh T, Phan H-P, Dao D V and Nguyen N-T 2018 Highly sensitive 3C-SiC on glass based thermal flow sensor realized using MEMS technology *Sens. Actuators* A **279** 293–305

Balakrishnan V, Phan H-P, Dinh T, Dao D V and Nguyen N-T 2017 Thermal flow sensors for harsh environments *Sensors* **17** 1–31

Beker L, Maralani A, Lin L W and Pisano A P 2018 Modeling, fabrication, and characterization of SiC concentrically matched differential capacitance output pressure sensors *Sens. Actuators* A **273** 293–302

Buchberger A, Coclite A M and Bergmann A 2018 Fast humidity sensors for harsh environment *Proc.* **2** 1–5

Buchberger A, Peterka S, Coclite A M and Bergmann A 2019 Fast optical humidity sensor based on hydrogel thin film expansion for harsh environment *Sensors* **19** 1–11

Chen R, Yan A, Wang Q and Chen K P 2014 Fiber-optic flow sensors for high-temperature environment operation up to 800 °C *Opt. Lett.* **39** 3966–9

De Luca A, Pathirana V, Ali S Z and Udrea F 2013 Silicon on insulator thermodiode with extremely wide working temperature range *17th Int. Conf. on Solid-State Sensors, Actuators and Microsystems (Transducers & Eurosensors XXVII) (Barcelona, 16–20 June, 2013)* 4

De Luca A, Pathirana V, Ali S Z, Dragomirescu D and Udrea F 2015 Experimental, analytical and numerical investigation of non-linearity of SOI diode temperature sensors at extreme temperatures *Sens. Actuators* A **222** 31–8

Fahrner W R, Job R and Werner M 2001 Sensors and smart electronics in harsh environment applications *Microsyst. Technol.* **7** 138–44

Fricke S, Friedberger A, Seidel H and Schmid U 2012 A robust pressure sensor for harsh environmental applications *Sens. Actuators* A **184** 16–21

Fu M, Zhang J, Jin Y, Zhao Y, Huang S and Guo C F 2020 A highly sensitive, reliable, and high-temperature-resistant flexible pressure sensor based on ceramic nanofibers *Adv. Sci.* **7** 1–8

Habisreuther T, Elsmann T, Pan Z, Graf A, Willsch R and Schmidt M A 2015 Sapphire fiber Bragg gratings for high temperature and dynamic temperature diagnostics *Appl. Therm. Eng.* **91** 860–5

Jiang X, Kim K, Zhang S, Johnson J and Salazar G 2014 High-temperature piezoelectric sensing *Sensors* **14** 144–69

Jin S 2011 Silicon carbide pressure sensors for high temperature applications *PhD Thesis* (Cleveland, OH: Department of Materials Science and Engineering, Case Western Reserve University)

Jin S, Rajgopal S and Mehregany M 2011 Silicon carbide pressure sensor for high temperature and high-pressure applications: Influence of substrate material on performance *16th Int. Solid-State Sensors, Actuators and Microsystems Conf. (Beijing, 5–9 June, 2011)* 2026–9

Jin S, Rajgopal S and Mehregany M 2012 Characterization of poly-SiC pressure sensors for high temperature and high-pressure applications *Materials Science Forum* vol 717–20 (Switzerland: Trans Tech Publications Ltd) pp 1211–4

Kim K, Zhang S, Huang W, Yu F and Jiang X 2011 $YCa_4O(BO_3)_3$ (YCOB) high temperature vibration sensor *J. Appl. Phys.* **109** 126103-1–3

Kim K, Zhang S, Salazar G and Jiang X 2012 Design, fabrication and characterization of high temperature piezoelectric vibration sensor using YCOB crystals *Sens. Actuators* A **178** 40–8

Lari A, Khodadadi A and Mortazavi Y 2008 Temperature-independent TiO_2–ZrO_2 oxygen lambda sensor *SENSORS, 2008 IEEE* 26–29 *(Lecce, October)* 839–42

Lee D-H, Hong H-K, Park C-K, Kim G H, Jeon Y-S and Bu J U 2001 A micromachined robust humidity sensor for harsh environment applications *14th IEEE Int. Conf. on Micro Electro Mechanical Systems (Cat. No. 01CH37090) (Interlaken, 25 January)* 558–61

Liang T, Li W, Lei C, Li Y, Li Z and Xiong J 2022 All-SiC fiber-optic sensor based on direct wafer bonding for high temperature pressure sensing *Photonic Sens.* **12** 130–9

Marsi N, Majlis B Y, Hamzah A A and Yasin F M 2015 Development of high temperature resistant of 500 °C employing silicon carbide (3C-SiC) based MEMS pressure sensor *Microsyst. Technol.* **21** 319–30

Nguyen T K, Phan H P, Dinh T, Dowling K M, Foisal A R M and Senesky D G *et al* 2018 Highly sensitive 4H-SiC pressure sensor at cryogenic and elevated temperatures *Mater. Des.* **156** 441–5

Plaza J A, Llobera A, Dominguez C, Esteve J, Salinas I, Garcia J and Berganzo J 2004 BESOI-based integrated optical silicon accelerometer *J. Microelectromech. Syst.* **13** 355–64

Ren J, Ward M, Kinnell P, Craddock R and Wei X 2016 Plastic deformation of micromachined silicon diaphragms with a sealed cavity at high temperatures *Sensors* **16** 1–12

Romanosky R R and Maley S M 2013 Harsh environment sensor development for advanced energy systems *Proc. SPIE Micro- and Nanotechnology Sensors, Systems, and Applications V, 87250H, Micro- and Nanotechnology Sensors, Systems, and Applications V, 7250H* 8725*(30 May 2013)*

Santra S, Guha P K, Ali S Z, Haneef I, Udrea F and Gardener J W 2008 SOI diode temperature sensor operated at ultrahigh temperatures—a critical analysis *SENSORS 2008, IEEE (Lecce, Italy, 26–29 October)* 78–81

Santra S, Udrea F, Guha P K, Ali S Z and Haneef I 2010 Ultra-high temperature (>300 °C) suspended thermodiode in SOI CMOS technology *Microelectron. J.* **41** 540–6

Sasago Y, Nakamura H, Odaka T, Isobe A, Komatsu S, Nakamura Y and Yamawaki T *et al* 2020 SiC-FET gas sensor for detecting sub-ppm gas concentrations *Adv. Sci. Technol. Eng. Syst. J.* **5** 151–8

Tan Q *et al* 2014 A harsh environment-oriented wireless passive temperature sensor realized by LTCC technology *Sensors* **14** 4154–66

Udrea F, Santra S and Gardner J W 2008 CMOS temperature sensors—concepts, state-of-the-art and prospects *Int. Semiconductor Conf. (Sinaia, Romania, 13–15 October)* 31–40

Wall B, Gruenwald R, Klein M and Bruckner G 2015 A 600 °C wireless and passive temperature sensor based on langasite SAW-resonators *Proc. Sensors AMA Conf.* 2015 *(Nürnberg, 19–21 May 2015)* 390–5

Wang Y, Ma L, Li W, Li W and Liu X 2020 A high-temperature mixed potential CO gas sensor for *in situ* combustion control *J. Mater. Chem.* A **8** 20101–10

Wieczorek G, Schellin B, Obermeier E, Fagnani G and Drera L 2007 SiC based pressure sensor for high-temperature environments *IEEE Sens.* **28** 748–51

Yao Z, Liang T, Jia P, Hong Y, Qi L, Lei C, Zhang B and Xiong J 2016 A high-temperature piezoresistive pressure sensor with an integrated signal-conditioning circuit *Sensors (Basel)* **16** 913

Zhai Y, Li H, Tao Z, Cao X, Yang C, Che Z and Xu T 2022a Design, fabrication and test of a bulk SiC MEMS accelerometer *Microelectron. Eng.* **260** 111793

Zhai Y, Li H, Tao Z, Yang C, Cao X, Che Z and Xu T 2022b Simulation analysis and fabrication of a silicon carbide-based piezoresistive accelerometer *J. Phys.: Conf. Ser.* **2246** 012007

Zhang S, Jiang X, Lapsley M, Moses P and Shrout T R 2010 Piezoelectric accelerometers for ultrahigh temperature application *Appl. Phys. Lett.* **96** 013506

IOP Publishing

Extreme-Temperature and Harsh-Environment Electronics (Second Edition)
Physics, technology and applications
Vinod Kumar Khanna

Chapter 23

Adapting medical implant electronics to human biological environments

When an implantable electronic device such as a cardiac pacemaker, a glucose biosensor for *in vivo* diabetes monitoring, or other similar gadget is introduced in the human body, the material implanted should not have any hazardous effects for the patient. Furthermore, the immune system immediately reacts to the foreign body for which biocompatibility of the material is essential to match it with the tissues and cells. Building electronic equipment for operation inside the body calls for proper material selection together with stringent quality control measures. The issues demanding special attention in developing implantable devices, associated strategies and precautions are discussed.

Implantable electronics devices such as cardiac pacemakers and defibrillators, neural stimulators, visual prosthesis and cochlear implants contain sophisticated electronic circuits which have to operate reliably inside the human body. In order to visualize the difficulties faced by the implanted devices in their functioning, we must know the biological environment, the abode of these devices. Preliminary ideas on humidity effects in chapter 15, waterproof electronics in chapter 16 and chemical corrosion prevention in chapter 17 will be helpful in understanding the contents of this chapter.

23.1 Environment inside the human body

23.1.1 Water in the body

A large fraction of the human body weight is ascribed to body fluids or water content. The water percentage varies with sex, e.g., it is 60%–67% for men and 52%–55% for women. There is a slight variation with age and the level of hydration. A 70 kg man contains 42%–47% (29.4–32.9 kg) water. The brain and the kidneys carry 80%–85% water, the heart and lungs have 75%–80% water, the muscles, liver and skin have 70%–75% water, the blood has 50% water, bones 20%–25% and teeth 8%–10% (MNT 2004–22).

The total quantity of water in the human body is subdivided into two compartments (Eliaz 2019):

(i) Extracellular fluid (ECF) compartment: It contains the water outside the cells.
(ii) Intracelluar fluid (ICF) compartment: It contains the water inside the cells.

The subdivision ratio is ECF:ICF=1:2 so that for the 70 kg man, the ECF has 14–15 L and ICF has 28–30 L.

$$\text{ECF volume} = \text{Interstitial fluid volume} + \text{Intravascular fluid volume} \tag{23.1}$$

Interstitial fluid is the fluid outside the blood vessels while intravascular fluid is the fluid inside the blood vessels.

$$\text{Interstitial fluid volume: Intravascular fluid volume} = 3:1 \tag{23.2}$$

$$\begin{aligned}\text{Interstitial fluid volume} &= \text{Lymphatic fluid volume}\\ &\quad + \text{Transcellular fluid volume}\end{aligned} \tag{23.3}$$

Lymphatic fluid is the fluid flowing in the lymphatic system, an extensive drainage network of delicate tubes throughout the body helping to maintain the balance of body fluids and protecting against infections. Examples of transcellular fluid, the fluid filling the epithelial-lined spaces, are the joint fluid and cerebrospinal fluid. Epithelium is the tissue covering the hollow organs and cavities of the body.

$$\text{Lymphatic fluid volume: Transcellular fluid volume} = 2:1 \tag{23.4}$$

$$\text{Intravascular fluid volume} = \text{Venous volume} + \text{Arterial volume} \tag{23.5}$$

23.1.2 Electrolytes in the body

The electrolytes participate in metabolism. They control the potentials of cell membranes. They also determine the osmolarity of fluids in the body.

Main cations: H^+, Na^+, K^+, Mg^{2+} and Ca^{2+}
Chief anions: OH^-, HCO_3^-, Cl^-, SO_4^{2-} and PO_4^{3-}

Through a mechanism called homeostasis, the body keeps the amount and distribution of fluids and electrolytes within the normal range and maintains them at constant levels.

23.2 Essential properties of packaging materials for reliable functioning of implanted medical electronic devices

The package is the housing of electronics serving to provide the correct environment for its operation inside the body. The quality and reliability of the package entirely depends upon the materials used for making it, and their proper integration and bonding. These are the materials which directly interface with the human body, not the materials used in the fabriaction of electronics. Therefore, great emphasis is laid on the properties of the package material. Any arbitrarily chosen material is unacceptable.

23.2.1 Hermeticity

As we have seen, the implanted electronic devices must work in the watery, corrosive environment inside the human body to satisfactorily discharge the assigned responsibilities. Therefore, they must be packaged inside an enclosure to provide an unperturbed ambience suitable for their working. At the same time, the package should prevent any toxic materials in the implant from leaching out into the body and harming it. To meet this goal, the package must be undoubtedly made of a benign material. However, eligibility for use as a packaging material is subject to possession of some essential properties (figure 23.1). The packaging material must:

(i) neither permit any substance in the body to enter inside it and disturb the functioning of the implanted electronics,
(ii) nor permit any material in the implanted electronics circuit to leak out into the body and impact human health.

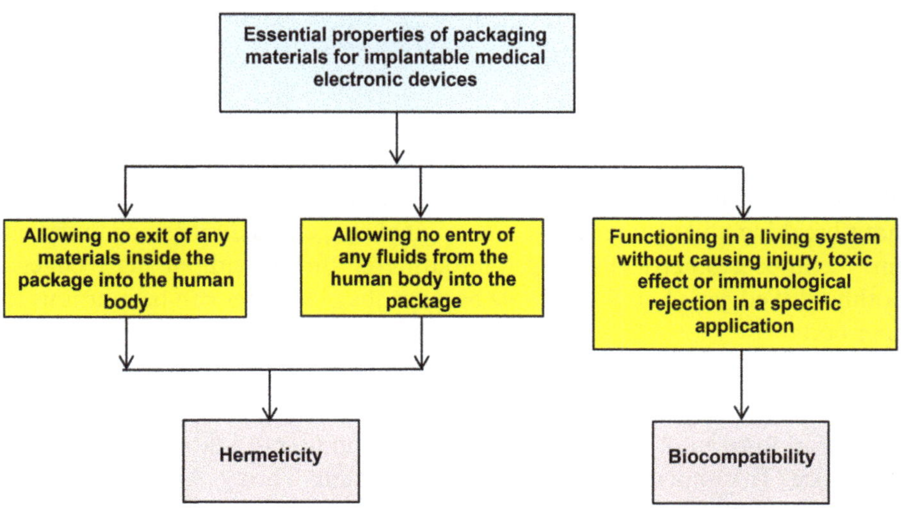

Figure 23.1. Properties of materials for packaging medical implants.

Conditions (i) and (ii) can be combined and placed under a single head, the hermeticity of the package, a term describing airtight or gastight sealing. Imperveousness to gas flow must be maintained throughout the lifespan of the implant securing both the electronics and the person in whose body the implant is placed.

23.2.2 Biocompatibility

A material fulfilling requirements (i) and (ii) must conform compulsorily to a fundamental requirement of acceptance of the package material by the body. The immunological responses and repair mechanisms of the body are extremely intricate. Insertion of a package of a particular material can produce deleterious effects such as irritation to the skin in the implanted area, formation of blood clots, development of chronic inflammation or production of substances that are toxic to the cells. We say that the material is rejected by the body.

The ability of a material to reside in the living body without producing any adverse effects on the host body is referred to as its biocompatibility. It is related to the manner in which the living body responds towards the insertion of a foreign object inside it, namely the foreign body reaction by the recipient of the implant. Biocompatibility is not an absolute concept. No material is biocompatible under all circumstances. So, biocompatibility can be stated with reference to spectific situations making it meaningful only in a contextual background. A biomaterial is a substance, either biological or synthetically produced, which can interface and interact with biological systems, and therefore can be introduced in the body for a diagnostic/therapeutic role, for augmentation of body function or for replaceement of an organ for restoration of function/healing caused by injury or disease. A non-biocompatible material can play havoc inside the body instead of providing the needed remedy by the implant.

Thus, the criteria of hermeticity and biocompatibility of the material used for packaging the implant must be fulfilled to the utmost extent achievable.

23.2.3 Mechanical flexibility

The expectations from materials do not end with hermeticity and biocompatibility. The wish list of implant designers is longer. A domineering demand is mechanical flexibility of the implantable device. A flexible material can be stretched or bent without breaking and also returns to its original shape and dimensions under elastic strain. It can be geometrically shaped into different forms without sharp edges. The smoothness of edges is highly desirable because sharp features can be injurious to the human body.

23.2.4 Weight

The material to be used for packaging implants should not add much to the weight of the device. The implant has to stay inside the human body. A heavy implant will cause more inconvenience to the patient than the relief brought by the concerned therapy. So, the patient will feel uncomfortable and disturbed by it.

23.2.5 Internal outgassing

Many materials are used in making the internal structure of the implants. These include materials that undergo outgassing during which any absorbed gas or liquid trapped inside the material is released whenever there are changes in temperature. The internal outgassing releases contaminants which spread inside the implant. It increases the vapor pressure and moisture content inside the package. The outgassing products impact the performance of electronic devices and circuits inside the package, and may lead to failure of the implantable electronics in the long run, thus reducing its operating lifetime.

23.2.6 Radio frequency transparency

Often the controlling circuit of implantable electronics devices is kept outside the body. In this arrangement, the implanted device has to be connected to the external circuit. This connection is done either with wires or without wires. The wireless connection is better because it avoids the risk of infections to the patient at places where wires enter the body. Moreover, wiring is inconvenient to the patient as it affects freedom of movement. For implants working wirelessly, such materials have to be selected which are transparent to electromagnetic waves. We know that highly conducting surfaces reflect electromagnetic waves and interfere with radio frequency (RF) signals. In fact, conducting materials are used for RF shielding.

23.2.7 Heat generation minimization

If the package of an implanted device comprises mainly conductive elements and the device is wirelessly operated, eddy currents will be induced in the conductive body of the package. These currents circulating in the body of the package will produce heat and lead to a local rise of temperature. As the temperature of the package material should not rise by more than 2 °C above the normal human body temperature (37 °C), a mandatory restriction is that the material used must not allow any temeperature excursion beyond this limit.

23.2.8 Thermal expansion coefficients matching

During the fabrication of some implants, materials with dissimilar properties are bonded together by high-temperature processes, e.g., brazing or soldering. In such cases, the materials joined should have matching coefficients of thermal expansion. Otherwise, a package which is hermetic in the beginning may lose its hermeticity afterwards because whenever the implant experiences cycles of ascending and descending temperatures, cracks are initiated at the interfaces of materials having mismatched thermal expansion coefficients. These cracks propagate through the package to the human body environment. The leakage of body fluids into the implant causes it to malfunction.

23.2.9 Ease of processing

The material should be easily processable so that device manufacturing is simplified. Many materials are amenable to processing using established thin-film and semiconductor microelectromechanical systems (MEMS) fabrication technologies. Such materials are obviously favored.

23.2.10 Other properties

A material that has qualified because of its hermeticity, biocompatibility and other desirable traits still faces the constraints imposed by cost and ease of manufacturing. In some cases, optical transparency is also necessary.

23.3 Studying biological response vis-à-vis material properties

As we are dealing with a biological environment, the first step is to explore the foreign body reaction triggered by insertion of a material inside the human body. The foreign body reactions cannot be reduced to zero but can be minimized by proper choice of materials. Let us first examine the biocompatibility issue.

In the second step, we will like to know about the biomaterials which have been found to be suitable for packaging/making implants. These materials differ in their hermeticity. There are arguments supporting a material due to its qualities and also for opposing the material owing to its drawbacks. These for and against features must be carefully weighed in material selection for implanted devices.

23.4 Foreign body reaction to implanted biomaterials

The human body is bestowed with the outstanding ability to differentiate between itself and an outside element. The moment a foreign object is put inside the body, a series of responses are activated leading eventually to the encapsulation of the object and its seclusion from the body.

23.4.1 Post implantation acute and chronic inflammation phases

During implantation of the medical electronics device shown in figure 23.2(a), a surgical incision in the human body is necessary. So, tissues in the implant area are damaged and blood is extravasted from the wounds. The human body responds to the injury caused by implantation through the development of inflammation which is subdivided into acute and chronic phases. For an implant package made of biocomptible materials, the acute inflammation phase resolves in less than 1 week. Chronic inflammation phase takes no longer than 2 weeks. Moreover, it is confined to the implant area. Acute inflammation stretching past 3 weeks suggests an infection.

23.4.2 Stages of inflammatory response

Commencing immediately after implantation, the inflammatory response comprises a series of events focused at the dilution, neutralization or separation of the affected area of the human body containing harmful outside materials from the rest of the

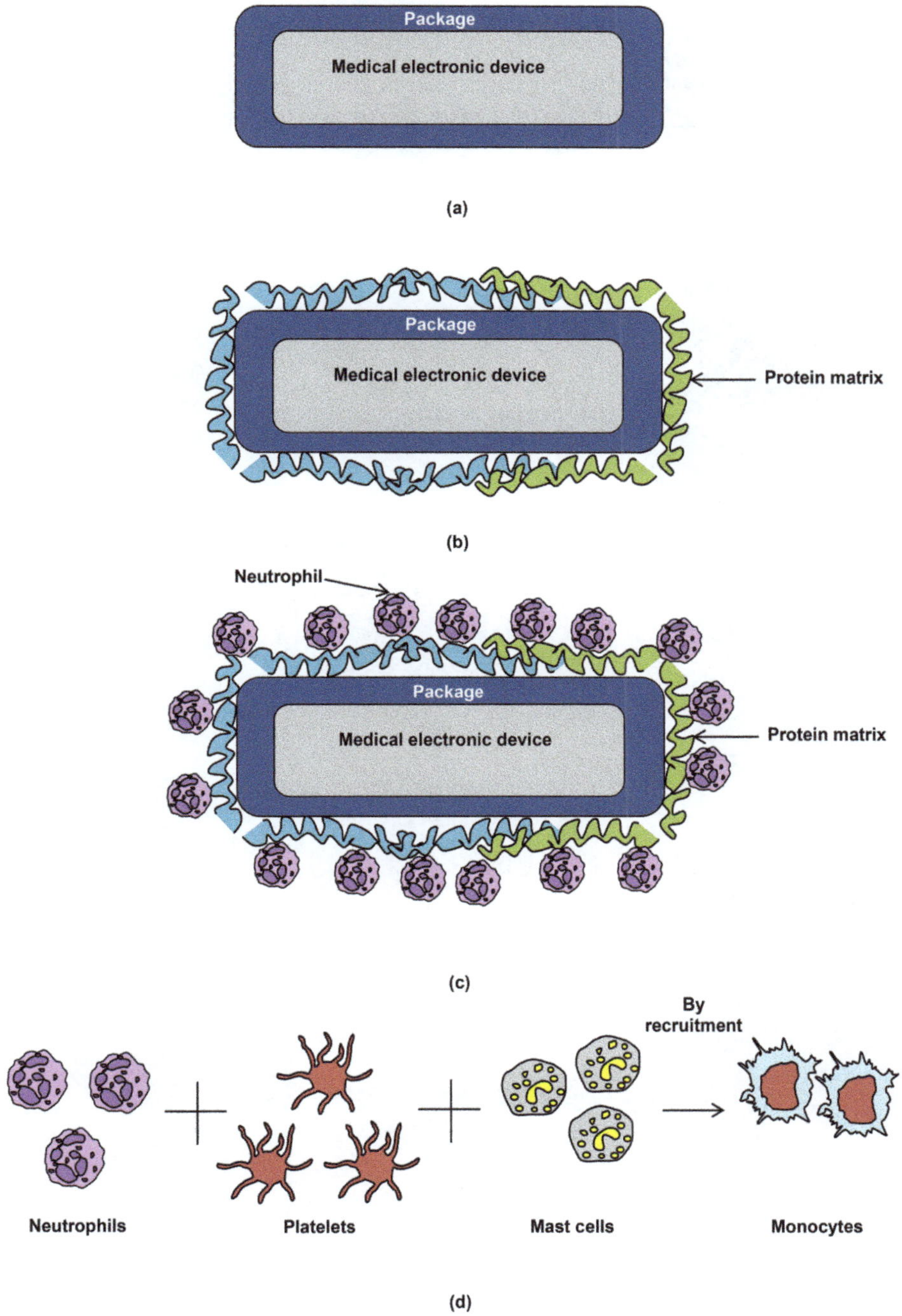

Figure 23.2. Steps in foreign body reaction: (a) the implanted device, (b) binding of protein to implant surface, (c) recruitment of neutrophils, (d) recruitment of monocytes, (e) production of macrophages and foreign body giant cells, (f) engulfing of the implant by macrophages, (g) covering of the implant with the protein matrix and FBGCs, and (h) encapsulation of the implant by fibrous tissue.

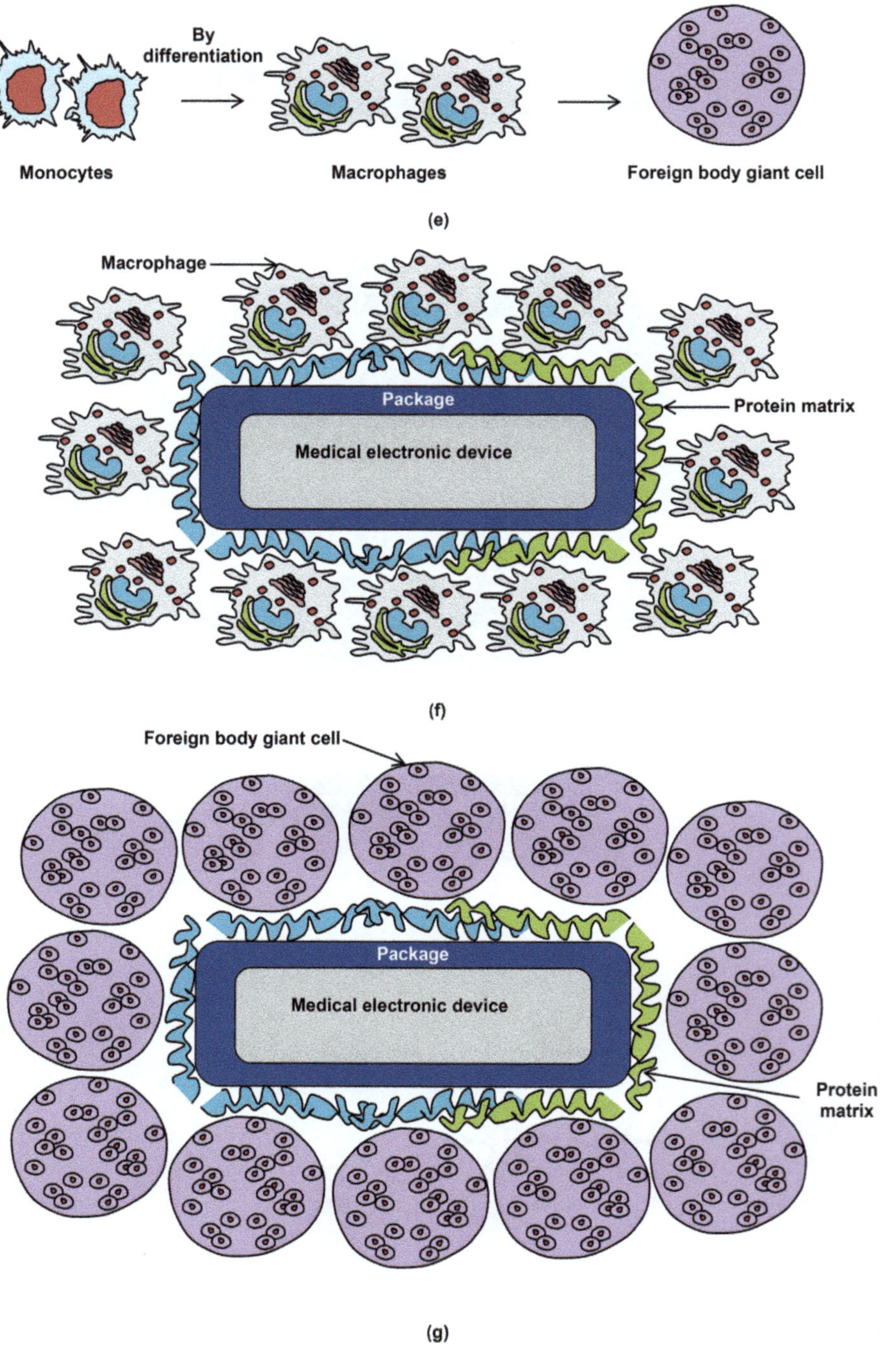

Figure 23.2. (Continued.)

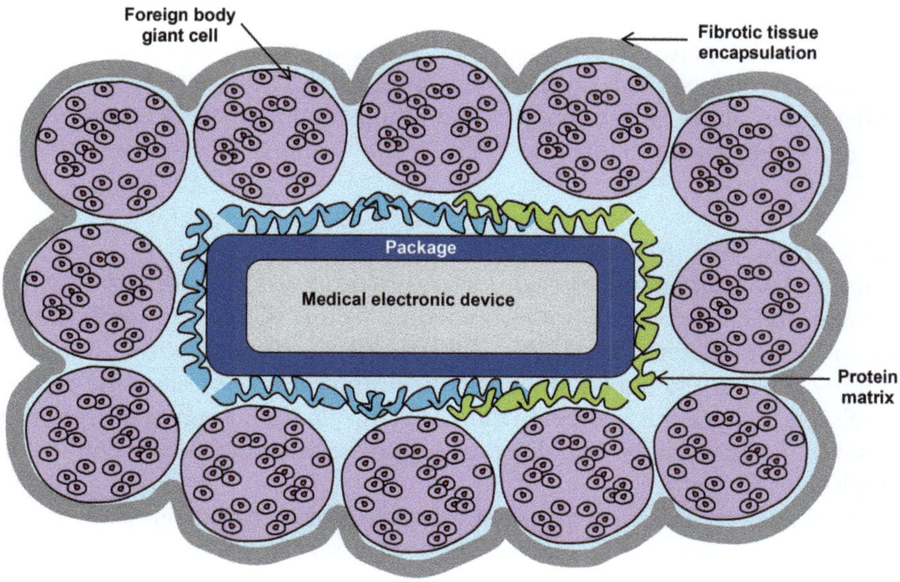

(h)

Figure 23.2. (Continued.)

body. Although the aim of inflammatory response is to protect the human body from the outside object, its influence on the stability and operation of the implant is a cause of anxiety.

23.4.2.1 Protein binding

Proteins such as albumin and fibrinogen in the extravasted blood are absorbed on the surface of implanted device. This protein absorption occurs within a few seconds of the implantation. The protein binding is shown in figure 23.2(b).

Consequent to protein absorption, a provisional matrix of proteins builds up over the implant. Through this matrix, the cells in the region interact with the implant surface. Slowly the smaller proteins are substituted by larger proteins via a process known as Vroman effect.

23.4.2.2 Neutrophils recruitment

As already mentioned, acute inflammation of the human body portion surrounding the implant ensues. The cytokines-rich matrix formed contains chemo-attractants (Sheikh *et al* 2015). These chemo-attractants can recruit cells of the immune systems in the site. So, within a few minutes of implantation, neutrophils move to the injured tissue area (figure 23.2(c)). They stick to the provisional matrix. The acute inflammation is characterized by the presence of neutrophils and the inflammatory response is mediated by degranulation of mast cells together with release of histamine and adsorption of fibrinogen. Interleukin-3 (IL-3) and interleukin-4 (IL-4) released from the mast cells control the degree of the forthcoming stages.

23.4.2.3 Monocytes recruitment

The neutrophils, platelets and mast cells recruit monocytes (figure 23.2(d)). Monocytes are created from myeloid progenitor cells. The myeloid progenitor cells engender mono blasts, pro-monocytes and ultimately monocytes. From the monocytes monophages and foreign body giant cells (FBGCs) are produced, figure 23.2(e).

23.4.2.4 Monocyte differentiation into macrophages

The monocytes differentiate into macrophages (Carnicer-Lombarte *et al* 2021). Cell differentiation is induced by colony-stimulating factors (CSFs). Stromal cells in the blood and tissues secrete these CSFs. The macrophages undergo multiplication. In less than 2 days of the implantation, the neutrophil layer vanishes and a layer of macrophages occupies its place (figure 23.2(f)). The macrophages layer is self-sustainable. It proliferates and liberates chemo-attractants which employ more macrophages into service.

Macrophages are classified as: classically activated or M1 macrophages, wound healing or reparative or M2 macrophages, and regulatory macrophages Mregs. M1 macrophages defend the host by secreting IL-12 and IL-23, nitric oxide synthase and pro-inflammatory cytokines. M2 macrophages are more concerned with repair than defence mechanisms. Regulatory macrophages moderate the immune response. They restrict inflammation by secreting anti-inflammatory cytokine IL-10.

The macrophages adhere to the surface of the implant through integrins, a group of transmembrane proteins which specifically bind with the proteins already present on the implant's surface. The $\alpha M\beta 2$ integrins are specifically bound to the serum proteins such as fibronectin and fibrinogen on the surface of the implant. After the macrophages have adhered to the implant surface, they experience cytoskeletal remodeling.

23.4.2.5 Phagocytosis attack by the macrophages and end to implant activity

The macrophages flatten out and surround the implant in an effort to cause phagocytosis and extend structures known as pedosomes which specialize in proteolysis. If the implant is large in size, the macrophages may be unsuccessful in phagocytosis. Then the degrading enzymes and reactive oxygen released break down the implant into smaller pieces that can be easily phagocytosed. The attack by macrophages poses a serious threat to stability of the implant because it can produce cracks from which toxic materials in the implant can ooze out. It can also bring implant function to a halt. If the phagocytosis is completed without any harmful effect, normalcy of the tissue is restored. Some implantable devices are designed to gainfully utilize this implant degradation process to destroy the implant after it has accomplished its intended therapeutical task.

23.4.2.6 Formation of foreign body giant cells (FBGCs)

Fusion of monocyte membranes produces FBGCs. Critical factors influencing membrane fusion are adhesion proteins and lipid rafts. Actin rearrangement is

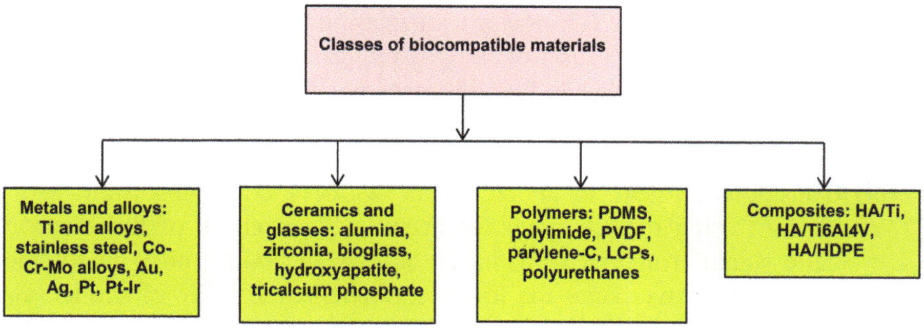

Figure 23.3. Classification of biocompatible materials with a few examples of each class.

also important. The implant is now covered with the protein matrix and FBGCs (figure 23.2(g)).

23.4.2.7 Indefinitely staying implant

When the implant has to reside in the human body for a long duration of time, a fibrotic process sets in during which the implant is encapsulated by a layer of fibrous tissue (figure 23.2(h)). This fibrotic tissue isolates the implant from its surroundings and renders the delivery of any therapy very difficult.

23.5 Biomaterials for implants

Material for implant packaging must be chosen to lessen the negative foreign body reaction while maintaining the designed functionality. According to their chemical composition and properties, biomaterials can be subdivided into four categories: metals, ceramics, polymers and composites (figure 23.3).

23.5.1 Metals

23.5.1.1 Advantages

(i) They provide good hermeticity. Very low moisture permeability of metals has set an industrial benchmark for implantable electronics.
(ii) A limited number of metals show good biocompatibility.
(iii) Degree of outgassing from metals is low.
(iv) Ductility (ability to be drawn into wires) of metals is good.
(v) Resistance of metals to wear (material loss by mechanical action) and impact is high.

23.5.1.2 Drawbacks

(i) Mechanical flexibility of metals is poor. Properties of metals are vastly different from those of biological tissues. The hard metallic boundaries and corners of metallic packages cause bleeding wounds.
(ii) RF transparency of metals is poor because of high electrical conductivity. As a consequence, the RF signal is attenuated by the walls of the metallic

package. Hence the RF power transmission through the metal is reduced. To avoid this diminution of signal strength, the antenna of the implant is often placed outside the metallic package. But this increases the overall size of the implant. So, the metal packaging becomes arduous in both ways.
(iii) Considerable heat is generated in wireless communication with the implant using metallic package because the strength of the electromagnetic signal from the transmitter outside the body must be sufficiently increased to overcome the losses suffered in the body of the metal. On using a high energy electromagnetic signal, large eddy currents are produced in the metal causing appreciable heating. This excessive heating effect is most worrying.
(iv) Optical transparency of a metallic package is poor.
(v) The metallic package for implant becomes inconveniently heavy.
(vi) Cost of metallic packaging is high.
(vii) Fabrication of a metallic package is difficult.
(viii) Resistance of metals to corrosion in biological environment is low.

23.5.2 Ceramics

23.5.2.1 Advantages

(i) Hermeticity of ceramics is medium. Ceramics are used to make hermetic feedthroughs for routing electrical signals into and out of the implant housing, e.g., the alumina–titanium feedthrough of the cardiac pacemaker.
(ii) Ceramic materials displaying good biocompatibility are referred to as bioceramics. These materials do not cause any allergic reactions in the patient. No cytotoxicity is observed in the implant recipient.
(iii) Outgassing of ceramics is low.
(iv) Corrosion resistance of ceramics is high.
(v) Ceramics are good dielectrics.
(vi) RF transparency of ceramics is good due to their low conductivity. Alumina ceramic is commonly used for packaging implants working by radio communication.
(vii) Heat is not produced in ceramic packages.

23.5.2.2 Drawbacks

(i) Mechanical flexibility of ceramics is poor like that of metals. They are brittle materials.
(ii) They are opaque to light.
(iii) Increase in weight of implant by ceramic packaging is medium.
(iv) Cost of ceramic material is high.
(v) Processing and manufacturing processes for ceramics are difficult. Packages with complex geometrical shapes cannot be made using ceramics.
(vi) Impact strength of ceramics is low.

23.5.3 Polymers

23.5.3.1 Advantages

(i) Many polymers are biocompatible.
(ii) Mechanical flexibility of polymers is excellent.
(iii) RF and optical transparencies of polymers are good.
(iv) Implants made by using polymers are light in weight. Hence they do not put any extra burden on the patient.
(v) Polymers are relatively inexpensive.
(vi) Polymers films can be easily deposited, often by spin coating method making processing simple and easy.

23.5.3.2 Drawbacks

(i) Most of the polymers are poor in hermeticity, and are permeable to gases, albeit to different extents. The penetration of water vapor through a polymer coating can play havoc on the protected electronic device. In presence of condensed water, any ions released from the devices triggers a corrosion mechanism of the polymer film further aggravating the moisture attack. A short circuiting of the devices takes place leading to failure.
(ii) Micro-cracks may appear in the polymer coating. The cracks are initiated by the mechanical strain originating from differential expansion and contraction of the polymer over an inelastic substrate. Through these micro-cracks, there is an outflow of toxic materials from the electronic devices into the human body causing cell death and intensifying foreign body response. The toxic materials may exceed the tolerable limits of the human body and may be extremely harmful to patients.
(iii) Polymers show a high degree of outgassing.

Note: The advantages of polymers are utilized in situations where neither metal packaging can be used owing to difficulties in molding and welding, nor ceramic packaging is possible due to sintering challenges. These circumstances arise when making implants of complex shapes with large number of input/output pins. An example of implants of this type is the retinal implant for blindness.

23.5.4 Composites

These are heterogeneous combinations of two or more distinct materials in which the combining materials are separated on a dimensional range larger than atomic scale. In these combinations, one material called the reinforcing material is embedded in the other material known as the matrix, e.g., a dental composite consisting of an acrylic polymer strengthened with ceramic particles. The materials are combined to produce a material in which the constituent materials work synergistically displaying superior properties over singular materials.

23.5.4.1 Advantages

Composites offer unique opportunities to overcome the undesirable properties of individual materials. An individual material may suffer from disadvantages, e.g., ceramics have low fracture strengths, but their combinations with other materials, e.g., metals, yields materials with improved fracture strength. So composite materials are designed to fulfil special requirements.

23.5.4.2 Drawbacks

(i) Design of implants using composite materials is far more complex than with monolithic materials owing to the large number of variables involved.
(ii) Clinical data on the long-term reliability of composites is lacking.
(iii) Interactions among different materials in composites are not completely understood.
(iv) Standards for biocompatibility and fatigue of composites are insufficient and unsatisfactory (Salernitano and Migliaresi 2003).

23.6 Metallic biomaterials

23.6.1 Titanium (Ti) and its alloys

Titanium shows a high degree of biocompatibility owing to its bio-inertness, resistance to corrosion by body fluids and hard tissue compatibility with respect to bone bonding and bone formation (Hanawa 2019). The structural and functional connection between the living bone and the dead medical implant material is known as osteointegration.

Although the mechanism of excellent biocompatibility of titanium is still unclear, it is known that a strongly-adherent, insoluble and chemically impermeable oxide film forms on the surface of titanium in the presence of oxygen. It is retained at the pH value inside the body. TiO_2 has a low tendency to form ions in an aqueous environment. It exhibits a low reactivity with macromolecules (Sidambe 2014). So, the surface-covering film on titanium passivates and protects the underlying titanium from the unfavorable conditions inside the human body. Moreover, the surface oxide of titanium does not denature proteins providing capacity for osteointegration.

Surface modifications of titanium can help to create an ideal cellular response. Porous and roughened surfaces accelerate osteointegration. Surface texture, hydrophobicity, and steric hindrance are the prominent surface features that influence biocompatibility. Therefore, commercially pure titanium (CP-Ti) and titanium alloys have found widespread applications in medical implants from head-to-toe regions, e.g., pacemakers, knee and hip joints, dental and orthopedic implants, etc.

CP-Ti is available in four grades: CP-Ti Grade 1, CP-Ti Grade 2, CP-Ti Grade 3 and CP-Ti Grade 4. The oxygen content increases from 0.18% in Grade 1 to 0.4% in Grade 4. The mechanical strength increases from 170 MPa for Grade 1 to 480 MPa for Grade 4.

Ti–6Al–4V is an alloy containing 90% Ti, 6% aluminum and 4% vanadium. It is an (α+β) alloy meaning that it is a mixture of α and β phases of titanium.

The α-phase of titanium has a hexagonal close-packed crystalline structure. Its β-phase has a body-centered cubic structure. The (α+β) alloys are stronger than pure alloys. Ti–6Al–4V alloy has a strength of 850 MPa. It displays excellent resistance to corrosion but releases aluminum and vanadium, which can be biologically problematic (Nicholson 2020).

23.6.2 Cobalt–chromium alloys

Co–Cr alloys, e.g., ASTM F75 (cast Co–28Cr–6Mo), ASTM F799 (wrought Co–28Cr–6Mo), ASTM F90 (wrought Co–20 Cr–15W–10Ni), ASTM F562 (wrought Co–35Ni–20 Cr–10Mo), are used in dental, orthopedic and cardiovascular implants (Mani 2016). ASTM stands for American Society for Testing and Materials.

They have a high corrosion, wear and fatigue resistance (Narushima *et al* 2015). The corrosion-resistant behavior of Co–Cr alloys within the human body arises from the spontaneously created passive Cr_2O_3 film on Cr.

23.6.3 Stainless steels

Steel grades are designated by SAE (Society of Automotive Engineers) International.

The most common steel is SAE 304 and the second most common is SAE 316L. The SAE 316L is a grade of SAE 316 with extra low carbon content.

SAE 304: 18–20 Cr, 8–10.6 Ni, 2 Mn, 0.75 Si, 0.1 N, 0.08C, 0.045 P, 0.03 S, Balance Fe

SAE 316L: 16–18 Cr, 10–14 Ni, 2–3 Mo, 2 Mn, 0.75 Si, 0.1 N, 0.045 P, 0.03 S, 0.03 C, Balance Fe

Corrosion resistance of stainless steels originates from the presence of 12%–13% Cr in them (Ortiz *et al* 2011). A stable, transparent, uniform thin film of Cr_2O_3 formed on Cr acts as an inert layer shielding the surface of steel and preventing the spreading of oxygen into the alloy. When mechanically or chemically damaged, this film is self-repaired in the presence of oxygen.

Inclusion of small amounts of Cu and Mo stabilizes Cr and reduces its vulnerability to pitting corrosion. SAE 316L steel is a Mo-bearing grade. Mo imparts to it a higher resistance to pitting and crevice corrosion. It is especially recommended in an environment containing chlorides. However, it is more expensive than SAE 304 steel.

23.7 Ceramic biomaterials

23.7.1 Classes of ceramics

Looking at their biological behavior, ceramics are subdivided into three classes: bioactive, inert and resorbable. Hydroxyapatite and bio glass are bioactive ceramics, alumina and zirconia belong to the inert group of ceramics while calcium phosphate is of resorbable type (absorbing again). Hydroxyapatite is the mineral form of calcium apatite. Its formula is $Ca_{10}(PO_4)_6(OH)_2$. Bio glass comprises a group of bioactive glasses: SiO_2, Na_2O, CaO and P_2O_5. Alumina (Al_2O_3) ceramic is an excellent insulator. It shows high mechanical strength and hardness. It is well known for its physical and chemical stability. It is the most widely used oxide ceramic. Zirconia (ZrO_2) offers high hardness, fracture toughness, wear and corrosion resistance. Calcium phosphate $Ca_3(PO_4)_2$ ceramic is used for regeneration of bone tissue for applications in orthopedics and dentistry to restore bones and teeth.

23.7.2 Processing of ceramics

Raw materials for ceramics are derived natural sources such as clay minerals, quartz sand, silicates, flint, etc in powder form (figure 23.4). Either wet forming or dry forming technique is used to process the powder.

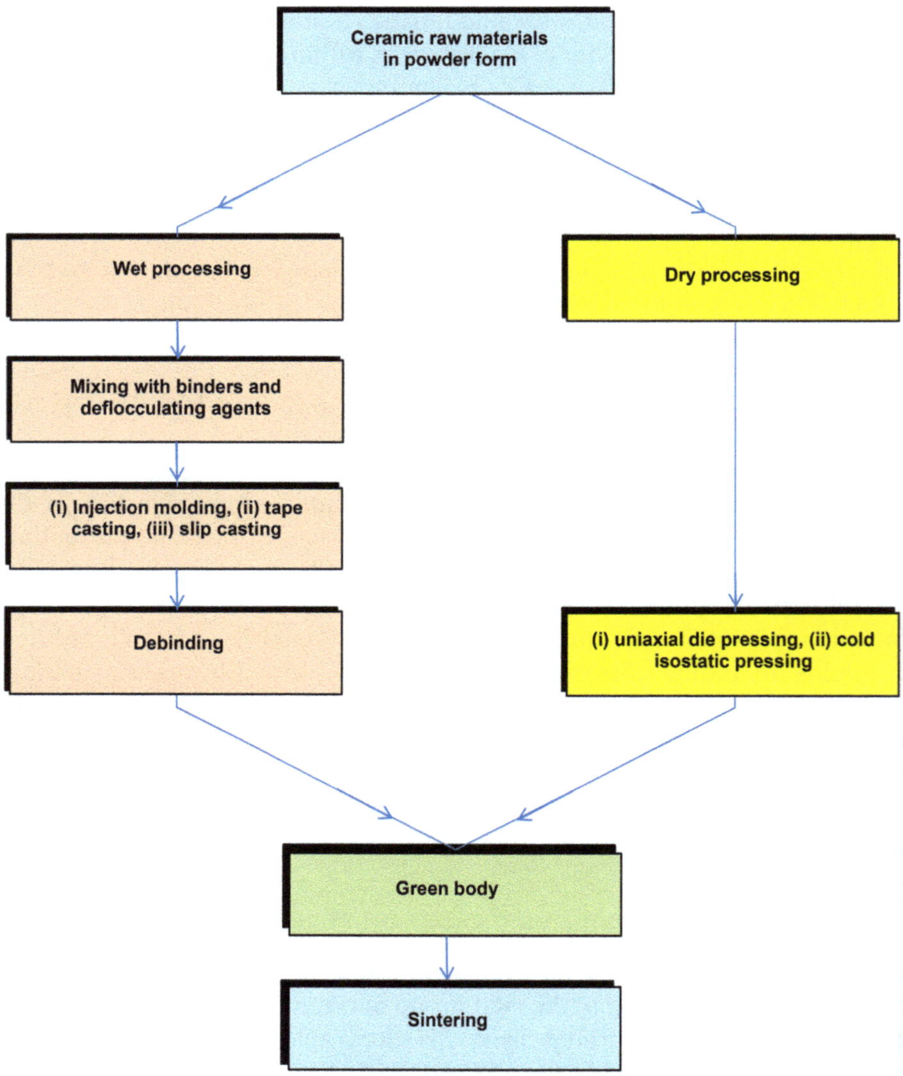

Figure 23.4. Flowchart for processing of ceramics.

In wet forming, the powder is mixed with binders and deflocculating agents. The latter prevent the suspended particles from coalescing into aggregates. The slurry obtained by mixing is put under any one of the following processes for consolidation:

(i) injection molding during which the slurry is mechanically forced into molds having the shapes of the objects to be realized;
(ii) tape casting (doctor blading or knife coating) in which the slurry is spread to form a thin layer on a flat surface, and dried; or
(iii) slip casting in which the slurry, also called slip is poured into a porous plaster mold.

Capillary forcers suck the water from the slurry into the mold. The particles deposit in a dense layer on the inner walls of the mold. When the required thickness of particles has been deposited, the remaining slurry is drained out leaving an object in a free-standing shape. The mold is opened, the object formed is taken out and dried. Note that injection molding applies pressure to inject the material into mold while slip casting uses the force of gravity for the same task. Molding is usually done with plastics. Casting is done with metals.

In all cases, de-binding is done by burning out the binders.

In dry forming, a slurry is not made. The powder is taken in dry form and put into any one of the following consolidation processes:

(i) uniaxial die pressing in which the powder is compacted by pressing along a single axis with a punch, mechanically or hydraulically, in a repetitive cycle, to form a rigid die;
(ii) cold isostatic pressing in which the powder is sealed in an elastomer mold. The mold is placed in a pressure chamber. A liquid is pumped into the pressure chamber. The mold is pressed uniformly in all directions by application of liquid pressure. According to Pascal's law, the pressure applied to a fluid enclosed in a container is transmitted to every point of the fluid and walls of the container without any change in magnitude.

The shaped solid resulting from wet or dry forming is called the green body. The word 'green' is used for pre-sintered material in the sense that it is native or fresh.

The final process is sintering of the ceramic. This is done at a high temperature above 1000 °C. During sintering, the individual particles enlarge and blend together to form a densified solid.

23.7.3 Making hermetic ceramic feedthroughs by conventional brazing

The principal application of ceramics in medical implants is for making electrical feedthroughs for cardiac pacemakers, cardioverter defibrillators, and neural electrodes. A feedthrough transports electrical energy from a circuit on one side of a surface to that on its opposite side without any leakage, e.g., a vacuum feedthrough. A ceramic-to-metal feedthrough can withstand thermal cycles and shocks as well

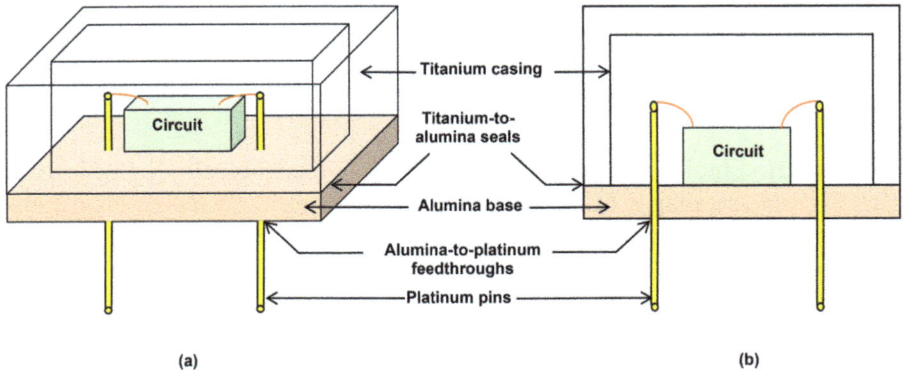

Figure 23.5. A biomedical implant with ceramic-to-metal (alumina-to-platinum) feedthroughs. The implant has a titanium casing with titanium-to-alumina seals.

as pressure variations. It is better than a glass-to-metal feedthrough as glass is more prone to breakage under vibration and acceleration conditions. A common ceramic-to-metal feedthrough in implantable electronics is one in which platinum pins pass through a ceramic substrate (figure 23.5).

The most widely used process for joining ceramic with metal is brazing, a technique in which the two materials are placed together and joined by melting and flowing a material called the braze or filler material at a high temperature around 450 °C; the joining temperature is lower than the melting point of the two materials taken. Materials with melting points above 450 °C are called brazes, those melting below 450 °C are known as solders. Brazes are alloys such as Ag–Cu, Au–Ni, Ni–Cr and Pd-based alloys.

The braze alloys are of two types: passive and active. For brazing with passive brazes, the ceramic must be pre-metallized by electroplating, physical or chemical vapor deposition (CVD) to form a Mo, Mn or W film. This film serves to improve the wetting of the surface. An extensively used method for brazing to alumina is the Mo–Mn method. In this method, a mixture of glass, Mo and Mn phases is applied to the alumina surface by painting or screen printing, and the alumina is heated in a wet H_2 atmosphere to 1500 °C. Brazing is done after coating the metallized alumina surface with a Ni film for wetting.

Noble metal brazing does not generally use Cu or Ni. It is usually done with Ag or Pt or their alloys. Sometimes Pd and Au are used. This brazing is done in air or in an environment rich in oxygen. The noble metal oxides formed during brazing bond with ceramic substrates, in particular with oxide ceramics.

The active braze alloy contains a metal such as Ti or Zr to promote wetting of the ceramic surface, requiring no prior metal coating. Removal of the pre-metallization step makes this process easier in implementation. For active braze alloys, a vacuum furnace (10^{-5} torr) is necessary. The Ti–Cu–Ni braze alloy is used for alumina–titanium bonding. As already said, Ti plays the role of increasing the wetting. As the use of Cu in the Ti–Cu–Ni alloy is disfavored from a toxicity standpoint, the Cu-free

Ti–Ni brazes have been found to be more suitable for medical implants, and are therefore used in hermetically sealing ZrO_2 to Ti (Jiang *et al* 2005).

To make an array of feedthroughs, the metal pins are brazed into vias formed in a ceramic substrate by laser machining. This is a well-established technology for hermetic sealing. It suffers from the limitation that the spacing between the pins called the pitch is too large (400–500 μm) so that the number of pins for interfacing with the nervous system is too small. The pitch achieved is much larger than the pad spacings in integrated circuits.

Clearly, a miniature-size implant with a high density of pins inserted at a small pitch~200 μm cannot be fabricated by this method. Moreover, the number of pins cannot be raised by increasing the size of the implant because then it will become too bulky and voluminous to be put inside the human body. This restriction on pins imposes limits on the performance of the implant. A cochlear implant with fewer pins will provide lower fidelity. A retinal prosthesis with small number of pins will yield poor image resolution.

23.7.4 Making ceramic feedthroughs using extruded metal vias

High pin density feedthroughs were fabricated by an extruded metal via process (Shah *et al* 2012). In this process (figure 23.6), an array of vias is formed on a 250 μm thick alumina substrate at a pitch of 200 μm using CO_2 laser. The vias have diameters of 53 μm on the laser beam-exit side. The diameter values are slightly larger on the side of laser beam entry. So, the vias have a tapered profile. The substrates are conformally coated with a Ti/Au film in a sputtering chamber. Lithography is performed on both sides. Patterning is done on one side for fixing the electronics. It is done on the other side for attaching the microelectrode array. Gold studs of larger size than vias are placed over the vias using an ultrasonic ball bonder. The gold studs are alloyed with Ti/Au film. The extrusion of studs together with ultrasonic energy, high temperature and force gives good sealing. For planarity of the substrates, the gold studs are coined by thermo-compression on a flip-chip bonder. This coining also improves the hermeticity of feedthroughs. Hermeticity is tested by helium leak test. The leak rate is 9×10^{-10} Torr L s^{-1}.

Using gold stud bumps, vertical feedthroughs were sealed in alumina substrates at densities of 1600 cm^{-2} with all manufacturing done at temperatures <420 °C. Water leak rates were found satisfactory for packages of volume <2 cm^3 and feedthrough count > 50 for an implantation period of 50 years (Langenmair *et al* 2018).

23.8 Polymeric biomaterials

23.8.1 PDMS (polydimethylsiloxane)

PDMS, $(C_2H_6OSi)_n$, is a silicone elastomeric substance (Volkov 2013). The silicone is a silicon-based polymer. An elastomer is a viscoelastic material with rubbery appearance. Hence PDMS is often called 'silicone rubber'. At low temperatures it behaves as an elastic solid and at high temperatures as a viscous liquid with unusual rheological or flow properties.

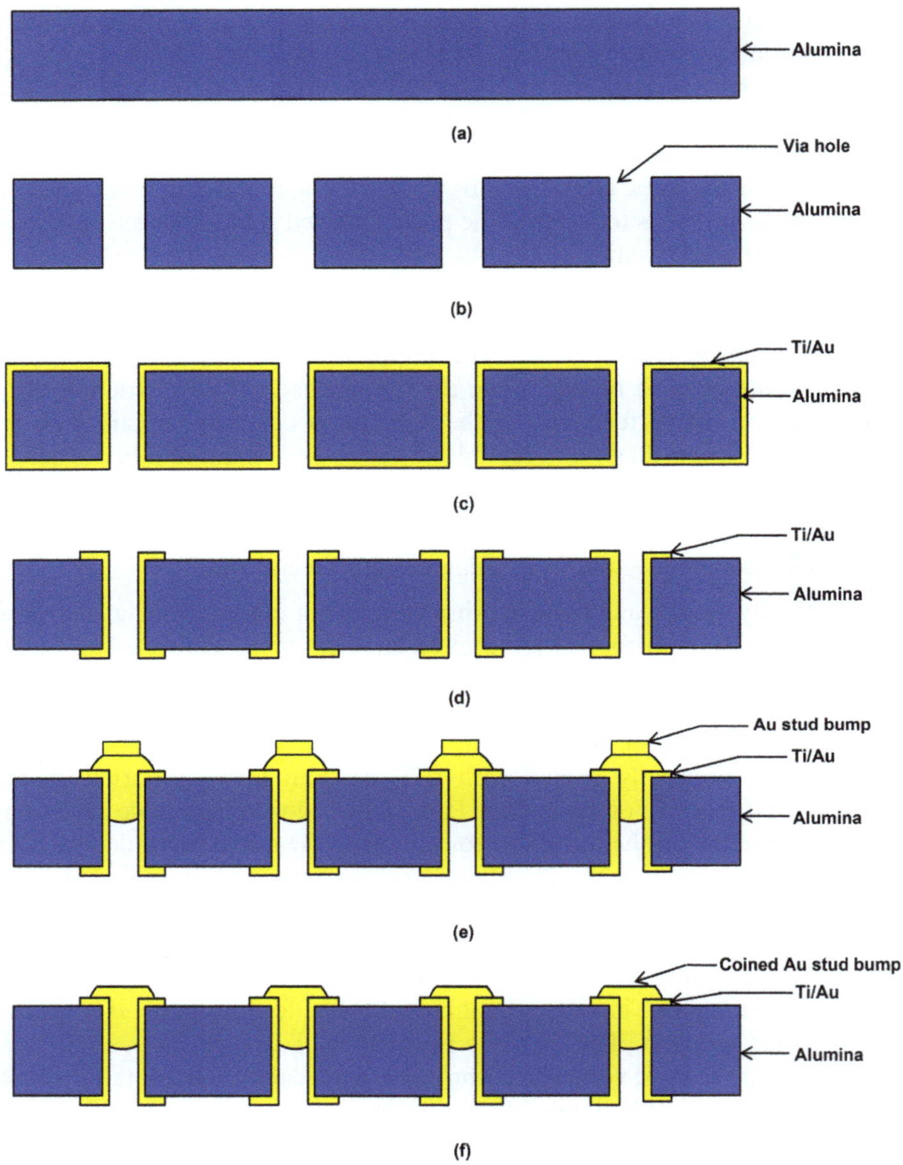

Figure 23.6. Fabrication of high-density alumina-to-gold feedthroughs showing: (a) alumina substrate, (b) vias machining in the substrate, (c) conformal coating of the substrate containing vias with Ti/Au film, (d) lithography on both sides of the substrate and patterning, (e) placing gold studs over the vias, extrusion of metal studs by bumping and their alloying with Ti/Au film, and (f) coining of studs by thermo-compression in a flip-chip bonder.

Advantages:

(i) PDMS is a transparent and chemically inert material with high UV and thermal resistance.

(ii) It offers long-term stability in a biological environment displaying high resistance to biodegradation.

(iii) It is easily processed, can be deposited by spin coating or drop casting in a broad thickness range spanning from nano to micrometers, and can be molded into different geometrical shapes.
(iv) It has a high dielectric strength ~14 MV m^{-1},
(v) It has a low glass transition temperature $T_g = -150$ °C.

Disadvantages:
(i) A PDMS package is soft and fragile because of the low tensile and shear strengths of this material.
(ii) Its high permeability to gases and vapors makes it unsuitable for hermetic sealing. A multilayer package avoids through-hole defects by employing individually deposited and cured layers which are subjected to pressure for densification to reduce gas permeability.
(iii) Variation of dielectric constant of PDMS from 2.3 to 3.8 in the frequency range 0–50 Hz precludes its utilization as a substrate for telemetric structures.

23.8.2 Polyimide

Polyimide resin, $C_{35}H_{28}N_2O_7$, is a polymer of imide monomers (Hicyilmaz and Bedeloglu 2021). It is available in the form of free-standing sheets or as a liquid. Polyimide in liquid phase is deposited on a substrate by spin coating, and is hardened by curing at 400 °C.

Advantages:
(i) Polymide shows high electrical resistivity = 1.5×10^{17} Ω-cm and high dielectric strength = 303 kV mm^{-1} with a dielectric constant = 3.4 at 1 kHz for a 25 μm thick film.
(ii) It has high mechanical flexibility and reliability.
(iii) It has high thermal stability and chemical resistance favoring its use as a reliable insulator on metal films at the 37 °C temperature of the human body.

Disadvantages:
(i) Polymide has a high moisture absorption rate ~2%–4%.
(ii) It suffers from lack of stiffness due to which a polyimide packaged implant easily buckles when inserted in the body.

23.8.3 PVDF (polyvinylidene fluoride)

PVDF, $-(C_2H_2F_2)_n-$, is a fluorocarbon-based semicrystalline, thermoplastic homopolymer of the vinylidene monomer (Saxena and Shukla 2021). It shows pyroelectric and piezoelectric properties. It can be processed by melt casting, solution casting, spin coating and film extrusion.

Advantages:
(i) PVDF shows high abrasion, creep and fatigue resistance.
(ii) It is chemically resistive to most solvents and acids.

(iii) It displays low water absorption.
(iv) It has a high dielectric strength ~1.4–110 kV m^{-1}.
(v) It produces minimal cellular response in contact with human tissue.

Disadvantages:
(i) PVDF has low thermal stability.
(ii) It gives poor adhesion with other materials.
(iii) It is incapable of forming smooth films.
(iv) Large changes in relative permittivity of PVDF from 6 to 10 between 1 kHz and 1 MHz are undesirable.

23.8.4 Parylene-C

Recall discussions on parylene coatings in section. 16.3. Parylene is the generic name of a polymer series. Parylene-C, $C_{16}H_{14}Cl_2$, is chlorinated poly(para-xylylene) (Golda-Cepa *et al* 2020).

Advantages:
(i) Parylene-C has low permeability to moisture and gases~0.06%–0.1%.
(ii) It offers possibility to deposit high-purity, pinhole-free coatings by CVD on substrates maintained at room temperature.
(iii) It has the ability to form ultrathin, hydrophobic, optically transparent, uniform conformal coatings on complex shapes wrapping on every corner, preserving correct angles in small areas, and adding no weight to components.
(iv) It has good temperature and UV stability.
(v) It has high tensile and yield strength.
(vi) It shows anti-corrosion properties.

Disadvantages:
(i) Parylene-C is resistant to solvents at room temperature causing difficulty in its removal if reworking is necessary.
(ii) It has weak adhesion with metals, particularly the noble metals.
(iii) Its use gives limited throughput because CVD is a time-consuming process restricted to small batches.

23.8.5 Liquid crystal polymers (LCPs)

LCPs are 4-hydroxybenzoic acid and related monomers-based aromatic polyesters straddling the boundary between ordered crystalline solids and dis-ordered amorphous liquids.

Advantages:
(i) LCPs provide low moisture absorption rate (0.04%) which is better than those of parylene-C and polyimide.
(ii) Gas permeability of LCPs is not influenced by the humidity or temperature of the surroundings.

(iii) Low dielectric constant (3) and low dissipation factor (0.002–0.0045) of LCPs make them useful for implants using wireless communication.
(iv) They have high mechanical strength, heat and chemical resistance, and small coefficient of thermal expansion.
(v) LCPs possess flexibility and show compatibility with semiconductor processing.
(vi) They offer possibility of layer-to-layer lamination with fusion bonding by application of heat and pressure.

Disadvantages:
(i) LCPs have poor adhesion with metals and flexible substrates.
(ii) Requirement of a high temperature ~285 °C for bonding of LCPs is inconvenient.
(iii) LCPs are not photo-patternable materials.

23.8.6 Thermoplastic polyurethane (TPU)

As its name indicates, it softens when heated and hardens on cooling. It is an elastomeric material with rubbery-like properties. It is a polymer composed of many organic units joined by carbamate or urethane molecules. Its chemical formula is $C_{17}H_{16}N_2O_4$ (Kim and Liu 2018).

Advantages:
(i) TPU gives high durability and flexibility.
(ii) It has a high tensile strength.
(iii) It has good ability to bear loads.
(iv) It can be solution-coated or vacuum-formed enabling various fabrication processes.

Disadvantages:
(i) Some grades of TPU have small shelf life.
(ii) TPU is not as cost-effective as alternative materials.

23.9 Composite biomaterials

These materials are classified in accordance with the matrix material used in the composite.

23.9.1 Metal matrix composites

Hydroxyapatite reinforced Ti–6Al–4V (HA/Ti–6Al–4V) improves biocompatibility and wear resistance of titanium which shows poor wear resistance. Also, bioinert Ti surface slows down osteointegration (Avila *et al* 2021).

23.9.2 Ceramic matrix composites

Mechanical properties of HA are inferior to bone posing risks of breakage at load-bearing sites. Properties of HA are controlled by incorporating small amounts of bioactive glass serving as a source of ions. Osteoblastic performance and antibacterial activities of HA are improved by making samarium (Sm^{3+})-doped glass-reinforced hydroxyapatite composites using 2.5% (w/w) of the P_2O_5 glass to 97.5% (w/w) of HA. The resulting material acts as a bioactive bone substitute (Morais *et al* 2014).

23.9.3 Polymer matrix composites

The elastic modulus of polyetheretherketone (PEEK) being appreciably lower than that of bone, stiffness of PEEK is too low to bear the stress of orthopedic implants. Therefore, carbon fiber reinforced (CF-PEEK) composites formed by adding high modulus carbon fibers to PEEK are used for these implants. CF-PEEK has similar modulus to bone and can endure the strain of prolonged fatigue (Liao *et al* 2020).

23.10 Implantable microelectrode arrays for neuroprosthetics

A microelectrode array is a bidirectional interface with the brain through which sensory signals are recorded and stimulation signals are applied (figure 23.7). It is called a neural implant or brain implant. Brain regions that have become dysfunctional because of a head injury or a paralysis stroke can be activated with this implant.

The Utah neural microelectrode array (Campbell *et al* 1990) is a 10 × 10 array consisting of P^+ silicon microneedles in an N-type wafer. The microneedles are fabricated by bulk micromachining of silicon (figure 23.8). The length of each needle is 1500 μm, width of the needle base is 80 μm, and the needle-to-needle spacing is

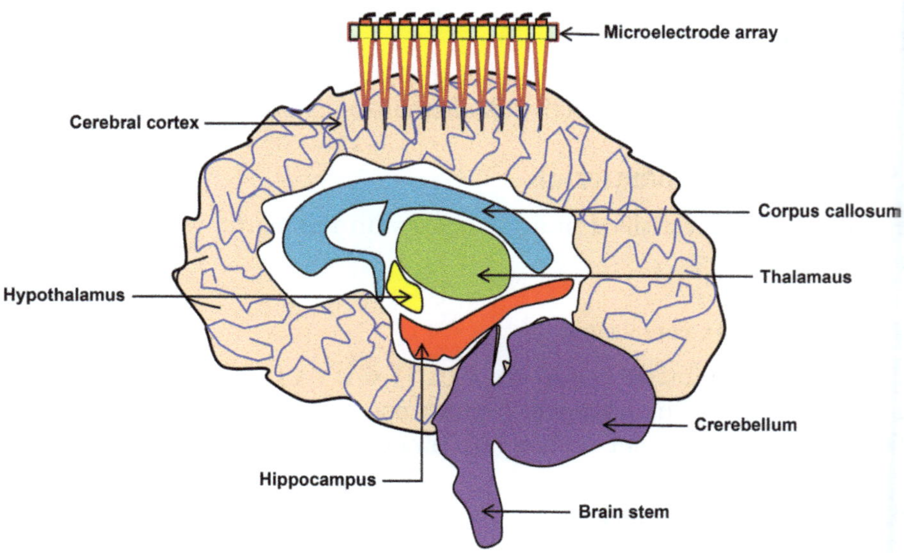

Figure 23.7. Microelectrode array showing microelectrodes implanted in the cortex.

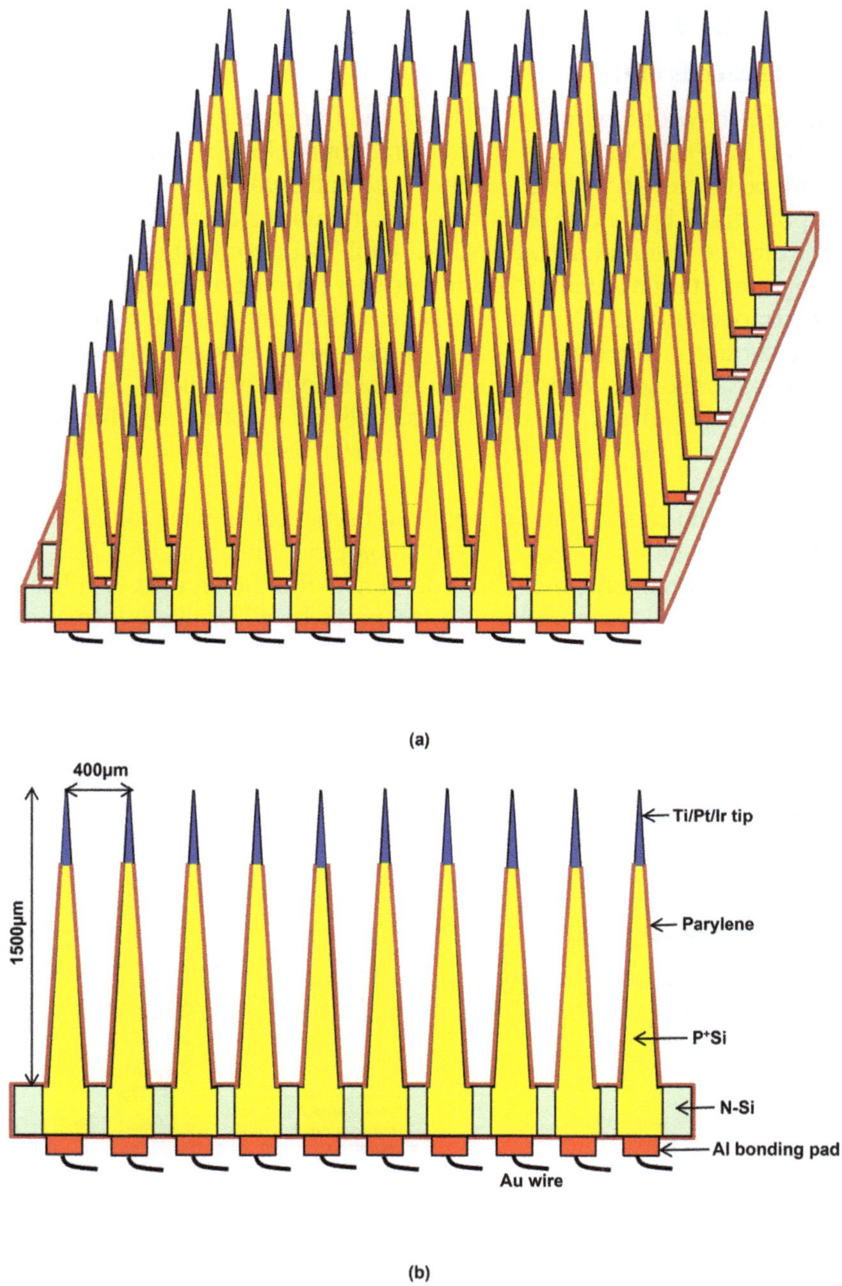

Figure 23.8. The Utah microelectrode array: (a) 3D drawing and (b) cross-sectional diagram.

400 μm. The tip of the needle is coated with Ti/Pt/Ir film by sputtering. The array is encapsulated with parylene-C deposited by CVD. Reactive ion etching is done to remove the parylene-C from the tips of silicon needles. Al pads are formed on the backsides of the needles. Gold wires are bonded to the pads for connections.

23.11 Optrode array with integrated LEDs

23.11.1 Applications of the array

It is used for electrocorticography (ECOG), and in optogenetics (Kwon *et al* 2013a, 2013b, 2014, 2015). ECOG, also called intracranial electroencephalography, is an invasive procedure of recording the electrical potentials from the cerebral cortex (cortical potentials) by placing electrodes directly on the cortical surface instead of doing it from the external surface of the skull as in conventional EEG (electroencephalogram). It is implemented during surgery (intraoperative ECOG) or outside surgery (extraoperative ECOG) after incision to implant the electrode grid subdurally on the surface of the brain. Using this method epileptogenic tissue is localized enabling neurosurgeons to identify the source of seizures in epilepsy patients who are resistant to epileptic drugs. The ECOG signals have better signal-to-noise ratio, spatial and spectral resolution than EEG signals because of absence of shielding effect of the skull.

Optogenetics, the combination of optics and genetical engineering, is a method of optical neuromodulation to monitor the biological functions of a cell or group of cells, thereby enabling the manipulation of activities of neurons or other types of cells with light. Genetic engineering is a technique by which the information in the genetic code of a living being is altered by addition of a new piece of code that allows neurons to make light-responsive proteins known as opsins. As cells can be turned on or off using light, responses of pain or fear can be activated or blocked.

Optogenetics has advanced the understanding of the behavior of neural circuits in various conditions, e.g., mood and obsessive-compulsive disorders, Parkinson's disease, etc. It offers an opportunity for treatment of inherited degenerations of retina.

23.11.2 Working of the array

Figure 23.9 illustrates the working of the array (Lee *et al* 2014). It performs two functions: optical and electrical. The optical function is to shine light on selected

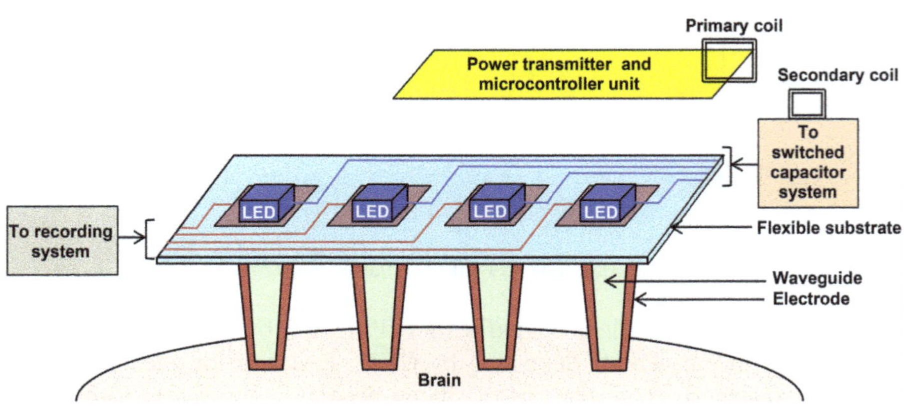

Figure 23.9. Conceptual diagram of optrode array.

brain sites. The electrical function is to measure the electrical signal generated in the illuminated site to investigate the responses of different regions to light activation. For convenience of mounting on the curved surfaces, the array is made on a flexible substrate.

For electrical function, LEDs are attached on the top surface of the substrate. Below each LED there is a trapezium-shaped waveguide made of a polymer to carry the light from the LED to the brain site under study. The LEDs are controlled by an SCS (switched capacitor system) powered inductively through a coil. The coil on the SCS is the secondary coil while the primary coil along with the transmitter is placed outside the body.

For electrical studies, the surface of the waveguide is coated with a metal. The tip of the waveguide touches the surface of the part of the brain over which the assembly is implanted. The current signals received from the brain sites are carried by the metal layer covering the waveguide and sent to a recording system.

23.11.3 Fabrication of the array

The array fabrication is subdivided into process stages that are enumerated in subsections below.

23.11.3.1 Fabrication of PDMS stamp

A 3″ glass wafer (figure 23.10(a)) is coated with a 30 μm thick SU-8 photoresist layer (figure 23.10(b)). The photoresist is pre-baked and lithography is performed using a mask. This mask is designed in such a way that after the photoresist has been developed, photoresist blocks of the same size as LEDs are formed at the exact distances at which the LEDs will be mounted. After post-baking the photoresist, a mold is formed (figure 23.10(c)). This mold is used to make the PDMS stamp. For making the stamp, PDMS is poured in the mold. PDMS flows and fills the gaps between the photoresist blocks (figure 23.10(d)). It is cured to make it hard. After hardening, the PDMS layer is carefully peeled off. This is the stamp defining the locations of LEDs (figure 23.10(e)).

23.11.3.2 Mounting of LEDs on a parylene-C support layer

A 3″ glass wafer is taken (figure 23.11(a)). Parylene-C film of thickness 5 μm is deposited over the glass wafer (figure 23.11(b)). The parylene-C film is coated with a 0.3 μm thick copper film (figure 23.11(c)). Lithography is done to define the pattern of contacts and interconnections on the copper film. The two contacts for each LED are shown (figure 23.11(d)). The exposed copper contact areas are covered with solder (figure 23.11(e)).

The separately formed PDMS stamp is taken. LEDs are fitted into the cavities in the stamp (figure 23.11(f)). The stamp with the LEDs fixed inside the cavities is turned upside down and brought over the structure with copper contacts formed in figure 23.11(g). The stamp is carefully lowered over the structure. Then the stamp and the structure are pressed together (figure 23.11(h)). After pressing,

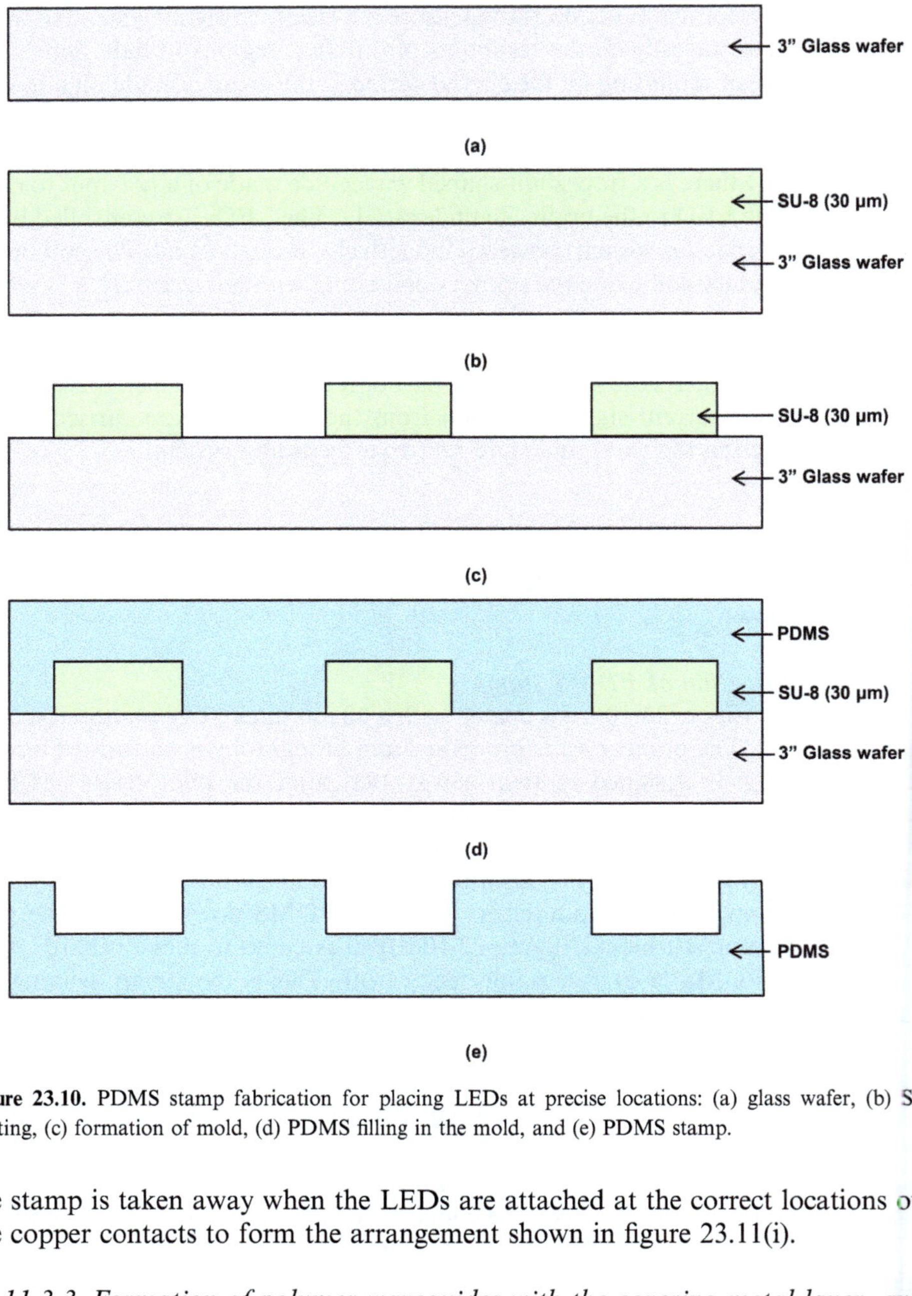

Figure 23.10. PDMS stamp fabrication for placing LEDs at precise locations: (a) glass wafer, (b) SU-8 coating, (c) formation of mold, (d) PDMS filling in the mold, and (e) PDMS stamp.

the stamp is taken away when the LEDs are attached at the correct locations over the copper contacts to form the arrangement shown in figure 23.11(i).

23.11.3.3 Formation of polymer waveguides with the covering metal layer, and assembly of the optrode array

On a 3″ glass wafer (figure 23.12(a)), SU-8 photoresist is coated. The thickness of SU-8 layer is 50 µm (figure 23.12(b)). The photoresist is pre-baked, exposed with a mask, developed and post-baked. The pattern obtained contains blocks of SU-8 at pre-decided locations. Over the patterned SU-8 layer, PDMS is poured and cured

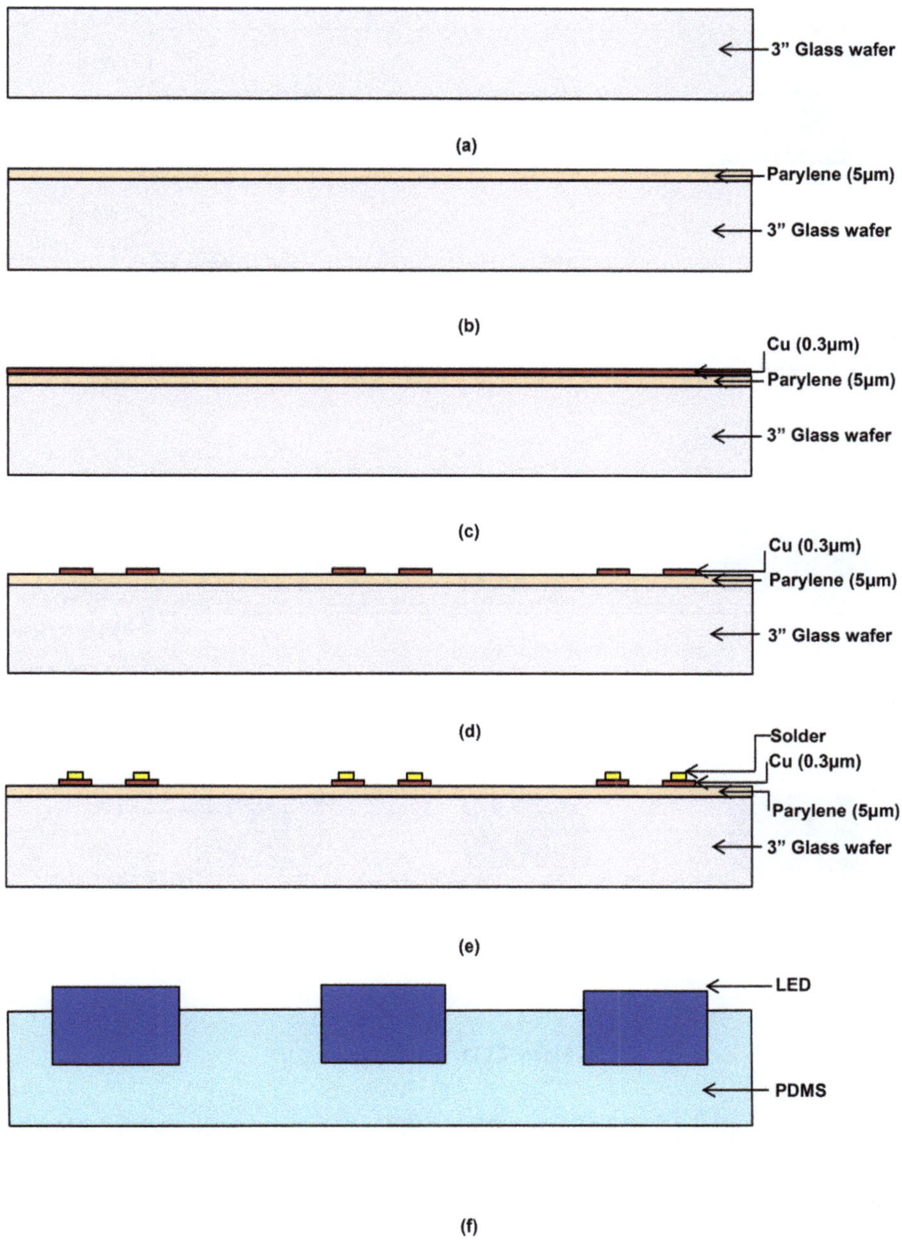

Figure 23.11. Process of LED mounting on parylene-C substrate: (a) glass wafer, (b) parylene-C deposition, (c) Cu film coating, (d) lithography for Cu contacts patterning, (e) covering exposed Cu with solder, (f) fitting LEDs in PDMS stamp, (g) bringing inverted stamp over the structure with Cu contacts, (h) pressing the stamp against the structure, and (i) withdrawal of stamp.

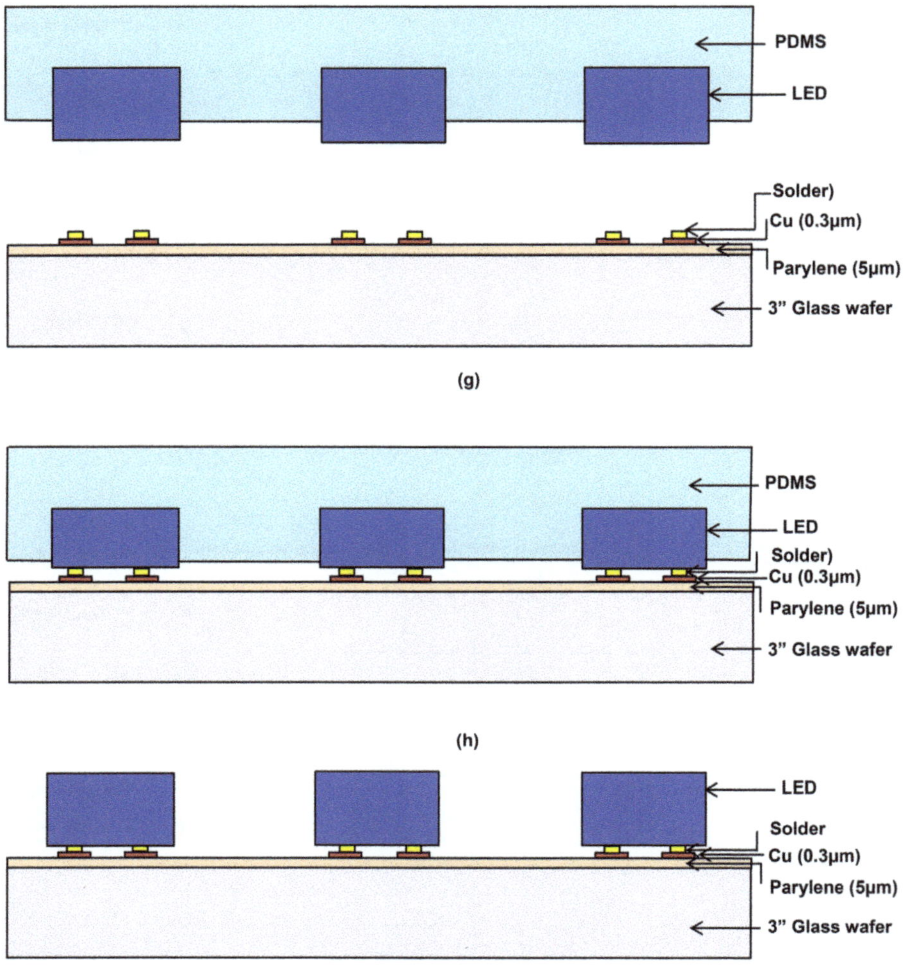

Figure 23.11. (Continued.)

for hardening (figure 23.12(c)). The PDMS is coated with photoresist (figure 23.12(d)). The photoresist is patterned by lithography. The pattern contains exposed regions of PDMS bounded by photoresist protected regions (figure 23.12(e)). The patterned structure is subjected to oxygen plasma treatment. During plasma treatment, the exposed regions of PDMS are converted from hydrophobic to hydrophilic nature (figure 23.12(f)). Photoresist is removed (figure 23.12(g)). A small quantity of SU-8 is dropped over the PDMS using a nanopipette forming a hemispherical dome-shaped film covering the treated PDMS (figure 23.12(h)). Droplet backside exposure (DBE) is done using a mask (figure 23.12(i)) forming microneedles of different lengths (figure 23.12(j)). DBE is a process used for making slanted microneedles in which the length of microneedles

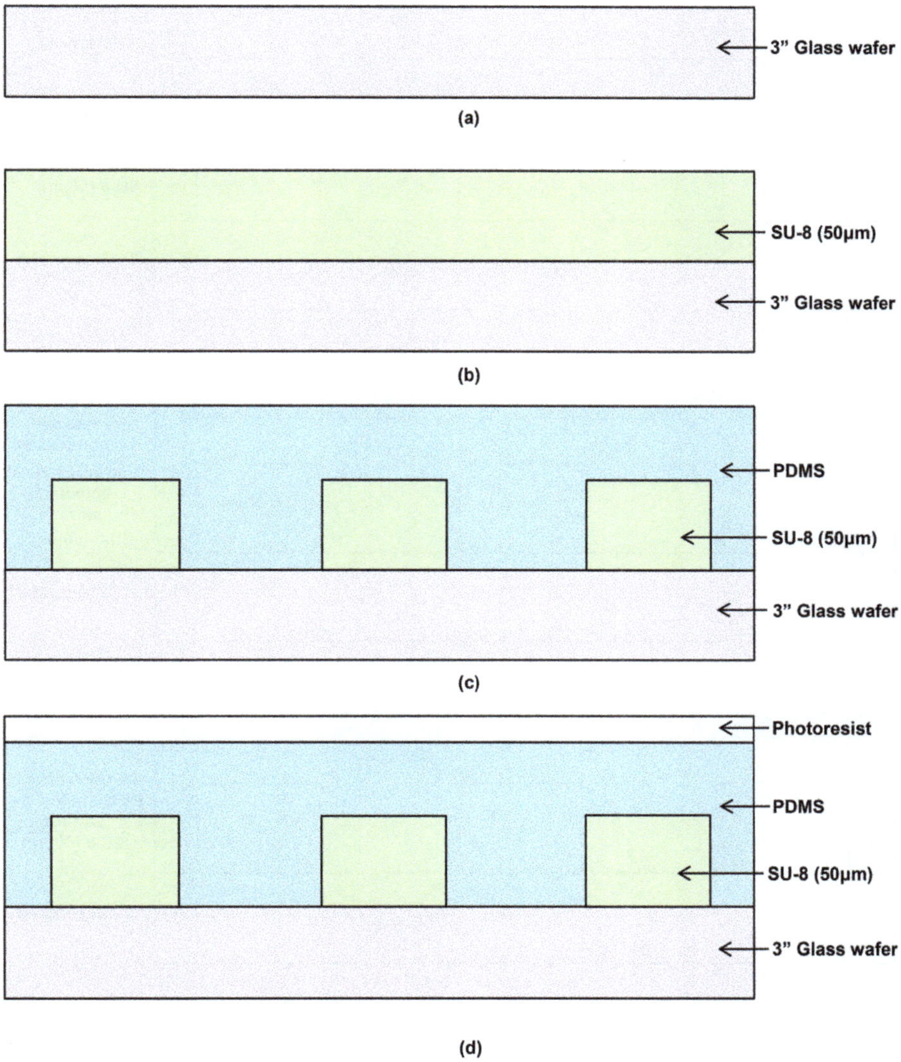

Figure 23.12. Process steps for making polymer waveguides and combining their separately fabricated parts together to complete the array: (a) glass wafer, (b) SU-8 coating, (c) SU-8 patterning and pouring PDMS on patterned SU-8, (d) photoresist coating, (e) photoresist patterning, (f) oxygen plasma treatment, (g) removal of photoresist, (h) dropping small quantity of SU-8, (i) droplet backside exposure, (j) microneedle formation, (k) developing SU-8 and oxygen plasma treatment, (l) etching waveguide tip, depositing parylene-C and removing it from waveguide tip, (m) separating PDMS from glass, (n) SU-8 removal, (o) aligning LEDs with cavities in PDMS, (p) fixing LEDs in PDMS cavities, and (q) removing the glass wafer.

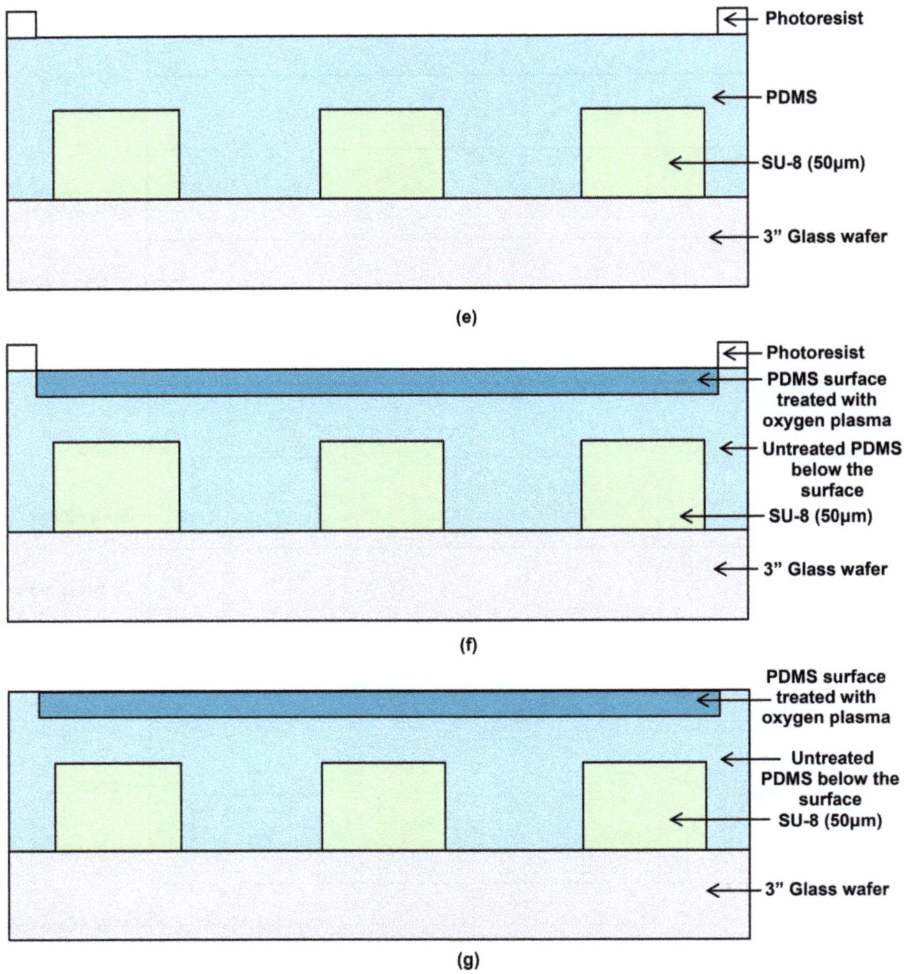

Figure 23.12. (Continued.)

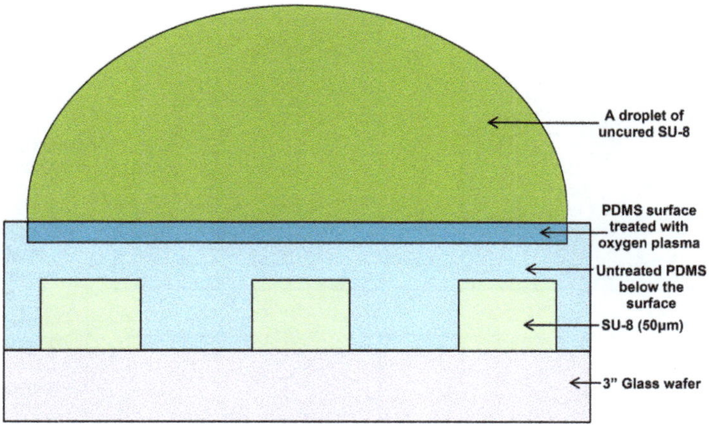

(h)

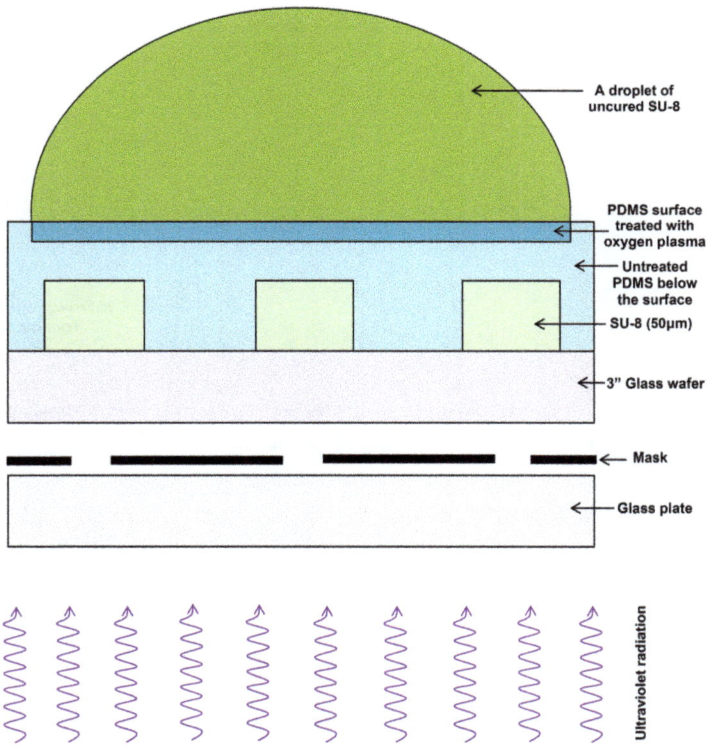

(i)

Figure 23.12. (Continued.)

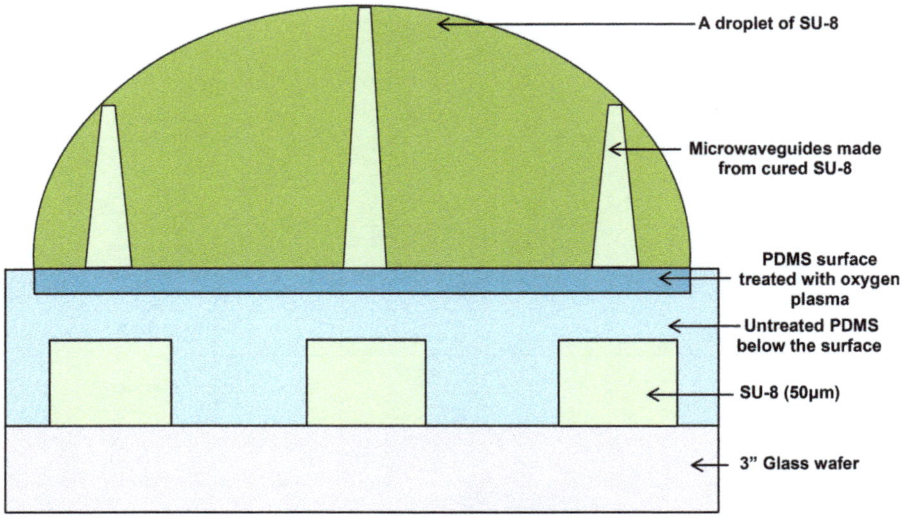

(j)

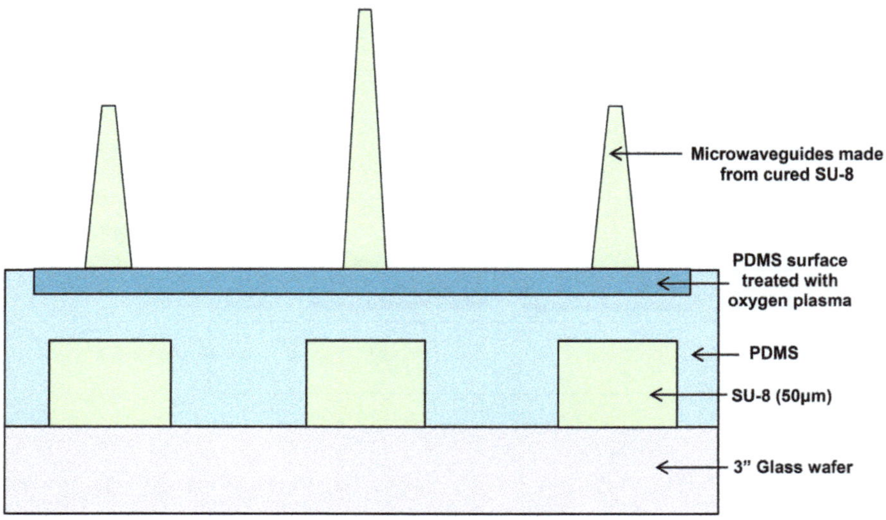

(k)

Figure 23.12. (Continued.)

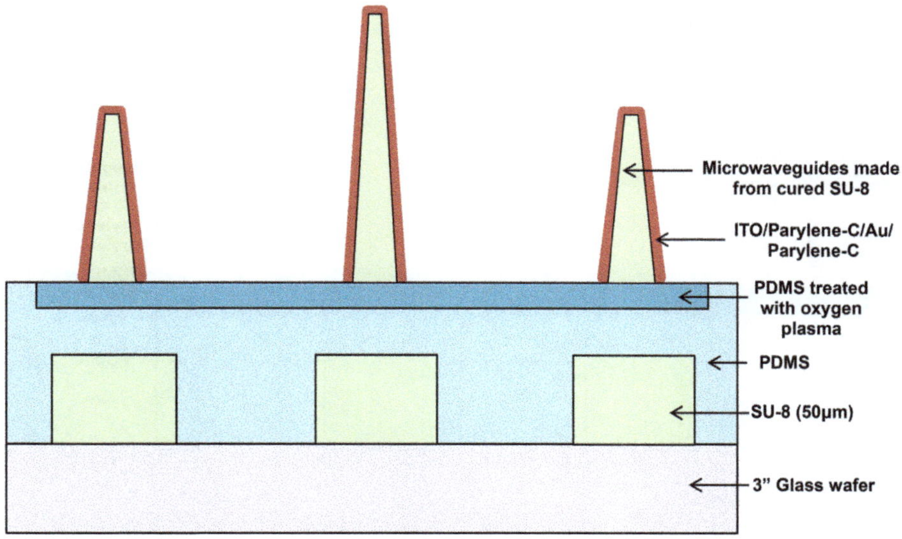

(l)

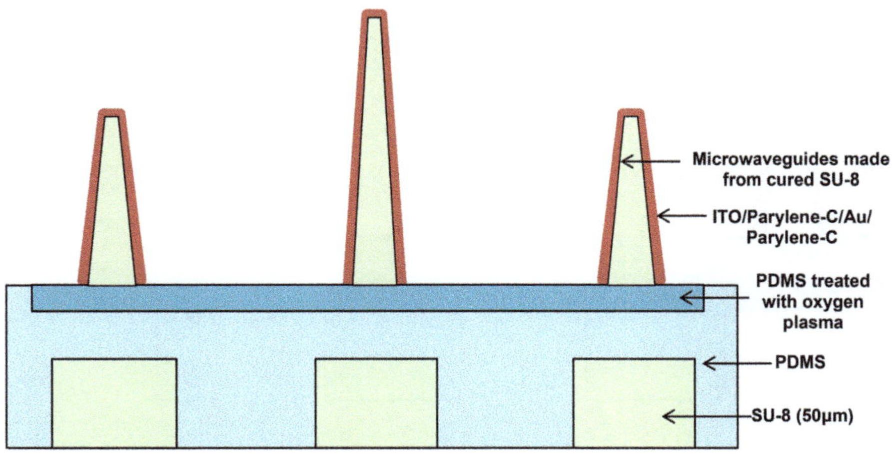

(m)

Figure 23.12. (Continued.)

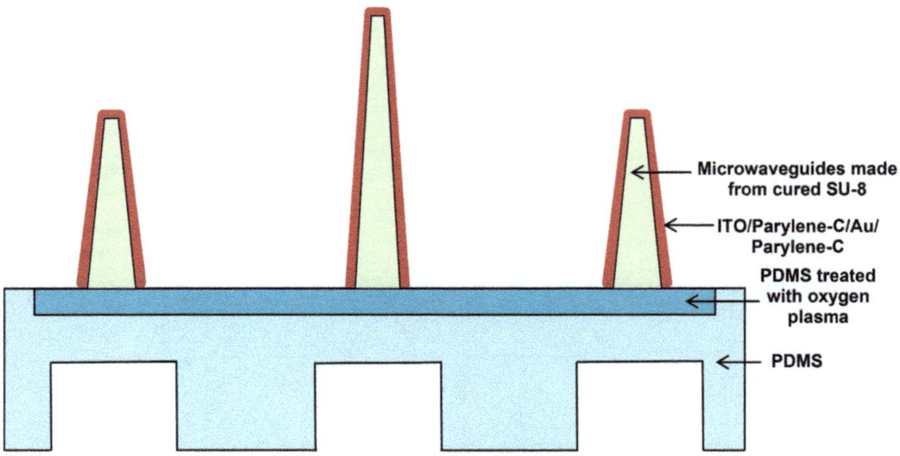

(n)

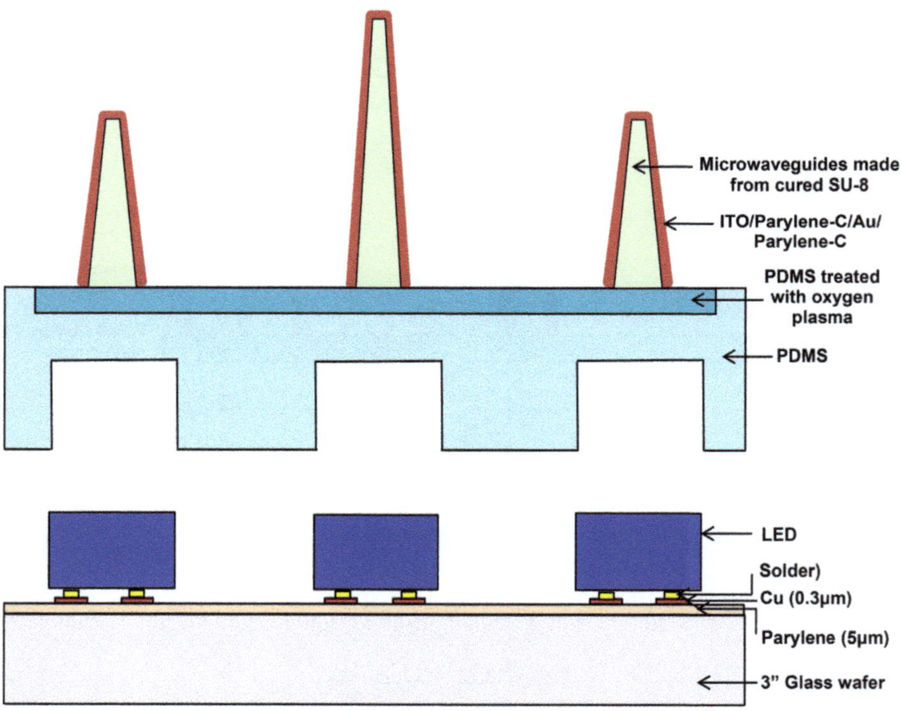

(o)

Figure 23.12. (Continued.)

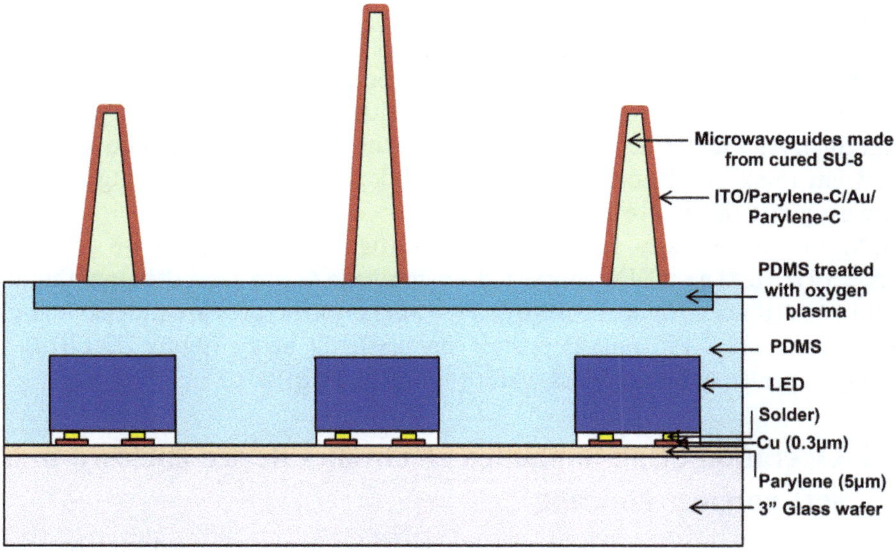

(p)

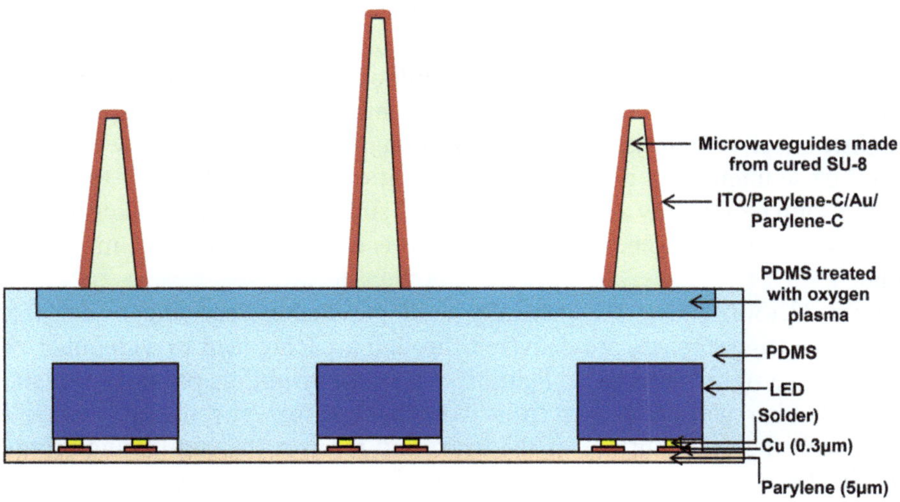

(q)

Figure 23.12. (Continued.)

is determined by the volume of SU-8 droplet. Neither specialized equipment nor intricate fabrication methods are needed in this process.

SU-8 is developed and polished with oxygen plasma (figure 23.12(k)). 0.1 μm thick ITO is DC sputtered, followed by deposition of 5 μm thick parylene-C, and 0.2 μm thick Au. Subsequent steps are: etching the waveguide tip to allow passage of light, 5 μm thick parylene-C protective layer deposition and removing parylene-C from waveguide tip (figure 23.12(l)).

PDMS is taken away from glass wafer (figure 23.12(m)). SU-8 is removed (figure 23.12(n)). The LEDs supported on parylene-C are placed below the above structure and aligned (figure 23.12(o)). Two aligned structures are pressed and glued together with SU-8 (50 μm) deposited on an LED array (figure 23.12(p)). The assembly is peeled off from glass wafer (figure 23.12(q)).

23.12 Operation of an implanted electronics device enclosed in a soft polymer covering

A pacemaker for the heart powered by the triboelectricity produced by cardiac motion itself cannot be enclosed in a brittle metallic package. Its packaging must be flexible enough so that the feeble movements received from the heart are able to produce sufficient electric current to drive the device. Let us probe into the packaging of the triboelectric nanogenerator of a self-powered symbiotic pacemaker.

Ouyang *et al* (2019) demonstrated a symbiotic pacemaker in the large animal model (5 kg porcine). The symbiotic pacemaker concept is inspired by the phenomenon of symbiosis in biology, meaning living together of two organisms by their close interaction and mutual co-operation. Here the body of the animal is both the source of energy and the recipient of stimulus. The pacemaker is powered by a piezoelectric nanogenerator which harvests energy from cardiac motion of the animal.

Figure 23.13(a) shows the generally used pacemaker receiving power from a battery. The trigger pulses are delivered through a pacing lead to a chamber of the heart, here the right ventricle. In figure 23.13(b), the symbiotic pacemaker is shown. This pacemaker gets its power from a nanogenerator. The nanogenerator is in contact with the outer surface of the heart and receives the stimulating vibrations from the heart muscle movements. It produces electricity by utilizing contact electrification and electrostatic induction. It feeds the generated charges to a power management unit (PMU). In the beginning the PMU is switched off. The charges produced in the nanogenerator are stored in the capacitor of the PMU. Afterwards, the PMU is turned on using a magnet as a wireless trigger. Then the pacemaker starts supplying pacing pulses to the heart.

The nanogenerator must be structurally stable. Additionally, its working should not be hindered by body fluids. So it must be safely packaged in such a manner that the cardiac motion is properly transferred to it. Obviously, as already mentioned, the nanogenerator cannot be packaged in a rigid and heavy metal casing. Soft, lightweight and flexible polymeric materials are required. A double-layer protection is

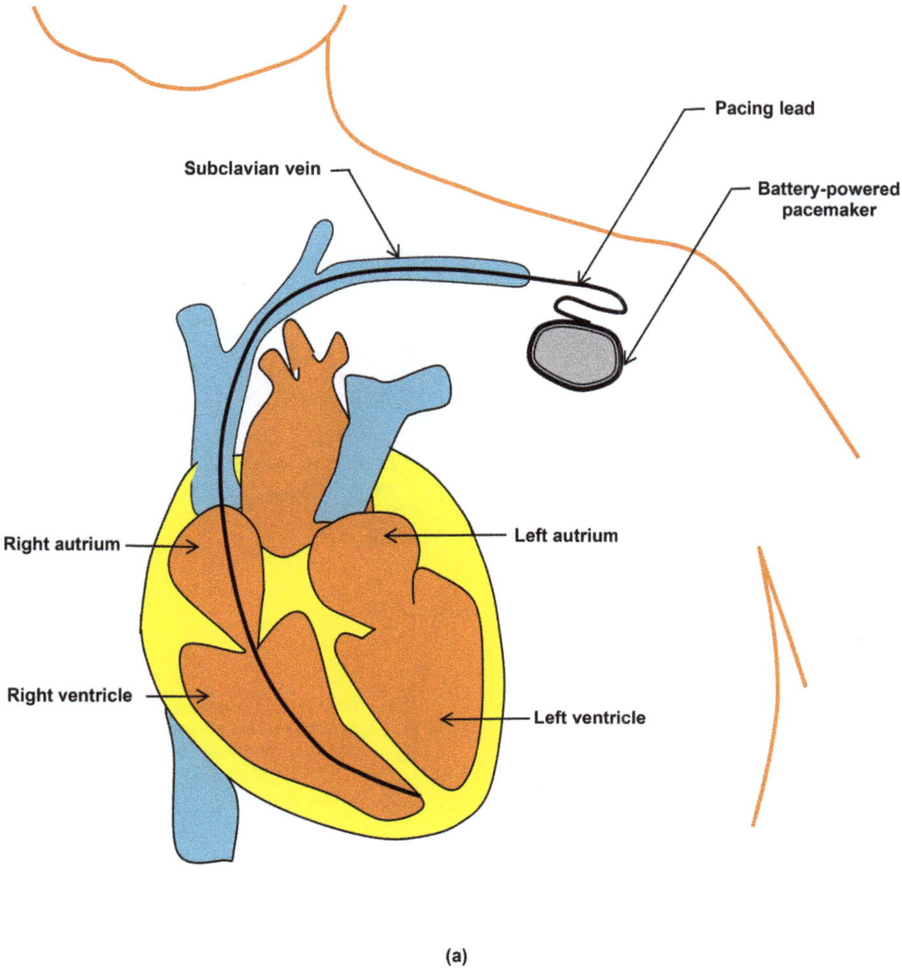

Figure 23.13. Cardiac pacemakers: (a) the common pacemaker with a battery and (b) the symbiotic device without battery.

therefore used. The first layer is a Teflon film. For the second layer, PDMS is mixed with the curing agent in the ratio 10:1 and deposited by spin coating over the Teflon film. The PDMS layer is hardened at 80 °C for 1 h.

For *in vivo* energy harvesting and arryhmia correcting experiments, the nanogenerator was implanted between the heart and the pericardium of anesthetized porcine. It was confirmed that the pacing stimuli can correct sinus arrythmia.

The open-circuit voltage of the nanogenerator reaches up to 65.2 V. The energy harvested in each cardiac cycle is 0.495 μJ. For comparison, the endocardial pacing threshold energy is 0.377 μJ for man.

Accelerated fatigue test is performed by subjecting the nanogenerator immersed in PBS solution to a periodic vibration for 10^8 cycles. The output voltage is 93 V,

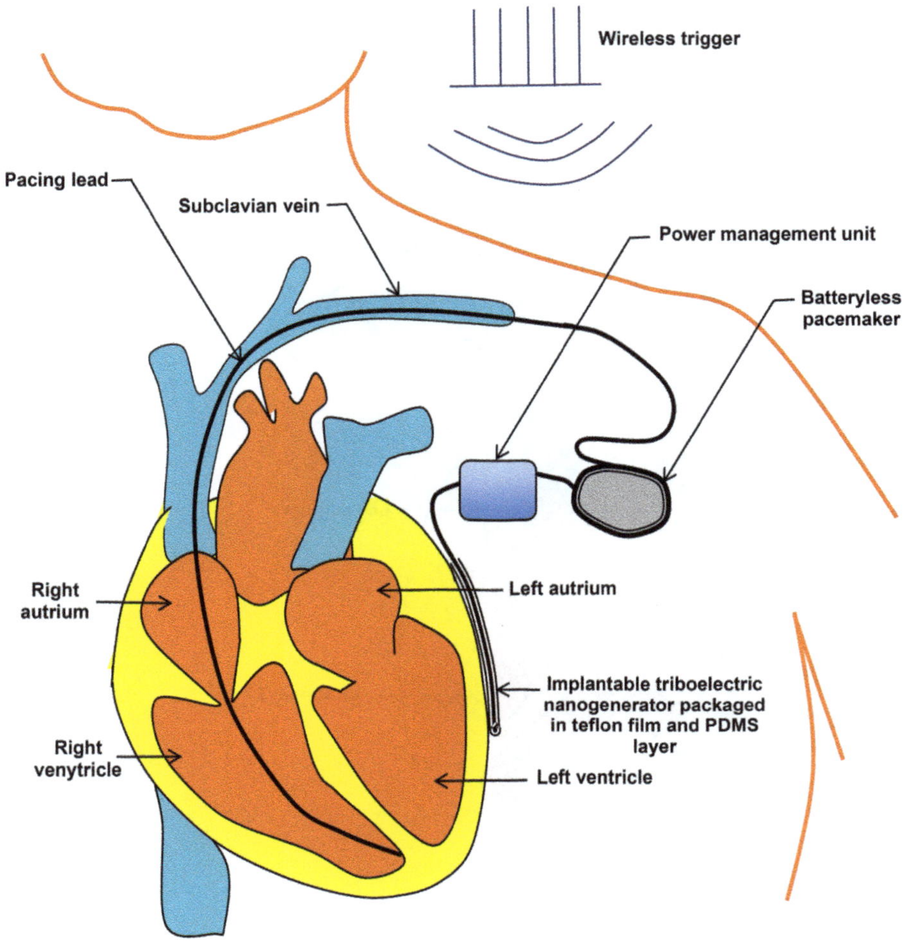

(b)

Figure 23.13. (Continued.)

transferred charge is 47 nC and the current is 9 μA, whereas the same parameters for testing in dry conditions are 91 V, 45 nC and 8 μA. The test shows that water cannot penetrate inside the nanogenerator. There are no deleterious effects of wet environment on the nanogenerator performance (Ouyang *et al* 2019).

23.13 Anti-foreign body reaction (FBR) techniques for domestication/mitigation of FBR to implants

After all the considerations regarding biomaterials are duly accounted for during construction of the implant, it is worthwhile to know additional methods that can be applied for its further amelioration. Techniques have been evolved to minimize the severity of FBR. Promising anti-FBR approaches are given in the ensuing subsections (Veiseh and Vegas 2019).

23.13.1 Optimization of size, shape and texture of the implant

The implant should be as small as possible to deliver the intended therapy. Bigger implants cause severe wounds in larger areas during insertion leading to difficulty in healing and more intense FBR. The shape of the implant should be such that there are no sharp edges. Implants with pointed geometrical features have a tendency to cause inflammation. The nature of surface of the implant, whether rough or smooth, impacts its biocompatibility. Smooth surfaces show better performance than rough ones owing to the differences in behaviors of morphages and immune cells towards these surfaces.

23.13.2 Drug co-delivery

Implants are specially engineered to release drugs slowly over extended periods of time to prevent their rejection by the human body. Let us cite some examples of the drugs used. Corticosteriods, e.g., dexamethasone are steroid hormones having anti-inflammatory, and immunosuppressive effects. Tyrosine kinase inhibitors are pharmacological agents disrupting the pathways for transmission of signals of protein kinases. All these drugs are found to reduce fibrosis and hinder rejection of the implant. A non-steroidal, immunosuppressant drug pirfenidone used in treating idiopathic pulmonary fibrosis is also found to decrease the collagen protein content around the implant, and hence thickness of layer covering it. Chemokines, a family of signalling proteins, also play a vital role in the migration and positioning of immune cells in the tissues, and in healing of wounds.

23.13.3 Using bioresorbable materials for building implants

A bioresorbable or biodegradable material is one which is naturally absorbed by the body gradually producing benign materials. Thus the implant is eliminated after it has performed its designed task and there is no need to extract it by surgery. This offers a route to overcome FBR and get rid of inflammation accompanying implantation. Additionally, the implant is destroyed after its use without any surgical intervention.

23.13.4 Using zwitterionic materials

These are a class of materials containing equivalent cationic and anionic groups, and showing zero overall electrical charge (He *et al* 2016). They are resistant to absorption of proteins and adhesion of cells, thereby inhibiting the formation of biofilms on surfaces. This property endows them with antifouling abilities for implants. Commonly used zwitterionic materials are: sulfobetaine methacrylate (SBMA), carboxybetaine methacrylate (CBMA) and 2-methacryloyloxyethyl phosphorylcholine (MPC) polymers (Kim *et al* 2019).

23.14 Sensors working in biological environments

Medical implant devices can continue to work in the living body by hiding inside a biomaterial package. Some sensors too can continue to work in a same fashion by covering with a film which blocks any fluids. But some sensors must be in direct contact with the environment under study because they will acquire the stimulus to create the requisite electrical signal through this interaction. So, the active surface of the sensor dealing with the biological medium is exposed to it. It is under constant attack by biofluids whereby sensitivity of the device may gradually fall in the course of time. Clearly, the protective coating is no longer hermetic. Instead, it must prevent the sensor from damage while allowing the interaction of the sensor with body fluids for which the signal is to be generated by the sensor.

23.14.1 Sensors which can work by indirect interaction through shielding films

To illustrate this class of sensors, we will explain the working of a sensor for intraocular pressure measurement. This sensor can operate without loss of sensitivity in the intended application. So, let us look into the packaging of a wireless, implantable intraocular pressure measuring system.

Polymer-based packaging approach helps in reducing the form factor and weight of the medical implant. These are highly desirable features when the implant is to be fitted in a small volume. A miniaturized system developed by Agarwal *et al* (2018) is tested by implanting in a porcine eye between sclera and conjunctiva, and the variations of intraocular pressure are studied (figure 23.14(a)).

The system consists of three components (figure 23.14(b)):

(i) a pressure sensor with digital readout,
(ii) charge storage and decoupling capacitors, and
(iii) a CMOS IC chip supporting wireless power/data telemetry and a serial I^2C communication interface with the sensor.

The implant also has an RF coil. The components are mounted on a flexible polymide substrate. The packaged pressure sensor is shown in figure 23.14(c). For packaging of the sensor, the roof of the plastic housing of the sensor is removed to obtain a sensor placed in an open-faced enclosure. The pressure sensor is submerged in silicone oil, and silicone oil is coated with parylene-C film *in situ* by CVD. The reason for preferring paraylene-C is that the CVD process provides a thin, uniform conformal coating over silicone oil. The thin, conformal coating is essential in order that the pressure sensitivity of the device is not lowered by the coating. A silicone chamber and a flexible silicon tube are glued to the PCB. The chamber covers the pressure sensor.

The IC chip is coated with a UV-curable epoxy. All the components are covered with silicone. Note the small space above the pressure sensor through which the intraocular fluid can interact with the parylene-on-oil layer on the pressure sensor through a cannula.

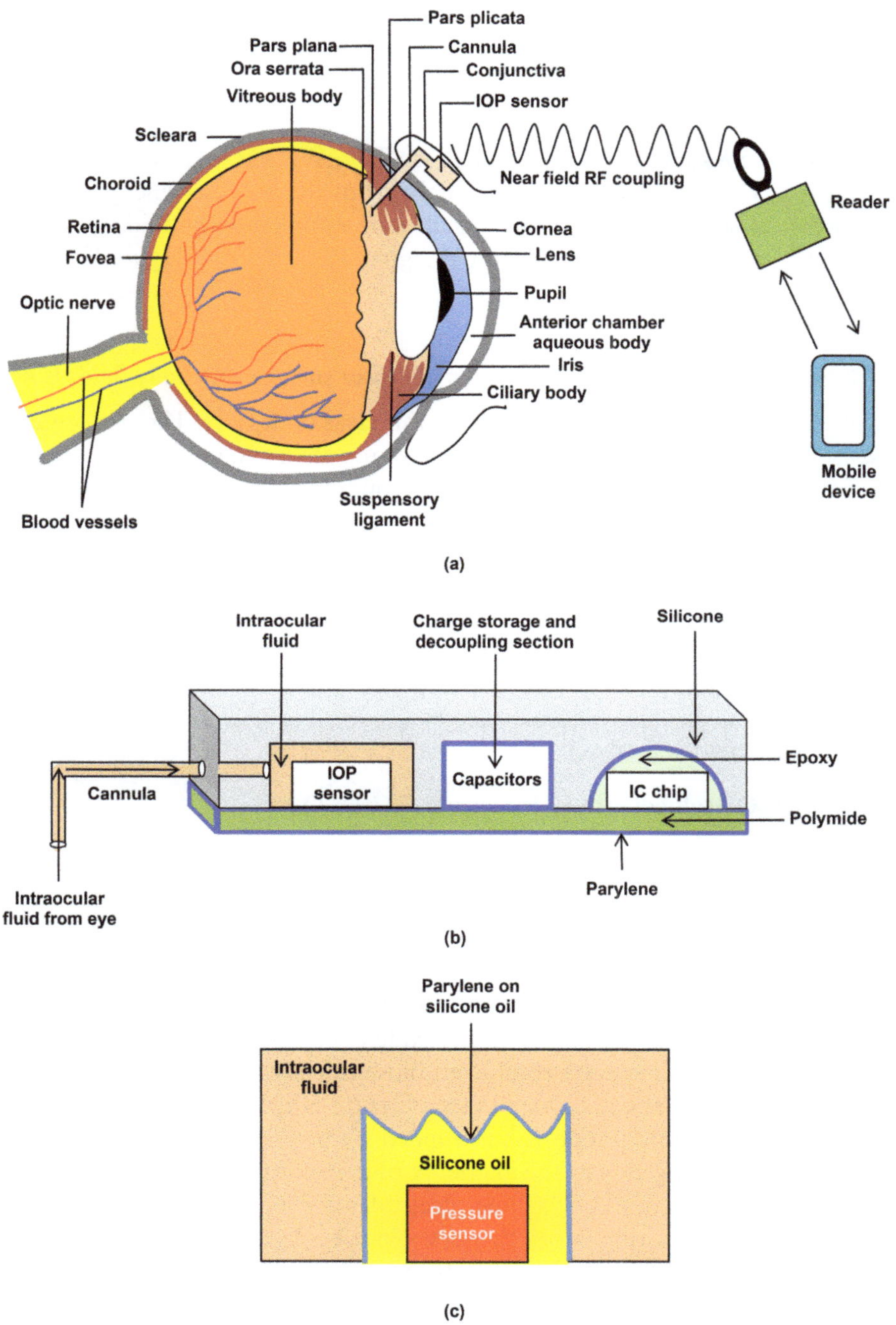

Figure 23.14. The intraocular pressure sensor implant: (a) the human eye showing its various parts, and the location where the pressure measuring system will be placed between the sclera and conjunctiva if tested in humans (as done in animal experiments), (b) the components of the pressure measuring system, and (c) the details of the pressure sensor for detecting pressure changes via the force acting on the parylene-on-oil film.

The RF coil in the implant receives power through near-field RF coupling at 915 MHz from a reader circuit. This coil is also used to transmit the measurement data to the reader circuit by RF backscattering. The reader relays the data to a mobile device.

The power consumed by the implant is 9.7 μW. The pressure sensitivity is 0.17 mm Hg. To assess the long-term reliability of the polymer encapsulation, the sensor was kept soaked in 0.9% saline solution resembling body fluid for 4 months. During this period, the drift in the output signal was <0.5 mm Hg at 27 °C–38 °C (Agarwal *et al* 2018).

23.14.2 Sensors in which direct interaction of sensor surface with body fluids is needed

We will talk about glucose biosensor under this class of devices. The performance of this sensor is degraded by exposure to its working environment.

Diabetes mellitus is a chronic health condition caused by production of insufficient insulin hormone by the pancreas or the inability of utilization of insulin produced. For efficient management of diabetes, it is essential to prevent hypo- (low blood sugar) or hyperglycaemia (high blood sugar) conditions. This is done by continuous glucose monitoring with an implanted biosensor. A widely used glucose biosensor is the amperometric biosensor using glucose oxidase (Gox) enzyme to catalyze the reaction between glucose and the dissolved oxygen in the glucose specimen solution. In this reaction, one gas (oxygen) is consumed and another gas (hydrogen peroxide) is evolved:

$$\begin{aligned} C_6H_{12}O_6 \text{ (glucose)} + O_2\text{(oxygen)} \rightarrow\ & C_6H_{10}O_6 \\ & \text{(glucono} - \text{delta} - \text{lactone)} \\ & + H_2O_2\text{(hydrogen peroxide)}, \\ & \text{in presence of Gox} \end{aligned} \tag{23.6}$$

So, measurement of glucose concentration can be done from decrease in oxygen concentration or increase in hydrogen peroxide concentration.

A Clark oxygen electrode-based glucose biosensor is shown in figure 23.15 at the top side. A platinum disc can be seen in the diagram of the sensor. This disc acts as a cathode. The platinum cathode is surrounded by a silver ring coated with silver chloride. The silver ring acts as the anode. A potential difference of 0.6 V is applied between cathode and anode. The two electrodes are immersed in saturated KCl solution, which transports charge carriers. The Ag/AgCl electrode together with KCl solution constitutes a reference electrode system. The electrodes and the KCl solution are kept inside a container. The bottom of this container is made of a membrane, which is permeable to oxygen. This means that the measured current will be due to oxygen reduction only according to the reaction:

$$O_2 + 4H^+ + 4e^- \rightarrow 2H_2O, \quad -0.6 \text{ V at Pt cathode} \tag{23.7}$$

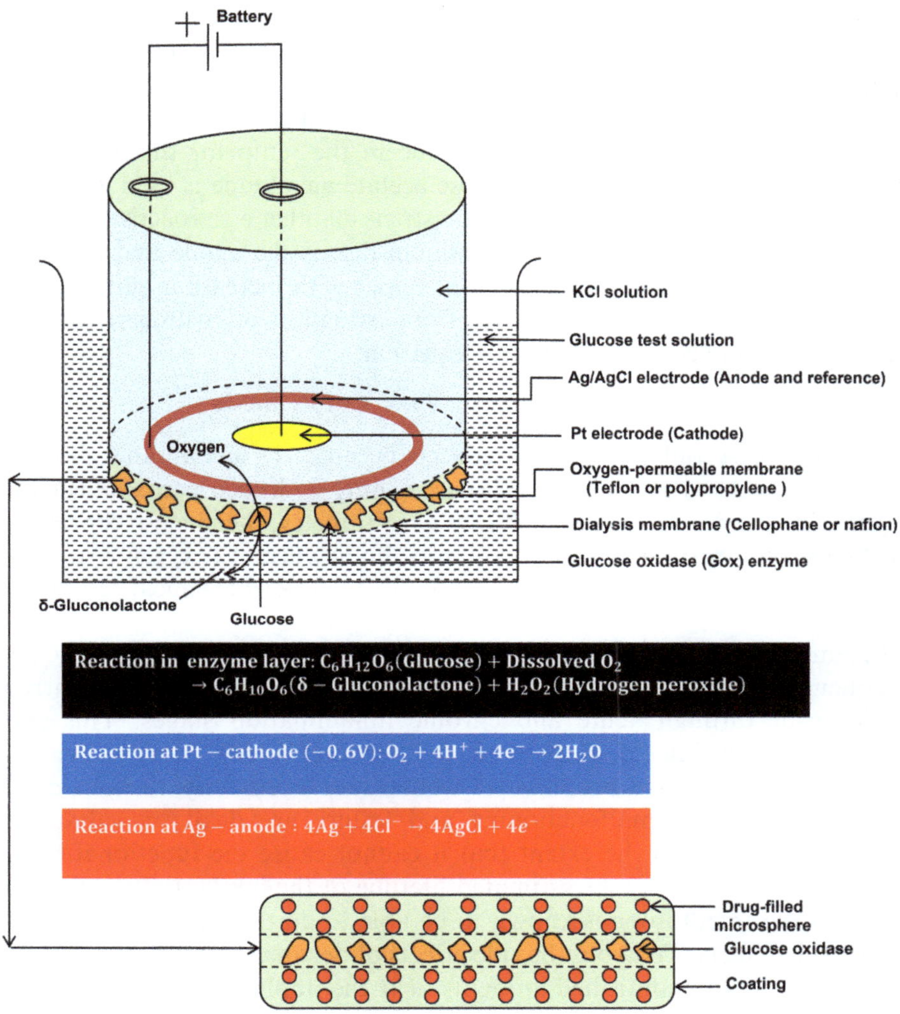

Figure 23.15. Amperometric glucose biosensor. The construction of the sensor is shown in the upper part of the diagram while the lower part of the diagram shows the use of anti-FBR coating.

The glucose oxidase enzyme is kept between the oxygen-permeable membrane and an underlying dialysis membrane. The specimen glucose solution is placed below the dialysis membrane, which allows the glucose to pass through it. In the presence of the enzyme, the glucose reacts with dissolved oxygen in the specimen solution. The products of this reaction are δ-gluconolactone and hydrogen peroxide. It must be noted here that the reaction of glucose with dissolved oxygen depletes the dissolved oxygen level. The more the glucose present in the specimen solution, the greater is the quantity of oxygen consumed. So the concentration of glucose is measured from the decrease in oxygen level. The decrease in oxygen level itself is obtained by measuring the current, which is due to oxygen reduction at the platinum cathode because oxygen alone is allowed to pass through the oxygen-permeable membrane. As the oxygen content falls, the current decreases. A fall in current indicates a rise in glucose concentration.

In an alternative arrangement, glucose concentration is measured from increase in hydrogen peroxide level. Changes are made in the setup for this measurement. Instead of a Teflon membrane, a cellulose acetate membrane is used which allows hydrogen peroxide to pass through it but restricts all other electroactive species. The polarity of electrodes is reversed. A platinum disc is the anode and a silver ring serves as the cathode. Here the current increases with increase in glucose concentration. The increase in current arises from oxidation of hydrogen peroxide at platinum anode in accordance with the equation:

$$H_2O_2 \rightarrow 2H^{+} + O_2 + 2e^{-}, \quad +0.6\ \text{V at Pt anode} \tag{23.8}$$

Thus, in one case only oxygen is reduced because the membrane allows only oxygen to pass through it. In the other case, only hydrogen peroxide is oxidized because of restriction to oxygen. Thus, in amperometry, the glucose concentration is evaluated by measuring the decrease in cathodic current (oxygen reduction) or increase in anodic current (H_2O_2 oxidation) of the electrochemical cell at a fixed potential.

The implanted biosensor cannot escape from the effects of FBR. As a consequence of FBR, a fibrous capsule built up of proteins and cells surrounds the biosensor through acute and chronic inflammation stages. This capsule opposes the diffusion of glucose to the sensing surface due to which the sensitivity of the device progressively falls, and eventually the device stops working. This device malfunctioning and failure are dealt by subduing the FBR and limiting it to the extent that it cannot cause the functionality of the sensor to deteriorate over the estimated lifetime of the implant. As already said, the steroid dexamethosone is known to be able to do this task. However, it has systemic side effects. To avoid the side effects of dexamethasone, this medication must be delivered in controlled amounts near the implant. As microspheres are used in drug delivery, they are viable options to achieve patient compliance. Preferred material for making microspheres is poly(lactic-co-glycolic) acid (PLGA) because its degradation occurs slowly with good reproducibility. A dexametahsone/PLGA system could provide release of dexamethasone for 1 month duration, which is extendable beyond this period by adding poly-ethylene glycol (PEG) (Hickey *et al* 2002).

In another study, PLGA microsphere/poly(vinyl alcohol) (PVA) hydrogel composites were used as drug-eluting coatings for glucose biosensors (Wang *et al* 2013). The composite forms a uniform coating around the sensor (figure 23.15 on the bottom side). In this scheme, the tissue-modifying drug is slowly released by the PLGA microsphere. At the same time, rapid diffusion of glucose to the sensor is allowed by the hydrogel. No inflammation or encapsulation was observed in tissues coated with composites containing dexamethasone over 1 month. In the absence of dexamethasone, these effects were significant and pronounced.

Further studies (Wang *et al* 2015) revealed that:

(i) the initial trauma suffered during implantation of the sensor determines the degree of acute inflammation while the size of the implanted sensor controls the creation of fibrous capsule,
(ii) a dexamethasone concentration of 150 mg microspheres mL^{-1} hydrogel is sufficient for FBR blockage in large sensors of size 0.75 mm × 0.75 mm × 9 mm, whereas in smaller sensors of size 0.5 mm × 0.5 mm × 5 mm, the required concentration was 75 mg mL^{-1}.
(iii) a small-size sensor covered with a composite drug-eluting coating shows satisfactory long-term performance.

Effect of zwitterionic material coating on implanted sensor behavior has also been investigated (Xie *et al* 2018). The signals received from implanted glucose sensors for continuous glucose monitoring are found to be noisy during the initial period after implantation requiring repeated calibration of the sensor. The noise is considerably reduced when the sensor is coated with a zwitterionic polymer found through combinatorial chemistry. It was confirmed that the immune responses are annuled by the coating, resulting in improved performance of the glucose meter.

23.15 Discussion and conclusions

Packaging of medical implants is a crucial process in which considerations must be made for biocompatibility of the package material to minimize the foreign body response together with its hermeticity to secure the electronic circuitry of the implant from human body fluids and above all, to keep the patient safe from hazardous materials leaking into the body. A wide range of biomaterials falling under metals, ceramics, polymers and composites classes are available. Properties of these materials must be carefully studied, and their relative advantages and disadvantages must be understood. By applying this knowledge, safe implantable devices can be developed.

The relentless miniaturization of implants poses new challenges to packaging engineers. Downsizing of implants along with integration of polymers in implant packaging can give low-cost, light-weight devices. The controlled release of anti-FBR agents must be fully utilized. Progress in these technologies will lead to new diagnostic and therapeutical methods revolutionizing healthcare.

Review exercises

23.1. What is the percentage of water content in the human body? How many kilograms of water are present in the body of a man weighing 70 kg?

23.2. How much water, expressed as percentages, is present in human: (a) brain, (b) lungs, (c) blood, (d) bones and (e) teeth?

23.3. Into what fluid compartments is the water in the body subdivided? What is the ratio of this subdivision?

23.4. Distinguish between interstitial and intravascular fluids? What is the ratio of these fluids in the human body?
23.5. Explain the terms: lymphatic and transcellular fluids? What is the sum of volumes of these two fluids called? What is the ratio of these volumes?
23.6. What roles are played by electrolytes in the body? Name two important cations and two anions in the human body. What is homeostasis?
23.7. Why is a great emphasis laid on the materials used for packaging medical implants? When looking for selection of implant packaging materials, name six essential properties that the chosen material must possess.
23.8. What is hermeticity? Explain its significance in the context of packaging of medical implants.
23.9. What material property will you check for compliance before saying that a chosen material is biocompatible? Is biocompatibility 'absolute'? Define the term 'biomaterial'.
23.10. What are your views on the mechanical flexibility of a material used for packaging an implantable device? Why will a rigid material be harmful?
23.11. Will you accept a material that releases gases for making the internal components of an implanted device? What will happen if such a material is used?
23.12. Can the material used for packaging a medical implant be opaque to electromagnetic waves? When can we use conductive materials for implant packaging? When are insulating materials used?
23.13. Large eddy currents are produced in the body of a metallic implant package when it is operated. How does this happen and in what way are eddy currents harmful?
23.14. Why is matching of thermal expansion coefficients essential for any two materials joined together to produce a package for a medical implant? How will materials with unmatched coefficients impact the performance of the implant?
23.15. What benefits accrue if the materials chosen for implant packaging are processable by standard semiconductor techniques?
23.16. An implant package is made of a biocompatible material. How long does the acute inflammation take to resolve? In what duration does the chronic inflammation phase subside?
23.17. Name two proteins in the blood extravasated during implant insertion which are absorbed on the surface of the implant? What is the process of competitive absorption of blood serum proteins on a surface called?
23.18. Into what types are macrophages classified? What functions do these types of macrophages carry out?
23.19. Name the proteins that bind the macrophages to the surface of the implant. In what way does the phagocytosis attack by macrophages threaten the stability of the implant?

23.20. By fusion of what membranes are FBGC produced? What are the factors influencing this fusion?
23.21. What biological process is set up for a permanently residing implant in the body? What does this process try to achieve?
23.22. Into how many classes are biomaterials divided? Are all metals biocompatible? Do metals introduce any outgassing issues in medical implants?
23.23. Are metals transparent to radio frequency waves? Why does packaging of an implant become complex if wireless communication is necessary?
23.24. What is the reason of generation of heat in an implant packaged in a metallic body and working through a wireless link?
23.25. Are sharp corners and boundaries of metallic packages acceptable for implantable electronics?
23.26. Do metallic packages provide hermeticity? Are metals susceptible to corrosion in body fluids? Yes/No.
23.27. What are bioceramics? Comment (yes/no) on whether bioceramics can be used for packaging implants on the basis of their hermeticity, outgassing, RF response and corrosion behavior.
23.28. Is heat produced in the alumina package of an implant when operated wirelessly?
23.29. What are the limitations of ceramics as packaging materials for implants?
23.30. What qualities speak in favor of polymers for packaging medical implants? What shortcomings can lead to rejection of polymers for packaging?
23.31. How does the penetration of moisture through the polymer coating of an implanted device affect the performance of the device?
23.32. Give an example of an implant in which neither metallic nor ceramic packaging is suitable but polymers are preferred.
23.33. Do polymers show good hermeticity and outgassing properties for medical implant packaging?
23.34. What is a composite material? Why did the demand for composite materials arise for medical implants? What is the special feature that these materials are able to provide but which single materials are unable to offer?
23.35. Highlight two disadvantages of using composite materials for packaging implants.
23.36. What is osteointegration? Doses titanium display good osteointegration behavior?
23.37. What roles does the surface oxide film formed on titanium play in its acceptance for medical devices?
23.38. What prominent qualities have favored the widespread use of titanium in medical implants?

23.39. What are the four grades in which CP-Ti is available? How does the oxygen content vary from grade to grade? Are the four grades equal in mechanical strength?
23.40. Write the percentage of titanium in Ti–6Al–4V alloy. What is the meaning of saying that it is an (α+β) alloy? What is the problematic issue with this alloy?
23.41. For what medical applications are Co–Cr alloys used? What are their beneficial properties? What property does the spontaneously produced chromium oxide impart to Co–Cr alloys?
23.42. Which international society designates steel grades? What are the most common and second most common steel grades? How much iron is present in these grades in percentage values?
23.43. Presence of which metal in stainless steels lends a corrosion-resistant property to them. Explain the mechanism by which steel becomes corrosion resistant due to this metal?
23.44. What are the three classes of ceramics? Give an example of each of the three classes?
23.45. What properties make alumina the most widely used oxide ceramic? Under which class of ceramics does hydroxyapatite fall?
23.46. Differentiate between the wet forming and dry forming processes used in ceramic processing. What is meant by green body? Why is it said to be green?
23.47. (a) Are doctor blading, knife coating and tape casting different processes?
(b) Both injection molding and slip casting are mold-based processes? What is the difference between them?
23.48. Does the dry forming process for ceramics require a slurry to be made? How is uniaxial die pressing carried out?
23.49. How is Pascal's law applied when the elastomer mold is pressed during cold isostatic pressing?
23.50. At what temperature is the green body formed during ceramic processing? What happens during sintering?
23.51. What is the main application of ceramics in medical implants? Which feedthrough is better: ceramic-to-metal or glass-to-metal? Why?
23.52. Name and illustrate with a diagram a commonly used ceramic-to-metal feedthrough in medical implants.
23.53. What is the most commonly used process for joining a metal to ceramic called? Describe this process.
23.54. Does 'a braze alloy' mean the same thing as 'a solder'? If not, what is the difference?
23.55. What is the difference between passive and active braze alloys? Which of the two kinds of alloys requires a pre-deposited metal layer and for what purpose?
23.56. How is brazing done with alumina by the Mo–Mn method?

23.57. What metals or alloys are used in noble metal brazing? What happens to the noble metal oxides formed during brazing?
23.58. Why is active brazing easier to implement than passive brazing? What braze alloy is used in alumina–titanium bonding? Why is the use of copper in this alloy disapproved?
23.59. What is the limitation of brazing technique for making an array of closely-spaced ceramic-to-metal feedthroughs in alumina? How will this limitation affect the performance of a retinal prosthesis made by brazing?
23.60. Name and describe a process for making high pin density feedthroughs using alumina substrate.
23.61. What is the process used for drilling vias in alumina substrate? How are extruded metal vias formed for making feedthroughs?
23.62. What is meant by the terms: (a) gold stud, (b) stud bumping, and (c) coining?
23.63. What is an elastomeric substance? Is PDMS elastomeric?
23.64. By what methods can PDMS be deposited on a substrate? Is it moldable?
23.65. What does glass transition temperature of a material signify? What is the glass transition temperature of PDMS? Will you call it low or high? Explain.
23.66. Why is hermetic sealing with PDMS difficult? Comment on the suitability of PDMS as a substrate for telemetric structures.
23.67. Is polyimide available in liquid form? Is it available as free-standing sheets?
23.68. Which one of the following properties of polyimide is not high: electrical resistivity, dielectric strength, mechanical flexibility, thermal stability, stiffness, chemical resistance, rate of moisture absorption.
23.69. Name four ways of processing PVDF. Is it thermoplastic? Does it exhibit piezoelectric and pyroelectric behaviors?
23.70. Enlist four advantages and four disadvantages of PVDF as a packaging material for medical implants.
23.71. Does PVDF show a constant dielectric constant value in the frequency range 1 KHz to 1 MHz?
23.72. Elaborate the advantages of using parylene as a coating material for depositing thin films during medical implant packaging. Are these films stable under temperature variation and UV exposure?
23.73. Which of the three polymers has the lowest moisture absorption rate: (a) polyimide, (b) LCP and (c) parylene-C?
23.74. What properties make LCPs attractive as packaging materials for implants based on wireless communication?
23.75. What advantages are offered by LCPs as a processing material? Are LCPs photo-patternable?
23.76. Can polyurethane be vacuum-formed? Comment on its tensile strength.

23.77. In the composite material HA/Ti–6Al–4V, what useful property is provided by hydroxyapatite which Ti–6Al–4V lacks?

23.78. What properties of hydroxyapatite are improved by incorporating small amounts of bioactive glass? How does the bioactive glass help HA?

23.79. Why does PEEK need to be reinforced with carbon fibers for using it in orthopedic implants? Which two materials have similar elastic moduli: bone, PEEK and CF-PEEK?

23.80. What is the implication of the statement that a microelectrode array provides a bidirectional interface with the brain?

23.81. Explain with a cross-sectional diagram the operation of the Utah microelectrode array? What material is used for making the needles of the array? What metals are used for making electrode tips? With what material is the array encapsulated?

23.82. Is electrocorticography a non-invasive medical procedure? Is electroencephalography invasive? Which gives a high signal-to-noise ratio: EEG or ECOG? Why? For what application do neurosurgeons use ECOG?

23.83. Elucidate the roles of optics and genetic engineering in optogenetics. What medical opportunities are offered by optogenetics?

23.84. Explain with the help of a diagram the operation of an optrode array clearly elaborating its optical and electrical functions. Is the waveguide for light transmission made of a metal or a polymer? How are the LEDs powered?

23.85. Why is PDMS stamp necessary for making the optrode array? How is it made?

23.86. How are LEDs mounted on a parylene-C film? Describe the processes of making polymer waveguides and assembling the optrode array.

23.87. How is the length of the microneedles controlled in droplet backside exposure process? Does this process require a specialized costly machine?

23.88. What will happen if a cardiac pacemaker powered by triboelectricity produced by motion of heart is packaged in a brittle metal package? By which natural phenomenon is the idea of symbiotic pacemaker inspired?

23.89. Draw diagrams of battery-powered and self-powered cardiac pacemakers. Point out the differences between the two types of pacemakers.

23.90. Argue in favor of using flexible materials in packaging the nanogenerator of a symbiotic pacemaker. Of what materials is its two-layer protection made?

23.91. What test is performed to confirm that the nanogenerator can perform satisfactorily in the wet environment of the human body?

23.92. A large-size medical implant with rough surface and sharp edges is less prone to FBR. Right or wrong.

23.93. Name a few drugs, which can be slowly released from medical implants to counter fibrosis.
23.94. Name a non-steroidal drug which is found to decrease the thickness of bio-layer formed on the implant during FBR.
23.95. What is a bioresorbable material? How is it helpful in overcoming FBR?
23.96. What properties bestow antifouling abilities on zwitterionic materials? Give two examples of these materials.
23.97. Medical electronics circuits work undisturbed within the package. But can the sensors do so? Name a sensor whose surface indirectly interfaces with the biological environment and one whose surface directly interfaces with this environment.
23.98. Give two main reasons for using polymers in the packaging of an implantable intraocular pressure measuring system? What are the three principal components of this system?
23.99. Draw the diagram of the pressure sensor used in the intraocular pressure monitoring system? In which oil is the pressure sensor submerged? By what film is this oil coated? Why should this film be very thin?
23.100. What material interfaces with the intraocular fluid and transfers pressure changes to the sensor? What is the material used for the transmission medium?
23.101. How is an intraocular pressure sensor implant powered? How is measured data obtained?
23.102. What test is performed to assess reliability of the polymer packaging used for an intraocular pressure measurement system?
23.103. Diabetes occurs because the pancreas does not produce enough .... hormone. What is the difference between hypo- and hyperglycaemia? Why is continuous glucose monitoring necessary for diabetes management?
23.104. Explain the working principle of a glucose biosensor using amperometry. Why is it necessary to use glucose oxidase enzyme?
23.105. Describe the two approaches applied to measure glucose concentration by an amperometric biosensor based on oxygen consumption and hydrogen peroxide liberation. What gas permeable membranes are used in the two cases? Write down the equations for oxygen reduction and increase in hydrogen peroxide concentration.
23.106. How does FBR hamper the functioning of a glucose biosensor? How does this obstruction show up in the measured response of the biosensor?
23.107. Which steroidal drug is used to prevent the degradation of the implanted glucose biosensor? What steps are taken to deliver this drug in controlled amounts near the implant?

23.108. What material is used for making microspheres for delivering anti-inflammatory drug near the implant? Why is this material chosen for this task?

23.109. Describe the use of the PLGA microsphere/PVA hydrogel composites for inhibiting FBR in a glucose biosensor.

23.110. How does a zwitterionic material coating on a glucose biosensor influence the performance of the sensor? What is the mechanism by which this coating works?

23.111. Do larger glucose biosensors require a higher dexamethasone concentration for preventing FBR?

References

Agarwal A, Shapero A, Rodger D, Humayun M, Tai Y-C and Emami A 2018 A wireless, low-drift, implantable intraocular pressure sensor with parylene-on-oil encapsulation *IEEE Custom Integrated Circuits Conf. (CICC) (San Diego, CA, 8–11 April 2018)* 1–4

Avila J D, Stenberg K, Bose S and Bandyopadhyay A 2021 Hydroxyapatite reinforced Ti6Al4V composites for load-bearing implants *Acta Biomater.* **123** 379–92

Campbell P K, Jones K E and Normann R A 1990 A 100 electrode intracortical array: structural variability *Biomed. Sci. Instrum.* **26** 161–5

Carnicer-Lombarte A, Chen S-T, Malliaras G G and Barone D G 2021 Foreign body reaction to implanted biomaterials and its impact in nerve neuroprosthetics *Front. Bioeng. Biotechnol.* **9** 1–22

Eliaz N 2019 Corrosion of metallic biomaterials: a review *Materials* **12** 1–91

Golda-Cepa M, Engvall K, Hakkarainen M and Kotarba A 2020 Recent progress on parylene C polymer for biomedical applications: a review *Prog. Org. Coat.* **140** 105493

Hanawa T 2019 Titanium–tissue interface reaction and its control with surface treatment *Front. Bioeng. Biotechnol.* **7** 170

He M, Gao K, Zhou L, Jiao Z, Wu M, Cao J, You X, Cai Z, Su Y and Jiang Z 2016 Zwitterionic materials for antifouling membrane surface construction *Acta Biomater.* **40** 142–52

Hickey T, Kreutzer D, Burgess D J and Moussy F 2002 Dexamethasone/PLGA microspheres for continuous delivery of an anti-inflammatory drug for implantable medical devices *Biomaterials* **23** 1649–56

Hicyilmaz A S and Bedeloglu A C 2021 Applications of polyimide coatings: a review *SN Appl. Sci.* **3** 22

Jiang G, Mishler D, Davis R, Mobley J P and Schulman J H 2005 Zirconia to Ti-6Al-4V braze joint for implantable biomedical device *J. Biomed. Mater. Res. Part* B **72** 316–21

Kim D, Lee M-J, Kim J-Y, Lee D, Kwon J-S and Choi S-H 2019 Incorporation of zwitterionic materials into light-curable fluoride varnish for biofilm inhibition and caries prevention *Sci. Rep.* **9** 10

Kim S and Liu S 2018 Smart and biostable polyurethanes for long-term implants *ACS Biomater. Sci. Eng.* **4** 1479–90

Kwon K Y, Sirowatka B, Weber A and Li W 2013a Opto-μECoG array: a hybrid neural interface with transparent μECoG electrode array and integrated LEDs for optogenetics *IEEE Trans. Biomed. Circuits Syst.* **7** 593–600

Kwon K Y, Bi X and Li W 2013b Droplet backside exposure for making slanted SU-8 microneedles *35th Annual Int. Conf. of the IEEE Engineering in Medicine and Biology Society (EMBC) (Osaka, 3–7 July)* 767–70

Kwon K Y, Lee H-M, Ghovanloo M, Weber A and Li W 2014 A wireless slanted optrode array with integrated micro LEDs for optogenetics *IEEE 27th Int. Conf. on Micro Electro Mechanical Systems (MEMS) (San Francisco, CA, 26–30 January)* 813–6

Kwon K Y, Lee H M, Ghovanloo M, Weber A and Li W 2015 Design, fabrication, and packaging of an integrated, wirelessly-powered optrode array for optogenetics application *Front. Syst. Neurosci.* **9** 69

Langenmair M, Martens J, Gierthmuehlen M, Plachta D T T and Stieglitz T 2018 Low temperature approach for high density electrical feedthroughs for neural implants using maskless fabrication techniques *40th Annual Int. Conf. of the IEEE Engineering in Medicine and Biology Society (EMBC) (Honolulu, HI, 18–21 July)* 2933–6

Lee H-M, Kwon K-Y, Li W and Ghovanloo M 2014 A wireless implantable switched-capacitor based optogenetic stimulating system *36th Annual Int. Conf. of the IEEE Engineering in Medicine and Biology Society (Chicago, IL, 26–30 August)* 878–81

Liao C, Li Y and Tjong S C 2020 Polyetheretherketone and its composites for bone replacement and regeneration *Polymers (Basel)* **12** 2858

Mani G 2016 Metallic biomaterials: cobalt–chromium alloys *Handbook of Biomaterial Properties* ed W Murphy, J Black and G Hastings (New York: Springer) ch 1b pp 159–66

MNT 2004–22 What is the average percentage of water in the human body *Medical News Today* https://medicalnewstoday.com/articles/what-percentage-of-the-human-body-is-water#where-in-the-body

Morais D S, Coelho J, Ferraz M P, Gomes P S, Fernandes M H and Hussain N S *et al* 2014 Samarium doped glass-reinforced hydroxyapatite with enhanced osteoblastic performance and antibacterial properties for bone tissue regeneration *J. Mater. Chem.* B **2** 5872–81

Narushima T, Ueda K and Alfirano 2015 Co–Cr alloys as effective metallic biomaterials *Advances in Metallic Biomaterials: Tissues, Materials and Biological Reactions* ed M Niinomi, T Narushima and M Nakai (Heidelberg: Springer) ch 7 pp 157–78

Nicholson J W 2020 Titanium alloys for dental implants: a review *Prosthesis* **2** 100–16

Ortiz A J, Fernández E, Vicente A, Calvo J L and Ortiz C 2011 Metallic ions released from stainless steel, nickel-free, and titanium orthodontic alloys: toxicity and DNA damage *Am. J. Orthod. Dentofacial Orthop.* **140** e115–22

Ouyang H, Liu Z, Li N, Shi B, Zou Y, Xie F, Ma Y, Li Z and Li H *et al* 2019 Symbiotic cardiac pacemaker *Nat. Commun.* **10** 1821

Salernitano E and Migliaresi C 2003 Composite materials for biomedical applications: a review *J. Appl. Biomater. Biomech.* **1** 3–18

Saxena P and Shukla P 2021 A comprehensive review on fundamental properties and applications of poly(vinylidene fluoride) (PVDF) *Adv. Compos. Hybrid Mater.* **4** 8–26

Shah K G, Delima T, Felix S, Sheth H, Tolosa V, Tooker A and Pannu S S 2012 High-density, bio-compatible, and hermetic electrical feedthroughs using extruded metal vias *Hilton Head Workshop 2012: A Solid-State Sensors, Actuators and Microsystems Workshop (Hilton Head Island, SC, June 2–6)* 4

Sheikh Z, Brooks P J, Barzilay O, Fine N and Glogauer M 2015 Macrophages, foreign body giant cells and their response to implantable biomaterials *Materials* **8** 5671–701

Sidambe A T 2014 Biocompatibility of advanced manufactured titanium implants—a review *Materials* **7** 8168–88

Veiseh O and Vegas A J 2019 Domesticating the foreign body response: recent advances and applications *Adv. Drug Delivery Rev.* **144** 148–61

Volkov A 2013 Polydimethylsiloxane (PDMS) *Encyclopedia of Membranes* ed E Drioli and L Giorno (Heidelberg: Springer) pp 1–2

Wang Y, Papadimitrakopoulos F and Burgess D J 2013 Polymeric 'smart' coatings to prevent foreign body response to implantable biosensors *J. Controlled Release* **169** 341–7

Wang Y, Vaddiraju S, Gu B, Papadimitrakopoulos F and Burgess D J 2015 Foreign body reaction to implantable biosensors: effects of tissue trauma and implant size *J. Diabetes Sci. Technol.* **9** 966–77

Xie X, Doloff J C, Yesilyurt V, Sadraei A, McGarrigle J J, Omami M and Veiseh O *et al* 2018 Reduction of measurement noise in a continuous glucose monitor by coating the sensor with a zwitterionic polymer *Nat. Biomed. Eng.* **2** 894–906

Chapter 24

Meeting the challenges faced by electronics in unfavorable space environments

Besides the high-impulse vibrations, wide temperature fluctuations, spacecraft charging effects, tin whisker growth concerns, atomic oxygen erosion, and radiation exposure, electronic equipment in the high-vacuum space environment of satellites and spacecraft face difficulties arising from release of gases trapped inside the material through outgassing. Techniques to overcome the manifold problems encountered in space missions are described. Sources and effects of vibration are mentioned, and protection and dampening methods of vibration are given. Necessity of thermal control is highlighted. Passive and active control approaches are discussed. Attention is drawn to the surface and internal charging effects in spacecraft, and appropriate counter-measures are explained. Ways to avoid electrical shorting and arcing due to tin whisker growth are outlined. Atomic oxygen in space affects polymers, metals, glasses and thermal coatings calling for provisions to be made for fighting against its attack. For protection from radiation in space, gamma ray and neutron radiation shielding materials are used, and conformal coatings are specially adapted. Effects of outgassing from various materials in the spacecraft in space are minimized by using molecular absorber pucks and coatings. Earnest action against different types of threats to electronics in space increases its operational lifetime and reduces chances of failures, paving the pathway to successful missions.

The electronic equipment onboard a space vehicle is the center of its computational intelligence, telecommunication system and control instrumentation.

doi:10.1088/978-0-7503-5072-3ch24

Therefore, it must be safeguarded against any dangers which either impede or stop its operation because partial or total failure of electronics is tantamount to partial or total failure of the mission. To protect the equipment, we must clearly know the exact nature of threats. Only then can useful solutions be suggested.

From the launch of the vehicle to its travel through space and return of the module to Earth, there are a series of threats, the extent of each of which is related to the particular situation. Vibrations at the time of launch, during flight and at the time of re-entry into the Earth's atmosphere must be carefully addressed and properly dealt with. The gases outgassed in the vacuum of space from various materials used in construction of the spacecraft may deposit as condensates on electronic devices, circuit boards and components such as charge-coupled device (CCD) cameras causing deleterious effects. Gaseous effusion from parts of a satellite on nearby electronic circuits frequently bring about premature failure and degrade reliability of devices. Similar remarks apply to other threats.

A few of the numerous challenges confronting space flights will be briefly explained and corrective measures or remedies taken will be described. Looking back, the vibration effects on electronics were described in chapter 20. Chapters 18 and 19 surveyed the influence of cosmic radiation in outer space on spaceship onboard electronics and the corrective measures undertaken. The reader may recall the basics of vibration-tolerant systems in chapter 20, radiation effects on electronics in chapter 18, and radiation hardening of electronics in chapter 19 to supplement this chapter.

24.1 The challenge of vibrations and shocks

24.1.1 Sources of vibrations in space vehicles

The principal sources of vibrations and shock are:

(i) rocket noise,
(ii) direct vibration excitation,
(iii) aerodynamic excitation,
(iv) turbulence environment, and
(v) pyrotechnic shock.

Powerful acoustic waves are emitted when rockets thunder at 180–200 dB. The noise created by rocket motor exhaust during launch is a broad-spectrum noise including audible (20 Hz–20 kHz) and subaudible (<20 Hz) ranges. It encompasses all types of noise from low frequency to mid frequency (5 Hz–40 kHz). Normal human conversations take place at 60 dB. A soft whispering sound is 30 dB. City traffic noise is 80–85 dB. Sound above 120 dB immediately harms the ears. Figure 24.1 presents a comparison of sound levels from various sources on the decibel scale to give an idea of the loudness level during rocket launches with respect to the commonly experienced sounds in everyday life.

Direct vibration excitations are post-lift off vibrations arising from instabilities in combustion in rocket engines. Low-frequency oscillations are called chugging while high-frequency oscillations are known as screaming.

Aerodynamic excitations depend on aerodynamic design of the vehicle and its flight profile.

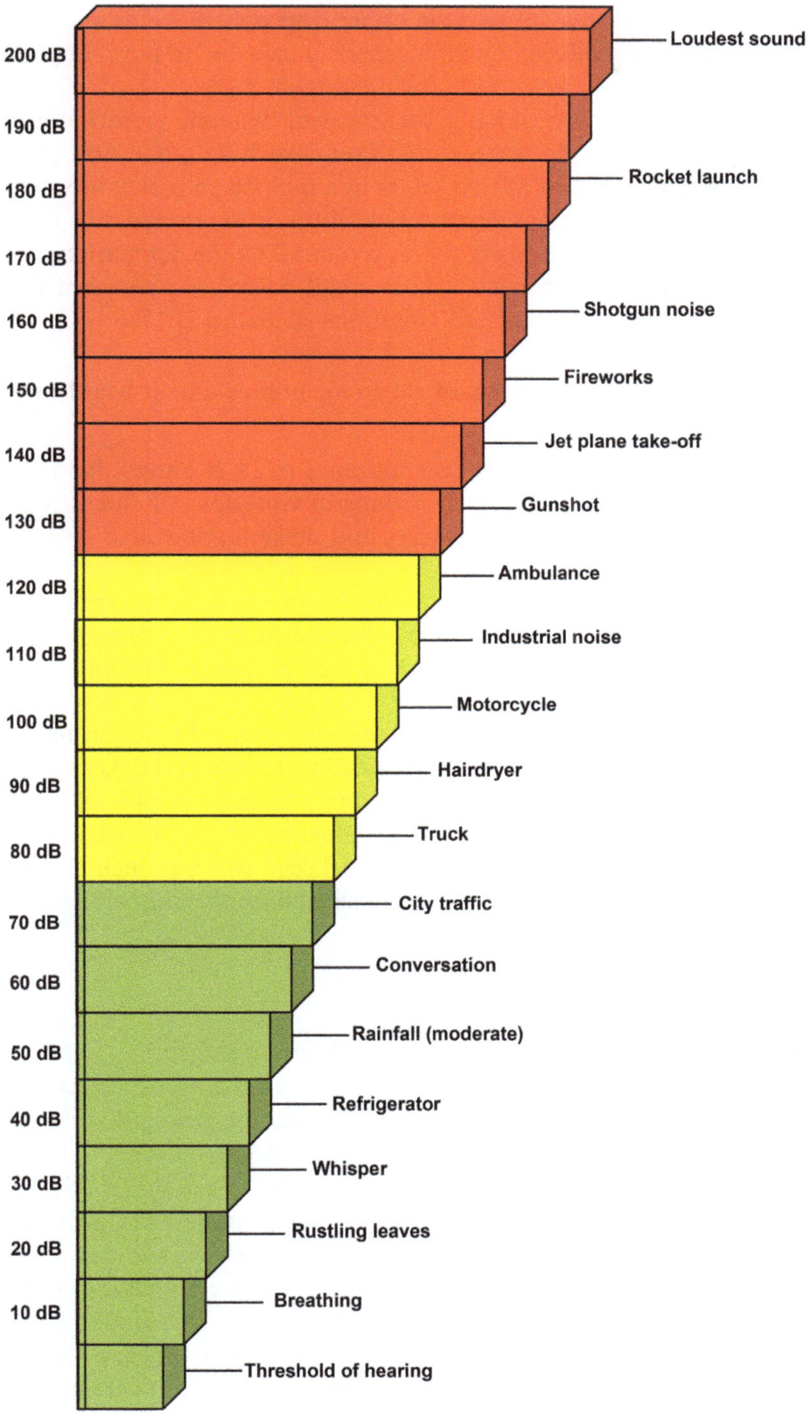

Figure 24.1. Chart displaying different sources of sound on the decibel scale.

Turbulence environments are faced in upper troposphere winds, e.g., at the time of re-entry. Forces are produced by atmospheric winds due to the change in path of the flight.

Pyrotechnic shock or pyro shock is the transient dynamic vibration experienced when an explosion occurs on a structure. Events such as separation of a satellite from parent spacecraft or that of two stages in a multistage rocket are synonymous to the occurrence of detonations. Pyrotechnic shock is the response of the structure to the high-frequency, high-intensity waves produced by the detonation. Pyro shock may rupture electronic components whose small parts have resonant frequencies matching with those of shock-created vibrations (Leonard 2017).

24.1.2 Effects of vibrations on onboard electronic printed-circuit board assemblies (PCBAs)

To start discussions in this section, remembrance of PCB knowledge is necessary. Figure 24.2 will be helpful in this study. Owing to vibrations, failures are noticed as breakage of solder joints or solder tracks, and delamination of PCBs. Vibration-prompted failures are:

(i) Mechanical deformation: It refers to change in shape and size and is said to be elastic if temporary and plastic when permanent. Copper foils, resin and other materials in PCBs differing in their ability to withstand stress bring about deformation.

(ii) Brittle fracture: It happens at the soldered joints due to tensile stresses.

(iii) Warpage: It is twisting or bending caused by unbalanced forces due to weights of components or heat sinks, and thermal expansion during soldering.

(iv) Fatigue and thermal failure: It is the initiation of cracks such as those due to soldering of low thermal expansion coefficient solder to high thermal expansion coefficient boards. Thermal coefficient mismatching is the main cause.

24.1.3 Protection of PCB from vibration

Precautions necessary are (Bhalerao 2021):

(i) Vibrations should be considered in PCB design stage.

 (a) Copper-clad laminates (figure 24.3) consisting of a base reinforcing material with a sheet of copper pasted on one/both sides should be used for PCBs as they provide better peel strength.

 (b) The structure of the board including the layers, distribution of copper and placement of components should preferably be symmetrical. Twists and wraps in the circuit board are thus avoided.

 (c) Wide conductors draw solder from pads. A conductor moving to a via connecting an inner power plane takes away heat from the pad or lead of the component at the time of soldering. So insufficient heat is absorbed by the solder resulting in cold solder joint error. The wetting is poor, and the solder connection has a porous greyish appearance. For such connections, the product (conductor length × conductor width) should be 1 mm^2 and pad-to-track ratio = 1.5/1.

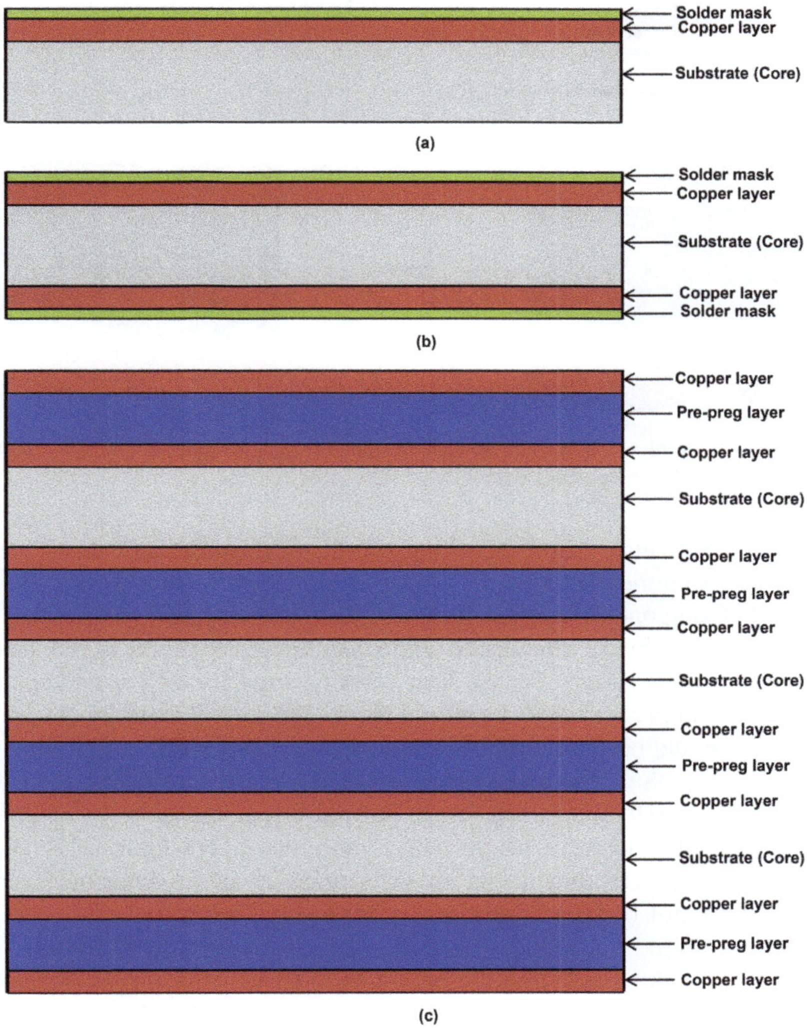

Figure 24.2. Recollection of PCB technology: (a) single-layer (or single-side), (b) double-layer (or two sides) and (c) multilayer PCBs. Pre-preg is a partially cured, reinforcing, dielectric material made of glass fiber weave impregnated with a synthetic resin. Partial drying makes it malleable so that it can be formed into sheets. It is used as a binding material between a PCB core and copper foil, and between two cores. The core of the PCB is a fully cured rigid base made of fully cured flame retardant FR-4 grade material laminated with copper foils on either side.

(d) Plated-through-holes (PTH) PCBs can be used. However, neither rivet nor eyelet (a small hollow Cu tube inserted on one side of PCB followed by pressing the rivets with a rivet tool to deform and connect the two sides), nor standoff (a threaded separator of specified length to raise one part in assembly above another part) is allowed.

(e) Due to poor fastening and non-performance in outgassing, solder masks (polymer layers applied on PCBs to prevent copper from

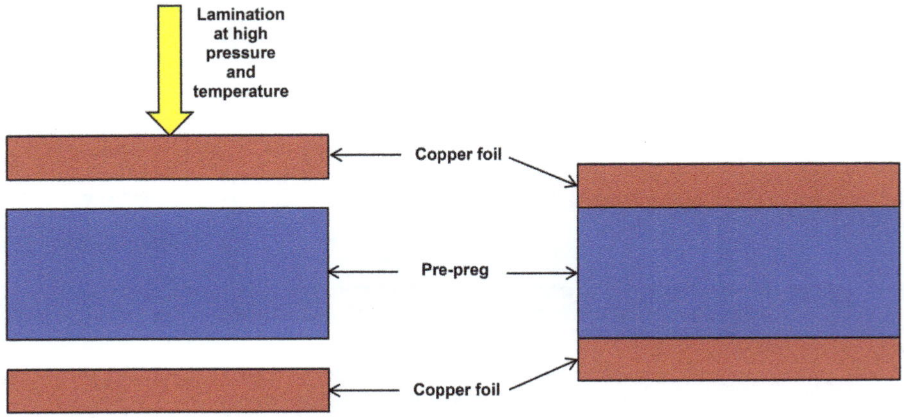

Figure 24.3. Formation of a copper-clad laminate by applying high pressure at a high temperature to a sandwich of copper foil/pre-preg/copper foil.

oxidizing, short-circuiting and environmental dust) are not recommended.

(f) Component derating: In space, there are only two mechanisms of heat transference: conduction and radiation. Convection is not possible in space vacuum. So, the heat carrying capacity of a conductor is lower than prescribed for terrestrial work. The current capacity of components should be derated by 66% to account for difference in heat flow process.

(g) Vias: Satellite and spacecraft PCBs require through-hole vias (with hole extending from top to bottom layer) and buried vias (located in the inner layer with no path to an outer layer), figure 24.4. Blind vias (hole extending from an exterior layer to an interior layer which cannot be seen after lamination) are not permitted.

Blind via reduces parasitic capacitance produced by the barrel of the plated hole and the planes through which it passes (Knack 2020). However, it suffers from reliability problems. During soldering a component to a pad with a blind via, air is trapped under the hole in the blind via when the solder paste is applied to the pad. This small air bubble weakens the joint and can open the circuit during thermal cycling. Other issues with blind vias are related to difficulties in plating of hole when the hole is deeper compared to diameter and stopping the drilling at the planned layer. Blind vias can cease to function from deformation of copper barrel due to non-uniformity of weaving style of glass cloth within the laminate.

(h) Critical wiring should be routed on the same layer with minimal via use.

(i) Variable components: Variable capacitors, inductors or trim pots should be avoided.

(j) Conformal coatings should be applied on exterior surfaces of electronic devices for avoidance of electrostatic discharge (ESD) and Paschen's discharge. A conformal coating is a non-conductive

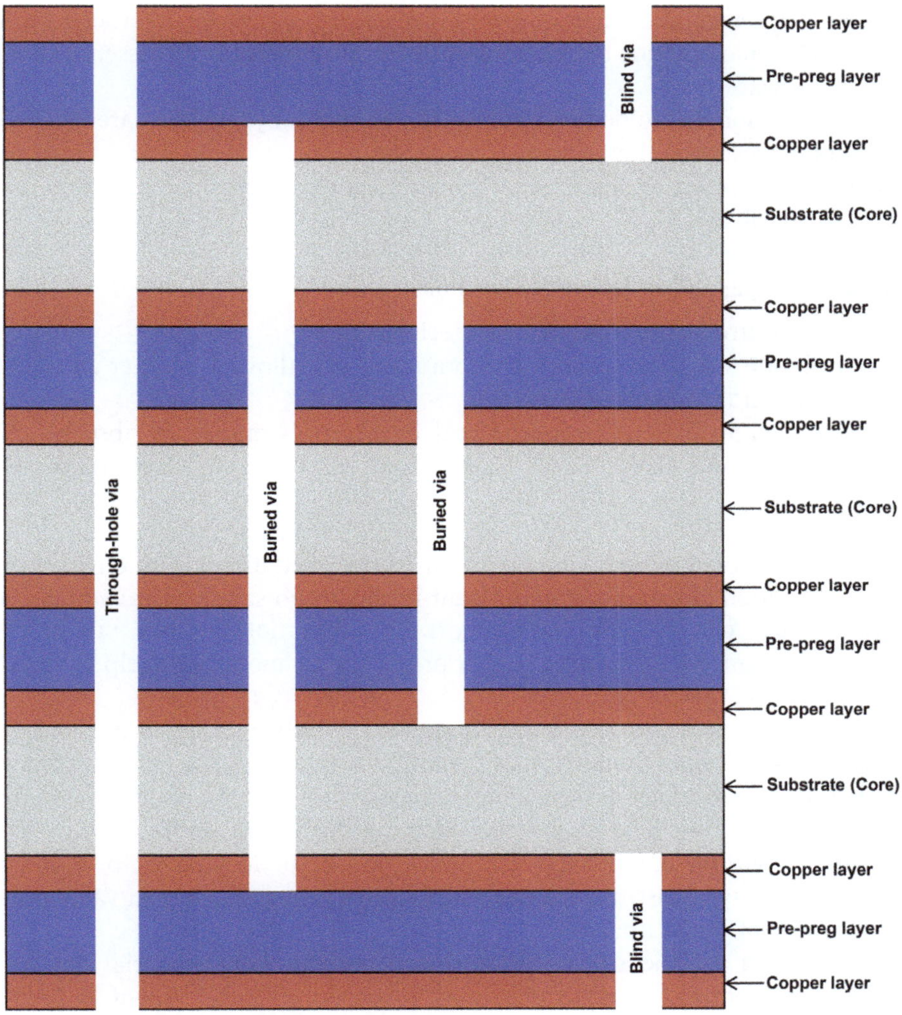

Figure 24.4. Through-hole, buried and blind vias in a PCB.

polymeric film consisting of acrylics, polyurethanes, epoxy, silicones, or a combination of resins, conforming to the surface of the PCB. ESD is the discharge of static electricity between two charged objects coming into contact, e.g., human body and an electronic device. Paschen's discharge is the electric discharge between two electrically charged surfaces taking place at a voltage, which for a given gas is a function of the gas pressure and intervening distance between the surfaces, in accordance with Paschen's law.

(ii) Proper mounting of components should be checked by extensive analysis for compliance of specifications by the manufacturer. Non-complying and defective components should be removed.

(iii) Component and interconnection reliability assurance is crucial because electronics cannot be repaired in space. Reliability is the essence of space hardware.
(iv) Perfection in soldering should be the goal by taking care that good practices are followed.
(v) Adhesives act as shock absorbers and should receive due attention during assembly.

24.1.4 Dampening and isolation of vibrations

Vibration is controlled by applying two techniques:
(i) isolation where the energy of vibration is not allowed to enter an object; it blocks transmission of vibration.
(ii) dampening where the mechanical energy of vibration is absorbed by a material and dissipated as heat so that less energy reaches the protected structure.

The two terms 'isolation' and 'dampening' are often used mutually and reciprocally.

Foam is a good absorber but is deficient in rigidity to support heavy machinery, whereas rubber can provide the support but does not dampen vibrations as effectively as foam. Metal springs, rubber or pneumatic mountings help in vibration isolation.

Vibration dampers are viscoelastic materials. Viscoelasticity is a property of materials to exhibit both fluidic (viscosity) and solid (elasticity) behavior. A damping coefficient is defined to act as a numerical representation of the ability of a material to dampen vibrations. The higher the damping coefficient, more effective is the material in vibration dampening. Tan delta (tan δ) or loss factor of a material measures its ability to disperse the energy of vibration as heat. A high value of tan δ is an indicator of better ability.

Sorbothane, a proprietary, polyether-based, polyurethane material has a high damping coefficient to attenuate vibrations. It is an ultra-soft, resilient and durable material which can withstand wide temperature fluctuations and can be cast into a variety of shapes. It has a long fatigue life, low creep rate, and also non-skid properties. It combines vibration isolation with dampening. It has been used to vibration isolate cameras and pumps on the International Space Station (Vibration Solutions 2022).

24.2 The challenge of temperature excursions beyond safe limits

24.2.1 Need of thermal control on space vehicles

Let us get a feeling of the likely temperatures expected in space. The temperature of space in the Universe is 2.725 48 ± 0.000 57 K (Fixsen 2009). It is the temperature of cosmic background radiation. This radiation is spread throughout the Universe. Recall from section 2.3.3. The surface temperature of the Moon increases significantly from just before dawn to that in the afternoon. Temperature measurements at the site of Apollo 15 (26°N, 3.6°E) on the Moon, recorded a minimum temperature = 92 K = −181 °C and maximum temperature = 374 K = 101 °C yielding a difference = 374 – 92 = 282 K (Heiken *et al* 1991). In the absence of thermal control, the temperature of an orbiting space station is +121 °C on the sun-facing side and −157 °C on the dark- or sun-shaded side (NASA Science 2022).

A thermal control system maintains the temperatures of different zones in the spacecraft within acceptable levels, either passively or actively. Typical permitted temperatures during flight are: electronics: −40 °C to +50 °C and batteries: −20 °C to +30 °C.

24.2.2 Passive thermal control

Here no powered equipment is used. As the spacecraft has limited power, so low-cost passive systems requiring little or no maintenance are preferable.

(i) Paints, coatings, surface finishes, or adhesive tapes: White paint has low absorptivity and low emissivity of sunlight. So, it is suitable for outer surfaces requiring less absorption of sunlight. Black paint has high absorptivity and high emissivity of sunlight. It is used on inner surfaces of a spacecraft for providing necessary heat exchanges. Adhesive tapes are easy to apply and remove, and can be used to make midway changes/corrections in assembly after it has reached an advanced stage (NASA 7.0 Thermal Control 2021).

(ii) Multilayer insulation (MLI) or super-insulation blankets: Optimum temperatures for electronics and batteries are maintained with the help of this insulation (figure 24.5). It comprises several thin sheets of heat-reflective material separated by heat-conduction insulating air spaces to jointly prevent loss of heat by both radiation and conduction. Each

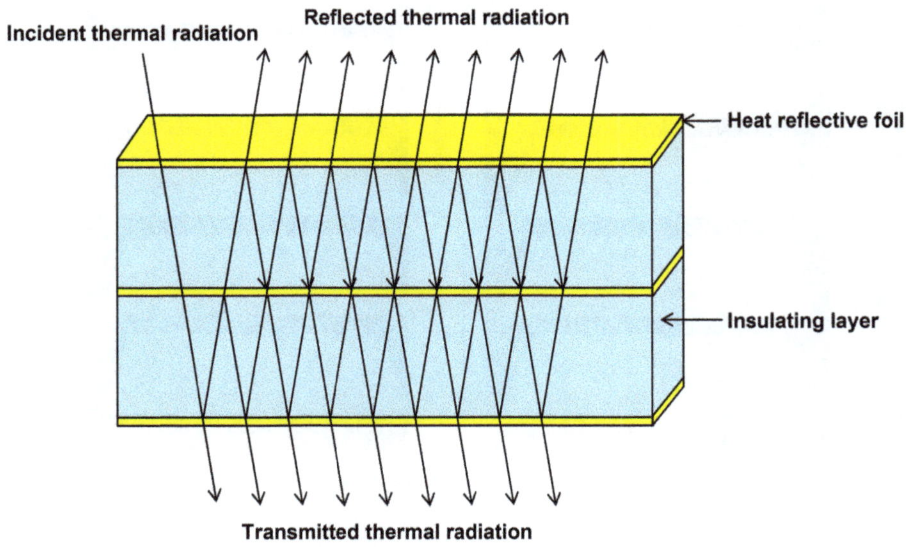

Figure 24.5. Multilayer insulation principle.

heat-reflective layer is a 0.5 mm thick polyester sheet coated with an Al, Cu or Au thin film. Between each heat-reflective layer and the next layer is an openly weaved fiberglass or ceramic threads layer serving as heat-conduction insulating air space. A space suit may contain a succession of 12 alternatively placed heat-reflective layers/conduction-insulating air spaces. Thus, a heat impervious system is built which can be easily draped and shaped as desired. Loss of heat by radiation is avoided because each heat-reflective layer sends back 95%–98% of incident IR radiation to its predecessor. Heat loss by conduction is avoided because of air pockets in the heat-conduction insulating air space (Sigma Technologies 2016).

(iii) Thermal straps: A thermal strap, also called a heat strap or thermal braid is a flexible, lightweight link for conduction of heat. It is connected between two locations, a heat source and heat sink. It consists of a copper cable or copper foil/aluminum foil/pyrolytic graphite sheet with end fittings, e.g., brackets/lugs.

(iv) Thermal interface material: This is a material such as paste or grease or a thin sheet sandwiched between two surfaces for heat transfer.

(v) Sunshields or sunshades: They are made of a material with low absorptivity of sunlight to obstruct the sunlight falling on the spacecraft by blocking the view of the Sun.

(vi) Thermal louvers: These consist of an array of blades (figure 24.6). The blades are controlled by bimetallic actuators. The actuators are adjusted

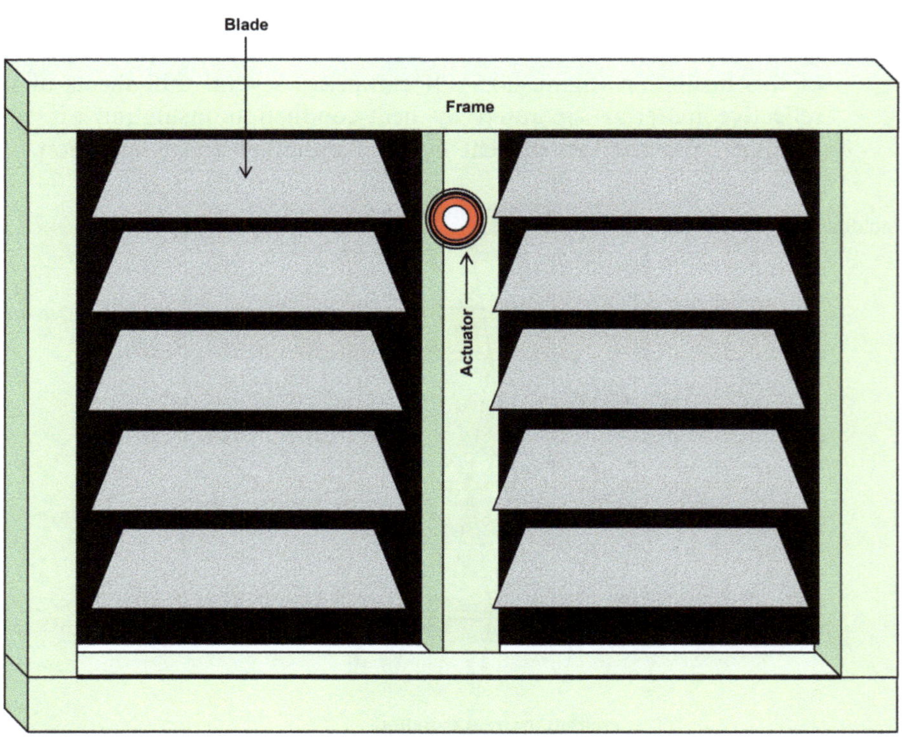

Figure 24.6. Thermal louvers.

for a defined temperature range of the heat radiating surface of a spacecraft. When the temperature rises above the set point, the actuators open the blades allowing the heat to escape from the radiating surface. Conversely, when the temperature falls, the blades are closed by the actuators. Thus, a shield is produced by which the heat losses from the radiating surface are decreased (Muraoka *et al* 2001).

(vii) Deployable radiators: A radiator is a surface dedicated to dissipation of excess heat. A deployable radiator (figure 24.7) is one which is deployed when it is necessary to dissipate excess heat and stowed otherwise. A deployable radiator is fabricated with an environment-adaptive function in the form of a reversible thermal panel (RTP) using graphite sheets of high thermal conductivity in the reversible fin and a shape memory alloy in the deployment/stowing actuator (Nagano *et al* 2006). The radiator and absorber sides of the radiator/absorber reversible fin have opposite

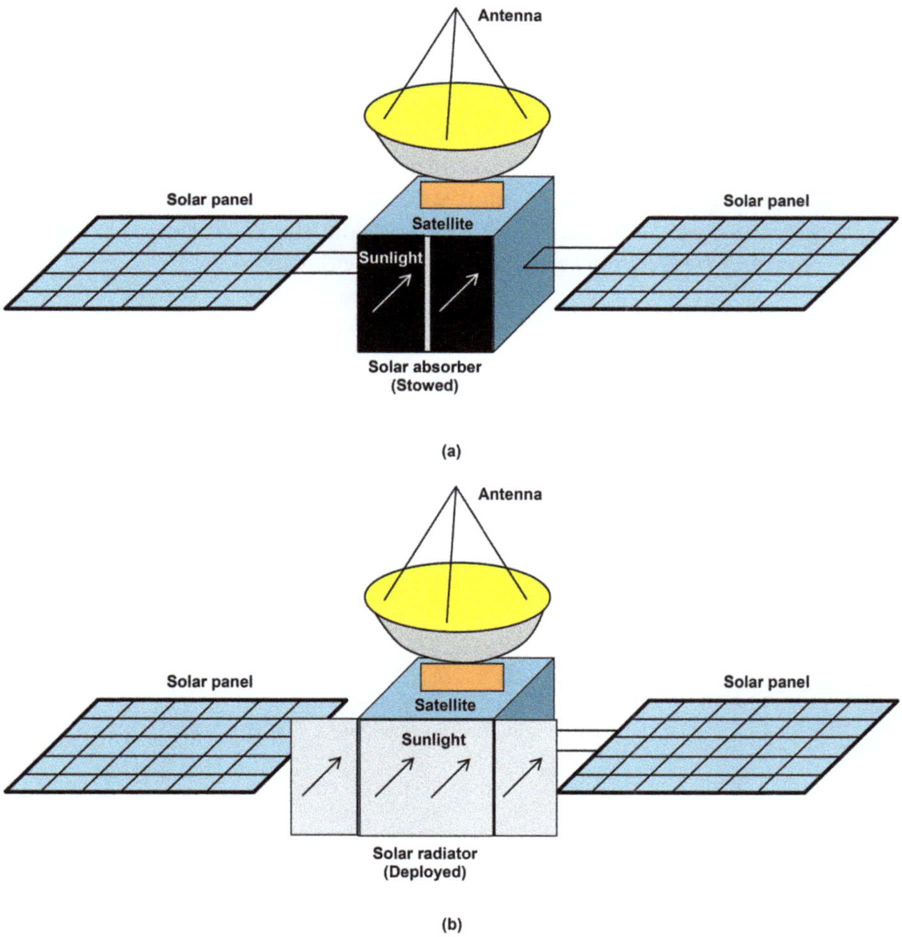

Figure 24.7. Deployable radiator: (a) stowed and (b) deployed conditions.

thermo-optical properties. The radiator side has low solar absorptance/hemispherical emittance while the absorber side has high solar absorptance/hemispherical emittance. The RTP changes its role from a solar radiator to a solar absorber by deploying/stowing the radiator/absorber reversible fin according to environmental changes and heat dissipation requirements.

(viii) Heat pipes: A heat pipe (figure 24.8) consists of a vacuum tight envelope in which lies a wick structure and the working fluid. It transports heat from a hot end to a cold end until equalization of temperatures of the two ends. Let us call the hot end the evaporator and the cold end the condenser, and suppose the wick is drenched in a liquid. At the evaporator hot end side, the liquid is evaporated and the vapor moves to the cold condenser end. At the condenser end, the vapor cools down into a liquid. The condensed liquid moves along the wick to the hot evaporator end via capillary action where it is evaporated again and the same process is repeated. This evaporation–condensation cycle involving phase changes and the circulatory flow of vapor and liquid phases continues until the temperature of the evaporator end = temperature of the condenser end (Advanced Cooling Technologies 2022).

(ix) Phase-change material (PCM) devices: A phase-change material stores heat as latent heat. There are two latent heat values associated with changes in phase of water, one with respect to steam, the latent heat of vaporization (540 cal g^{-1}) and another relative to ice, the latent heat of freezing (80 cal g^{-1}). By absorbing heat at phase transition, the PCM provides cooling. By releasing heat at phase transition, it provides heating.

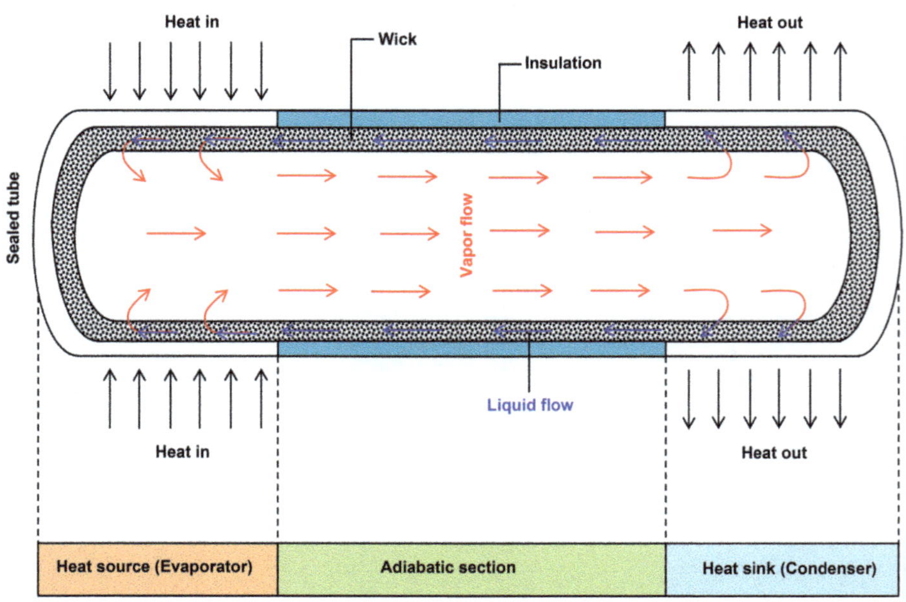

Figure 24.8. Heat transfer in a heat pipe.

The PCM is used in a reversible, closed loop system to construct a thermal storage unit (figure 24.9). For this purpose, the PCM is enclosed in a metal housing. The housing is attached to a source of heat, which is here the hot electronic circuit. On receiving heat from the source, the housing is warmed. The PCM absorbs heat and changes phase from liquid to vapor. The vapor gets cooled on passing through a pipe surrounded by a finned heat exchanger. PCM releases heat to transform from vapor back to liquid. As the energy of heat source (the electronic circuit) is consumed during vaporization of liquid, the housing and the attached heat source (the circuit) cools. A thermal storage unit is used with components undergoing repeated temperature cycling. The unit has no moving parts. Stability in controlling temperature is achieved (Collette *et al* 2011).

(x) Thermal switches: A thermal switch is a device that controls heat transference between a heat producing component and a heat sink, e.g., a radiator. It is an electromechanical switch in the form of bimetallic disk with different coefficients of thermal expansion. It is pre-calibrated for snapping at a particular temperature.

A low-mass, two-phase thermal switch (TPTS) to maintain the temperature of electronics assembly against variations in thermal loads and environments is shown in figure 24.10. The switch is passively activated by the temperature of the electronics assembly. It consists of a sealed bellows made of metal. The bellows contains a wick structure and the working fluid. It is enclosed in a hermetic housing, with one end of the bellows fastened to the housing and the opposite end to a heat sink. The housing

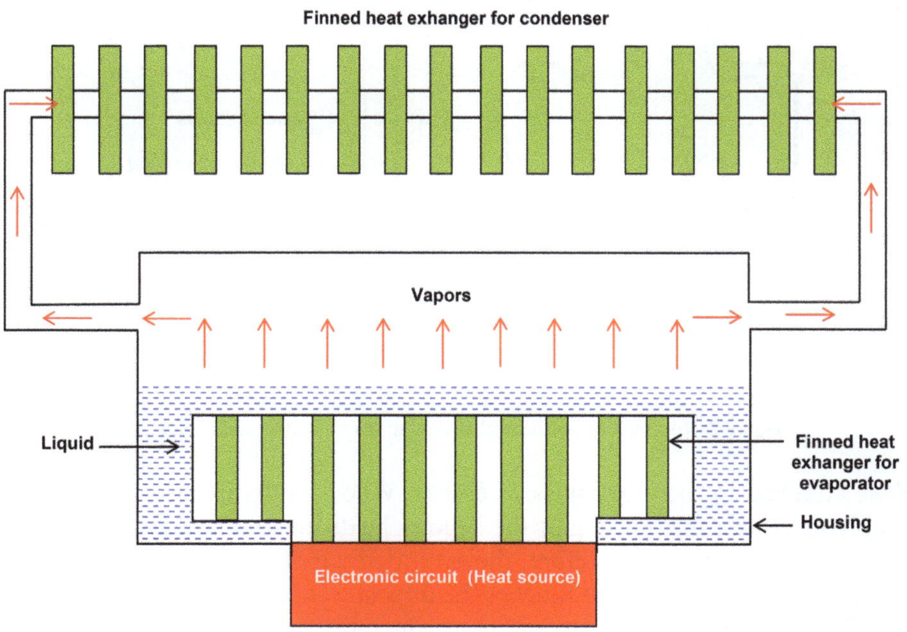

Figure 24.9. Phase-change cooling system.

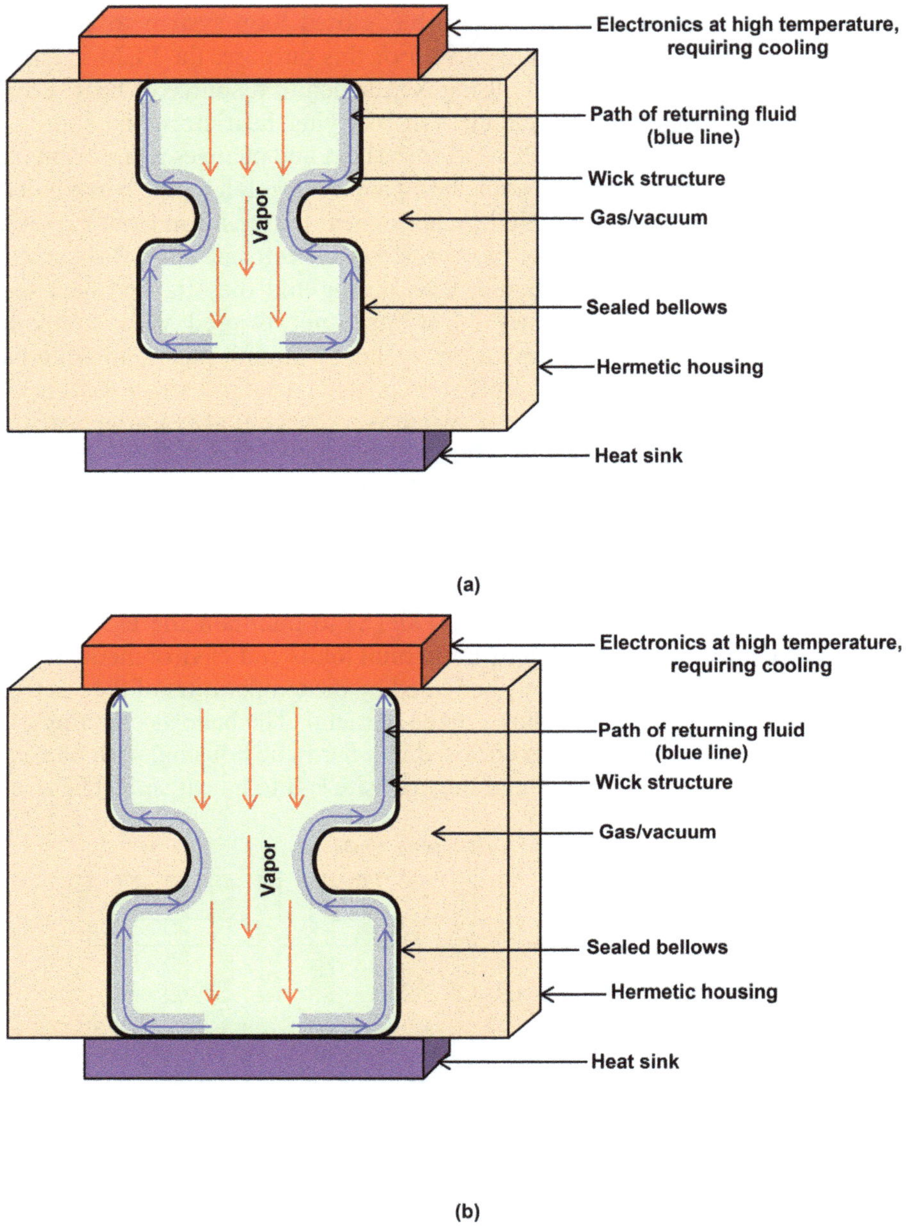

Figure 24.10. Two-phase thermal switch in: (a) off state and (b) on-state.

itself is attached to the electronics assembly. When the temperature of electronics assembly increases, the housing of the bellows heats up so that the working fluid in the bellows is vaporized. Therefore, the saturated vapor pressure inside the bellows rises. The bellows expands and touches the opposite end of the housing joined to the heat sink. The saturated vapor pressure at which the contact is established between the bellows and the heat sink is the set point of the switch (NASA Tech Briefs 2016).

24.2.3 Active thermal control

When the heat loads are high, or the temperature tolerance limits are stringent, active control systems working on input power are used.

(i) Electrical resistance heaters: The type of heater widely used in spacecraft is the patch heater (figure 24.11(a)). The heater is fabricated in various shapes, sizes and wattages. The heater is used with a temperature sensor to control battery temperatures within narrow limits.

In this heater, the heating element is encapsulated between insulating polyimide or kapton sheets. The heating element is a 25 μm thick metal foil, usually a nickel base alloy. It is laminated to a kapton substrate. A resistance pattern is formed on the metal foil by chemical etching. The other side of the foil is also laminated with kapton. Lead wires are fixed to the heater and covered with epoxy (Tempco 2021–2).

Another kind of heater is the cartridge heater (figure 24.11(b)) consisting of a wound resistor in a cylindrical metal enclosure.

(ii) Cryocoolers: A cryocooler is a small refrigerator to reach temperatures <100 K. The term 'cryocooler' is used for small cooling systems working at low

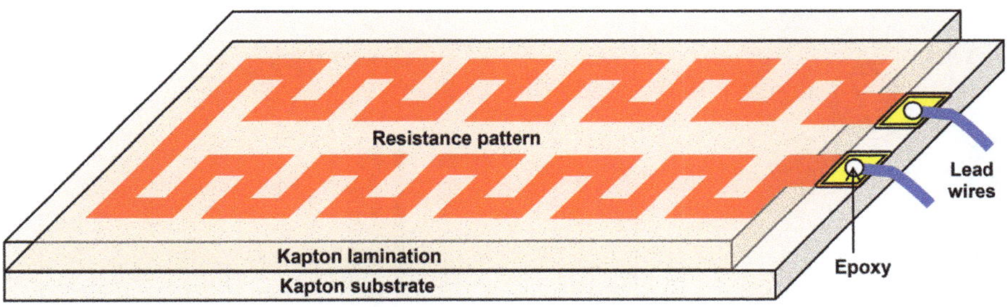

(a)

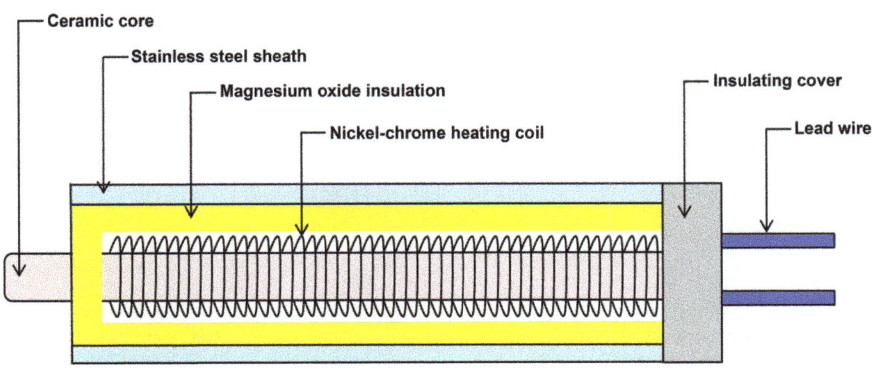

(b)

Figure 24.11. Heaters: (a) patch and (b) cartridge.

input powers <20 kW, with some systems requiring ~2–3 W. Five types of cryocoolers are used for different applications. These are: Joule–Thomson, Brayton, Stirling, Gifford–McMahon, and pulse tube cryocoolers. They differ in their principles of operation. A cryocooler may be suitable for a given set of applications but unsuitable for others. So, one or two types of cryocoolers match with the requirements of each application. Naturally, these are the types preferred for that application (Radebaugh 2009).

Joule–Thomson and Brayton cryocoolers work on recuperative cycles, whereas the Stirling, pulse tube and Gifford–McMahon cryocoolers work on regenerative cycles.

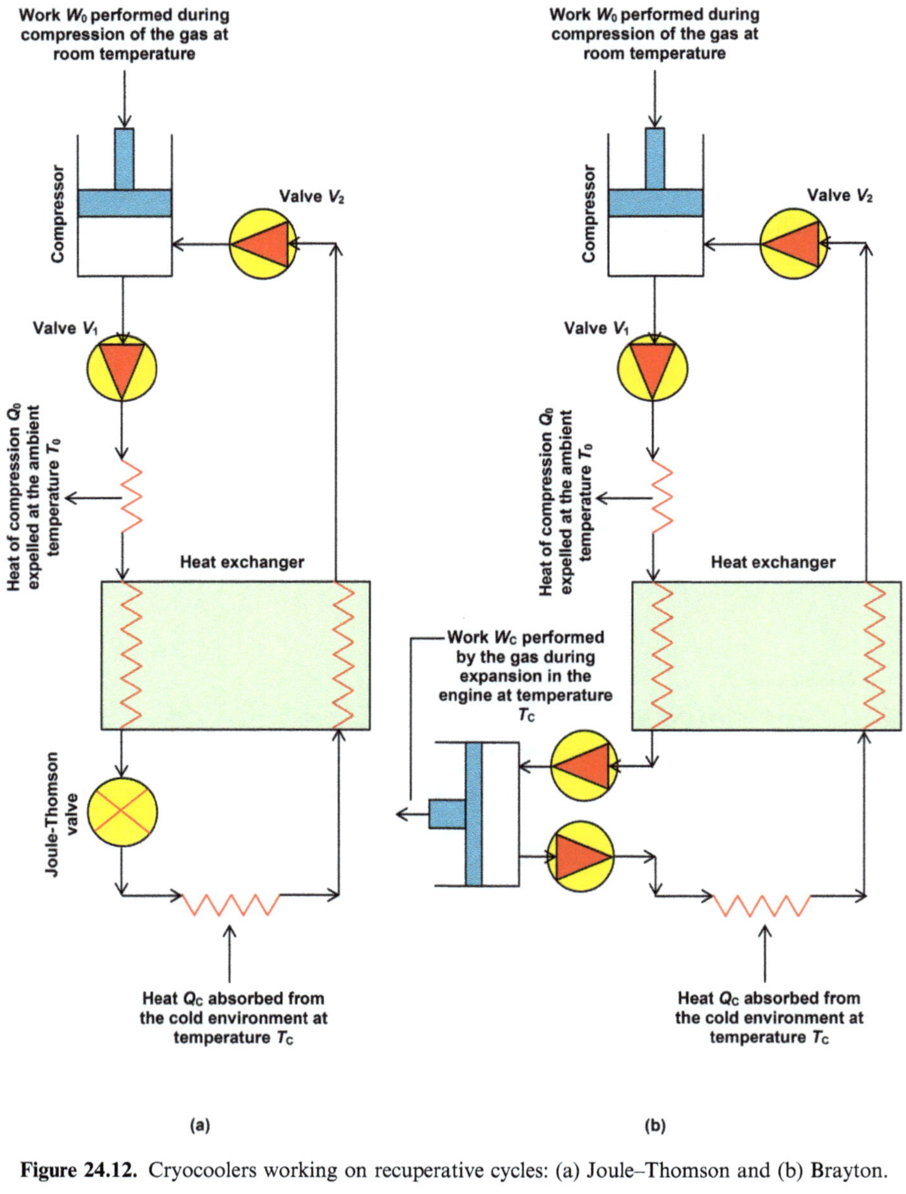

Figure 24.12. Cryocoolers working on recuperative cycles: (a) Joule–Thomson and (b) Brayton.

Figure 24.12 shows the recuperative cycle cryocoolers, Joule–Thomson cooler in figure 24.12(a), and the Brayton cooler in figure 24.12(b). In figure 24.13, the regenerative-cycle cryocoolers are shown, the Stirling cooler in figure 24.13(a), the pulse-tube cooler in figure 24.13(b) and Gifford–McMahon cooler in figure 24.13(c).

A key component of all refrigerators is the heat exchanger. This component performs the task of transference of heat from one fluid to another without mixing the fluids. The heat exchanger for the recuperative cycles is called the recuperative heat exchanger; it is abbreviated as 'recuperator'. In the same way, the heat exchanger for the regenerative cycles is known as the regenerative heat exchanger which is shortened to 'regenerator'.

In the recuperative cycle cryocoolers, the recuperator has two separate channels. One channel is for high-pressure fluid. The other channel is for low-pressure fluid.

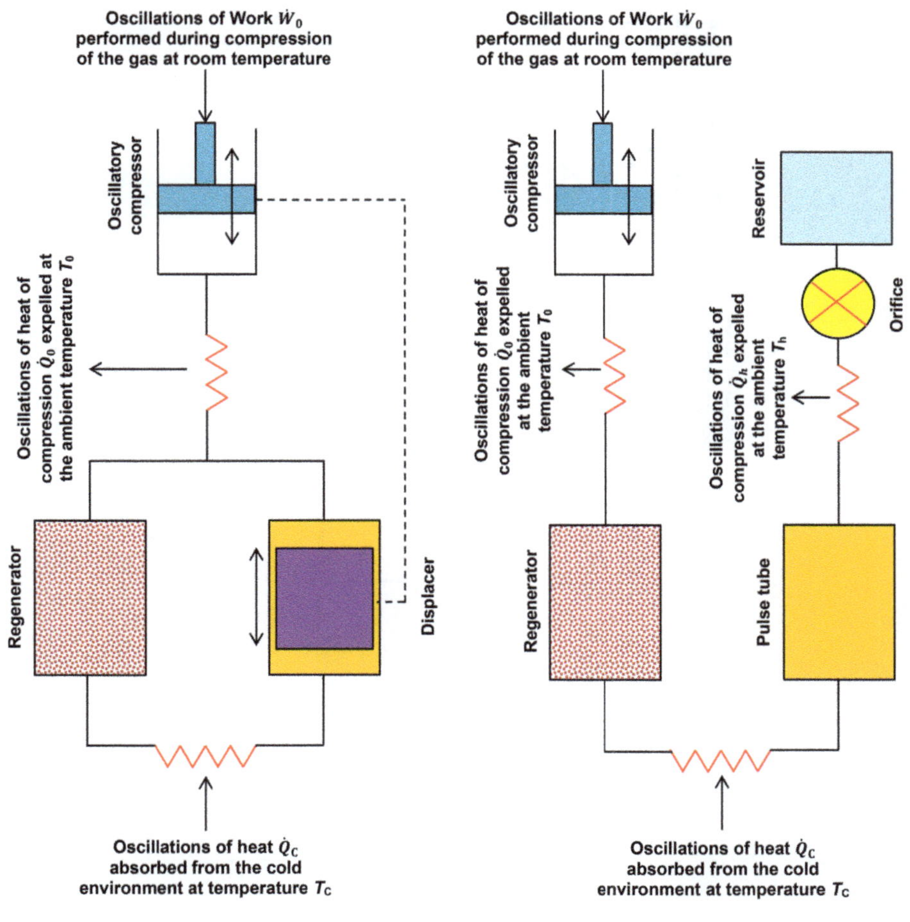

Figure 24.13. Cryocoolers working on regenerative cycles: (a) Stirling, (b) pulse tube and (c) Gifford–McMahon.

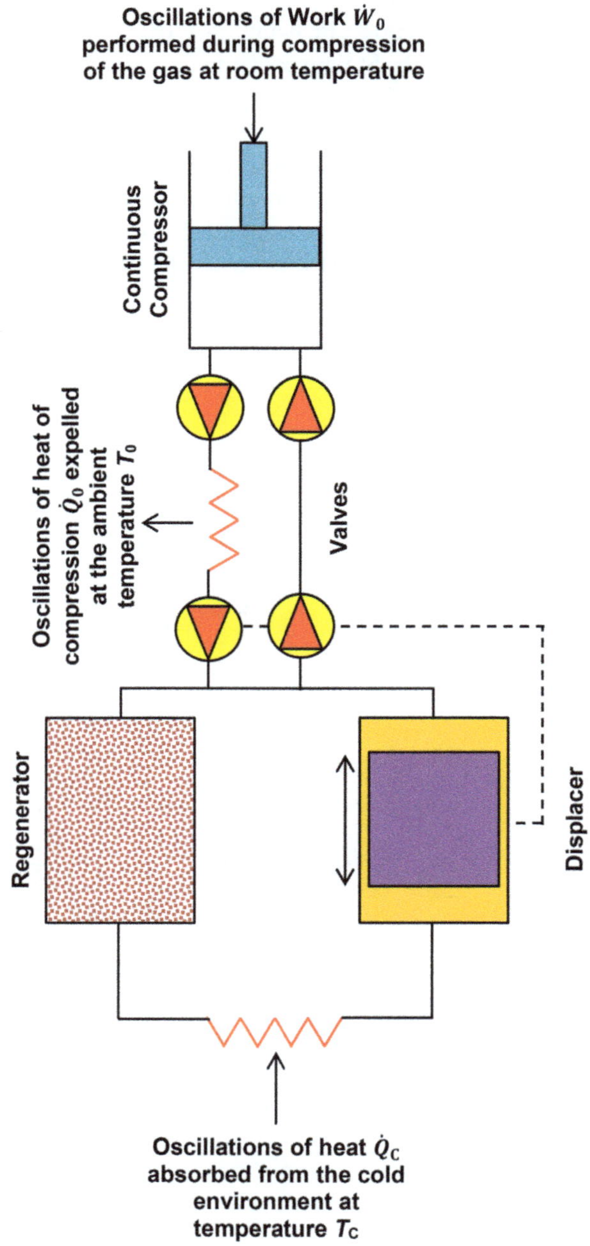

(c)

Figure 24.13. (Continued.)

The working fluid flows steadily in one direction at fixed high and low pressures. The high-pressure fluid goes to the high-pressure channel of the recuperator and the low-pressure fluid goes to its low-pressure channel. In the Joule–Thomson cycle, the low temperature expansion takes place through an orifice, capillary, or valve. The working fluid is liquefied at the cold end at 4.2 K for helium. In the case of hydrogen the temperature is 20 K; for nitrogen it is 77 K. In the Brayton cycle, the low temperature expansion occurs with an expansion engine.

The regenerative-cycle cryocoolers work like alternating current systems with oscillating flows and oscillating pressures. Unlike the recuperator having two flow channels, the regenerator has only one flow channel. During every half cycle, the flow in the channel changes direction. In the first half-cycle, the inflowing hot stream is precooled by transferring heat to the regenerator matrix. The matrix has sufficient capacity for storage of heat during the half cycle. In the second half cycle, the matrix transfers the stored heat to the returning cold stream.

Refrigeration takes place at proper phase angles between flow and pressure at the cold end with maximum cooling taking place under the condition that flow and pressure are in phase at the cold end. The working fluid is helium. Heating takes place with increasing pressure and cooling with decreasing pressure. Typical frequency ~1–2 Hz for Gifford–McMahon (GM) cryocoolers and some pulse tube cryocoolers. It is ~30–60 Hz for Stirling cryocoolers and some pulse tube cryocoolers. Gifford–McMahon (GM) cryocoolers and GM-pulse tube cryocoolers attain the lowest temperatures ~3 K. The Stirling cryocoolers and Stirling pulse tube cryocoolers are generally used for temperatures >20 K.

In Stirling and Gifford–McMahon cycles, phase modulation between pressure and flow waves is essential. The phase angle is adjusted by the phase of the displacer relative to the piston. Another function of the displacer is to recoup the expansion work at the cold end. The pulse tube cryocooler generally has no displacer. So, this expansion work is lost.

In the Stirling cycle, the oscillating pressure is produced by a valveless compressor acting as a pressure oscillator. In the Gifford–McMahon cycle, it is produced with the help of valves. The valves switch the cold head between a source at low and high pressure. The low- and high-pressure sources are generated using a compressor having inlet and outlet valves.

In the schematic diagrams of cryocoolers, W_0 is the work done at room temperature during compression of the gas. The heat of compression Q_0 is expelled at the ambient temperature T_0. The heat absorbed from the cold environment at temperature T_C is denoted by Q_C. In the case of the Brayton cooler, the gas performs work W_C during expansion in the engine at temperature T_C.

(iii) Thermoelectric coolers: A thermoelectric cooler is a small-size solid-state heat pump. Being a solid-state technology, chances of failure are minimal. It does not need any refrigerant such as chlorofluorocarbons. It is good for localized cooling.

The thermoelectric cooler works on the principle of Peltier effect, viz., the cooling of one junction and heating of other junction when current flows through a circuit comprising two dissimilar conductors or semiconductors, e.g., in a circuit made by connecting a battery at the two ends of a bismuth wire with two pieces of copper wire, temperature increases at the junction where current moves from copper to bismuth but decreases at the junction where current moves from bismuth to copper. The hot junction is connected to a heat sink.

Peltier effect is the inverse of Seebeck effect, the production of an electromotive force and hence a current when a temperature difference is maintained between the two junctions of dissimilar conductors forming a loop. Thermocouples work on Seebeck effect.

In its simplest form, a thermoelectric cooler (figure 24.14) consists of a P-type semiconductor and an N-type semiconductor joined by a metal interconnection, e.g., a copper layer (Samsuddin *et al* 2013). The P-type and N-type semiconductors

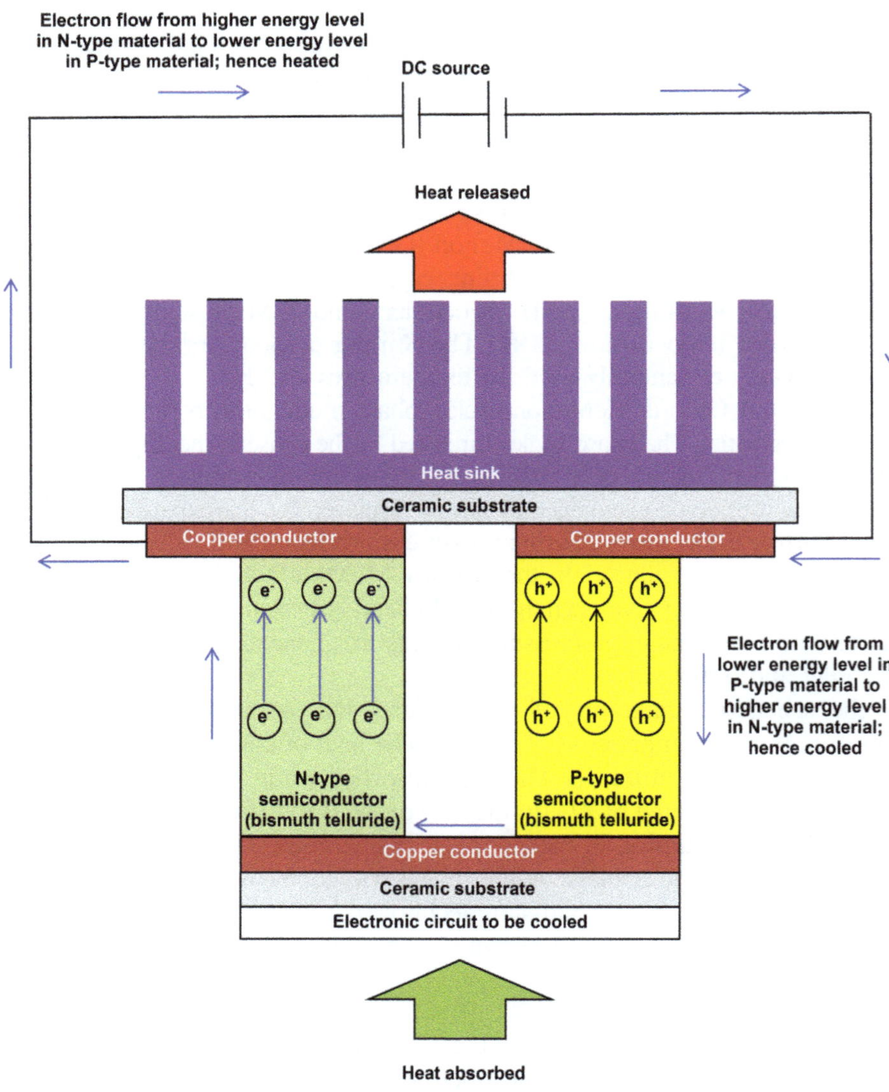

Figure 24.14. Diagram illustrating the principle of cooling by a thermoelectric cooler.

together constitute a thermocouple. The semiconductor used is bismuth telluride (Bi_2Te_3). The thermoelectric cooler is operated by applying a DC source. Electrons flow from the lower energy level in P-type semiconductor to higher energy level in N-type material. In this process of electron flow from lower to higher energy level, heat is absorbed, and P-type/N-type junction is cooled. Electrons continue their motion and flow from the higher energy level in the N-type semiconductor to lower energy level in P-type semiconductor. In this process of electron flow from higher to lower energy level, heat is released to the environment and N-type/P-type junction is heated.

Practical thermoelectric coolers are fabricated as modules comprising single-stage or multistage assemblies. A single stage contains several couples connected electrically in series and thermally in parallel (figure 24.15). All the cold junctions are soldered to metal interconnection on one ceramic plate. All the hot junctions are soldered to the metal interconnection on the other ceramic plate. A multilayer

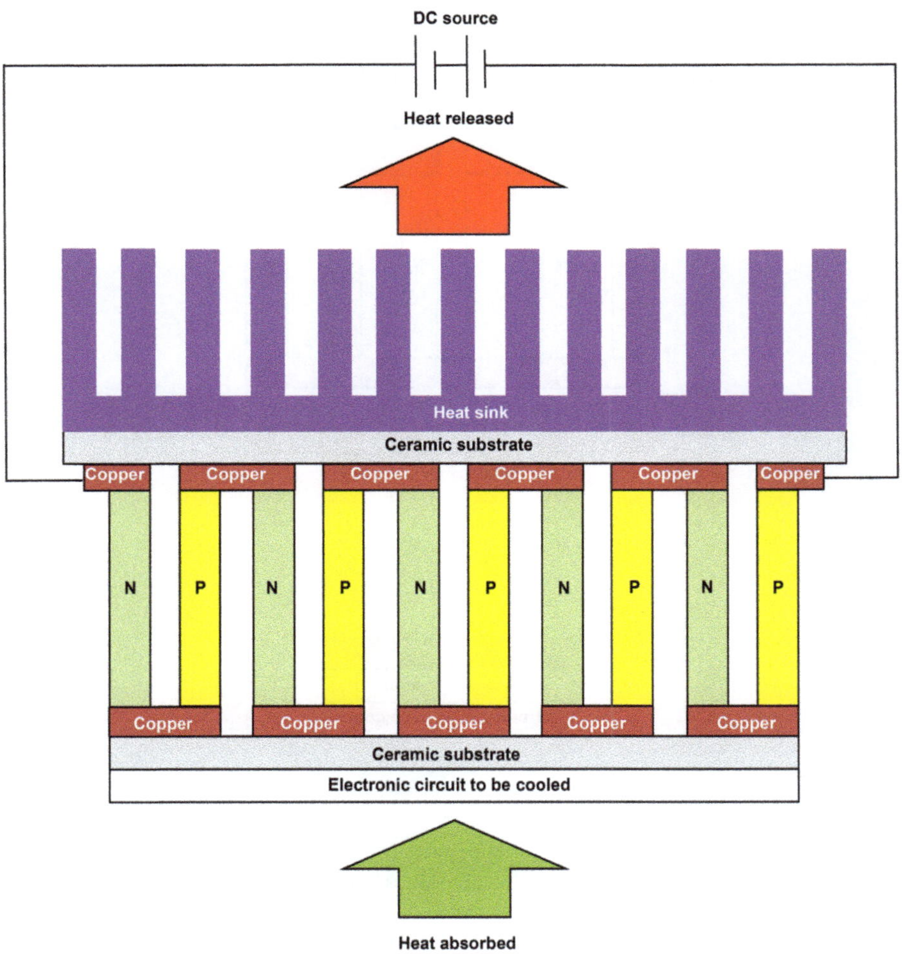

Figure 24.15. A single-stage thermoelectric cooler comprising several couples.

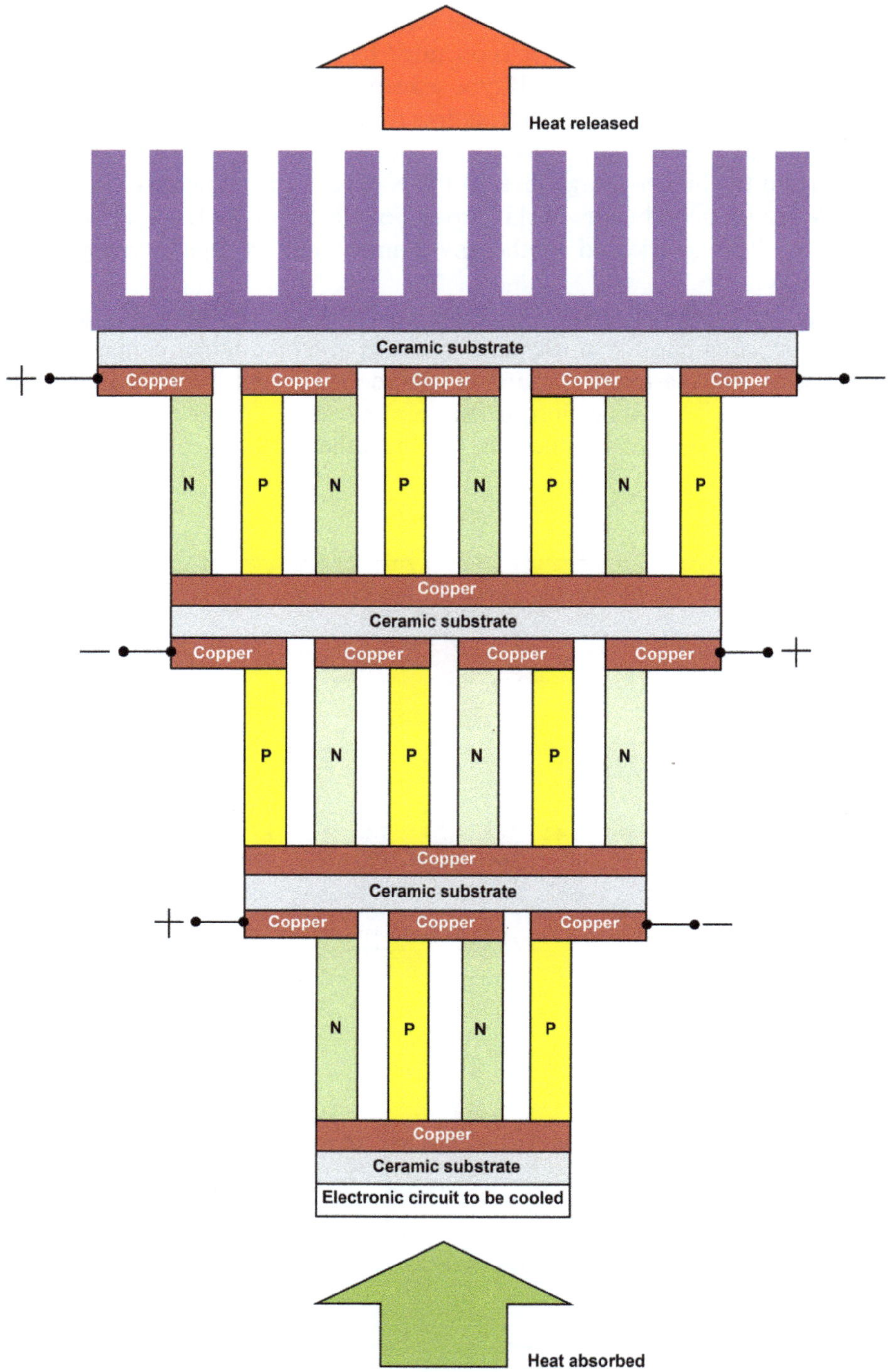

Figure 24.16. A three-stage thermoelectric cooler.

thermoelectric cooler (figure 24.16) is a layered structure comprising layers of single-stage coolers. It is a 2-stage, 3-stage, …, n-stage cooler accordingly as the number of stages is 2, 3, …, n.

A thermoelectric cooler can also be used as a thermoelectric generator via the Seebeck effect in which a temperature difference between the junctions creates a potential difference across them (von Lukowicz *et al* 2016).

(iv) Mechanically pumped fluid loops: A pumped fluid loop differs from the conventional thermal control in the respect that connection between the temperature-controlled components and heat radiating surface is conductive or radiative in the traditional approach, whereas it is convective in the fluid loops. A fluid flowing between the heat source and the heat sink convectively gathers heat from the heat source and again convectively dissipates it at the heat sink through a heat exchanger. The cooled fluid is returned to the heat source for continuation of cooling. The prime mover of the fluid is a mechanical pump, and the fluid is circulated through tubing. The fluid loop can cool several locations via convective cooling.

The pumped fluid loop can provide precision temperature control for electronics components with different power dissipation ratings. It can be automatically switched on/off. It is robust, predictable, versatile and offers flexibility in integration with any configuration. However, the system may break down partially or completely in the case of leakages or failure of the pump. Mechanically pumped fluid loop is not used in small spacecraft because it consumes more power.

24.3 The challenge of electrical charging of spacecraft

Spacecraft charging is a broad term encompassing the build-up of charges on the surface of the spacecraft or in its interior (Mikaelian 2001). It is subdivided under two headings:

(i) surface charging, and
(ii) internal charging.

24.3.1 Surface charging

Surface charging refers to charging confined to the surface of the spacecraft. It causes the development of an electrostatic potential difference between the surface of the spacecraft and the surrounding plasma environment. Surface charging is of two types: absolute and differential.

Absolute surface charging occurs when the complete spacecraft surface acquires an electrical potential with respect to its surroundings. If the spacecraft surface is metallic, the surface is an equipotential surface.

Differential surface charging takes place when a difference of potential is created between two portions of the spacecraft. This happens when the spacecraft surface consists of portions made from conducting and dielectric materials.

24.3.1.1 Negative charging of the spacecraft surface in the space plasma
99.999 of matter in the Universe is in the plasma state, the fourth state of matter consisting of ions and electrons. Plasma fills the Solar System, the magnetosphere of the Earth and other planets, the interstellar and intergalactic space. This plasma environment of space constitutes the space plasma (Lai 2012). A spacecraft is immersed in this space plasma (figure 24.17). Electrons are light particles whereas ions are heavier. So, electrons travel faster than ions. At any instant, the electron flux incident on the spacecraft from space plasma is much greater than the ion flux. So, the number of electrons on the spacecraft surface exceeds the number of positive ions. The presence of excess of negatively charged electrons on the spacecraft surface produces charge inequality. Consequently, the surface is at a negative potential relative to the neutral plasma ambience (Lai and Cahoy 2017, Lai and Pradipta 2022).

24.3.1.2 Differential charging of the spacecraft in sunlight
Figure 24.18 illustrates the difference in charging behavior of the shadowed spacecraft surface in shade and the sunlit surface facing the Sun. In the diagram the sunlit surface is shown as positively charged, as explained in the next paragraph. But we shall explain that the ultimate potential of this surface will also be negative as for the shadowed side.

The side of the spacecraft facing the Sun receives sunlight. This sunlight causes photoemission based on the photoelectric effect from the spacecraft surface. The electron loss from the surface of the spacecraft (including electrons released by photoemission plus electrons produced by secondary electron emission) surpasses the electron gain from the space plasma. It is expected that the sunlit face of the spacecraft will be positively charged. But this is not so (Lai and Tautz 2006). The reverse happens because the opposite face of the spacecraft not receiving sunlight loses an insignificant number of electrons by photoemission, and remains negatively charged. Due to the large population of negative electron charges on the shaded side, an electric field is set up on this side. This electric field wraps around the spacecraft to the unshaded side to create a potential barrier. In this potential barrier, the photoemitted electrons as well as the electrons liberated by secondary emission on the sunlit side remain entrapped (Lai and Cahoy 2015). As these electrons are prevented from escaping from the sunlit side, they maintain a negative charge condition on this side also. Thus, the potentials on the dark and bright sides of the spacecraft differ. The dark side of the spacecraft is at a high negative potential while the bright side is at a low negative potential. This is a differential charging situation.

24.3.1.3 Inimical effects of spacecraft surface charging
Spacecraft charging leads to undesirable electrostatic discharges. It generates unwanted electromagnetic fields. So, it has devastating effects on electronic communications, control, navigation and instrumentation systems on board. It vitiates any scientific measurements carried out on the space flight. It is correlated to spacecraft anomalies, which are unexpected incidents happening during operational lifetime of a spacecraft that deviate from its normal behavior. They range from

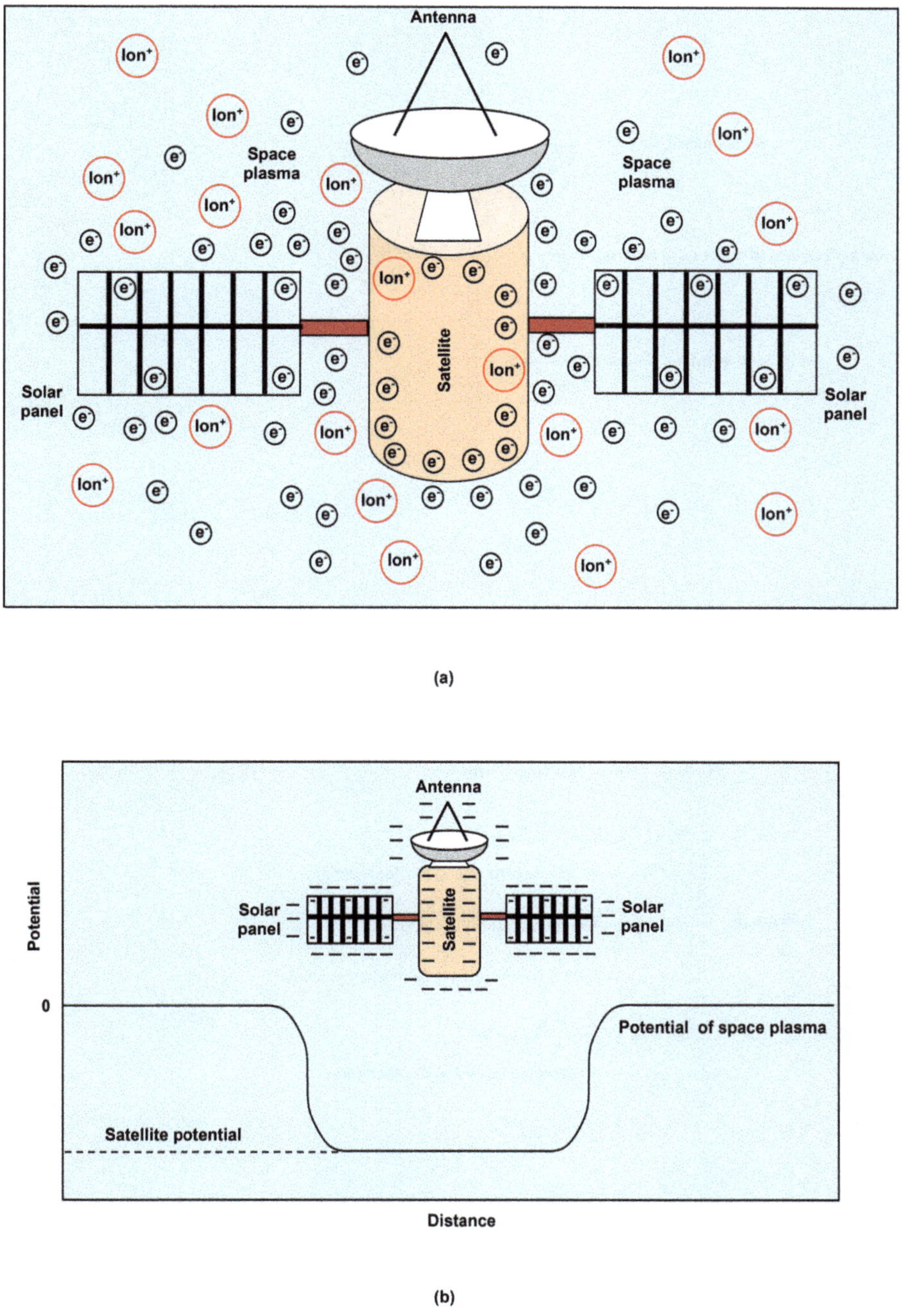

Figure 24.17. Spacecraft in space plasma: (a) charging, and (b) potential in plasma environment.

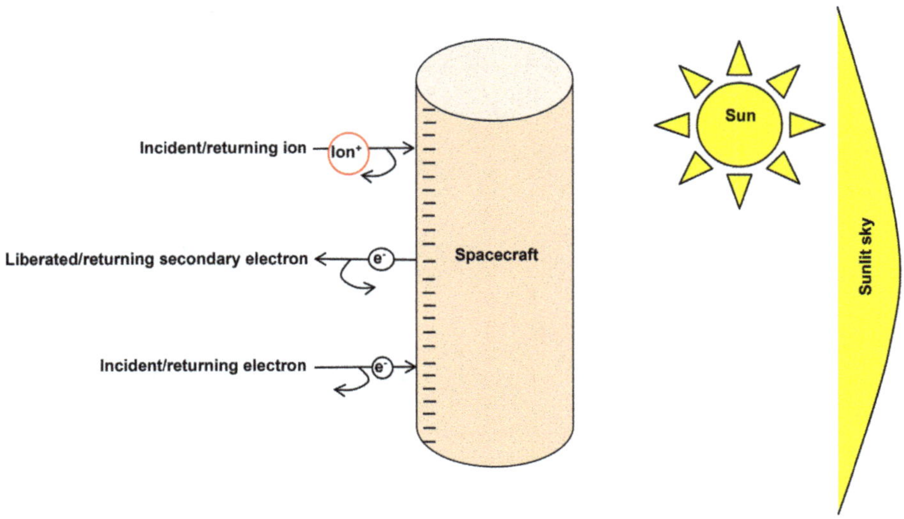

(a)

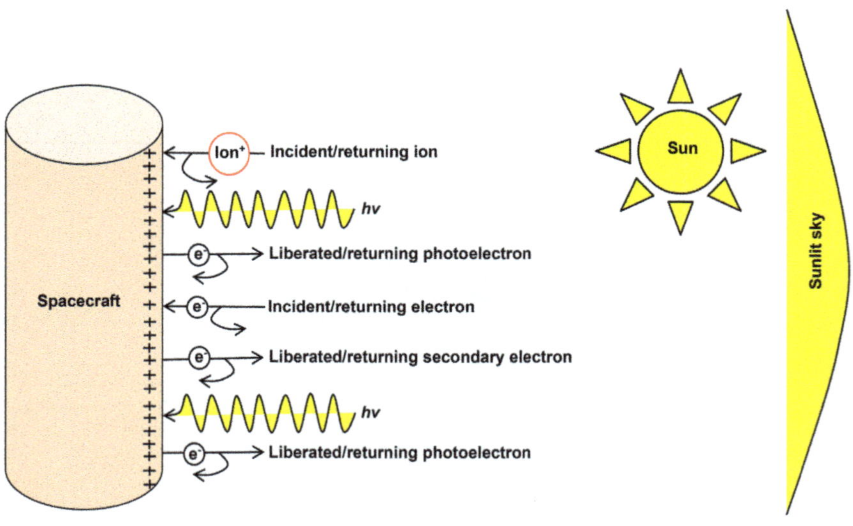

(b)

Figure 24.18. Particles and radiation striking the two surfaces of the spacecraft: (a) the surface away from the Sun and in the shadow, and (b) the surface facing the Sun and receiving sunlight. In (a), the plasma ions and electrons fall on the surface while in (b) the sunlight photons are included apart from plasma ions and electrons. In both cases, the returning or backscattered particles are shown. In (a) the secondary electrons are liberated from the surface while in (b) photoelectrons as well as secondary electrons are released from the surface.

temporary errors to loss of contact with the spacecraft. An anomaly database is maintained to empirically establish a causal relationship with unpredicted events. Obviously, spacecraft charging is a matter of utmost concern, and suitable measures must be devised to obliterate its impact.

24.3.1.4 Surface charge annihilation techniques

Charged-particle emission methods: One group of methods involves emission of charged particles from the surface of the spacecraft, viz. electrons, ions, and plasma.

(i) Electron emission: It can be stimulated from a surface by elevating its temperature; this method is called thermionic electron emission. It can also be induced by applying a high electrostatic field, which is known as field electron emission. Figure 24.19 conceptually shows how these phenomena can be applied to a spacecraft for electron removal.

Electron emission acts as a direct approach towards getting rid of excess electrons from the surface of a spacecraft. It is useful for metallic surfaces only. When the surface is a dielectric material, it will not work.

(ii) Low-energy ion emission: Ion emission is the release of ions from the surface of a solid. For ion emission, the force restraining the ion to the surface is overcome by heating (thermionic ion emission). Another way is to bombard the surface with energetic particles such as ions (ion–ion

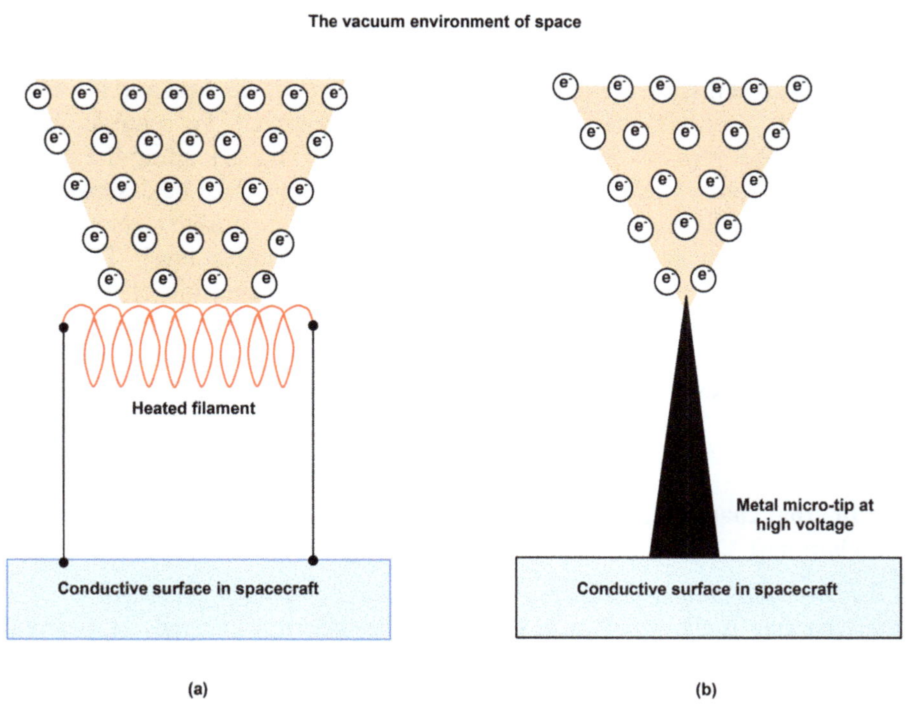

Figure 24.19. Excess electron removal from a spacecraft surface by: (a) thermionic emission from a heated filament and (b) high-field electron emission from a metal micro-tip at a high applied voltage.

emission) or electrons (electron–ion emission). The method easily applicable is chosen.

When low-energy positive ions are liberated from the negatively charged surface of the spacecraft, the returning ions are attracted towards the portions on the surface which have the highest electron densities, and so are most negatively charged (Lai 1989). In this way, potentials at different spots on the spacecraft surface are neutralized to counter differential charging.

(iii) Low-energy plasma emission: A plasma is produced by applying an electric and/or magnetic field across a dielectric gas, e.g., in a discharge tube. In a DC plasma, the gas is ionized. The applied electric field pulls the electrons towards the anode and positive ions move towards the cathode. Collisions between electrons and neutral gas atoms rapidly increase the number of charged particles in the plasma.

Surface coatings: Another group of methods relies on protection mechanisms based on covering the surface of the spacecraft with different types of coatings:

- Applying a partially conducting paint or a metallic wire mesh on the surface: To avoid creation of different potential points on the surface, an insulating surface is covered with a partially conducting coating, e.g., indium oxide zinc orthotitanate, alodyne, etc. A conductive wire mesh may also be fixed over the surface. The coating is better than the mesh because potential differences may be produced between the wires and surface. Nonetheless, the mesh may sometimes be more convenient.

  Plasma emission method synergizes the effects of electron and ion emission methods so that they act co-operatively together for charge removal. However, it requires a container containing neutral gas loading the spacecraft with additional wight. Further, control mechanisms must be designed to start a plasma shower when the potential reaches an intolerable level and end the cycle as soon as the potential reaches an acceptable level. These efforts introduce complexity and inconvenience in implementation of this scheme.
- Covering the spacecraft surface with a material with high secondary electron emission yield: This approach is effective when the primary electron energy is less than 1 keV. Above 1 keV, secondary electron emission decreases and becomes <1. Then the coating can no longer protect against charging.

24.3.2 Internal charging (deep dielectric charging or bulk charging or buried charging)

24.3.2.1 Internal charging causes

Internal charging is the build-up of charges inside dielectric materials or ungrounded metals enclosed by the spacecraft by electrons and ions penetrating through the shielding material inside the spacecraft, usually happening for electron energies in MeV range or above (Lai *et al* 2018). These high-energy electrons manage to reach great depths inside the dielectrics. Depending on the resistivity of the dielectric, they reside there for protracted time spans of hours, days and even months (figure 24.20). They produce internal electric fields. In figure 24.21, effects of penetration of both ions and electrons are examined.

Since internal electrostatic discharge is also possible through ungrounded conductors, charging threats do not take place in dielectrics only. So, the generic term 'internal charging' is more appropriate than 'deep dielectric charging' (NASA 2017).

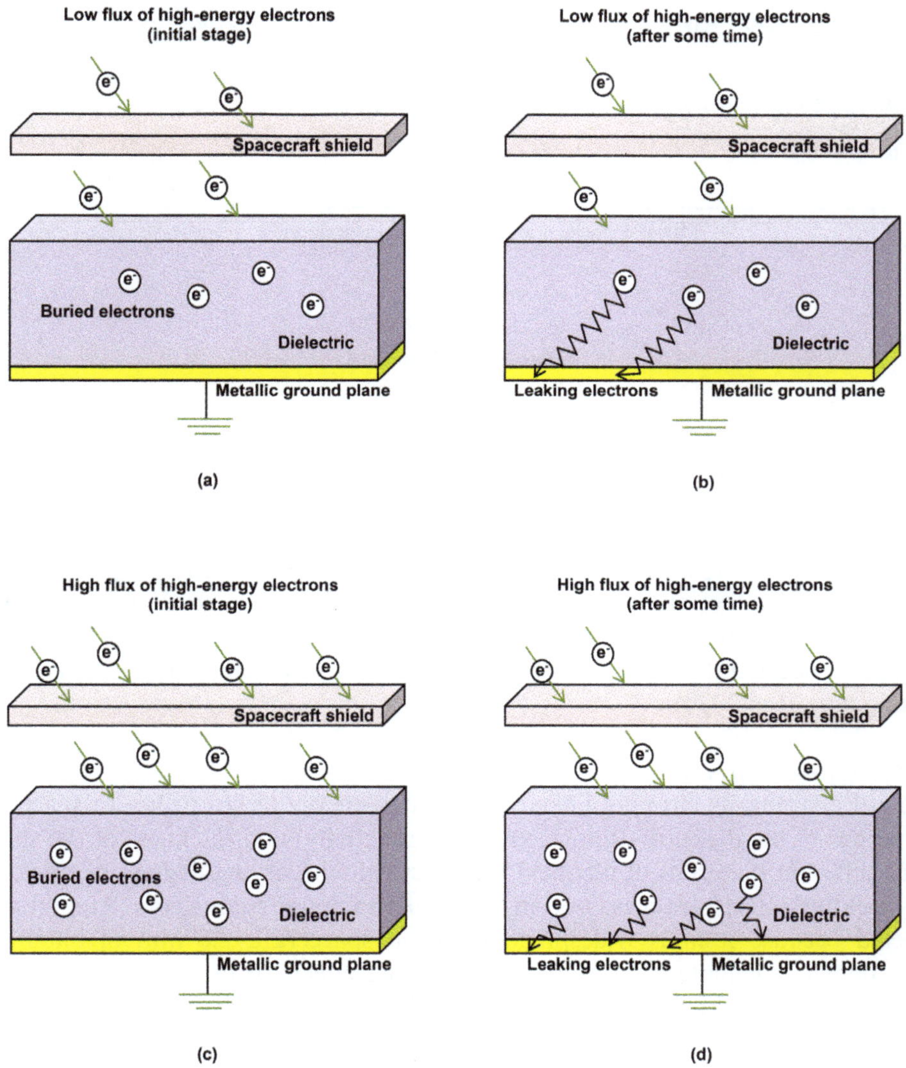

Figure 24.20. Deep dielectric charging with high-energy electrons showing a dielectric layer with grounded metal plane: (a) and (b) low flux of electrons in the initial stage when the electrons are buried in the dielectric and after some time when a fraction of the electrons manage to leak out to ground; (c) and (d) high flux of electrons in the beginning stage when the electrons get buried in the dielectric and after some time has elapsed with few electrons leaking out; (e) and (f) very high flux of electrons in the starting and later stages. In (f) many more electrons are incident than those able to leak out so that a huge electron population builds up. The spatial variation of electron density sets up a large electric field able to trigger an electrical discharge causing damage to the dielectric.

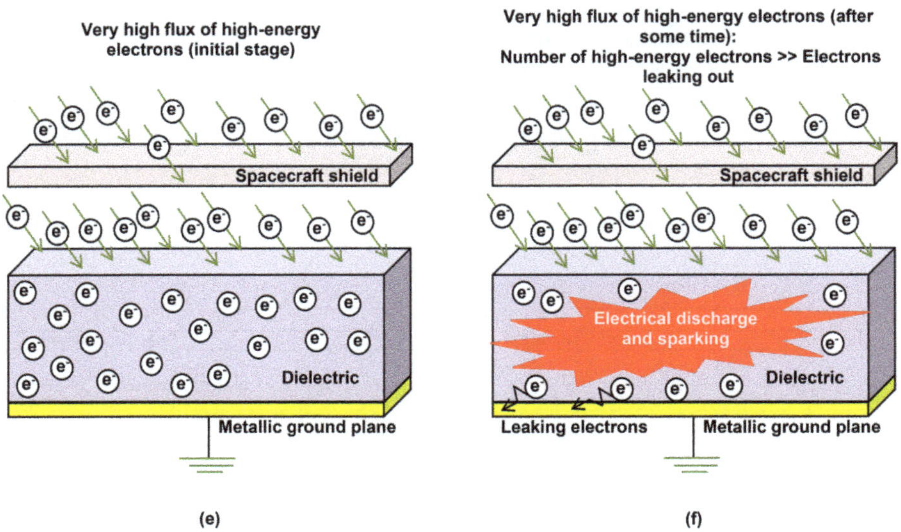

Figure 24.20. (Continued.)

24.3.2.2 Internal charging effects

Like surface charging, internal charging too can lead to spacecraft failures and anomalies. When the charge accumulation has reached a critical stage at which the resulting electric field exceeds the breakdown strength of the dielectric, arc discharge takes place. Such electrostatic discharges degrade the properties of the dielectric and affect signal measurements. Twenty five percent of the reported spacecraft anomalies and debacles arise from internal charging effects. So internal charging necessitates careful consideration of electronic devices in space (Xiangqian *et al* 2020).

24.3.2.3 Shielding the electron flux

Internal charging is strongly linked to the high-energy electron flux in space, the properties of the dielectric (thickness and conductivity) and thickness of the shield. Since different materials of identical thickness exhibit varying degrees of shielding, the shielding effects are measured in terms of aluminum equivalents. Aluminum is selected because it finds widespread usage in spacecraft. 2.8 mm of aluminum equivalent is adequate to stop internal charging effects in GEO (geosynchronous equatorial orbit). Empirically it has been found that a perpendicularly striking electron flux lower than 0.1 pA cm^{-2} is unable to cause any ill effects, which is therefore suggested as a safe flux (Xiangqian *et al* 2017).

Dielectrics such as PTFE (polytetrafluoroethylene) and FR-4 (flame retardant glass-reinforced epoxy laminate material) differ in their electrical conductivities. The higher the conductivity, the lower is the probability of discharge. So, shielding thickness required to decrease the risk of charging varies with the dielectric, which should be kept in mind by spacecraft engineers (Xiangqian *et al* 2020).

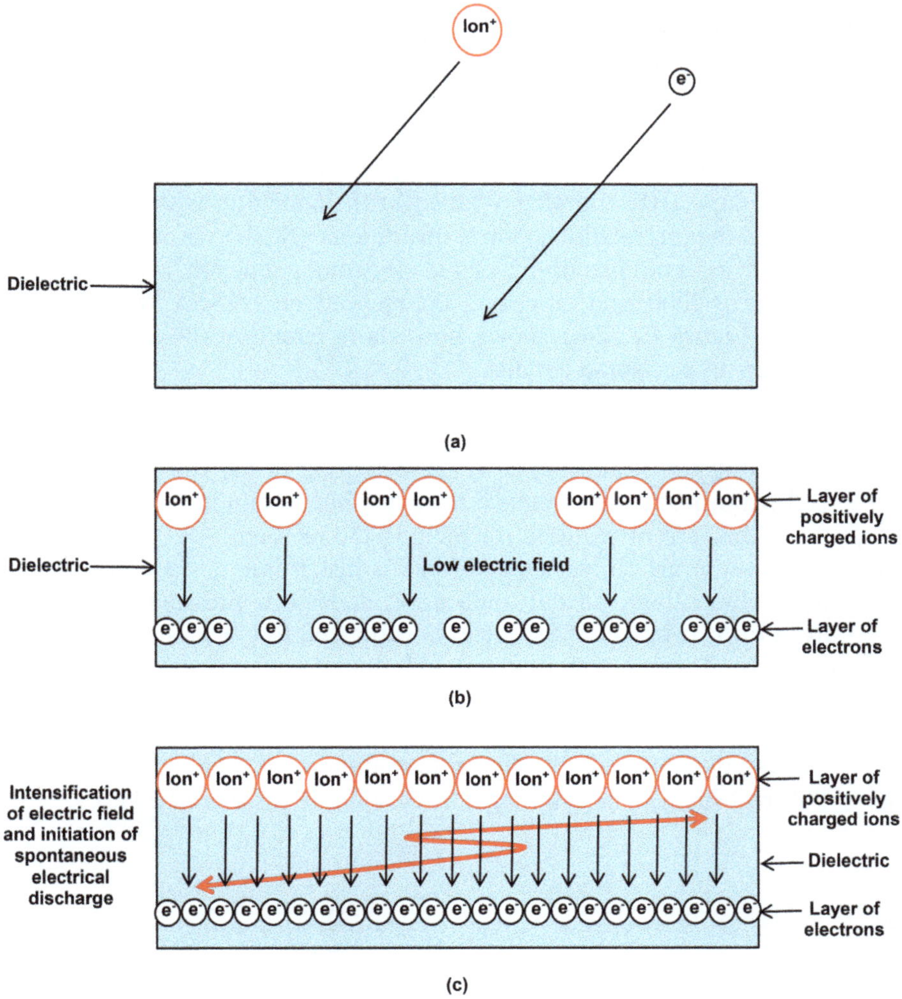

Figure 24.21. Penetration of positive ions and electrons in a dielectric: (a) initiation of the penetration, (b) penetration due to a small flux of ions and electrons, and (c) penetration with a large flux of ions and electrons. The ions being heavier and slower than electrons, their penetration is restricted to shallow depths in the dielectric. The electrons being lighter and faster particles can reach deep down in the dielectric. So, a layer of positively charged ions builds up at a small depth below the surface of the dielectric. Similarly, a layer of electrons is formed at a large depth under the dielectric surface. Due to these positively and negatively charged layers, an electric field is established. In (b) this field is moderate, while in (c) it becomes strong enough to activate a discharge.

24.4 The challenge of tin whisker growth

24.4.1 Tin whiskers

Tin whiskers were fleetingly discussed in section 17.6. These are hair-like, spider-web-thin, electrically conductive crystalline protrusions from pure tin, cadmium, or zinc surfaces (Zhao *et al* 2017). They can be described as spontaneous filamentary metallurgical outgrowths.

24.4.2 Risks to electronic circuits

(i) Short circuits: Thin strands sprouting from metal finishes often break away from the base metal causing intermittent or permanent electrical short circuits by bridging closely-spaced circuit components at different voltages, e.g. the non-NASA on-orbit commercial satellites Solidaridad 1, Galaxy VII, Galaxy IIIR completely failed in their mission with loss of primary and redundant satellite control processors (SCPs) through tin whisker-induced short circuits on tin-plated electromagnetic relays in August 2000, November 2000, and January 2006, respectively (NASA Whisker Failures 2009). Figure 24.22(a) shows how short circuits can be caused by tin whiskers in electronic circuits.

The created short circuits are either:

(a) Permanent short circuits or glitches if the current flowing in the whisker is inadequate to cause melting of the whisker, or

(b) transient short circuits if the available current in the whisker exceeds the current required for melting the whisker.

(ii) Metal vapor arcs in vacuum: When a whisker initiates a short circuit in a high-voltage, high-current application under low pressure, a metal vapor arc is formed in which the metal is vaporized into a plasma of conductive tin ions. This plasma can carry hundreds of amperes of current. A destructive arc is sustained by consuming tin from the surrounding plated area (figure 24.22(b)). This arc can persist for a long duration until consumption of available tin or termination by a circuit protection fuse or circuit breaker or external extinguishing (Brusse *et al* 2002).

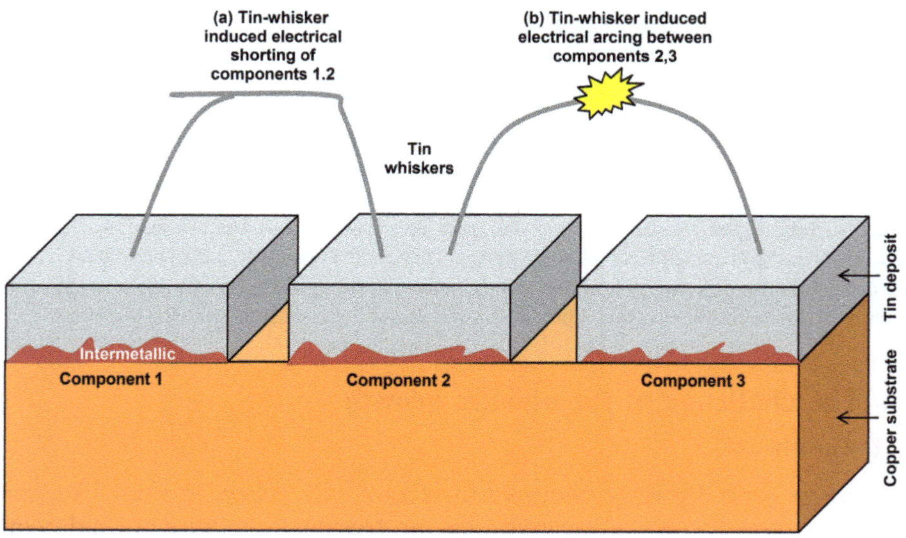

Figure 24.22. Effects of tin whiskers on electronics: (a) creation of an electrical short circuit and (b) production of arcing between two tin whisker ends.

(iii) Whisker debris: Whiskers can break away from their parent substrate. The scattered whisker debris creates mechanical issues in microelectromechanical systems and optical devices and slip rings.

24.4.3 Theories of whisker growth

The whiskers are supposed to grow as a mechanism of relief of residual compressive stresses within electroplated tin films introduced:

(i) by plating chemistry, or
(ii) via formation of intermetallics during diffusion of substrate into tin (figure 24.23) or diffusion in the opposite direction tending to change the lattice spacing in the plating, or
(iii) by mechanical stresses during bending, stretching, tightening of screws, or
(iv) through scratches or nicks in the plating, or
(v) by mismatching of thermal expansion coefficients of the plating and substrate under temperature variations.

24.4.4 Methods to reduce whisker growth

This can be accomplished:

(i) By avoiding tin-plated components wherever alternatives exist.
(ii) By adding lead to the tin coating, an unappealing method because lead must be avoided for RoHS (restriction of hazardous substances)-compliance.
(iii) By replating areas susceptible to whisker formation with nickel or tin/lead.
(iv) By using reflowed tin (reheating the tin-electroplated component above the melting point of tin, 231.9 °C and cooling it to release internal stresses).

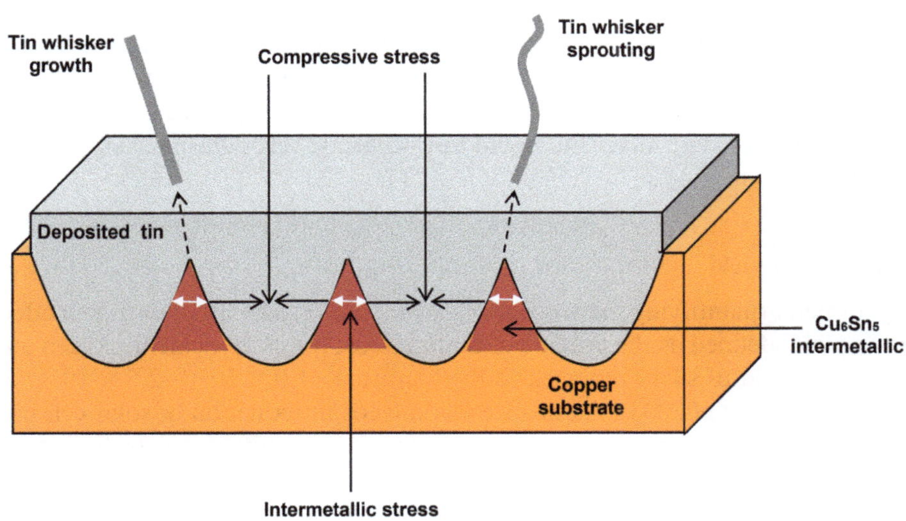

Figure 24.23. Mechanism of tin whisker growth. Cu_6Sn_5 intermetallic formed from 6Cu + 5Sn occupies more volume than 6Cu + 5Sn producing compressive stress at the Cu_6Sn_5 interface with Sn.

(v) By using hot tin dip coating (immersion of the component into a bath of pure molten tin at a temperature >232 °C). This method completely reflows the tin plating but success is limited by the fact that it is often difficult to reflow the entire surface.

(vi) By conformally coating all whisker prone surfaces or encapsulating with foam to prevent short circuits. Conformal coating/foam restrain arcing by restricting the availability of tin. Coating materials used are urethane and urethane acrylate. Acrylic, silicone, and parylene-C are also applied. 50–75 µm thick arathane 5750-A/B LV, a two-part, translucent urethane coating, is used to insulate electronic components and PCBs. The type of foam and foam density decide its effectiveness.

(vii) By interposing physical barriers, e.g., insulating washers, ceramic spacers, etc.

Figure 24.24 shows how chances of electrical shorting are reduced by conformally coating electronic circuits.

24.5 The challenge of erosion of spacecraft materials by atomic oxygen

24.5.1 Crippling effects of atomic oxygen on space missions

Atomic oxygen, a principal neutral species in low earth orbits (LEOs) environment is known to jeopardize space missions by damaging spacecraft materials (Verker *et al* 2020). A communication satellite, Intelsat VI, was abandoned on March 14, 1990, in LEO due to failure of proper separation of third stage. Investigations attributed the failure to degradation of silver foil interconnects in the solar cell array of the satellite that were exposed to atomic oxygen flux (Koontz *et al* 1991). From study of the effects of atomic oxygen on boron nitride, silicon nitride and solar cell interconnects in Intelsat VI satellite in LEOs, it was found that silver oxidizes to form high-resistance silver oxide. Atomic oxygen permeates through boron nitride coated on silver but could not penetrate through silver nitride (Cross *et al* 1991).

24.5.2 Erosion yield

The parameter quantifying the susceptibility of a material to atomic oxygen effect is 'erosion yield' defined as the volume of material lost for each incident oxygen atom. The erosion yield of silver reported in the literature ~1.0×10^{-24}–2.5×10^{-23} cm^3/atom (Gouzman *et al* 2003). As silver erosion leads to reduction of solar cell power output and premature termination of the space mission, suitable protective coatings for silver are necessary. Sol–gel method was applied for depositing SiO_2 coatings on silver foils. The erosion yield of the coatings tested by ground-based AO simulator was 2×10^{-26} cm^3/atom, which was significantly lower than that of the uncoated foils (Duo *et al* 2012).

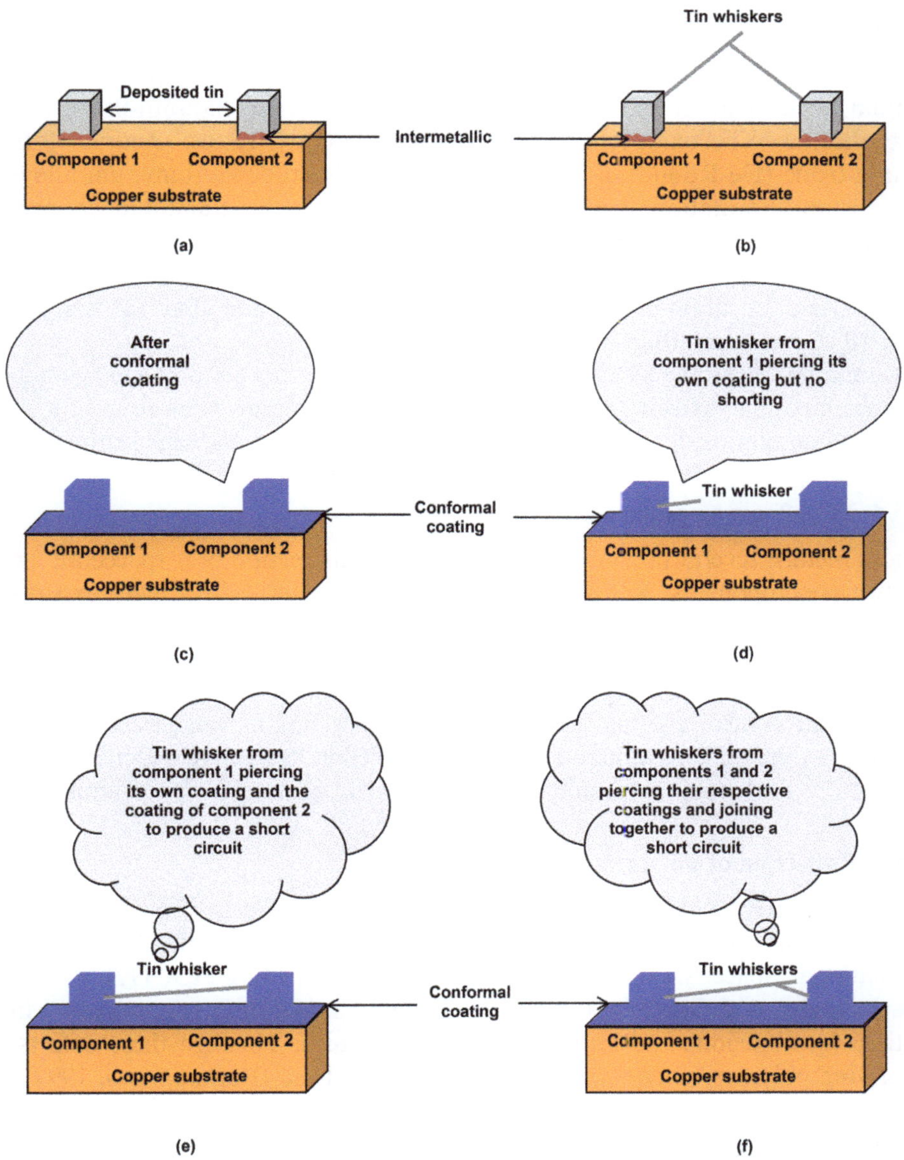

Figure 24.24. Prevention of tin whisker shorting by conformal coating of electronic components: (a) non-coated components, (b) electrical short circuit produced by tin whiskers ejecting from two nearby components 1 and 2, and joining together, (c) coated components, (d) a tin whisker from a coated component breaking through its coating and coming out but still not long enough to cause shorting between components 1 and 2, (e) shorting produced by a tin whisker ejecting from component 1 and growing sufficiently long to pierce the coating of the component 2, and (f) shorting caused by a tin whisker from component 1 piercing its coating and connecting with another tin whisker which has come out of its coating of component 2. Chances (e) and (f) are considerably less than when the circuits were left uncoated as in (b) because coatings of both components 1 and 2 must be pierced through. When coating of component 1 or component 2 only is pierced, there will be no shorting. This shows the effectiveness of coating in mitigating this kind of shorts.

24.5.3 AO effects on metals

No macroscopic changes are observed in metals with the exception of silver and osmium, which cannot be used in uncoated form. Silver oxidation follows a linear parabolic law. The oxide film breaks due to thermal stresses. Osmium forms a volatile oxide (De Rooij 2010). Aluminum is slowly eroded. Aluminum alloys are AO-resistant. Formation of an adherent oxide layer protects aluminum, copper, and steel. Gold and platinum display high resistance to corrosion by atomic oxygen. Thin Au films are deposited on exposed surfaces for protection against AO. The erosion yield of titanium is 3.9×10^{-27} cm^3/atom while that for tungsten is 4.4×10^{-28} cm^3/atom (Samwel 2014).

Aluminum, silver and silver alloys used in solar reflectors are protected with glassy ceramics. Erosion yields in the range $0.4–2.3 \times 10^{-28}$ cm^3/atom are found for silicon oxides, aluminum oxides and magnesium fluoride (Dooling and Finckenor 1999).

24.5.4 AO effects on polymers

Besides metals, several polymers are used in the construction of a spacecraft and its electronics components. Various materials such as material in a flight experiment, material from a repair mission or material in a laboratory-simulated LEO environment, were exposed to atomic oxygen (Dooling and Finckenor 1999). These embraced a wide selection including organic materials, metallic, non-metallic, optical and thermal control coatings. Polymers and other organic materials were found to be most vulnerable to atomic oxygen effects. Teflon has an AO reaction efficiency of 5×10^{-26} cm^3/atom. Kapton has an erosion yield of 3×10^{-24} cm^3/atom.

24.5.5 Protection of polymers

Sputter-deposited metal oxide coatings, e.g., aluminum oxide, silicon dioxide etc are used for protecting polymers against AO attack. Solar array blankets on the International Space Station were coated with 130 nm thick SiO_x ($1.9 < x < 2.0$) film (Rutledge and Olle 1993). Scratches or pin window defects are created in the coatings by abrasions and impacts of micrometeroids. Through these defects, the underlying polymer, silicone and fiberglass, are oxidized (Forkapa *et al* 1999).

24.5.6 AO effects on glasses and thermal coatings

Most glasses used as optical materials are AO resistant. Long-term exposure of thermal coatings with organic binders must be avoided. After degradation of binders, pigment particles are left behind. These act as contaminants (Dooling and Finckenor 1999).

24.6 The challenge of radiation showers

24.6.1 Inapplicability of common shielding practices to electronics in space

The highly penetrating gamma and x-rays are stopped by building barriers of lead, concrete, or water. Radioactive materials are stored under water or kept in rooms with walls lined with lead or concrete. The heavy traditional radiation shields cannot be carried in space where weight reduction is an essential prerequisite. So, when thinking about electronics shielding in space flights, the Earth-based practices in shielding need to be revised in view of constraints imposed by requirements of being light in weight, small in volume and effective in cost. One cannot think of constructing a heavy concrete wall in a spacecraft.

24.6.2 Gamma ray shielding materials

Gamma ray shielding is done using materials containing high atomic number Z and high volumetric mass density elements (De Vanzo and Hayes 2020). Lead, tungsten, cast iron, steel and concrete are commonly used.

24.6.3 Neutron radiation shielding materials

Opposite to gamma ray shielding, materials for neutron shielding contain low atomic number elements to slow down the neutrons. Hydrogen being highly efficient against neutrons, neutron shielding materials such as polymers, composites, etc, invariably contain hydrogen. Low atomic number elements do not act as shields against electromagnetic radiation. This happens because the probability of interaction of the material with photons is proportional to Z^h where $h \geqslant 2$.

24.6.4 Adapting conformal coatings for shielding electronics in space

24.6.4.1 Protection from neutrons by hydrogenous component of coatings

A large number of low atomic number coatings known as conformal coatings are used to protect electronics from humidity, temperature and vibration. These conformal coatings contain elements such as carbon, hydrogen and oxygen. Neutrons can easily transfer their energy to the hydrogenous component of the conformal coatings. So, these low Z coatings are useful as shields against neutrons.

24.6.4.2 Modification of coatings for protection from gamma rays

From the earlier remark that gamma ray shielding requires high atomic number elements, it is evident that the conformal coatings can be adopted for gamma radiation shielding by incorporating high atomic number elements in their structure. Thus, the polymer matrix of conformal coating, already ready for neutron protection, is modified to include protection against gamma rays.

24.6.4.3 Strategies devised for protection from (neutron + gamma) mixed radiation environments

How to combine the low Z and high Z elements in a common matrix? It has been found that metal oxide polymer composites can be employed for this purpose. The polymer, as we already know, is effective against neutrons. What metal oxides can be used? High thermal conductivity metal oxides such as tungsten oxide and gadolinium oxide are suitable for this task. Additionally, they also improve the mechanical properties of the composite.

24.6.4.4 The extra benefit of lead avoidance

The most advantageous feature of polymer composites containing metal oxides is the elimination of lead, much dreaded because of its toxicity. At low energies, tungsten, bismuth, and gadolinium oxides have been found to be comparable to lead in reduction of x-ray exposure.

24.6.4.5 Shielding effectiveness dependence on metal oxide percentage, and compromising solutions

Expectedly, the gamma ray shielding effectiveness increases with the percentage of metal oxide particles included in the polymer matrix. However, a high metal oxide content is accompanied by a decrease in neutron shielding effectiveness. A trade-off between gamma and neutron shielding effectiveness may be necessary.

24.6.4.6 Acrylics as preferred materials for coatings

As acrylic conformal coatings are easy to apply, inclusion of metal oxide particulate in these coatings is a viable way to provide shielding against gamma radiation as well as neutrons for electronic systems onboard a spacecraft.

24.6.4.7 Modeling of composites

De Vanzo and Hayes (2020) performed modeling of metal oxide impregnated acrylic conformal coatings on silicon-based electronics. Their simulations showed that the gamma radiation decreases by 300 times at energies <10 keV. Neutron displacement damage falls up to 225% across the Watt spectrum. Watt distribution function is a mathematical function representing the energy distribution of fission neutrons released when a heavy nucleus, e.g., 236U produced by 235U + n (neutron), undergoes fission.

24.7 The challenge of outgassing in vacuum environment of space

24.7.1 Outgassing sources and mechanisms

Several materials used in the construction of a spacecraft are outgassing sources, to a greater or smaller degree. Adhesives, tapes, plastics, epoxies, potting compounds, and lubricants are the likely sources of outgassing.

Outgassing is the release of gases or vapors from a solid (figure 24.25). These are the gases which were absorbed, dissolved or trapped inside the solid. The extent of outgassing is a function of temperature and vacuum level besides the material properties. One mechanism of outgassing is thermal desorption which is detachment of molecules from the surface on receiving energy greater than the energy binding them to the surface. Another mechanism is sublimation which is the direct conversion of a material from solid to gaseous phase without passing through the liquid phase.

24.7.2 Effects of outgassing

Outgassing affects the operation of the spacecraft in two ways. First, the outgassing material itself may suffer from changes in its properties so that it is unable to

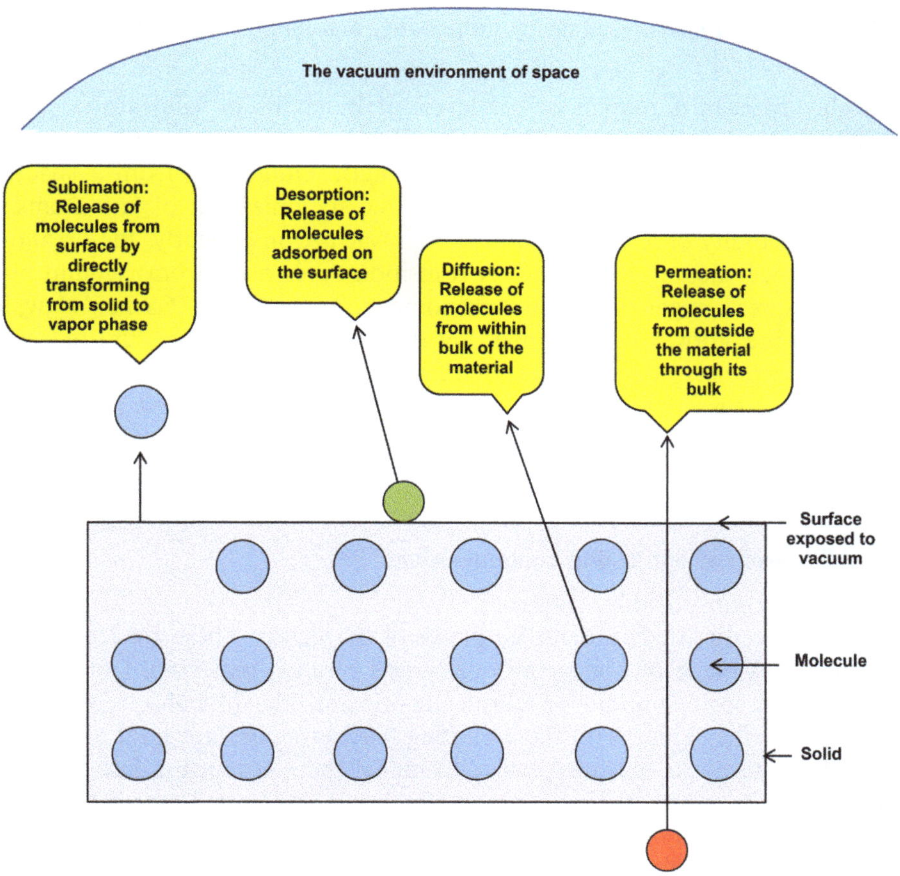

Figure 24.25. Main sources contributing to outgassing from a surface in vacuum.

perform the intended function. Second, the gases or vapors liberated during outgassing contaminate the spacecraft environment. So, the spacecraft is contaminated by the outgassing of its own materials in space.

In a spacecraft, optical lenses and mirrors, laser guns, solar cell arrays, sensors and thermal control coatings are sufferers of outgassing-induced molecular contamination. Cryogenic systems will also malfunction. The outgassed materials from a location may deposit on adjoining electronic and optical components impacting their operation, e.g., solid particles condensing on the solar panel may reduce its transparency thereby blocking sunlight (Jiao *et al* 2019). Photovoltaic efficiency of the solar panel will decline. The charge-coupled device sensor of an onboard camera may be obscured by the condensate impairing its picture quality. Sensing and imaging performance worsens. With degradation of thermal coatings, the control on absorptivity and emissivity of sunlight is poor. So, the temperature of the spacecraft varies affecting temperature-sensitive electronics and instruments. Insulating or conducting films may deposit on electronic devices causing discharges or short circuits.

24.7.3 Lowering of space vacuum by outgassing, and hampering of high-voltage operations

Present-day spacecraft require power levels of hundreds of kilowatts. Operating potentials range up to hundreds of kilovolts (Huang *et al* 2015). A perfect vacuum is the ideal insulator with highest dielectric strength. Outgassing from a spacecraft surface lowers the vacuum level in space. Due to ionization of gas atoms, the insulating capability of space vacuum is diminished. Consequently, the breakdown voltage of space vacuum is lowered. A transition takes place from vacuum breakdown to low-pressure gas breakdown phenomena. According to Paschen's law, low-pressure gas breakdown is determined by the product of the pressure of gas, the spacing between electrodes and the condition of electrode surface. In high-voltage solar arrays, outgassing initiates plasma close to the metal–dielectric junctions, and leads to inception of arcing. Delimiting the influence of outgassing from spacecraft materials on the lowering of space vacuum is therefore mandatory.

24.7.4 Alleviation of outgassing contamination

24.7.4.1 Molecular adsorber pucks

Adsorber pucks are fitted at identified places of spacecraft. These are zeolite-coated cordierite pucks made of honeycomb patterned ceramic materials. Cordierite is a blue magnesium iron aluminium cyclosilicate mineral. The difficulty faced in the use of pucks is that they along with the mounting hardware consume a lot of spacecraft space. While the pucks are inexpensive to make, their supporting fixtures are not cost-effective. Moreover, they add to the total and distributed mass of the spacecraft. Hence, an easily-installable alternative occupying small space and of less weight is needed, which is fulfilled by molecular adsorber coating. This coating is developed at Goddard Space Flight Center (GSFC) of NASA.

24.7.4.2 Molecular adsorber coating (MAC)

MAC is better than pucks from size, weight, and cost viewpoints. It is a sprayable paint which does not require any supporting hardware. This coating is a mixture of zerolite with an inorganic binder material Ludox® colloidal silica, a dispersion of silica particles in water with a slender particle size distribution in the low nanometer size range (Grace 2022).

The MAC may comprise zeolite (0.2–0.4 mass fraction), binder (0.4–0.8 mass fraction) and water (0.0–0.3 mass fraction) (Straka *et al* 2012). The zeolite is added to the binder with continuous stirring until it acquires a homogeneous milky appearance. A requisite quantity of distilled water is poured for adjusting viscosity. The MAC is taken in a spray gun. A primer based on zinc oxide/potassium silicate is applied to the surface to be coated. This primer acts as an intermediary to improve adhesion between the surface and MAC. The MAC is sprayed on the primer covered surface followed by drying in air and baking at 60 °C–120 °C for 2 h. For preventing the on-orbit molecular contamination from affecting electronics, inside/outside of all electronic boxes are completely protected with MAC. Coating is also applied on the internal surfaces of instruments and baffles as well as internal structural walls of spacecraft buses.

The coating is highly permeable and porous. In this coating, zerolite is a white pigment. The Ludox® binder is transparent. Zeolite is an aluminosilicate mineral with the general formula $M_xAl_xSi_{1-x}O_2{\cdot}yH_2O$ where M = metal or H^+ ion, $0<x<1$, y is the number of water molecules. Zeolite belongs to the group of solids called molecular sieves. The cavities in its nanotextured structure are the sites where the contaminant molecules are captured and trapped. The large surface-area-to-volume ratio of zeolite maximizes trapping efficiency (Abraham *et al* 2012).

Ground investigations have been carried out on MAC in simulated spaceflight conditions. These conditions represent the vacuum environment in space and the planet Mars. MAC has been found to be effective in absorption of high molecular weight plasticizers such as DOP (dioctyl phthalate) (Abraham and Jallice 2018).

24.7.4.3 Black MAC

The aforementioned sprayable paint contains inorganic white materials made from zeolite. Some applications require low reflectivity properties, which are achieved with black coatings. Such black coatings are used on interior surfaces in a spacecraft for controlling stray light. The stray light reduces image quality by impairing the contrast, and increases background noise. Black coatings are also used for absorbing light within optical paths between cavities of cameras, telescopes and other systems. For these applications, molecular absorption coatings have been developed using black pigments providing similar adsorptive properties to the white coatings (Abraham *et al* 2014).

24.8 Discussion and conclusions

Imagine the operation of electronics in the serene and friendly terrestrial atmosphere. Contemplate the operation of the same electronics in the hazardous space environment. There is an enormous difference between the two environments. This difference needs to be emphasized, appreciated and taken into account when sending electronics into space where there are little or no chances of repair and maintenance.

Review exercises

24.1. Why is failure of electronic equipment onboard a space craft equivalent to failure of the mission?

24.2. What are decibel levels of rocket noise and normal sound audible to human ears? At what intensity on the decibel scale does sound harm human hearing?

24.3. Name the five sources of noise in space vehicles. What is meant by pyrotechnic shock? What is its impact on electronic components?

24.4. What type of mechanical deformation is observed in PCBs due to vibrations? What is its main cause?

24.5. What are the reasons for brittle fracture and warpage produced on vibration of a PCB?

24.6. What is the foremost factor responsible for fatigue and thermal failure of PCBs on vibration?
24.7. Why is use of copper-clad laminates preferred?
24.8. Why should solder masks not be used in PCBs for space applications?
24.9. What is meant by eyelet and standoff in a PCB?
24.10. Why is heat-carrying capacity of conductor lower in space than on Earth? How is the lower heat capacity taken into account during choosing components for PCBs?
24.11. What are through-hole, buried and blind vias in a PCB?
24.12. What is the advantage of a blind via? What are its drawbacks?
24.13. What is a conformal coating? Explain electrostatic discharge and Paschen's discharge.
24.14. What are the two techniques applied to control vibrations?
24.15. Distinguish between foam and rubber in reference to control of vibrations.
24.16. What is Sorbothane? What are its merits to weaken vibrations?
24.17. What is the temperature of space in Universe? How much is the difference in the temperature on lunar surface just before dawn and in the afternoon?
24.18. If thermal control is not done, what will be the temperatures of an orbiting space station on the sun-facing side and dark side?
24.19. Within what ranges should the temperature be maintained during a space flight for normal functioning of electronics and batteries?
24.20. What is the difference between black paint and white paint regarding absorptivity and emissivity of sunlight? For which surfaces of a spacecraft is black paint better for heat exchange and for which surface is white paint superior?
24.21. What is multilayer insulation? How does it provide a heat impervious system?
24.22. What is a thermal strap? Explain its use.
24.23. How does a thermal louver produce a shield by which heat losses from a radiating surface are decreased?
24.24. Explain with a diagram the function of a deployable radiator. How does it change its role from a solar absorber to a solar radiator?
24.25. Draw the diagram of a heat pipe and explain how it transports heat between two locations?
24.26. In what form does a phase-change material store thermal energy? How is it used to build a thermal storage unit?
24.27. What is a thermal switch? Draw the diagram of a two-phase thermal switch and explain its operation.
24.28. In what respects are passive thermal control systems more suitable for spacecraft than active thermal control systems?
24.29. Differentiate between patch and cartridge heaters used in a spacecraft.
24.30. What is a cryocooler? Name the five types of cryocoolers.

24.31. Write the names of the cryocooler which work on: (a) recuperative cycles and (b) regenerative cycles.
24.32. What is the role of heat exchanger in a cryocooler? What is the heat exchanger for recuperative cycles called? By what name is the heat exchanger for regenerative cycles addressed?
24.33. How many channels does a recuperator have? How many channels are there in a regenerator?
24.34. How does low-temperature expansion take place in a Joule–Thomson cycle?
24.35. What is the typical frequency for: (a) Gifford–McMahon (GM) cryocoolers and (b) Stirling cryocoolers? For attaining what temperatures are these cryocoolers used?
24.36. Explain the working of: (a) recuperative cycle cryocoolers, and (b) regenerative-cycle cryocoolers.
24.37. On what principle does a thermoelectric cooler work? Does it require a refrigerant for cooling?
24.38. What is the difference between Peltier and Seebeck effects?
24.39. Can a thermoelectric cooler be used as a thermoelectric generator?
24.40. In a pumped fluid loop, will it be correct to say that the connection between temperature-controlled components and heat radiative surface is radiative? Is such a system useful for a small spacecraft?
24.41. What are the two headings under which spacecraft charging is subdivided? How does absolute surface charging of a spacecraft differ from differential surface charging?
24.42. Why is the surface of spacecraft negatively charged with respect to the surrounding plasma environment?
24.43. Electrons are released from the sunlit side of the spacecraft by photoelectric emission. So, this side should become positively charged. Why does this side remain negatively charged?
24.44. How do the electrons on the shaded side of a spacecraft create an electric field which wraps around to the sunlit side to produce a barrier to photoemitted electrons from this side preventing them from escaping? Explain why the sunlit and shaded sides are at different potentials of the same polarity?
24.45. What are the undesirable effects of spacecraft charging on electronics? Why is it necessary to maintain a spacecraft anomaly database?
24.46. What are names of the phenomena in which electrons are released from a surface by: (a) heating and (b) applying an electric field?
24.47. What is the direct method of getting rid of excess electrons from the surface of a spacecraft? Will this method work for a dielectric surface?
24.48. Explain the meaning of the following terms: (a) thermionic ion emission, (b) ion–ion emission, (c) electron–ion emission and (d) photoemission.
24.49. How are the potentials at different points on the surface of a spacecraft equalized during release of low-energy ions?

24.50. What are the practical difficulties faced in utilizing low-energy plasma emission for removing the excess charge from the surface of a spacecraft?
24.51. Plasma emission is a combination of electron emission with ion emission. Explain.
24.52. How does a partially conductive surface coating or mesh of wire help in avoiding spacecraft charging?
24.53. Why is the application of a surface coating on a spacecraft ineffective in preventing charging when the primary electron energy exceeds 1 keV.
24.54. At what range of incident electron energies does internal charging of a spacecraft become worrisome?
24.55. How is internal charging of a spacecraft caused? Is it restricted to dielectrics only? Which term is more correct: internal charging or deep dielectric charging?
24.56. How does internal charging cause spacecraft failures and anomalies?
24.57. Can internal charging in a spacecraft cause an arc discharge?
24.58. Why are shielding effects to electron flux measured in terms of aluminum equivalents? What magnitude of electron flux is considered a safe limit?
24.59. Is shielding thickness required to decrease the risk of charging the same for all dielectrics? If not, why?
24.60. Name three metals from the surfaces of which tin whiskers grow? How do these whiskers cause short circuits?
24.61. How do tin whiskers set up metal vapor arcs in vacuum.? What is whisker debris? What issues does it create?
24.62. What are the suggested mechanisms of whisker growth?
24.63. How does reflowed tin or hot tin dip coating help to reduce tin whisker growth?
24.64. Give some examples of conformal coatings/foam materials used to restrain whisker effects.
24.65. What metals are used during replating areas of circuits for protection against tin whisker growth? Is it advisable to add lead to tin coating for reducing whiskers?
24.66. How are silver foil interconnects degraded by atomic oxygen? Can atomic oxygen penetrate to reach silver when it is coated with: (a) boron nitride, (b) silver nitride?
24.67. Define erosion yield of a material. What is its unit?
24.68. How does erosion of silver cause the performance of solar cells to deteriorate? Can sol–gel SiO_2 coating protect solar cells again erosion by atomic oxygen? Compare its erosion yield with that of silver?
24.69. What is the effect of atomic oxygen on osmium? Can it be used in uncoated form?
24.70. How are aluminum, copper and steel affected by atomic oxygen?
24.71. Name two metals which are highly resistant to corrosion by atomic oxygen?

24.72. Which of these two metals have a smaller erosion yield: (a) titanium, (b) tungsten?
24.73. Can deposition of gold film on exposed metal surfaces protect them from atomic oxygen?
24.74. Are polymers and organic materials vulnerable to atomic oxygen effects? Is the use of sputter-deposited metal oxide coatings helpful in protection? What happens when micrometeoroids strike against the coatings in space?
24.75. What are the effects of atomic oxygen on glasses and thermal coatings?
24.76. Can a wall of concrete stop gamma and x-rays? Can water be used as a barrier to these radiations?
24.77. Why are radiation shielding practices used on Earth not adoptable in a spacecraft?
24.78. What are the conflicting requirements of materials effective in shielding gamma rays and neutron radiation?
24.79. Gamma ray shielding uses low atomic number and low-density materials. Right or wrong?
24.80. Neutron radiation shielding requires low atomic number materials. Correct or incorrect?
24.81. Do low atomic number materials act as shields against electromagnetic radiation? Yes or no?
24.82. Can conformal coatings containing carbon, hydrogen and oxygen be used to shield electronics against neutrons?
24.83. Which component of a conformal coating (carbon, hydrogen, or oxygen) provides the neutron shielding capability to it?
24.84. What modification is necessary in a conformal coating containing carbon, hydrogen and oxygen in order that it can be used as a shield against gamma rays?
24.85. Name two metal oxides which are used in metal oxide polymer composites designed to act as shields against both gamma rays and neutrons.
24.86. Are tungsten and gadolinium oxides effective in reducing low energy x-ray exposure?
24.87. If the metal oxide percentage becomes very high in the metal oxide polymer composite, what is the effect on the neutron shielding capability of the composite?
24.88. A metal oxide polymer composite is largely made of polymer with very little metal oxide content. Comment on its gamma ray shielding capacity.
24.89. Does inclusion of metal oxide particulate in acrylics yield an effective coating solution for protection from gamma rays and neutrons?
24.90. What does Watt distribution function represent?
24.91. What is meant by outgassing? What are the mechanisms of outgassing? Give five examples of likely outgassing sources in a spacecraft.

24.92. What are the effects of outgassing on optical components, thermal control coatings, solar cell arrays and sensors of the spacecraft?
24.93. Explain the effect of outgassing on: (a) the vacuum level in space and (b) high voltage operations carried out in spacecraft.
24.94. Suggest two ways to capture the contaminants released during outgassing in space.
24.95. What are molecular adsorber pucks made of? Do they need any supporting fixture?
24.96. What are molecular adsorber coatings composed of? In what respects do molecular adsorber coatings provide a better solution to the outgassing problem than molecular adsorber pucks?
24.97. Write the general formula of zeolite. How does it trap the gases liberated during outgassing?

References

Abraham N S, Hasegawa M M and Straka S A 2012 Development and testing of molecular adsorber coatings *Proc. SPIE, Optical System Contamination: Effects, Measurements, and Control 2012* **8492** 849203

Abraham N S, Hasegawa M M, Straka S A, Carosso N E and Egges J 2014 Black molecular adsorber coatings for spaceflight applications *Proc. SPIE, Systems Contamination: Prediction, Measurement, and Control 2014* **9196** 91960F

Abraham N S and Jallice D E 2018 Preliminary testing of NASA's molecular adsorber coating technology for future missions to Mars *Proc. SPIE, Systems Contamination: Prediction, Control, and Performance 2018* **10748** 107480E

Advanced Cooling Technologies 2022 *Heat pipe resources: an overview of heat pipe technology* https://1-act.com/resources/heat-pipe-resources/#:~:text=A%20heat%20pipe%20is%20a,wick%20in%20the%20evaporator%20section

Bhalerao R 2021 How vibrations in space vehicles affect PCBA *Sierra Circuits* https://protoexpress.com/blog/vibrations-space-vehicles-affect-pcba (accessed 28 March 2022)

Brusse J A, Ewell G J and Siplon J P 2002 Tin whiskers: attributes and mitigation *CARTS 2002: 22nd Capacitor and Resistor Technology Symp. (25–29 March 2002)* 67–80

Collette J P, Rochus P, Peyrou-Lauga R, Pin O, Nutal N, Larnicol M and Crahay J 2011 IAC-11.C2.8.1 – phase change material device for spacecraft thermal control *Proc. 62nd Int. Astronautical Congress 2011 (IAC 2011) 7 (Cape Town, 3–7 October 2011)* (Paris: The International Astronautical Federation) pp 6020–31

Cross J B, Koontz S L and Lan E H 1991 Atomic oxygen interaction with spacecraft materials: relationship between orbital and ground-based testing for materials certification *Report* LA-UR-91-3052; CONF-9109285-1ON: DE92000190 Los Alamos National Laboratory

De Rooij A 2010 Corrosion in space *Encyclopedia of Aerospace Engineering* ed R Blockley and W Shyy (New York: Wiley)

Dooling D and Finckenor M M 1999 Material selection guidelines to limit atomic oxygen effects on aircraft surfaces *NAS/TP-1999-209260*

Duo S W, Song M M, Liu T Z and Li M S 2012 SiO_2 coatings prepared by sol–gel process protecting silver from atomic oxygen erosion *Appl. Mech. Mater.* **121–126** 3044–7

De Vanzo M and Hayes R B 2020 Ionizing radiation shielding properties of metal oxide impregnated conformal coatings *Radiat. Phys. Chem.* **171** 108685

Fixsen D J 2009 The temperature of the cosmic microwave background *Astrophys. J.* **707** 916–20

Forkapa M J, Banks B, Rutledge S K, Ma D H and Sechkar E 1999 Atomic oxygen durability testing of an international space station solar array validation coupon *Space Technology Proc.: Protection of Materials and Structures from the Low Earth Orbit Space Environment, Proc. of ICPMSE-3, 3rd Int. Space Conf. (Toronto, April 25–26, 1996)* J I Kleiman and R C Tennyson (Dordrecht: Springer) pp 9–30

Gouzman I, Grossman E, Murat M, Noter Y, Saar N, Zilberman G, Minton T K, Garton D J, Buczala D and Brunsvold A 2003 A study of atomic oxygen interactions with protected silver surfaces *Proc. 9th Int. Symp. on Materials in a Space Environment (Noordwijk, 16–20 June (ESA SP-540, September)* 487–92

Grace 2022 *Ludox® colloidal silica* https://grace.com/products/ludox/ (accessed 28 March 2022)

Heiken G H, Vaniman D T and French B M 1991 *Lunar Sourcebook: A User's Guide to the Moon* (Cambridge: Cambridge University Press) p 34

Huang J, Liu G, Wang Z, Jiang L, Yuan X, Liu Y, Deng J and Yi Z 2015 High voltage breakdown induced by outgassing of space materials *AIP Adv.* **5** 037119-1–8

Jiao Z, Jiang L, Sun J, Huang J and Zhu Y 2019 Outgassing environment of spacecraft: an overview *IOP Conf. Ser.: Mater. Sci. Eng.* **611** 012071

Knack K 2020 *Blind and buried vias—what are they and how are they used?* https://resources.altium.com/p/blind-and-buried-vias-what-are-they-and-how-are-they-used (accessed 29 March 2022)

Koontz S L, Cross J B, Hoffbauer M A and Kirkendahl T D 1991 Atomic oxygen degradation of Intelsat 4-type solar array interconnects: laboratory investigations *NASA Technical Memorandum, Report No. TM 102175* National Aeronautics and Space Administration (NASA) i–v 1–19

Lai S T 1989 An overview of electron and ion beam effects in charging and discharging to spacecraft *IEEE Trans. Nucl. Sci.* **36** 2027–32

Lai S T 2003 A critical overview on spacecraft charging mitigation methods *IEEE Trans. Plasma Sci.* **31** 1118–24

Lai S T and Tautz M F 2006 Aspects of spacecraft charging in sunlight *IEEE Trans. Plasma Sci.* **34** 2053–61

Lai S T 2012 *Fundamentals of Spacecraft Charging: Spacecraft Interactions with Space Plasmas* (Princeton, NJ: Princeton University Press) pp 1–10

Lai S T and Cahoy K 2015 Trapped photoelectrons during spacecraft charging in sunlight *IEEE Trans. Plasma Sci.* **43** 2856–60

Lai S T and Cahoy K 2017 Spacecraft charging *Encyclopedia of Plasma Technology* ed J L Shohet vol II (Boca Raton, FL: CRC Press) pp 1352–66

Lai S T, Cahoy K, Lohmeyer W, Carlton A, Aniceto R and Minow J 2018 Chapter 16: Deep dielectric charging and spacecraft anomalies *Extreme Events in Geospace: Origins, Predictability, and Consequences* ed N Buzulukova (Amsterdam: Elsevier) pp 419–32

Lai S T and Pradipta R 2022 *The Physics of Satellite Charging: Causes, Effects, and Applications* (Boca Raton, FL: CRC Press) pp 1–10

Leonard C 2017 Challenges for electronic circuits in space applications, part 1 *MPD Microwave Product Digest* https://mpdigest.com/2017/06/22/challenges-for-electronic-circuits-in-space-applications-part-1/

Mikaelian T 2001 *Spacecraft charging and hazards to electronics in space* (York: York University) pp 1–28 arXiv:0906.3884, Cornell University, 2009

Muraoka I, De Sousa F L, Ramos F M and Parisotto W R 2001 Numerical and experimental investigation of thermal louvers for space applications *J. Braz. Soc. Mech. Sci. Eng.* **23**

Nagano H, Nagasak Y and Ohnishi A 2006 A simple deployable radiator with autonomous thermal control function *J. Thermophys. Heat Transf.* **20** 856–64

NASA Whisker Failures 2009 https://nepp.nasa.gov/whisker/failures/index.htm#satellite (accessed 29 March 2022)

NASA Tech Briefs 2016 *Two-phase thermal switch* https://techbriefs.com/component/content/article/tb/pub/briefs/mechanics-and-machinery/24538 (accessed 31 March 2022)

NASA 2017 Mitigating in-space charging effects—a guideline *NASA-HDBK-4002A w/CHANGE 1* revalidated 2017-10-19, p 13

NASA 7.0 Thermal Control 2021 https://nasa.gov/smallsat-institute/sst-soa/thermal-control (accessed 29 March 2022)

NASA Science 2022 *Staying cool on the ISS* https://science.nasa.gov/science-news/science-at-nasa/2001/ast21mar_1 (accessed 1 April 2022)

Radebaugh R 2009 Cryocoolers: the state of the art and recent developments *J. Phys.: Condens. Matter* **21** 164219

Rutledge S and Olle R 1993 Space station freedom solar array blanket cover lay atomic oxygen durability testing *38th Int. SAMPE Symp. (Anaheim, CA, May 10–13)*

Samsuddin N I, Ahmad S, Hasbullah N F and Toha S F 2013 Parametric system identification of thermoelectric cooler for single photon avalanche diode application *Res. J. Appl. Sci. Eng. Technol.* **6** 712–9

Samwel S W 2014 Low earth orbital atomic oxygen erosion effect on spacecraft materials *Space Res. J.* **7** 1–13

Sigma Technologies 2016 *Multilayer insulation—what makes it different* https://sigmalabs.com/blog/multi-layer-insulation-what-makes-it-different.html (accessed 30 March 2022)

Straka S A, Peters W C, Triolo J J, Hasegawa M M, Novo-Gradac K J, Miller C B and Abraham N S 2012 Molecular adsorber coating *US Patent* US20120034384A1

Tempco 2021–2 *Kapton® heaters* https://tempco.com/Products/Electric-Heaters-and-Elements/Flexible-Heaters/Kapton-Heaters.htm (accessed 30 March 2022)

Verker R, Bolker A, Carmiel Y, Gouzman I, Grossman E, Minton T K and Remaury S 2020 Ground testing of an on-orbit atomic oxygen flux and ionizing radiation dose sensor based on material degradation by the space environment *Acta Astronaut.* **173** 333–43

Vibration Solutions 2022 What is Sorbotane? https://vibrationsolution.com/pages/what-is-sorbotane (accessed 30 March 2022)

von Lukowicz M, Abbe E, Schmiel T and Tajmar M 2016 Thermoelectric generators on satellites —an approach for waste heat recovery in space *Energies* **9** 541

Xiangqian Y *et al* 2017 Mitigating deep dielectric charging effects in space *IEEE Trans. Nucl. Sci.* **64** 2822–8

Xiangqian Y, Siyu S, Hongfei C, Qiugang Z, Hong Z, Weihong S and Ao C 2020 Mitigating deep dielectric charging effects at the orbits of Jovian planets *Trans. Nanjing Univ. Aeronaut. Astronaut.* **37** 804–15

Zhao H, Liu Q-b, Lan Y-p, Wang H and Yao D-w 2017 The application and research status of tin whisker formation in electric usage *AIP Conf. Proc.* **1839** 020003-1–10

IOP Publishing

Extreme-Temperature and Harsh-Environment Electronics (Second Edition)
Physics, technology and applications
Vinod Kumar Khanna

Chapter 25

Electronics jamming counteraction and cybersecurity assurance in adversary environments

Most electronic equipment that works well in friendly circumstances may not be able to perform reliably in hostile conditions. During electronic warfare, jamming of electronics is a tool used by an enemy to cripple the defence services. Electromagnetic signals are used to disrupt or block communication services. This is intentional electromagnetic interference unlike that produced unknowingly by faulty electronic design without caring for electromagnetic compatibility. Electronics communication must therefore be designed bearing in mind that it may be required to work dependably in antagonistic situations. No less threatening is the hacking of electronics. Protecting electronics against malicious attacks of hackers calls for building cybersecurity in electronics systems by strengthening software and hardware facilities. In this chapter, different types of jamming and hacking of electronics are surveyed, and countermeasures evolved for making electronics sufficiently sturdy to withstand these dangers are described.

Two familiar techniques, namely jamming and hacking, can be applied with malicious intentions to make various electronic equipment, instrumentation, computing, control and communication systems function in a haywire or erratic manner, and even become non-functional. Warring groups exploit these techniques to weaken their opponents. In this chapter, the types of attacks launched on electronics by these approaches will be described, and the schemes devised for stalling these attempts will be explained.

doi:10.1088/978-0-7503-5072-3ch25

25.1 A jamming attack

The wireless medium is a shared medium. The shared nature of this medium makes it easy for adversaries to obstruct legitimate users from using this medium through disruption of either transmission or reception services. Figure 25.1 shows a typical jamming environment in which a jammer of power P_J is transmitting signals on the same frequency f at which a transmitter of power P_T is communicating with a receiver. P_J is greater than P_T so that the communication of signals between the transmitter and the receiver is blocked. This denial-of-service attack is known as a jamming attack. The jamming attack is successful if the signal-to-noise ratio at the receiver decreases to a value less than 1.

25.2 Types of jamming and jammers

25.2.1 Classification by type of jamming signal used

According to Mpitziopoulos *et al* (2009), jamming is classified as (figure 25.2):

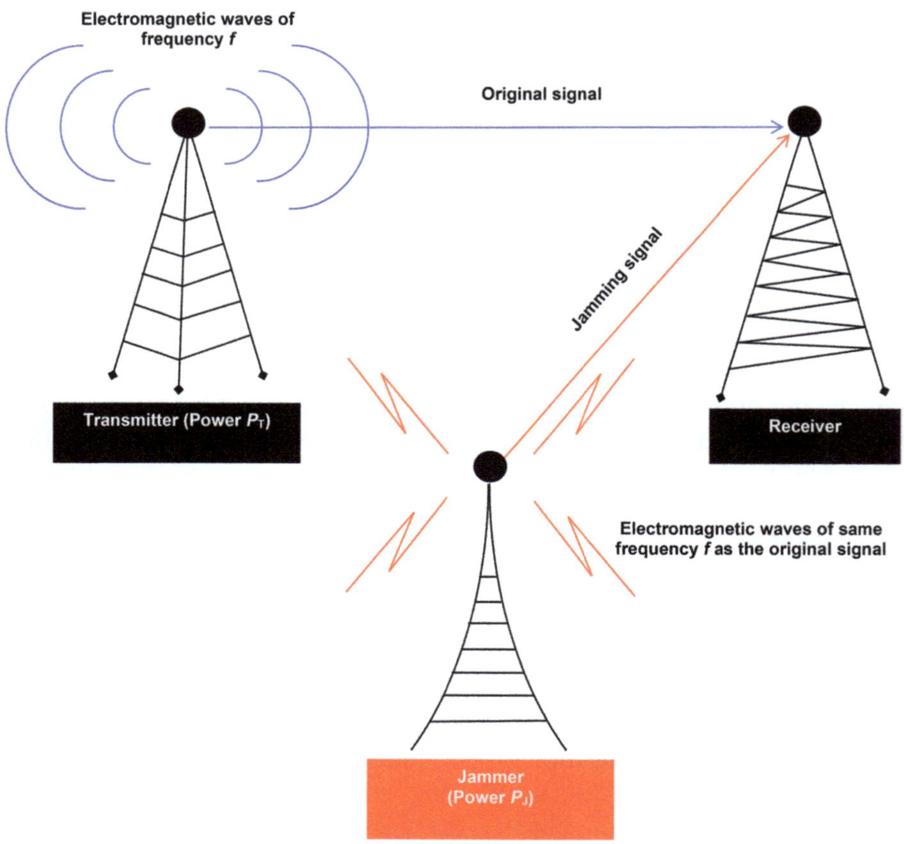

Figure 25.1. A jamming environment.

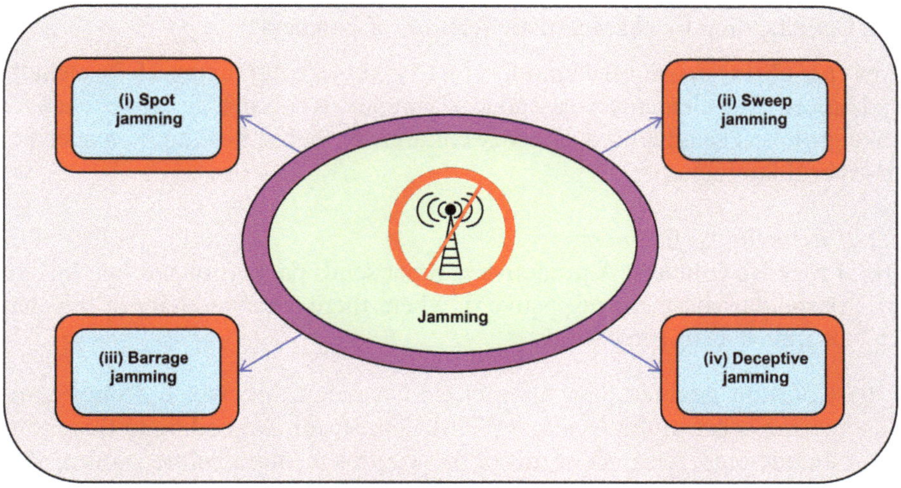

Figure 25.2. Types of jamming with reference to the jamming signal used.

(i) Spot jamming: A powerful jamming signal having sufficiently high power to supersede the signal of the target transmitter is applied at the same frequency as that of the target transmitter using the same modulation technique as the target transmitter. Since the jamming signal has only one frequency, only a single frequency is jammed and the target transmitter can easily avoid jamming by changing its frequency.

(ii) Sweep jamming: The jamming signal continuously shifts at full power from one frequency to another over a range of frequencies so that signals of different frequencies in this range are jammed. However, the jamming signal is applied for a limited time duration at one frequency. So, for any given frequency, jamming occurs only for the limited time at which the jamming signal is applied at that particular frequency, and not during the remaining period, when the jamming signal moves to other frequencies.

(iii) Barrage jamming: Instead of shifting from one frequency to another, the jamming signal moves from one range of frequencies to another. So, jamming takes place over a range of frequencies. As the power of the jamming signal is distributed over several frequencies in the range, jamming at any one specified frequency is weaker than if the power was concentrated at a single frequency.

(iv) Deceptive jamming: Legitimate packets of fake useless data are transmitted instead of random bits to make other nodes believe that genuine transmission is taking place. The other nodes remain in receiving states during the period of this transmission of fake data, and so are jammed. This type of jamming cannot be easily detected as it is a fake transmission disguised as authentic.

25.2.2 Classification by characteristic features of jammers

Grover *et al* (2014) classify the jammers into two categories (figure 25.3): elementary and advanced. The elementary category of jammers is subdivided into proactive and reactive groups. The advanced category contains the subsets of function-specific and smart-hybrid jammers.

25.2.2.1 Elementary jammers

(a) **Proactive jammers:** A proactive jammer sends packets or random bits in one particular channel irrespective of where there is any activity in the channel or not. It is of three kinds:

(i) Constant jammer: This jammer emits random bits which do not obey the carrier sense multiple access (CSMA) protocol. According to this protocol, a node must sense the status of the wireless medium before transmission. A frame is transmitted only under the condition that the node is idle for a distributed coordination function interframe space (DIFS) interval. The transmission is deferred if the channel is busy during this period. Jamming is achieved by keeping the wireless medium continuously busy so that the inter-communication among nodes cannot take place.
(ii) Random jammer: Jamming is done by intermittently transmitting random or regular bits. It alternates between sleeping and jamming modes.
(iii) Deceptive jammer: This done by emitting regular packets in place of random bits as outlined above in section 25.2.1(iv), giving the impression that legitimate transmission is in progress.

(b) **Reactive jammers:** The reactive jammer sends jamming signals only when the network is active. It is subdivided into:

(i) Reactive request-to-send/clear-to-send jammer: It jams the network when a request-to-send message is sensed. Since the request-to-send packet from a sender is distorted, the clear-to-send reply is not returned from the receiver.
(ii) Reactive data/acknowledgement jammer: Network jamming is done by corrupting the transmission of data or acknowledgement packets.

25.2.2.2 Advanced jammers

(a) **Function-specific jammers:** A function-specific jammer works in reference to a pre-determined function. Its subdivisions include:

(i) Follow-on jammer: It hops over all the channels quickly jamming each channel for a short time. It is effective against frequency-hopping spread spectrum anti-jamming technique with slow hopping.
(ii) Channel-hopping jammer: Hopping proactively between different channels, it can jam multiple channels simultaneously.

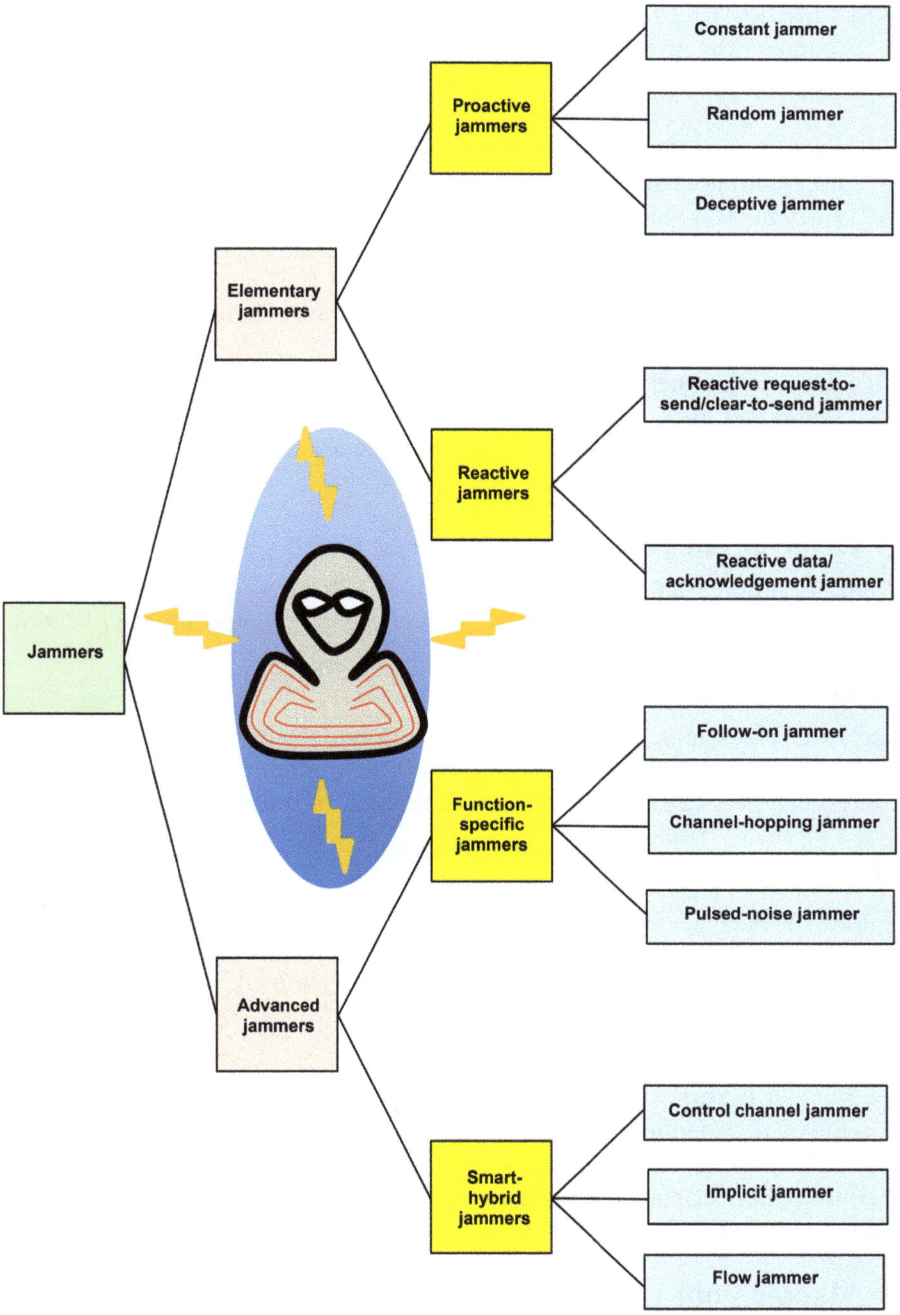

Figure 25.3. Classes of jammers from the viewpoint of their features.

(iii) Pulsed-noise jammer: It can change channels. Also, it can jam on different bandwidths at different time periods. Simultaneous jamming of multiple channels can be done.

(b) **Smart-hybrid jammers:** The smart-hybrid jammer is called by this name because it is designed to be effective in jamming and at the same time power efficient. It is said to be hybrid because of possibility of implementation as both proactive and reactive jammers. Under this class fall:

(i) Control channel jammer: It aims at the control channel of a multi-channel network.
(ii) Implicit jammer: Besides target disabling, it denies service at other nodes of the target also.
(iii) Flow jammer: Traffic flow in the network is decreased using multiple jammers.

25.3 Detection of jamming attacks

This entails differentiating between legitimate and adversarial causes of loss of connectivity by utilizing statistical data on various parameters (Xu *et al* 2006).

25.3.1 From signal strength

The presence of a jammer impacts the distribution of signal strength. So, a jamming attack can be detected by comparing the average strength of the signal with respect to a threshold level calculated from ambient levels of noise.

25.3.2 From carrier sensing time

During jamming, the channel appears busy to the source. Hence a legitimate source is prevented from sending out packets. Carrier sensing time therefore becomes a decisive parameter to ascertain whether a device is jammed.

25.3.3 From packet delivery ratio (PDR)

As the jammer corrupts transmission of signals in the network, the power delivery ratio is lowered by jamming. Hence PDR is a suitable parameter to determine the presence of a jammer.

A combination of statistics such as signal strength with PDR can discriminate between a jammed and congested network.

25.4 Mapping out jammed area and planning the defense strategy against jamming

After detection of jamming, the jammed area must be carefully mapped out. Wood *et al* (2003) detected jamming by monitoring utilization of the channel. When the channel utility falls below a threshold value, the network is deemed to be jammed. Once jamming is detected, the jammed nodes broadcast JAMMED messages announcing that they have been jammed. These messages are not received by adjoining jammed nodes but the non-jammed neighbors on the boundary of the jammed region get the messages initiating a mapping procedure. They exchange and combine information about which nodes are jammed, and what are their locations, thus mapping out the territory of the jammed area.

25.5 Approaches to overcome jamming

Two different but complementary approaches are applied to overcome jamming:

25.5.1 Retreating away from the jammer

This retreat is done in two ways:

(i) Spatial retreat: Knowledge of the jammed area is applied to relocation of jammed nodes away from the jammed area. This is possible for mobile networks such as mobile sensor networks.
(ii) Spectral evasion: The frequency of transmission is changed to a frequency different from the jammed frequency.

25.5.2 Resource adjustment to actively compete with the jammer

The power level and coding of the communication are altered to provide effective communication in the presence of a jamming signal.

25.5.3 Adopting jamming-resistant communication techniques

Spread spectrum methods are used.

25.6 Retreating methods

25.6.1 Spatial retreat

Spatial retreats become complex if the jammer itself is mobile. By moving to a new position, the mobile jammer makes relocation of large chunks of original network nodes necessary. With enormous relocation, the network may lose its uniformity. It may even be partitioned. Hence the spatial retreat scheme consists of two phases:

(i) Escape phase: In this phase, the nodes in the jammed region move to non-jammed areas without losing connection with the remaining network.
(ii) Reconstruction phase: In this phase, the nodes migrate to new positions in such a manner that coverage of the network is uniform, and the jammer cannot partition the network.

25.6.2 Channel surfing

Channel surfing is moving from a channel of one frequency to a channel of another frequency. It seems to resemble frequency-hopping modulation. But it is dissimilar to frequency-hopping modulation in which the carrier frequency is continually changing. During channel surfing, the change in frequency does not take place in this fashion. It happens on demand.

It is evident that the switching of several devices to new channels must be reliably coordinated. In this coordination, difficulties are expected in distributed computing asynchrony, latency and scalability. These difficulties are avoided by a harmonized switching of channels and multiplexing of the spectra. In such switching, the complete network changes to the new channel. As soon as a node detects that it is jammed, it changes its channel. At the same time, it sends beacons to other nodes declaring that it is now on the new channel. Nodes at the boundary of the jammed area are not jammed but their neighbors inside the boundary are jammed. The boundary nodes find that their neighboring nodes are absent on the original channel. They probe the new channel to find whether their neighboring nodes are present there. When a node detects beacons on the new channel, it immediately returns to the original channel and broadcasts a message to all the nodes of the network to switch to the new channel.

The latency in channel switching depends on the size of the network. As the network enlarges, the latency incurred during channel switching becomes significant. This increased latency may result in a situation in which some nodes are in the new channel while other nodes are still in the original channel. Such an instability can be prevented by making only nodes in the jammed area switch channels. Nodes on the boundary of the jammed area can act as relay nodes between various spectral regions.

25.7 Competition method: regulation of transmitted power and error correcting code

Here the nodes adjust the power of their communication and their coding in order that the packets are reliably transmitted and received. A high-power signal is more resistant to jamming because the power of the jamming signal must exceed the power of the jammed signal. An increase in power of transmission raises the signal-to-noise ratio. Therefore, capability to change the power of the transmitter is essential so that suitable power can be applied depending on conditions. However, an increase in power level loads the power budget. Further, the area of coverage of the transmission increases which leads to unintentionally interfering with regular radio transmissions.

When a stronger error correcting code is used, the probability of successful decoding of the packets by the receiver is enhanced. But this is achieved at the cost of a lower information rate and hence throughput. Low density parity coding (LDPC) provides a high throughput attaining the theoretical Shannon limit. The complexity of decoding is also low (Pirayesh and Zeng 2022).

Clearly, power control and code adaptation techniques will not be able to counter constant jamming at high power. Protocols must be built to adapt the codes in response to the severity of threat perception.

25.8 Jamming-resistant spread-spectrum communication systems

Spread-spectrum systems comprise a class of systems that were developed with the specific objective of preventing jamming (Ström *et al* 2002). In order to jam a signal, the adversary must know that a signal is being transmitted. Only after the adversary knows about the transmission can a jamming signal be sent to disrupt the transmission. So, the first precaution is that the transmitted signal must be such that an adversary fails to detect it and be able to know about the transmission. The second precaution is that even if the adversary is able to know about transmission, the signal being sent should be difficult to jam, which is possible if the signal has an in-built anti-jamming property.

In a spread-spectrum system, the precautions are taken care of by hiding the signal in a larger bandwidth than is necessary for transmission. Figure 25.4 shows how the bandwidth is increased at the transmitter to conceal the signal and decreased at the receiver to recover it. As can be seen, spreading decreases the spectral density of the signal but its total energy remains the same. However, it is

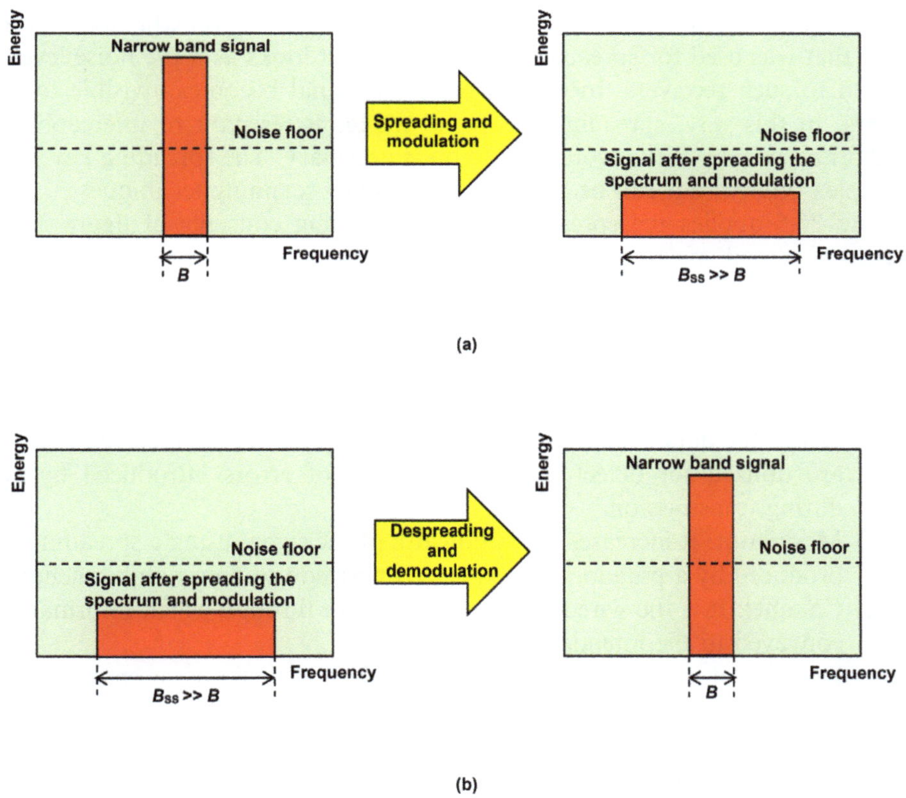

Figure 25.4. Changing the bandwidth of the signal: (a) spreading operation at the transmitter stretches the bandwidth of the narrowband signal from B to B_{SS} which is much larger than B, (b) despreading operation at the receiver restores the bandwidth of the spread signal B_{SS} to the original bandwidth B of the narrowband signal.

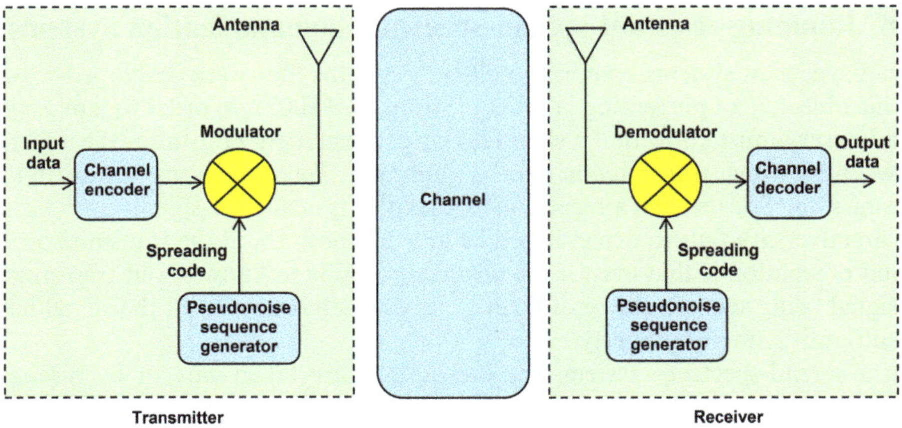

Figure 25.5. General model of spread system communication showing its three constituent parts, the transmitter, channel and receiver.

distributed over a wider frequency range. The signal energy can lie below the noise floor. So, it appears as a noise or interfering signal to any receiver who does not have the key that was used for spreading the signal, i.e., it looks as if the noise level has increased to such receivers. In other words, the signal becomes invisible to these receivers. In this way, spreading the signal makes it resistant to interception by unauthorised people. But a word of caution is necessary. The spreading key should be complex otherwise it can be decoded by applying scanning techniques.

Figure 25.5 applies the spreading and despreading concepts of figure 25.4 to construct a spread spectrum communication system. The system consists of three parts: (i) transmitter, and (ii) receiver connected via (iii) the channel. From the transmitter to the receiver, the data signal passes successively through the stages:

(i) Channel encoder: It is a device to translate data into a code to provide protection against errors. Encoding is done by adding controlled redundant bits to the data. These are known as parity bits. At the receiver, these bits are utilized for detection and correction of errors introduced by noise during transmission.
(ii) Modulator: It increases the bandwidth of the signal using a spreading code produced by a pseudonoise sequence or pseudorandom number generator.
(iii) Channel: It is the wired or wireless medium through which information is conveyed to the intended audience,
(iv) Demodulator. It despreads the received data using the same spreading code that was used at the transmitter, and
(v) Channel decoder: It restores the binary data into original form without the parity bits, the error detection bits that were added to the data block.

There are many ways of concealing the signal but the two most common techniques are frequency-hopping spread system (FHSS) and direct sequence spread spectrum (DSSS), which we will study in the coming subsections. We will also study a hybrid FHSS/DHSS scheme.

25.8.1 Frequency-hopping spread spectrum (FHSS)

25.8.1.1 The frequency-hopping concept

Taking the narrowband signal to be transmitted, the frequency of the carrier is switched among many channels extending over a large bandwidth. This frequency switching is done in a random manner. The transmission time comprises a sequence of cycles. Each cycle is made up of a number of hopping periods t_H. These hopping periods are dwell times of frequency of transmission at a particular frequency. Figure 25.6 shows typical cycles of frequency hopping.

Although the transmission is done using $M = 2^k$ different carrier frequencies forming 2^k channels, in any hop period t_H, only one frequency is used while other frequencies are idle. The IEEE 802.11 standard uses a 300 ms interval during which a few bits are transmitted in an encoded format with the help of an analog-to-digital encoding scheme, e.g., frequency-shift keying (FSK) or binary phase-shift keying (BPSK). The signal thus obtained is centered on a particular frequency.

The sequence of channels is laid out by a spreading code supplied by a frequency synthesizer on receipt of a k-bit pattern from a pseudorandom noise bit generator. A pseudorandom number is a sequence of numbers satisfying one or more statistical tests of randomness but derived from a known starting point called the seed by a

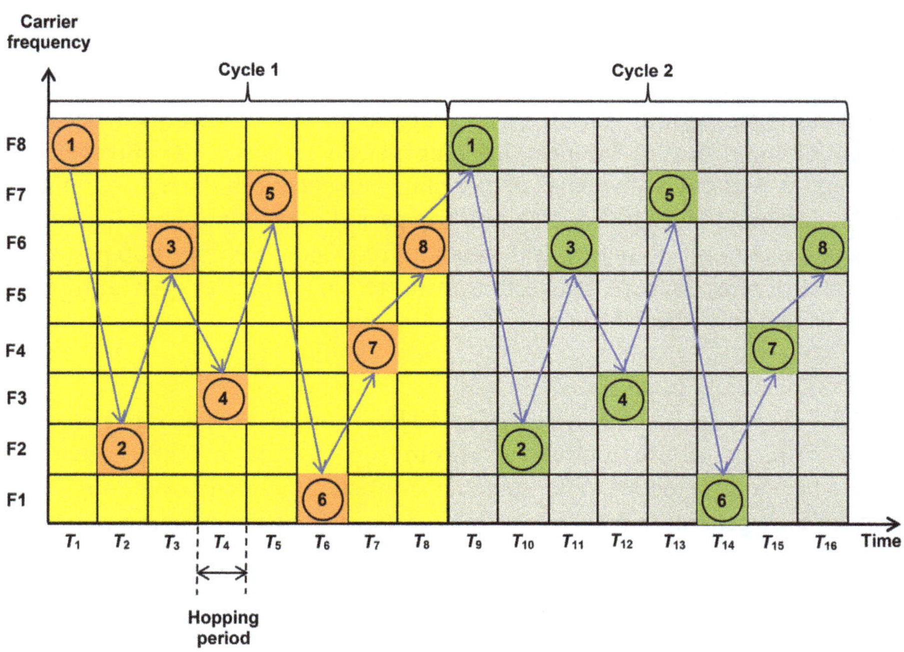

Figure 25.6. Frequency-hopping signal showing how frequency changes from one hopping period to the next in a particular cycle 1 when 8 carrier frequencies F8, F2, F6, F3, F7, F1, F4, F6 are used. The same sequence of frequency jumping is following in the cycle 2 and in further cycles onwards.

definite mathematical procedure, the algorithm. A pseudorandom number generator is the algorithm mapping the seed to the longer pseudorandom string. It applies mathematical formulae to produce random numbers. The k-bit pattern is supplied to a channel frequency table in which the frequencies are allocated to various k-bit patterns. For $k = 3$, there are 8 k-bit patterns: 000, 001, 010, 011, 100, 101, 110, 111. The corresponding frequencies are denoted as F1, F2, F3, F4, F5, F6, F7, F8. Whenever, any k-bit pattern enters the table from the pseudorandom noise bit generator, the frequency assigned to that pattern is sent to the frequency synthesizer. At each consecutive hop period, a different frequency value is chosen. The channels are sequenced in accordance with a spreading code prescribed by the pseudorandom noise bit generator. The signal representing the spreading code is called the chipping signal $c(t)$ because it contains a large number of bits which are known as chips. The transmitter and the receiver share this code between them. So, synchronization is preserved between them during tuning to the sequence of channels.

The bandwidth B_{FHSS} determined by the spacing between the M carrier frequencies and the width of the channels is much larger than the bandwidth B of the original signal.

$$B_{FHSS} \gg B \tag{25.1}$$

At any instant of time, only a small percentage of the bandwidth is exploited. Exactly what percentage is used and in what manner is confidential. As the frequency switching algorithm is a secret shared only between the transmitter and the receiver, the adversary is unable to know in which portion of the bandwidth the signal is being transmitted rendering it demanding, though not impossible, for the adversary to jam the signal. If an intruder trying to intercept the signal accesses it during a hopping period, only a small fraction of data is accessed because the frequency changes during the next hopping period to which the intruder cannot quickly adapt without knowledge of hopping sequence. Similarly, a jammer may succeed in jamming the signal during one hop period but will not be able to do so for long enough due to the change in transmission frequency from one hop period to the next. Thus, privacy is kept between transmitter and receiver. An anti-jamming property is also furnished to them.

25.8.1.2 Signal flow in an FHSS system

Figure 25.7 shows the block diagram of an FHSS transceiver. The input binary data enter an FSK or BPSK modulator. From the modulator, they move to the frequency-hopping spreader. The spreader consists of a pseudorandom noise bit generator. This generator produces a k-bit pattern for each hopping period t_{H}. It supplies the k-bit patterns to a channel frequency table. From the table, the frequency for the obtained k-bit pattern for one hop period is sent to the frequency synthesizer. The same process is repeated for the ensuing hop period, and so on to get a cycle of the frequency hopping, and then further cycles of hopping. These carrier frequencies are fed by the frequency synthesizer to the mixer. Thus, the two input signals to the mixer are the encoded binary data from the input port supplied

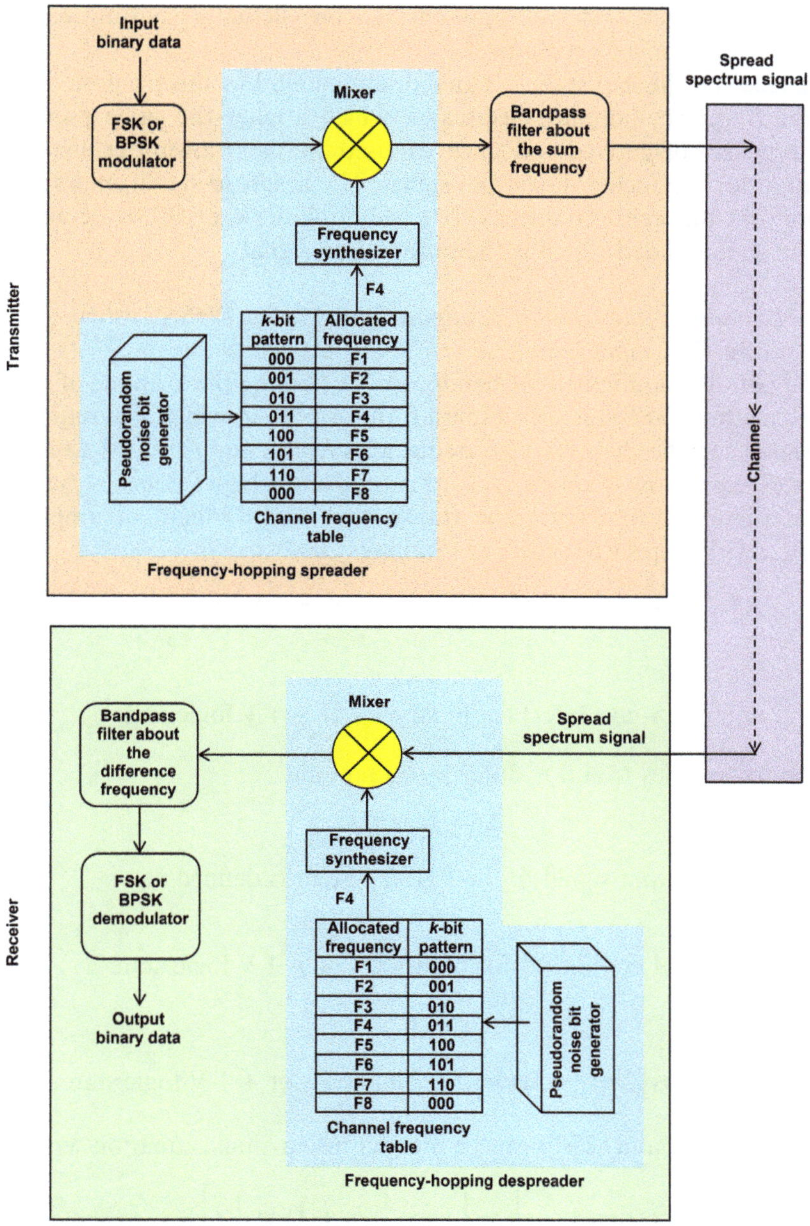

Figure 25.7. Layout of frequency-hopping spread spectrum system subdivided into transmitter, channel and receiver blocks.

by the FSK or BPSK modulator, and the carrier frequencies. In the mixer, the two signals combine together to yield a spread signal. This spread signal is passed through a bandpass filter about the sum frequency. The reason for needing this bandpass filter and an identical filter in the receiver will be explained in further discussion. The filtered spread signal is the one actually transmitted. It passes through the communication channel.

In the receiver, the first stage is signal despreading. For this purpose, the signal is fed to the frequency-hopping despreader which applies the same pseudorandom noise bit generator-derived sequence as used in the transmitter to extract the information for producing the output signal. The despread signal enters a bandpass filter about the difference frequency. After filtering, FSK or BPSK demodulation is done. The demodulated signal is the output data signal.

25.8.1.3 Examining the need of bandpass filters in an FHSS system using FSK modulation

Let us explain the requirement of bandpass filters taking the example of modulation by BFSK method. BFSK is the simplest form of FSK in which the frequency of the carrier signal switches between two frequencies $f_1 = f_0$ and $f_2 = f_0 + \Delta f$ for binary 0 and 1 states, e.g., a low frequency f_1 for binary 0 and a high frequency f_2 for binary 1 condition (figure 25.8). Hence the transmitted digital signal of amplitude A is represented with respect to time t as (Stallings 2007)

$$s_d(t) = A\cos(2\pi f_1 t) \text{ for binary 0, or } -1 \text{ V logic state} \quad (25.2)$$

and

$$s_d(t) = A\cos(2\pi f_2 t) \text{ for binary 1, or } +1 \text{ V logic state} \quad (25.3)$$

where the frequencies f_1 and f_2 differ by an amount

$$\Delta f = f_2 - f_1 \quad (25.4)$$

On this basis, the input signal to the FHSS system is defined as

$$s_d(t) = A\cos\{2\pi f_0 t\} \text{ for binary 0, or } -1 \text{ V logic state} \quad (25.5)$$

and

$$s_d(t) = A\cos\{2\pi(f_0 + \Delta f)t\} \text{ for binary 1, or } +1 \text{ V logic state} \quad (25.6)$$

Equations (25.5) and (25.6) can be merged into a single equation written as

$$s_d(t) = A\cos\left[2\pi\left\{f_0 + \frac{1}{2}(b_i + 1)\Delta f\right\}t\right] \quad (25.7)$$

where b_i is the ith data bit.

For binary 0, $b_i = -1$, equation (25.7) is transformed to

$$s_d(t) = A\cos\left[2\pi\left\{f_0 + \frac{1}{2}(-1 + 1)\Delta f\right\}t\right] = A\cos(2\pi f_0 t) \quad (25.8)$$

which is the same as equation (25.5).

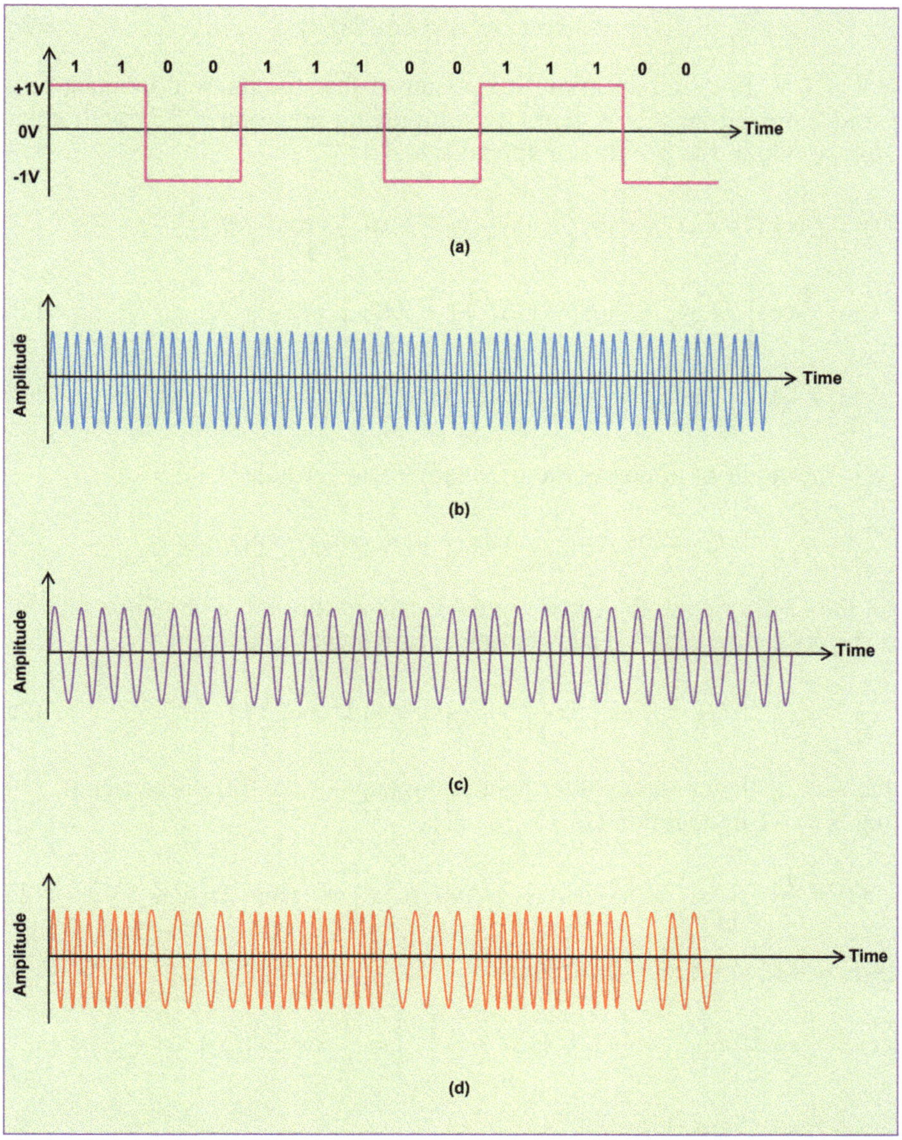

Figure 25.8. Frequency-shift keying modulation: (a) non-return-to-zero waveform, (b) high frequency carrier wave, (c) low frequency carrier wave and (d) BFSK modulated wave.

For binary 1, $b_i = +1$, equation (25.7) changes to

$$s_d(t) = A\cos\left[2\pi\left\{f_0 + \frac{1}{2}(1+1)\Delta f\right\}t\right] = A\cos\left\{2\pi\left(f_0 + \Delta f\right)t\right\} \qquad (25.9)$$

which is equation (25.6).

Let f_i be the frequency of the spreading signal during ith hopping period. Then the spreading or chipping signal can be represented as

$$c(t) = A\cos(2\pi f_i t) = \cos(2\pi f_i t) \tag{25.10}$$

since $A = 1$ V. For simplification, it is assumed that the phase difference between data and chipping signals is zero. By multiplying equation (25.7) with equation (25.10), we obtain the product or spread signal as

$$\begin{aligned} p(t) = s_d(t)c(t) &= A\cos\left[2\pi\left\{f_0 + \frac{1}{2}(b_i + 1)\Delta f\right\}t\right]\cos(2\pi f_i t) \\ &= \frac{1}{2}\cos\left[2\pi\left\{f_0 + \frac{1}{2}(b_i + 1)\Delta f\right\}t + 2\pi f_i t\right] \\ &\quad + \frac{1}{2}\cos\left[2\pi\left\{f_0 + \frac{1}{2}(b_i + 1)\Delta f\right\}t - 2\pi f_i t\right] \end{aligned} \tag{25.11}$$

which is the result of applying the trigonometrical formula

$$\cos x \cos y = \frac{1}{2}\{\cos(x + y) + \cos(x - y)\} \tag{25.12}$$

To get the FHSS signal, the spread signal is passed through a bandpass filter which stops the difference signal. The resulting spread signal is given by

$$s(t) = \frac{1}{2}\cos\left[2\pi\left\{f_0 + \frac{1}{2}(b_i + 1)\Delta f + f_i\right\}t\right] \tag{25.13}$$

So, the task of the bandpass filter here is stoppage of the difference signal. Putting $b_i = -1$ in equation (25.13), we get

$$s(t) = \frac{1}{2}\cos\left[2\pi\left\{f_0 + \frac{1}{2}(-1 + 1)\Delta f + f_i\right\}t\right] = \frac{1}{2}\cos\{2\pi(f_0 + f_i)t\} \tag{25.14}$$

When $b_i = +1$, equation (25.13) gives

$$\begin{aligned} s(t) &= \frac{1}{2}\cos\left[2\pi\left\{f_0 + \frac{1}{2}(1 + 1)\Delta f + f_i\right\}t\right] = \frac{1}{2}\cos\{2\pi(f_0 + \Delta f + f_i)t\} \\ &= \frac{1}{2}\cos\{2\pi(f_0 + f_i + \Delta f)t\} \end{aligned} \tag{25.15}$$

Equation (25.13) represents the FHSS signal transmitted. When this signal reaches the receiver, it is multiplied by the same chipping signal $c(t)$ that was used at the transmitter, yielding a signal

$$r(t) = s(t)c(t) = \frac{1}{2}\cos\left[2\pi\left\{f_0 + \frac{1}{2}(b_i + 1)\Delta f + f_i\right\}t\right]\cos(2\pi f_i t) \tag{25.16}$$

On applying the trigonometrical formula of equation (25.12) to the signal $r(t)$ in equation (25.16) we get

$$
\begin{aligned}
r(t) &= \frac{1}{2} \times \frac{1}{2}\cos\left[2\pi\left\{f_0 + \frac{1}{2}(b_i + 1)\Delta f + f_i\right\}t + 2\pi f_i t\right] \\
&+\frac{1}{2} \times \frac{1}{2}\cos\left[2\pi\left\{f_0 + \frac{1}{2}(b_i + 1)\Delta f + f_i\right\}t - 2\pi f_i t\right] \\
&= \frac{1}{4}\cos\left[2\pi\left\{f_0 + \frac{1}{2}(b_i + 1)\Delta f + f_i + f_i\right\}t\right] \\
&+\frac{1}{4}\cos\left[2\pi\left\{f_0 + \frac{1}{2}(b_i + 1)\Delta f + f_i - f_i\right\}t\right] \\
&= \frac{1}{4}\cos\left[2\pi\left\{f_0 + \frac{1}{2}(b_i + 1)\Delta f + 2f_i\right\}t\right] \\
&+\frac{1}{4}\cos\left[2\pi\left\{f_0 + \frac{1}{2}(b_i + 1)\Delta f\right\}t\right]
\end{aligned}
\tag{25.17}
$$

In this equation, the first term arising from the sum frequency needs to be discarded because we wish to get a signal of the same form as given in equation (25.7) for $s_d(t)$. This is done by interposing a bandpass filter to block the sum frequency in the signal $r(t)$. After filtering, the final signal becomes

$$
r(t) = \frac{1}{4}\cos\left[2\pi\left\{f_0 + \frac{1}{2}(b_i + 1)\Delta f\right\}t\right] \tag{25.18}
$$

which is of the desired form defined in equation (25.7) at the beginning. So, the task of the bandpass filter here is to cast aside the component with sum frequency.

25.8.1.4 MFSK-based FHSS

For bandwidth efficiency, the MFSK (Multiple frequency-shift keying) modulation is preferred. It is a variant of FSK in which more than two frequencies are used. In this scheme, M different signal elements represented by M frequencies are used for encoding the digital input. Here $M = 2^L$ with L denoting the number of bits per signal element. For one signal element, the transmitted signal of amplitude A is expressed by the equation

$$
s_i(t) = A\cos(2\pi f_i t) \text{ where } 1 \leqslant i \leqslant M \tag{25.19}
$$

In this equation,

$$
f_i = f_c + (2i - 1 - M)f_d \tag{25.20}
$$

where f_c is the carrier frequency and f_d is the difference frequency. If T is the bit period, time taken by L bits in each signal element is LT. So time occupied by each signal element $LT = T_s$ (say), and the rate of repetition of signal elements = $1/T_s$. The minimum frequency separation between signal elements extends from $f_c + f_d$ to $f_c - f_d$, i.e., $2f_d$. Hence total frequency range of M signal elements = $2Mf_d$ which is the required bandwidth W_d of the input data signal giving

$$
W_d = 2Mf_d \tag{25.21}
$$

We can also write

$$2f_{\rm d} = \frac{1}{T_{\rm s}} \tag{25.22}$$

Equations (25.21) and (25.22) show that

$$W_{\rm d} = M \times 2f_{\rm d} = M \times \frac{1}{T_{\rm s}} = \frac{M}{T_{\rm s}} \tag{25.23}$$

FHSS is carried out by translation of MFSK signal to a new frequency in every 1/ $f_{\rm c} = T_{\rm c}$ seconds. This is done by modulation of the MFSK signal with the carrier signal of FHSS. Depending upon the relative values of $T_{\rm c}$ and $T_{\rm s}$, FHSS is classified as

$$T_{\rm c} \geqslant T_{\rm s} \rightarrow \text{Slow FHSS} \tag{25.24}$$

and

$$T_{\rm c} < T_{\rm s} \rightarrow \text{Fast FHSS} \tag{25.25}$$

25.8.1.5 Protection from jamming by FHSS

When FHSS uses 2^k different channels (distinguish from 2^L frequencies used for MFSK), the frequency hops among 2^k different values. The bandwidth of the spread spectrum is

$$W_{\rm s} = 2^k W_{\rm d} \tag{25.26}$$

Generally, the number of channels (2^k) is quite large. Hence

$$W_{\rm s} > > W_{\rm d} \tag{25.27}$$

Suppose a jammer of power $S_{\rm Jammer}$ is trying to jam an MFSK transmitter of bandwidth $W_{\rm d}$. FHSS with 2^k channels is applied. For effective jamming, all the 2^k channels must be jammed. Hence, the available jamming energy per channel is $S_{\rm Jammer}/2^k$, implying that the jamming energy decreases by a factor 2^k for any one channel. This reduction in jamming energy is observed as an increase in signal-to-noise ratio. It is known as the processing gain and is expressed as

$$G = 2^k \tag{25.28}$$

Applying equation (25.26), the gain is written in the form

$$G = \frac{W_{\rm s}}{W_{\rm d}} \tag{25.29}$$

25.8.2 Direct sequence spread spectrum (DSSS)

The power of the RF signal to be transmitted is spread over the entire bandwidth in a way that appears random to the jammer. This is done by multiplying the data in the RF carrier signal with a digital signal known as the pseudorandom noise signal which is a sequence of +1 and −1 values at a frequency or chip rate much higher than the frequency of the RF signal. It is said to be pseudo because it is not real Gaussian noise. 'Pseudo' implies its close or deceptive resemblance to noise. Thus, the RF signal is substituted by a signal of extremely wide bandwidth. The noise-like signals can easily escape detection by the adversary. On the receiver side, the original data is recovered by despreading, demodulation and filtering out the noise. Figure 25.9 shows the various signals involved in DSSS: input information signal, spreading signal and spread signal on the transmitter side and spread signal, spreading signal and output signal on the receiver side.

Note that the chip means a single bit of the pseudorandom noise sequence. Chip rate is the rate of sending the chips or bits.

25.8.2.1 DSSS using BPSK modulation

A common form of phase-shift keying is binary phase-shift keying in which the two binary states 0 and 1 are represented by two different phases (figure 25.10). If the amplitude of the carrier wave is A and its frequency is f_c, the BPSK signal is expressed by two equations

$$s_d(t) = A\cos(2\pi f_c t + \pi) = -A\cos(2\pi f_c t) \text{ for binary 0, or } -1 \text{ V logic state} \tag{25.30}$$

and

$$s_d(t) = A\cos(2\pi f_c t) \text{ for binary 1, or } +1 \text{ V logic state} \tag{25.31}$$

Let $d(t)$ be a discrete function which acquires the value −1 for one bit time if the corresponding bit in a stream of bits is 0, and takes the value +1 for one bit time when the corresponding bit in the bit stream is 1. In terms of this discrete function $d(t)$, equations (25.30) and (25.31) can be combined together and written as a single equation

$$s_d(t) = A\, d(t)\cos(2\pi f_c t) \tag{25.32}$$

To verify that this is so, we see that for binary 0, $d(t) = -1$, we have from equation (25.32)

$$s_d(t) = -A\cos(2\pi f_c t) \tag{25.33}$$

which is the same as equation (25.30).For binary 1, $d(t) = +1$, we get from equation (25.32)

$$s_d(t) = A\cos(2\pi f_c t) \tag{25.34}$$

which is equation (25.31).In the transmitter, the BPSK signal $s_d(t)$ is multiplied with the chipping signal $c(t)$ to get the spread signal $s(t)$ as

$$s(t) = s_d(t)c(t) = Ad(t)\cos(2\pi f_c t)c(t) \tag{25.35}$$

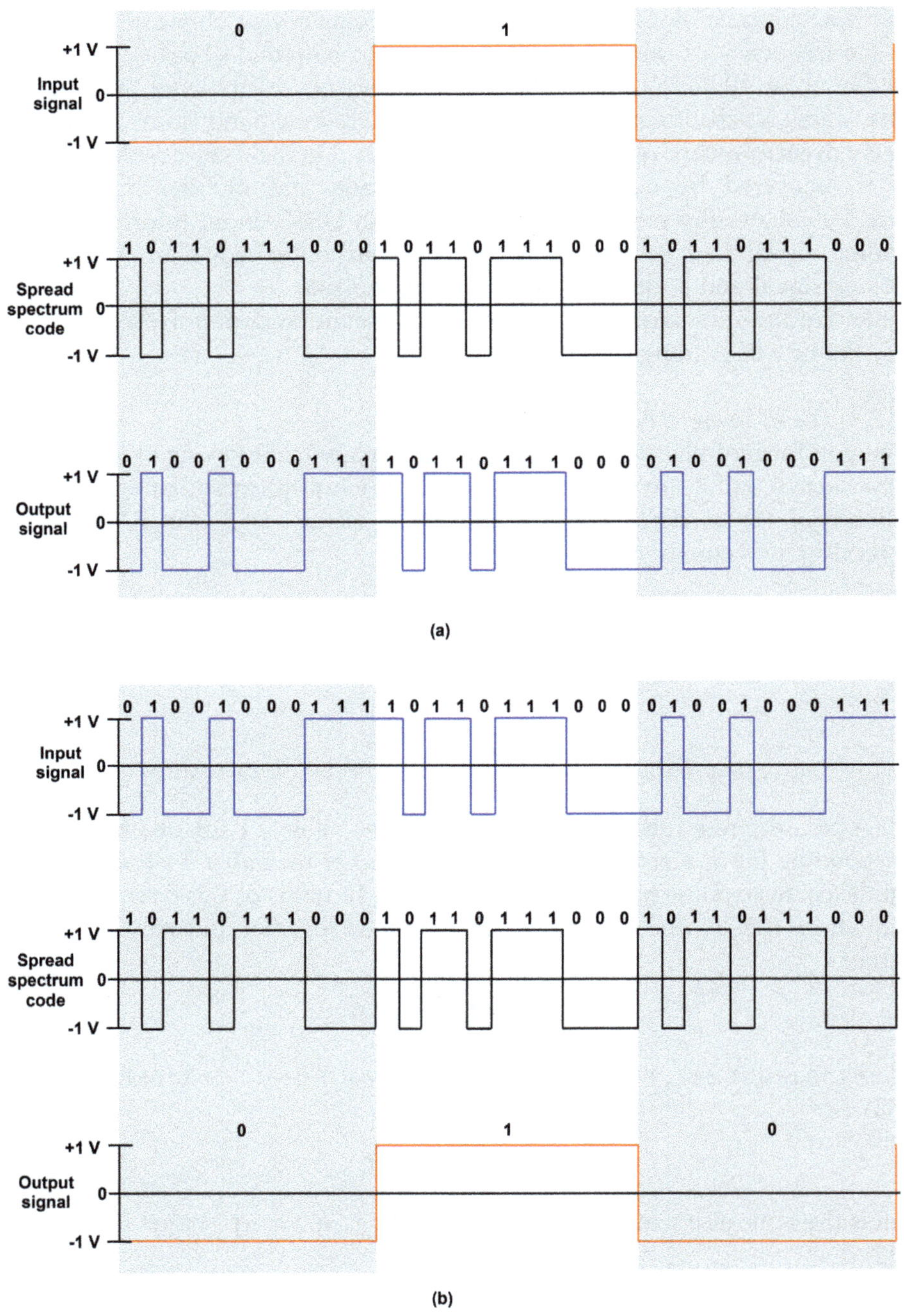

Figure 25.9. Signals and signal transformations in a DSSS system from transmitter to receiver: (a) transmitter side and (b) receiver side.

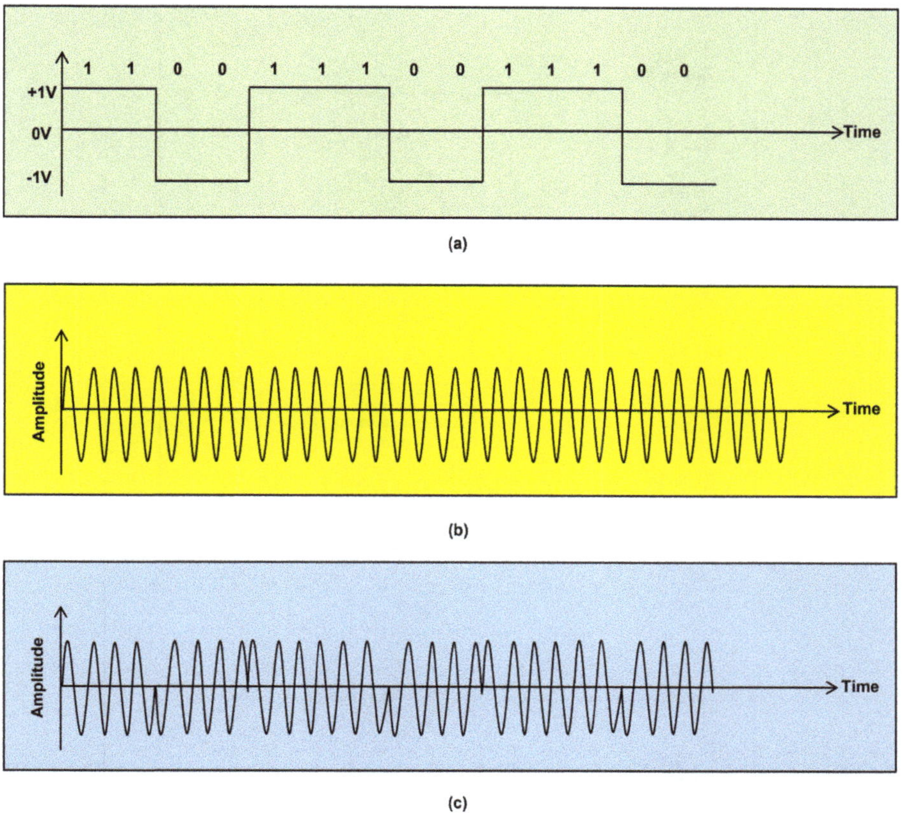

Figure 25.10. Phase-shift keying modulation: (a) non-return-to-zero waveform, (b) carrier wave and (c) BPSK modulated wave.

This spread signal $s(t)$ is transmitted to the receiver. In the receiver, it is multiplied by the same chipping signal $c(t)$ to get the signal

$$r(t) = s(t)c(t) = Ad(t)\cos(2\pi f_c t)c(t) \times c(t) \tag{25.36}$$

But

$$c(t) \times c(t) = c^2(t) = 1 \tag{25.37}$$

because $c(t) = +1$ or -1 and so squaring $c(t)$ will always yield 1. Hence equation (25.36) takes the form

$$r(t) = Ad(t)\cos(2\pi f_c t) \tag{25.38}$$

Comparing equations (25.38) and (25.32), it is found that

$$r(t) = s_d(t) \tag{25.39}$$

showing that the signal sent by the transmitter is recovered at the receiver. Figure 25.11 shows the operations carried out at the transmitter and receiver.

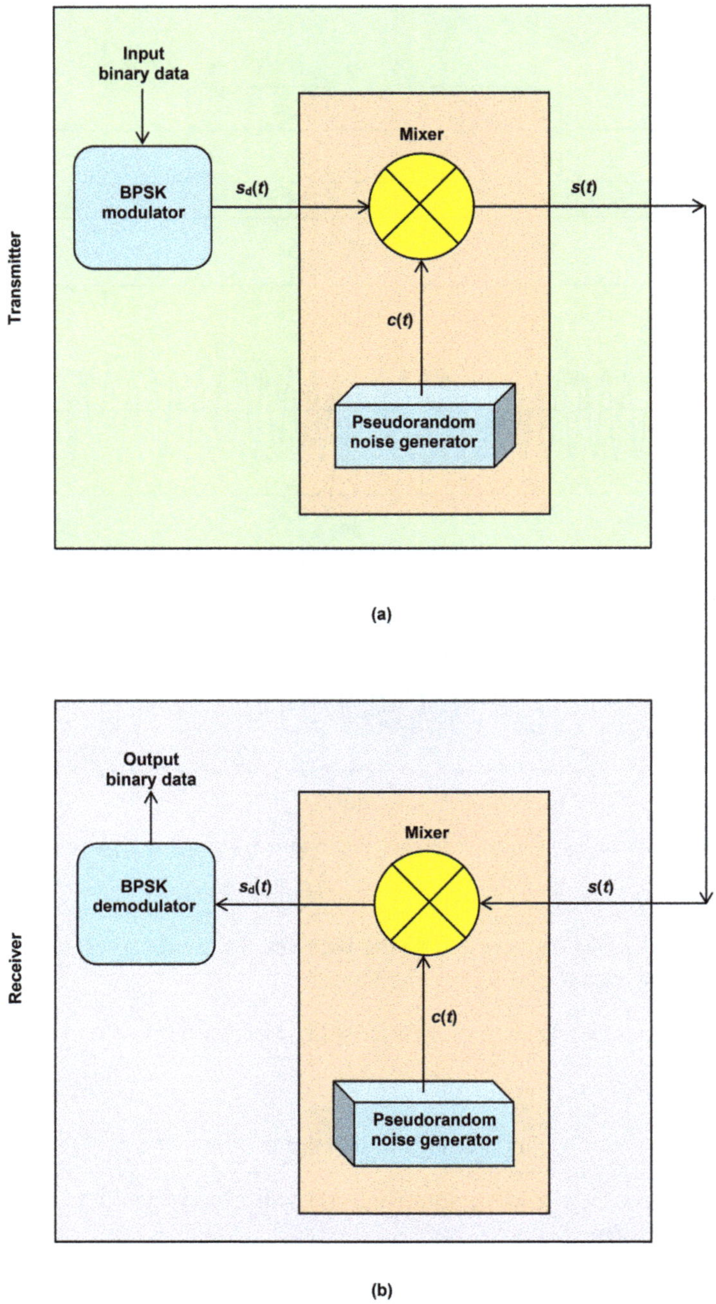

Figure 25.11. The stages of (a) signal modulation and spreading in a DSSS transmitter and (b) signal despreading and demodulation in a DSSS receiver.

25.8.2.2 *Inclusion of the effect of noise in the analysis and implementation of DSSS with BPSK scheme*

For implementation of DSSS, the BPSK is performed on the input information signal to produce the signal $s_d(t)$, which is sent to direct spectrum spreader consisting of a mixer fed with chipping signal $c(t)$ by a pseudorandom noise source. Thus, the spread spectrum signal $s(t)$ is generated: $s(t) = s_d(t)c(t)$. This spread signal is transmitted to the receiver. The signals at the different sections of DSSS are shown in figure 25.12.On the way, noise $n(t)$ enters the signal. In the simplified analysis done in section 25.8.2.1, the effect of noise on the transmission was not included. As the signal $s(t)$ flows from the transmitter to the channel, additive white noise $n(t)$ creeps in distorting the signal $s(t)$ so the that the signal at the receiver end is not $s(t)$ but

$$s_{\text{Corrected}}(t) = s(t) + n(t) = s_d(t)c(t) + n(t) \tag{25.40}$$

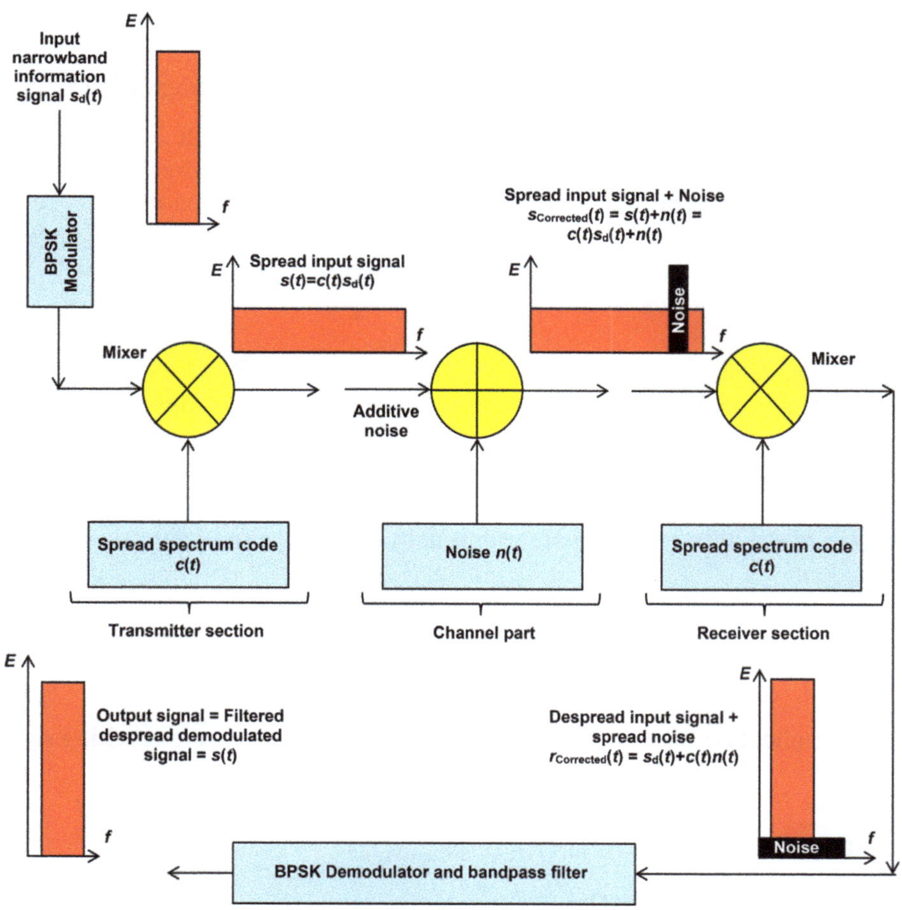

Figure 25.12. Direct sequence spread spectrum communication system.

because the noise is additive in nature. In the receiver, the signal $s_{\text{Corrected}}(t)$, i.e., the spread signal plus noise signal is multiplied by the same chipping signal $c(t)$. Hence the signal coming out from the receiver is

$$r_{\text{Corrected}}(t) = c(t)s_{\text{Corrected}}(t) = c(t)\{s_d(t)c(t) + n(t)\} = c^2(t)s_d(t) + c(t)n(t) \quad (25.41)$$

As already noted, $c^2(t) = 1$, Therefore, equation (25.41) reduces to

$$r_{\text{Corrected}}(t) = s_d(t) + c(t)n(t) \quad (25.42)$$

This equation shows that the signal obtained is the signal $s_d(t)$ plus spread-out noise signal $c(t)n(t)$. The spread-out noise signal is a wideband signal. This signal is subjected to BPSK demodulation and bandpass filtering to get the output signal $s(t)$. The demodulator includes a bandpass filter to filter out the noise. So, on passing the signal $r_{\text{Corrected}}(t)$ produced after multiplication of signal $s_{\text{Corrected}}(t)$ with $c(t)$ through a bandpass filter matched to the BPSK data, the original signal $s(t)$ sent by the transmitter is retrieved on the receiver side, and most of noise signal is attenuated.

25.8.2.3 Spectra in the DSSS system

Three different spectra are observed in this communication process (figure 25.13):

(i) the spectrum of the input information signal with data rate $1/T$ for a bit width T, and signal energy stretching over a width $2/T$ on the frequency axis;
(ii) the spectrum of the pseudorandom noise signal with data rate $1/T_c$ for a bit width T_c, and energy distributed over a much larger width $2/T_c$; and
(iii) the spectrum of the combined signal with bit rate of $(1/T + 1/T_c)$, and energy extending over $2(1/T+1/T_c)$.

25.8.2.4 Weakening of a jamming signal applied to DSSS

Consider a jamming signal $s_{\text{Jamming}}(t)$ with power S_{Jamming} applied to the DSSS system at its center frequency. Its impact on the DSSS can be examined by an approach similar to how the effect of noise was analyzed in section 25.8.2.2. In the presence of jamming, the signal at the receiver end is not $s_{\text{Corrected}}(t)$ given by equation (25.40) above but has an additional term due to the jamming signal. It is expressed as

$$\begin{aligned} s_{\text{Corrected-Jamming}}(t) &= s_{\text{Corrected}}(t) + s_{\text{Jamming}}(t) = s(t) + n(t) + s_{\text{Jamming}}(t) \\ &= s_d(t)c(t) + n(t) + s_{\text{Jamming}}(t) \end{aligned} \quad (25.43)$$

Upon multiplication by the chipping signal $c(t)$ in the receiver, the resulting signal is

$$\begin{aligned} r_{\text{Jamming}} &= c(t)s_{\text{Corrected-Jamming}}(t) = c(t)\left\{s_d(t)c(t) + n(t) + s_{\text{Jamming}}(t)\right\} \\ &= c^2(t)s_d(t) + c(t)n(t) + c(t)s_{\text{Jamming}}(t) \\ &= s_d(t) + c(t)n(t) + c(t)s_{\text{Jamming}}(t) \end{aligned} \quad (25.44)$$

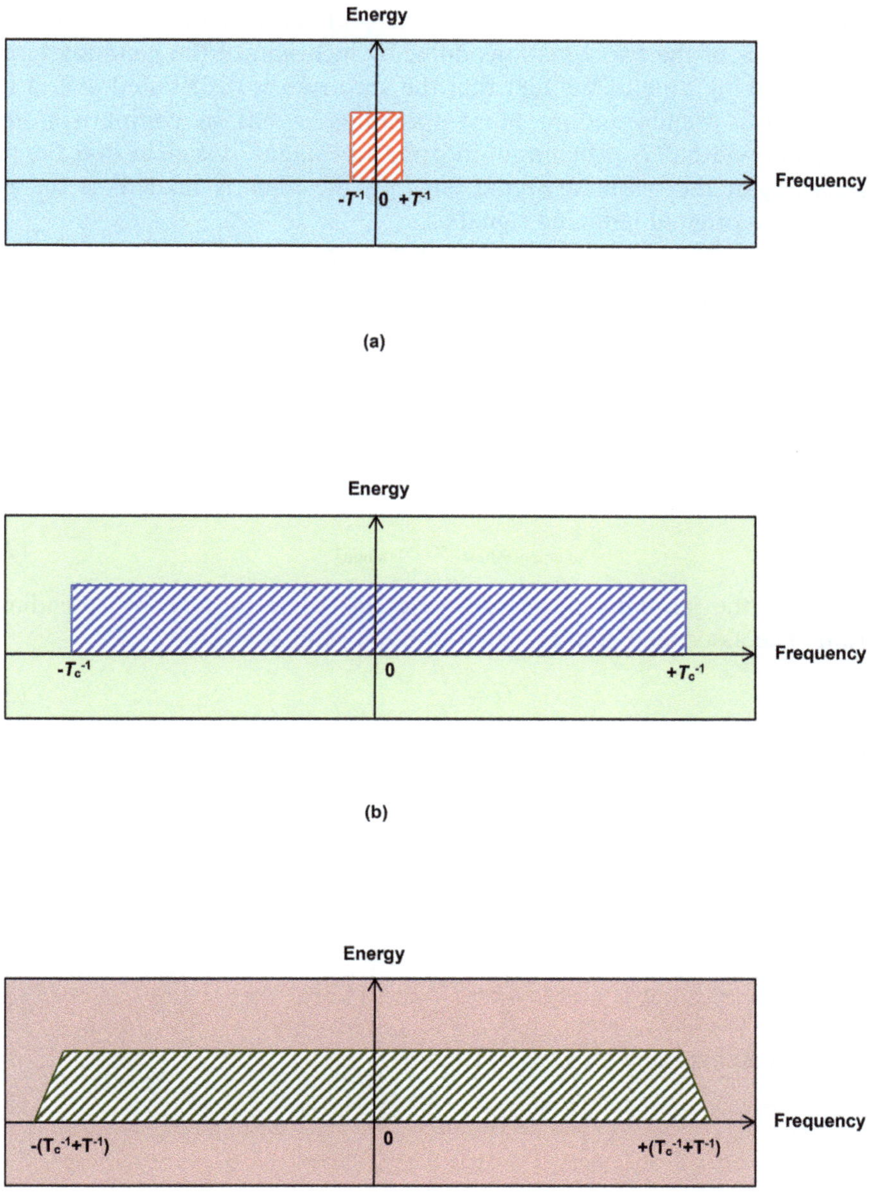

(c)

Figure 25.13. Spectra of (a) the input data signal, (b) the spreading signal and (c) the combined signal in a DSSS system.

by putting $c^2(t) = 1$. Looking at equations (25.42) and (25.44), it is noted that the right-hand sides of the two equations differ by inclusion of the jamming term $c(t)$ $s_{\text{Jamming}}(t)$. Taking note of the fact that the spectrum is distributed over a larger width $2/T_c$ for pseudorandom noise spreading signal in comparison to the much smaller width $2/T$ of the input information signal, it is clear that the power $S_{\text{Jamming(Filter)}}$ of the jamming signal entering the filter is related to the power S_{Jamming} of the original jamming signal as

$$\frac{S_{\text{Jamming(Filter)}}}{S_{\text{Jamming}}} = \frac{2/T}{2/T_c} = \frac{T_c}{T} \tag{25.45}$$

from which

$$S_{\text{Jamming(Filter)}} = S_{\text{Jamming}}\frac{T_c}{T} \tag{25.46}$$

Since

$$T_c \ll T \tag{25.47}$$

$$S_{\text{Jamming(Filter)}} \ll S_{\text{Jamming}} \tag{25.48}$$

meaning that the jamming power decreases in the ratio T_c/T by spreading the spectrum. The associated gain in signal-to-noise ratio is given by

$$G = \frac{T}{T_c} \tag{25.49}$$

If R is the bit rate in the input information signal and R_c the bit rate in the spread signal,

$$\frac{T}{T_c} = \frac{1/R}{1/R_c} = \frac{R_c}{R} \tag{25.50}$$

Further, if W is the bandwidth of the input information signal and W_s the bandwidth of the spread signal,

$$\frac{R_c}{R} \approx \frac{W_s}{W} \tag{25.51}$$

Combining together equations (25.49)–(25.51),

$$G = \frac{T}{T_c} = \frac{R_c}{R} \approx \frac{W_s}{W} \tag{25.52}$$

25.8.3 Hybrid FHSS/DSSS

It is a combination of FHSS with DSSS to increase the resistance to jamming beyond the capabilities of either of them when used alone.

25.9 Ethical hacking

25.9.1 The white hat hacker

After learning about jamming and its prevention techniques, let us move forward to ethical hacking. Hacking is an unlawful activity, a surreptitious attack seeking to gain unauthorized access to digital devices such as computers, smartphones and even entire networks with *mala fide* intentions to gather information, to corrupt systems or to seek financial gain. Ethical hacking is an authorized effort to investigate how an unauthorized person will be able to access a computer system (Harper *et al* 2022). An ethical hacker, also called a white hat hacker, is an expert in security of computers and networks who is hired by an organization to improve its security posture. With prior permission from the organization, and in a lawful and legitimate fashion, the ethical hacker duplicates the strategies of a malicious hacker on the target computer system to identify its vulnerabilities. Remediation advice is given to the organization to fix the critical issues demanding attention so that a harmful hacker may not be able to exploit them.

25.9.2 Phases of ethical hacking

Ethical hacking is done in five phases (figure 25.14). These phases are (Horvath 2022):

(i) **Reconnaissance:** In this phase, the available information about the system is collected including the security measures that are already applied. In passive reconnaissance, essential facts about the company, names of staff, email and IP addresses are gathered. In active reconnaissance, specific information about the system, server or application under consideration is sought.
(ii) **Scanning:** In this phase a vulnerability scan of the system is performed using automated tools for mapping the network to know about the firewall and routers.
(iii) **Gaining access:** This is done by attacking and infiltrating into the system with the help of the acquired information in first and second phases, its analysis and drawing a blueprint of the network.
(iv) **Maintaining access:** This is essential unless the goal of hacking is fulfilled. Trojans and backdoors are used in this phase.
(v) **Covering of tracks:** This is necessary in order that the system may not know that it was attacked. Entry of the hacker remains hidden by deleting all logs and any folders created by the hacker, together with uninstalling the applications and traces of any other activity carried out during hacking.

Malware is an important tool widely used for intruding into computers. So, we begin with malware. Then the different kinds of hacking attacks will be dealt with followed by hacking prevention techniques.

25.10 Malware (malicious software)

'Malware' is a blanket term for any intrusive software to damage a computer, network or server by infecting, stealing or by gaining remote control (figure 25.15). Malware manifests itself by causing aberrations in behavior of the computer such as its slowing down, loss of its storage capacity, changes in browser settings, inundation

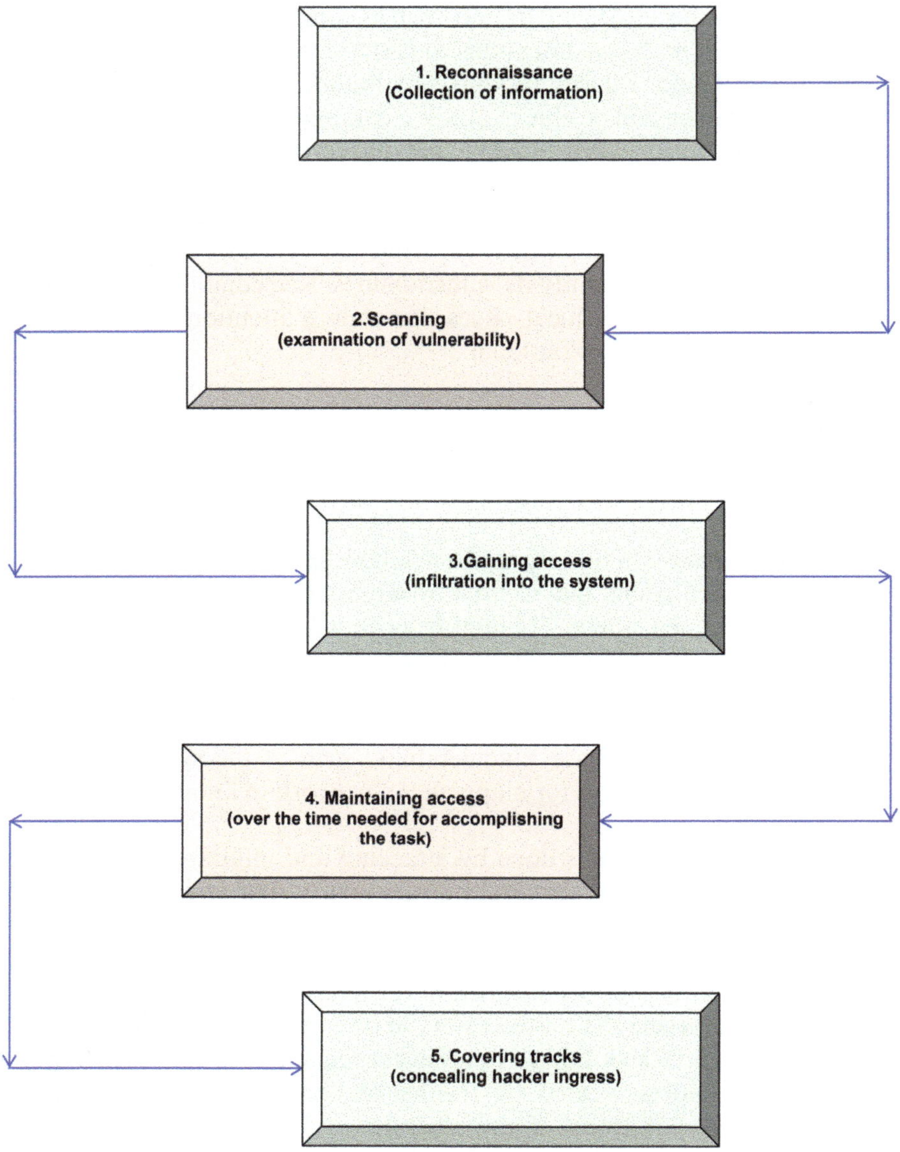

Figure 25.14. Phases of ethical hacking process.

of computer screen with annoying advertisements, etc. Malware for cybercrime automation is known as crimeware.

Common forms of malware include viruses, worms, trojan horses, wiper, spyware, ransomware, rootkit, keylogger, etc (Sikorsk and Honig 2012). Malware is removed by installing and running a cybersecurity program and changing all passwords.

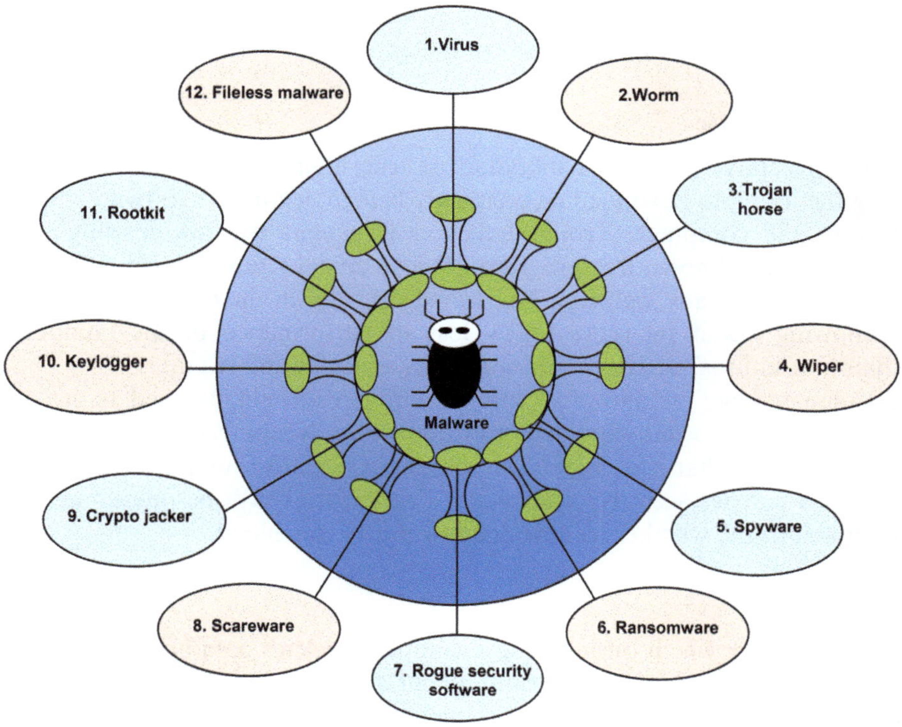

Figure 25.15. Family of malware.

25.10.1 Virus

The virus is a computer program which upon execution replicates itself by copying into other programs and inserts its own code, causing damage to software and data. It is different from a computer worm because it needs a host program to run and requires user action for transmission while the worm can work without a host program and can spread across systems and networks on its own. A virus is malware but all malware are not viruses. Based on the method used by a virus to infect a computer, the virus is referred to as a file, boot sector, macro or script virus.

25.10.2 Worm

This is a self-replicating program spreading to other computers on the network while remaining active on the infected computers. It is transmitted through weaknesses in software. A worm spreads through a file sent as an email attachment, via a link to web or link to ICQ (derived from 'I seek you') message, or via a peer-to-peer file sharing network. Often, a worm arrives as an attachment with email which when opened downloads the worm into the computer. The worm installs and starts working. The worm is a stand-alone program unlike a virus which requires an active host program to run. The worm modifies and deletes files. By making multiple copies of itself without human interaction, the worm exhausts the available hard drive space and bandwidth of the infected computer.

25.10.3 Trojan horse

This is a deceptive program concealing its true intent. It infects a computer by tricking the user into downloading seemingly benign, legitimate software, and takes control of the computer. Trojan infection is caught by downloading cracked applications or unknown free programs. It also comes from opening email attachments or visiting shady websites. The trojan can stay in the computer without the user knowing about its presence. A loss in performance of the computer or exhibition of unusual activities are symptoms of trojan infection.

A trojan scanner or malware-removal software should be used to get rid of trojans. Diagnostic scans should be run frequently. Keeping the operating system updated ensures that the user has the latest security updates. Updating of applications patches security weaknesses. One must always be questioning about email attachments, whether to download or not to download.

25.10.4 Wiper

This is a malware which on infecting a computer deletes data and software on its hard drive. The wiper targets important files of the computer which are crucial for its functioning. It deletes those files or overwrites corrupted data on these files. Sometimes it encrypts the files. A wiper attack is observed in the form of voluminous deletion of files. Immediately data should be backed up to another location without internet connection.

25.10.5 Spyware

This is spying software that collects information about a person or organization and sends it to a third party without consent. Spyware gains entry into the computer through drive-by-download or a pop-up window. It captures data such as passwords, usernames, credit card numbers, etc, through keystrokes, screen captures, etc. The stolen data are relayed to the spyware author.

Programs must be downloaded from trusted websites. Internet security with reliable anti-malware must be used.

25.10.6 Ransomware

This is a malware for denying access to users or an organization of the files on their computers by encrypting the files, and scaring or intimidating users into paying a ransom fee to obtain the decryption key. The ransomware gains access to a computer from a phishing email containing a link to a website providing a download. After entering the computer, the ransomware starts encrypting files. Once the encryption process is completed, a ransom is demanded, typically as an amount of cryptocurrency for reinstating the victim's access to files. As the attack generally begins with an email prompting the user to click on a link, training and education on cybersecurity are necessary. Further, data must be regularly backed up to avoid payment of ransom or loss of data. Patching is essential because the attacker looks for unpatched vulnerabilities. Unpatched systems are always prone to viral infections.

25.10.7 Rogue security software

This is a software misleading a user to think that his/her computer is virus infected. Mimicking a real anti-virus solution, it gives pop-up messages, alerting and convincing the user to pay for a virus removal tool. In fact, malware on the computer is the rogue security software itself. No data is stolen but the user is trapped into paying for a fake anti-virus tool. The problem may seem to be over temporarily after which payment is asked for another service. This software often comes with files downloaded from untrusted websites. Running valid anti-virus program on the computer provides protection against this malware.

25.10.8 Scareware

This is a malware which causes anxiety or threat perception in users to purchase unwanted software. Pop-ups appear on the user's screen claiming to have found some virus in the user's computer and inviting the user to download or purchase a quick-fix software. Instead of resolving the issue, this software steals the user's personal data.

Uniform resource locator (URL) filters compare network traffic against a database preventing users from accessing harmful sites. Pop-up ad blockers are programs for prohibiting pop-ups.

25.10.9 Crypto jacker

This is a malware to infect computers for malicious cryptocurrency mining without knowledge of the user. The computing power of the victim is used for generating cryptocurrency. Web browser downloads cause such attacks which compromise desktops, laptops and smartphones. When the user clicks on a malicious link, a crypto mining code is loaded into the user's computer. Alternatively, a JavaScript code is embedded in a web page. Through it, crypto mining is done on the computer of a visitor to the web page. A comprehensive cybersecurity program provides protection against crypto jacking.

25.10.10 Keylogger

The keylogger is a hardware or software to record the keys struck on the keyboard to covertly monitor the actions of a person (Malwarebytes 2022). A hardware keylogger is planted in the hardware of the victim's computer or may be physically lodged as a plug-in device on the user's unlocked computer in the absence of the user. A software keylogger gets inside a computer like other malware.

The keylogger can take screenshots, steal passwords and personal information, and transmit them to a remote computer. Keyloggers are not always used for nefarious work. They are used by organizations to solve technical problems in their computers or to monitor the activities of employees. Parents can use it to keep watch over their children using computers.

It is beneficial to be sceptical about downloading email attachments. Infection is preventable by always-on anti-virus programs.

25.10.11 Rootkit

This is a malware designed to get unauthorized access to a computer while itself remaining hidden. Most rootkits enter a computer by riding on trusted software or with virus infection. The controller of rootkit can remotely change the configurations of the host computer leading to strange behavior.

Security of the computer is made more effective by patching the operating system and applications against vulnerabilities. Virus definitions must be always kept up-to-date.

25.10.12 Fileless malware

This malware differs from traditional malware in the respect that the attacker is not required to get any executable file downloaded and installed in the targeted computer. Owing to its non-dependence on installed files, no footprint is left by this malware making it difficult to detect. Instead of installed files, the legitimate programs of the victimized computer are used to infect it. Nothing is written permanently on the hard disc drive of the computer. The fileless malware exists in the random-access memory. So, if a computer is rebooted, the attack of fileless malware must be re-launched.

The best solution to avert a fileless malware attack is to seal the entry points of this malware into the computer. Any unpatched applications must be patched up. Any software vulnerability issues must be resolved. Patching and updating the software of the computer is mandatory.

25.11 Hacking threats and attacks

25.11.1 Advanced persistent threat (APT)

This is a politically- or economically-motivated cyberattack in which the attacker gains and maintains access on a computer for an extended period of time without getting detected. The access is gained through social engineering, human intelligence or infiltration. During the dwell time, the attacker monitors, intercepts and relays sensitive data.

25.11.2 Arbitrary code execution (ACE)

This is an attack to run arbitrary commands or code on a target computer. It happens when an attacker spots a security flaw in software or hardware, and exploits the flaw to achieve administrative control of the device and run commands of attacker's choice. ACE exploit is a program that utilizes such a flaw. Remote code execution (RCE) deals with carrying out this kind of task over a network.

ACE is done through the instruction pointer in a running process. The pointer indicates the next instruction to be followed. So, control over instruction pointer is established by injecting a code and using a vulnerability to alter the instruction pointer such that it points towards the injected code instead of the next planned instruction. In this way, the computer is deceived into implementing the injected code. This deception is possible because the computer does not differentiate between data and code. In many computers, insertion of a no-execute bit makes this trespassing difficult.

25.11.3 Backdoor attack

This is an undocumented way of accessing a computer for stealing data, defacing a website or hijacking the server. The backdoor is a type of malware to bypass normal authentication for getting remote access to a computer and issuing commands.

Installation of a backdoor entails remote file inclusion (RFI). In an RFI attack (figure 25.16), the referencing function in an application is used for uploading malware from a distant URL (uniform resource locator). With the help of a scanner or search engine, the attacker identifies a website with weak security. The RFI vulnerability is then utilized to upload a malicious piece of code known as the backdoor shell. The outcomes are compromising of the website, hijacking of the server and stealing of data. Local file inclusion (LFI) is done in servers by a similar technique through web browsers using insecure local file upload functions. Weeding out the backdoor shell files is difficult as they are always masked with alias names. However, file inclusion risks can be minimized by properly validating and sanitizing the inputs. Nonetheless dedicated security is essential.

Note on hardware backdoor: Hardware backdoors include unwanted codes in the firmware of computer chips like a central processing unit (CPU) or network card inserted by a member of a design team or a third-party internet protocol (IP) provider for covertly bypassing normal authentication procedures (Waksman and Sethumadhavan 2011). They are highly challenging because they cannot be removed by anti-virus software and can evade disk encryption.

25.11.4 Code injection and cross-site scripting (XSS)

Code injection entails introduction of a code in a computer program to change the course of its execution. Lack of validation of input/output data, e.g., data format, allowed characters, etc, is responsible for such assaults. A bug created by processing invalid data is exploited. Code injection is used to change values in a database, install malware or gain higher access to resources.

Cross-site scripting is a client-side code injection attack in which malicious executable scripts are inserted into the codes of an otherwise benign and trusted website (PortSwigger 2022). 'Client side' refers to all things in a web application that are displayed or happen on an end-user device, including what the user sees and any actions which the application performs on the user's browser.

In an XSS attack, web pages or applications are used by the attacker to send a malicious code, often in the form of a segment of JavaScript, through which the user's

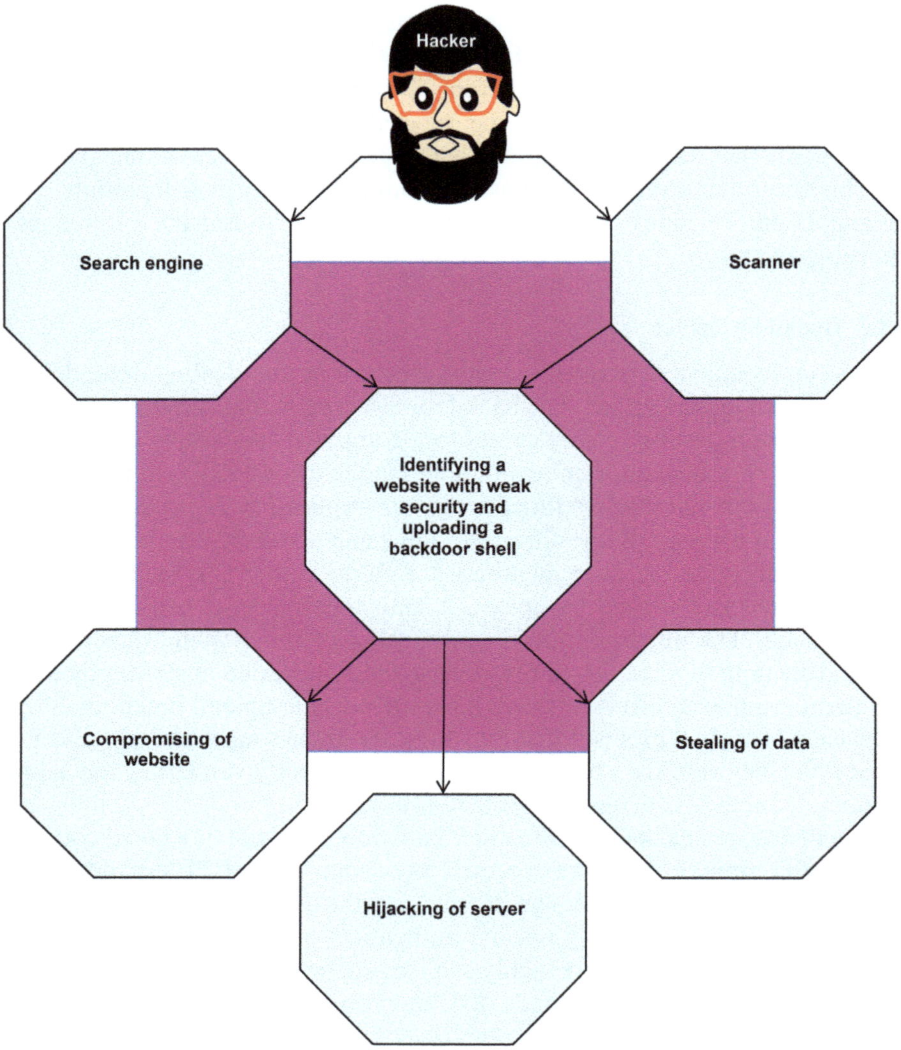

Figure 25.16. Flow of an RFI attack.

interaction with a vulnerable application is impaired. Damage caused by XSS ranges in severity from annoyance to the end user to complete compromise of an account.

Common XSS attacks are:

(i) **Reflected XSS:** In reflected XSS, the attacker's payload is carried by a request sent to a server on the web and is contained within the hypertext transfer protocol (HTTP) response. The victims are enticed into making requests to servers by sending malicious links to them. Emails or social media spread the attack. As the injected script is reflected off a web server, the attack is called reflected XSS. The attack carried out through a single request/response cycle is said to be non-persistent or type I XSS.

(ii) **Stored XSS:** In stored XSS, also known as persistent or type II XSS, the attacker's payload is injected into the chosen web application. The malicious code permanently resides in the application's database if it does not validate its input. Then spiteful script is entered in a user input field, e.g., comments on posts. When the infected web page is opened by a visitor, the attacker's payload appears in the unsuspecting user's browser who unknowingly becomes a prey on executing the malicious script.

(iii) **DOM-based XSS:** This occurs when a web application writes data supplied by the user to DOM, the document object model, an application programming interface for hypertext markup language (HTML) and extensible markup language (XML) documents. The application reads the data and transmits it to the browser of the user. Improper treatment of data enables the attacker to inject the malicious payload, which is stored in the DOM and is executed upon reading the data from DOM.

XSS is preventable by applying precautionary procedures:

(i) All inputs should be treated as if they are untrusted.

(ii) Defensive security measures such as escaping and encoding must be used. Escaping is a technique of securing output by stripping out incorrectly formed HTML or script tags.

(iii) Trusted sanitizers should be used to clean and parse HTML.

(iv) Setting HttpOnly flag for cookies makes them inaccessible via client-side JavaScript.

25.11.5 Drive-by-download and data breach

This is an authorized or unauthorized download happening during opening an attachment to an email or clicking a link to a website, known as a drive-by-download. A drive-by-download attack seeks to hijack, spy on, or incapacitate a user's device or damage data. The attack is initiated either through (Kaspersky 2022a):

(i) authorized downloads such as clicking on a deceptive link, or

(ii) unauthorized downloads where infection is caught during a visit to a website.

During attacks of category (i), a vector is created for delivering malware. The malware is installed on the user's device when the user interacts with the vector. With the help of malware, the attacker gets entry into the user's device.

During category (ii) attacks, a malicious component is plugged into a web page exploiting a security flaw. Upon triggering the component by a visiting user, the malware is downloaded into the user's device to start evil activities. Both zero-day exploits with unknown fixes and known exploits with un-installed solutions are utilized by the attacker.

For protection, endpoint users should install programs using admin accounts because admin privileges are required for downloads to get installed without the user's permission. The web browser and the operating system should always be kept updated. An ad-blocker is helpful because online ads are frequently used to load infections. While internet security software is essential on the user's device, visits to suspicious websites should be avoided. A website owner should invariably keep all components updated and remove the outdated ones. Admin accounts should have strong usernames and should use strong passwords. Software for website security is indispensable.

A data breach means a violation of security for viewing or stealing sensitive data. To prevent breaches, all systems in a network should use fully patched and updated software. It is better to be prudent and vigilant while giving information on a social media profile. Opening emails from unfamiliar senders must be avoided. A disaster recovery plan should be ready to combat breaches.

25.11.6 Denial-of-service (DoS) attack

In this attack (figure 25.17), a machine or network is overloaded to make it fail and become inaccessible to intended users by sending more traffic from a source or distribution of sources (distributed DoS). Main types of DoS attack are given in subsections below (Cloudflare 2022).

25.11.6.1 Buffer overflow attack

A buffer is a memory space where data is temporarily stored during transference from one location to another in a computer. When a program writes more data on a buffer exceeding its capacity, the extra data is overwritten in memory locations adjacent to the destination buffer. This situation is called buffer overflow.

In a buffer overflow attack, more traffic is sent to a network address than it has been designed to handle leading to buffer overflow. If an executable code is overwritten, the program behaves erratically. The attacker may introduce a different code. So, the execution path of a program may be changed resulting in damage to files and exposure of private information.

25.11.6.2 ICMP flood attack

Generally, internet control message protocol (ICMP) echo-request and eco-reply messages are used to ping a network. A ping (packet inter-network groper) is a program used to verify the existence of a destination IP address and acceptance of requests by it.

During the attack, also called a smurf attack or ping of death, the network is flooded with spoofed (hoaxed) request packets compelling it to use all its resources to reply with an equal number of reply packets, thereby becoming unavailable to normal traffic.

To prevent this attack, the ICMP functionality of the victim router or computer is disabled. The perimeter firewall is set to block pings. Outside attacks can be avoided but not those from inside the network. In another way to avert this attack,

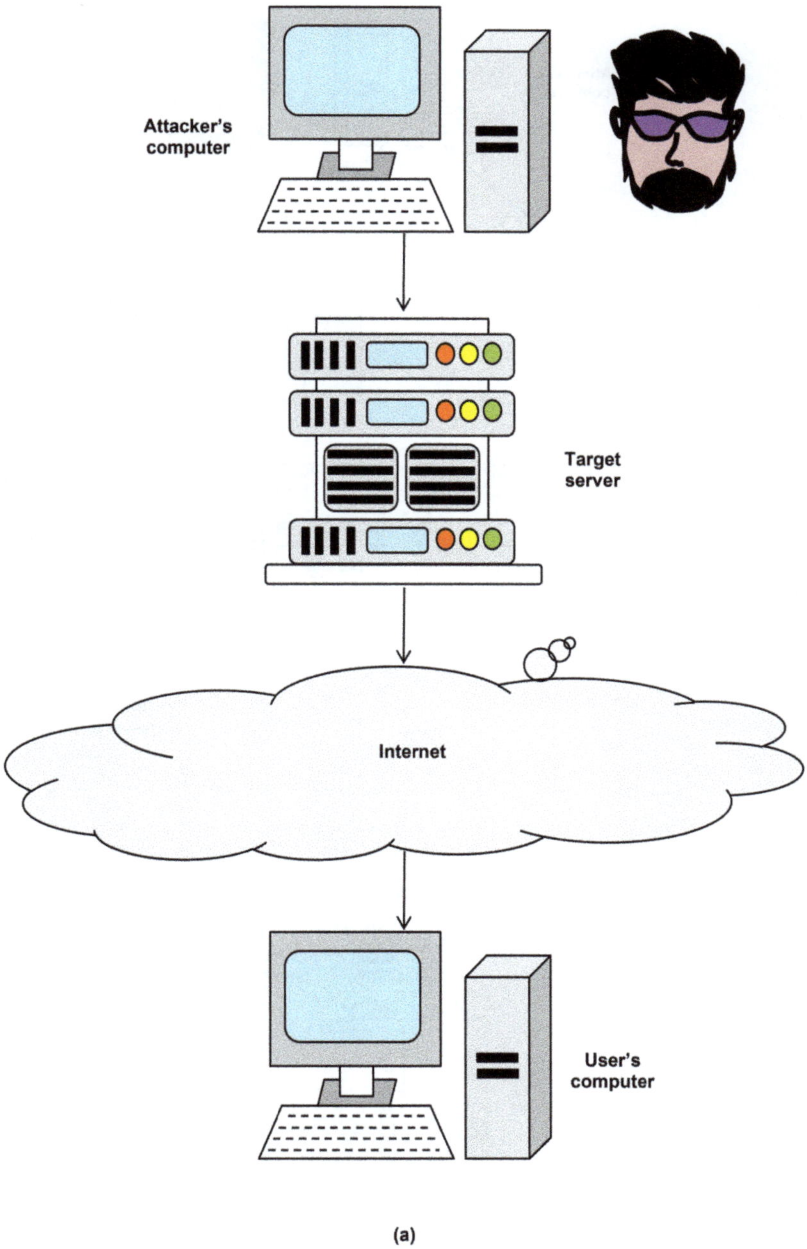

Figure 25.17. Types of DoS attack by: (a) a single computer and (b) a distribution of computers.

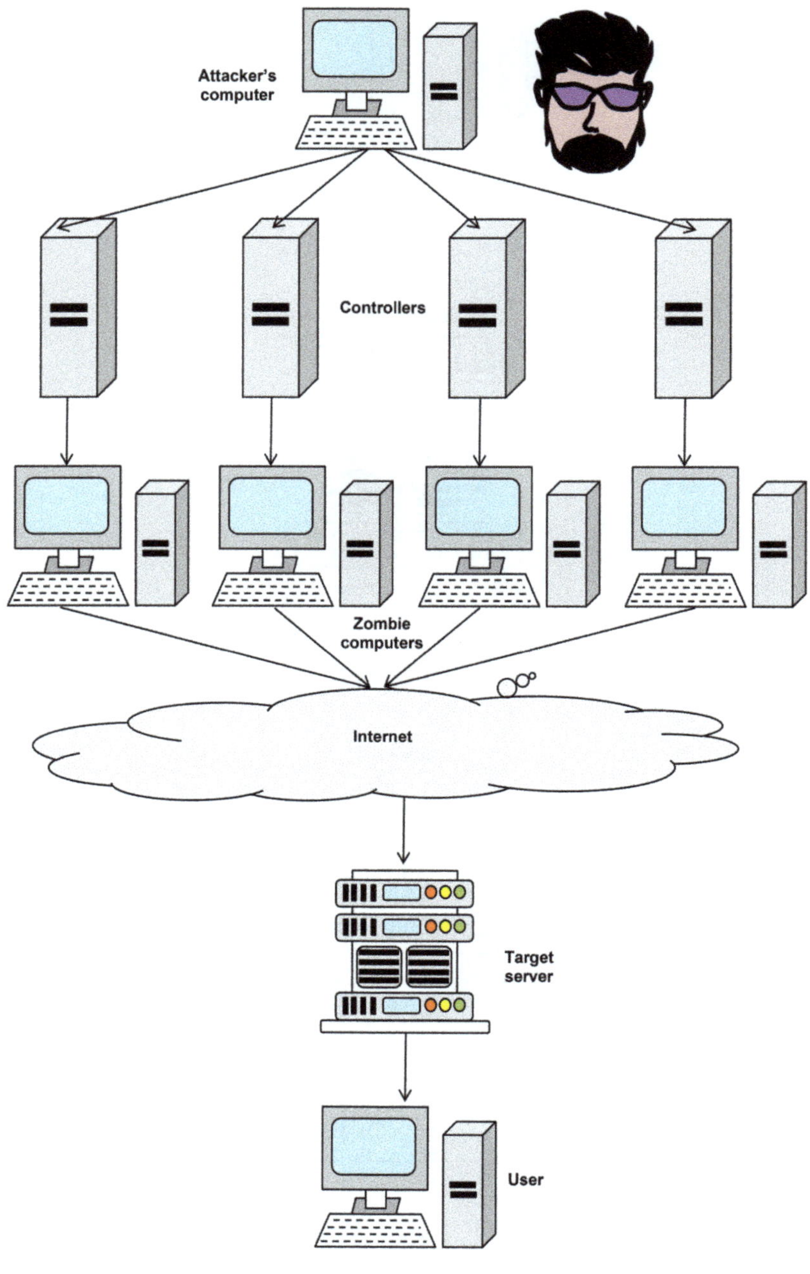

Figure 25.17. (Continued.)

thresholds for the allowed rate of receipt of packets of various forms of traffic are fixed in the default packet handling page. Traffic of the specified type will be dropped when the packet rate crosses the limit.

25.11.6.3 SYN flood attack

In this attack, also known as TCP SYN (Transmission control protocol Synchronize) flood or half-open attack, a server is showered with requests for connection. The corresponding acknowledgements are not replied to, resulting in a large number of open TCP connections. Consumption of resources of the server by these open connections crowds out the genuine network traffic. Neither new connections can be opened nor normal server function is provided to the already connected users.

25.11.6.4 Distributed denial-of-service (DDoS) attack

In place of one computer, many previously compromised and controlled computers called zombies are used by the attacker to send a large volume of requests to a network to put it out of action. Thus, it is a synchronized attack orchestrated by multiple computers on a single target.

An early sign of a DoS attack is obtained by carefully watching network traffic. A sudden increase in traffic or abnormal traffic pattern is a warning of this kind of attack. So, traffic reports must be constantly studied. Moreover, the anti-virus software must be kept updated. All security vulnerabilities must be patched and sealed.

25.11.7 Eavesdropping

This is a sniffing, snooping or sneaking act in which attackers secretly listen to the private network traffic between servers and mobile devices without consent by taking advantage of insecure connections (Alpine Security 2021). Data sent on open networks without using passwords or any form of encryption are liable to fall prey to eavesdropping. Weak passwords allow attackers to gain access into networks to intercept communication. The attacker inserts software in the network path using malware and clandestinely monitors the activity. In passive eavesdropping, the attacker observes the communication without interfering with it. In active eavesdropping, the attacker not only observes communication but also tries to modify it.

Data encryption must always be used in transit to thwart eavesdropping attacks.

25.11.8 Email spoofing

Here an email is sent from a fake address duping the recipient that it is from a trusted source, e.g., a colleague or branded company. To give an example, an email comes to a customer stating that the customer's account will be terminated soon. To continue the account, the customer must click a link and change the password. The customer is fooled into believing that the email has come from a correct source by including standard elements from the official website of the source. If the customer accepts to do as told, the attacker can access the account and indulge in crime.

Sender policy framework (SPF) is a protocol for authentication of emails. It provides a method for receiving email servers to verify whether an incoming email from a domain has been sent from an authorized source. It works using the established domain name system (DNS). For a given domain, a policy is published giving details of email servers which are permitted to send emails from that domain. It is called an SPF record and forms a part of the DNS record. On receipt of an email by a server, it looks at the rules for the return path known as the bounce domain in DNS. A comparison is made between the IP address of the email sender and the IP addresses mentioned in the SPF record. Then the rules laid out in the SPF record of the sending domain are applied to take a decision about the acceptance, rejection or flagging of the received email. In this way the SPF protects email senders/recipients from spoofed emails.

Be cautious of emails producing a feeling of danger or urgency. Copy and paste some portion of the text of a received email message into a search engine. Perhaps experiences and warnings of earlier deceived people are published on the internet.

25.11.9 Exploit

The exploit is a snippet of code or a specialized program infiltrating software/hardware by taking advantage of security flaws and causing installation of malware. It is a tool to leverage a security vulnerability. Exploit is not malware itself but a channel to deliver malware. In an exploit attack, the computer is slowed down or may suffer frequent freezes or crashes.

An exploit which has been investigated and documented by experts is called a known exploit. Patches have been created for fixing the responsible security flaws. An unknown or zero-day exploit is one which has not been identified and researched. As the name suggests, a zero-day exploit is created and launched on the same day on which a hacker discovers a vulnerability. The exploit developer has known about the vulnerability for ‘zero days’.

Vendors patch known bugs in software on a regular basis. To mitigate exploit attacks, any software patches and updates must be applied soon after release. The operating system must be kept up-to-date.

25.11.10 Malvertising

Online advertising for spreading malware is known as malvertising or malicious advertising. An infected ad from an advertising network is used to inject malicious codes into the user’s computer. Characteristically, the attacker breaches a third-party server and inserts malicious code in a display ad. Upon clicking by a user, the corrupted code is installed in the computer of the user. In general, the attacks are done using drive-by-download, a compelled redirection of the browser to a malicious site or by execution of Java or Flash script for displaying the advertised content. Besides keeping all software and extensions up-to-date and installing anti-virus software and blockers, one must not allow Java or Flash programs to run automatically when surfing the internet.

25.11.11 Social engineering

This is the tactic of psychological manipulation of people through human error for divulging confidential information such as passwords to gain control over their computers or for stealing financial data (Kaspersky 2022b). As an example, a password phishing email is sent by the attacker. The email looks as if it has come from an esteemed company. In the email, a link is provided to enter the password and banking information in the hope that a potential victim will fill in the requested data which will be utilized for malevolent activities. By training and education and keeping vigilant, people can stay protected. Two-factor authentication (2FA) is an identity and access management technique in which a user is granted access to resources and data of an application after furnishing two pieces of evidence to an authentication mechanism, e.g., a password as the first factor and finger print or facial scan as the second factor. Anti-virus software, email filters, and firewalls warn against suspicious objects/activities.

25.11.12 Phishing

Phishing is a social engineering attack in which a fraudulent message is sent to a person pretending to be sent from a legitimate institution, a trusted or reputed entity to trick the user into revealing personal information such as login credentials, credit card numbers, etc or to install malware in the user's computer. The best defence against phishing is alertness and caution about threatening calls for immediate action to click or open an attachment, on emails from new or uncommon senders, or email messages having spelling or grammatical errors because professional companies have editorial teams to ensure perfection.

25.11.13 Privilege escalation

This implies exploitation of a design flaw in an operating system, e.g., cross-site scripting, improper cookie handling or weak passwords, to gain elevated access to its normally protected resources. In horizontal privilege escalation, the attacker takes over the account of another user and uses the privileges granted to the user. In vertical privilege escalation, the attacker seeks more privileges from an already taken over and compromised account.

The attacker performs reconnaissance, changes security settings of privileges and loads more software into the targeted computer. Access is deepened and may lead to breach of data.

Anomalous activities on user accounts indicate privilege escalation. Hence behavioral analytics is a useful tool for detecting attempts of invasion.

Cross-site scripting and improper handling of cookies can be corrected by paying attention to programming. Password weakness can be avoided by user education.

25.11.14 Spamming

This means sending a large number of unsolicited messages, usually emails or texts, generally for advertising purpose, to several recipients. Spam is a junk email. Spam is a source of nuisance. It is sent because it offers a cheap way of promoting products. Spamming differs from phishing which is a serious offence because phishers want to capture sensitive personal information.

25.11.15 Zombie attacks

A zombie computer is a network-connected computer that has been compromised by a hacker, i.e., a hacker-grabbed computer, and that can be used by the hacker for malicious activities. A group of zombie computers is used to launch a DDoS attack on a website by saturating it with requests. Often the server crashes. Then the true users cannot avail of the facility. A different type of attack is made with pulsing zombie computers which attack target system at a low intensity so that the server is not disabled but is slowed down. These pulsing zombie attacks may extend over long periods, even years evading detection. The real cause of slowing down is a low-intensity DDoS attack but users may confuse it with other types of problems. Another way in which zombie computers are used is for sending spam. By this method, the attackers are difficult to detect. Also, bandwidth expenses of attackers are reduced because bandwidth costs are borne by the owners of the zombie computers, not by the attackers.

25.11.16 Botnet attacks

The botnet (from 'ro**bot** + **net**work') is a group of zombie computers, i.e., a hijacked computer network (bot network) under single-party control that is infected with malware and used for cybercrimes such as generating spam and phishing emails and performing DDoS attacks. The person controlling the bot is known as a bot herder. In the centralized client-server model, the full botnet is operated by a single command-and-control server. In the decentralized peer-to-peer model, the bots are not in direct contact with the command-and-control server. Instead, they share information and commands with each other.

A botnet is produced in three stages. In the first stage, a weakness is found in a website application or behavior of users. In the second stage, the chosen devices are infected with malware and control is set up over them. In stage 3, the bot herder mobilizes malicious activities.

Common protections against botnet attacks are periodic security awareness lectures, keeping software updated, regularly updating passwords especially the privacy and security options, and using a quality anti-virus program.

25.12 Defences against hacking

25.12.1 Access control software

This is a program for authentication and authorization to use resources of a computer. It regulates the granting of admittance to ensure correct utilization of resources in a computing environment. There are four types of access control:

(i) Discretionary access control (DAC): It is the least restrictive and extremely flexible kind of control in which the software and data are completely managed by individual end-users;
(ii) Managed/mandatory access control (MAC): It is the converse of DAC, and the most restrictive type in which the system owner only has control over permissions or privileges;
(iii) Role-based access control (RBAC): In this type of access control, the system administrator grants access keeping in view the role of the key holder in the organization; and
(iv) Role-based access control: In this kind of access control, access is given or denied according to a set of dynamic rules laid down by the system administrator.

25.12.2 Anti-keylogger

This is a program to detect the presence of any keylogger software on the computer. If found, the software is deleted or immobilized. Anti-keylogger scans are specially recommended on computers used for banking for protection of passwords, credit card numbers, etc. Further, anti-keylogger scanning of a computer should be frequently done to minimize the time duration during which a keylogger is present on a computer, e.g., if anti-keylogger scans are done every alternate day, the period of presence of a keylogger on the computer is restricted to at best a day.

25.12.3 Anti-malware

This is a program for real-time protection of a computer from malicious software by detecting and removing it. Malware is a broad term. Virus is a subset of malware. Therefore, anti-malware protection is more comprehensive than anti-virus software. An anti-virus program detects specific types of viruses from their known signatures. Anti-malware software works on non-signature-based anomalies. It has more advanced features and more extensive coverage than an anti-virus program. It scans the system thoroughly looking for spyware, adware, worms and spams. It performs behavior monitoring. Any malware is identified from its behavior and character, isolated and further analyzed, and prevented from being installed.

25.12.4 Anti-spyware software

This is a program to detect and remove any spyware software that is installed on a computer without the user's knowledge to gather information from the computer. Spyware may install more software on the computer and may also degrade its processing power, making anti-spyware necessary. The detection of spyware is done through rule-based techniques or from definition files for identification of spyware.

25.12.5 Anti-subversion software

This is a program to protect against any software performing unintended functions such as by altering the program code, e.g., malicious rules can be loaded by code tampering. Anti-subversion protection is provided either statically or dynamically. Static anti-subversion is done at the time of code construction by security auditing and code verification against various types of attacks. Dynamic anti-subversion is done during execution of the code by checking for unplanned behaviors of the program and protecting against subversion.

25.12.6 Anti-tampering software

This is a program to make modification of the computer software difficult. Tampering can be prevented externally through anti-malware or anti-virus scanners or internally by tamper-proofing defense mechanisms via runtime integrity checks or by obfuscation making codes difficult to understand. Often white-box cryptography is used in which the output generated from a given input by a standard cryptographic implementation is produced by a process which is internally different from the standard process. Hence cipher keys are not disclosed so that an attacker is put in misperception. Confidentiality of keys brings protection.

25.12.7 Anti-theft system

This is a program for tracing back a stolen or lost item, e.g., an RFID (radio-frequency identification) tag or GPS (global positioning system) locator. Electronic devices such as laptops and mobile phones have tracking software for phoning home to inform about their location. Phoning home is useful in getting back missing items. Security tags attached to commercial products work in combination with a surveillance system to prevent shoplifting. Certain items can be disabled in case of theft or loss, e.g., a bank card is disabled by a telephone call to the bank when it is stolen.

25.12.8 Cryptographic/encryption software

This is a program to conceal the content of a message by encoding it, e.g., by converting the information from plaintext to ciphertext. The code is decipherable only by authorized personnel who hold the key. Only these people can access the original plaintext. This means that encryption is a method to make a message unreadable by unauthorized parties and so irrelevant to them.

Encryption is of two types: symmetric and asymmetric (Kessler 2022). This classification is based on the number of keys used in transferring information, and the two classes are named accordingly.

Symmetric encryption uses a single key both for encryption and decryption (figure 25.18(a)). The plaintext is converted into ciphertext using a key K. The ciphertext is reconverted to plaintext with the same key K. As this key must be transmitted by the sender to the recipient to read the message, the transmission must be adequately secured otherwise it may be stolen by anyone on the way, the message may be read and misused.

Asymmetric encryption uses two keys, one key for encryption and another for decryption (figure 25.18(b)). One key is called public key. The other key is known as private key. The plaintext is transformed into ciphertext with public key K_{Public}. The ciphertext is converted back to plaintext with private key K_{Secret}. The public key

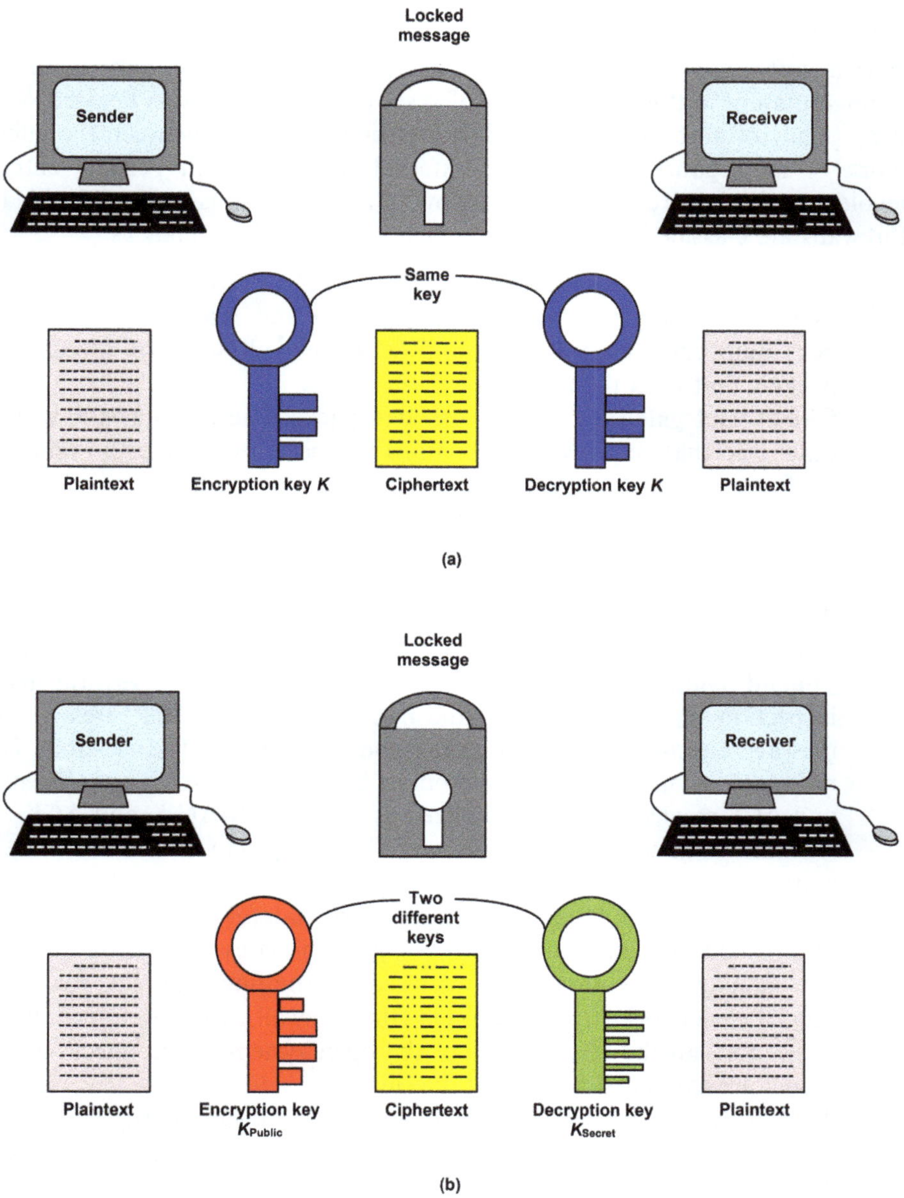

Figure 25.18. Types of encryptions: (a) symmetric and (b) asymmetric.

K_{Public} is either known to the general public or shared among lawful users. The private key K_{Secret} is kept strictly confidential. It is shared only between the dispatcher and the receiver. This makes the enforcement of rigorous security measures unnecessary. The relaxation provided makes data security easier.

25.12.9 Firewall

This is a software- or hardware-based device for network security controlling incoming/outgoing traffic on the basis of a set of security rules as decided on the security policy of an organization. It is a first line of defence between the internal network of an organization and the external internet which permits entry of benevolent traffic inwards and blocks inflow of any dangerous traffic (Liu 2011).

Firewalls are classified on the basis of their operational methods as:

(i) **Packet-filtering firewalls:** These are the most basic type of firewalls. They function by analyzing data packets for source and destination IP addresses, port numbers and protocols, and selectively allowing or blocking the packets that fulfil the checklist.

(ii) **Circuit-level gateways**: They work like packet-filtering firewalls with the difference that requested transactions are permitted while the remaining traffic is filtered out.

(iii) **Proxy firewalls:** They work at the application layer between two systems. Hence, they are called application-level gateways. They pretend to be original clients on a web server. After connection is made, the incoming data packets are examined and transferred to the original client only if found safe. Thus, additional protection is provided to the original client.

(iv) **Stateful multilayer inspection (SMLI) firewalls:** They perform packet inspection along with TCP handshake verification. TCP uses a three-message handshake for negotiating and beginning a TCP session. These messages are named as: SYNchronize, SYNchronize ACKnowledgement, and ACKnowledge, and symbolized as SYN, SYN-ACK, and ACK.

When a connection is made, the firewall produces a state table for storage of source and destination IP addresses and port numbers. Security rules for traffic regulation are formulated by stateful inspection technology, also called dynamic filtering. In this technology, the state of active connections is monitored. The acquired information is applied for permitting the network traffic through the firewall. In stateful multilayer inspection, all the seven layers of the open system interconnection (OSI) model are examined for packet filtering.

(v) **Next-generation firewalls:** They integrate multifarious security features and functionalities including deep packet inspection, stateful inspection, access control and blockage of application-layer attacks, which required different software in the past.

25.12.10 Intrusion detection system/intrusion prevention system (IDS/IPS)

IDS/IPS is a program or device to check malicious activity or violations of policy in a network, and take suitable necessary actions.

An IDS is a passive system for threat detection. When a threat is perceived, an alert is created. The IDS is signature or anomaly-based or a hybrid of both types. The signature-based IDS uses a library of known threats for threat identification. An anomaly-based IDS models normal behavior of a system and looks at nonconformities with this model for threat recognition. A hybrid IDS combines both techniques.

Unlike the IDS, an IPS is an active system which possesses IDS capabilities of identifying threats using signatures, anomalies or both, and also the necessary corrective capability to prevent occurrence of intrusion.

25.12.11 Sandbox

This is an isolated virtual environment program allowing one to run the software without affecting the remaining system. The sandbox is a safe testing environment mimicking the end-user environment to provide its accurate representation. Inside the sandbox, a new programme can be run and observed to see whether it has any harmful effects. During the running of this program, the remaining system remains shielded from any mishappening in the sandbox. Sandboxing is particularly effective against zero-day threats which have not been seen before and which do not match with the signatures of known malware; hence they remain undetected by traditional signature matching.

25.12.12 Security information and event management (SIEM)

This is a program to automate collection of logging data from firewalls, anti-virus, intrusion detection software, etc into a central repository to analyze the trends in data and provide alerts in real time. Artificial intelligence is used to automate the manual processes for detecting threats. Any anomalies in behavior are exposed. Event correlation is an important part of SIEM. It is done by data analytics to identify and understand complex patterns in data. As the manual workflows for detailed analysis are offloaded, the mean time for threats detection and also that for responding to them are drastically reduced. Pre-planned correlation rules are applied to produce alerts for initiating actions before major threats develop.

SIEM has become an indispensable tool for organizations that have to fulfil strict regulatory compliance. Automatic data collection and analysis minimizes time and efforts in compliance reporting. Aided by artificial intelligence (AI)-driven automation, SIEM enables quick recognition of threats and provides auditing of regulatory compliance, thereby increasing the efficiency of the organization in the detection of advanced and hitherto unknown threats, and lowering security risks.

25.12.13 Software patch

This is a quick-repair program to resolve functionality problems, add new features or improve security. A patch means a set of changes to a computer program for improving its performance. Security patches, feature updates and bug fixes are the three common types of patches. From the security viewpoint, a patch may be looked upon as a software or an operating system update for enhancement of security features. Updates must be downloaded only from trusted websites of vendors.

25.12.14 Vulnerability management software

This is a program to identify and mitigate security vulnerabilities. A vulnerability is a weakness that can be exploited by a threat. Vulnerability management is an ongoing or continuing process of identification and remediation of vulnerabilities at different points of a system. It differs from vulnerability assessment which is one-time evaluation of a system's security. A good vulnerability management program utilizes knowledge about threat intelligence and business operations for prioritization of security risks and their mitigation.

25.12.15 Packet sniffer

Also known as a packet analyzer, this is a hardware or software device to monitor the flow of streams of data between any two points, e. g., between two computers or between a computer and network. It is used by organizations for tracking the use of network by staff. It captures information about the website visited by a user, the website contents, and the files downloaded. It also scans incoming data from a network to know about particular elements in malicious codes. In this way, it prevents the computer from being infected with viruses. A packet sniffer works in an unfiltered or filtered mode. In the unfiltered mode, all packets are stored in a memory disc to be examined later. In the filtered mode, only packets containing specific data elements are allowed.

A packet sniffer allows a network administrator to monitor data traffic and interpret it for diagnosing performance problems. A hacker uses the sniffer for unauthorized monitoring of network activity.

25.12.16 Public key infrastructure services

Simple passwords may not be sufficient to confirm identity of individuals or systems. More rigorous proofs are essential for protection of people, devices and data in internet activities such as banking, e-commerce and emails. Public key infrastructure (PKI) shown in figure 25.19, is a combination of hardware, software, policies and processes to manage identity and security in electronic transfer of information. It provides a framework for executing various security tasks such as authentication of different parties taking part in electronic business transactions, authorization of access to systems, encryption of contents of messages and verification of the authors of messages. It oversees the issue of digital certificates for protection of data and provision of reliable end-to-end communications (Buchmann *et al* 2013).

The enabling technologies for PKI are public key or digital certificates and public key encryption. A public key is a cryptographic code i.e., a large numerical value used for encryption of data. A public key certificate is an electronic document. It contains information about the key, its owner and a digital signature of the issuer. The digital signature is the mathematical scheme for establishing authenticity of a digital document. The issuer is the entity or establishment which has verified the contents of the certificate.

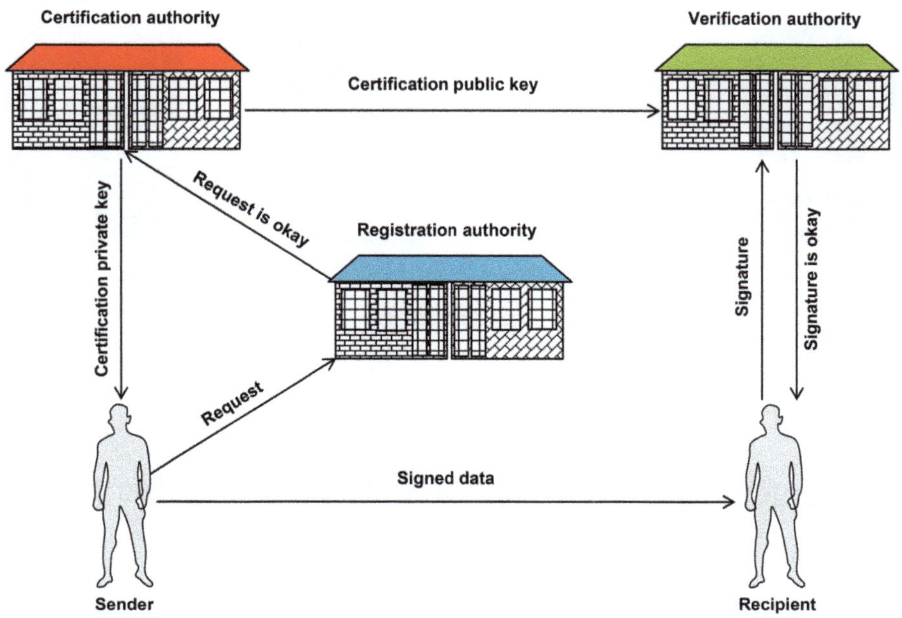

Figure 25.19. Public key infrastructure.

Public key encryption facilitates PKI through a pair of keys, which are separate but related to each other. One key is the public key used for message encryption. The other key is a private key used for message decryption. The message can be decrypted only by the intended user who holds the private secret key. The organization acting as the prime governing body for digital certificates is known as the certificate authority (CA). It is the certificate issuer referred to above.

25.12.17 Managed detection and response (MDR) services

MDR services (figure 25.20) combine technology with human expertise to search for threats and respond to them, thereby providing the functions of a modern security operations center (MSOC) to a customer (Blokdyk 2021). The threats are remotely watched using endpoint detection and response (EDR) tools. The endpoint can be a server, a workstation or a mobile device. An EDR tool is a technology platform for quick investigation and termination of attacks at endpoints. For threat detection, data about user logins and process execution at endpoints are collected, correlated and analyzed using data analytics and contextual information to find out anomalies in behavioral patterns. The anomalies point towards suspicious activities. The suspicious activities are flagged to alert security analysts and to initiate responses automatically, e.g., an endpoint may be temporarily obstructed for preventing the spreading of malware in the network. Remedial actions are planned and executed for threat containment at endpoints.

It is desired that an EDR tool should be easily integrated with existing security because all threats are not covered by end point security. Moreover, artificial

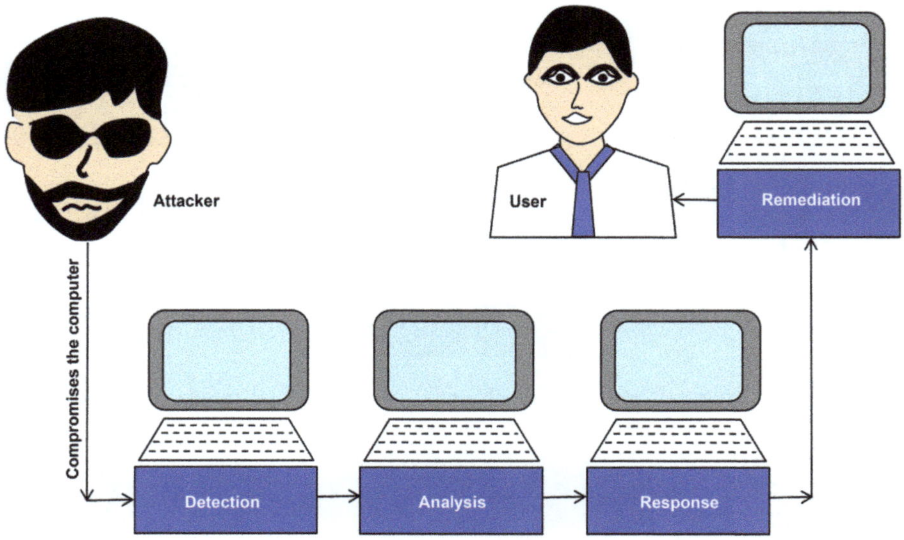

Figure 25.20. Management, detection and response system.

intelligence and machine learning are applied for automation of threat detection and alert creation. Both historical and current data must be used in order that a comprehensive picture of an incident is drawn. Forensic capabilities can aid in threat tracking. Thus, a threat hunting, monitoring and responding cyber service is built up.

25.12.18 Vulnerability assessment and penetration testing (VAPT) tools

These are tools that help in identifying the vulnerabilities in a network or web application such as absence of a functionality or presence of a malicious code that an attacker may utilize for breaching security (Rahalkar 2018). Code analysis is performed statically as well as dynamically.

A VAPT tool (figure 25.21) attacks the system from within the network and from outside the network in the same manner as a potential hacker will do (Wilhelm 2013). In this simulated cyberattack, all sensitive data locations, key applications, network connections and access points are included. Exploitable flaws and their severity are exposed. If a security breach can be made as proven by unauthorized access, the weakness is removed. Again, the VAPT tool attacks the system to find whether the system has been corrected. The corrections and testing are done till confidence in system security is achieved

25.13 Discussion and conclusions

Jamming can look as innocent as noise. It can become so severe as to completely block communication. So, the types of jamming and the jammers used must be systematically studied and kept in mind to cope with any situation. Crucial services

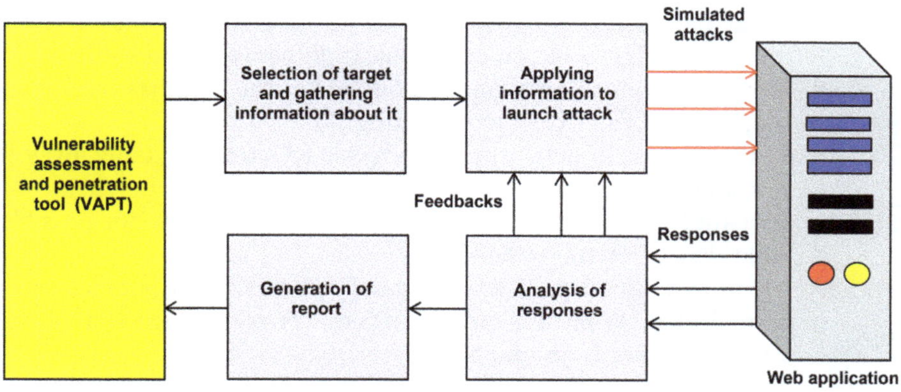

Figure 25.21. Penetration testing.

wherever high financial risks are involved or human lives are in danger must always be built using jamming-resistant techniques.

Due to the proliferation of electronic computers and networking, and the complexity and sophistication of computer systems and networks, cybersecurity has become a subject of paramount importance. Cybersecurity professionals are needed the world over to protect the vast amount of information storage and use at stake together with the associated financial concerns. Cybercrimes such as malware attacks, phishing, unauthorized data access and financial frauds in online banking are fairly common. Therefore, it is necessary to educate users to provide awareness and inculcate habits such as changing passwords at regular intervals, avoid visiting phishing websites and take extreme caution in opening links or downloading email attachments.

Review exercises

25.1. Explain the meaning of jamming in electronics communication. How is a jamming attack carried out?

25.2. Name the different types of jamming and mention principal features of each type.

25.3. How does barrage jamming differ from sweep jamming?

25.4. What is a proactive jammer? Explain the operation of its constant, random and deceptive sub-classes.

25.5. When does a reactive jammer send signals? How is jamming achieved by: (a) a reactive request-to-send/clear-to-send jammer, (b) reactive data/acknowledgement jammer?

25.6. What is a function-specific jammer? Name a sub-type of this jammer, which is effective in jamming frequency-hopping spread spectrum communication with slow hopping.

25.7. Why is a smart-hybrid jammer said to be 'smart'? Why is it called a hybrid jammer?

25.8. What parameters will you measure to ascertain whether a jamming attack is taking place on communication between two stations?
25.9. How is a jammed area mapped out? After the area has been mapped out, how do you overcome jamming?
25.10. For what type of networks is it possible to retreat spatially away from the jammed area to avoid getting jammed?
25.11. If spatially retreating away from the jammed area is not possible, in what way is retreating done?
25.12. Mention a situation in which spatial retreat from a jammer becomes complicated. What are the two phases through which spatial retreat from a jammer is accomplished?
25.13. What is the meaning of channel surfing? Is it the same as frequency-hopping modulation or dissimilar to it?
25.14. What difficulties are faced during switching of several devices to new channels? How are these difficulties avoided?
25.15. How is latency in channel switching related to size of the network? What kind of instability is created in the network by this latency?
25.16. How do regulation of transmitter power and error correcting code help in avoiding jamming? What features of communication are negatively affected by this method?
25.17. Name a communication system specially developed for preventing jamming.
25.18. For jamming a communication, the jammer must know that a communication is taking place. Even after the jammer knows that communication is in progress and tries its disruption, the task must be formidable for the jammer. Give your arguments to justify that the spread system communication meets both these requirements.
25.19. Does the total energy of the signal change when it is spread to occupy a large bandwidth?
25.20. Why does a spread spectrum communication signal look like a noise signal?
25.21. Why should the spreading key be complex in spread spectrum communication?
25.22. What are the functions of channel encoder and modulator in a spread spectrum communication system?
25.23. What are the roles of demodulator and channel decoder in a spread spectrum communication system?
25.24. What are two popular methods of concealing the signal used in spread spectrum communication?
25.25. What is meant by hopping period in a transmission cycle of FHSS communication? Are all the frequencies used in any given hop period?
25.26. What is: (a) a pseudorandom number, and (b) a pseudorandom number generator?

25.27. How do the pseudorandom number generator, channel frequency table and frequency synthesizer work in coordination to define a sequence of channels according to a spreading code?
25.28. What is a chipping signal? Why is it called by this name?
25.29. Explain why a frequency-hopping signal is difficult to intercept and jam.
25.30. Draw the structural diagram of a frequency-hopping spreader showing its constituent parts, and explain how signal spreading is done?
25.31. With the help of a diagram, explain the operation of a frequency-hopping despreader.
25.32. Explain with a block diagram the working of a frequency-hopping spread spectrum communication system.
25.33. What is the full form of BFSK? Draw an input binary sequence and explain how is BFSK done showing the waveforms of the carrier and BFSK modulated signals.
25.34. In an FHSS transmitter, the spread signal is passed through a bandpass filter to stop the difference frequency. Explain why this filter is necessary taking a carrier signal whose frequency switches between f_0 and $f_0+\Delta f$.
25.35. Why is a bandpass filter necessary to stop the sum frequency in an FHSS receiver? At which stage is this filtering done?
25.36. In a BFSK modulation scheme, the transmitted signal is given by

$$s_d(t) = A\cos(2\pi f_1 t) \text{ for binary 0} \tag{25.53}$$

and

$$s_d(t) = A\cos(2\pi f_2 t) \text{ for binary 1 .} \tag{25.54}$$

If

$$f_2 - f_1 = \Delta f \tag{25.55}$$

can you write a single equation to represent the two separate equations for binary 0 and 1 states?
25.37. (a) In an FHSS system, the transmitted BFSK digital signal is given by

$$s_d(t) = A\cos\{2\pi(f_0 + \Delta f)t\} \tag{25.56}$$

before spreading. The frequencies f_0 and $f_0+\Delta f$ correspond to binary sates 0 and 1. If a chipping signal of frequency f_i

$$c(t) = \cos(2\pi f_i t) \tag{25.57}$$

is applied to this signal, show that the product signal is expressed as

$$p(t) = \frac{1}{2}\cos\left[2\pi\left\{f_0 + \frac{1}{2}(b_i + 1)\Delta f\right\}t + 2\pi f_i t\right] + \frac{1}{2}\cos\left[2\pi\left\{f_0 + \frac{1}{2}(b_i + 1)\Delta f\right\}t - 2\pi f_i t\right] \quad (25.58)$$

consisting of two terms: a term containing the sum frequency and another term with difference frequency. Removing the difference frequency term by filtering, write the equation for transmitted spread spectrum signal.

(b) The received spread spectrum signal is multiplied by the chipping signal $c(t)$. Show that the despread signal is

$$r(t) = \frac{1}{4}\cos\left[2\pi\left\{f_0 + \frac{1}{2}(b_i + 1)\Delta f + 2f_i\right\}t\right] + \frac{1}{4}\cos\left[2\pi\left\{f_0 + \frac{1}{2}(b_i + 1)\Delta f\right\}t\right] \quad (25.59)$$

25.38. What is MFSK? Why is it preferred over BFSK?

25.39. Depending on the relative magnitudes of frequency-hopping rate and the information bit rate, state the condition in which FHSS is said to be slow. When is it treated as fast?

25.40. If a jammer of power S_{Jammer} is trying to jam an MFSK transmitter using FHSS with 2^k channels, what is the available jamming energy per channel? What is the processing gain?

25.41. Both FHSS and DSSS transmit information in the form of a wide bandwidth signal. What is the difference between their ways of information transmission?

25.42. In DSSS, the sequence of input data is spread over the full bandwidth of the spread spectrum signal, while in FHSS the data sequence is spread over small frequency slots of the spread spectrum signal bandwidth. Explain.

25.43. What is the most elementary form of PSK? What is its principle? How is it different from the digital modulation based on frequency of the carrier wave?

25.44. Consider a BPSK signal

$$s(t) = Ad(t)\cos(2\pi f_c t) \quad (25.60)$$

where the symbols have their usual significance. Before transmission, it is multiplied by a chipping signal $c(t)$. Again, after reception at the receiver, it is multiplied by $c(t)$. What is the final signal obtained? Why?

25.45. A spread signal $s(t)$ is sent from a DSSS transmitter. On the way to receiver, a noise $n(t)$ is added to this signal. What happens to noise when the signal $s(t)+n(t)$ is multiplied by chipping signal $c(t)$ at the receiver? How is this noise removed?

25.46. In a DSSS system:
   (a) What is the frequency range over which energy of the information signal extends if the data rate is $1/T$?
   (b) What is the frequency range of energy of the spread signal if the pseudorandom noise signal has a data rate of $1/T_c$?
   (c) What is the frequency range of energy of (information bit + spread spectrum) signal?

25.47. A jamming signal is applied to a DSSS system. Analyze its effect on the performance of the system and show that the jamming signal at the receiver is reduced in the ratio of data rates of pseudorandom noise and information signals. What is the resulting processing gain?

25.48. Hacking is an illegal activity. When is it said to be ethical? What is the intent of ethical hacking? What is an ethical hacker called?

25.49. What are the five phases of ethical hacking? Outline the work done in each phase.

25.50. What is a malware? How does it show its presence in the computer? Write the names of ten common types of malware.

25.51. Define a virus in relation to computers. What harm does it cause to the computer? How is it different from a computer worm?

25.52. What is a computer worm? How does it spread? What damage does it inflict on the computer? Elaborate the statement, 'the worm is a standalone program'.

25.53. Why is a Trojan horse called by this name? How is this malware similar to the famed giant wooden Trojan horse used by Greek soldiers to deceive the Trojans and capture the city of Troy?

25.54. How is a Trojan horse infection caught? How can protection from this malware be achieved?

25.55. What does wiper malware do on infecting a computer? What steps should be immediately taken when one observes a massive deletion of files of the system?

25.56. What is spyware malware? How does it gain entry into the computer? What does it do when it is inside the computer?

25.57. What does ransomware malware start doing on entering a computer? When does it ask for ransom? What should be done to avoid ransomware attacks?

25.58. How does rogue security software mislead users? Does it steal data? What does it advise the user to pay for?

25.59. What kind of threat perception does scareware malware build up? How do URL filters and pop-up blockers help in protecting from it?

25.60. What malware is used for mining cryptocurrency? How does this malware attack happen?

25.61. What is a keylogger? What is it used for? Is it software or hardware, or can it be in either form?

25.62. How is a hardware keylogger implanted in a computer?
25.63. What harmful activities does a keylogger perform during its stay on the computer? Can a keylogger be used for beneficial activity?
25.64. How does rootkit malware enter a computer? What facility does a rootkit provide to its controller?
25.65. How does fileless malware differ from traditional malware? Why does this malware leave no footprint?
25.66. In which type of memory does fileless malware sit? Is it written on the hard disc drive? Will a fileless malware attack work if the computer is rebooted or it needs to be re-launched?
25.67. What is APT? What happens during an APT attack?
25.68. How is a computer deceived to execute an arbitrary code by gaining control over the instruction pointer? How is an ACE attack countered?
25.69. Is backdoor a malware? What is a backdoor attack on a computer?
25.70. What is an RFI attack? What are its outcomes? Why is it difficult to remove backdoor shell files?
25.71. What is the meaning of a hardware backdoor? At what stage is it inserted? How is this insertion done? Can it be got rid of with the help of software?
25.72. What factors encourage the launching of a code injection attack on a computer? What harm can code injection do to the computer?
25.73. What is done during cross-site scripting? How does it impact interaction of the user with a trusted website?
25.74. Distinguish between type I and type II XSS attacks.
25.75. Why is type I XSS called reflected XSS? How are users urged to make requests to the server?
25.76. Why is type II XSS known as stored XSS? How does the attacker's payload appear on the browser of a user in type II XSS?
25.77. Which is non-persistent, Type I or type II XSS?
25.78. When does DOM-based XSS occur? How is the data transmitted to a user's browser? How does this data enable the attacker to inject a malicious payload? When is this payload executed?
25.79. What is meant by escaping HTML?
25.80. How is a drive-by-download attack initiated through authorized downloads? How does it start with unauthorized downloads? What types of programs should end-users install for protection? Are ad blockers helpful?
25.81. What is a data breach? How can a data breach be prevented?
25.82. Explain with diagrams DoS and DDoS attacks. What are zombie computers?
25.83. What is a buffer in computer memory? What is meant by buffer overflow? How is a buffer flow attack done?
25.84. What is ping software? What ICMP messages are used to ping a network?

25.85. How does the network become unavailable to normal traffic during an ICMP flood attack? Suggest ways to prevent ICMP flood attacks.
25.86. How is SYN flood attack carried out by showering a server with requests but not replying with acknowledgements?
25.87. Can carefully watching network traffic indicate a DoS attack? How?
25.88. What is eavesdropping? What kind of data is prone to eavesdropping? How is eavesdropping done? Can data encryption thwart this attack?
25.89. What is email spoofing? How does the attacker dupe the user by email spoofing?
25.90. What is the protocol for email authentication using the domain name system called? Explain how it protects the people communicating through emails from spoofed emails.
25.91. What information does an SPF record contain and how is this information applied to infer whether an email is from a genuine source?
25.92. How does an exploit infiltrate in a computer? Is it a malware? Can it deliver malware?
25.93. What is a zero-day exploit? What does 'zero-day' mean in reference to an exploit?
25.94. What is malvertising? How is it done? What precaution must be taken during internet browsing to avoid malvertising?
25.95. What is the name of a cyberattack that applies psychological manipulation of people for disclosure of secret information? Give an example.
25.96. What is two-factor authentication? How is it done? How does it protect against a social engineering attack?
25.97. Why is phishing said to be a social engineering attack? What defensive action is recommended against phishing?
25.98. What kind of design flaw in the computer operating system encourages an attacker to get privilege escalation? Distinguish between horizontal and vertical privilege escalations.
25.99. What are indications of privilege escalation on a computer? Name a useful tool for its detection.
25.100. What is spamming? How does it differ from phishing?
25.101. Which attack is launched by using a group of zombie computers? What is the result of this attack?
25.102. Name a cyberattack carried out by using pulsing zombie computers. What confusion is created by this attack?
25.103. Who bears the bandwidth costs in an attack made with zombie computers, the attacker or the owners of zombie computers?
25.104. What are the two words from which the name 'botnet' is derived? What is the person controlling the botnet called?
25.105. What are centralized and decentralized models of botnet? Describe the three stages of production of a botnet.
25.106. What is access control software? Describe the four types of access control provided by the software.

25.107. Which is the most flexible type of access control provided by access control software? Name the most restrictive type of access control.
25.108. What does an anti-logger do? For what type of computer applications are anti-logger scans essential? How often should anti-logger scanning of computers be done?
25.109. Which computer protection is more comprehensive? Anti-virus or anti-malware protection.
25.110. What does anti-spyware software do? How is spyware detected?
25.111. What is the function of anti-subversion software? Distinguish between static and dynamic anti-subversion.
25.112. What is software tampering? How is it prevented? How does white-box cryptography maintain confidentiality?
25.113. What is an anti-theft system? Name an item which can be disabled when lost or stolen.
25.114. What is encryption of data? How does it make a message unreadable to unauthorized parties?
25.115. Explain the difference between symmetric and asymmetric encryption of text using diagrams.
25.116. Why should a symmetrically encrypted message be strongly secured while there is no such requirement for an asymmetrically encrypted message?
25.117. Is firewall a software or hardware device? How does it act as a first line of defense between the internal network and external internet of a company?
25.118. What are the most basic firewalls? What parameters do they analyze to allow entry of network traffic in the computer?
25.119. What feature distinguishes circuit-level gateways from packet-filtering firewalls?
25.120. Why are proxy firewalls called by this name? How do they provide additional protection to the client?
25.121. What are the three messages used in a TCP handshake for negation and starting of a TCP session?
25.122. What is multilayer stateful inspection? How does a multilayer stateful firewall work?
25.123. What features and functionalities are incorporated in next-generation firewalls? How do these firewalls differ from traditional firewalls?
25.124. Which of the two systems (intrusion detection system, intrusion prevention system) is an active system and which one is passive?
25.125. What does an intrusion detection system do? What are the three types in which this system is built?
25.126. Point out the difference between the intrusion detection and intrusion prevention systems.
25.127. What is a sandbox? How does the use of sandbox provide security to the computer? Against what type of threats is the sandbox especially useful?

25.128. What tasks are performed by a security information and event management system? Why is this system an essential tool for an organization? How does this system reduce the mean time for detecting threats and responding to them?
25.129. What is a software patch and why is it needed? What are the three common types of patches?
25.130. What is security vulnerability? What does security management software do? How does its function differ from that of security assessment software?
25.131. Is packet sniffer a software or hardware? What does it do?
25.132. What information does a packet sniffer capture? How is this information helpful to an organization?
25.133. Compare the filtered and unfiltered modes of operation of a packet sniffer.
25.134. For what different purposes do a network administrator and a hacker use a packet sniffer?
25.135. What is PKI? For what tasks does it provide a framework? What activities does it govern?
25.136. Name the technologies which have enabled public key infrastructure. What is the governing body for digital certificates called?
25.137. Explain the terms: (a) public key, (b) public key certificate and (c) public key encryption.
25.138. Why can a PKI message be decrypted by the intended user only?
25.139. How do managed detection and response services provide the facilities of a modern security operations center?
25.140. What is an endpoint detection and response tool? How is threat detection carried out? What happens when suspicious activity is noticed?
25.141. What are the desirable features expected in an EDR tool to build a cyber service for hunting, monitoring and responding to threats?
25.142. Explain with a diagram how a vulnerability assessment and penetration tool works and helps in improving the security of a system.

References

Alpine Security 2021 The 8 most common cyber attacks and how to stop them https://alpinesecurity.com/blog/the-8-most-common-cyber-attacks-and-how-to-stop-them/

Blokdyk G 2021 *Managed Detection and Response: A Complete Guide* (Brendale, Australia: 5STARCooks)

Buchmann J A, Karatsiolis E and Wiesmaier A 2013 *Introduction to Public Key Infrastructures* (Berlin: Springer)

Cloudflare 2022 What is a denial-of-service attack? https://cloudflare.com/en-gb/learning/ddos/glossary/denial-of-service/

Grover K, Lim A and Yang Q 2014 Jamming and anti-jamming techniques in wireless networks: a survey *Int. J. Ad Hoc Ubiquitous Comput.* **17** 197–215

Harper A, Linn R, Sims S, Baucom M, Tajeda H, Fernandez D and Frost M 2022 *Gray Hat Hacking: The Ethical Hacker's Handbook* (New York: Mc-Graw Hill)

Horvath I 2022 The five phases of ethical hacking, Invensis, Inc https://invensislearning.com/blog/phases-of-ethical-hacking/

Kaspersky 2022a What is a drive-by-download? https://kaspersky.com/resource-center/definitions/drive-by-download

Kaspersky 2022b What is social engineering? https://kaspersky.co.in/resource-center/definitions/what-is-social-engineering

Kessler 2022 *An overview of cryptography* https://garykessler.net/library/crypto.html

Liu A X 2011 *Firewall Design and Analysis* (Singapore: World Scientific)

Malwarebytes 2022 Keylogger https://malwarebytes.com/keylogger

Mpitziopoulos A, Gavalas D, Konstantopoulos C and Pantziou G 2009 A survey on jamming attacks and countermeasures in WSNs *IEEE Commun. Surv. Tutor.* **11** 42–56

Pirayesh H and Zeng H 2022 Jamming attacks and anti-jamming strategies in wireless networks: a comprehensive survey *IEEE Commun. Surv. Tutor.* **24** 767–809

PortSwigger 2022 Cross-site scripting https://portswigger.net/web-security/cross-site-scripting

Rahalkar S 2018 *Network Vulnerability Assessment: Identify Security Loopholes in Your Network's Infrastructure* (Birmingham: Packt)

Sikorski M and Honig A 2012 *Practical Malware Analysis: The Hands-On Guide to Dissecting Malicious Software* (San Francisco, CA: No Starch Press) pp 1–5

Stallings W 2007 *Data and Computer Communications* (Hoboken, NJ: Pearson-Prentice Hall) pp 276–87

Ström E, Ottosson T and Svensson A 2002 An introduction to spread spectrum systems *Technical Report* No. R016/2002pp Department of Signals and Systems, Gothenburg Chalmers University of Technology, Göteborg 1–28

Waksman A and Sethumadhavan S 2011 Silencing hardware backdoors *2011 IEEE Symp. on Security and Privacy (Oakland, CA, 22–25 May, 2011)* 49–63

Wilhelm T 2013 *Professional Penetration Testing: Creating and Learning in a Hacking Lab* (Amsterdam: Syngress: Imprint of Elsevier)

Wood A D, Stankovic J A and Son S H 2003 JAM: a jammed-area mapping service for sensor networks, RTSS 2003 *24th IEEE Real-Time Systems Symp. (Cancun, 5 December 2003)* (Piscataway, NJ: IEEE) pp 286–97

Xu W, Ma K, Trappe W and Zhang Y 2006 Jamming sensor networks: attack and defense strategies *IEEE Netw.* **20** 41–7

www.ingramcontent.com/pod-product-compliance
Ingram Content Group UK Ltd.
Pitfield, Milton Keynes, MK11 3LW, UK
UKHW051911100926
8556IPUK00005B/850

9 780750 350709